THE
INTERNATIONAL SERIES
OF
MONOGRAPHS ON PHYSICS

GENERAL EDITORS

W. MARSHALL   D. H. WILKINSON

H. S. W. MASSEY, E. H. S. BURHOP
AND H. B. GILBODY

ELECTRONIC AND IONIC IMPACT PHENOMENA

SECOND EDITION
IN FOUR VOLUMES

VOLUME III

Slow Collisions of Heavy Particles

BY H. S. W. MASSEY

OXFORD
AT THE CLARENDON PRESS
1971

Oxford University Press, Ely House, London W. 1

GLASGOW NEW YORK TORONTO MELBOURNE WELLINGTON
CAPE TOWN SALISBURY IBADAN NAIROBI DAR ES SALAAM LUSAKA ADDIS ABABA
BOMBAY CALCUTTA MADRAS KARACHI LAHORE DACCA
KUALA LUMPUR SINGAPORE HONG KONG TOKYO

© OXFORD UNIVERSITY PRESS 1971

PRINTED IN GREAT BRITAIN

PREFACE TO THE SECOND EDITION

THE immense growth of the subject since the first edition was produced has raised many problems in connection with a new edition. Apart from the sheer bulk of new material, the interconnections between different parts of the subject have become very complex, while the sophistication of both theoretical and experimental techniques has greatly increased.

It became clear at the outset that it was no longer possible to attempt a nearly comprehensive treatment. As sources of data on cross-sections, reaction rates, etc., for use in various applications are now available and are becoming more comprehensive, it seemed that in the new edition the emphasis should be on describing and discussing experimental and theoretical techniques, and the interpretation of the results obtained by their use, rather than on compilation of data. Even so, a greater selectivity among the wide range of available material has been essential. Within these limitations the level of the treatment has been maintained roughly as in the first edition, although some allowance has been made for the general increase in the level of sophistication.

When all these considerations were taken into account it became clear that the new edition would be between four and five times larger than the first. To make practicable the completion of the task of writing so much against the rate of production of new results, it was decided to omit any discussion of phenomena occurring at surfaces (Chapters V and IX of the first edition) and to present the new edition in four volumes, the correspondence with the first edition being as follows:

Second edition	First edition
Volume I	Chapters I, II, III
Volume II	Chapters IV, VI
Volume III	Chapter VII
Volume IV	Chapters VIII, X

In this way Volumes I and II deal with electron impact phenomena: Volume I with electron–atom collisions and Volume II with electron–molecule collisions. Volume II also includes a detailed discussion of photo-ionization and photodetachment which did not appear in the first edition.

Volumes III and IV deal, in general, with collisions involving heavy particles. Thus, Volume III is concerned with thermal collisions involving neutral and ionized atoms and molecules, Volume IV with higher energy collisions of this kind. In addition, recombination is included in this volume as well as a description of collision processes involving slow positrons and muons, which did not appear in the first edition.

Because of the complicated mesh of cross-connections many difficult decisions had to be made as to the place at which a new technique should be described in detail. Usually it was decided to do this in relation to one of the major applications of the technique rather than attempting, in a wholly artificial way, to avoid forward references at all stages.

In covering such a wide field in physics, and indeed also in chemistry, acute difficulties of notation are bound to arise. The symbol k for wave number is now so universally used in collision theory that we have been so impious as to use κ instead of k for Boltzmann's constant. Unfortunately, k is also used very widely by physical chemists to denote a rate constant. In some places we have adhered to this but elsewhere, to avoid confusion with wave numbers, we have substituted a less familiar symbol. Another unfamiliar usage we have employed is that of f for oscillator strength to distinguish it from f for scattered amplitude. Again we have been unfashionable in using F instead of E for electric field strength because of risk of confusion with E for energy.

No attempt has been made to adopt a set of symbols of universal application throughout the book, although we have stuck grimly to k for wave number and Q for cross-section as well as to e, h, κ and c.

We have tried not to be too pedantic in choice of units though admitting to a predilection for eV as against kcal/mole. When dealing with phenomena of strongly chemical interest we have at times used kcal/mole, but always with the value in eV in brackets.

The penetration of the work into chemistry, though perhaps occurring on a wider front, is no deeper than before. The deciding factor has always been the complexity of the molecules involved in the reactions under consideration.

In order to complete a volume it was necessary, at a certain stage, to close the books, as it were, and turn a blind eye to new results coming in after a certain date—unless, of course, they rendered incorrect anything already written. The closing date for Volumes I and II was roughly early 1967, for Volume III mid-1968, and for Volume IV about one year later. Notes on later advances over the whole field will be included at the end of Volume IV.

London
February 1969

H. S. W. M.
E. H. S. B.
H. B. G.

PREFACE TO THE FIRST EDITION

There are very many directions in which research in physics and related subjects depends on a knowledge of the rates of collision processes which occur between electrons, ions, and neutral atoms and molecules. This has become increasingly apparent in recent times in connection with developments involving electric discharges in gases, atmospheric physics, and astrophysics. Apart from this the subject is of great intrinsic interest, playing a leading part in the establishment of quantum theory and including many aspects of fundamental importance in the theory of atomic structure. It therefore seems appropriate to describe the present state of knowledge of the subject and this we have attempted to do in the present work.

We have set ourselves the task of describing the experimental techniques employed and the results obtained for the different kinds of collision phenomena which we have considered within the scope of the book. While no attempt has been made to provide at all times the detailed mathematical theory which may be appropriate for the interpretation of the phenomena, wherever possible the observations have been considered against the available theoretical background, results obtained by theory have been included, and a physical account of the different theories has been given. In some cases, not covered in *The theory of atomic collisions*, a more detailed description for a particular theory has been provided. At all times the aim has been to give a balanced view of the subject, from both the theoretical and experimental standpoints, bringing out as clearly as possible the well-established principles which emerge and the obscurities and uncertainties, many as they are, which still remain.

It was inevitable that some rigid principles of exclusion had to be practised in selecting from the great wealth of available material. It was first decided that phenomena involving the collisions of particles with high energies would not be considered, and that other phenomena associated with the properties of atomic nuclei such as the behaviour of slow neutrons would also be excluded. It was also natural to regard work on chemical kinetics as such, although clearly involving atomic collision phenomena, as outside the scope of the book, but certain of the more fundamental aspects are included. Phenomena involving neutral atoms or molecules only have otherwise been included on an equal footing with those involving ions or electrons. A further extensive class of phenomena have been excluded by avoiding any discussion of collision processes occurring within solids or liquids, confining the work to processes occurring in the gas phase or at a gas–solid interface. Among the latter phenomena electron diffraction at a solid surface has been rather arbitrarily excluded as it is a subject already adequately dealt with in other texts. Secondary electron emission and related effects are, however, included.

By limiting the scope of the book in this way it has just been possible to provide a fairly comprehensive account of the subjects involved. It is perhaps too much to hope that even within these limitations nothing of importance has been missed, but it is believed that the account given is fairly complete. Extensive tables of observed and theoretical data have been given throughout for reference purposes and the extent to which the data given are likely to be reliable has been indicated. Every effort has been made to provide a connected and systematic account but it is inevitable that there will be differences of opinion as to the relative weight given to the various parts of the subject and to the different contributions which have been made to it.

We are particularly indebted to Professor D. R. Bates for reading and criticizing much of the manuscript and for many valuable suggestions. Dr. R. A. Buckingham has also assisted us very much in this direction while Dr. Abdelnabi has checked some of the proofs. We also wish to express our appreciation of the remarkable way in which the Oxford University Press maintained the high standard of their work under the present difficult circumstances.

H. S. W. M.
E. H. S. B.

London
August 1951

ACKNOWLEDGEMENTS
VOLUME III

IT is again a pleasure to thank Mrs. J. Lawson for the skilful assistance so freely given in preparing diagrams and tables, obtaining and checking references, preparing the author index, and in many other ways. In this she has been helped by her husband, not only in carrying out electronic computation in connection with the work but in many other directions also. As before, Mrs. M. Harding has dealt with her customary cheerfulness and skill with the big bulk of badly written manuscript and has helped in many other details of checking, etc.

We are much indebted to Dr. A. Omont and Dr. S. Zienau for checking certain of the proofs and making valuable suggestions. Professor E. W. McDaniel, Professor J. B. Hasted, Dr. E. E. Ferguson, Dr. R. A. Young, and Dr. D. M. Hunten have provided us with material in advance of publication. We have benefited by discussions with Dr. S. Corrigan on molecular beams, with Drs. M. Bloom and R. F. Snider on nuclear magnetic relaxation, and with Professor Hasted on a number of subjects.

As with the earlier volumes, the work done by the Clarendon Press and by the printers has been first class, and again we have benefited from the skill shown by all concerned with checking the material at each stage.

CONTENTS

16. COLLISIONS BETWEEN ATOMS AND MOLECULES UNDER GAS-KINETIC CONDITIONS—ELASTIC COLLISIONS

1. Introduction—classification of possibilities	1295
2. General nature of the interaction between atoms	1298
3. Elastic collisions of gas atoms—general discussion	1301
3.1. Cross-sections effective in scattering, viscosity, and diffusion	1301
4. Quantal and classical cross-sections for rigid spherical atoms	1304
4.1. The effect of symmetry	1308
5. Cross-sections for extended range interactions	1309
5.1. Classical formulae	1309
5.1.1. Orbiting	1310
5.1.2. The rainbow angle	1312
5.1.3. Glory scattering	1312
5.1.4. Example	1312
5.1.5. Diffusion and viscosity coefficients	1315
5.1.6. Small-angle scattering	1315
5.1.7. A classical 'total' cross-section	1317
5.2. Relation of classical to quantal formulae	1317
5.3. Semi-classical approximation	1319
5.3.1. Glory interference effects	1320
5.3.2. Interference effects near the rainbow angle	1321
5.4. Quantum theory of scattering by long-range potentials	1322
5.4.1. The total cross-section for potentials varying as r^{-s}	1323
5.4.2. Small-angle scattering for potentials varying as r^{-s}	1325
5.4.3. The effect of glory scattering	1326
5.4.4. Orbiting under quantum conditions	1329
6. Collision parameters for analytically represented interactions	1333
6.1. Analytical representations of interatomic interactions	1333
6.2. Reduced parameters when the semi-classical approximation is valid	1335
6.3. Cross-sections in terms of reduced parameters	1337
6.3.1. The low- and high-velocity regions	1338
6.4. Dependence of the rainbow angle on interaction parameters	1339
6.5. Conditions for orbiting	1340
6.6. Determination of potential parameters from experimental data	1340
7. The measurement of total cross-sections by molecular-ray methods	1342
7.1. The production and detection of molecular beams	1342
7.2. The single-beam method for measuring total cross-sections	1346

7.2.1. The principle of the method	1346
7.2.2. Angular resolution	1349
7.2.3. Typical experimental arrangements	1351
7.2.3.1. The use of velocity selectors	1355
7.3. The crossed-beam method for measuring total cross-sections	1361
8. The measurement of differential cross-sections	1366
9. Discussion of experimental results for collisions of alkali-metal atoms with rare-gas and mercury atoms	1369
9.1. The classical approximation and its failure at very small scattering angles	1369
9.2. The velocity variation of the total cross-section	1373
9.2.1. The asymptotic form of the interaction	1375
9.2.2. The 'glory' oscillations	1377
9.3. Absolute measurement of cross-sections—determination of the van der Waals constant C	1381
9.3.1. Comparison with calculated values	1386
9.4. Measurements of differential cross-sections	1387
9.4.1. Analysis of observations of rainbow effects	1387
9.4.2. Further experiments at high resolution	1389
9.5. Self-consistency of data derived from different types of experiment	1396
10. The interaction between helium atoms at large and intermediate separations	1399
10.1. The low-temperature evidence ($T < 20\ °K$)	1400
10.2. Results from observations of transport properties and second virial coefficients for $T > 20\ °K$	1408
10.3. Measurement of the total cross-section for He–He collisions	1411
11. Interactions between other pairs of rare-gas atoms	1413
12. Collisions involving atoms which may interact in more than one way	1418
12.1. Atoms both in ^2S states—collisions between hydrogen atoms	1418
12.1.1. Formulae for the cross-sections	1418
12.1.2. The H–H interactions	1419
12.1.3. The total elastic cross-sections for H–H collisions	1421
12.1.4. The viscosity of atomic hydrogen	1424
12.2. Atoms both in ^2S states—collisions between alkali-metal atoms	1429
12.2.1. Experimental method	1429
12.2.2. Analysis of data	1431
12.3. The viscosity and thermal conductivity of atomic oxygen	1433
12.4. Observation of the anisotropy of van der Waals forces	1436
12.4.1. The magnitude of the effects expected	1437
12.4.2. The experimental method	1439
12.4.3. The observed results	1442

13.	Collisions involving molecules	1444
	13.1. Special features associated with such collisions	1444
	13.2. Collisions of K and Cs atoms with various molecules	1446
	13.3. Collisions between dipolar molecules and other (non-polar) molecules	1449
	13.4. The H–H_2 interaction	1452
	13.5. The H_2–H_2 interaction	1453
	13.5.1. The mean interaction—evidence from transport and total cross-sections	1454
	13.5.2. The anisotropic component of the interaction	1457
	13.6. Interaction of H_2 molecules with rare-gas atoms	1461

17. COLLISIONS BETWEEN ATOMS AND MOLECULES UNDER GAS-KINETIC CONDITIONS—COLLISIONS INVOLVING EXCITATION OF VIBRATION AND ROTATION—REACTIVE COLLISIONS BETWEEN MOLECULAR BEAMS

1. Introductory remarks concerning the probability of vibrational and rotational transitions in gas-kinetic collisions	1463
2. Experimental methods for studying the probability of vibrational transitions occurring on impact between molecules	1466
2.1. The dispersion and absorption of high-frequency sound	1466
2.1.1. Variation of dispersion and absorption with pressure, temperature, and impurity content	1470
2.1.2. Methods for measuring the velocity and absorption of high-frequency sound waves	1470
2.1.2.1. The ultrasonic interferometer	1470
2.1.2.2. Further methods for measurement of sound absorption	1473
2.1.3. Some notes on the derivation of relaxation times from measurements of ultrasonic dispersion and absorption	1475
2.2. Shock-wave investigations	1476
2.2.1. Principles of the shock tube and its application to relaxation measurements	1476
2.2.2. Measurement techniques	1483
2.2.3. Optical techniques for temperature measurement	1487
2.3. The effect of persistence of vibration in gas dynamics	1491
2.4. Spectroscopic methods for measurement of vibrational relaxation time for molecules in ground electronic states	1496
2.4.1. The quenching of infra-red fluorescence	1497
2.4.2. The use of flash spectroscopy	1502
2.5. Spectroscopic methods for measurement of vibrational relaxation times for molecules in excited electronic states	1504
2.5.1. The use of laser-induced fluorescence	1508
2.6. The spectrophone	1511

CONTENTS

3. Results of experimental observations of vibrational relaxation times — 1518
   - 3.1. Relaxation times for low-lying vibrational levels — 1518
     - 3.1.1. Pure diatomic gases — 1518
     - 3.1.2. Mixtures of diatomic molecular gases with rare gases — 1522
     - 3.1.3. Pure polyatomic gases and mixtures with rare gases — 1523
     - 3.1.4. The effect of diatomic and polyatomic impurities on the vibrational relaxation times of molecular gases — 1532
   - 3.2. Relaxation times for higher vibrational levels — 1537
   - 3.3. Summarizing remarks — 1538
4. Theoretical discussion of vibrational relaxation — 1540
   - 4.1. Head-on collisions of atoms with diatomic molecules — 1540
     - 4.1.1. Application of distorted-wave method with exponential interaction — 1540
     - 4.1.2. Temperature dependence of deactivation probability — 1543
     - 4.1.3. Semi-classical method — 1544
   - 4.2. Head-on collisions between diatomic molecules, with exponential interaction—vibrational transfer — 1545
   - 4.3. Use of a more realistic interaction — 1547
   - 4.4. Three-dimensional theory of vibrational deactivation — 1550
     - 4.4.1. Semi-empirical treatment — 1550
     - 4.4.2. Application of the close-coupling (truncated eigenfunction) method — 1553
   - 4.5. Elementary theory of vibration–rotation transfer — 1561
   - 4.6. Summarizing remarks on the theory of vibrational relaxation — 1563
5. The excitation and deactivation of molecular rotation on impact between molecules — 1564
   - 5.1. Observed rotational relaxation in H_2 and D_2 — 1565
   - 5.2. Theoretical calculation of rotational relaxation times for H_2, D_2, and HD — 1566
   - 5.3. Rotational excitation of H_2 and of D_2 by collision with H and with He atoms — 1576
   - 5.4. Rotational excitation of heavier molecules — 1577
     - 5.4.1. Experimental evidence from acoustic and shock-wave observations — 1577
     - 5.4.2. Rotational relaxation and thermal conductivity of gases — 1579
       - 5.4.2.1. Rotational relaxation from measurements of thermal transpiration — 1582
     - 5.4.3. Spectroscopic evidence — 1583
   - 5.5. Rotational excitation and molecular beam experiments — 1588
     - 5.5.1. Theoretical considerations — 1588
     - 5.5.2. Experiments on the scattering of beams of polar molecules in selected rotational states — 1593
       - 5.5.2.1. Dependence of elastic cross-section on molecular orientation — 1595
       - 5.5.2.2. Collisions involving rotational transitions — 1599

CONTENTS

5.6. Theoretical discussion of collisions between polar molecules	1606
5.6.1. Comparison with observation	1612
5.7. Resonant transfer of rotational energy and the thermal conductivity of polar gases	1615
5.8. Collisions involving change of molecular orientation only	1616
6. Reactive collisions between molecular beams	1622
6.1. Introduction	1622
6.2. The determination of the total reaction cross-section—relation to non-reactive scattering—complex phase shifts	1624
6.3. Experimental analysis of angular, recoil, and internal energy distribution in the reaction $K + Br_2 \rightarrow KBr + Br$	1634
6.4. Results for reactions of alkali atoms with other halide molecules	1643
6.5. The use of ionization detectors in the study of reactions between crossed molecular beams—the reaction of D with Br_2	1646
6.6. Concluding remarks	1648

18. COLLISIONS BETWEEN ATOMS (AND/OR MOLECULES) UNDER GAS-KINETIC CONDITIONS—COLLISIONS INVOLVING ELECTRONIC TRANSITIONS

1. Introduction	1651
2. Quenching of radiation	1653
2.1. Quenching experiments with resonance radiation	1653
2.1.1. Quenching experiments with flame gases	1656
2.1.2. Quenching cross-sections from lifetime measurements	1662
2.1.3. Quenching cross-sections from the rate of decay of imprisoned resonance radiation	1663
2.2. Use of flash photolysis and time-resolved spectroscopy	1666
2.3. Deactivation of atoms excited by shock waves	1670
2.4. Deactivation of excited atoms produced by optical dissociation	1675
2.5. Deactivation of $Hg(6^3P_1)$ and $Hg(6^3P_0)$	1676
2.6. Deactivation of $Hg(7^3S_1)$ and $Hg(6^3D_1)$	1691
2.7. Preliminary discussion of results on quenching cross-sections	1692
3. Polarization of resonance radiation—collisions which change the orientation and alignment of the total electronic angular momentum	1696
3.1. Introduction—the disorientation and disalignment cross-sections	1696
3.2. Principles of the method of measurement of disorientation and disalignment cross-sections for atoms with no nuclear spin using Hanle effect	1700
3.3. Experimental measurement of cross-sections	1703
3.4. Double resonance method	1711
3.5. Depolarization of $Hg(6^3P_2)$	1715
3.6. Depolarization of atoms excited by step-wise excitation	1717
3.7. Effect of nuclear spin	1719

CONTENTS

- 3.8. The measurement of cross-sections for self-depolarization — 1722
- 3.9. Depolarizing cross-sections for neon — 1725
- 3.10. Depolarization of rubidium and caesium resonance radiation — 1729
- 3.11. Results of measurements of depolarization cross-sections — 1731
- 4. Sensitized fluorescence — 1731
  - 4.1. Sensitized fluorescence and fine-structure transitions — 1734
    - 4.1.1. Collision-induced transitions between $^2P_{\frac{1}{2}}$ and $^2P_{\frac{3}{2}}$ states of alkali-metal atoms — 1734
    - 4.1.2. The excitation of fine-structure transitions in other atoms — 1739
    - 4.1.3. Cross-sections for 'mixing' collisions between the fine-structure levels of $He(2^3P)$ — 1740
  - 4.2. Sensitized fluorescence and collisions involving transfer of excitation — 1743
    - 4.2.1. Introduction — 1743
    - 4.2.2. Experimental methods and results—energy resonance — 1744
      - 4.2.2.1. Mercury-sensitized sodium fluorescence — 1744
      - 4.2.2.2. Mercury-sensitized indium fluorescence — 1749
      - 4.2.2.3. Mercury-sensitized thallium fluorescence — 1750
      - 4.2.2.4. The enhancement of spark lines — 1755
    - 4.2.3. Wigner's spin-conservation rule — 1756
      - 4.2.3.1. Transfer collisions in helium—apparent breakdown of Wigner's rule — 1757
      - 4.2.3.2. Transfer collisions in helium—importance of transfer through F states — 1758
      - 4.2.3.3. Transfer collisions in helium—application of time-resolved spectroscopy — 1761
- 5. Collisions involving metastable atoms — 1767
  - 5.1. Total and differential cross-sections for collisions of metastable atoms with other atoms and molecules — 1769
    - 5.1.1. Measurement of total cross-sections — 1769
    - 5.1.2. Discussion of results — 1770
    - 5.1.3. Measurement of differential cross-sections — 1774
  - 5.2. Measurement of the rates of processes which destroy metastable atoms — 1776
    - 5.2.1. Introduction — 1776
    - 5.2.2. Experimental analysis of loss processes for metastable atoms in discharge afterglows — 1780
      - 5.2.2.1. Helium — 1780
      - 5.2.2.2. Neon — 1783
      - 5.2.2.3. Argon — 1792
      - 5.2.2.4. Effect of impurities — 1792
      - 5.2.2.5. Excitation transfer between metastable helium and normal neon atoms — 1796
    - 5.2.3. Metastable atoms in flowing afterglows — 1799

5.2.4. Data about metastable atoms from analysis of electron decay rates in discharge afterglows	1801
5.2.4.1. Helium and neon	1801
5.2.4.2. Mercury	1806
5.2.4.3. Effect of impurities	1809
5.3. Further study of Penning ionization by metastable atoms	1810
5.3.1. Enhancement by metastable atoms of ionization produced by alpha particles	1810
5.3.2. Measurement of cross-sections using metastable atom beams	1815
5.3.3. The energy distribution of the electrons produced by Penning ionization	1817
5.3.4. Discussion of results	1821
5.4. Measurement of the cross-sections for transfer of excitation between metastable and normal helium atoms	1822
5.4.1. Lower limits from beam experiments	1822
5.4.2. Optical-pumping method	1823
5.5. Reaction rates involving metastable oxygen atoms	1828
5.5.1. Introduction	1828
5.5.2. Experimental methods and results for quenching of $O(^1S)$	1831
5.5.3. Excitation of $O(^1S)$ in three-body recombination	1838
5.5.4. The quenching of metastable $O(^1D)$	1839
6. Collisions involving exchange of electron spin	1841
6.1. Introduction	1841
6.2. Methods for measuring spin-exchange cross-sections	1844
6.2.1. From the relaxation time of a hyperfine population	1844
6.2.2. Use of optical pumping to Zeeman levels	1851
6.2.3. Nuclear magnetic-resonance method	1854
6.2.4. Stimulated emission method	1858
6.2.5. Use of the hydrogen maser	1861
6.2.6. Results of observations of spin-exchange cross-sections	1864
7. Spin-reversal collisions	1864
8. Collisions involving excited atoms—discussion and theoretical interpretation of results	1868
8.1. Collisions in which exact resonance applies ($\Delta E = 0$)—excitation transfer	1869
8.1.1. Symmetrical and accidental resonance	1869
8.1.2. The case of symmetrical resonance	1870
8.1.3. Application to collisions of metastable helium atoms in helium	1876
8.1.3.1. The deactivation of $He(2^1S)$ in collisions with normal He atoms	1879
8.1.4. Application to transfer of excitation on impact between hydrogen atoms	1881

- 8.2. Collisions in which $\Delta E = 0$—spin-exchange collisions — 1884
  - 8.2.1. Spin-exchange collisions between H atoms — 1885
  - 8.2.2. Spin-exchange collisions between alkali-metal atoms — 1888
  - 8.2.3. Spin-exchange collisions between H and O atoms — 1890
- 8.3. Collisions in which $\Delta E = 0$—collisions producing depolarization — 1892
  - 8.3.1. Collisions between atoms of the same electronic structure — 1894
  - 8.3.2. Collisions between atoms with different electronic structure — 1901
  - 8.3.3. Numerical values and comparison with observation — 1902
    - 8.3.3.1. Self-depolarization of $^3P_1$ levels — 1902
    - 8.3.3.2. Depolarization of $^3P_1$ levels in collisions with foreign atoms — 1905
  - 8.3.4. Depolarization of states of integral J other than $^3P_1$ — 1906
    - 8.3.4.1. Hg and Cd(6^3S_1) — 1907
    - 8.3.4.2. Hg(6^1P_1) — 1907
    - 8.3.4.3. Hg(6^3P_2) and Cd(5^3P_2) — 1908
    - 8.3.4.4. Hg(6^3D) and Hg(6^1D) — 1908
  - 8.3.5. Inclusion of nuclear spin — 1909
  - 8.3.6. Impact disorientation and disalignment of atoms in excited states with $J = \tfrac{1}{2}$ or $\tfrac{3}{2}$ — 1912
- 8.4. Collisions in which the resonance is imperfect ($\Delta E \neq 0$) — 1914
  - 8.4.1. The crossing-point case — 1915
  - 8.4.2. The case of no crossing point — 1918
  - 8.4.3. Unsymmetrical or accidental resonance — 1923
  - 8.4.4. Summarizing remarks — 1924
  - 8.4.5. Collisions in which transitions occur between fine-structure levels — 1925
- 8.5. Collisions in which electron spin-flip occurs — 1928
- 8.6. Remarks on collisions involving molecules in which electronic transitions occur — 1929

19. COLLISIONS UNDER GAS-KINETIC CONDITIONS—IONIC MOBILITIES AND IONIC REACTIONS

- 1. Introduction — 1932
- 2. The behaviour of positive ions in gases—the mobility of positive ions — 1933
  - 2.1. Introductory remarks — 1933
  - 2.2. Techniques for measuring mobilities of unclustered positive ions — 1936
    - 2.2.1. The electrical-shutter method — 1936
    - 2.2.2. Hornbeck's method — 1938
    - 2.2.3. The pulse method of Biondi and Chanin — 1942
    - 2.2.4. Parallel-plate method — 1944
    - 2.2.5. Mobilities derived from electron decay measurements in afterglows — 1951
    - 2.2.6. Cyclotron-resonance method — 1954

CONTENTS

2.3. Observed mobilities for the alkali-metal ions in rare gases	1957
2.4. The theory of the mobility of ions in non-reacting gases	1958
2.5. Mobilities of alkali-metal ions in molecular gases	1964
2.6. Mobilities of NO^+ ions in He, Ar, H_2, and N_2	1965
3. Reactions involving positive ions	1966
3.1. Cluster formation	1966
3.1.1. The clustering of water molecules	1966
3.1.2. The appearance of clustered alkali ions in pure rare gases	1968
3.1.3. Theory of ion clustering	1968
3.2. The mobilities of ions in their parent gases—the effect of charge exchange and molecular ion cluster formation	1972
3.2.1. The mobility of helium ions in helium—historical account	1972
3.2.2. The mobility of helium ions in helium—experimental techniques and results at 300° K—drift-tube measurements	1975
3.2.3. The mobility of helium ions in helium—experimental techniques and results at 300° K—afterglow observations	1982
3.2.4. The mobility of helium ions in helium—experimental techniques and results at 300° K—determination of atomic to molecular ion conversion rate by measurements of the optical emission from an afterglow	1987
3.2.5. The mobility of helium ions in helium—experimental results at temperatures below 300° K	1990
3.2.6. The mobilities of other rare-gas ions in their parent gases	1994
3.2.7. The mobility and reactions of Hg^+ ions in mercury vapour	1996
3.3. The theory of the effect of charge transfer on the mobilities of ions in their parent gases	1997
3.3.1. Applications to H^+ in H and D^+ in D	1999
3.3.2. Application to He^+	2000
3.3.3. Application to other atomic ions in their parent gases	2001
3.3.4. Mobilities of diatomic ions in their parent (atomic) gases	2002
3.4. Mobilities of positive ions in their parent (diatomic) gases—ionic reactions in general—experimental methods	2003
3.4.1. Use of a mass spectrograph without a drift tube	2004
3.4.2. The associated drift tube and mass-spectrometer technique	2012
3.4.3. Drift tube combined with inlet and exit mass analysis	2015
3.4.4. Static afterglow studies using mass spectrographs	2018
3.4.5. The flowing-afterglow method	2021
3.5. Results of mobility and ionic reaction observations	2028
3.5.1. Hydrogen ions in hydrogen	2028
3.5.2. Nitrogen ions in nitrogen	2036
3.5.3. Ions in oxygen and in helium–oxygen mixtures	2043
3.5.4. Ions in nitrogen–helium mixtures	2051
3.5.5. Other reactions of He^+ ions	2052
3.5.6. Ions in nitrogen–oxygen mixtures	2052

3.5.7. Reactions of nitrogen and of oxygen ions with nitric oxide   2060
3.5.8. Summary of results for reactions of helium, nitrogen, and oxygen ions with N_2, O_2, NO, N, and O   2061
3.5.9. Reactions of nitrogen and of oxygen ions with CO and CO_2 2062
3.5.10. Reactions of C^+, CO^+, and CO_2^+ with O_2 and CO_2 2063
3.5.11. Reactions of Ar^+ ions   2063

4. Reactions involving negative ions   2064
 4.1. Mobility of negative ions in oxygen   2065
 4.2. Detachment of electrons from negative ions in collisions with gas molecules   2083
  4.2.1. Introduction   2083
  4.2.2. Adaptation of the pulse method   2085
   4.2.2.1. Analysis of the low-pressure data in O_2 2088
   4.2.2.2. Analysis of the high-pressure data in O_2 2091
   4.2.2.3. Experimental results in O_2 and their interpretation   2092
   4.2.2.4. Attachment and detachment rates in oxygen at high F/p—detachment from O^- 2096
  4.2.3. The rates of associative detachment reactions   2101
 4.3. Charge transfer reactions involving negative ions at thermal energies   2105
 4.4. The formation of complex negative ions   2106
  4.4.1. Negative ions in CO_2 and in O_2–CO_2 mixtures   2106
  4.4.2. Ions in H_2O and in O_2–H_2O mixtures   2110
  4.4.3. Ions in O_2–N_2 mixtures   2113
 4.5. Notes on the theory of detachment reactions   2113

AUTHOR INDEX   *1*

SUBJECT INDEX   *12*

16

COLLISIONS BETWEEN ATOMS UNDER GAS-KINETIC CONDITIONS—ELASTIC COLLISIONS

1. Introduction—classification of possibilities

WE now begin the consideration of impact phenomena in which both colliding systems are of atomic mass. In dealing with electron impacts it was convenient to discuss collisions involving electrons of homogeneous energy before going on to consider phenomena in which electron swarms are concerned. We adopt the opposite order for atomic collisions. This is partly because of the rich variety of phenomena involved and partly because advantage may be taken of the highly developed gas-kinetic theory.

Collisions between atomic or molecular systems include a number of additional possibilities which do not arise in electron impacts. In general, we still have the distinction between elastic and inelastic collisions, according as energy is or is not exchanged between the relative translational motion and internal motion of either or both colliding systems. However, inelastic collisions between atoms or molecules may involve a change of the electronic, vibrational, and/or rotational states of either or both colliding systems. The latter possibility, that the internal motion of both may change, does not occur with electron impacts.

We may classify the possibilities according to the scheme in Table 16.1. The colliding systems are denoted by A and B or, if the molecular character needs to be emphasized, by AC and BD. Electronic excitation is denoted by a ' as in A', that of vibration and/or rotation by a *, and the change of kinetic energy of relative translation, or energy discrepancy, by ΔE. Some physical phenomena in which the particular kind of collision is important under gas-kinetic conditions are also listed in the table. It will be appreciated that this can only be so if the kinetic energy of relative motion is increased by the collision, or if it is decreased by an amount comparable with the temperature energy $\frac{3}{2}\kappa T$.

There is one point about the classification which might be raised. If in the transfer collisions (k) or (m) the translational energy change is

TABLE 16.1

Summary of atomic and molecular collision processes

General nature of collision	Detailed description	Symbolic representation	Examples of physical phenomena in which effective
Elastic		$A+B \to A+B$	(a) Viscosity, diffusion, and thermal diffusion of gases. Mobility of positive ions in gases
Inelastic, in which one colliding system does not change its state of internal motion	Electronic excitation and de-excitation involving:		
	Change of term or configuration	$A+B' \to A+B+\Delta E$	(b) Quenching of resonance radiation. Deactivation of metastable atoms by atom impact
	Change of magnitude of total electronic angular momentum only		(c) Sensitized fluorescence—optical pumping
	Change of orientation of total electronic angular momentum only		(d) Depolarization of resonance radiation—line widths in double resonance
	Spin exchange only		(e) Line widths in electron and nuclear magnetic resonance
	Spin reversal		(f) Depolarization of oriented alkali metal atoms by foreign gases
	Ionization	$A+B \to A+B^+ +e - \Delta E$	(g)
	Detachment	$A+B^- \to A+B+e - \Delta E$	(h)
	Excitation of vibration and/or rotation	$A+BD \to A+BD^* - \Delta E$	(i) Dispersion of high-frequency sound. Rate of unimolecular reactions
	Electronic and vibrational excitation	$A+BD \to A+BD^* - \Delta E$	(j)

TABLE 16.1 (cont.)

General nature of collision	Detailed description	Symbolic representation	Examples of physical phenomena in which effective
Inelastic, in which both colliding systems change their state of internal motion	Transfer of electronic excitation	$A' + B \rightarrow A + B' \pm \Delta E$	(k) Quenching of resonance fluorescence. Sensitized fluorescence
		$A' + B \rightarrow A + B'^+ + e \pm \Delta E$	(l) Excitation of spark lines by foreign gases
	Charge transfer	$A^+ + B \rightarrow A + B^+ \pm \Delta E$	(m)
		$A^+ + B \rightarrow A + B'^+ \pm \Delta E$	(n) Excitation of spark lines by foreign gases
		$A^+ + B^- \rightarrow A' + B' \pm \Delta E$	(o) Mutual neutralization of positive and negative ions
	Transfer of vibration or rotation	$AC^* + BD \rightarrow AC + BD^* \pm \Delta E$	(p)
	Interchange of electronic and vibrational excitation	$AC' + BD \rightarrow AC + BD^* \pm \Delta E$	(q) Photochemical phenomena
		$A' + BD \rightarrow A + B + D \pm \Delta E$	(r) Photochemical phenomena
		$A' + BD \rightarrow AB + D \pm \Delta E$	(s) Associative ionization. The inverse process is one of electron-ion recombination
		$A' + B' \rightarrow AB^+ + e \pm \Delta E$	(t)
	Atomic rearrangement collisions	$AC + BD \rightarrow AB + CD \pm \Delta E$	(u) Chemical reactions
	Ionic reactions	$A^\pm + BC \rightarrow AB^\pm + C \pm \Delta E$	(v)
	Associative detachment	$A + B^- \rightarrow AB + e \pm \Delta E$	(w)
	Mutual neutralization	$A^+ + B^- \rightarrow A' + B'' \pm \Delta E$	(x)

zero, should they be regarded as elastic or inelastic? If the atoms A and B are identical there is no means of determining experimentally whether there has been any transfer of excitation or of charge. It is strictly correct then, to consider the transfer processes as contributing to the elastic collisions. In dealing, for example, with the diffusion of metastable He atoms or of He^+ in He, this must be allowed for (see Chap. 18, § 5.2 and Chap. 19, § 3.2). If the atoms are not identical but isotopic the energy discrepancy may be very small, but it is finite and the energy transfer process may be distinguished. In this case it is best to consider the process as it really is, an inelastic one.

We now proceed to separate discussion of the different physical phenomena in their relation to the particular collision processes. In this chapter we deal with elastic collisions between neutral systems. Collisions involving excitation of molecular vibration and rotation form the subject of the next chapter, which also includes a brief discussion of chemical reactions as studied by crossed molecular beam techniques. The following chapter deals with collisions, again between neutral systems, in which electronic transitions occur. Collisions in which one of the systems is charged form the subject matter of Chapter 19. This concludes the discussion of collisions under gas-kinetic conditions except for those parts of Chapter 20, in the next volume, which deal with recombination between ions.

2. General nature of the interaction between atoms

We shall give here a brief summary of the general nature of atomic interactions as indicated from theory as this is important for the interpretation of many collision phenomena with which we are concerned in this chapter.

We consider first two atoms in their ground states with completely filled outer shells, such as two rare-gas atoms. At large separations r an attractive interaction,

$$V(r) \sim -C/r^6, \tag{1}$$

exists which is usually known as the van der Waals attraction. This arises from a dynamic polarization of one atom by the other, the electrons tending to revolve in their orbits in the respective atoms in such a relative phase that their average separation over a period of time is as large as possible. Various approximate theoretical methods† have been given for the estimation of C for a given pair of atoms, in terms of such

---

† MOTT, N. F. and SNEDDON, I. N., *Wave mechanics and its applications*, § 32 (Clarendon Press, Oxford, 1948); see also § 9.3.1 of this chapter.

properties as their polarization. These methods are not very accurate, particularly for complex atoms.

The expression (1) is really the first term in an asymptotic series and, as the atoms approach, additional attractive terms varying as r^{-8}, r^{-10}, etc., should be included. It is doubtful how useful the addition of such terms is, because, for values of r at which they are becoming appreciable, the chemical (or intrinsic) forces are also coming into play. These forces are strongly repulsive when the atoms have completely filled outer shells and fall off exponentially with distance. They arise from a combination of direct Coulomb forces, averaged over the charge distribution of the atoms, with forces arising from the exchange of electrons between the atoms† (see Chap. 12, § 1).

The combination of the van der Waals and chemical forces gives an interaction of the shape illustrated in Fig. 16.1. Gas-kinetic studies at ordinary temperatures yield information mainly about the depth and shape of the attractive potential where the minimum energy is of the order κT. They can also give information about the slope of the repulsion. Detailed prediction of atomic interactions at separations effective in gas-kinetic collisions has been confined mainly to the simplest atoms, such as helium, and even then it is not certain how the different contributions to the interaction should be combined at separations near the potential minimum. For trial purposes, in attempting to derive the interaction from experimental results, an interaction of the form

$$\lambda r^{-s} - \mu r^{-t} \qquad (2)$$

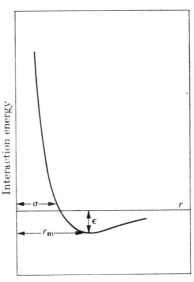

Fig. 16.1. Variation of the interaction energy with distance r between the nuclei of two atoms, each of which has a completely filled outer shell.

is often used, the first term representing the chemical repulsion, the second the van der Waals attraction. The four parameters λ, μ, s, and t can then be chosen to give the best agreement with observed data. The repulsive term may be replaced by the exponential form $Ae^{-r/a}$ which

† Ibid., § 33.

receives some justification from the quantum theory of the valence force. Examples of the use of various empirical expressions which give the correct general shape for the interaction will be given below (see §§ 6.1, 10, 12, and 13).

The only essential modification which is introduced if one of the atoms is replaced by an ion such as Li$^+$ is the addition of an attraction of the form

$$V \sim -\tfrac{1}{2}\alpha\epsilon^2/r^4, \tag{3}$$

due to the polarization of the atoms by the ionic charge, α being the polarizability of the atom. α is given in terms of the dielectric constant K of the gas formed by the atoms at n.t.p. by the relation

$$\alpha = (K-1)/8\pi N, \tag{4}$$

where N is the number, $2\cdot 7 \times 10^{19}$, of atoms per cm^3 under these conditions. Some uncertainty arises in these cases as to when the polarization saturates, i.e. at what values of r the interaction (3) ceases to increase as r is reduced. If this does not occur before the chemical repulsion dominates the situation there is no difficulty but, if the polarization saturates at greater distances, the addition of the term (3) may seriously overestimate the depth of the potential minimum.

For two rare-gas atoms, with completely filled outer shells, i.e. atoms in ^1S states, there is a unique interaction of the form we have described. The same holds for interaction between rare-gas atoms and other atoms in ^2S or ^3S states except in the latter cases when the atoms are of the same kind (as for example He1^1S and He2^3S).

In all other cases the interaction will not be unique, depending on the relative orientation of spin and of orbital angular momentum in the interacting atoms (see Chap. 12, Tables 12.3 and 12.4). Thus two atoms in ^2S states can interact in quite different ways depending on whether the unpaired electron spins in each atom are parallel or antiparallel. In particular, for two H atoms the triplet interaction is of the same form as that between rare-gas atoms but the singlet one, in which the electron spins are antiparallel, exhibits a strong chemical attraction. As a result of this the interaction, while of the same general shape as in Fig. 16.1, has an attraction with depth 4·4 eV as compared with a fraction of an electron-volt. The separation r_m at the minimum is also considerably reduced. Chemical attraction of this kind arises in many other cases in which the interaction is not unique. In addition, if the atoms are not both in S states, the van der Waals interaction will depend on the orientation of the angular momentum in either or both atoms.

Finally, we must draw attention to the special features associated with the interaction of an atom with an ion, or excited atom, of its own kind, for example of He with He⁺ or of He with He'. In such cases there are two possible interactions, one of which arises when the electron distribution is symmetrical, the other when it is antisymmetrical with respect to the nuclei. These two interactions may be represented for many purposes in the form

$$V = V_0(r) \pm U(r), \qquad (5)$$

the alternative signs for $U(r)$ corresponding to the two cases. $V_0(r)$ has much the same form as the interaction between normal atoms. In general, $U(r)$ is of such magnitude as to produce, in one of the two cases, an attractive energy of a few electron-volts, much greater than that due to the van der Waals forces. For an ion and an atom, both in their ground states, $U(r)$ falls off exponentially with r, though more slowly than the repulsive part of $V_0(r)$. An example is provided by the interactions between H and H⁺ illustrated in Chapter 13, Fig. 13.39. Important cases where this effect must be taken into account are discussed in Chapter 19, §§ 3.2 and 3.3. On the other hand, for a normal and excited atom $U(r)$ falls off as r^{-p} if the transition from the excited to the ground state is associated with an electric $(p-1)$-pole moment. Thus for a normal He atom and one in a p state $U(r)$ falls off as r^{-3}, both interactions being thus of very long range. If the transition is associated with no moment, i.e. is of s–s type, $U(r)$ again falls off exponentially though more slowly than $V_0(r)$. The same applies if the transition involves electron exchange between the two atoms, for example in the interaction between a normal and a metastable helium atom in the $2\,^3S$ state. The matter is further discussed in Chap. 18, § 8.1, in which the existence of the two interactions is related to the phenomena of transfer of excitation or of charge on collision between a normal and excited atom.

The above considerations apply strictly only if one of the atoms or ions concerned possesses a completely filled outer shell of electrons. In many of the cases that we shall be discussing below this condition is fulfilled. If it is not, more possibilities arise than we have discussed above, including in particular the occurrence of attractive chemical forces between neutral atoms (see § 12).

3. Elastic collisions of gas atoms—general discussion

3.1. *Cross-sections effective in scattering, viscosity, and diffusion*

The definitions of total and differential elastic collision cross-sections given in Chapter 1, § 2 for electron impacts apply equally well to collisions between atoms or molecules. It is necessary only to allow for the fact that the colliding systems are now of comparable mass. The motion of the centre of mass must be separated out so that the angle of scattering θ is the angle the direction of relative motion is turned through by the collision.

In dealing with electron collisions we introduced, in addition to the differential and total cross-sections $I(\theta)$ and Q_0, a momentum transfer or loss cross-section defined by

$$Q_d = 2\pi \int_0^\pi (1-\cos\theta) I(\theta) \sin\theta \, d\theta. \tag{6}$$

This cross-section is the appropriate one for dealing with electron diffusion phenomena. In the same way the diffusion coefficient for a mixture of gases depends on the momentum-transfer cross-section defined by (6), θ having the same significance as above.

According to the theory of Chapman† and Enskog‡ the coefficient of diffusion is given by

$$D = \frac{3\pi^{\frac{1}{2}}}{16(n_1+n_2)P_{12}^d} \left\{ \frac{2(M_1+M_2)}{M_1 M_2} \kappa T \right\}^{\frac{7}{2}} (1-\epsilon_0)^{-1}, \tag{7}$$

where

$$P_{12}^d = \int_0^\infty v^5 Q_d(1,2) \exp\left\{-\frac{1}{2} \frac{M_1 M_2 v^2}{\kappa T(M_1+M_2)}\right\} dv. \tag{8}$$

M_1, M_2 are the masses of the atoms of the respective gases, and n_1, n_2 their concentrations in number of atoms/cm³. $Q_d(1,2)$ is the cross-section (6) for collision of an atom of each kind with relative velocity v. T is the absolute temperature. ϵ_0 is a correction depending on the nature of the collisions and the concentrations n_1, n_2. It is never greater than 0·136 and can generally be ignored.

If the diffusion cross-section is constant or is replaced by a suitable mean value Q_d the formula (7) gives

$$D = \frac{3}{8}\left(\frac{\pi\kappa T(M_1+M_2)}{2M_1 M_2}\right)^{\frac{1}{2}} \frac{1}{(n_1+n_2)Q_d}.$$

Since the mean relative velocity of two gas molecules is given by

$$\bar{v} = \left\{\frac{8\kappa T(M_1+M_2)}{\pi M_1 M_2}\right\}^{\frac{1}{2}}$$

we have

$$D = \frac{3\pi}{32} \frac{\bar{v}}{(n_1+n_2)Q_d}.$$

† CHAPMAN, S., *Phil. Trans. R. Soc.* **A216** (1916) 279; **217** (1916) 115.
‡ ENSKOG, D., Inaugural dissertation (Uppsala, 1917).

When $n_1 \ll n_2$, as for diffusion of electrons in a gas, we have

$$D = \frac{3\pi}{32} \frac{\bar{v}}{n_2 Q_d} \simeq \frac{\bar{v}}{3n_2 Q_d},$$

which is the expression used in Chapter 5, § 2.1 (9).

The mobility μ of an ion in a pure gas at specified temperature and pressure is given by the relation

$$\mu = eD/\kappa T, \tag{9}$$

where D is given by the formula (7). It is thus determined, as for electrons, by the momentum-transfer cross-section for collisions between an ion and a gas atom.

A further cross-section Q_η arises in the theory of the viscosity η and heat conductivity σ of gases.† According to the theory of Chapman‡ and Enskog§ the coefficient of viscosity η of a gas at temperature T is given by

$$\eta = \frac{20\kappa^3 T^3}{M^2} \left(\frac{4\pi\kappa T}{M}\right)^{\frac{3}{2}} \frac{1+\epsilon}{\pi R_{11}^\eta}, \tag{10}$$

where M is the mass of a gas atom and ϵ is a correction which is never greater than 0·017. R_{11}^η is given by

$$R_{11}^\eta = \tfrac{1}{2} \int_0^\infty v^7 Q_\eta \exp\{-Mv^2/4\kappa T\} \, dv, \tag{11}$$

where

$$Q_\eta = 2\pi \int_0^\pi I(\theta) \sin^3\theta \, d\theta. \tag{12}$$

This cross-section gives greater weight to large angle deviations than Q_d, as would be expected from the following considerations. The greater the rate at which energy is equalized by sharing in collisions the smaller the viscosity and heat conductivity. If energy is equalized after impact of two similar molecules the angle of scattering is 90°. Collisions for which $\theta \simeq 90°$ are therefore more effective in stifling conductivity than any others. Hence the weighting by $\sin^2\theta$ in (12). On the other hand, diffusion is most retarded by collisions in which back-scattering occurs, i.e. $\theta \simeq 180°$. This is allowed for in (6) by the weighting factor $\sin^2 \tfrac{1}{2}\theta$.

† These are connected, for monatomic gases, by the relation

$$\sigma = (1+\delta)\tfrac{5}{2} c_v \, \eta, \tag{13}$$

where δ is a small correcting term and c_v is the specific heat at constant volume (see KENNARD, E. H., *Kinetic theory of gases*, p. 179 (McGraw-Hill, New York, 1938)).
‡ CHAPMAN, S., *Phil. Trans. R. Soc.* A**216** (1916) 279; **217** (1916) 115.
§ ENSKOG, D., Inaugural dissertation (Uppsala, 1917).

Another transport phenomenon of interest is that of thermal diffusion.† Under the influence of a temperature gradient a composition gradient is set up in a gaseous mixture. Thus for a binary mixture, if n, n_1 are the respective concentrations of all molecules and of those of species 1 only, then

$$\operatorname{grad}(n_1/n) = -k_T \operatorname{grad}(\ln T), \tag{14 a}$$

where T is the temperature and k_T is a coefficient, dependent on the temperature and on the nature of the gases, which is known as the *thermal diffusion ratio*. It depends on the diffusion and viscosity cross-sections for collisions between molecules of the same, as well as different, species in the mixture. The temperature dependence of k_T is mainly determined by that of $(6c-5)$ where

$$c = P_{12}^{\mathrm{d}}/R_{12}^{\mathrm{d}}, \tag{14 b}$$

P_{12}^{d} being given by (8) and R_{12}^{d} by (11) with Q_η replaced by Q_{d} and M by $M_1 M_2/(M_1+M_2)$.

Even with an extended range of interaction between gas atoms the cross-sections Q_{d} and Q_η have finite values on the classical as well as on the quantum theory. This is because of the reduced emphasis on small angle deviations. In most gas-kinetic phenomena classical methods give results of sufficient accuracy in the theory of viscosity and diffusion. The only exceptions are the lightest gases at very low temperatures. On the other hand, the total elastic cross-section Q_0 is *never* given correctly on classical theory. For an extended range of interaction it gives an infinite value whereas the quantum theory normally gives a finite value. Even for rigid spherical atoms of radius $\tfrac{1}{2}a$ for which the classical value of Q_0 is πa^2, the quantum theory gives a different value for *all* velocities of relative motion. Thus for all collisions between atoms Q_0 can only be calculated from quantum theory. We shall now examine these questions in a little more detail.

4. Quantal and classical cross-sections for rigid spherical atoms

The collision between two rigid spherical atoms of radius $\tfrac{1}{2}a$ and of mass M may be regarded in the same way as the collision of a particle of mass $\tfrac{1}{2}M$ with an infinitely massive rigid spherical obstacle of radius a. For such a collision the interaction energy is given by

$$\left.\begin{array}{ll} V = 0 & (r > a) \\ \to \infty & (r \leqslant a) \end{array}\right\}. \tag{15}$$

† CHAPMAN, S., *Phil. Trans. R. Soc.* A**217** (1916) 115.

The classical cross-section is πa^2. The quantal formula is the same as that given in Chapter 6, § 3.1:

$$Q_0 = \frac{4\pi}{k^2} \sum_{l=0}^{\infty} (2l+1)\sin^2\eta_l. \tag{16}$$

The wave number $k = \frac{1}{2}Mv/\hbar$, where v is the relative velocity of the colliding atoms. η_l is the phase shift produced in the de Broglie waves for the relative motion associated with angular momentum $\{l(l+1)\}^{\frac{1}{2}}\hbar$.

The discussion of the formula (16) follows on similar lines to that of Chapter 6, § 3.4. We first suppose that the relative velocity is so low that ka is very small. Under these conditions those waves for which $l > 0$ will hardly be affected by the inaccessible region $r < a$. The wave with $l = 0$ is a sine wave which, because of this inaccessibility, must vanish at $r = a$ instead of at $r = 0$, i.e. a phase change $\eta_0 = -ka$ is produced. Hence, in the limit $k \to 0$,

$$Q_0 \to \frac{4\pi}{k^2} k^2 a^2 = 4\pi a^2, \tag{17}$$

four times the classical value.

A difference between classical and quantal values is to be expected when the wavelength λ is long compared with the dimensions of the scattering obstacle. However, it is not difficult to show that, even in the limit of very short wavelengths, the quantum value does not tend to πa^2 but to twice that value.

If λ is large all the phase shifts η_l for which $ka > l$ will be large. We may break up the sum in (16) into two parts so that

$$Q_0 = \frac{4\pi}{k^2} \left(\sum_{l=0}^{L} + \sum_{L}^{\infty} \right) (2l+1)\sin^2\eta_l, \tag{18}$$

with $L \simeq ka$. The second sum may be neglected because the phases η_l are very small for $ka < L$. In the first, $\sin^2\eta_l$ will fluctuate rapidly with l as, for $l < L$, $\eta_l \gg \pi$. It is therefore a good approximation to replace $\sin^2\eta_l$ by its mean value of $\frac{1}{2}$ and write

$$Q_0 \simeq \frac{1}{2} \frac{4\pi}{k^2} \sum_0^L (2l+1)$$

$$\simeq 2\pi L^2/k^2$$

$$= 2\pi a^2. \tag{19}$$

A more accurate analysis confirms this result.

In order to analyse the matter further it is instructive to examine, under the same conditions, the difference between the classical and quantal forms for the differential cross-sections. With classical mechanics the probability of scattering per unit solid angle is independent of angle in this case, so that

$$I(\theta) = \tfrac{1}{4}a^2. \tag{20}$$

The quantum expression for $I(\theta)$ has been given in Chapter 6, § 3.2, and is

$$I(\theta) = \frac{1}{4k^2}\left|\sum\{\exp(2i\eta_l)-1\}(2l+1)P_l(\cos\theta)\right|^2. \tag{21}$$

At $\theta = 0$, $P_l(\cos\theta) = 1$ independent of l, so

$$I(0) = \frac{1}{4k^2}\left\{\left|\sum(2l+1)(\cos 2\eta_l-1)\right|^2 + \left|\sum(2l+1)\sin 2\eta_l\right|^2\right\}.$$

We may now adopt the same procedure as for Q_0 by summing up to $l = L$ and ignoring all higher terms. Furthermore, because of interference, the second sum will be negligible compared with the first: $\sin 2\eta_l$ will fluctuate violently in sign whereas $\cos 2\eta_l - 1$ is one-signed and has a mean value of -1. This gives (see Chap. 8 (22))

$$I(0) \simeq L^4/4k^2 = \tfrac{1}{4}k^2a^4 = kQ_0/4\pi. \tag{22}$$

This is k^2a^2 times greater than the classical value at that angle, the ratio tending to infinity as $\lambda \to 0$. The large value arises from the coherence of the harmonics which all have the same sign at $\theta = 0$ (see Fig. 6.1).

As θ increases first P_L and then successively lower-order harmonics change sign and produce interference so that the value of $I(\theta)$ rapidly falls. For large values of θ it may be shown† that the classical expression (20) for $I(\theta)$ holds. As θ decreases this will remain approximately true until the coherence effect becomes pronounced. This can be taken as the angle for which P_L has its first zero, i.e. approximately $\pi/L = \pi/ka$. This gives the schematic form of $I(\theta)$, according to quantum theory, which is illustrated in Fig. 16.2 (b), namely

$$\begin{aligned}I(\theta) &= \tfrac{1}{4}a^2\{k^2a^2-(k^2a^2-1)(ka\theta/\pi)\} & (\theta < \pi/ka) \\ &= \tfrac{1}{4}a^2 & (\theta > \pi/ka)\end{aligned}. \tag{23}$$

With this rough approximation we have, for large ka,

$$2\pi\int_{\pi/ka}^{\pi} I(\theta)\sin\theta\,d\theta \simeq \pi a^2, \qquad 2\pi\int_0^{\pi/ka} I(\theta)\sin\theta\,d\theta \simeq \pi a^2,$$

† MOTT, N. F. and MASSEY, H. S. W., *The theory of atomic collisions*, 3rd edn, chap. v (Clarendon Press, Oxford, 1965).

indicating that the doubling of the classical cross-section comes from the increased scattering at small angles. As λ decreases the angular range over which the classical formula is valid increases, but the quantum excess over the classical value at very small angles also increases so that the cross-section remains doubled.

A more accurate analysis confirms this description. Fig. 16.2 (a) illustrates the function $I(\theta)$ calculated by Massey and Mohr† without approximation for the case $ka = 20$, showing that the schematic representation (23) reproduces the main features.

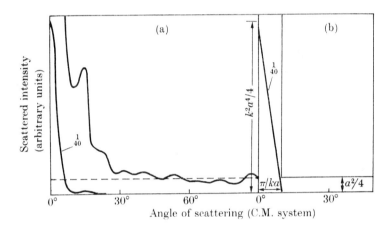

Fig. 16.2. Angular distributions of atoms for the rigid sphere model ($ka = 20$). (a) --- classical distribution; —— distribution calculated by quantum theory. (b) Schematic angular distribution.

The doubling of the total cross-section by the wave theory was first pointed out by Massey and Mohr† and a detailed discussion in terms of diffraction theory has been given by Wergeland.‡

As the diffusion and viscosity cross-sections Q_d and Q_η attach much less weight to small-angle deviations it is not surprising that in the short wavelength limit the quantal and classical formulae for these cross-sections agree. Fig. 16.3 illustrates the variation of Q, Q_d, and Q_η with relative velocity according to the quantum theory. The temperature scales indicate the range in which quantum effects are likely to be important for helium. They are calculated for an atomic radius

$$\tfrac{1}{2}a = 1{\cdot}05 \text{ Å}.$$

† MASSEY, H. S. W. and MOHR, C. B. O., *Proc. R. Soc.* A**141** (1933) 434.
‡ WERGELAND, A., *Skr. norske Vidensk-Akad.* 1945, no. 9.

For Q_η and Q_d the corresponding relative velocity is that which is most effective in determining viscosity and diffusion phenomena at the temperatures indicated while for Q it is equal to the r.m.s. velocity at the temperature. It will be seen that, below 80 °K, the quantum theory must be used. For heavier gases, on the other hand, quantum effects are negligible at temperatures down to the liquefying points.

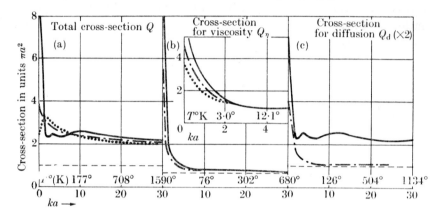

FIG. 16.3. Cross-sections calculated using the rigid sphere model. (a) Total collision cross-section, Q. (b) Cross-section for viscosity, Q_η. (c) Cross-section for diffusion, Q_d ($\times 2$): --- classical; —·—·— quantum theory for dissimilar atoms; ——— quantum theory for identical atoms obeying the Bose–Einstein statistics; ···· quantum theory for identical atoms obeying Fermi–Dirac statistics.

4.1. *The effect of symmetry*

In considering collisions of similar atoms a further quantal effect is introduced. This arises from the need to satisfy certain symmetry requirements.† Thus, if the atoms obey Bose–Einstein statistics, the wave function representing the relative motion must be unchanged when the atoms are interchanged. This requires that, in the formula for the differential cross-section, all harmonics of odd order must be excluded. This will be so for normal helium atoms with nuclei ⁴He. On the other hand, for collisions between helium atoms possessing ³He nuclei, which obey Fermi–Dirac statistics, a different modification is introduced. For this case

$$I(\theta) = \tfrac{3}{4}I_a(\theta) + \tfrac{1}{4}I_s(\theta), \tag{24}$$

where I_a includes only the odd-order harmonics, I_s only the even order.

The effect of these symmetry requirements on the cross-sections Q_η, Q_d, and Q, for the rigid sphere case, is illustrated in Fig. 16.3.

† For a detailed discussion see MOTT, N. F. and MASSEY, H. S. W., *The theory of atomic collisions*, 3rd edn, chap. xi (Clarendon Press, Oxford, 1965).

5. Cross-sections for extended range interactions

For actual atoms the interaction energy has not the form (15) but is finite at all finite values of r. As, however, it exhibits a steeply rising repulsion for $r < r_0$ (see Fig. 16.1), a number of the general conclusions derived from the study of the rigid-sphere model remain true. Thus the classical theory may be used for the discussion of viscosity and diffusion except for helium at low temperatures. It is never valid for the determination of total collision cross-sections. In fact with the form (2) for V no classical value exists. The quantum theory predicts an angular distribution of the form (23) where the range a is of the order r_0. This may be made use of in designing experiments to measure Q as it makes possible an estimate of the angular resolving power required to give a result effectively independent of the geometry of the apparatus (see § 7.2).

5.1. *Classical formulae*

Let p be the impact parameter corresponding to a particular classical orbit of relative motion of the colliding systems. The classical differential cross-section for collisions in which the impact parameter lies between p and $p+\mathrm{d}p$ is $2\pi p\,\mathrm{d}p$, so the usual differential cross-section $I_{\mathrm{cl}}(\theta)\,\mathrm{d}\omega$ for collision in which the direction of relative motion is deflected through an angle θ into the solid angle $\mathrm{d}\omega$ is given by

$$2\pi I_{\mathrm{cl}}(\theta)\sin\theta\,\mathrm{d}\theta = 2\pi p\frac{\mathrm{d}p}{\mathrm{d}\theta}\mathrm{d}\theta, \tag{25}$$

so that

$$I_{\mathrm{cl}}(\theta) = \frac{p}{\sin\theta}\frac{\mathrm{d}p}{\mathrm{d}\theta}. \tag{26}$$

We must therefore examine the relation between p and θ.

According to classical orbit theory the angle of deflexion ϑ is given in terms of the impact parameter p by

$$\vartheta = \pi - 2\int_{r_0}^{\infty}\frac{\mathrm{d}r}{r\phi(r)}, \tag{27}$$

where

$$\phi(r) = \left\{\frac{r^2}{p^2} - 1 - \frac{r^2}{p^2 E}V(r)\right\}^{\frac{1}{2}}. \tag{28}$$

In this form E is the kinetic energy of relative motion before and after the collision. $V(r)$ is the interaction energy between the atoms when at a distance r apart and r_0 is the positive zero of the denominator, or the largest such zero if more than one exists. A proof of this result is given by Mott and Massey in *The theory of atomic collisions*, 3rd edn, chapter v.

Before proceeding further we note that (27) does not necessarily give the desired relation between θ and p. The angle of deflexion is not necessarily confined to values between 0 and π as is the angle of scattering. Furthermore, while the angle of deflexion ϑ is a single-valued function of p the reverse is not necessarily true. This may well occur even if ϑ lies between 0 and π. We now have the following cases to consider.

If p is a single-valued function of ϑ and $0 \leqslant \vartheta \leqslant \pi$ then (27) applies with

$$\theta = \pi - 2 \int_{r_0}^{\infty} \frac{\mathrm{d}r}{r\phi(r)}. \tag{29}$$

If, for a given value of ϑ, $0 \leqslant \vartheta \leqslant \pi$, there exist several solutions $p_1, p_2, ..., p_s$ for p, then

$$I_{\mathrm{cl}}(\theta) = \frac{1}{\sin\theta}\left\{p_1\left(\frac{\mathrm{d}p}{\mathrm{d}\theta}\right)_1 + p_2\left(\frac{\mathrm{d}p}{\mathrm{d}\theta}\right)_2 + \ldots + p_s\left(\frac{\mathrm{d}p}{\mathrm{d}\theta}\right)_s\right\}. \tag{30}$$

This may be extended to cases in which, for $0 \leqslant |\vartheta| \leqslant \pi$, there exist solutions $p_1, p_2, ..., p_s$ to give

$$I_{\mathrm{cl}}(\theta) = \frac{1}{\sin\theta}\left\{p_1\left|\left(\frac{\mathrm{d}p}{\mathrm{d}\theta}\right)_1\right| + p_2\left|\left(\frac{\mathrm{d}p}{\mathrm{d}\theta}\right)_2\right| + \ldots + p_s\left|\left(\frac{\mathrm{d}p}{\mathrm{d}\theta}\right)_s\right|\right\}. \tag{31}$$

5.1.1. *Orbiting.* If $|\vartheta|$ exceeds π the relative position vector makes more than a complete revolution during the collision and orbiting is said to occur. To examine this possibility a little further it is convenient to work in terms of the classical relative angular momentum $J = Mvp$ where M is the reduced mass. We may write (27) in the form

$$\vartheta = \pi - \frac{2J}{Mv}\int_{r_0}^{\infty}\frac{\mathrm{d}r}{r^2(1-V_{\mathrm{eff}}/E)^{\frac{1}{2}}}, \tag{32}$$

where
$$V_{\mathrm{eff}} = V + J^2/2Mr^2 \tag{33}$$

and is the effective potential for the relative radial motion so that the radial velocity is given by

$$\tfrac{1}{2}Mv_r^2 = E - V_{\mathrm{eff}}. \tag{34}$$

If the conditions are such that v_r is small over an appreciable accessible region of r then there exists the possibility that the angular motion will cause the relative position vector to make more than one revolution during passage through this region of r. In particular, if at the closest distance of approach r_0 with relative angular momentum J,

$$(\mathrm{d}V_{\mathrm{eff}}/\mathrm{d}r)_{r=r_0} = 0 \tag{35}$$

then ϑ will be logarithmically infinite for a collision with this relative angular momentum.

Such a condition will never arise if V is a positive monotonic function of r, i.e. a monotonic repulsion. If, however, V is attractive the possibility exists. Thus Fig. 16.4 illustrates the variation of V_{eff} with J in such a case. For sufficiently small J, V_{eff} will exhibit a maximum at some value of r_0.

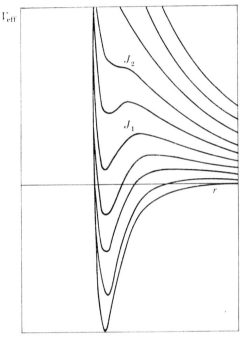

FIG. 16.4. Variation of the effective potential V_{eff} with r for a monotonic attractive field, for different values of the angular momentum J, which increases upwards.

The classical differential cross-section under these circumstances has been discussed by Ford and Wheeler† who show that the contribution from the neighbourhood of the singularity takes the form

$$I^0_{\text{cl}}(\theta) = J_1^2 (M^2 v^2 a \sin\theta)^{-1}\{e^{-(\theta+\vartheta_1)/a} + \tfrac{1}{2} e^{-(\theta+\vartheta_2)/2a} + \text{higher order terms}\}. \tag{36}$$

Here J_1 is the angular momentum for which (34) and (35) apply and

$$a = \left(\frac{2J_1^2}{Mr^4} \Big/ \frac{d^2 V_{\text{eff}}}{dr^2}\right)^{\frac{1}{2}}_{r=r_0}. \tag{37}$$

ϑ_1 and ϑ_2 are certain constants.

† FORD, K. W. and WHEELER, J. A., *Ann. Phys.* **7** (1959) 259.

In many applications orbiting does not occur but care must always be taken to check this in any particular case.

5.1.2. *The rainbow angle.* It remains to discuss two other singularities that may arise. The first is known as the rainbow effect because of its relation, in semi-classical theory, to the formation of rainbows by light scattering. It arises when the slope $d\vartheta/dp$ of the deflexion function vanishes. This may well occur when $0 \leqslant |\vartheta| \leqslant \pi$. In such a case I_{cl} as given by (26) must tend to infinity at the rainbow angle $\theta_r = |\vartheta_r|$. If p_r is the corresponding value of the impact parameter then, near p_r, we may write

$$\vartheta(p) = \vartheta_r + \alpha(p-p_r)^2, \tag{38}$$

so that the contribution from the rainbow singularity to the classical differential cross-section is, near θ_r,

$$I_{cl}^r = \frac{p_r}{2 \sin\theta \{\alpha|\vartheta-\vartheta_r|\}^{\frac{1}{2}}} \quad (\theta < \theta_r),$$
$$= 0 \quad (\theta > \theta_r). \tag{39}$$

5.1.3. *Glory scattering.* The remaining singularity arises when the angle of deflexion vanishes or is a multiple of π. This is known as the glory effect because of its association with certain meteorological phenomena. If $|\vartheta| \to 0$ or $2s\pi$, where s is integral, the singularity is in the forward direction, if to an odd multiple of π in the backward direction. Near a forward glory, occurring for an impact parameter p_g, we may write

$$\vartheta(p) = 2s\pi + \beta(p-p_g). \tag{40}$$

This gives two contributions I_{cl}^g to I_{cl} corresponding to $\theta = \vartheta - 2s\pi$, for $\beta(p-p_g) > 0$, and $\theta = 2s\pi - \vartheta$, for $\beta(p-p_g) < 0$. For both

$$\left|\frac{dp}{d\theta}\right| = \frac{1}{|\beta|},$$

$$I_{cl}^g(\theta) = \frac{2}{\sin\theta} \frac{p_g}{|\beta|}. \tag{41}$$

A similar analysis applies to a backward glory.

5.1.4. *Example.* As an illustration consider the classical scattering corresponding to a deflexion function of the form shown in Fig. 16.5 (a) which is characteristic of collisions between gas atoms at ordinary temperatures with interaction potentials of the general form shown in Fig. 16.1.

It will be seen that glory and rainbow singularities occur for $p = p_g, p_r$ respectively and that for $\theta < |\vartheta_r|$, where ϑ_r is the rainbow angle, contri-

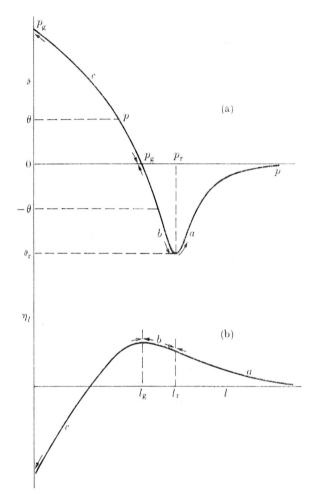

FIG. 16.5. (a) Typical shape of the deflexion function $\vartheta(p)$ for collisions between atoms interacting with potential energy functions of the general shape shown in Fig. 16.1. Branches a, b, c arise respectively from the long-range attraction, the attraction at medium ranges, and the short-range repulsion. p_g, p_r are the impact parameters associated with glory and rainbow singularities respectively. (b) Corresponding variation of the phase shift η_l with l $(= Mvp/\hbar)$.

butions to $I_{\rm cl}(\theta)$ come from three branches—p_r is a triple-valued function of θ in this range. This is more clearly seen in Fig. 16.6, which is a plot of p^2 as a function of $\cos\theta$. The three branches are indicated as a, b, and c respectively. They arise respectively from the long-range attraction, the attraction at medium ranges, and the short-range repulsion in terms of the general form of the interaction shown in Fig. 16.1.

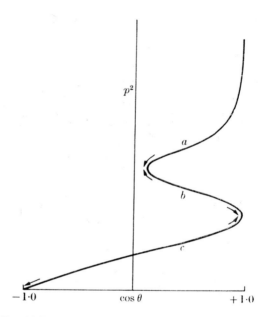

Fig. 16.6. Variation of p^2 with $\cos\theta$ corresponding to the deflexion function shown in Fig. 16.5.

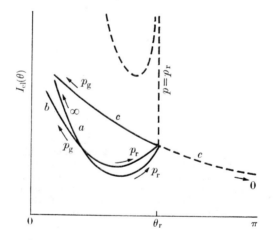

Fig. 16.7. --- Differential cross-section $I_{\text{cl}}(\theta)\,d\omega$ for collisions corresponding to the deflexion function shown in Fig. 16.5, according to the classical theory. ——— Contributions from the different branches a, b, c shown separately, the range of p being indicated in each case; as $\theta \to 0$ the contributions from c and b coalesce but the scattering is dominated by that from a.

Fig. 16.7 shows the contributions of the separate branches to I_{cl} obtained in terms of the formula (31). It is noteworthy that, near the glory singularity, b and c tend to coincidence, as do a and b near the rainbow singularity.

5.1.5. *Diffusion and viscosity coefficients.* Provided orbiting does not occur, classical formulae for the diffusion and viscosity coefficients D and η may be written down without difficulty, no special treatment being necessary for the other cases discussed above. We have

$$D = \frac{3}{64}\left\{\frac{\pi}{2}\frac{M_1 M_2}{(M_1+M_2)\kappa T}\right\}^{-\frac{1}{2}}\frac{(1-\epsilon_0)^{-1}}{J(n_1+n_2)}, \qquad (42)$$

$$\eta = \frac{5}{8}\left(\frac{M\kappa T}{\pi}\right)^{\frac{1}{2}}\frac{1+\epsilon}{I}, \qquad (43)$$

where

$$J = \int_0^\infty \int_0^\infty x^2 e^{-x} p \cos^2 \tfrac{1}{2}\alpha \, dp\, dx, \qquad (44\,\text{a})$$

$$I = \int_0^\infty \int_0^\infty x^3 e^{-x} p \sin^2 \alpha \, dp\, dx, \qquad (44\,\text{b})$$

$$\alpha = 2\int_{r_0}^\infty \frac{dr}{r\phi(r)}. \qquad (44\,\text{c})$$

ϵ_0, ϵ, are as in (7) and (10) respectively and ϕ as in (28).

5.1.6. *Small-angle scattering*

For large values of the impact parameter p the deflexion angle ϑ will be small in magnitude and may be obtained approximately as follows.† By means of the substitution $x = r_0/r$ the expression (27) for ϑ may be written

$$\vartheta = \pi - 2\int_1^0 \left[\frac{r_0^2}{p^2}\left\{1-\frac{V(r_0/x)}{E}\right\} - x^2\right]^{-\frac{1}{2}} dx \qquad (45\,\text{a})$$

with

$$1-\frac{V(r_0)}{E} = \frac{p^2}{r_0^2}. \qquad (45\,\text{b})$$

Eliminating p^2 from (45 a) by means of (45 b) we obtain

$$\vartheta = \pi - 2\int_1^0 \left[\left\{\frac{1-V(r_0/x)/E}{1-V(r_0)/E}\right\} - x^2\right]^{-\frac{1}{2}} dx.$$

† KENNARD, E. H., *Kinetic theory of gases*, p. 179 (McGraw-Hill, 1938).

When p is large $V(r_0/x)$ will be small throughout the range of integration so

$$\vartheta \simeq \pi - 2 \int_1^0 \left[\left\{1 - \frac{V(r_0/x)}{E}\right\}\left\{1 + \frac{V(r_0)}{E}\right\} - x^2\right]^{-\frac{1}{2}} dx$$

$$\simeq \pi - 2 \int_1^0 (1-x^2)^{-\frac{1}{2}}\left[1 + \frac{1}{2E}\left\{\frac{V(r_0/x) - V(r_0)}{1-x^2}\right\}\right] dx.$$

Since
$$\int_1^0 (1-x^2)^{-\frac{1}{2}} dx = \tfrac{1}{2}\pi,$$

$$\vartheta \simeq \frac{1}{E} \int_0^1 \frac{V(r_0) - V(r_0/x)}{(1-x^2)^{3/2}} dx. \qquad (46)$$

In particular, if
$$V(r) = -C_s r^{-s}, \qquad (47)$$

$$\vartheta = -\frac{C_s}{E} r_0^{-s} \int_0^1 \frac{1-x^s}{(1-x^2)^{3/2}} dx. \qquad (48)$$

If $|\vartheta| \ll 1$, $r_0 \simeq p$, so
$$\vartheta \simeq -\frac{C_s}{E} p^{-s} F(s), \qquad (49\,\text{a})$$

where
$$F(s) = \int_0^1 \frac{1-x^s}{(1-x^2)^{3/2}} dx. \qquad (49\,\text{b})$$

The small-angle classical differential cross-section is then given by
$$I_{\text{cl}} = g(s)(C_s/E)^{2/s}\theta^{-2(1+1/s)}, \qquad (50)$$
where
$$g(s) = s^{-1}\{F(s)\}^{2/s}.$$

We note that, since integration by parts gives

$$\int \frac{x^s}{(1-x^2)^{\frac{3}{2}}} dx = \frac{x^{s-1}}{(1-x^2)^{\frac{1}{2}}} - (s-1) \int \frac{x^{s-2}}{(1-x^2)^{\frac{1}{2}}} dx,$$

and
$$\int \frac{dx}{(1-x^2)^{3/2}} = \frac{x}{(1-x^2)^{\frac{1}{2}}},$$

$$F(s) = (s-1)f(s), \qquad (51)$$

where
$$f(s) = \int_0^1 \frac{x^{s-2}}{(1-x^2)^{\frac{1}{2}}} dx$$

$$= \int_0^{\frac{1}{2}\pi} \sin^{s-2}\theta\, d\theta$$

$$= \frac{(s-3)(s-5)\ldots 1}{(s-2)(s-4)\ldots 2}\frac{\pi}{2} \quad (s \text{ even} > 2) \qquad (52)$$

$$= \frac{(s-3)(s-5)\ldots 2}{(s-2)(s-4)\ldots 3} \quad (s \text{ odd} > 3).$$

The use of this formula in practice is limited by the fact that, for sufficiently small scattering angles, classical theory becomes invalid (cf. the rigid sphere case discussed in § 4). Thus the condition of validity of (50) is that $\theta \ll 1$ but $\theta > \theta_{\text{cl}}$ where θ_{cl} is such that for $\theta < \theta_{\text{cl}}$ the quantum formulation is essential. This will be further discussed in § 9.1.

5.1.7. *A classical 'total' cross-section.* Although the total elastic cross-section for a potential which is of unbounded range, such as (47), does not exist we may define an effective classical total cross-section

$$Q_{\text{cl}} = 2\pi \int_{\theta_0}^{\pi} I_{\text{cl}}(\theta) \sin\theta \, d\theta, \qquad (53\text{ a})$$

where θ_0 is a finite lower limit determined say by the angular resolution of the measuring equipment. Provided $\theta_0 > \theta_{\text{cl}}$ we may evaluate Q_{cl} using the small angle approximation (50). Thus for the potential (47)

$$Q_{\text{cl}} = \pi s g(s)(C_s/E)^{2/s} \theta_0^{-2/s}. \qquad (53\text{ b})$$

5.2. *Relation of classical to quantal formulae*

We now consider the scattering in terms of the usual quantal formulae for the differential and total cross-sections in terms of phase shifts. As for the rigid-sphere case discussed in § 4, the phase shifts η_l are large for $l < L$ where $L \gg 1$. Under these circumstances the 'classical' approximation for η_l given in Chapter 6, § 3.7 (58) is a good one. As modified by Langer it gives

$$\eta_l = \int_{r_0}^{\infty} \left\{ k^2 - \frac{2MV}{\hbar^2} - \frac{(l+\tfrac{1}{2})^2}{r^2} \right\}^{\tfrac{1}{2}} dr - \int_{r_1}^{\infty} \left\{ k^2 - \frac{(l+\tfrac{1}{2})^2}{r^2} \right\}^{\tfrac{1}{2}} dr, \qquad (54)$$

where k, the wave number of relative motion, equals $(2ME/\hbar^2)^{\tfrac{1}{2}}$ and r_0 and r_1 are the respective outermost zeros of the integrands. We note now that

$$\frac{\partial \eta_l}{\partial l} = \frac{\pi}{2} - (l+\tfrac{1}{2}) \int_{r_0}^{\infty} \frac{dr}{r^2 [k^2 - (2MV/\hbar^2) - \{(l+\tfrac{1}{2})/r\}^2]^{\tfrac{1}{2}}}. \qquad (55)$$

If we write
$$(l+\tfrac{1}{2})\hbar = J = Mvp, \qquad (56)$$

in the notation of § 5.1.1, then

$$\frac{\partial \eta_l}{\partial l} = \frac{\pi}{2} - \int_{r_0}^{\infty} \frac{dr}{r\phi(r)}, \quad \text{with } \phi(r) \text{ as in (28)},$$

$$= \tfrac{1}{2}\vartheta(p) \quad (\text{see (27)}), \qquad (57)$$

where ϑ is the classical angle of deflexion for an impact parameter p.

Now to obtain the classical approximation $I_{\rm cl}(\theta)\,\mathrm{d}\omega$ for the differential cross-section from the full quantum formula (21) we use the asymptotic form for $P_l(\cos\theta)$ for large l,

$$P_l(\cos\theta) \sim \left(\frac{2}{l\pi\sin\theta}\right)^{\frac{1}{2}} \sin\{(l+\tfrac{1}{2})\theta+\tfrac{1}{4}\pi\}, \tag{58}$$

which is valid provided $l\theta > 1$. Since

$$\sum_l (2l+1)P_l(\cos\theta) = 0, \quad \theta \neq 0,$$

we then have
$$f(\theta) \simeq \sum A(l)\{\exp B_+(l) - \exp B_-(l)\}, \tag{59}$$

where
$$A(l) = -\frac{1}{2k}(2l/\pi\sin\theta)^{\frac{1}{2}}, \tag{60 a}$$

$$B_\pm(l) = 2\eta_l \pm (l+\tfrac{1}{2})\theta \pm \tfrac{1}{4}\pi. \tag{60 b}$$

The only appreciable contribution to the sum of a large number of oscillating terms as in (59) comes from regions of l in which the phase of the oscillation is stationary. Thus if

$$\frac{\mathrm{d}B}{\mathrm{d}l} = 0, \quad \text{for} \quad l = l_0, \tag{61}$$

and vanishes for no other value of l, then

$$\sum A(l)\exp\{iB(l)\} \simeq A(l_0)\exp iB(l_0) \int_{-\infty}^{\infty} \exp\{i\beta(l-l_0)^2\}\,\mathrm{d}l, \tag{62}$$

where
$$\beta = \frac{1}{2}\left(\frac{\mathrm{d}^2 B}{\mathrm{d}l^2}\right)_{l=l_0}. \tag{63}$$

If $\mathrm{d}B/\mathrm{d}l = 0$ for more than one value of l, say $l = l_1, l_2, l_3, \ldots$, then on the right-hand side of (62) we have a sum of terms of the same form as in (62) and with l_0 successively taking the values l_1, l_2, l_3, etc.

For a region of stationary phase in either of the sums occurring in (59)

$$2\frac{\partial \eta_l}{\partial l} \pm \theta = 0, \tag{64}$$

the choice of sign being such as to maintain the convention that $\theta > 0$. Considering again the case in which there is only one solution $l = l_0$ of (61) we see that

$$\left(\frac{\mathrm{d}^2 B}{\mathrm{d}l^2}\right)_{l=l_0} = 2\left(\frac{\partial^2 \eta_l}{\partial l^2}\right)_{l=l_0} = \frac{\partial \theta}{\partial l}, \tag{65}$$

so
$$f(\theta) = |f|e^{iC(l_0)}, \tag{66}$$

where
$$|f| = \left\{\frac{l_0}{k^2 \sin\theta}\left(\frac{\partial l}{\partial \theta}\right)_{l=l_0}\right\}^{\frac{1}{2}} \tag{67}$$

and $C(l_0)$ is the phase.

Writing now
$$(l+\tfrac{1}{2})\hbar = J$$
$$= Mvp,$$
and remembering that $l_0 \gg 1$, we have
$$|f(\theta)|^2 = \frac{p}{\sin\theta}\frac{\mathrm{d}p}{\mathrm{d}\theta} = I_{\mathrm{cl}}(\theta). \tag{68}$$

5.3. *Semi-classical approximation*†

The phase of $f(\theta)$, while unimportant in this case, is important if there are several values of l for which (61) is satisfied. We then find the total scattered amplitude is
$$f(\theta) = |f_1|e^{iC(l_1)} + |f_2|e^{iC(l_2)} + \ldots \tag{69}$$
and
$$|f(\theta)|^2 = |f_1|^2 + |f_2|^2 + \ldots + 2|f_1||f_2|\cos\{C(l_1) - C(l_2)\} + \ldots \tag{70}$$
$$= I_{\mathrm{cl}} + \text{oscillating interference terms}.$$

This is known as the semi-classical approximation. It reduces to the classical result if the angular resolution of the observing equipment is insufficient to resolve the oscillations.

The condition of validity of the semi-classical approximation for scattering at a given angle θ is that the values of l for which
$$\left|\frac{\partial \eta_l}{\partial l}\right| = \tfrac{1}{2}\theta \tag{71}$$
should all be much greater than unity and η_l should also be large compared with unity for these values of l. If there is only one value of l for which (71) applies, the classical differential cross-section will be a very good approximation, but if there are several such values of l it will only be accurate on the average because of the presence of the oscillatory interference terms.

To examine these oscillations further we note that, to include all the various possibilities of sign,
$$C(l) = 2\eta_l - 2(l+\tfrac{1}{2})\frac{\partial \eta}{\partial l} - \frac{\pi}{4}\left(2 - \frac{\partial^2\eta/\partial l^2}{|\partial^2\eta/\partial l^2|} - \frac{\partial \eta/\partial l}{|\partial \eta/\partial l|}\right). \tag{72}$$

Consider now the semi-classical theory applied to a classical deflexion function of the form shown in Fig. 16.5 (a). Remembering the correspondence (56) we see that, for $\theta < \theta_r$, there will be three values of l for which (71) holds. These give the separate contributions from the branches a, b, and c shown in Fig. 16.7. Fig. 16.5 (b) shows the way the

† The semi-classical treatment in this section is due to FORD, K. W. and WHEELER, J. A., *Ann. Phys.* **7** (1959) 259.

phase η_l varies with l corresponding to (71). This may be obtained at once, using (57) and (56), which give

$$\eta_l = \tfrac{1}{2}k \int_\infty^p \vartheta(p)\,\mathrm{d}p. \tag{73}$$

It is of exactly the form expected for scattering by a potential of the form shown in Fig. 16.1. Thus, following the discussion in Chapter 6, § 3.4, the main contribution to the phase shift η_l comes from separations r such that $kr \simeq l$. For large l, r falls in the region of the long-range attraction giving rise to a positive phase shift just as in electron scattering by atoms. As l decreases, r falls in a region of increasing attraction until the potential minimum is reached. For the corresponding value of l, η_l is a maximum which we shall refer to as η_m. As l decreases further η_l decreases steadily to become negative when r falls within the repulsive region—a repulsion increases the local wavelength so that the number of zeros of the radial wave-function is reduced, not increased as with an attraction.

We have now

$$f_a = k^{-1}(l_a/2\eta_a'' \sin\theta)^{\frac{1}{2}} \exp\{i(2\eta_a + l_a\theta - \tfrac{1}{2}\pi)\}, \tag{74 a}$$

$$f_b = k^{-1}(l_b/2|\eta_b''|\sin\theta)^{\frac{1}{2}} \exp\{i(2\eta_b + l_b\theta - \pi)\}, \tag{74 b}$$

$$f_c = k^{-1}(l_c/2|\eta_c''|\sin\theta)^{\frac{1}{2}} \exp\{i(2\eta_c - l_c\theta - \tfrac{1}{2}\pi)\}, \tag{74 c}$$

where η_a, η_b, η_c are written for η_{l_a}, η_{l_b}, η_{l_c} respectively and $\eta'' = \partial^2\eta/\partial l^2$.

5.3.1. *Glory interference effects.* Special interest attaches to the limiting behaviour at small angles. In such cases

$$l_b \to l_c = l_\mathrm{g}, \quad \eta_b \to \eta_c = \eta_\mathrm{m}, \quad \eta_b'' \to \eta_c'' = \eta_\mathrm{g}'', \quad \text{say.} \tag{75}$$

Hence

$$f_b + f_c \to k^{-1}(l_\mathrm{g}/2|\eta_\mathrm{g}''|\theta)^{\frac{1}{2}}\exp\{i(2\eta_\mathrm{m} - 3\pi/4)\}[\exp\{i(l_\mathrm{g}\theta - \tfrac{1}{4}\pi)\} +$$
$$+ \exp\{-i(l_\mathrm{g}\theta - \tfrac{1}{4}\pi)\}]$$
$$= 2k^{-1}(l_\mathrm{g}/2|\eta_\mathrm{g}''|\theta)^{\frac{1}{2}}\exp\{i(2\eta_\mathrm{m} - 3\pi/4)\}\cos(l_\mathrm{g}\theta - \tfrac{1}{4}\pi)$$
$$= 2(I_\mathrm{cl}^\mathrm{g})^{\frac{1}{2}}\exp(iC_\mathrm{g})\cos(l_\mathrm{g}\theta - \tfrac{1}{4}\pi), \tag{76}$$

where $C_\mathrm{g} = 2\eta_\mathrm{m} - 3\pi/4$.

The magnitude of the contribution from branch a will clearly be much larger because along it $\mathrm{d}p/\mathrm{d}(\cos\theta)$ is very large. We may therefore write for the total scattered amplitude

$$f(\theta) \to f_a\{1 + (f_b + f_c)/f_a\}$$
$$= (I_\mathrm{cl}^a)^{\frac{1}{2}}\exp(iC_a)[1 + 2(I_\mathrm{cl}^\mathrm{g}/I_\mathrm{cl}^a)^{\frac{1}{2}}\exp\{i(C_\mathrm{g} - C_a)\}\cos(l_\mathrm{g}\theta - \tfrac{1}{4}\pi)], \tag{77}$$

so that
$$I(\theta) = I_{\text{cl}}^a(\theta)[1+\{H(\theta)\}^2+2H(\theta)\cos C(\theta)], \tag{78}$$
where
$$H(\theta) = 2(I_{\text{cl}}/I_{\text{cl}}^a)^{\frac{1}{2}}\cos(l_g\theta - \tfrac{1}{4}\pi), \tag{79}$$
$$C(\theta) = 2(\eta_m-\eta_a)-l_a\theta-\tfrac{1}{4}\pi. \tag{80}$$

The oscillatory term arising from the product $H(\theta)\cos C(\theta)$ is a sum of two stationary oscillations in θ, one of period $2\pi/|l_a-l_g|$ and one of period $2\pi/|l_a+l_g|$.

5.3.2. *Interference effects near the rainbow angle.* In the neighbourhood of the rainbow angle the contributions from branches a and b tend to coalesce and the stationary phase treatment of p needs modification because $\eta_a'' \to \eta_b'' \to 0$. To examine the interference effects that occur in this region we write for the angle of deflexion

$$\vartheta = \vartheta_{\text{r}}+q(l-l_{\text{r}})^2, \tag{81}$$

where l_{r}, the value of l at the rainbow angle, $= Mvp_{\text{r}}/\hbar$. Thus $q = \hbar^2\alpha/M^2v^2$ where α is as in (38).

To avoid complications about signs we discuss the case illustrated in Fig. 16.5 (a) in which $\theta = -\vartheta$ near the rainbow angle. It follows from (73) that

$$\eta = \eta_{\text{r}}-\tfrac{1}{2}\theta_{\text{r}}(l-l_{\text{r}})+\tfrac{1}{6}q(l-l_{\text{r}})^3, \tag{82}$$

so that, in (60 b), near l_{r},

$$B(l) = 2\eta_{\text{r}}-\theta(l-l_{\text{r}})+\tfrac{1}{3}q(l-l_{\text{r}})^3+l\theta+\tfrac{1}{4}\pi. \tag{83}$$

Hence at θ, near θ_{r},

$$\sum A(l)\exp iB(l) = A(l_{\text{r}})\exp\{iB(l_{\text{r}})\}\int_{-\infty}^{\infty}\exp\{\tfrac{1}{3}q(l-l_{\text{r}})^3+(\theta-\theta_{\text{r}})(l-l_{\text{r}})\}\,dl$$
$$= \frac{2\pi}{k}\left(\frac{l_{\text{r}}}{2\pi\sin\theta}\right)^{\frac{1}{2}}e^{iC_{\text{r}}}q^{-1/3}\text{Ai}\{q^{-1/3}(\theta-\theta_{\text{r}})\}, \tag{84}$$

where
$$\text{Ai}(x) = \frac{1}{2\pi}\int_{-\infty}^{\infty}\exp i\{xz+\tfrac{1}{3}z^3\}\,dz \tag{85}$$

is the Airy function that has the general form shown in Fig. 16.8 and

$$C_{\text{r}} = 2\eta_{\text{r}}+l_{\text{r}}\theta-\frac{3\pi}{4}. \tag{86}$$

It can be seen from Fig. 16.8 that the rainbow singularity in the classical differential cross-section is no longer present. Instead the semi-classical rainbow amplitude has a maximum at an angle a little smaller than θ_{r}.

The full semi-classical amplitude in the case we have been discussing is obtained by adding (84) to f_c of (74 c), giving

$$I_{\text{sc}}(\theta) = I_{\text{cl}}^c(\theta)+I_{\text{sc}}^{\text{r}}(\theta)+2\{I_{\text{cl}}^c(\theta)\}^{\frac{1}{2}}\{I_{\text{sc}}^{\text{r}}(\theta)\}^{\frac{1}{2}}\cos\{C_{\text{r}}(\theta)-C_c(\theta)\}, \tag{87}$$

where
$$I_{\text{sc}}^{\text{r}} = \frac{2\pi}{k^2}\frac{l_{\text{r}}}{\sin\theta}q^{-2/3}\text{Ai}^2\{q^{-1/3}(\theta-\theta_{\text{r}})\}. \tag{88}$$

This shows the presence of interference oscillations of period
$$\simeq 2\pi/|C_r - C_c| = 2\pi/|l_r + l_c|.$$
If the scattering is observed with insufficient resolution to distinguish these oscillations the variation of the scattered intensity at the rainbow angle will follow the shape of $[\mathrm{Ai}\{q^{-1/3}(\theta - \theta_r)\}]^2$. Thus it will have a broad maximum near $q^{-\frac{1}{3}}(\theta_r - \theta) = 1$ with width of order $3q^{\frac{1}{3}}$ and subsidiary maxima at $q^{-\frac{1}{3}}(\theta_r - \theta) = 3\cdot 2$, etc.

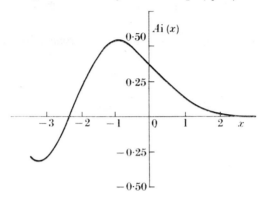

FIG. 16.8. The Airy function Ai(x).

Fig. 16.9 illustrates, for a particular representative case,† the type of angular distribution of scattered intensity that arises. The pattern results from the rapid modulation of a background intensity which varies near the rainbow angle in the way expected of the Airy function. It will be seen that the background exhibits a slow oscillation as predicted from the form of that function but the location of subsidiary (or supernumerary) rainbow maxima is not very close to that given from it. This is not surprising since the form (88) will only hold close to the rainbow angle. The subsidiary maxima correspond to the supernumerary rainbows in the optical case.

5.4. *Quantum theory of scattering by long-range potentials*

The total elastic cross-section Q is given by
$$Q = \frac{4\pi}{k^2} \sum (2l+1)\sin^2\eta_l,$$
where k is the wave number of relative motion. For all cases of potential importance the phase shifts η_l converge slowly with l so that a large number of terms contribute to the right-hand sum. Thus we may write
$$Q = \frac{8\pi}{k^2} \int_0^\infty l \sin^2\eta_l \, dl. \tag{89}$$

† MASON, E. A. and MONCHICK, L., *J. chem. Phys.* **41** (1964) 2221.

Furthermore, η_l is related to the classical deflexion angle ϑ by

$$\frac{\partial \eta_l}{\partial l} = \tfrac{1}{2}\vartheta(l), \tag{90}$$

so
$$\eta_l \simeq \tfrac{1}{2}\int_\infty \vartheta\,\mathrm{d}l, \qquad \text{where } l \simeq kp. \tag{91}$$

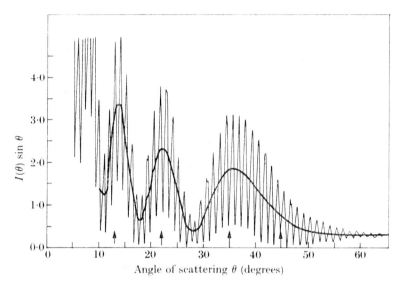

Fig. 16.9. Illustrating the behaviour of the differential cross-section per unit angle near a rainbow singularity, calculated by Mason and Monchick for a (12,6) interaction with $r_\mathrm{m} = 4\cdot 8$ Å and $K = 2\cdot 8$ (see § 6.1). The arrows denote the classical rainbow angle (45°), the first maximum (35°), the first minimum (22°), and the second maximum of the Airy function.

5.4.1. *The total cross-section for potentials varying as* r^{-s}. When l is large, ϑ and hence η_l will be small. For the case of an interaction

$$V(r) = -C_s r^{-s}, \tag{92}$$

it has been shown in § 5.1.6 that, when l is large, since $l \simeq kp$,

$$\vartheta \simeq -2(s-1)A_s l^{-s}, \tag{93}$$

where
$$A_s = Mf(s)C_s \hbar^{-2} k^{s-2}, \tag{94}$$

with $f(s)$ as in (52). Hence for large l, using (91),

$$\eta_l \simeq A_s l^{-s+1}. \tag{95}$$

As l decreases η_l will increase until it becomes comparable with and even much greater than unity. The approximation (95) is no longer valid under these conditions, say for $l < L$. To evaluate (89) approximately two procedures may be used.

In the first, due to Massey and Mohr,† we write

$$Q = \frac{8\pi}{k^2}\left[\int_0^L l\sin^2\eta_l\,dl + A_s^2\int_L^\infty l^{-2s+3}\,dl\right], \qquad (96)$$

the approximation $\sin^2\eta_l \simeq \eta_l^2$ being a good one when $l > L$. For $l < L$, $\sin\eta_l$ will oscillate rapidly with l because η_l will pass through several multiples of π as $l \to 0$. If it is assumed that the phase of η_l fluctuates at random we may then replace $\sin^2\eta_l$ by a mean value of $\tfrac{1}{2}$ to give

$$Q \simeq \frac{2\pi}{k^2}L^2 + \frac{4\pi}{k^2}\frac{A_s^2}{s-2}L^{-2s+4}. \qquad (97)$$

The choice of L is somewhat arbitrary but the value of Q does not depend very strongly on it. If we take L as the value of l for which η_l as given by (95) has the value x, of order unity, then

$$L \simeq (x/A_s)^{1/(1-s)} \qquad (98)$$

and

$$Q \simeq \frac{2\pi}{k^2}x^{2/(1-s)}A_s^{2/(s-1)}\left(1+\frac{2}{s-2}x^2\right)$$
$$= 2\pi\{2MC_s f(s)/k\hbar^2\}^{2/(s-1)}B_s(x), \qquad (99)$$

where

$$B_s(x) = x^{2/(1-s)}\left(1+\frac{2}{s-2}x^2\right). \qquad (100)$$

Alternatively, we may write

$$Q = \gamma_{\text{MM}}(s)\{C_s/\hbar v\}^{2/(s-1)}, \qquad (101\text{ a})$$

where, taking $x = \tfrac{1}{2}$ as was done by Massey and Mohr,

$$\gamma_{\text{MM}} = \pi\left(\frac{2s-3}{s-2}\right)\{2f(s)\}^{2/(s-1)}. \qquad (101\text{ b})$$

The second method, due to Landau and Lifshitz,‡ does not divide the range of integration but substitutes directly the approximation (95) for η_l in (89) and evaluates the resulting integral directly. Using the result that

$$\int_0^\infty x^{-z}\cos bx\,dx = \tfrac{1}{2}\pi b^{z-1}\sec(\tfrac{1}{2}\pi z)/\Gamma(z), \qquad (102)$$

they find

$$Q = \gamma_{\text{LL}}(s)\{C_s/\hbar v\}^{2/(s-1)}, \qquad (103\text{ a})$$

where

$$\gamma_{\text{LL}} = \pi^2\{2f(s)^{2/(s-1)}\}\operatorname{cosec}\left(\frac{\pi}{s-1}\right)\Big/\Gamma\left(\frac{2}{s-1}\right). \qquad (103\text{ b})$$

† Massey, H. S. W. and Mohr, C. B. O., *Proc. R. Soc.* **A144** (1934) 188.
‡ Landau, L. D. and Lifshitz, E. M., *Quantum mechanics*, p. 916 (Pergamon Press, 1959).

The same result follows by using a semi-classical approximation due to Schiff† according to which

$$Q \simeq 4 \int_{-\infty}^{\infty}\int_{-\infty}^{\infty} \sin^2\left\{\frac{1}{2\hbar v} \int_{-\infty}^{\infty} V(x,y,z)\,dz\right\} dx\,dy, \qquad (104)$$

where V is the interaction energy and v the velocity of relative motion. When V is spherically symmetrical (104) is equivalent to the approximation of Landau and Lifshitz but it may also be used when V is not spherically symmetrical (see § 12.4.1).

The expressions (101 a) and (103 a) differ only in the numerical factors $\gamma(s)$. The total elastic cross-section varies as $v^{-2/(s-1)}$ or $E^{-1/(s-1)}$ and is proportional to the $2/(s-1)$th power of the force constant C_s.

Values of γ_{MM} and γ_{LL} are compared for various values of s in Table 16.2.

TABLE 16.2

s	4	6	7	8	12
γ_{MM}	10·613	7·547	7·002	6·771	6·296
γ_{LL}	11·373	8·083	7·529	7·185	6·584

For the important practical case $s = 6$, γ_{MM} would agree with γ_{LL} if the value of x in (100) were taken to be 0·56 instead of 0·5.

Comparison with exact computations for realistic potentials shows that γ_{LL} gives cross-sections very close to the exact values.

5.4.2. Small-angle scattering for potentials varying as r^{-s}‡

The differential cross-section at small angles may also be calculated by similar approximate methods. For such angles we may write

$$P_l(\cos\theta) \simeq J_0(l\theta) = (1 - \tfrac{1}{4}l^2\theta^2 + \ldots), \qquad (105)$$

so that
$$I(\theta) = \frac{1}{4k^2}\left|\sum(2l+1)(e^{2i\eta_l}-1)P_l(\cos\theta)\right|^2 \simeq f_i^2 + f_r^2, \qquad (106)$$

where
$$f_i = \frac{2}{k}\int l\sin^2\eta_l(1-\tfrac{1}{4}l^2\theta^2)\,dl, \qquad (107\text{ a})$$

$$f_r = \frac{1}{k}\int l\sin 2\eta_l(1-\tfrac{1}{4}l^2\theta^2)\,dl. \qquad (107\text{ b})$$

---

† SCHIFF, L. I., *Phys. Rev.* **103** (1956), 443.
‡ PAULY, H., *Z. Phys.* **157** (1959), 54; HELBING, R. and PAULY, H., *Z. Phys.* **179** (1964) 16; MASON, E. A., VANDERSLICE, J. T., and RAW, C. J. G., *J. chem. Phys.* **40** (1964) 2153.

These expressions may be evaluated by either of the approximate methods used for calculating the total cross-section. Following through the Landau–Lifshitz procedure we find

$$f_\text{i} = (kQ/4\pi)\{1 - (k^2 Q/16\pi) g_1(s)\theta^2\}, \tag{108 a}$$

$$f_\text{r} = (kQ/4\pi)\tan\{\pi/(s-1)\}\{1 - (k^2 Q/16\pi)g_2(s)\theta^2\}, \tag{108 b}$$

where
$$g_1(s) = \pi^{-1}\tan\{\pi/(s-1)\}[\Gamma\{2/(s-1)\}]^2/\Gamma\{4/(s-1)\}, \tag{109 a}$$

$$g_2(s) = g_1(s)\tan\{2\pi/(s-1)\}\cot\{\pi/(s-1)\}. \tag{109 b}$$

Q is the total elastic cross-section given by (103) and we note that (108 a) is consistent with the theorem proved in Chapter 8, § 2.2.2 that the imaginary part of the forward scattered amplitude $= kQ/4\pi$.

The differential cross-section $I(\theta)\,d\omega$ is now given, for small angles θ, by

$$I(\theta) = (kQ/4\pi)^2[1+\tan^2\{\pi/(s-1)\}]\{1-\tfrac{1}{2}g_2(s)(k^2Q/8\pi)\theta^2\}, \tag{110}$$

where $\tan\{\pi/(s-1)\}$ is the ratio of the real to the imaginary part of the forward scattered amplitude. It decreases as s increases so that, in the limit of the rigid spherical case $s \to \infty$, it vanishes.

The classical formula (50) will be valid only when $\theta > \theta_\text{cl}$ where

$$\theta_\text{cl} \simeq k^{-1}(Q/\pi)^{\frac{1}{2}}. \tag{111}$$

For $\theta < \theta_\text{cl}$ the true differential cross-section falls below the classical value to reach a finite limit when $\theta \to 0$. This is illustrated in Fig. 16.10.

5.4.3. *The effect of glory scattering.*† So far we have assumed that no complications arise due to the presence of any glory scattering which contributes to the intensity at small angles (see Fig. 16.7). For realistic potentials, glory effects will certainly arise. Because of the strong repulsive interaction at small distances the phase shifts η_l will all pass through positive maxima (see Fig. 16.5 (b)) as l decreases and then decrease to reach large negative values as $l \to 0$. Because of the relation (90) the maxima will occur where $\vartheta_l = 0$, i.e. at a glory singularity. Thus η_l is a maximum where $l = l_\text{g}$, in the notation of § 5.3.1.

The region of the maximum will be one of stationary phase for η_l so that the assumption of random-phase variations when $\eta_l \gg 1$, which is made in deriving (101, 103), will not be valid for all $l < L$.

To allow for this Düren and Pauly‡ broke up the range of integration in (89) into four ranges:

$$\text{I},\ 0 \leqslant l < l_\text{g}-\tfrac{1}{2}\Delta l; \quad \text{II},\ l_\text{g}-\tfrac{1}{2}\Delta l < l < l_\text{g}+\tfrac{1}{2}\Delta l;$$
$$\text{III},\ l_\text{g}+\tfrac{1}{2}\Delta l < l < L; \quad \text{IV},\ l > L. \tag{112}$$

Here l_g is the value of l for which $\partial\eta/\partial l = 0$ and Δl is the range of l over which the phase is effectively stationary.

---

† BERNSTEIN, R. B., *J. chem. Phys.* **37** (1962) 1880; ibid. **38** (1963) 2599.
‡ DÜREN, R. and PAULY, H., *Z. Phys.* **175** (1963) 227.

16.5 COLLISIONS BETWEEN ATOMS

For l lying between $l_g \pm \tfrac{1}{2}\Delta l$, η_l may be written approximately as

$$\eta_l = \eta_m + \kappa_1(l-l_g)^2, \qquad (113)$$

where η_m is the maximum value of η_l as a function of l. In other ranges the same approximations may be made as in the original random-phase analysis of Massey and Mohr.

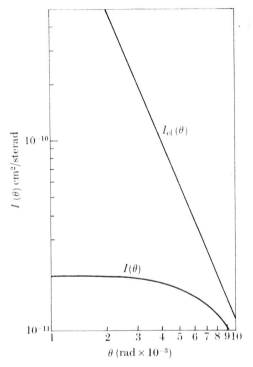

Fig. 16.10. Illustrating the deviation of the quantal from classical scattering at small angles for an inverse sixth-power attraction $-C/r^6$. The results given apply roughly to the case of collisions between two argon atoms at room temperature.

It is then found that
$$Q = Q_{\text{MM}} + \Delta Q, \qquad (114\text{ a})$$

where
$$\Delta Q = -\pi(4l_g/k^2)(\pi/4\kappa_1)^{\frac{1}{2}}\{\cos 2\eta_m C(\tfrac{1}{2}\kappa_1 \Delta l^2) + \sin 2\eta_m S(\tfrac{1}{2}\kappa_1 \Delta l^2)\}. \quad (114\text{ b})$$

C and S are the usual Fresnel integrals defined by

$$C(x) = (2/\pi)^{\frac{1}{2}}\int_0^{x^{\frac{1}{2}}} \cos t^2\,dt, \qquad S(x) = (2/\pi)^{\frac{1}{2}}\int_0^{x^{\frac{1}{2}}} \sin t^2\,dt.$$

No serious error is made by making $\Delta l \to \infty$ in (114 b), giving

$$\Delta Q = -\pi(4l_g/k^2)(\pi/\kappa_1)^{\frac{1}{2}}\cos(2\eta_m - \tfrac{1}{4}\pi). \qquad (115)$$

The same result may be obtained by using the relation between the total cross-section and the imaginary part of the forward scattered amplitude (Chap. 8, § 2.2.2). We have for this amplitude

$$f_i(0) = f_i^0(0) + f_i^g(0)$$
$$= (kQ_{MM}/4\pi) + f_i^g(0), \tag{116}$$

where f^g is the amplitude due to the glory scattering.

To calculate $f^g(0)$ we must modify the analysis of § 5.3.1, which is not valid when $l_g \theta < 1$. In place of (58) we use

$$P_l(\cos\theta) \simeq J_0(l\theta), \quad l\theta < 1, \tag{117}$$

and the approximation (113) for η_l, to give

$$f_g = \frac{e^{2i\eta_m}}{4ik} \int_0^\infty e^{2i\kappa_1(l-l_g)^2} l\, J_0(l\theta)\, dl. \tag{118}$$

Using the integral representation

$$J_0(x) = \frac{1}{2\pi} \int_0^{2\pi} e^{ix\cos\phi}\, d\phi,$$

and replacing the lower limit of integration in (118) by $-\infty$, which is valid provided $\kappa_1 l_g^2 \gg 1$, we find

$$f_g = (l_g/k)\left(\frac{2\pi}{-\kappa_1}\right)^{\frac{1}{2}} e^{i(2\eta_m - \frac{1}{4}\pi)} J_0(l_g \sin\theta), \tag{119}$$

giving

$$f_g(0) = (l_g/k)\left(\frac{2\pi}{-\kappa_1}\right)^{\frac{1}{2}} e^{i(2\eta_m - \frac{1}{4}\pi)}. \tag{120}$$

We then have

$$f_i(0) = \frac{k}{4\pi}(Q_{MM} + \Delta Q), \tag{121 a}$$

where

$$\Delta Q = -\frac{4\pi l_g}{k^2}\left(\frac{\pi}{\kappa_1}\right)^{\frac{1}{2}} \cos(2\eta_m - \tfrac{1}{4}\pi), \tag{121 b}$$

in agreement with (115).

The effect of the glory is thus to produce an oscillatory deviation from the monotonic variation of the total cross-section with energy given by (115). This is because η_m will in general decrease steadily with increasing energy from a value at very low energy which is a large multiple of π. Fig. 16.11 illustrates this behaviour.

The number of oscillations about the mean variation of Q with relative energy provides a lower limit on the number of bound states of the diatomic system which can arise due to the attractive interaction at long range.

As the relative wave number k decreases, l_g decreases so that in the limit of vanishing k, $l_g \to 0$. The phase shift η_0 in this limit tends to $n\pi$ where n is the number of diatomic bound states with zero relative angular momentum (see Chap. 6, § 3.6). The number of oscillations observed in a plot of Q against $1/k$ will be equal to the number of multiples of π contained in η_0 as $l_g \to 0$, and hence to n. This argument would be exact if the semi-classical approximation were valid down to $k = 0$. Nevertheless, it can be said that the number of oscillations observed, which in practice will be for not too small k, will never be less than n.

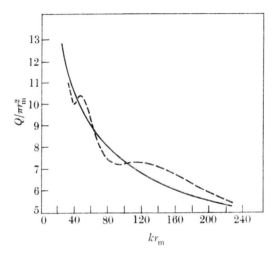

Fig. 16.11. Variation of the total cross-section with relative wave number k calculated for a $(12, 6)$ interaction. r_m is the parameter defined in § 6.1. —— mean cross-section $\bar{Q}$. --- total cross-section $Q = \bar{Q} + \Delta Q$.

5.4.4. *Orbiting under quantum conditions.* An orbiting singularity arises in the limit when the rainbow angle tends to infinity. The condition that must be satisfied at such a singularity has been discussed in § 5.1.1 and a formula, (36), given for the classical differential cross-section in the neighbourhood of the singularity. Unlike the glory and rainbow singularities, a semi-classical approximation, based on the approximate expression (54) for the phase shift, cannot be derived when an orbiting singularity is present. Thus, referring to Fig. 16.4, suppose that orbiting occurs when $J = J_1$, so that

$$V_{\text{eff}}(r_1) = E_1, \qquad \left(\frac{dV_{\text{eff}}}{dr}\right)_{r=r_1} = 0 \quad \text{at } J = J_1,$$

E_1 being the kinetic energy of relative motion. As the angular momentum decreases below some value $J > J_1$ the classical turning point moves gradually in until when $J = J_1$ it is at r_1. A further decrease of J, however, causes the turning point to make a sudden jump inwards to a considerably smaller value of r, because the maximum in $V_{\text{eff}}(r)$ is now less than E_1. According to (54) this sudden jump produces a discontinuity in the phase shift η_l considered as a function of l where $(l+\tfrac{1}{2})\hbar \simeq J$. A similar discontinuity arises if we consider not the variation of η_l with l for fixed E but with E for fixed l. No such discontinuity can occur in the actual phase shift and it is removed when due allowance is made for quantum-mechanical tunnelling through the potential barrier presented by V_{eff} for $J < J_2$ in Fig. 16.4. In terms of the approximation (54) for the phase shift this means that it must be modified to include a contribution from the classically inaccessible region.

A further important influence of this region arises from the presence of virtual energy levels. Thus consider as in Fig. 16.12 (a) the variation of V_{eff} with r for a fixed value of J (or l) $< J_2$ so that an inner attractive well exists. If this well is deep and/or wide enough there will exist a number of energy levels within it which, for $E > 0$, will not be stable because of leakage through the barrier. These are virtual levels (see Chap. 9, § 9) which, because of their finite lifetime, are broadened so that, if Γ is the half-breadth, the lifetime τ is given by

$$\Gamma = \hbar/\tau. \qquad (122)$$

In general, the deeper the level the greater the lifetime τ, as a greater thickness of barrier has to be penetrated. Now consider the behaviour of the phase shift η_l as a function of energy E. As $E \to 0$, $\eta_l \to s\pi$, where s is the number of *real* energy levels ($\tau \to \infty$) associated with V_{eff} for $J = (l+\tfrac{1}{2})\hbar$. η_l will vary gradually as E increases until it approaches the energy $E^{(1)}$ of the first virtual level. In the limit in which the breadth $\Gamma^{(1)}$ of this level tends to zero, η_l will jump discontinuously by π at $E = E^{(1)}$. Because of the finiteness of $\Gamma^{(1)}$ the change by π will not be discontinuous but will be more or less sharp depending on whether τ is large or small. Similar behaviour will occur as the energy passes through that of other virtual levels so that the variation of η_l with E takes the form shown in Fig. 16.12 (b). As far as the total elastic cross-section Q is concerned typical resonance effects (see Chap. 9, § 9) will occur over an energy range of order Γ at an energy corresponding to a virtual level. In general, the width Γ will be so small that the energy resolution available in practice will not be high enough to observe these effects but, if the virtual level occurs near the top of a barrier, the corresponding resonance may be observed. A detailed discussion for the specific case of the interaction of two H atoms is given in § 12.1.3 (cf. Figs. 16.60 and 16.12 (b)).

In most cases of practical significance the barrier maximum in V_{eff} occurs at large separations where the interatomic interaction is an attraction of the form
$$V(r) = -C_s r^{-s},$$

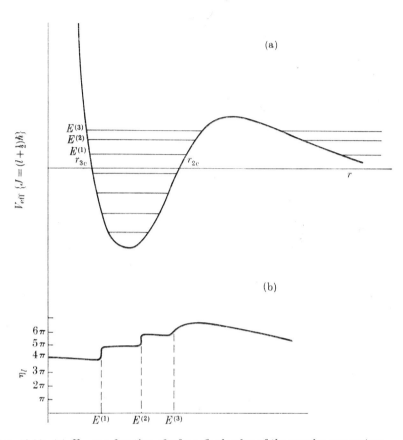

FIG. 16.12. (a) V_{eff} as a function of r for a final value of the angular momentum
$$J\{=(l+\tfrac{1}{2})\hbar\} < J_2 \text{ in Fig. 16.4.}$$
(b) Variation of η_l with kinetic energy E of relative motion, the number of real energy levels in V_{eff} being 4.

so, following (33),
$$V_{\text{eff}} \simeq -C_s r^{-s} + l(l+1)\hbar^2/2Mr^2.$$

The value, r_m, of r at the maximum is given by
$$sC_s r_m^{-s-1} - l(l+1)\hbar^2/Mr_m^3 = 0,$$
giving
$$r_m = \{sMC_s/\hbar^2 l(l+1)\}^{1/(s-2)}.$$

The energy $V_{\text{eff}}(r_m)$, $= E_m$, at the barrier peak is therefore

$$V_{\text{eff}}(r_m, l) = \alpha_s l(l+1)^{s/(s-2)}, \qquad (123\text{ a})$$

where
$$\alpha_s = \{\hbar^{2s}/(sM)^s C_s^2\}^{1/(s-2)}(\tfrac{1}{2}s - 1). \qquad (123\text{ b})$$

Fig. 16.13 shows a typical plot of $V_{\text{eff}}(r_m, l)$ against $l(l+1)$. We may also indicate on the same diagram the distribution of virtual levels that must

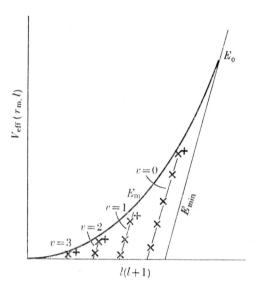

Fig. 16.13. Typical plots of E_m, the height of the maximum and $E_{\min}$, the height of the minimum, of V_{eff} as a function of $l(l+1)$. The distribution of virtual levels (denoted by ✗) is also shown. A + indicates a level that may be of sufficiently long life for observation through a resonance effect in collision experiments.

fall below the plotted curve. These may be classified just as the vibrational levels of a diatomic molecule. Thus the deepest level in the well is designated as a $v = 0$ level, the next deepest $v = 1$, and so on. For realistic potentials the levels for different l and fixed v fall on straight lines as indicated in Fig. 16.13 and shown in Fig. 16.60 for the actual case of H–H interaction. In Fig. 16.13 the levels that are likely to be of greatest width are indicated. In all cases they are close to the barrier maximum.†

Diagrams of the form shown in Fig. 16.13 are often used in the discussion of predissociation of diatomic molecules due to high rotation.

† See BERNSTEIN, R. B., *Phys. Rev. Lett.* **16** (1966) 385.

In such cases the atoms are initially bound in the virtual state and transitions from these states are observed spectroscopically. With this technique the virtual states most readily detected are those of long lifetime so that collision experiments which reveal, through the observation of resonance, the existence of short-lived virtual states, will provide complementary information.

We include also in Fig. 16.13 a typical plot of the energy at the bottom of the potential well, when $\geqslant 0$, as a function of $l(l+1)$. This will, at some value E_0 of E, join the plot of $V_{\text{eff}}(r_m, l)$. For $E > E_0$ orbiting can no longer occur but only rainbow scattering.

The simple approximation (54) cannot include effects due to virtual levels because it ignores all contributions from inside the classical closest distance of approach. On the other hand, a similar approximation applied within the two inner classical turning points can be used to obtain an approximation to the energies of the virtual levels provided they do not lie too close to the barrier maximum. For such a level, in the notation of (54),

$$\int_{r_{3c}}^{r_{2c}} \left\{ k^2 - \frac{2MV}{\hbar^2} - \frac{(l+\tfrac{1}{2})^2}{r^2} \right\}^{\tfrac{1}{2}} \, dr = (n+\tfrac{1}{2})\pi \quad (n = 0, 1, 2, \ldots), \quad (124)$$

r_{2c} and r_{3c} being the two inner zeros of the integrand (see Fig. 16.12 (a)). Further more elaborate approximations have been discussed by Ford, Hill, Wakano, and Wheeler† and by Curtiss and Powers.‡

We have concentrated attention on the variation of η_l with E for fixed l but a similar discussion can be followed through in describing the variation of η_l with l for fixed E. A typical example, when orbiting occurs ($E < E_0$) is illustrated in Fig. 16.14 showing the deflexion function ϑ as well as η_l.

6. Collision parameters for analytically represented interactions

6.1. *Analytical representations of interatomic interactions*

We shall now illustrate the results of the preceding sections by describing results of calculations carried out accurately for interactions of realistic form as in Fig. 16.1. At the same time we shall discuss ways of determining the characteristics of the interaction from observed total and differential elastic cross-sections.

---

† Ford, K. W., Hill, D. L., Wakano, M., and Wheeler, J. A., *Ann. Phys.* **7** (1959) 239.
‡ Curtiss, C. F. and Powers, R. S., *J. chem. Phys.* **40** (1964) 2145.

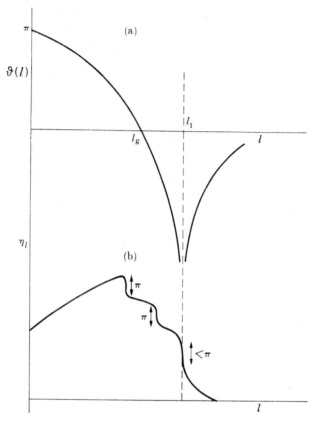

FIG. 16.14. (a) Typical deflexion function ϑ when orbiting occurs at $l = l_1$. (b) Corresponding behaviour of η_l as a function of l.

If one of the atoms is in a ^1S state the interaction will be unique and, if the atoms are all neutral, have the asymptotic form $-C/r^6$ but the short-range repulsion does not obey any such simple law. For the purpose of analysing experimental data on total and differential elastic cross-sections we start from analytical representations of the potential which are as simple as possible. It is convenient to write any such potential in terms of dimensionless variables in the form

$$V = \epsilon f(r/r_m), \tag{125}$$

where ϵ is the depth of the attractive potential well and r_m is the nuclear separation at the potential minimum (see Fig. 16.1).

Three forms for V have proved very useful. The simplest is the so-called Lennard–Jones potential

$$f(x) = \{6/(n-6)\}x^{-n} - \{n/(n-6)\}x^{-6}, \tag{126}$$

where n is > 6 and is often chosen to be 12, a value that often gives good results.

This potential for $n = 12$ has been generalized by Kihara to the form

$$f(x) = \left(\frac{1-\alpha}{x-\alpha}\right)^{12} - 2\left(\frac{1-\alpha}{x-\alpha}\right)^{6}. \tag{127}$$

The third form, introduced by Buckingham, takes account of the theoretical form for the repulsive potential which falls off exponentially with distance by writing

$$f(x) = \left(\frac{6}{\alpha-6}\right)\exp\{-\alpha(x-1)\} - \frac{\alpha}{\alpha-6}x^{-6}. \tag{128}$$

We shall refer to these as the $(n, 6)$, $(12, \alpha, 6)$, and $(\exp \alpha, 6)$ potentials respectively. The constant C in the van der Waals' attraction is given by

$$C = \epsilon r_m^6 \left\{\frac{n}{n-6}, \, 2(1-\alpha)^6, \, \frac{\alpha}{\alpha-6}\right\} \tag{129 a}$$

for the respective cases. It is also sometimes important to know the separation σ at which the potential vanishes (see Fig. 16.1). It is given respectively by

$$\sigma = r_m \left\{\left(\frac{6}{n}\right)^{1/(n-6)}, \, \alpha + 0\cdot 926/(1-\alpha), \, \beta\right\}, \tag{129 b}$$

where β is the solution of

$$6e^{-\alpha(x-1)} = \alpha x^{-6}. \tag{129 c}$$

Finally, the reduced curvature κ of the potential at the minimum is given for the respective cases by

$$\kappa = \left\{6n, \, \frac{72}{(1-\alpha)^2}, \, 6\alpha\left(\frac{\alpha-7}{\alpha-6}\right)\right\}. \tag{130}$$

6.2. *Reduced parameters when the semi-classical approximation is valid*

With the form (125) for V, the Schrödinger equation for relative motion with wave number k and angular momentum $\{l(l+1)\}^{\frac{1}{2}}\hbar$ is

$$\frac{d^2 G_l}{dr^2} + \left\{k^2 - \frac{l(l+1)}{r^2} - \frac{2M\epsilon}{\hbar^2} f\left(\frac{r}{r_m}\right)\right\} G_l = 0.$$

The substitution $x = r/r_m$ reduces this to

$$\frac{d^2 G_l}{dx^2} + \left\{k^2 r_m^2 - \frac{l(l+1)}{x^2} - \frac{2M\epsilon r_m^2}{\hbar^2} f(x)\right\} G_l = 0. \tag{131}$$

This shows that the phase shift $\eta_l(kr_m)$ is a function only of $B = 2M\epsilon r_m^2/\hbar^2$ which tends to ∞ in the classical limit and is often referred to as the quantum parameter. Alternatively, the further substitution $y = xkr_m$ leads to

$$\frac{d^2 G_l}{dy^2} + \left\{1 - \frac{l(l+1)}{y^2} - \frac{\epsilon}{E}f\left(\frac{y}{A}\right)\right\}G_l = 0, \qquad (132)$$

where $A = kr_m$ and $E = k^2\hbar^2/2M$. This shows that the phase shift η_l is a function A and K ($= E/\epsilon$, $= A^2/B$).

If the semi-classical approximation is valid a simpler dependence of the phase shifts on the reduced variables can be derived. Thus, turning to (27), we see that the classical deflexion function is given by

$$\vartheta = \pi - 2\int_{x_0}^{\infty} x^{-1}\left\{\frac{x^2}{p^{*2}} - 1 - \frac{x^2}{p^{*2}K}f(x)\right\}^{-\frac{1}{2}} dx, \qquad (133\text{ a})$$

where

$$p^* = p/r_m. \qquad (133\text{ b})$$

Since

$$\frac{\partial \eta_l}{\partial l} = \frac{\partial \eta}{k\,\partial p} = \tfrac{1}{2}\vartheta(p^*, K), \qquad (134)$$

we have

$$\frac{\partial \eta^*}{\partial p^*} = \tfrac{1}{2}\vartheta(p^*, K), \qquad (135)$$

where

$$\eta^* = \eta/kr_m = \eta/A. \qquad (136)$$

Under these conditions η^* is a function only of p^*, K. To retain contact with the angular momentum quantum number we can express this by saying that $\eta_{l^*}^*$ is a function of K only, in this approximation, l^* being a reduced angular-momentum quantum number given by

$$l^* = (l+\tfrac{1}{2})/A. \qquad (137)$$

Under thermal energy conditions, for interactions between gas atoms, the semi-classical approximation is a good one. Bernstein† has examined its validity in some detail for the (12, 6) interaction. For values of $B = 2M\epsilon r_m^2/\hbar^2$ ranging from 62·5 to 375, K from 0·2 to 3·2, and l from 0 to 30, the semi-classical approximation to η_l given by (134) differed from the phase shift calculated by direct solution of the equation (131) by no more than 0·03 radians.

† BERNSTEIN, R. B., J. chem. Phys. 36 (1962) 1403.

6.3. *Cross-sections in terms of reduced parameters*

In terms of the reduced phase shifts for the (12, α, 6) potential we have, for the total elastic cross-section,

$$Q = \bar{Q} + \Delta Q, \qquad (138\text{ a})$$

where

$$\frac{\bar{Q}}{\pi r_m^2} = 2 \cdot 573 \left\{ \frac{A}{K}(1-\alpha)^6 \right\}^{2/5}, \qquad (138\text{ b})$$

$$\frac{\Delta Q}{\pi r_m^2} = 4p_m^* \left\{ \frac{4\pi}{-A(\partial \vartheta/\partial p)_m^*} \right\}^{\frac{1}{2}} \cos(2A\eta_m^* - \tfrac{1}{4}\pi). \qquad (138\text{ c})$$

p_m^*, η_m^*, and $\partial \vartheta/\partial p^*$ all refer to values at a maximum of η^* considered as a function of p^*.

Düren and Pauly have given tables of these quantities as functions of K for $\alpha = -0\cdot3\,(0\cdot1)\,+0\cdot5$.

The location of the extrema are given by

$$A\eta_m^* = \eta_m = (N - \tfrac{3}{8})\pi \qquad (N = 1, 2, \ldots) \qquad (139)$$

Bernstein† has considered the form taken by this relation for the $(12, 6) \equiv (12, 0, 6)$ potential when K is large. Under these conditions the classical deflexion function, given by (133), takes the form, in the notation of (49), (132), and (133),

$$\vartheta = \frac{1}{K}\{F(12)p^{*-12} - 2F(6)p^{*-6}\}, \qquad (140)$$

and

$$\eta^* = \tfrac{1}{2}\int_{p^*}^{\infty} \vartheta \, \mathrm{d}p^* = -\frac{1}{2K}\left\{\frac{F(12)}{11}p^{*-11} - \frac{2F(6)}{5}p^{*-5}\right\}. \qquad (141)$$

Hence

$$p_m^* = \{F(12)/2F(6)\}^{1/6}$$
$$= 0\cdot 947, \qquad (142)$$

and

$$\eta_m^* = K^{-1}\{6F(6)/55\}p_m^{*-5}$$
$$= 0\cdot 134\pi K^{-1}. \qquad (143)$$

In general, for not too small K, we may write for this potential

$$\eta_m^* = 0\cdot 134\pi\, K^{-1}\{1 - c_1 K^{-1/2} + c_2 K^{-1} + \ldots\}. \qquad (144)$$

From the accurately calculated values c_1 is found to be $\simeq 0\cdot 25$. Reverting to the actual phase shift η_m we have then

$$\eta_m \simeq 0\cdot 268\pi(\epsilon r_m/\hbar v)\{1 - 0\cdot 25(\epsilon/E)^{\frac{1}{2}} + \ldots\}, \qquad (145)$$

and extremes of ΔQ will occur when

$$N - \tfrac{3}{8} = 0\cdot 268(\epsilon r_m/\hbar v)\{1 - 0\cdot 25(\epsilon/E)^{\frac{1}{2}} + \ldots\}. \qquad (146)$$

† BERNSTEIN, R. B., *J. chem. Phys.* **37** (1962) 1880; ibid. **38** (1963) 2599.

This expansion will presumably converge provided $E/\epsilon > 0.06$. Taking into account the first two terms only gives η_m^* correct to 0.10 per cent even for K as low as 0.2 and is therefore very useful in the analysis of experimental data.

For $(12, \alpha, 6)$ and $(\exp \alpha, 6)$ potentials with $\alpha \neq 0$ no such simple expressions are available. Nevertheless, for a given value of α, η_m^* is a function only of $1/K$ for these potentials also. Writing

$$\eta_m^* = \sum a_n K^{-(n+1)/2}, \qquad (147)$$

Bernstein and O'Brien† have evaluated a_1, a_2, a_3, and a_4 numerically for all three sets of potentials for different values of n and α. They find that a_2, a_3, and a_4 are practically independent of the form of the potential but that a_1 is largely determined by the reduced curvature κ of the potential at the separation r_m (see (130)). Some typical results are given in Table 16.3.

TABLE 16.3

Coefficients a_n in the expansion of η_m^* in powers of $K^{\frac{1}{2}}$

Reduced curvature κ	a_1 Type of potential		
	$(n, 6)$	$(12, \alpha, 6)$	$(\exp \alpha, 6)$
48	0.14700	1.4742	0.4521
72	0.4216	0.4216	0.4146
96	0.3963	0.3884	0.3926
120	0.3804	0.3647	0.3780
	$a_2 = 2.0 \times 10^{-3}$	$a_3 = -1.90 \times 10^{-1}$	$a_4 = 8.8 \times 10^{-2}$

6.3.1. *The low- and high-velocity regions.* It follows from (146) that if the interaction is of (12, 6) type no oscillations will appear in the total cross-section as a function of relative velocity when

$$\epsilon r_m/\hbar v < 1.$$

Also from (141) it may be seen that when

$$\epsilon r_m/\hbar v \simeq 1$$

the cross-section given by (103) for the r^{-12} repulsion only is equal to that for the r^{-6} term only.

We may therefore distinguish two regions of relative velocity, a low one in which $v < \epsilon r_m/\hbar$ and a high one in which the inequality is reversed. In the former region the cross-section shows oscillatory behaviour as a

† BERNSTEIN, R. B. and O'BRIEN, T. J. P., *J. chem. Phys.* **46** (1967) 1208.

function of v and the 'mean' cross-section varies as $v^{-2/5}$. No oscillations occur in the high-energy region and the variation with velocity is much slower, as $v^{-2/11}$.

Although these considerations have been arrived at for the (12, 6) potential specifically, they apply in general to interactions of similar type. The distinction between high and low velocities in this context is often very useful. For interaction between light atoms ϵr_m is usually much smaller than for heavy atoms so that, for the same relative velocity, collisions between the former may fall in the high velocity region and for the latter in the low velocity region.

6.4. *Dependence of the rainbow angle on interaction parameters*

It is also useful to consider the dependence of the rainbow angle on the parameters and shape of the interaction potential. In the limit of high energies, for the (12, 6) potential we have (see (140))

$$\vartheta = \frac{1}{K}\{F(12)p^{*-12} - 2F(6)p^{*-6}\},$$

so

$$\frac{\partial \vartheta}{\partial p^*} = -\frac{12}{K}\{F(12)p^{*-13} - F(6)p^{*-7}\}. \tag{148}$$

Hence, for a rainbow angle,

$$F(12)p^{*-13} - F(6)p^{*-7} = 0, \tag{149}$$

so

$$p^* = \left\{\frac{F(12)}{F(6)}\right\}^{1/6} \tag{150}$$

and

$$\vartheta_r = -\frac{1}{K}\frac{\{F(6)\}^2}{F(12)} \tag{151}$$

$$= -\frac{2 \cdot 05}{K}. \tag{152}$$

Under these conditions the rainbow angle depends only on K and not at all on r_m. This result is generally true for realistic atomic interactions as pointed out by Mason†—the rainbow angle is determined mainly by the ratio ϵ/E and only to a small extent on the potential shape and parameters. Fig. 16.15 illustrates this by showing ϑ_r calculated numerically as a function of K for three potentials (12, 6),‡ (exp 15, 6),† and (exp 12, 6).† Tables of ϑ_r for Kihara potentials have also been given by Schlier.‡

† MASON, E. A., *J. chem. Phys.* **26** (1957) 667.
‡ SCHLIER, C., *Z. Phys.* **173** (1963) 352.

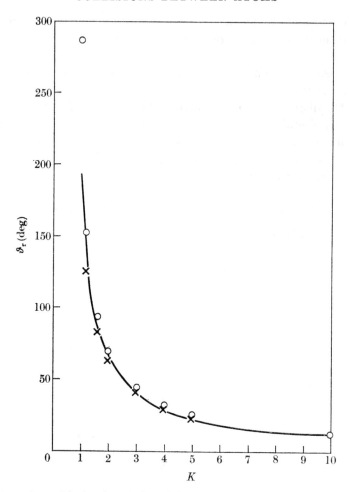

Fig. 16.15. Calculated variation of the rainbow angle ϑ_r with the parameter K for potentials of the form —— (12, 6); × (exp 12, 6); ○ (exp 15, 6).

6.5. *Conditions for orbiting*

For (12, 6) potentials the condition for orbiting to occur (see § 5.4.4 and particularly Fig. 16.13) is that the parameter $K \leqslant 0.8$. The critical reduced angular momentum $J_1^* = p^* K^{\frac{1}{2}}$, below which the reduced effective potential V_{eff}/ϵ shows a maximum, is 1·453.

6.6. *Determination of potential parameters from experimental data*

Measurements may be made, using techniques described in § 7, of differential and total elastic cross-sections as a function of a mean or even of a definite relative energy of collision. The differential cross-sections may be observed over a wide angular range with a resolution which continually improves with advances in technique.

The first step is to verify that the interaction behaves asymptotically as r^{-6}. This may be done by observing either or all of the following.

(a) The classical differential cross-section $I_{cl}\,d\omega$ at small angles θ which are nevertheless $> \theta_{cl}$, where θ_{cl} is the smallest scattering angle for which the classical approximation is valid. According to (50) I_{cl} should vary as $\theta^{-7/3}$.

(b) The total elastic cross-section Q as a function of relative velocity. As described in § 5.4.3, Q can be written as

$$Q = \bar{Q} + \Delta Q, \tag{153}$$

where $\bar{Q}$ should vary as $v^{-2/5}$ while ΔQ is an oscillatory function of v. To separate $\bar{Q}$ and ΔQ, the observed value of $\log Q$ may be plotted against $\log v$ so that a mean curve may be drawn through the oscillations, giving $\log \bar{Q}$.

(c) The classical 'total' cross-section Q_{cl} defined as in § 5.1.7, as a function of relative velocity. If the minimum scattering angle observed is $> \theta_{cl}$ then Q_{cl} should vary as $v^{-1/6}$.

Having verified the asymptotic form of the interaction, the van der Waals' constant $C = \epsilon r_m^6$ may be obtained by absolute measurement of the total elastic cross-section at a definite relative velocity or of the classical differential cross-section at a definite relative velocity and scattering angle. ϵr_m may be determined from the positions of the extrema in the oscillatory part ΔQ of the total cross-section as explained in § 6.3. ϵ itself may be determined from the values of the rainbow angles observed at different relative velocities (see § 6.4).

A lower limit to the number of bound diatomic states that exist is given by the number N of extrema observed in ΔQ as a function of v. Information about r_m may be obtained from the wavelength of the oscillations in the differential cross-sections in the neighbourhood of a rainbow angle or elsewhere. Thus the angular width between successive maxima at not too large scattering angles will be of the order kr_m, where k is the relative wave number.

Finally, some information about the curvature of the interaction function near the potential energy minimum may be obtained if the location of the extrema in ΔQ are determined with sufficient precision.

Having obtained broad features of the interaction between a particular pair of gas atoms it is then possible to refine the model by trial-and-error methods.

We shall discuss the information obtained about these interaction parameters from experimental measurements on various interatomic scattering differential and total cross-sections in §§ 9–13 after first describing the experimental techniques employed.

7. The measurement of total cross-sections by molecular-ray methods

The study of the scattering of atoms and molecules with gas-kinetic velocities may be carried out by methods which are the same in principle as those used for studying the total cross-sections for scattering of electrons. The electron beam must be replaced by a beam of gas atoms. It is necessary, therefore, to develop techniques for the production of such beams as well as methods for measuring the strength of the beam at any point. The beam must, of course, be undirectional and the atoms composing it must not collide with each other. The conditions are therefore those of effusive flow of a gas. A molecular beam of this sort will differ from the electron beams of Ramsauer's experiments (Chap. 1, § 4.1) in that the atoms within it will possess a Maxwellian distribution of velocities instead of a single definite velocity. Although a great deal can be done despite this limitation, methods of obtaining beams of neutral particles with homogeneous velocities in the thermal range have been devised and are now coming into use to an increasing extent.

The first experimental production of a molecular beam was carried out by Dunoyer† in 1911 but the full development of the technique was carried out at Hamburg by Stern and his collaborators over the period from 1923 to 1933. It is not our intention here to describe the full technical details of molecular-ray experiments. An excellent account has been given by Ramsey in *Molecular beams* (Clarendon Press, Oxford, 1956). We shall, however, summarize the types of source and methods of detection before describing the most detailed applications of the method which have yet been made to the study of atomic collisions.

7.1. *The production and detection of molecular beams*

A typical source of a molecular beam consists of a cavity containing a suitable pressure of the gas or vapour concerned which communicates with an evacuated chamber through a small orifice or slit (see Fig. 16.16 (a)). We shall refer to the cavity as the oven because, for dealing

---

† DUNOYER, L., *C.r. hebd. Séanc. Acad. Sci., Paris* **152** (1911) 592; *Radium, Paris* **8** (1911) 142.

with substances that have very low vapour pressures at ordinary temperatures, it must be heated so as to produce the working pressure. In general, this pressure is such that the mean free path λ of the molecules in the oven is comparable with or greater than the diameter d of the orifice so that the flow through the orifice is effusive and not hydrodynamic. Otherwise there is a tendency for so-called cloud formation to

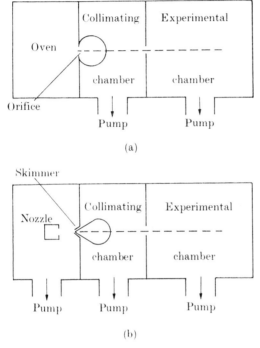

Fig. 16.16. (a) Schematic arrangement of an effusive oven beam. (b) Schematic arrangement of a supersonic nozzle beam.

occur on the high-vacuum side of the orifice, with consequent loss of intensity by scattering.

One of the most important requirements of a source is that it should provide a beam of maximum intensity because geometrical collimation, velocity selection, selection of internal molecular states and detection all involve reduction of effective intensity which may be very serious. The aim is to design the outflow system from the oven so as to maximize the flow in the desired direction for a given total flow. The amount of the latter which can be tolerated depends on the available pumping speed, including the possibility of condensation for vapours.

Increased directivity in the flow from the oven can be achieved in several ways. Thus, for effusive flow, the flux of molecules received through a circular orifice, of radius d, normal to its surface, at a distance l from it, is given by

$$I_0 = \tfrac{1}{4}n\bar{v}d^2/l^2, \tag{154}$$

n being the concentration and $\bar{v}$ the mean velocity of the molecules in the oven. For a given vapour the free path λ is given by $1/nQ$, where Q is an effective collision cross-section. Since $d \simeq \lambda$ we have

$$I_0 = \tfrac{1}{4}\bar{v}d/Ql^2. \tag{155}$$

It follows from this that the intensity from a linear array of N orifices of radius d will be the same as that from a single orifice of radius Nd, provided the oven pressure is N times greater. Such a linear array is equivalent to a slit of long dimension Nd and width d. A stack of M such slits will yield an intensity M times greater than the single circular orifice of radius Nd. If δ is the maximum aperture dimension tolerable in relation to the geometry of the experiment we must have $Nd \leqslant \delta$ and $Md \leqslant \delta$. The gain in beam intensity is then comparable with δ/d. In practice δ is of order 1 cm for l of order 1 m and the gain is about $1/2d$, where d is the slit width in centimetres.

An improvement in directivity is obtained if the plane aperture is replaced by a channel of length comparable with the free path λ. Thus, if the channel is a capillary of length L and radius a, the axial flow remains the same as in (154) but the total flow is $4a/3L$ times smaller than from the plane aperture. A many-channel array therefore presents considerable advantages. Since λ must be comparable with L, and $a \ll L$ it is necessary to use a densely packed array of very small diameter channels as λ is small at the working gas pressures involved.

A further technique, first suggested by Kantrowitz and Grey,† is to replace the effusive flow by a supersonic jet. In such a case the orifice in the effusive system is replaced by a skimmer (see Fig. 16.16 (b)). Because of the high axial flow speed there is increased flux along the beam direction for a given total flow. For some purposes there is the additional advantage that the gas is cooled by passage through the supersonic nozzle so that the velocity spread in the beam is reduced.

The methods of detection of slow molecular beams may be summarized as follows.

(a) *Condensation*, in which the molecular beam is condensed on a cooled surface and the beam is detected by the changed appearance of

---

† KANTROWITZ, A. and GREY, J., *Rev. scient. Instrum.* **22** (1951) 328.

the surface, this change being sometimes enhanced by chemical means. This method of detection is qualitative, although attempts have been made to use it as the basis of a quantitative method by estimating the thickness of the deposit either by weighing or by optical means.

(b) *Chemical targets*, in which some specific type of molecular or atomic species can be detected by chemical action. The most widely known example of this method of detection is that of atomic hydrogen by a target of MoO_3, giving a blue MoO_2 trace.

(c) *Manometric method*, in which the molecular beam enters a space closed except for the orifice by which the beam enters. The pressure in the space builds up until the rate of effusion of gas out through the orifice is equal to the rate at which gas is brought in by the beam. The pressure can then be measured by any suitable type of ion gauge.

(d) *Surface ionization detector*, available particularly for neutral beams of alkali atoms (see also Chap. 17, § 6), which strike a suitable hot filament, undergo surface ionization, and are measured by the ion current between the hot surface and a suitable detector held at a negative potential with respect to it. When applicable this is by far the most convenient and accurate method.

(e) *Electron impact ionization detector*. This method is available for use with any neutral beams and is particularly suitable for measurement of relative intensities. After passage through the detector slit the beam is ionized by a stream of electrons emitted from an oxide-coated cathode. The ions are extracted from this region electrostatically and analysed in a mass-spectrograph. To achieve high sensitivity it is necessary to reduce the background signal by chopping the neutral beam and using phase-sensitive detection with narrow-band amplification in much the same way as described, for example, in Chapter 1, § 4.2, Chapter 3, § 2.3, and Chapter 15, § 2.

In Fig. 16.17 we illustrate a typical arrangement of a detector of this sort used by Beier[†] for studying the scattering of helium beams by rare gases. Mass analysis is carried out by a Wien filter and the ion current measured by an open photomultiplier (Du Mont Type SP 187). The dimensions of the detector are as indicated. The slit in the ion extraction electrode Z is of dimension 0.4×2.5 mm and the entrance slit F to the filter 0.2×3 mm. Applied voltages, to extract the ions in the direction of the beam, to direct them on to the entrance slit of the filter and to accelerate them into the photomultiplier, are as indicated. To minimize production of ions from background gas the ionizing electron beam is of

† BEIER, H. J., *Z. Phys.* **196** (1966) 185.

rectangular cross-section of the same breadth as that of the atomic beam (0·1 × 2·5 mm). The Wien filter was chosen as mass analyser because of its high transmission and comparative simplicity. High resolving power is not required. For application of phase-sensitive detection the atomic beam is chopped at 65 c/s.

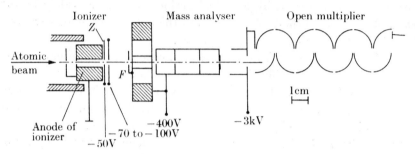

FIG. 16.17. Schematic arrangement of the impact ionization detector used by Beier.

With this arrangement the minimum particle densities (n_m) and currents (i_m) that can be detected for different rare-gas beams are as given in Table 16.4.

TABLE 16.4

Gas	He	Ne	Ar	Kr	Xe
n_m (atoms/cm³)	$6·3 \times 10^3$	$2·8 \times 10^5$	$4·4 \times 10^4$	$4·0 \times 10^3$	$1·5 \times 10^3$
i_m (atoms/s)	$1·9 \times 10^6$	$3·9 \times 10^7$	$4·3 \times 10^6$	$2·7 \times 10^5$	$3·3 \times 10^4$

Applications of this detector are described in §§ 10.3 and 11.

For further information regarding these methods reference should be made to the book by Ramsey (loc. cit).

7.2. *The single beam method for measuring total cross-sections*

7.2.1. *The principle of the method.* To determine the total cross-section for collisions between two kinds of gas atoms A and B a narrow molecular beam of the atoms A may be fired through the gas of atoms B contained in a small scattering chamber. The intensity of the beam is measured after passage through the chamber both when the gas B is absent (I_0) and when N_B atoms/cm³ are present (I). From the ratio I/I_0 of these intensities the average cross-section may be derived as follows.

Let the masses of the atoms A and B be M_A and M_B and the temperatures of the corresponding gases be T_A and T_B, respectively. The chance that an atom A, moving with velocity v, should pass through the chamber of length l is given by

$$P(v) = \exp(-l/\lambda_v). \qquad (156)$$

16.7 COLLISIONS BETWEEN ATOMS

The free path λ_v may be calculated as follows. The number of collisions suffered per second by an atom A with gas molecules moving with velocity u in a direction making an angle θ with that of the beam is given by

$$\nu(u,\theta)\sin\theta\, du d\theta = N_B Q(V) V f_B(u,\theta)\sin\theta\, du d\theta, \qquad (157)$$

where

$$V^2 = u^2 + v^2 - 2uv\cos\theta, \qquad (158)$$

and $f_B(u,\theta)\sin\theta\, du d\theta$ is the fraction of gas molecules with velocities between u and $u+du$ moving in directions making angles between θ and $\theta+d\theta$ with the atomic beam. N_B is the number of scattering molecules per unit volume and $Q(V)$ is the total cross-section for collisions between an atom A and a gas atom B in which the relative velocity is V. Since the time taken for an atom to traverse a distance δl is $\delta l/v$ the number of collisions suffered by the atom in traversing this distance is

$$\delta l/\lambda_v = \frac{\delta l}{v} \iint \nu(u,\theta)\sin\theta\, du d\theta. \qquad (159)$$

We thus have

$$\frac{1}{\lambda_v} = \frac{N_B}{v} \int_0^\infty \int_0^\pi V Q(V) f_B(u,\theta)\sin\theta\, du d\theta. \qquad (160)$$

$f_B(u,\theta)$ is given by the usual Maxwell–Boltzmann distribution function

$$f_B(u,\theta) = \frac{2}{\pi^{\frac{1}{2}}} \frac{1}{\alpha_B^3} u^2 \exp(-u^2/\alpha_B^2), \qquad (161)$$

where

$$\alpha_B = (2\kappa T_B/M_B)^{\frac{1}{2}} = v_{mB}. \qquad (162)$$

Taking $Q(V)$ to have the form

$$Q(V) = A_s V^{-2/(s-1)}, \qquad (163\ a)$$

we find

$$\lambda_v^{-1} = N_B Q(v) F(s, v/v_{mB}), \qquad (163\ b)$$

where

$$F(s,x) = 2\pi^{-\frac{1}{2}} e^{-x^2} x^{(4-2s)/(s-1)} \int_0^\infty y^{(2s-4)/(s-1)} e^{-y^2}\sinh 2xy\, dy. \qquad (164)$$

Thus the effective cross-section $Q_{\text{eff}}(v)$, which is such that

$$\lambda_v^{-1} = N_B Q_{\text{eff}}(v), \qquad (165)$$

is given by

$$Q_{\text{eff}} = Q(v) F(s, v/v_{mB}). \qquad (166)$$

In the special case in which $s \to \infty$, corresponding to a constant cross-section,

$$F(\infty, x) = \pi^{-\frac{1}{2}} \psi(x)/x^2, \qquad (167)$$

where

$$\psi(x) = x e^{-x^2} + (2x^2+1) \int_0^x e^{-z^2}\, dz. \qquad (168)$$

(163) then reduces to the well-known formula in the kinetic theory of gases.†

The most important practical case is that in which $s = 6$ and in Table 16.5, $F(6,x)$, as well as $F(\infty,x)$, is tabulated.‡

The formula (166) gives all that is necessary for the analysis of experimental results if a beam of homogeneous velocity is used. In many experiments, however, a thermal beam will be used and it is necessary to average over the velocity

† KENNARD, E. H., *Kinetic theory of gases* (McGraw-Hill, 1938).
‡ BERKLING, K., HELBING, R., KRAMER, K., PAULY, H., SCHLIER, C., and TOSCHEK, P., *Z. Phys.* **166** (1962) 406.

distribution of the atoms A as well. Because the probability of an atom emerging from the exit slit of the oven source is proportional to its velocity, this distribution function will be given by

$$f_A \, dv = (2/v_{mA}^4)v^3 \exp(-v^2/v_{mA}^2) \, dv, \tag{169}$$

where $v_{mA} = (2\kappa T_A/M_A)^{\frac{1}{2}}$. The ratio I/I_0 of transmitted to incident intensity is then obtained by averaging (156) over f_A to give

$$I/I_0 = \overline{P(v)} = 2v_{mA}^{-4} \int_0^\infty \exp[-\{xN_B A_s F(s, v/v_{mB})/v^{2/(s-1)}\} - v^2/v_{mA}^2] \, dv. \tag{170}$$

TABLE 16.5

The functions $F(6, x)$ and $F(\infty, x)$

x	$F(6, x)$	$F(\infty, x)$	x	$F(6, x)$	$F(\infty, x)$	x	$F(6, x)$	$F(\infty, x)$
0·30	2·2027	3·8731	1·30	1·1452	1·2904	3·00	1·0269	1·0556
0·40	1·8782	2·9690	1·40	1·1256	1·2520	3·50	1·0197	1·0409
0·50	1·6699	2·4403	1·60	1·0963	1·1943	4·00	1·0151	1·0312
0·70	1·4212	1·8632	1·80	1·0759	1·1540	4·50	1·0119	1·0247
0·90	1·2827	1·5677	2·00	1·0612	1·1249	5·00	1·0096	1·0200
1·10	1·1988	1·3969	2·50	1·0389	1·0800	∞	1·000	

The determination of A_s from this formula is rather tedious and a simpler procedure may usually be adopted without serious error. This consists in writing

$$I/I_0 = \exp\{-lN_B \overline{Q}\}, \tag{171}$$

where

$$\overline{Q} = \int Q_{\text{eff}}(v) f_A(v) \, dv$$

$$= \int Q(v) F(s, v/v_{mB}) f_A(v) \, dv$$

$$= Q(v_{mA}) G(s, v_{mA}/v_{mB}), \tag{172 a}$$

where

$$G(s, y) = 2 \int_0^\infty F(s, xy) x^{(3s-5)/(s-1)} e^{-x^2} \, dx. \tag{172 b}$$

This may be expressed in closed form as

$$G(s, y) = \Gamma\left(\frac{2s-3}{s-1}\right)\left(1 + \frac{1}{y^2}\right)^{(s-3)/2(s-1)} \tag{173}$$

and is tabulated in Table 16.6 for $s = 6$ and $s \to \infty$.

TABLE 16.6

The functions $G(6, y)$, $G(\infty, y)$, $G'(6, y)$, $G'(\infty, y)$

y	$G(6, y)$	$G(\infty, y)$	$G'(6, y)$	$G'(\infty, y)$	y	$G(6, y)$	$G(\infty, y)$	$G'(6, y)$	$G'(\infty, y)$
0·30	1·9683	3·4801	2·4850	4·3692	1·8	1·0097	1·1440	1·1627	1·2399
0·40	1·6874	2·6926	2·1180	3·3478	2·0	0·9959	1·1180	1·1391	1·1944
0·50	1·5095	2·2361	1·8817	2·7491	2·5	0·9738	1·0770	1·0996	1·1337
0·70	1·3002	1·7438	1·5967	2·0917	3·0	0·9613	1·0541	1·0761	1·0957
0·90	1·1855	1·4948	1·4340	1·7498	3·5	0·9536	1·0401	1·0621	1·0720
1·10	1·1159	1·3515	1·3312	1·5473	4·0	0·9485	1·0308	1·0508	1·0559
1·30	1·0708	1·2616	1·2619	1·4170	4·5	0·9450	1·0245	1·0435	1·0448
1·40	1·0540	1·2289	1·2355	1·3686	5·0	0·9424	1·0198	1·0381	1·0366
1·60	1·0282	1·1792	1·1938	1·2940	∞	0·9314	1·0000	1·0000	1·0000

A further important point that must not be overlooked is that the sensitivity of the detector system may depend on the velocity of the beam atoms. Thus, with the electron-impact ionization detector, the sensitivity will be proportional to the time spent by an atom in the ionizing electron beam and so to $1/v$. Hence, if such a detector is used, the extra factor v in (169) as compared with the Maxwell–Boltzmann distribution is exactly cancelled out and instead of (169) we must take

$$f_A \, dv = \frac{2}{\pi^{\frac{1}{2}}} \frac{1}{v_{mA}^3} v^2 \exp(-v^2/v_{mA}^2) \, dv. \tag{174}$$

(172 a) still applies but with G replaced by a different function G' also tabulated in Table 16.6.

On the other hand the detecting efficiency of the surface ionization detector is independent of v and (169) applies.

In many experiments phase-sensitive modulation detection is used to reduce the ratio of background noise to wanted signal. When this is done allowance must be made for the velocity dependence which is thereby introduced into the detection sensitivity. In a system of this kind the output signal is proportional to the cosine of the phase angle between the input signal and an arbitrary reference signal. Thus, if the beam is chopped at a frequency ν and traverses a distance d between the beam chopper and the detector the output signal includes the factor

$$\cos\{(2\pi\nu/vd)+\phi\}, \tag{175}$$

where ϕ is an arbitrary phase angle depending on the setting of the phase-shifter dial.

It must be appreciated that analysis of data in terms of the formulae we have discussed will only be really accurate if the elastic cross-section is a monotonic function of v described by (163 a). In fact it will exhibit the undulating behaviour discussed in § 5.4.3. Also the assumption that $2/(s-1)$ in (163) is $2/5$ will not give good results for collisions between light atoms for which the depth of the attractive well may be small compared with κT at ordinary temperatures. For such atoms $2/(s-1)$ is more nearly $2/11$ if the potential is of (12, 6) type.

The accuracy of the cross-section derived naturally depends on accurate knowledge of the density of the scattering gas. Earlier measurements were carried out in ignorance of the 'pumping-effect' error (Chap. 3, § 2.1.4) in absolute pressure measurements made with McLeod gauges, so that the densities measured in many cases were in error by ±30 per cent or so. This error depends on the gauge temperature, the nature of the gas, the diameter of the connecting tube between the gauge and the cold trap, and the gas pressure.

7.2.2. *Angular resolution.* It is important in planning experiments of this kind to ensure that the angular resolving power is great enough.

As a guide for this purpose the schematic angular distribution, illustrated in Fig. 16.2 (b) for the collisions of rigid spheres, may be used. For the parameter a the sum of the gas-kinetic radii of the colliding atoms (determined from viscosity or other phenomena) may be substituted or, with possibly greater accuracy, $\sqrt{2}$ times this sum. It is then possible to determine the minimum angle of deviation θ_0 which is to be counted as a collision in the actual apparatus, so that the ratio

$$\int_{\theta_0}^{\pi} I(\theta)\sin\theta \, d\theta \Big/ \int_{0}^{\pi} I(\theta)\sin\theta \, d\theta$$

is as near unity as is desired. If ϑ_0 is the minimum angle of deviation in degrees, measured in the laboratory system, so that the error in the determination of the cross-section should not exceed 10 per cent,

$$\vartheta_0 \simeq 277/a(MT)^{\frac{1}{2}},$$

where M is the atomic mass of the incident atom, T the equivalent temperature (in °K) of the incident particles, and a is measured in Å. Table 16.7 shows ϑ_0 for a number of different incident particle energies for some cases of scattering that have been investigated.

TABLE 16.7

Angular resolution in apparatus for the measurement of total collision cross-section

ϑ_0 = minimum angle of deviation to be counted as a collision (measured in degrees in the laboratory system), if the error in the total cross-section is not to exceed 10 per cent.

Target atom	Energy of incident particle (lab. system) (eV)	ϑ_0 (degrees) (lab. system) Incident atom							
		He	Ne	Ar	Li	Na	K	Rb	Cs
He	0·0255 (= 300 °K)	3·6	—	—	1·5	0·73	0·50	0·36	0·27
	0·0862 (= 1000 °K)	2·0	—	—	0·80	0·40	0·27	0·20	0·15
	1 (= 11 600 °K)	0·59	—	—	0·23	0·12	0·08	0·06	0·04
Ne	0·0255 (= 300 °K)	—	1·4	—	1·4	0·62	0·40	0·27	0·19
	0·0862 (= 1000 °K)	—	0·75	—	0·75	0·34	0·22	0·15	0·11
	1 (= 11 600 °K)	—	0·22	—	0·22	0·10	0·06	0·04	0·03
Ar	0·0255 (= 300 °K)	—	—	0·70	0·87	0·41	0·27	0·19	0·14
	0·0862 (= 1000 °K)	—	—	0·30	0·48	0·23	0·15	0·10	0·08
	1 (= 11 600 °K)	—	—	0·11	0·14	0·07	0·04	0·03	0·02

Formulae to correct observed total cross-sections for imperfect angular resolution have been derived[†] for cases in which the effective interaction varies as r^{-6}. If Q_{ex} is the observed and Q the true cross-section then

$$Q = Q_{\text{ex}}/R(\rho),$$

† PAULY, H., Z. Phys. **157** (1959) 54; VON BUSCH, F., ibid. **193** (1966) 412; CROSS, R. J., GISLASON, E. A., and HERSCHBACH, D. R., J. chem. Phys. **45** (1966) 3582.

where ρ is a resolution parameter defined by

$$\rho = (k_i Q^{\frac{1}{2}} \pi) w/l.$$

w is the half-width at half-height of the beam profile in the plane of the detector, l the distance from scattering chamber to detector, and k_i the wave number of the motion of the particles in the incident beam (not that of relative motion of beam and target atoms). $R(\rho)$ is given by

$$\begin{aligned} R(\rho) &= 0\cdot6218 - 0\cdot4246\rho + 0\cdot3752\, \exp(-0\cdot8129\rho^2) + \\ &\quad +0\cdot3817\rho\, \mathrm{erf}(0\cdot9016\rho) \quad (\rho < 1\cdot592), \\ &= 0\cdot0844 + 0\cdot5804\rho^{-1/3} \quad (\rho > 1\cdot592). \end{aligned} \quad (176)$$

It is convenient to express the angular resolving power of any equipment in some agreed form so that relative performance may be judged effectively. Kusch[†] has proposed that the resolution be defined in terms of the minimum scattering angle for which the efficiency of detection of scattering is 50 per cent and this has been universally adopted. This angle is approximately equal to $1\cdot5w/l$.

7.2.3. *Typical experimental arrangements.* Because of the convenience and accuracy of the hot-wire method of detection the earliest and much of the later work has been concerned with the study of total cross-sections for collisions of alkali-metal atoms with gas atoms and molecules of various kinds.

The earliest work, that of Broadway,[‡] was concerned with the scattering of a beam of sodium atoms by a beam of mercury atoms. It was directed towards the establishment of the quantal result that

$$\lim_{\theta \to 0} I(\theta)$$

is finite and for this purpose the scattering between angles of $0\cdot2$ and $1°$ was studied. Evidence in favour of the finite limit was found.

The first detailed study was made by Rosin and Rabi,[§] who measured the cross-sections for collisions between the alkali and rare-gas atoms and provided strong evidence of the correctness of the quantum viewpoint. The apparatus they used is illustrated diagrammatically in Fig. 16.18. A molecular beam of alkali atoms issued from the oven 1 through the aperture and fore slit A and, passing through the image slit B, entered the scattering volume 2 containing the rare gas that was admitted through the inlet C. That part of the beam that had not suffered collision in the scattering volume passed out through D into a highly evacuated

[†] Kusch, P., *J. chem. Phys.* **40** (1964) 1.
[‡] Broadway, L. F., *Proc. R. Soc.* **A141** (1933) 626.
[§] Rosin, S., and Rabi, I., *Phys. Rev.* **48** (1935) 373.

region 3. The intensity of this beam could then be surveyed by means of a fine tungsten filament E which could be rotated by means of a ground glass joint so as to traverse the beam. The pressure in the scattering chamber was measured by a McLeod gauge connected to the outlet F.

The slit B was 0·01 mm wide and only 0·05 mm long to reduce scattering within it. The beam was limited to a height of 0·25 mm. The length of path in the scattering chamber was 2 mm and the beam passed into

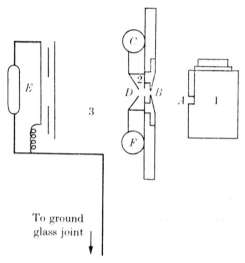

Fig. 16.18. Apparatus used by Rosin and Rabi for measuring total collision cross-sections for impacts of alkali metal and rare-gas atoms.

the detecting chamber through a channel, of 1 mm length, of a shape specially designed to offer high resistance to gas flow. A tungsten filament of 0·02 mm diameter, subtending an angle of 1·7′ at the scattering chamber, constituted the detecting wire.

To make the measurements it was only necessary to sweep the detecting filament through the beam in the absence of the scattering gas and repeat the observations with the gas present at a measured pressure. The important result emerged that, although the magnitude of the beam was reduced by the scattering gas, its shape was practically unaltered. This is illustrated for a typical case, the scattering of sodium by argon, in Fig. 16.19. The significance of this absence of broadening is that the number of collisions involving a deviation through less than a few minutes of arc is a small fraction of the total number of collisions. This is in agreement with the quantum viewpoint but in complete contradiction to the classical theory, according to which at least 50 per cent of the total scattering would take place between 1 and 10′. It indicates further that the angular resolving power of the apparatus is sufficiently high to give accurate values of the total cross-sections. In deriving the cross-sections in the manner explained in § 7.2.1 correction was made for the effect of the gas-pressure gradient in the channel between scattering and detecting chambers. The final values obtained are given in Table 16.10.

Similar experiments (K in He and Ar) were carried out by Rosenberg† a few years later. After a considerable interval (nearly twenty years) this work was resumed by Rothe and Bernstein‡ using equipment very similar in principle. Fig. 16.20 illustrates the general arrangement of their apparatus. C is a section of the bulkhead separating the oven and detector chambers which were separately pumped to pressures of

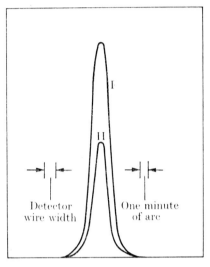

FIG. 16.19. Shape of the beam of alkali-metal atoms in the experiments of Rosin and Rabi. I, before absorption. II, after absorption.

5×10^{-7} and 1×10^{-7} torr respectively. The oven and collimating slits B and D respectively were each of width $0\cdot0025 \pm 0\cdot001$ cm. The inlet and outlet channels of the scattering chamber were each 1·27 cm long, 0·38 cm high, and 0·028 cm wide. As the central cavity was 2·53 cm long the effective scattering path was taken to be

$$1\cdot27 + 2\cdot53 + 1\cdot27 \times \tfrac{1}{2} = 4\cdot44 \text{ cm.}$$

The distances BD and DI were 11·12 and 19·68 cm respectively.

With this geometry the angle subtended at the mid-point of the scattering path by the detector wire (0·0025 cm in diameter) was about 30″. With a beam intensity of 6×10^8 atoms/s ion currents of order 10^{-10} A were recorded by the detector.

Pressure in the scattering chamber was measured by a Knudsen gauge calibrated with a McLeod gauge. The working pressure in the scattering chamber ranged between 1×10^{-6} and 2×10^{-4} torr.

† ROSENBERG, P., *Phys. Rev.* **55** (1939) 1267.
‡ ROTHE, E. W. and BERNSTEIN, R. B., *J. chem. Phys.* **31** (1959) 1619.

In all these, and many later, experiments a thermal beam was used. The first measurements that used beams of homogeneous velocity were those of Estermann, Foner, and Stern.† These experiments were designed to investigate the deflexion of a beam of caesium atoms due to gravity and had a very high angular resolution. As far as total cross-section measurements were concerned they essentially employed gravity as a

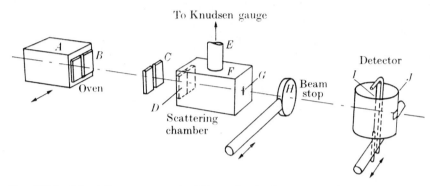

FIG. 16.20. Schematic arrangement of the molecular beam scattering experiments of Rothe and Bernstein.

velocity separator. Fig. 16.21 illustrates the principle of the method and the important dimensions. For the main purpose of the work the scattering chamber C was not included. Caesium atoms evaporated from the oven O were collimated by a fore slit S_1 and collimating slit S_2 each of 0·02 mm width. The beam was detected by a tungsten wire detector D of 0·02 mm diameter which could be moved parallel to itself in a vertical plane perpendicular to the undeviated beam. Owing to the long path length between S_2 and D an atom with the most probable velocity corresponding to a temperature of 450 °K suffered a downward deflexion due to gravity of 0·174 mm before arriving at the detector. Hence, as the detector traversed a vertical plane, the intensity of the beam observed varied in a manner determined by the velocity distribution of the atoms in the beam issuing from S_2. The current observed at a particular setting of the detector was thus a measure of the flux of atoms with velocity lying within narrow limits.

In the first experiments of Estermann, Simpson, and Stern‡ the velocity distribution of the atoms was studied. To measure the effective cross-sections, or more correctly the free path λ_v of (165) for a fixed value of v, it was only necessary to introduce the scattering chamber C, the

† ESTERMANN, I., FONER, S. N., and STERN, O., Phys. Rev. 71 (1947) 250.
‡ ESTERMANN, I., SIMPSON, O. C., and STERN, O., ibid. 238.

centre of which was 85 cm from S_2. A change in direction of the velocity of a beam atom through an angle $0\cdot02/850$ radians, $\simeq 5''$, prevented it from passing through so it would be regarded as scattered. This gave a very high resolving power which should have been completely adequate. The measurement of I/I_0 was then carried out in the usual way with different pressures of gas in C for different settings of the detector D

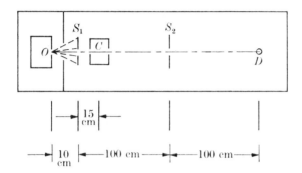

FIG. 16.21. Apparatus used by Estermann, Foner, and Stern for measuring total collision cross-sections for impacts of caesium atoms with rare-gas atoms.

corresponding to different velocities v of the beam atoms. This was done for collisions of the caesium atoms with helium, nitrogen, and caesium. The value obtained at the most probable velocity, for helium, is large. It is about $2\frac{1}{2}$ times greater than that found by other investigators (see Table 16.10).

For Cs–Cs collisions a special scattering chamber of monel metal was used and the pressure determined by an auxiliary detector which measured the efflux of vapour through a side slit and a collimating slit. The cross-section found at the most probable velocity is $2\cdot35\times10^{-14}$ cm².

7.2.3.1. *The use of velocity selectors.* A number of years elapsed before a programme of measurements using velocity-selected beams was fully initiated. These have been based mainly on the rotating slotted disc (Fizeau) type of velocity-selector but in some experiments† a Stern–Gerlach magnet has been used.

In general the velocity resolution attainable with magnetic methods of velocity selection is much poorer than with mechanical selectors. However, for experiments in which the most probable velocity in the

---

† COHEN, V. W. and ELLET, A., *Phys. Rev.* **52** (1937) 502; LULLA, K., BROWN, H. H., and BEDERSEN, B., ibid. **136A** (1964) 1233; FLUENDY, M. A. D., MARTIN, R. M., MUSCHLITZ, E. E., and HERSCHBACH, D. R., *J. chem. Phys.* **46** (1967) 2172.

beam before selection is higher than 9×10^5 cm s^{-1} it is difficult to use a mechanical selector.

As an example of a typical mechanical velocity-selector we give some details of the six-disc instrument designed by Hostettler and Bernstein.† The aim was to obtain performance, as regards high transmission and elimination of velocity 'sidebands', comparable with that obtained with selectors consisting of cylinders cut with many helical grooves, which have been extensively used,‡ both for obtaining molecular and neutron beams of homogeneous energy, but which are mechanically more difficult to construct. This was achieved by using relatively thick discs suitably placed. Additional advantages are the lightness of the rotor, which can therefore be run at relatively high speeds, ease of fabrication, the discs having straight perpendicular slots, and applicability to permanent gases as well as vapours (the molecules of the wrong speed are removed mainly by collision with the face of a disc rather than the side wall of a groove).

To discuss the design details it is convenient to consider the surface of the rotor as unrolled in a plane as shown in Fig. 16.22 (a). In the selector, molecules of the desired velocity are transmitted through those slots of the first and last discs which are rotated with respect to each other through an angle ϕ. On the plane diagram of Fig. 16.22 (a) the trajectories of molecules are straight lines, the cotangents of whose inclinations to the planes of the discs are proportional to the molecular speed. The intermediate discs are placed so as to interrupt as far as possible all molecules that do not have the selected speed.

The geometry of the rotor is determined by the radius r of each disc, the disc thickness d, the length L of the rotor, the slot width l_1 and the wall thickness l_2 between slots, as well as the angle ϕ. It is convenient in discussing performance to introduce the derived quantities

$$\beta = d/L, \quad \gamma = l_1/r\phi \quad \text{and} \quad \eta = l_1/(l_1+l_2). \tag{177}$$

For ease of assembly the slots in the end discs should, in a projection such as that of Fig. 16.22 (b), either coincide or fall mostly between each other so that

$$\phi r = n(l_1+l_2)/2,$$

where n is an integer. This means that

$$\gamma = 2\eta/n. \tag{178}$$

Similarly, ease of alignment limits the possible positions of the intermediate discs to $n-1$. With n chosen as 15 and $\eta = \frac{1}{2}(l_1+l_2)$ it was found by trial and error that, to eliminate all velocity side-bands, at least four intermediate discs are required, one of which is placed at a position such that L/n is half integral. The chosen positions are shown in Fig. 16.22 (a).

If v_0 is the velocity to be selected, the rotor must be run at an angular velocity ω such that

$$v_0/L = \omega/\phi. \tag{179}$$

---

† HOSTETTLER, H. U. and BERNSTEIN, R. B., *Rev. scient. Instrum.* **31** (1960) 872.
‡ MILLER, R. C. and KUSCH, P., *Phys. Rev.* **99** (1955) 1314; DASH, J. G. and SOMMERS, H. S., *Rev. scient. Instrum.* **24** (1953) 91; GREENE, E. F., ROBERTS, R. W., and ROSS, J., *J. chem. Phys.* **32** (1960) 940.

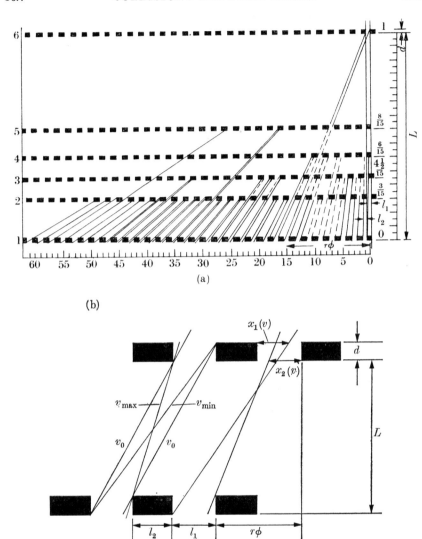

FIG. 16.22. (a) Illustrating the geometry of the slotted-disc velocity selector of Hostettler and Bernstein. The diagram is obtained by supposing the surface of the rotor unrolled on to a plane. (b) Geometry relating to the determination of the resolution and transmission of a slotted-disc velocity selector.

Referring to Fig. 16.22 (b), if $v_{\max}$, $v_{\min}$ are the maximum and minimum transmitted velocities respectively,

$$v_{\max}/(L-d) = \omega/(\phi - l_1/r), \tag{180 a}$$
$$v_{\min}/(L+d) = \omega/(\phi + l_1/r), \tag{180 b}$$

giving
$$v_{\max} = v_0(1-\beta)/(1-\gamma),$$
$$v_{\min} = v_0(1+\beta)/(1+\gamma).$$

The transmission of the rotor is given by

$$T = \int_{v_{\min}}^{v_{\max}} I(v)B(v)\,dv,$$

where $B(v)$ is the transmission for velocity v and $I(v)$ the velocity distribution function of the incident beam. Since $v \simeq v_0$ we may write

$$T \simeq I(v_0) \int_{v_{\min}}^{v_{\max}} B(v)\,dv. \tag{181}$$

To calculate $B(v)$ we see from Fig. 16.22 (b) that

$$B(v) = 1 - x_1/l_1 \quad (v < v_0),$$
$$= 1 - x_2/l_1 \quad (v > v_0),$$

whereas in (180 a) and (180 b)

$$v/(L+d) = \omega/(\phi + x_1/r), \qquad v/(L-d) = \omega/(\phi - x_2/r).$$

Eliminating x_1 and x_2, using (177), (178), and (179), gives

$$B(v) = \eta[1 - \{(1+\beta)(v_0/v) - 1\}/\gamma] \quad (v_0 > v > v_{\min})$$
$$= \eta[1 + \{(1-\beta)(v_0/v) - 1\}/\gamma] \quad (v_{\max} > v > v_0)$$
$$= 0 \quad \text{(otherwise)}. \tag{182}$$

Substitution in (181) gives

$$T = I(v_0)\eta v_0 \gamma^{-1}[(1+\beta)\ln\{(1+\beta)/(1+\gamma)\} + (1-\beta)\ln\{(1-\beta)/(1-\gamma)\}], \tag{183}$$

which, for $\beta \ll 1, \gamma \ll 1$, reduces to

$$T = G v_0 I(v_0),$$

where $G = \eta\gamma(1-\beta/\gamma)^2$, and is purely geometrical. The effective fractional time the selector is open is given by

$$\eta' = \eta(1-\beta/\gamma). \tag{184}$$

The function $B(v)$ in (182) is nearly triangular in form so the resolution R, defined by the width at half-intensity, is given by

$$R = (v_{\max} - v_{\min})/2v_0$$
$$= (\gamma - \beta)/(1-\gamma^2)$$
$$\simeq \gamma - \beta. \tag{185}$$

In practice correction must be made for the fact that the incident beam will not be perfectly collimated† and for imperfect alignment of the beam axis along the rotor shaft.

A selector of this kind constructed by Hostettler and Bernstein had the following dimensions.

No. of slots per disc 278, radial length of slots 0·8 cm, slot width $l = 0{\cdot}0813$ cm, average wall thickness l_2 between slots 0·0905 cm, $r = 7{\cdot}6$ cm, $d = 0{\cdot}1628$ cm, $L = 9{\cdot}997$ cm, $\phi = 0{\cdot}1695$ rad, $\gamma = 0{\cdot}0631, \beta = 0{\cdot}01628, \eta = 0{\cdot}4732, G = 0{\cdot}0164, \eta' = 0{\cdot}35$, resolution $R = 0{\cdot}0468$.

The ratio v/ω is 6·177 if v is in cm s^{-1} and ω in rev min^{-1} so that with the highest obtainable rate of rotation (17 000 rev min^{-1}) $v_0 = 1{\cdot}05 \times 10^5$ cm s^{-1}.

† DASH, J. G. and SOMMERS, H. S., *Rev. scient. Instrum.* **24** (1953) 91.

The discs were made of Alcoa 2024–T3 aluminium alloy. Ball-bearings enclosed in the motor were greased before assembly with Mil G 3278 grease and those for the disc assembly lubricated regularly with DC 702 silicone pump fluid. This permitted operation at pressures below 3×10^{-7} torr.

Rothe, Rol, Trujillo, and Neynaber† in their experiments on the velocity variation of total cross-sections for Li and K atoms in xenon used a velocity-selector of this type and also a phase-sensitive modulation

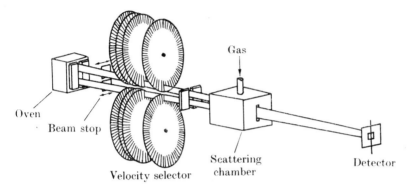

Fig. 16.23. Schematic arrangement of the molecular beam scattering experiments of von Busch, Strunck, and Schlier.

detector. Apart from the velocity-selector the apparatus was very similar to that of Rothe and Bernstein referred to on p. 1353. The detector wire, instead of being of pure tungsten, was an alloy of 92 per cent platinum with 8 per cent tungsten, which was found to give rise to less noise in detection. For measurement of the intensity of the lithium beam the wire was continuously sprayed with oxygen. Typical ion currents recorded were about 3×10^{-12} A.

As an example of an elaborate experiment using velocity selection we describe the arrangement used by von Busch, Strunck, and Schlier‡ to observe with precision the dependence of the total cross-section on relative velocity for collisions between alkali metal and rare-gas atoms. For this purpose they used two velocity-selected beams of alkali metal atoms and compared the absorptions suffered by each beam after passage through the same gas chamber. Fig. 16.23 illustrates this arrangement diagrammatically. The two beams issue from the same oven, pass through separate slotted-disc velocity selectors and thence through the same scattering gas chamber to be measured by the same detector. If

† ROTHE, E. W., ROL, P. K., TRUJILLO, S. M., and NEYNABER, R. H., *Phys. Rev.* **128** (1962) 659.
‡ VON BUSCH, F., STRUNCK, H. J., and SCHLIER, C., *Z. Phys.* **199** (1967) 518.

$I_p(v_1)$ is the intensity measured at this detector for the beam of velocity v_1 when the chamber pressure is p then by measuring $I_p(v_1)$, $I_{p'}(v_1)$, $I_p(v_2)$, $I_{p'}(v_2)$ we have, for the ratio of the effective cross-sections Q_{eff} when the beam velocities are v_1, v_2 respectively,

$$Q_{\text{eff}}(v_1)/Q_{\text{eff}}(v_2) = 1 + \ln\left\{\frac{I_p(v_1)I_{p'}(v_1)}{I_p(v_2)I_{p'}(v_2)}\right\} \bigg/ \ln\{I_p(v_2)/I_{p'}(v_2)\},$$

and it is not necessary to know the pressures p, p' or the effective path length within the scattering chamber, provided the geometry of the chamber is the same for both beams.

The principal dimensions of the system are as follows:

Width of oven and collimator slots	50 μm
Width of detector slit	100 μm
Distance from oven to collimator	53·8 cm
Distance from oven to detector	104·2 cm
Distance from centre of gas chamber to detector	38·9 cm
Length of gas chamber	9·5 cm
Dimension of beam entrance slit to chamber	1·0 × 0·04 cm
Dimension of beam exit slit from chamber	1·0 × 0·9 cm
Velocity resolution of selector	0·042

The surface ionization detector was a wire of 92 per cent Pt and 8 per cent W, 100 μm thick and the measured ion current between 10^{-11} and 10^{-13} A. An account of the results obtained with this apparatus is given in § 9.2.2.

TABLE 16.8

The functions $F_1(6, x)$, $F_1(\infty, x)$, $G_1(6, x)$, $G_1(\infty, x)$, $G_1'(6, x)$, $G_1'(\infty, x)$

x	0·30	0·40	0·50	0·70	0·90	1·10	1·30	1·40	1·60
$F_1(6, x)$	2·2284	1·9120	1·7104	1·4704	1·3351	1·2506	1·1941	1·1726	1·1390
$F_1(\infty, x)$	3·9196	3·0251	2·5033	1·9332	1·6382	1·4640	1·3522	1·3108	1·2469
$G_1(6, x)$	1·9968	1·7227	1·5493	1·3441	1·2289	1·1568	1·1084	1·0899	1·0607
$G_1(\infty, x)$	3·5333	2·7532	2·3005	1·8090	1·5560	1·4069	1·3110	1·2754	1·2203
$G_1'(6, x)$	2·5109	2·1508	1·9198	1·6409	1·4799	1·3765	1·3053	1·2776	1·2332
$G_1'(\infty, x)$	4·4169	3·4053	2·8095	2·1558	1·8125	1·6063	1·4715	1·4207	1·3415

x	1·8	2·0	2·5	3·0	3·5	4·0	4·5	5·0	∞
$F_1(6, x)$	1·1141	1·0952	1·0641	1·0460	1·0345	1·0267	1·0214	1·0174	1·0000
$F_1(\infty, x)$	1·2007	1·1662	1·1104	1·0784	1·0587	1·0452	1·0360	1·0293	1·0000
$G_1(6, x)$	1·0389	1·0222	0·9941	0·9773	0·9665	0·9590	0·9537	0·9498	0·9314
$G_1(\infty, x)$	1·1803	1·1503	1·1013	1·0728	1·0549	1·0427	1·0342	1·0280	1·0000
$G_1'(6, x)$	1·1994	1·1731	1·1276	1·0994	1·0806	1·0673	1·0576	1·0504	1·0000
$G_1'(\infty, x)$	1·2830	1·2386	1·1648	1·1208	1·0926	1·0730	1·0592	1·0489	1·0000

An interesting application of the magnetic method has been made by Fluendy, Martin, Muschlitz, and Herschbach† in the course of experiments on the scattering of H atoms by various gases. Here the magnetic analyser served a dual role of separating H atoms from H_2 molecules and of velocity selection—when the oven source was at a temperature above 2500 °K the most probable velocity of the H atoms exceeded 9×10^5 cm s^{-1} which is too high for convenient use of mechanical methods.

Fig. 16.24 (a) illustrates the general arrangement of the apparatus. Dissociation of H_2 was produced thermally in a tungsten tube furnace heated to temperatures between 2700 and 3000 °K. With a H_2 pressure of about 0·5 torr, 60 per cent dissociation was achieved. The mixed beam emerging from the furnace was modulated at about 10 c/s by a rotating disc, H, which also interrupted a light beam so as to provide a signal for phase-sensitive detection. The mixed beam could also be interrupted by a mechanical flag, G. Velocity selection and separation of H from H_2 was performed by the inhomogeneous deflecting magnet, J. The exit slits from the magnet were offset from the axis of the incident beam so as to transmit only one of the spin states. The position of the slit could be adjusted in increments of 0·0064 mm. After passage through the exit slit the H atom beam passed into a region within a cold shield, R, through the scattering chamber, M, to be detected bolometrically by the heat of recombination on a thin platinum strip (10^{-3} mm thick, 0·25 mm wide, 6·5 mm long, and of resistance 40 Ω). The maximum sensitivity of this detector was about 10^{10}–10^{11} atoms cm^{-2} s^{-1}. With the offset exit slit most of the spurious modulated signal due to infra-red radiation from the furnace was removed except for that reflected from the magnetic pole which was between 10 and 30 per cent of the maximum H atom signal. This was consistent to about 5 per cent. It was measured with zero field and subtracted. Under the conditions of the experiments the velocity resolution was about 27 per cent. Fig. 16.24 (b) shows the beam geometry and the slit widths. The angular resolution, in the laboratory system, according to the Kusch criterion, was 4·8′. Results obtained with this equipment are discussed in § 12.1.3.

7.3. *The crossed-beam method for measuring total cross-sections*

It is possible to determine total cross-sections by observing the attenuation of one beam by passage of a second beam across it. There is no difficulty in modifying the analysis of § 7.2.1 to apply to this case.

† FLUENDY, M. H. D., MARTIN, R. M., MUSCHLITZ, E. E., and HERSCHBACH, D. R., *J. chem. Phys.* **46** (1967) 2172.

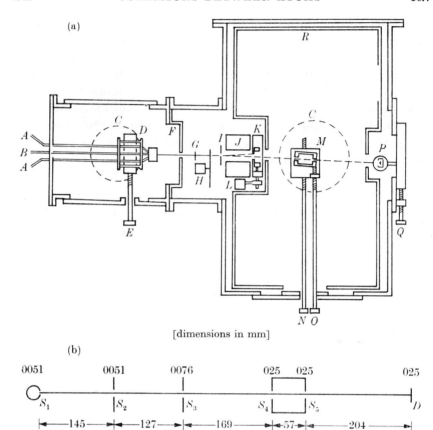

Fig. 16.24. (a) Arrangement of apparatus used by Fluendy, Martin, Muschlitz, and Herschbach for observing the scattering of H atoms in various gases: A, copper tubing ($\frac{1}{2}$ in diam.) supplying electric current and cooling water to tungsten oven; B, H_2 supply tube; C, oil diffusion pumps; D, hydrogen oven and supporting framework; E, traverse control for aligning the oven; F, water-cooled radiation shield separating chambers; G, beam flag; H, chopper wheel and motor; I, magnet entrance slit; J, deflecting magnet; K, magnet exit slit; L, motor for traversing magnet exit slit; M, scattering chamber; N, scattering chamber traverse control; O, scattering chamber rotation control; P, bolometer detector; Q, bolometer traverse control; R, cold shields. (b) Geometry of the beam used in the apparatus. S_1, oven slit; S_2 and S_3, entrance and exit slits for the magnet; S_4 and S_5, slits for scattering chamber; D, detector filament.

Thus, if the primary beam is velocity-selected, with velocity v, the ratio of transmitted, I, to incident intensity I_0 is given by

$$\ln(I/I_0) = \overline{lN_\mathrm{B}}\, Q(v) F_1(s,\, v/v_\mathrm{mB}) \qquad (186)$$

in the notation of § 7.2.1. This differs from the single-beam case in two respects. Thus $\overline{lN_\mathrm{B}}$ represents the mean value of this product over the region of interaction of the two beams and is less readily determined than in single-beam experiments. F_1 differs from F in (164) because of the unidirectional velocities possessed by the scattering atoms. When the beams intersect at 90°, which is usual in experiments of this kind,

$$F_1(s,x) = 4\pi^{-\frac{1}{2}}x^{-2\alpha} \int_0^\infty (x^2+z^2)^\alpha z^2 e^{-z^2}\,dz,$$

with $\alpha = (s-3)/2(s-1)$, and is tabulated for $s = 6$ and $s \to \infty$ in Table 16.8.†

Similarly, if the primary beam is not velocity-selected we have, again in the notation of § 7.2.1,

$$\ln(I/I_0) = \overline{l N_B}\, Q(v_{mA}) G_1(s, v_{mA}/v_{mB}), \qquad (187)$$

if the detector sensitivity is independent of beam velocity. If it is proportional to the beam concentration, as in the impact ionization detector, then G_1 in (187) is replaced by G'_1. Both G_1 and G'_1 are tabulated in Table 16.8 for $s = 6$ and $s \to \infty$.

One difficulty associated with the use of crossed-beam methods for determining total cross-sections is the relatively small attenuation produced which emphasizes the importance of reducing background signals. This may be done by chopping one of the beams mechanically and using phase-sensitive detectors as in crossed beam experiments involving charged particles (see, for example, Chap. 1, § 4.2, Chap. 3, § 2.3, Chap. 4, § 3, and Chap. 15, § 2). The further difficulty already referred to, is that of defining the scattering region both geometrically and in terms of the concentration of scattering atoms. Special attention must be paid to geometrical definition of the scattering beam and to the uniformity of atom concentration over the cross-section. On the other hand, crossed-beam methods offer greater flexibility in choice of scattering atom (the applications referred to above in earlier chapters have all been made to measure cross-sections for atoms or ions that are not attainable in bulk). Furthermore, it is possible to avoid the use of a McLeod gauge for pressure measurement as the concentration of scattering atoms in the secondary beam can, in principle, be determined by flow-rate measurements.

As an example of equipment in which a velocity-selected primary beam is used we take that constructed by Beck and Loesch‡ to determine the velocity dependence of the total cross-sections for collisions of K with Hg, Xe, Kr, and Ar atoms and of Cs with Hg. In all cases the

† BERKLING, K., HELBING, R., KRAMER, K., PAULY, H., SCHLIER, C., and TOSCHEK, P., loc. cit., p. 1347.
‡ BECK, D. and LOESCH, H. J., *Z. Phys.* **195** (1966) 444.

alkali-metal atom beam was the primary velocity-selected one. No attempt was made to determine absolute cross-sections so that, provided the beam geometry and concentration remained constant, there was no need to evaluate $\overline{N_\text{B}} l$ in (186).

The dimensions of the apparatus are shown in the schematic diagram of Fig. 16.25. The velocity selector, of slotted-disc type, gave a velocity width at half intensity of 3 per cent.

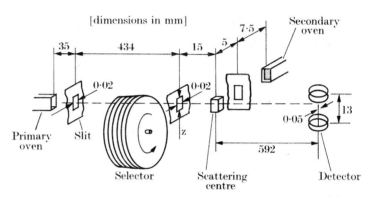

Fig. 16.25. Schematic arrangement of the molecular beam scattering experiments of Beck and Loesch.

The secondary beams were passed out of the secondary oven through a many-channel opening. For the rare gases the pressure in the secondary oven was between 0·2 and 0·5 torr while the temperature of the front chamber was 77 °K for Ar and Kr and 105 °K for Xe. The beam traversed about 2 mm before reaching the centre of the scattering region. It was not collimated and measurements made with a potassium beam at the same oven pressure indicated that its half breadth was about 30°. The beam was condensed on a liquid helium-cooled surface opposite the oven.

For the mercury beam the pressure in the secondary oven was much higher (between 5 and 10 torr), the temperature of the front chamber of the oven being 473 °K. The beam passed through a collimator before reaching the scattering region so that its half breadth was about 22°. Because of the high oven-pressure the mean velocity of the mercury atoms did not correspond exactly with that calculated from the oven temperature but measurements with a potassium beam at comparable pressure showed that the two probably did not differ by more than 20 per cent.

The angular resolving power, according to the Kusch criterion, was 30″ of arc.

We shall discuss the analysis of the observations obtained with this apparatus, which were directed towards the investigation of the undulatory effects in the cross-section due to glory interference, in § 9.2.2.

Landorf and Mueller† have constructed a crossed-beam apparatus for total cross-section measurements in which both primary and secondary beams are of permanent gases so that detection is carried out by the electron impact ionization technique. With this arrangement, which was designed to measure absolute cross-sections, the primary beam can be of either gas. When the masses of the two atoms involved are very different the angular resolution in the centre of mass system depends markedly on which beam is primary so that, if the results obtained are the same in both cases, there is a strong presumption that the angular resolution is adequate.

Fig. 16.26 illustrates the general arrangement of the apparatus. The principal dimensions are as follows.

	Primary beam	Scattering beam
Distance source aperture to collimator aperture (cm)	10·20	1·00
Distance collimator aperture to beam intersection (cm)	1·91	0·30
Distance collimator aperture to detector aperture (cm)	17·11	15·50
Source aperture (cm)	0·103 (diam.)	0·997 × 0·1028 (width × height)
Collimator aperture (cm)	0·103 (diam.)	0·996 × 0·1028 (width × height)
Detector aperture (cm)	0·051$_5$ (diam.)	
Intersecting region (cm)	0·103 (diam.)	0·992 × 0·103 (width × height)

The detector could be rotated about the intersection of the beam axes. It subtended an angle of 11·3′ at the beam intersection. Beam stops were provided, that for the primary beam being placed between the electron and the beam intersection and that for the scattering beam in the collimator region. These stops were important in determining the importance of background effects due to scattering of the primary beam in the collimator and detector regions. It was found that the attenuation from this latter source was about 8 per cent of that arising from scattering in the intersecting beam region. While that for the collector region was larger it did not depend on whether the scattering beam stop was in place or not and so did not affect the determination of the attenuation by the scattering beam.

† LANDORF, R. W. and MUELLER, C. R., *J. chem. Phys.* **45** (1966) 240.

The density of scattering atoms in the intersecting region was determined from the flow rate of the scattering gas from the storage reservoir and from the geometry of the source and collimator apertures. With primary beams scattered by He beams the flow rates used ranged from

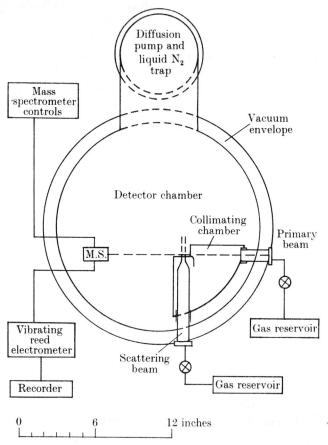

Fig. 16.26. Schematic arrangement of the molecular beam scattering experiments of Landorf and Mueller.

0·33 to 1·98 torr/min and the attenuation ratio I/I_0 from 0·972 to 0·859. Interchanging the beam flow rates from 0·08 to 0·135 torr/min gave attenuation ratios ranging from 0·985 to 0·975. Results obtained with this apparatus are discussed in § 11.

8. The measurement of differential cross-sections

Experiments to measure differential cross-sections consist essentially in crossing two beams A and B, usually at right angles, and measuring the intensity say of A as a function of the angle of scattering θ_A measured relative to the direction of the beam A. Analysis of the results is complicated by the need to transform from this laboratory reference system to the centre of mass (C.M.) system in which the angle of scattering is θ. If both beams are thermal with a considerable velocity spread the analysis becomes extremely complicated. Ultimately, to avoid this, experiments must be carried out with both velocity-selected beams A and B. In default of this one beam of relatively light atoms can be velocity-selected while the other consists of heavy atoms, preferably at a low temperature. The mean velocity of the latter can then be small compared with the homogeneous selected velocity of the atoms in the other beam.

If $\mathbf{v}_A$ and $\mathbf{v}_B$ are the velocities of the atoms in two beams intersecting at right angles, measured in the laboratory system, $\mathbf{w}_A$, $\mathbf{w}_B$ the corresponding velocities in the centre of mass system, the initial relative velocity $\mathbf{v}_r$ is given by

$$\mathbf{v}_r = \mathbf{v}_A - \mathbf{v}_B = \mathbf{w}_A - \mathbf{w}_B. \tag{188 a}$$

After an elastic collision the corresponding quantities will be $\mathbf{v}'_A$, $\mathbf{v}'_B$, $\mathbf{w}'_A$, $\mathbf{w}'_B$, $\mathbf{v}'_r$, where

$$v'_r = v_r, \quad w'_A = w_A, \quad w'_B = w_B. \tag{188 b}$$

The angles of scattering in the laboratory and centre of mass system are given by

$$\cos\theta_A = \frac{\mathbf{v}_A \cdot \mathbf{v}'_A}{v_A v'_A}, \quad \cos\theta = \frac{\mathbf{w}_A \cdot \mathbf{w}'_A}{w_A^2}. \tag{189}$$

We must specify also the angle ϕ_A between the plane of $\mathbf{v}_A$, $\mathbf{v}'_A$ and that of $\mathbf{v}_A$, $\mathbf{v}_B$. Since the velocity $\mathbf{v}_c$ of the centre of mass is unchanged in the collision and

$$\mathbf{w}_A = \mathbf{v}_A - \mathbf{v}_c, \quad \mathbf{w}'_A = \mathbf{v}'_A - \mathbf{v}_c,$$

then

$$\mathbf{w}_A - \mathbf{w}'_A = \mathbf{v}_A - \mathbf{v}'_A. \tag{190}$$

The geometry of the system is as shown in Fig. 16.27.

It follows from (189) that

$$\cos\theta = (w_A^2 + w'^2_A - v_A^2 - v'^2_A + 2v_A v'_A \cos\theta_A)/2w_A^2. \tag{191}$$

Also

$$w_A = M_B v_r/(M_A + M_B), \tag{192 a}$$

$$v_r = (v_A^2 + v_B^2)^{\frac{1}{2}}, \tag{192 b}$$

$$v_c = (M_A^2 v_A^2 + M_B^2 v_B^2)^{\frac{1}{2}}/(M_A + M_B), \tag{192 c}$$

$$v'_A = v_c \cos\gamma \pm \{w_A^2 - v_c^2 \sin^2\gamma\}^{\frac{1}{2}}, \tag{192 d}$$

$$\cos\gamma = \cos\theta_A \cos\alpha - \sin\theta_A \sin\alpha \sin\phi_A, \tag{192 e}$$

$$\tan\alpha = M_B v_B/M_A v_A. \tag{192 f}$$

From these relations θ may be related to θ_A and ϕ_A.

There still remains the transformation from the solid angle element in one system to that in the other. Thus if $I(\theta_A)\,d\omega_A$ is the intensity scattered through θ_A into the solid angle $d\omega_A$, then in the centre of mass system

$$I(\theta) = I(\theta_A)\frac{d\omega_A}{d\omega}. \tag{193}$$

Referring to Fig. 16.27 and remembering the definition of solid angle we see that

$$\frac{d\omega_A}{d\omega} = \left(\frac{w'_A}{v'_A}\right)^2 |\cos \zeta'_A|$$

$$= \frac{w'_A}{2v'^3_A}(w'^2_A + v'^2_A - v^2_c). \tag{194}$$

It is apparent from these formulae how complicated the analysis would become if the velocity distribution of the atoms in even one of the beams were taken into account.

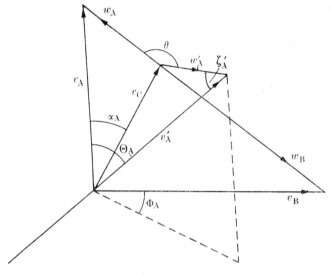

Fig. 16.27

In a typical experiment, such as that carried out by Morse and Bernstein[†] for the scattering of atomic beams of K and Cs by Hg, the primary beam was collimated, modulated at 25 c/s with a chopper and velocity selected with a triangular velocity distribution of half width 4·7 per cent of the nominal value (see p. 1358). The speed of the alkali atoms could be varied between 185 and 1000 m/s.

The secondary beam, crossing the primary at 90° was a collimated mercury beam with an average speed of 235 m/s.

The detector was a platinum–tungsten ribbon surface ionizer followed by an electron multiplier, cathode follower, narrow band amplifier, phase-sensitive rectifier, and recorder.

In the angular distribution measurements the detector was held fixed and the complete assembly of primary and secondary ovens, etc., could

† MORSE, F. A. and BERNSTEIN, R. B., *J. chem. Phys.* **37** (1962) 2019.

be rotated as a whole through an angle ranging from 30° on one side of the primary beam to 60° on the other. For the primary beam the half widths of the source and collimator slit were 0·0152 and 0·0190 cm respectively, the corresponding values for the secondary beam being 0·159 cm for each slit. The half-width of the detector was 0·0254 cm and its distance from the collimating slit was 29·35 cm. With this geometry the Kusch resolution angle (see p. 1351) was 14′.

The intensity of the mercury beam was maintained constant to better than 1 per cent by controlling the mercury oven temperature to within $\pm 0 \cdot 2$ °C. Constancy of the primary beam was checked at frequent intervals by direct measurement with the detector set at zero scattering angle.

Similar experiments will be briefly described in connection with results obtained, in § 9.4.

9. Discussion of experimental results for collisions of alkali-metal atoms with rare-gas and mercury atoms

9.1. *The classical approximation and its failure at very small scattering angles*

As discussed in § 5.1.6 the classical formula for the scattering by an inverse-power interaction must fail at sufficiently small angles so that the scattered intensity tends to a finite limit in the forward direction and the total cross-section exists. Indirect evidence that this occurs was provided from measurements of the total cross-section, which showed that the observed value was independent of the angular resolution in the apparatus provided this was smaller than a certain value, $\theta_{\min}$ say. However, much more definite results were obtained by Helbing and Pauly,† who measured the scattered intensity as a function of angle of scattering in the laboratory system for collisions of potassium atoms with argon and with xenon atoms. This they did by observing the scattering of an alkali metal atom beam by a beam of rare-gas atoms intersecting it at 90°.

The surface ionization detector wire was 8 μm thick so that it subtended an angle of 5″ at the beam intersection. Allowing for other geometrical features, scattering from the primary beam through 15″ of arc could be detected. To observe the variation of scattered intensity with scattering angle θ_A in the laboratory system the detector wire could be moved parallel to itself in a direction parallel to the axis of the scattering

† HELBING, R. and PAULY, H., *Z. Phys.* **179** (1964) 16.

beam. To examine the effect of the finite length of the detector measurements were made with two lengths of wire (1·8 and 14·2 mm).

For small angle collisions, $\theta_A \ll 1$, the relations between the scattering angles (θ, ϕ) in the C.M. and laboratory system (θ_A, ϕ_A) for beams intersecting at 90° take the form

$$\theta = \frac{v_A}{v_r}\left(1+\frac{M_A}{M_B}\right)\theta_A, \qquad (195\,a)$$

$$\tan|\phi| = \frac{v_A}{v_r}\tan|\phi_A|, \qquad (195\,b)$$

$$\frac{d\omega}{d\omega_A} = \frac{v_A}{v_r}\left(1+\frac{M_A}{M_B}\right)^2. \qquad (195\,c)$$

Here, as in § 8, v_A is the velocity of the primary beam, v_r the relative velocity of two colliding atoms, while M_A, M_B are the masses of the atoms in the primary and scattering beams respectively.

The classical differential cross-section at small angles in the centre of mass system is given by (50) as

$$I_{cl}(\theta)\,d\omega = \frac{1}{s}\{(s-1)f(s)\}^{2/s}\theta^{-2(1+1/s)}(C_s/E)^{2/s}\,d\omega, \qquad (196)$$

for an interaction of the form $\qquad V = -C_s/r^s,$

$f(s)$ being as defined in (52) and E being the relative kinetic energy

$$\tfrac{1}{2}\{M_A M_B/(M_A+M_B)\}v_r^2.$$

This is not valid for $\theta < \theta_{cl}$, where θ_{cl} is given by (111), and in the limit in which $\theta \to 0$ (see (110))
$$I(\theta) \to (kQ/4\pi)^2\{1+\tan^2\pi/(s-1)\}, \qquad (197)$$

where Q is the total elastic cross-section given by (103) and k is the wave number of relative motion. A convenient extrapolation to join (196) and (197) is to write

$$I(\theta) = I_{cl}[1-\exp\{-(\theta/\theta_0)^{2(1+1/s)}\}], \qquad (198)$$

where $\qquad \theta_0^{-2(s+1)/s} = \left(\frac{kQ}{4\pi}\right)^2[1+\tan^2\{\pi/(s-1)\}](C_s/E)^{-2/s}s\{(s-1)f(s)\}^{-2/s}. \qquad (199)$

Using the form (103) for Q we have

$$\theta_0 = k^{-1}(\pi/Q)^{\frac{1}{2}}\chi$$

where, in the notation of (103),

$$\chi = (s-1)^{1/(s+1)}(2\gamma/\pi)^{(1-s)/2(s+1)}(2f)^{(s-1)/s(s+1)}[16/s\{1+\tan^2\pi/(s-1)\}]^{s/2(s+1)},$$

and is not very different from θ_{cl}.

Assuming the form (198) the relations (195 a), (195 b), and (195 c) show that in the laboratory system the differential cross-section

$$I(\theta_A)\,d\omega_A = \frac{1}{s}\{(s-1)f(s)\}^{2/s}(C_s/E_A)^{2/s}\left(\frac{v_A}{v_r}\right)^{(2-s)/s}\theta_A^{-2(s+1)/s}\times$$
$$\times[1-\exp\{-(\theta_A/\theta_{0A})^{2(s+1)/s}]\,d\omega_A, \qquad (200)$$

where $\qquad \theta_{0A} = k_A^{-1}(\pi/Q)^{\frac{1}{2}}\chi. \qquad (201)$

Using (200), Helbing and Pauly derived the appropriate form for the observed differential cross-section in the laboratory system $J(\theta_A)\,d\omega_A$ allowing for the velocity distribution in the two beams. They found

$$J(\theta_A) = \frac{1}{s}\{(s-1)f(s)\}^{2/s}(C_s/E_{Am})^{2/s}\,\theta_A^{-2(s+1)/s}\,H(s,y,\theta_A)/Q(v_{Am}), \qquad (202)$$

where $E_{Am} = \tfrac{1}{2}M_A v_{Am}^2,$

v_{Am} being the most probable velocity of atoms of the primary beam. $H(s,y,\theta_A)$ is a complicated function of s, θ_A and y, the ratio v_{Am}/v_{Bm} of the most probable velocities in the primary and scattering beams. If $y \gg 1$,

$$H(s,y,\theta) = \left\{G'\!\left(\frac{s+2}{2-s},y\right)\!\bigg/G'(s,y)\right\}\Gamma\!\left\{2-\frac{s-2}{s(s-1)}\right\} \times$$
$$\times\left[1-\left\{1+\left(\frac{\theta_A}{\theta_{omA}}\right)^{\!2(s+1)/s}\right\}^{-2+(s-2)/s(s-1)}\right], \qquad (203)$$

$G'(s,y)$ being the function defined in § 7.2, p. 1349, and θ_{omA} is the value of θ_{0A} (see (201)) for the most probable velocity v_{mA}.

It follows that in the classical region $J(\theta_A)$ will vary as $\theta_A^{-2(s+1)/s}$, i.e. as $\theta_A^{-7/3}$ in the important practical case of $s = 6$, when $\theta_A > \theta_{mA}$.

There remains only to allow for the finite height of the detector ($2h_0$) and of the primary beam ($2z_0$). If cartesian coordinates ξ, η are chosen in the plane of movement of the detector, with origin at the point of intersection with the axis of the primary beam, we have, for small angle scattering,

$$\theta_A = (\xi^2+\eta^2)^{\frac{1}{2}}/D, \qquad d\omega_A = d\xi d\eta/D^2, \qquad (204)$$

where D is the distance from the intersection of the beam axes to the plane of the detector. Helbing and Pauly then show that when $h_0 \gg z_0$ the observed signal

$$J^*(\xi)\,d\xi$$

integrated over the beam and detector heights, varies as $\xi^{-(s+2)/s}$ in the classical region defined by
$$\xi > \theta_{omA}D.$$

For $\xi \to 0$ it tends to a finite limit. The noteworthy feature is now the dependence of the signal on $\xi^{-(s+2)/s}$ instead of $\xi^{-2(s+1)/s}$ as it would be if h_0 and $z_0 \to 0$. Thus with $s = 6$ and a detector wire for which $h_0 \gg z_0$ the signal should vary as $\xi^{-4/3}$ instead of $\xi^{-7/3}$.

Fig. 16.28 compares results obtained by Helbing and Pauly for J^* for K–Ar collisions using the two detector wires. For the longer one (14·2 mm) the variation of J^* with ξ or θ_A is exactly as predicted. Thus, for $\theta_A > 6'$, J^* varies as $\theta_A^{-4/3}$, which is manifest as a linear variation on the log-log plot of Fig. 16.28 with slope 4/3. With the short detector (1·8 mm) the slope in the classical region is greater corresponding to a variation as $\theta_A^{-1\cdot73}$.

Fig. 16.29 shows a comparison between the observed and calculated forms for $J^*(\xi)$ for K–Xe collisions, assuming $s = 6$ and the form (198) for $I(\theta)$. The agreement is very good, showing clearly both the predicted

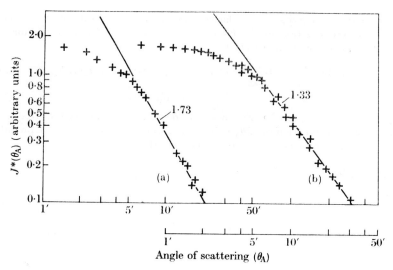

FIG. 16.28. Dependence of the small angle scattering, in the laboratory system, of K atoms by Ar atoms, on the length of the detector wire as observed by Helbing and Pauly. (a) Detector length 1·8 mm: + observed, —— for variation as $\theta_A^{-1\cdot73}$. (b) Detector length 14·2 mm: + observed, —— for variation as $\theta_A^{-4/3}$.

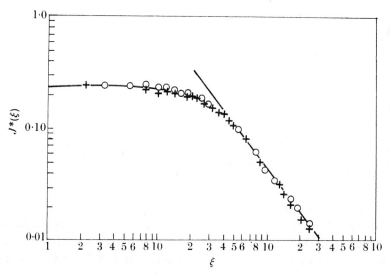

FIG. 16.29. Comparison of observed and calculated differential cross-sections in the laboratory system for the scattering of K atoms by Xe atoms. +, ○ experimental signals $J^*(\xi)$ (see text). —— calculated from (202), normalized to agree in absolute magnitude.

classical region and the constancy of the cross-section near the zero angle limit. It also establishes the validity of the assumption that $s = 6$.

From observations of this type it is possible to determine the critical angle θ_{0mA} and thence the total cross-section Q from (201). Comparison of results obtained in this way with those from other types of experiments will be discussed in § 9.3 (see Tables 16.10 and 16.11).

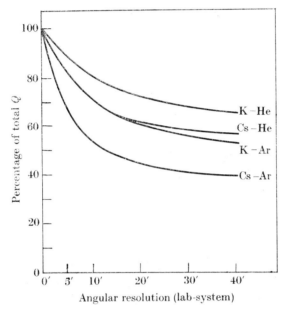

FIG. 16.30. Dependence of the apparent 'total' cross-section on the angular resolution (in the laboratory system) for collisions of K and Cs atoms with rare-gas atoms.

Helbing and Pauly, from their observations of small-angle scattering, were able to derive information about the angular resolving power necessary in an apparatus designed for the measurement of total cross-sections to any desired accuracy. Fig. 16.30 shows, from their results, the fraction of the true total cross-section for collisions of K and Cs with He and Ar atoms which could be measured with equipment of different angular resolution in the laboratory system. The theoretical estimates of Table 16.7 are in general agreement with these results as are those from equations (176).

9.2. *The velocity variation of the total cross-section*

In § 6.3.1 a distinction is made between low- and high-velocity regions as far as the form of the variation of the total cross-section with relative

velocity is concerned. If ϵ and r_m are the usual potential parameters introduced in § 6.1 then the relative velocity v is regarded as low in this sense if
$$v \ll \epsilon r_\mathrm{m}/\hbar.$$
We shall in most cases discuss here collisions for which this condition is satisfied, i.e. the low-velocity type. Under these conditions the variation of the cross-section takes the form of a gradual decrease with velocity on which are superimposed oscillatory variations. The total cross-section can thus be written in the form
$$Q = \bar{Q} + \Delta Q, \qquad (205)$$
where $\bar{Q}$ represents the gradually varying background and ΔQ the oscillatory part.

$\bar{Q}$ is determined by the long-range attraction and if this varies as r^{-s} for large r, $\bar{Q}$ varies as $v^{-2/(s-1)}$ (see § 5.4.1). In practice we expect $s = 6$ so $\bar{Q}$ should vary as $v^{-0.4}$.

The wavelength of the oscillations in ΔQ which arise from glory interference (see § 5.4.3) varies as v^{-1}. To display ΔQ it is convenient to plot $Qv^{0.4}$ as a function of v^{-1}. According to the analysis of § 6.3 the extrema are located at values v_e of v for which
$$N - \tfrac{3}{8} = (D\epsilon r_\mathrm{m}/\hbar v_\mathrm{e})\{1 - F(\epsilon/E)^{\frac{1}{2}} + ...\}, \qquad (206)$$
the maxima occurring at integral, and the minima at half-integral, values of N.

D is a constant whose value depends weakly on the shape of the interaction assumed and is largely determined by the 'curvature' of the interaction at the minimum (see p. 1335). For the (12, 6) potential it is 0·268. E is the energy of relative motion. F is a constant which also depends very weakly on the shape of the interactions assumed.

In view of (206) it is convenient to plot the extrema on an N, v^{-1} diagram. Although N is not at first determined absolutely this may be done by requiring that extrapolation to $v_\mathrm{e}^{-1} \to 0$ gives $N = 3/8$. From the slope of the plot for large v_r, $D\epsilon r_\mathrm{m}/\hbar$ may be obtained and hence, apart from a small uncertainty depending on the detailed shape of the interaction, ϵr_m.

The first experiments, in which the variation with velocity of the absorption of a velocity-selected alkali-metal atomic beam by rare gas contained in a scattering chamber was observed, were directed in the first instance towards the study of the background $\bar{Q}$ so that the index s involved in the asymptotic form of the interaction could be determined. With increased velocity resolution it has been possible in later work to make quantitative studies also of ΔQ.

9.2.1. *The asymptotic form of the interaction.* We consider first the evidence supporting the theoretical expectation that $s = 6$. Fig. 16.31 shows results obtained by Rothe, Rol, Trujillo, and Neynaber† for K–Xe collisions. The K beam was velocity-selected, using a selector

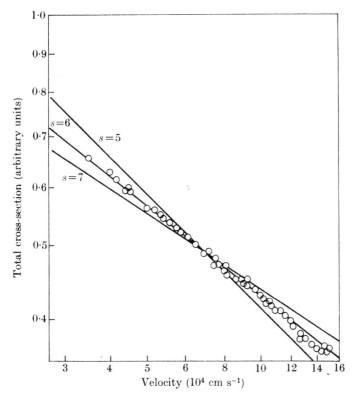

Fig. 16.31. Variation with relative velocity of the total cross-section for collision of K atoms with Xe atoms. ○ experimental points. ——— theoretical variation for $s = 5, 6, 7$, as indicated.

of the type described in § 7.2.2, while the absorbing Xe atoms were at room temperature. The cross-section $Q(v)$ was determined from the formula (166). $Q(v)$ is plotted in Fig. 16.31 on a log-log scale as a function of v, the velocity of the K atoms. Theoretically this plot should be a straight line of slope $2/(s-1) = 0.4$. Plots expected when $s = 5, 6$, and 7 are shown in Fig. 16.31 and it is clear that the experimental results lie very close to that with $s = 6$.

† ROTHE, E. W., ROL, P. K., TRUJILLO, S. M., and NEYNABER, R. H., *Phys. Rev.* **128** (1962) 659.

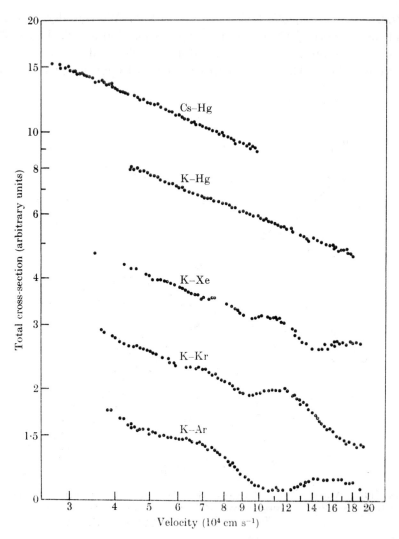

Fig. 16.32. Total cross-sections observed by Beck and Loesch as a function of relative velocity for collisions of alkali-metal atoms with rare-gas and with mercury atoms, as indicated.

Fig. 16.32 shows results obtained in later experiments by Beck and Loesch,† who observed the scattering of a primary beam of velocity-selected alkali-metal atoms by an intersecting beam of rare-gas or mercury atoms using the technique described in § 7.3. For Cs–Hg and K–Hg collisions ΔQ is small and the observed points lie closely on a

† loc. cit., p. 1363.

straight line of slope 0·4. Although for K–Xe, K–Kr, and K–Ar, ΔQ is important it is still apparent that the mean curves through the oscillations have much the same slope. In other experiments, directed towards the study of ΔQ, an analysis has usually been based on the assumption that $\bar{Q}$ varies as $v^{-0.4}$ and the self-consistency of the results obtained provides further support for the inverse sixth-power dependence of the van der Waals interaction at large separations.

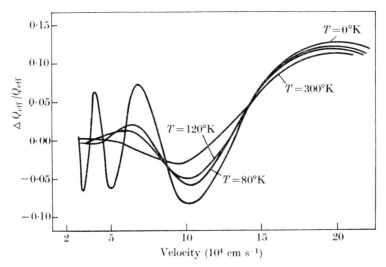

Fig. 16.33. Illustrating the dependence of the fraction $\Delta Q_{\text{eff}}/Q_{\text{eff}}$ (see text) for K–Ar collisions, on the temperature of the argon in the scattering chamber. The potassium beam is velocity-selected and $\Delta Q_{\text{eff}}/Q_{\text{eff}}$ is shown as a function of this velocity for values of the argon temperature T indicated. The curves are calculated, assuming a (12, 6) interaction with $\epsilon = 8\cdot 6 \times 10^{-15}$ erg, $r_{\text{m}} = 5\cdot 46 \times 10^{-8}$ cm.

9.2.2. *The 'glory' oscillations.* Turning now to the analysis of experiments concerned particularly with ΔQ we note that, while the location of the extrema in terms of relative velocity is not very sensitive to the temperature of the target atoms, the amplitude of the oscillations depends quite strongly on it. This is illustrated in Fig. 16.33 which gives results calculated for a typical (12, 6) potential, roughly representative of K–Ar interaction, with

$$\epsilon = 8\cdot 6 \times 10^{-15} \text{ erg}, \quad r_{\text{m}} = 5\cdot 46 \times 10^{-8} \text{ cm}.$$

Direct observation, without allowance for the velocity distribution of the target atoms, yields the effective cross-section Q_{eff} defined in § 7.2.1. We may write this in the form (205), i.e.

$$Q_{\text{eff}} = \bar{Q}_{\text{eff}} + \Delta Q_{\text{eff}}. \tag{207}$$

In Fig. 16.33 the ratio $\Delta Q_{\text{eff}}/\bar{Q}_{\text{eff}}$ is calculated for the assumed potential, taking the target atoms to be Ar atoms at different temperatures.

Although it is in principle possible to unfold the Maxwell distribution of the target atoms from ΔQ_{eff} this must be much less direct than for $\bar{Q}$. If an attempt is made to use the observed amplitudes to derive information about the interaction it is best to start from an assumed interaction, calculate Q and thence Q_{eff} and ΔQ_{eff} and compare with observations, proceeding thenceforward by further trial and error methods.

The first observations which exhibited oscillatory behaviour due to glory interference were those of Rothe, Rol, Trujillo, and Neynaber† for Li–Xe and K–Xe collisions. Some of their results are shown in Fig. 16.34. Later observations of high precision have been obtained by von Busch, Strunck, and Schlier,‡ using the comparison technique described in § 7.2.3.1 and applied to absorption of K and of Cs velocity selected beams by Ar, Kr, or Xe atoms in a scattering chamber. Another set of precise observations have been made by Beck and Loesch,§ using the crossed-beam technique described in § 7.3 and applied to scattering of primary velocity-selected beams of potassium atoms by secondary beams of mercury and of rare-gas atoms.

In presenting these results it is convenient to work in terms of

$$\frac{\Delta Q_{\text{eff}}}{\bar{Q}_{\text{eff}}} = \frac{Q_{\text{eff}}}{\bar{Q}_{\text{eff}}} - 1. \tag{208}$$

For K–Kr collisions results are available from three separate experiments and these are compared in Fig. 16.35. In all cases the temperature of the Kr atoms was 77 °K. Quite good agreement is apparent, particularly as regards the location of extrema.

Fig. 16.36 shows plots obtained by von Busch, Strunck, and Schlier and by Beck and Loesch for the determination of the relation between N and v_e^{-1} in (206). These plots are of the predicted form and from them $D\epsilon r_m/\hbar$ may be determined as described above.

Table 16.9 gives values obtained in this way for ϵr_m for the interaction of alkali-metal atoms with rare-gas and with mercury atoms, assuming in all cases a (12, 6) interaction.

As discussed on p. 1328 the number of extrema is a lower limit to the number of bound diatomic states that exist. Observed numbers of extrema are given in Table 16.9 showing, for example, that the van der Waals attraction between K and Hg is strong enough to support at least sixteen bound states.

† loc. cit., p. 1359.   ‡ loc. cit., p. 1359.   § loc. cit., p. 1363.

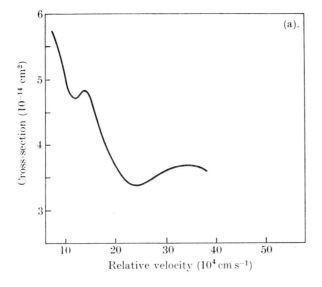

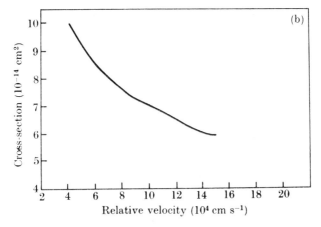

FIG. 16.34. Variation of total cross-section with relative velocity for Li–Xe and K–Xe collisions as observed by Rothe, Rol, Trujillo, and Neynaber. (a) ^{7}Li–Xe, (b) K–Xe.

Finally, we give in Fig. 16.37 results of direct measurement of $Q_{\text{eff}}(v)$ for K–He and K–Ne collisions. In these cases no oscillations appear in the cross-section–velocity curves. This may be due to the effective relative-velocity v_{eff} being substantially greater than v because of the high mean-velocity of the lighter Ne and He atoms. As ϵr_m will be smaller for these interactions $\epsilon r_m/\hbar v_{\text{eff}}$ may well be < 1, so that the collisions no longer fall in the low-velocity region.

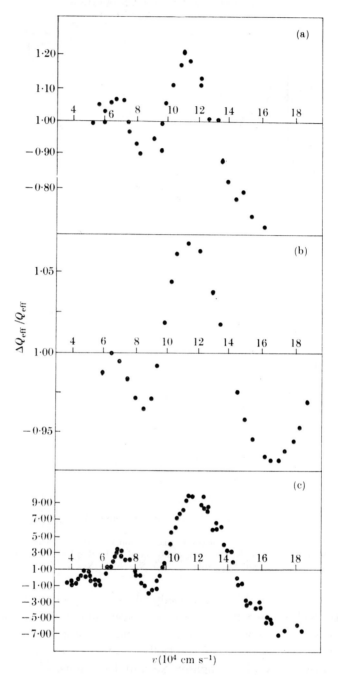

FIG. 16.35. Comparison of the velocity variation $\Delta Q_{\text{eff}}/Q_{\text{eff}}$ for K–Kr collisions as observed by different experimenters. (a) Rothe, Neynaber, Scott, Trujillo, and Rol. (b) von Busch, Strunck, and Schlier. (c) Beck and Loesch.

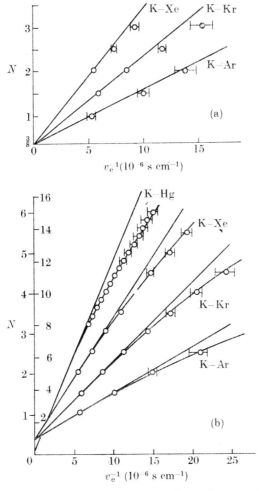

FIG. 16.36. Plots of observed extrema in total cross-sections against v_e^{-1}. (a) Observed by von Busch, Strunck, and Schlier. (b) Observed by Beck and Loesch. The inner scale of ordinates refers to the K–Hg case.

9.3. *Absolute measurement of cross-sections—determination of the van der Waals constant C*

So far we have discussed the results of observations of small-angle scattering and of the velocity dependence of the total cross-section in which relative intensity measurements were made but no absolute determinations. These sufficed to establish the inverse sixth-power dependence of the long-range interaction and, through the observations of the locations of glory extrema, the value essentially of ϵr_m. To determine

TABLE 16.9

Values of ϵr_m (in 10^{-22} erg cm) and lower limit N_l to numbers of bound states for the interaction of alkali metal atoms with rare-gas atoms and with mercury atoms, obtained from observation of glory interference effects

Alkali-metal atom	Ar $\epsilon r_m(N_l)$	Kr $\epsilon r_m(N_l)$	Xe $\epsilon r_m(N_l)$	Hg $\epsilon r_m(N_l)$	Reference
Li	4·3	6·1	10·6	—	a
	—	—	—	5·89	b
Na	4·8(2)	7·95(3)	10·8(3)	—	c
K	4·5(3)	7·6(5)	11·9(5)	45·6(16)	d
	4·8₅(3)	7·6₅(3)	11·7₅(3)	—	c
	—	7·6₅	—	—	e
Rb	—	7·9(3)	—	—	c
Cs	—	8·2₅(3)	—	—	c

a. ROTHE, E. W., ROL, P. K., and BERNSTEIN, R. B., *Phys. Rev.* **130** (1963) 2333.
b. ROTHE, E. W. and VENEKLASEN, L. H., *J. chem. Phys.* **46** (1967) 1208.
c. VON BUSCH, F., STRUNCK, H. J., and SCHLIER, C., *Z. Phys.* **199** (1967) 518.
d. BECK, D. and LOESCH, H. J., ibid. **195** (1966) 444.
e. ROTHE, E. W., NEYNABER, R. H., SCOTT, B. W., TRUJILLO, S. M., and ROL, P. K., *J. chem. Phys.* **39** (1963) 493.

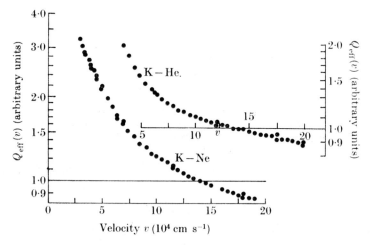

FIG. 16.37. Variation of the total cross-section for K–He and K–Ne collisions with velocity v of the potassium beam, as observed by von Busch, Strunck, and Schlier.

the strength of the interaction and, in particular, the van der Waals constant C, a knowledge of the absolute value of the cross-section at some relative velocity is essential. Thus, if $\bar{Q}(v)$ is the background cross-section (see (205)) when the relative velocity is v, C is given (see (103)) by

$$C = \hbar v(\bar{Q}/\gamma_{\mathrm{LL}})^{5/2} = \hbar v(\bar{Q}/8 \cdot 083)^{5/2}. \tag{209}$$

In many of the earlier experiments the absorption of a thermal instead of a velocity-selected beam was studied in which case the formula (172 a) is applicable but, in carrying out the analysis, the velocity variation of Q was often entirely ignored and Q_{eff} defined by (166) and (167) alone derived. This gives less definite results, because Q_{eff} is an ill-defined average over the velocity distribution of the atoms in the beam and in the target gas. It was usual to regard it as the cross-section for a collision in which the relative velocity has its average value.

Another uncertainty was introduced at first because of the need to determine whether the constant γ in (209) is best given by the Massey–Mohr (γ_{MM}) or Landau–Lifshitz (γ_{LL}) approximation. Although these differ by only 7 per cent the divergence in the derived values of the van der Waals constant is nearly 18 per cent. As explained on p. 1325, exact numerical calculation favours the Landau–Lifshitz value, which is now used throughout.

An uncertainty of a different kind was present in many of the earlier observations because of the pumping error in McLeod-gauge calibrations. In 1959 Rothe and Bernstein† made a great number of measurements of the absorption of thermal potassium and caesium beams in various gases. Pressure measurements were made with a Knudsen gauge calibrated against a McLeod gauge.

Table 16.10 reproduces their results for the rare gases, given relative to results for argon. Fortunately, although the absolute measurements obtained from this calibration are subject to the pumping error the relative measurements made with the Knudsen gauge are unaffected. Values of Q_0 in brackets were obtained by Mais (a),‡ Rosin and Rabi (b),§ Rosenberg (c),∥ and Helbing and Pauly (d)†† and it will be seen that they agree quite well. As the angular resolution used was rather different in these experiments it seems likely that the full total cross-section was measured.

† ROTHE, E. W. and BERNSTEIN, R. B., *J. chem. Phys.* **31** (1959) 1619.
‡ MAIS, W., *Phys. Rev.* **45** (1934) 773.
§ loc. cit., p. 1351.
∥ ROSENBERG, P., *Phys. Rev.* **55** (1939) 1267.
†† loc. cit., p. 1369.

The importance of the pumping error was pointed out independently at much the same time by Bennewitz and Dohmann[†] and by Rothe and Neynaber.[‡] The latter authors then repeated earlier measurements made by Rothe, Rol, and Bernstein[§] of the absorption of velocity-selected K and Li beams in helium, argon, and krypton. They found that the pumping error led to underestimation of the pressure by factors

TABLE 16.10

Cross-sections for interaction of potassium and caesium with rare gases relative to values for interaction with argon

	Mean relative velocity (10^5 m/min)	Relative Q_0
(a) Potassium with		
Helium	1·396	0·292 (0·283 (a), 0·284 (b), 0·291 (c))
Neon	0·826	0·400 (0·45 (a), 0·45 (b))
Argon	0·727	(1·00)
Krypton	0·669	1·26
Xenon	0·649	1·45 (1·5 (d))
(b) Caesium with		
Helium	1·297	0·310 (0·28 (b))
Neon	0·641	0·440 (0·50 (b))
Argon	0·506	(1·00)
Krypton	0·416	1·28
Xenon	0·382	1·58

of 1·07, 1·24, and 1·31 for helium, neon, and argon respectively. For K–Kr collisions the derived value of the van der Waals constant came out to be
$$C = 5 \cdot 5 \times 10^{-58} \text{ erg cm}^6.$$

Using this the value for other potassium–rare-gas interactions may be derived immediately from Table 16.10. Furthermore, from the apparent absolute values assumed by Rothe and Bernstein for K–Ar collisions the correction to their absolute measurements for caesium–rare-gas collisions can be derived.

Table 16.11 gives the van der Waals constant derived in this way for Na, K, and Cs interactions with rare-gas atoms. Values for Li obtained from the observations of Rothe, Rol, and Bernstein as corrected by Rothe and Neynaber are also included. Results are also given for K–Ar

[†] BENNEWITZ, H. G. and DOHMANN, H. D., *Z. Phys.* **182** (1965) 524.
[‡] ROTHE, E. W. and NEYNABER, R. H., *J. chem. Phys.* **42** (1965) 3306.
[§] ROTHE, E. W., ROL, P. K., and BERNSTEIN, R. B., *Phys. Rev.* **130** (1963) 2333.

and K–Xe, obtained by Florin,† using a velocity selected beam and with due precautions to avoid pumping error in pressure measurements, and by Helbing and Pauly.‡

TABLE 16.11

van der Waals constants for interactions of Li, Na, K, *and* Cs *atoms with rare-gas atoms*

	van der Waals constant C (10^{-58} erg cm^6)			
	Obs.	Calc.		
		(a)	(b)	(c)
Li				
Ar	2·1	1·63	1·9	1·8
Kr	2·7	2·68	2·9	2·6
Na				
Ar	2·0	1·82	1·97	
K				
Ar	3·1 (2·8) [4·1]	2·58	2·65	2·2
Kr	5·5	3·85	4·0	3·4
Xe	6·9$_5$ (7·4) [10·7]	6·05	6·4$_5$	5·3
Cs				
Ar	3·2	3·16	3·0$_5$	2·5
Kr	4·8	4·98	4·5$_5$	3·7
Xe	7·5	7·95	7·4	5·9

Observed results are those of Rothe, Rol, and Bernstein as corrected by Rothe and Neynaber (loc. cit., p. 1384) for pumping error in pressure measurements, except for those bracketed ( ) due to Florin, and [ ] due to Helbing and Pauly.‡
(a) Calculated by Dalgarno and Davison using (214).
(b) Calculated by Dalgarno and Kingston (loc. cit., p. 1387) using (212).
(c) Calculated from Slater–Kirkwood formula (218) with N_a, N_b replaced by n_a, n_b.

We also give in Table 16.11 values for the van der Waals constant for Na–Ar interactions derived from observations made by Berkling, Schlier, and Toschek§ using the apparatus described in § 12.4.2, which was primarily constructed for the study of anisotropic terms in van der Waals interactions. They measured the appropriate cross-section, using a velocity-selected sodium beam, relative to the cross-section for K–Ar collisions. Assuming the van der Waals constant for K–Ar interaction to be as found by Rothe and Bernstein,∥ it is then possible to derive the corresponding values for Na–Ar, assuming that both cases are in the low-velocity region (see § 6.3.1) and that the cross-section varies as $(C/v)^{2/5}$.

† FLORIN, Dissertation, Bonn, 1964.
‡ loc. cit., p. 1369.
§ BERKLING, K., SCHLIER, C., and TOSCHEK, P., *Z. Phys.* **168** (1962) 81.
∥ loc. cit., p. 1383.

9.3.1. Comparison with calculated values.

Theoretical expressions for the van der Waals constant C_{ab} for interaction between atoms A and B may be obtained by variational methods or by second-order perturbation theory. For all but the simplest atoms the latter method must be relied upon, and attention is paid to obtaining formulae that involve as far as possible atomic quantities that may be obtained from experiment. Thus Margenau[†] showed that, according to second-order perturbation theory, in atomic units,

$$C_{ab} = \tfrac{3}{2} \sum_{m \neq 0} \sum_{n \neq 0} f^a_m f^b_n / (\epsilon^a_0 - \epsilon^a_m)(\epsilon^b_0 - \epsilon^b_n)(\epsilon^a_0 + \epsilon^b_0 - \epsilon^a_m - \epsilon^b_n). \tag{210}$$

$\epsilon^a_0, \ldots, \epsilon^a_m, \ldots, \epsilon^b_0, \ldots, \epsilon^b_n, \ldots$ are the allowed energy levels for the respective atoms and f^a_m, f^b_n are the corresponding oscillator strengths for transitions from the ground states to the mth and nth excited states, respectively, as defined in Chapter 7, § 5.2.1. In many cases information about the oscillator strengths is available from other experiments—thus measurements of photo-ionization cross-sections gives the oscillator strengths for transitions to the continuum. Incomplete sets of experimental values may be completed to a good approximation by use of sum rules (Chap. 7, § 5.2.2).

If $\epsilon^b_0 - \epsilon^b_n \gg \epsilon^a_0 - \epsilon^a_m$ for all m and n which contribute appreciably to the sum in (210) use may be made of the formula

$$\alpha_b(0) = \sum_{n \neq 0} f^b_n / (\epsilon^b_0 - \epsilon^b_m)^2, \tag{211}$$

where $\alpha_b(0)$ is the static polarizability of atom B, to reduce (210) to

$$C_{ab} = \tfrac{3}{4} \alpha_b \sum_{m \neq 0} f^a_m / (\epsilon^a_0 - \epsilon^a_m). \tag{212}$$

This condition is well satisfied for the heavier alkali metals interacting with rare gases as $f^a_m \simeq 1$ for transitions to the first excited states and $\simeq 0$ for all other transitions that do not involve inner shell excitation. The large denominators associated with the latter render their contributions negligible.

An alternative form to (210) is often useful. This gives[‡]

$$C_{ab} = (3/\pi) \int_0^\infty \alpha_a(u) \alpha_b(u) \, du, \tag{213}$$

where

$$\alpha_a(u) = \sum_{m \neq 0} f^a_m / \{(\epsilon^b_0 - \epsilon^a_m)^2 + u^2\} \tag{214}$$

[†] MARGENAU, H., *Rev. mod. Phys.* **11** (1939) 1.
[‡] DALGARNO, A. and DAVISON, W. D., *Advances in Atomic and Mol. Physics* **2** (1966) 1.

and similarly for $\alpha_b(u)$. $\alpha_a(0)$ and $\alpha_b(0)$ are thus the static polarizabilities of the respective atoms.

A useful approximation follows if we write

$$\alpha(u) = a/(b^2+u^2) \tag{215}$$

and choose a and b so that the expression is exactly correct for $u = 0$ and for $u \to \infty$. Since

$$\lim_{u\to\infty} \alpha(u) = u^{-2} \sum f_m^a = Nu^{-2}, \tag{216}$$

where N is the number of atomic electrons, we have

$$\alpha(u) = \frac{N}{\{N/\alpha(0)\}+u^2} \tag{217}$$

so that, on substitution in (213) we obtain

$$C_{ab} = \frac{3}{2} \frac{\alpha_a(0)\alpha_b(0)}{\{\alpha_a(0)/N_a\}^{\frac{1}{2}}+\{\alpha_b(0)/N_b\}^{\frac{1}{2}}}. \tag{218}$$

This simpler formula was derived many years before by Slater and Kirkwood,† using a different method. It tends to overestimate C_{ab} because the contributions to the summation from the inner shells are too large. An empirical modification which gives better results involves the replacement of N_a, N_b by n_a, n_b, the number of outer-shell electrons in the respective atoms.

Values of C_{ab} calculated in three different ways are compared with the observed values in Table 16.11. Of the three calculated sets of values the most accurate are those of Dalgarno and Davison‡ who used (214) with the f-values determined from the most recent optical (including photo-ionization) measurements. Earlier calculations by Dalgarno and Kingston§ were based on (212) with the polarizabilities of the rare-gas atom determined from refractive-index measurements and less accurate values for the oscillator strengths for the alkali-metal atoms. The least accurate are those listed under (c) which were obtained from the Slater–Kirkwood formula (218) with N_a, N_b replaced by n_a, n_b. On the whole, the agreement between the more accurate theoretical values and the observed is not unsatisfactory. It seems also that the simpler Slater–Kirkwood formula gives reasonable results.

9.4. *Measurements of differential cross-sections*

9.4.1. *Analysis of observations of rainbow effects.* As discussed in § 6.4 the rainbow angle θ_r depends mainly on K the reduced energy of relative

---
† SLATER, J. C. and KIRKWOOD, J. G., *Phys. Rev.* **37** (1931) 682.   ‡ loc. cit., p. 1386.
§ DALGARNO, A. and KINGSTON, A. E., *Proc. phys. Soc.* **73** (1959) 455.

motion E/ϵ where ϵ is the depth of the attractive potential well. In fact, to quite a good approximation, θ_r varies as $1/K$, the constant of proportionality being 2·05 for a (12, 6) potential.

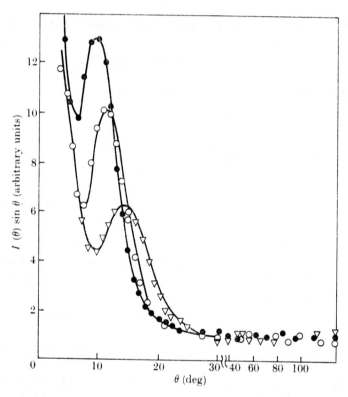

FIG. 16.38. Differential cross-section per unit angle (in the C.M. system) for K–Kr collisions as observed by Beck. In all cases the temperature of the secondary (Kr) beam is 180 °K. The K beam was velocity-selected, the velocities in the three cases shown being as follows: ● 6·61 × 10⁴ cm s⁻¹; ○ 6·09 × 10⁴ cm s⁻¹; ▽ 5·41 × 10⁴ cm s⁻¹.

Fig. 16.38 illustrates some of the first observations of the rainbow effect, in the scattering of potassium by krypton. The observations are those of Beck† carried out with a velocity-selected potassium beam, the angular resolution being of the order 30′. The three curves refer to different values of a nominal relative collision energy defined by

$$E = \frac{1}{2}\frac{M_A M_B}{M_A + M_B}(v_A^2 + 2\kappa T_B/M_B), \qquad (219)$$

† BECK, D., J. chem. Phys. **37** (1962) 2884.

where v_A is the velocity of the potassium beam, T_B the temperature of the rare-gas source, and M_A, M_B the respective masses of the potassium and rare-gas atoms. It will be seen that, as expected, the rainbow maximum moves to smaller angles as the relative energy decreases.

If all collisions took place at the same relative energy, the semi-classical theory outlined on p. 1322 shows that the rainbow angle θ_r is to be taken as the angle at which the intensity on the high-angle side of the rainbow maximum falls to 44 per cent of the value at the maximum. Beck showed that the energy spread due to the thermal character of the krypton beam had little effect on the location of θ_r. To isolate the contribution from the rainbow effect the large-angle scattering was extrapolated linearly through the rainbow region and then subtracted off. Having obtained θ_r the depth ϵ of the potential minimum was determined by assuming interactions of (exp α, 6) type with $\alpha = 12, 13, 14$, and 15. The derived values of ϵ are given in Table 16.12.

It is also possible to obtain a good estimate of the interaction parameter r_m from the width of the rainbow peak. According to (88) the semi-classical contribution $I_{sc}^r \, d\omega$ to the differential cross-section arising from rainbow scattering is given by

$$I_{sc}^r = \frac{2\pi}{k^2} \frac{l_r}{\sin\theta} q^{-2/3} [\mathrm{Ai}\{q^{-1/3}(\theta-\theta_r)\}]^2.$$

Here k is the wave number of relative motion, l_r is the angular momentum quantum number corresponding to the rainbow angle, and q is such that, near the rainbow angle,

$$\theta = \theta_r + q(l-l_r)^2.$$

By fitting the shape of the observed minimum q may be found. For an (exp α, 6) potential we may write for the deflexion function near the rainbow

$$\vartheta = \vartheta_r + q^*(\alpha)(p^*-p_r^*)^2.$$

Here p^* is the reduced impact parameter, p/r_m, so

$$l = kr_m p^*,$$

and

$$q = q^*/k^2 r_m^2.$$

If q^* is known from the calculated deflexion function r_m may be obtained.

The values obtained in this way by Beck are given in Table 16.12. They are subject to a somewhat greater uncertainty than those for ϵ because they are much more sensitive to the spread of relative impact energy.

9.4.2. *Further experiments at high resolution.* In the early experiments little detail is apparent in the observed angular distributions, apart from the rainbow maximum. Observation of interference effects has proved possible in later work largely because of improved resolution in velocity selection. Thus in Fig. 16.39 we show differential cross-sections for unit angle $I(\theta)\sin\theta$ in the C.M. system for Na–Hg and K–Hg collisions obtained by von Hundhausen and Pauly,[†] who observed

TABLE 16.12

Values of ϵ and r_m for the K–Kr interaction derived from analysis of observed rainbow scattering assuming (exp α, 6) interactions

α	12	13	14	15
ϵ (10^{-14} erg)	1·24	1·22	1·20	1·18
r_m (Å)	5·9	6·3	6·8	7·3

the scattering of velocity-selected alkali-metal atomic beams by a thermal mercury beam intersecting at 90°. Because of the relatively large mass of the mercury atom the thermal spread of velocity in the target beam is not serious, particularly for Na–Hg collisions. Thus considerable detail can be seen for these collisions, less detail for those between K and Hg. Fig. 16.40 shows the variation of the rainbow angle θ_r with the relative kinetic energy. (In the log-log scale of the diagram the variation is linear with slope -1 which corresponds with theoretical expectation (see (152)).

Values for the parameters ϵ and r_m for Na–Hg, K–Hg, Rb–Hg, Na–Xe, and K–Xe interactions derived from these observations on the assumption of (12, 6) and (8, 6) potentials are given in Table 16.13. For K–Hg the agreement with the values of ϵ derived by Morse and Bernstein[‡] from earlier rainbow-scattering observations is good. It is even better with the value derived from later experiments of Beck, Dummel, and Henkel.[§] A detailed analysis of the rainbow scattering data for K–Hg has been given by Beck and Loesch.[||]

Even more detail was observed by Barwig, Buck, von Hundhausen, and Pauly[††] for Na–Kr and Na–Xe collisions. Again the sodium atom beam was velocity-selected while the temperature of the oven source of the rare-gas beam could be controlled between liquid air and room temperature. Fig. 16.41 shows results obtained for the differential cross-section

[†] VON HUNDHAUSEN, E. and PAULY, H., *Z. Naturf.* **19a** (1964) 810.
[‡] loc. cit., p. 1368.
[§] BECK, D., DUMMEL, H., and HENKEL, U., *Z. Phys.* **185** (1965) 19.
[||] BECK, D. and LOESCH, H. J., ibid. **196** (1966) 66.
[††] BARWIG, P., BUCK, U., VON HUNDHAUSEN, E., and PAULY, H., ibid. **196** (1966) 343.

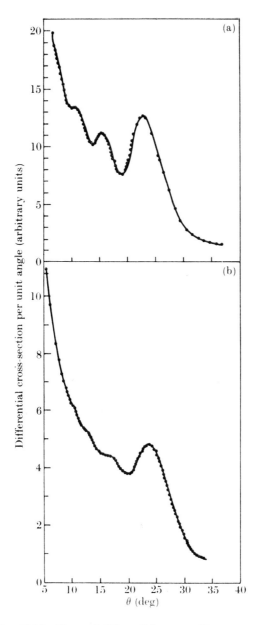

FIG. 16.39. Observed differential cross-sections per unit angle (in the C.M. system) for scattering (a) of Na by Hg with relative velocity $1\cdot 476 \times 10^5$ cm s^{-1}, (b) of K by Hg, the K atoms having the same mean energy as the Na atoms in (a).

per unit angle for Na–Kr collisions from three different relative velocities, the Kr oven temperature being 77 °K. Interference effects in the neighbourhood of the rainbow maximum are well resolved in each case. Fig. 16.42 shows similar detail for Na–Xe in which interference effects at small angles are displayed by plotting $I(\theta)\theta^{7/3}$ (see (196)) as a function of θ, in the centre of mass system. These observations were taken with

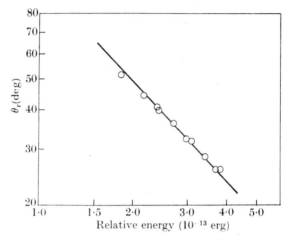

FIG. 16.40. Observed variation of rainbow angle θ_r with relative kinetic energy for Na–Hg collisions. ○ observed points. —— best theoretical fit using (12, 6) potential.

the oven temperature 133 °K. When this temperature was increased to 295 °K the interference effects, though still detectable, were much less apparent. It was also found that the resolution depended on the intensity of the secondary beam and that it was an advantage to maintain this intensity below the level required to give 50 per cent absorption of the beam.

Detailed analysis of these data was carried out using a potential of (8, 6) form which of all $(n, 6)$ potentials gave the best fit to the observed data. Figs. 16.43 and 16.44 illustrate the remarkably close agreement achieved by appropriate choice of parameters ϵ, r_m with this potential. These are given in Table 16.13. Despite the good agreement, the derived potential does not give very well either the location of the extrema or the amplitude of the oscillations observed by von Busch, Strunck, and Schlier in their study of the glory effect (see p. 1378). It seems that the functional form of the assumed potential is too simple to include all the observable features. Further evidence concerning this matter is discussed in § 9.5.

Finally, we refer to the observations of Groblicki and Bernstein[†] directed at the study of interference detail rather than rainbow effects. For this purpose they observed the scattering of a velocity-selected beam

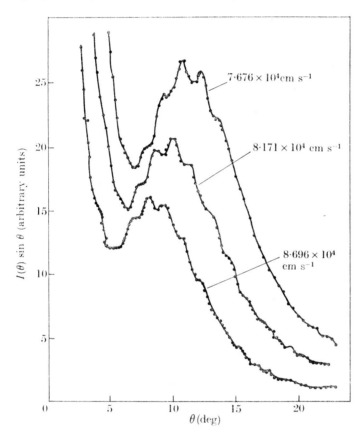

FIG. 16.41. Observed differential cross-section per unit angle (in the C.M. system) for scattering of Na by Kr with relative velocities as indicated.

of lithium atoms by a thermal mercury beam intersecting at 90°, the angular resolving power being 15' according to the Kusch criterion. Measurements were made over the angular range (in the C.M. system) from 0·5° to between 20° and 30°.

At angles of scattering θ (in the C.M. system) of less than 2° the scattered intensity varies as $\theta^{-7/3}$ as expected from the classical theory (see pp. 1316 and 1370). Fig. 16.45 (a) illustrates a typical angular distribution (in the lab. system) over the whole observed angular range. The curve

[†] GROBLICKI, P. J. and BERNSTEIN, R. B., J. chem. Phys. **42** (1965) 2295.

shown was obtained using a beam of ^{6}Li atoms. It was verified that if the ^{6}Li is replaced by the heavier isotope ^{7}Li the same scattering pattern is obtained for the same selected velocity of the lithium beam. The

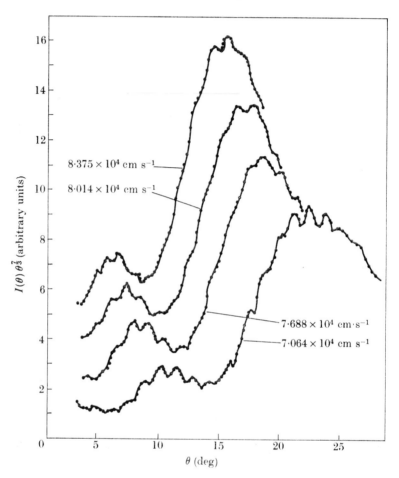

FIG. 16.42. The product $\theta^{7/3}I(\theta)$, where $I(\theta)\,d\omega$ is the differential cross-section in the C.M. system, for scattering of Na by Xe, as observed by Barwig et al.

quantum oscillations are brought out more clearly if $\theta^{7/3}I(\theta)$, instead of $I(\theta)$ as usual, is plotted as a function of θ. This may be seen by reference to Fig. 16.45(b).

To analyse the data interactions of (12, 6) type were assumed. Calculations were carried out for reasonable values of the parameter $B = 2M\epsilon r_m^2/\hbar^2$ where M is the reduced mass. Having chosen B, phase shifts $\eta_l(A)$ were calculated for various values of $A = kr_m$ (see (136))

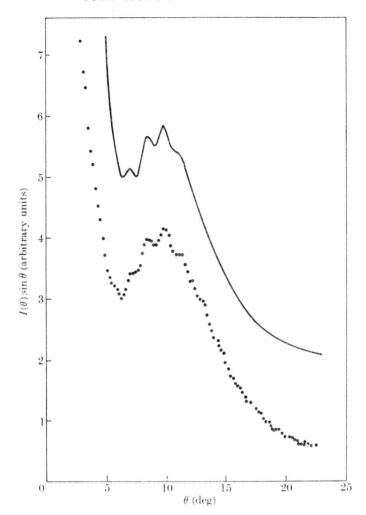

FIG. 16.43. Differential cross-section per unit angle for scattering of Na by Kr atoms with relative velocity $8 \cdot 171 \times 10^4$ cm s^{-1}. ● observed. ——— calculated assuming an (8, 6) potential with an optimized choice of the parameters ϵ and r_m.

and hence the angular distribution. For given B the value of r_m could then be chosen to give the best fit to the observations. This was repeated for other choices of B with the hope of selecting the best pair of values of ϵr_m^2 and of r_m. The accuracy of the observations was hardly adequate to carry out this programme. However, Fig. 16.46 shows the best fit obtained with the location of extrema in the angular distributions, taking $B = 1000$ and $r_m = 2 \cdot 83$ Å. The derived value of ϵ is $7 \cdot 8 \times 10^{-14}$ ergs, which is not unreasonable.

9.5. *Self-consistency of data derived from different types of experiment*

We have discussed the analysis of experimental results on glory and rainbow scattering in terms of the parameters ϵ and r_m. It has already

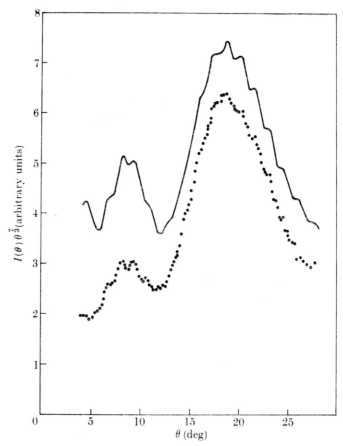

Fig. 16.44. Product $\theta^{7/3}I(\theta)$ for scattering of Na by Xe atoms with relative velocity $7\cdot68\times10^4$ cm s^{-1}. ● observed. —— calculated, assuming an (8, 6) potential with optimized choice of the parameters ϵ and r_m.

been noted that the $(n, 6)$ potential, which gives the best fit to the high-resolution rainbow data with $n = 8$, does not give a good fit to the glory observations for the two cases of Na–Kr and Na–Xe interactions. Part of this discrepancy may well be due to the oversimplified form of interaction assumed and indicates that the accuracy of observation is now high enough to make it worth while to seek to obtain more than just the two parameters ϵ and r_m for an assumed simple potential of $(n, 6)$ or similar type.

The glory and rainbow data for Na–Xe, K–Xe, and K–Hg have all been analysed on the assumption of a (12, 6) potential. As the glory data give ϵr_m and the rainbow ϵ and r_m separately a simple consistency check is possible for these cases. This may be done by referring to Table 16.14, which compares the product ϵr_m derived from the two sets of data.

TABLE 16.13

Values of the parameters ϵ and r_m derived from analysis of rainbow-scattering data obtained with high angular resolution, assuming $(n, 6)$ interactions with $n = 8$ and $n = 12$

Interacting pair	$n = 8$		$n = 12$	
	ϵ (10^{-14} erg)	r_m (Å)	ϵ (10^{-14} erg)	r_m (Å)
Na–Kr	1·39†	4·98†		
Na–Xe	2·10†	4·93†		
	1·93‡	5·2‡	1·72‡	6·8‡
K–Xe	1·80‡	7·3‡	1·52‡	9·4‡
Na–Hg	9·18‡	4·1‡	8·14‡	5·2‡
K–Hg	8·69‡	4·2‡	7·73‡	5·3‡
			7·43§	
			7·63‖	
Rb–Hg	8·14‡	4·4‡	7·11‡	5·5‡

† Observed by Barwig, Buck, von Hundhausen, and Pauly.
‡ Observed by von Hundhausen and Pauly.
§ Observed by Morse and Bernstein.
‖ Observed by Beck, Dummel, and Henkel.

There is a considerable difference for K–Xe but for the other two cases the agreement is within 10 per cent. It could presumably be improved by working with an $(n, 6)$ potential with $n < 12$. Unfortunately in no other cases have both glory and rainbow data been analysed with the same $(n, 6)$ potential.

A further consistency check is possible in principle from the values of the van der Waals constant C derived from absolute measurement of cross-sections. Thus, for an $(n, 6)$ interaction,

$$C = \{n/(n-6)\}\epsilon r_m^6, \tag{220}$$

so we can compare values of C derived directly with those derived from (220). However, at the time of writing, the only case which may be checked is that of K–Xe. Taking ϵ and r_m as given by the rainbow data (Table 16.13) for $n = 12$ we find $C = 21 \cdot 0 \times 10^{-57}$ erg cm^6 as against about 7×10^{-58} erg cm^6 from total cross-section data (Table 16.11).

Remembering that r_m is poorly determined from the rainbow data we may write alternatively

$$C = \{n/(n-6)\}(\epsilon r_m)^6/\epsilon^5, \tag{221}$$

using ϵr_m as given from the glory data and only ϵ from the rainbow data.

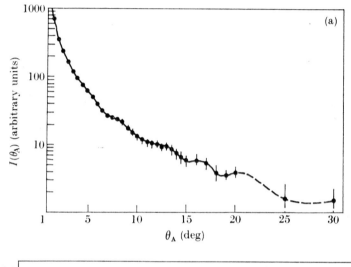

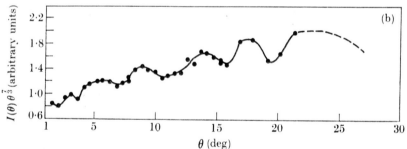

Fig. 16.45. (a) Variation of differential cross-section $I(\theta_A)$ per unit solid angle for Li–Hg collisions in the laboratory system, observed by Groblicki and Bernstein. (b) As for (a) but plotting $\theta^{7/3}I(\theta)$ instead of $I(\theta)$, where $I(\theta)$ and θ are now in the C.M. system.

This gives $C = 7 \times 10^{-57}$ ergs, still considerably larger than that derived directly.

It seems that, at the time of writing, the experimental techniques are developed to a stage that establishes the validity of the theoretical considerations discussed in §§ 5, 6. Further increase in precision of measurement is required to make possible a determination of the interatomic interactions in more detail than can be represented by simple potentials involving two parameters only.

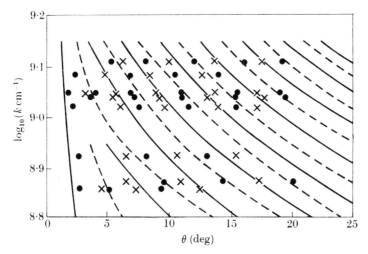

Fig. 16.46. Comparison of observed and calculated locations of maxima and minima in the differential cross-sections for Li–Hg collisions at different relative wave numbers. —— maxima, calculated; × maxima observed; – – – minima, calculated; ● minima observed.

TABLE 16.14

Comparison of ϵr_m values derived from analysis of glory and of rainbow scattering data in terms of a (12, 6) potential

Interaction	ϵr_m (erg cm × 10^{22})	
	From glory observations	From rainbow observations
Na–Xe	10·8	11·7
K–Xe	11·9	14·3
K–Hg	45·6	41

10. The interaction between helium atoms at large and intermediate separations

A great deal of work has been devoted to the accurate determination of the interaction energy of two helium atoms at separations of gas-kinetic importance. This is partly because of the relatively large magnitude of the quantal modifications in this case and partly because of the need for accurate knowledge of the interaction in the interpretation of the remarkable phenomena exhibited by liquid helium.

Most of the evidence about the interaction at large and intermediate separations, in which the detailed shape of the repulsive part of the interaction is not concerned, came at first from analysis of data on the viscosity, thermal conductivity, and second virial coefficient of helium (^{4}He) gas as a function of temperature in the range 0–1000 °K. Once it became possible to concentrate the ^{3}He isotope, experimental determination of the transport properties of ^{3}He and of ^{3}He–^{4}He mixtures enlarged the data that may be used to determine the basic interaction in the same range of atomic separations. It is only recently that useful data have been forthcoming on total collision cross-sections for He–He collisions at thermal energies.

We begin by examining the evidence in the low-temperature range 0–20 °K, in which quantum effects are particularly important, and then proceed to discuss the evidence from higher temperatures. Information available from analysis of observations at considerably higher relative energies must be utilized if the repulsive interaction is to be investigated at separations considerably less than r_m. Such observations have been made using monoenergetic atomic beams produced through charge transfer to He$^+$ beams and will be described and discussed in Chapter 22.

10.1. *The low-temperature evidence* ($T < 20$ °K)

The first quantal analysis of the viscosity observations was made in 1934 by Massey and Mohr[†] who showed that the approximate formula for the interaction,
$$V(r) = be^{-ar} - cr^{-6}, \tag{222}$$
where $b = 7 \cdot 7 \times 10^{-10}$ ergs, $a = 4 \cdot 6 \times 10^8$ cm^{-1}, and $c = 1 \cdot 47 \times 10^{-60}$ erg cm^6, given by Slater[‡] on theoretical grounds, while not correct, was a fair approximation. A number of attempts to improve on (222) have been made using the second virial coefficient as well as the viscosity and thermal conductivity data. At first the work was hampered by the paucity and lack of self-consistency of the observations. Thus the thermal conductivity results that were available until 1950 were not consistent with the viscosity results.[§] The most detailed analysis carried out before that time by Buckingham, Hamilton, and Massey[||] derived an interaction that was reasonably consistent with the thermal conductivity and virial coefficient data but not with the viscosity. This position

---

[†] MASSEY, H. S. W. and MOHR, C. B. O., *Proc. R. Soc.* A**144** (1934) 188.
[‡] SLATER, J. C., *Phys. Rev.* **32** (1928) 349.
[§] The thermal conductivity σ is related to the viscosity coefficient η by
$$\sigma = \frac{15}{4} \frac{\kappa}{M} \eta,$$
where κ is Boltzmann's constant and M the mass of an atom.
[||] BUCKINGHAM, R. A., HAMILTON, J., and MASSEY, H. S. W., *Proc. R. Soc.* A**179** (1941) 103.

was clarified when later measurements of the viscosity were made and found to agree quite well with values derived from thermal conductivity observations. Further tests of any assumed interaction became available when it was possible to obtain the light helium isotope ³He in quantities sufficient for measurement of its transport properties. As a result the position for temperatures below 20 °K is now fairly satisfactory.

The most extensive calculations have been carried out using two forms of interaction, one of the (12, 6) type and one a more elaborate one introduced by Buckingham and Corner,† which is a generalization of the $(\exp \alpha, 6)$ form to allow for the modification of the long-range attraction by the presence of a term falling off as r^{-8}. Their potential is thus
$$V = -\epsilon f(r/r_\mathrm{m}), \tag{223}$$

where
$$f(x) = f_1 x^{-6}(1+\beta x^{-2}) - f_2 \exp\{-\alpha(x-1)\}, \tag{224 a}$$

with
$$f_1 = \alpha/\{\alpha(1+\beta)-6-8\beta\}, \qquad f_2 = -1+(1+\beta)f_1. \tag{224 b}$$

For $x < 1$ it is modified by multiplying f_1 by a factor $\exp\{-4(1-x)^3/x^3\}$, which has the effect of quickly damping out the attractive terms as x decreases. In all calculations with this potential α and β were fixed as 13·5 and 0·2 respectively, the latter being roughly the value expected from the perturbation theory of the long-range interaction.

For both the interactions two parameters remain unfixed. From early evidence it seemed that the quantum parameter $B = 2M\epsilon r_\mathrm{m}^2/\hbar^2$, where M is the reduced mass, must be close to 7·29 and in the calculations carried out using the potential (223) B was fixed at this value, and the van der Waals constant C, $= f_1 \epsilon r_\mathrm{m}^6$, varied to give the best fit to the data. Fig. 16.47 (a) shows the comparison with the observed viscosity‡ and thermal conductivity§ data in the temperature range up to 30 °K or so, obtained with the two values 1·45 and 1·55 × 10⁻⁶⁰ erg cm⁶ for C. It will be seen that, if equal weight is given to data below 4 °K and between 15 and 25 °K, the most probable value for C is about 1·50 × 10⁻⁶⁰ erg cm⁶. This gives the values of ϵ and r_m in Table 16.16.

† BUCKINGHAM, R. A. and CORNER, J., ibid. 189 (1947) 118.
‡ VAN ITTERBEEK, A. and VAN PAEMEL, O., Physica, 's Grav. 7 (1940) 265; KEESOM, W. H. and KEESOM, P. H., ibid. 29; BECKER, E. W. and MISENTA, R., Z. Phys. 140 (1955) 535; BECKER, E. W., MISENTA, R., and SCHMEISSNER, F., ibid. 137 (1954) 126; COREMANS, J. M. J., VAN ITTERBEEK, A., BEENAKKER, J. J. M., KNAAP, H. F. P., and ZANDBERGEN, P., Physica, 's Grav. 24 (1958) 557.
§ UBBINK, J. B. and DE HAAS, W. J., ibid. 10 (1943) 465; FOKKENS, K., TACONIS, K. W., and DE BRUYN OUBOTER, R., Proc. 8th Int. Conf. Low Temp. Phys., p. 34 (Butterworths, London).

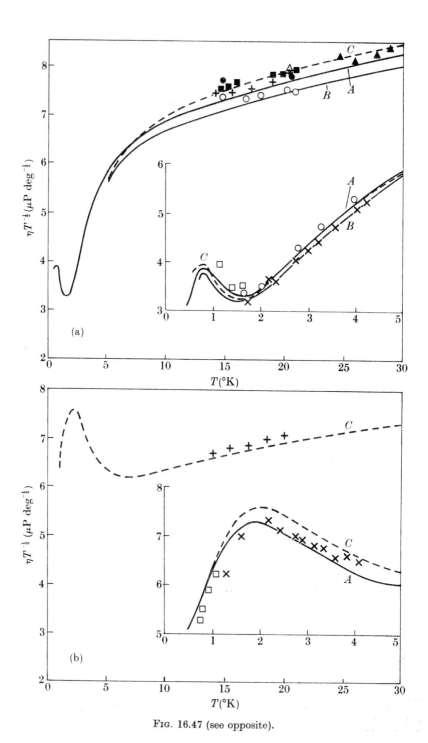

Fig. 16.47 (see opposite).

With the (12, 6) potential similar results† are obtained as may be seen by reference to Fig. 16.47, the corresponding values of ϵ and r_m being as given in Table 16.16. Fig. 16.48 (a) illustrates the comparison between the second virial coefficient as calculated‡ with the (12, 6) interaction and the observed data of Keller.§ It will be seen that, while the agreement is not bad, there are real discrepancies at the lower temperatures.

Turning now to the difference in behaviour to be expected between ^{3}He and ^{4}He we note that this will arise to a large degree, in this temperature range, from the different statistics obeyed by the two isotopes. The magnitude of this effect is clearly seen by reference to Fig. 16.49. This gives the total and viscosity cross-sections calculated‖ assuming an interaction 1·3 times the Slater interaction (222) which, while not accurate, is close enough for illustrative purposes. It will be seen that, at very low temperatures, there is a big difference between results expected when the atoms obey different statistics—the cross-sections are much smaller when the atoms obey the Fermi–Dirac rather than the Bose–Einstein statistics. This is manifest in the results shown in

---

† KELLER, W. E., *Phys. Rev.* **105** (1957) 41; MONCHICK, L., MASON, E. A., MUNN, R. J., and SMITH, F. J., ibid. **139** (1965) A1076.
‡ Using (12, 6) interaction, DE BOER, J. and MICHELS, A., *Physica, 's Grav.* **6** (1939) 409.
§ KELLER, W. E., *Phys. Rev.* **97** (1955) 1.
‖ BUCKINGHAM, R. A. and MASSEY, H. S. W., *Proc. R. Soc.* **A168** (1938) 378.

FIG. 16.47. (a) The viscosity η of helium (^{4}He) at low temperatures T °K,

observed directly
- ● van Itterbeek and van Paemel;
- ■ Keesom and Keesom,
- × Becker, Misenta, and Schmeissner,
- + Becker and Misenta,
- ▲ Coremans, van Itterbeek, Beenakker, Knaap, and Zandbergen,

derived from thermal conductivity measurements
- ○ Ubbink and de Haas,
- □ Fokkens, Taconis, and de Bruyn Ouboter.

calculated
- —— A, with interaction (223), $B^2 = 7·29$, $C = 1·45 \times 10^{-60}$ erg cm^6,
- —— B with interaction (223), $B^2 = 7·29$, $C = 1·55 \times 10^{-60}$ erg cm^6,
- --- C with (12, 6) interaction, parameters as in Table 16.16

(b) The viscosity η of helium (^{3}He) at low temperatures,

observed directly
- × Becker, Misenta, and Schmeissner,
- + Becker and Misenta,

derived from thermal conductivity measurements
- □ Fokkens, Taconis, and de Bruyn Ouboter.

calculated
- —— A with interaction (223), parameters as in Table 16.16.
- --- C with (12, 6) interaction, parameters as in Table 16.16.

Fig. 16.47 (b), in which observed† and calculated‡ values of the viscosity and thermal conductivity of ³He are illustrated as functions of temperature in the range 0–30 °K. Comparison with the corresponding results for ⁴He (Fig. 16.47 (a)) shows that at low temperatures the

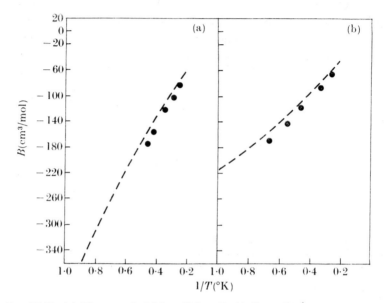

FIG. 16.48. (a) The second virial coefficient B of helium (⁴He) at low temperatures T °K. ● observed (Keller); --- calculated with (12, 6) interaction, parameters as in Table 16.16. (b) The second virial coefficient of helium (³He) at low temperatures. ● observed (Keller). --- calculated with (12, 6) interaction, parameters as in Table 16.16.

viscosity of ³He is considerably larger than that for ⁴He due to the small viscosity cross-section. It will be seen also that both assumed interactions give quite good agreement with the observations.

Fig. 16.48 (b) shows the observed and calculated values of the second virial coefficient of ³He.§ The situation is similar to that for ⁴He in that the agreement, while not unsatisfactory, is not complete as the deviations

---

† Viscosity: BECKER, E. W. and MISENTA, R., *Z. Phys.* **140** (1955) 535; BECKER, E. W., MISENTA, R., and SCHMEISSNER, F., ibid. **137** (1954) 126.
Thermal conductivity: FOKKENS, K., TACONIS, K. W., and DE BRUYN OUBOTER, R., loc. cit.

‡ With interaction (223): BUCKINGHAM, R. A. and SCRIVEN, R. A., *Proc. phys. Soc.* **65** (1952) 376; with (12, 6) interaction: KELLER, W. E., *Phys. Rev.* **105** (1957) 41; MONCHICK, L., MASON, E. A., MUNN, R. J., and SMITH, F. J., ibid. **139** (1965) A1076.

§ Experimental: KELLER, W. E., *Phys. Rev.* **98** (1955) 1571. Calculated: DE BOER, J., VAN KRANENDONK, J., and COMPAAN, K., ibid. **76** (1949) 1728; KILPATRICK, J. E., KELLER, W. E., HAMMEL, E. F., and METROPOLIS, N., ibid. **94** (1954) 1103.

seem to be outside likely experimental error. There is no doubt, however, that the difference between the behaviour of the two isotopes in this respect, as for the viscosity, is closely of the form expected from the theory.

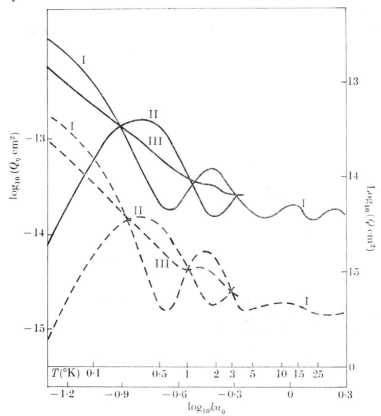

Fig. 16.49. Illustrating the effect of symmetry properties on the total and viscosity cross-sections for an interaction roughly representative of that between He atoms. —— total cross-section Q; - - - viscosity cross-section Q_η; I, Bose–Einstein statistics; II, Fermi–Dirac statistics; III, without allowance for symmetry properties.

Using the interactions discussed above, the coefficient of diffusion D_{12} of ^{3}He in ^{4}He may be calculated in terms of the appropriate diffusion cross-section (7) to obtain the results shown† in Fig. 16.50. Bendt‡ has measured D_{12} for mixtures with average composition 7·94 mole per

† With interaction (223): BUCKINGHAM, R. A. and SCRIVEN, R. A., loc. cit.; with (12, 6) interaction: MONCHICK, L., MASON, E. A., MUNN, R. J., and SMITH, F. J., loc. cit.
‡ BENDT, P. J., Phys. Rev. 110 (1958) 85.

cent ³He from 1·74 to 296 °K and his results are included in the figure. As with the temperature data, there is a very good agreement with both sets of calculated values.

It is also possible by the use of nuclear spin resonance techniques to determine experimentally the self-diffusion coefficient of ³He. Essentially the method consists in orienting ³He nuclei in a magnetic field

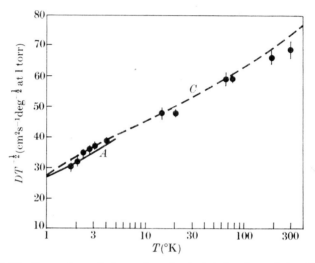

Fig. 16.50. Comparison of observed and calculated values of the diffusion coefficient D of ³He in ⁴He. Calculated —— A with interaction (223), parameters as in Table 16.16; ––– C with (12, 6) interaction, parameters as in Table 16.16. Observed ● Bendt.

and then removing the field. The relaxation time for disorientation is determined by the diffusion time of oriented nuclei to the containing walls as there is a negligibly small chance of spin flip in a collision within the gas (cf. Chap. 18, § 7). By measuring the relaxation time for gas contained in a vessel of suitable geometry the self-diffusion coefficient may be obtained. Fig. 16.51 compares results obtained† in this way with results of calculations‡ based on the (12, 6) interaction as above. The agreement is again very good.

Finally, we have the evidence from thermal diffusion. Fig. 16.52 compares observed§ and calculated‖ thermal diffusion ratios k_T (see (14 a))

† LUSZCZYNSKI, K., NORBERG, R. E., and OPFER, J. E., *Phys. Rev.* **128** (1962) 186.
‡ MONCHICK, L., MASON, E. A., MUNN, R. J., and SMITH, F. J., loc. cit., p. 1405.
§ WATSON, W. W., HOWARD, A. J., MILLER, N. E., and SHIFFRIN, R. M., *Z. Naturf.* **18a** (1963) 242; VAN DER VALK, F., Thesis, Amsterdam.
‖ With interaction (223): BUCKINGHAM, R. A. and SCRIVEN, R. A., loc. cit., p. 1404; with (12, 6) interaction: MONCHICK, L., MASON, E. A., MUNN, R. J., and SMITH F. J., loc. cit., p. 1405.

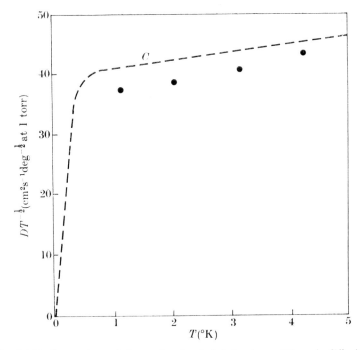

Fig. 16.51. Comparison of observed and calculated values of the spin diffusion coefficient D of ^{3}He. Calculated --- C with (12, 6) interaction, parameters as in Table 16.16. Observed ● Luszczynski, Norberg, and Opfer.

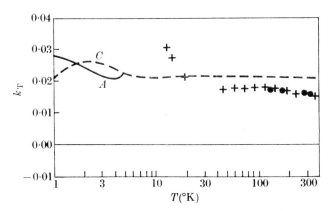

Fig. 16.52. Comparison of observed and calculated thermal diffusion ratios k_T for equimolar mixtures of ^{3}He and ^{4}He. Calculated: ——— A with interaction (223), --- C with (12, 6) interaction parameters as in Table 16.16. Observed: ● Watson, Howard, Miller, and Shiffrin; + van der Valk.

for equimolar mixtures of ^{3}He and ^{4}He. The agreement is only good down to temperatures of 20 °K, but below this temperature the observations are less reliable so the discrepancies are not serious.

10.2. *Results from observations of transport properties and second virial coefficients for $T > 20$ °K*

The gradual transition from a purely quantum to an effectively classical behaviour may be bridged by using expansions for the viscosity and diffusion cross-sections of the form†

$$Q_\eta = Q_\eta^{\text{cl}}\left\{1+\left(\frac{\epsilon}{B^2K}\right)q_\eta^{(1)}(K)+\left(\frac{\epsilon}{B^2K}\right)^2 q_\eta^{(2)}(K)+\ldots\right\},$$

$$Q_\text{d} = Q_\text{d}^{\text{cl}}\left\{1+\left(\frac{\epsilon}{B^2K}\right)q_\text{d}^{(1)}(K)+\left(\frac{\epsilon}{B^2K}\right)^2 q_\text{d}^{(2)}(K)+\ldots\right\}, \quad (225)$$

where $B = 2M\epsilon r_\text{m}^2/\hbar^2$ and K is the kinetic energy of relative motion in units ϵ. Q_η^{cl} and Q_d^{cl} are the respective cross-sections according to classical mechanics. These expansions in powers of $\hbar^4$ converge well when $K \gg 1$ (conventionally for $K > 6$). Explicit expressions have been obtained for $q_\eta^{(1)}$ and $q_\text{d}^{(1)}$.

As an indication of the way in which the transition to the classical formulae proceeds we give in Table 16.15 some typical values of Q_η^{cl}, Q_d^{cl} and of the correction terms ΔQ_η, ΔQ_d such that

$$Q_\eta = Q_\eta^{\text{cl}}+\Delta Q_\eta, \qquad Q_\text{d} = Q_\text{d}^{\text{cl}}+\Delta Q_\text{d}. \quad (226)$$

TABLE 16.15

Illustrating the importance of quantum effects on transport cross-sections at different reduced relative energies K of impact

K	6	8	10	15	20	30	40	50
$Q_\text{d}^{\text{cl}}/\pi r_\text{m}^2$	0·745	0·701	0·668	0·617	0·585	0·544	0·516	0·496
$\Delta Q_\text{d}/\pi r_\text{m}^2$	−0·047	−0·030	−0·021	−0·010	−0·005$_5$	−0·002$_5$	−0·001$_4$	−0·000$_9$
$Q_\eta^{\text{cl}}/\pi r_\text{m}^2$	0·916	0·848	0·802	0·736	0·698	0·650	0·620	0·598
$\Delta Q_\eta/\pi r_\text{m}^2$	−0·079	−0·069	−0·048	−0·023	−0·013	−0·006	−0·003	−0·002

These have been calculated using the interaction (223). As an indication of the relation of the results of Table 16.15 to the gas temperature we may take the equivalent reduced temperature $T^* = \kappa T/\epsilon$ as $\tfrac{2}{3}K$. Thus with ϵ/κ of the order 10 °K, $K = 50$ corresponds to T about 350 °K.

Using results such as in Table 16.15 and corresponding results calculated purely from classical theory for larger K, the variation of the

† DE BOER, J. and BIRD, R. B., *Physica 's Grav.* **20** (1954) 185.

viscosity of helium with temperature up to 1100 °K has been calculated by Buckingham, Davies (A. R.), and Davies (A. E.),† using the same two interactions of the form (223) as in the analysis of the data below 25 °K. Fig. 16.53 shows the comparison with the observed results.‡ Once again, an interaction with a value of C close to 1.50×10^{-60} erg cm^6 will give

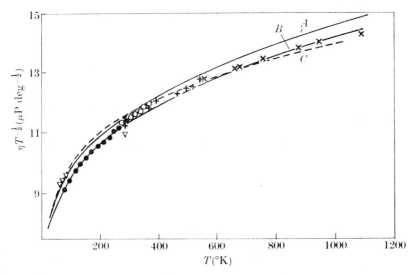

Fig. 16.53. The viscosity η of helium at normal to high temperatures T °K. Observed: × Trautz and Zink; ○ Wobser and Müller; + Trautz and Husseini; ● Johnston and Grilly; ∇ Rietveld, van Itterbeek, and van den Berg. Calculated: —— A with interaction (223) $B^2 = 7.29$, $C = 1.45 \times 10^{-60}$ erg cm^6; —— B with interaction (223) $B^2 = 7.29$, $C = 1.55 \times 10^{-60}$ erg cm^6; ---- C with (12, 6) interaction, parameters as in Table 16.16.

a fit that is as good as the spread of the observed data warrants. It is remarkable that a comparatively simple interaction gives good results over such a wide range of temperature. The lack of sensitivity to the details of the repulsive interaction is also clear from the fact that a single interaction of (12, 6) type is equally successful in representing the behaviour of the viscosity over this range. This may be seen by reference to Fig. 16.53, which gives the results for $T > 20$ °K, obtained by the (12, 6) type with parameters as in Table 16.16 which gives a good fit to the transport properties and second virial coefficient for $T < 20$ °K.

† loc. cit., p. 1454.
‡ TRAUTZ, M. and HUSSEINI, I., *Annln der Physik* **2** (1929) 733; TRAUTZ, M. and ZINK, R., ibid. **7** (1930) 427; JOHNSTON, H. L. and GRILLY, E. R., *J. phys. Chem.* **46** (1942) 948; WOBSER, R. and MÜLLER, F., *Kolloidbeihefte* **52** (1941) 165; RIETVELD, A. O., VAN ITTERBEEK, A., and VAN DEN BERG, G. J., *Physica, 's Grav.* **19** (1953) 517.

On the other hand, the observed virial coefficient for $T > 20$ °K, is not well reproduced with the same (12, 6) interaction but requires a considerably smaller value of ϵ as given in Table 16.16. This may be seen from the comparison with the observed results[†] shown in Fig. 16.54. It will be noted that the calculated values are quite sensitive to the choice of parameters. The interaction (223) gives good results over the restricted temperature range for which its predictions have been calculated.

TABLE 16.16

Values of ϵ and r_m which give a good fit to transport and second virial coefficient data for helium

Temperature range	Interaction form	Data fitted	(ϵ/κ) °K	r_m (Å)
$T < 20$ °K	(223)	Viscosity Thermal conductivity	9·97	2·974
	(12, 6)	Second virial coefficient	10·22	2·869
$T > 20$ °K	(223)	Viscosity	9·97	2·974
	(12, 6)	Viscosity	10·22	2·869
		Second virial coefficient	6·03	2·95

The value $1·50 \times 10^{-60}$ erg cm^6 derived for the van der Waals constant C is somewhat larger than the best theoretical values. Thus Davison,[‡] using an elaborate variational calculation, finds $1·396 \times 10^{-60}$ erg cm^6. Bruch and McGee[§] have therefore attempted to obtain an empirical interaction which not only agrees asymptotically with this theoretical value but is also in agreement with the best theoretical determinations of the asymptotic value of the dipole-quadrupole term.[‡] They were able to obtain a good fit with the observed second virial coefficient in the range 50–1000 °K with an interaction which at intermediate distances is characterized by the parameters $\epsilon/\kappa = 12·5$ °K, $r_m = 2·98 \times 10^{-8}$ cm, and a reduced curvature κ (see (130)) of 80. At the time of writing, the validity of this interaction for dealing with viscosity data at all temperatures and second virial coefficient data at low temperatures has not been checked.

[†] Boks, J. O. A. and Onnes, H. K., *Leiden Communs* **170a** (1924); Holborn, L. and Otto, J., *Z. Phys.* **10** (1922) 367; **23** (1924) 77; **33** (1925) 1; Nijhoff, G. P., Keesom, W. H., and Ilün, B., *Leiden Communs* **188c** (1927), Nijhoff, G. P. and Keesom, W. H., ibid. **188b** (1927); Michels, A. and Wouters, H., *Physica, 's Grav.* **8** (1941) 923; Schneider, W. G. and Duffie, J. A. H., *J. chem. Phys.* **17** (1949) 751; Yntema, J. L. and Schneider, W. G., ibid. **18** (1950) 641.
[‡] Davison, W. D., *Proc. phys. Soc.* **87** (1966) 133.
[§] Bruch, L. W. and McGee, I. J., *J. chem. Phys.* **46** (1967) 2959.

10.3. *Measurement of the total cross-section for* He–He *collisions*

Total cross-sections for collisions between helium atoms with relative velocities corresponding to ordinary temperatures have been measured by Harrison,[†] by Rothe and Neynaber,[‡] by Beier[§] and by Moore, Datz, and van der Valk.[‖] In all cases the technique involved observation of

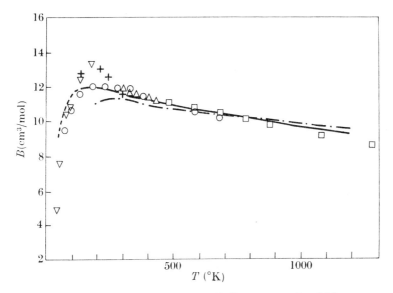

FIG. 16.54. The second virial coefficient of helium at normal to high temperatures. Calculated: ——— (12, 6) potential with $\epsilon/\kappa = 6\cdot03$; — · — · — (12, 6) potential with $\epsilon/\kappa = 10\cdot22$; – – – interaction (223). Observed: + Boks and Onnes; ▽ Nijhoff, Keesom, and Ilün; ○ Holborn and Otto; △ Michels and Wouters; □ Schneider, Duffie, and Yntema.

the absorption of a thermal helium beam by helium gas in a cell, the beam being chopped mechanically so that phase-sensitive detection could be used to discriminate against background. The intensity of the transmitted beam was monitored with an ionization detector. Table 16.17 summarizes some of the detailed features of the experiments. The arrangement of the detector used by Beier is illustrated in Fig. 16.17, while the experiment of Rothe and Neynaber is essentially similar to that described in § 9.3. In three sets of experiments special care was taken to render negligible the pumping error in the absolute pressure

---

[†] HARRISON, H., *J. chem. Phys.* **37** (1962) 1164.
[‡] ROTHE, E. W. and NEYNABER, R. H., ibid. **43** (1965) 4177.
[§] BEIER, H. J., *Z. Phys.* **196** (1966) 185.
[‖] MOORE, G. E., DATZ, S., and VAN DER VALK, F., *J. chem. Phys.* **46** (1967) 2012.

calibration, in which a McLeod gauge was employed. Thus Beier followed the procedure recommended by Meinke and Reich† of cooling (to 10 °C) the mercury in the reservoir of the McLeod. No precautions were taken by Harrison but it is possible that, as helium is the gas concerned, the error in his experiments was not large.†

TABLE 16.17

Details of equipment used for measuring total cross-sections for He–He *collisions*

Observers	Beam temp. (°K)	Scattering gas temp. (°K)	Modulation frequency (c/s)	Resolution
Harrison	83–2900	80	100	15′
Rothe and Neynaber	300	300	20	50″
Beier	4·8–300	300	65	2′
Moore, Datz, and van der Valk	77–600	liq. N_2	1080	15′

In these experiments the scattering takes place under high-velocity conditions (see § 6.3.1) because the van der Waals attraction between helium atoms is relatively so small. The observations were therefore analysed assuming that the cross-section is effectively independent of velocity (i.e. the function $G''_1(\infty, x)$ is used in the analysis of § 7.2.1). The results obtained are illustrated in Fig. 16.55.

It will be seen that Harrison's values are somewhat larger than those obtained in the other experiments. This may possibly be due to inaccuracy in the absolute pressure calibration. For comparison, cross-sections calculated by Bernstein and Morse‡ for the following interactions are also shown:

I, a (12, 6) potential with $\epsilon = 1·40 \times 10^{-15}$ ergs, $r_m = 2·973$ Å;

II, an (exp 12, 6) potential with the same values for ϵ and r_m.

It is clear that the cross-section is to a large extent determined by ϵ and r_m though there is some dependence on the shape. For both interactions the cross-sections are smaller than those observed by Harrison but, within the rather large experimental error, agree with those measured in the other three experiments. Since the cross-section depends largely on ϵ and r_m it is likely that this agreement would be

---

† MEINKE, C. and REICH, G., *Vacuum* **13** (1963) 579.
‡ BERNSTEIN, R. B. and MORSE, F. A., *J. chem. Phys.* **40** (1964) 917.

preserved if the Buckingham–Corner potential (223), which has the same values for these parameters, were used for the theoretical calculations.

At the time of writing the situation can be regarded as promising. It should not be long before useful discriminatory information about He–He interactions will be obtained from total cross-section measurements.

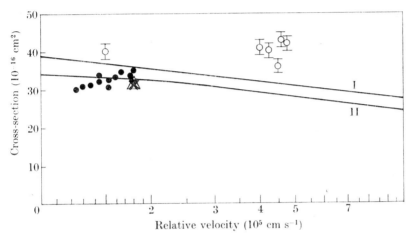

FIG. 16.55. Total cross-sections for He–He collisions. Observed by: ○ Harrison; ● Beier; × Rothe and Neynaber; △ Moore, Datz, and van der Valk. Calculated by Morse and Bernstein for: I, (12, 6) interaction; II, (exp 12, 6) interaction.

11. Interactions between other pairs of rare-gas atoms

Total cross-sections for the scattering of helium atoms by the other rare-gas atoms have been measured by Düren, Helbing, and Pauly† and by Beier,‡ using the single-beam technique without velocity analysis. Beier's equipment has already been described (see Fig. 16.17 and § 10.3 above). That of Düren et al. is similar in principle but with somewhat poorer angular resolution. In a later investigation a velocity selector was added and further measurements carried out by Düren, Feltgen, Gaide, Helbing, and Pauly.§ Landorf and Mueller∥ have applied the double beam method (see § 7.3) to He–Ar scattering as well as other cases. Attention should also be drawn to the earlier observations of Rothe,

---

† DÜREN, R., HELBING, R., and PAULY, H., Z. Phys. **188** (1965) 468.
‡ BEIER, H. J., ibid. **196** (1966) 185.
§ DÜREN, R., FELTGEN, R., GAIDE, W., HELBING, R., and PAULY, H., Phys. Lett. **18** (1965) 282.
∥ LANDORF, R. W. and MUELLER, C. R., ibid. **19** (1966) 658.

Marino, Neynaber, Rol, and Trujillo† by the single-beam method in which the absorption of rare-gas beams by argon was studied. This work was subject to the same pumping error as the similar work of Rothe and Bernstein for absorption of K and Cs beams but corrections may be

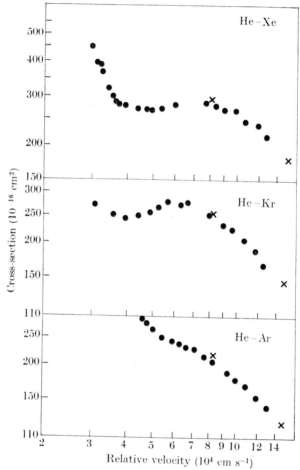

FIG. 16.56. Observed total cross-sections for collisions of He with A, Kr, and Xe atoms. ● Beier. × Düren, Helbing, and Pauly.

applied by comparison with later results of Rothe and Neynaber for Ar–Ar collisions in which the pumping error was not present.

There is very good agreement between the observations of Beier and of Düren, Helbing, and Pauly, as may be seen by the comparison shown in Fig. 16.56, in which the cross-sections are given as functions of the

† ROTHE, E. W., MARINO, L. L., NEYNABER, R. H., ROL, P. K., and TRUJILLO, S. M., Phys. Rev. **126** (1962) 578.

relative velocity of the colliding atoms. In fact, while the relative velocities given by Beier are deduced from an analysis of the type discussed in § 7.2.1 assuming an inverse sixth-power interaction, those for the results of Düren et al. have been calculated simply from the formula for the mean relative velocity

$$\bar{v}_\mathrm{r} = \bar{v}_\mathrm{B} + \bar{v}_\mathrm{S}^2/3\bar{v}_\mathrm{B}, \qquad (227)$$

where $\bar{v}_\mathrm{B} = (9\pi\kappa T_\mathrm{B}/8M_\mathrm{B})^{\frac{1}{2}}$, $\bar{v}_\mathrm{S} = (8\kappa T_\mathrm{S}/\pi M_\mathrm{S})^{\frac{1}{2}}$, T_B, T_S being the temperatures of the source oven and scattering chamber respectively and

TABLE 16.18

Potential parameters for interaction of He with Ne, Ar, Kr, and Xe atoms

	ϵ (10^{-14} erg)			r_m (10^{-8} cm)		
	I	II	III	I	II	III
He–Ne	0·232	0·264		3·05	3·00	
He–Ar	0·350	0·491	0·338	3·45	3·36	3·36
He–Kr	0·365	0·608		3·75	3·47	
He–Xe	0·385	0·667		4·05	3·72	

I, derived from total cross-section measurements of Düren et al. and of Beier.
II, derived by applying the mixture rules to parameters derived from analysis of measurements of the viscosities of the pure gases.
III, derived from analysis of mutual diffusion data.

M_B, M_S the respective masses of the beam and scattering atoms. There is also very good agreement with the observations made by Düren, Feltgen et al. using the velocity selector. The value found by Landorf and Mueller for He–Ar appears to be a little higher. On the other hand, the corrected value found by Rothe, Marino et al. is somewhat smaller.

Düren et al. analysed their data on the assumption of (12, 6) interactions and determined the parameters ϵ and r_m to give the best fit. Their results are given in Table 16.18.

For comparison values derived from analysis of mutual diffusion data for He–Ar† are included as well as others derived as follows. As a rough means for determination of the viscosity of a mixture of two gases A and B the parameters ϵ(AB), r_m(AB) for the interactions between A

† SRIVASTAVA, B. N. and SRIVASTAVA, K. P., *J. chem. Phys.* **30** (1959) 984; SRIVASTAVA, K. P., *Physica, s' Grav.* **25** (1959) 571.

and B are related to those $\epsilon(AA)$, $r_m(AA)$, etc., by the rules†

$$\epsilon(AB) = \{\epsilon(AB)\epsilon(BB)\}^{\frac{1}{2}}, \qquad r_m(AB) = \tfrac{1}{2}\{r_m(AA)+r_m(BB)\}. \quad (228)$$

Knowing $\epsilon(AA)$, etc. from analysis of the viscosity data for the pure gases $\epsilon(AB)$ and $r_m(AB)$ may then be obtained.

It will be seen that the agreement is not very satisfactory. A further check is to use the parameters to derive the van der Waals constant C_{AB} from the relation

$$C_{AB} = 2\epsilon(AB)\{r_m(AB)\}^6 \quad (229)$$

and compare it with calculated values. This comparison is given in Table 16.19. It is remarkable, though possibly fortuitous, that, although

TABLE 16.19

Comparison of observed and calculated values of the van der Waals constant for interaction between helium atoms and rare-gas atoms

	van der Waals constant (erg cm$^6 \times 10^{-60}$)					
	I	II	III	IV	V	VI
He–Ne	3·73	—	3·85	—	3·0	2·78
He–Ar	11·8	8·1	14·1	13·3	9·2	8·7
He–Kr	20·3	—	21·1	—	12·9	12·5
He–Xe	33·9	—	35·5	—	17·9	17·2

I, derived from analysis of observations of Düren et al. and of Beier.
II, derived from analysis of corrected observations of Rothe, Marino et al.
III, derived from analysis of mutual diffusion data.
IV, derived from use of mixture relations and analysis of viscosity data.
V, calculated by Kingston‡ using (212).
VI, calculated by Dalgarno and Davison§ using (213).

the parameters ϵ, r_m derived from the total cross-section and viscosity data are so different (see Table 16.18), the van der Waals constants are much closer. This is not surprising as the long-range interaction is likely to be the dominant factor, except possibly for He–Ne. On the other hand, there is very poor agreement with the calculated values which are all considerably smaller. It may be significant that the value for He–Ar derived from the corrected data of Rothe, Marino et al. agrees much

---

† HIRSCHFELDER, J. O., CURTISS, C. F., and BIRD, R. B., *Molecular theory of gases and liquids*, p. 567 (Wiley, New York, 1954).
‡ KINGSTON, A. E., *Phys. Rev.* **135** (1964) A1018.
§ DALGARNO, A. and DAVISON, W. D., loc. cit., p. 1386.

better but, for this case, the mutual diffusion data favour the larger values. The situation is clearly not yet satisfactory.

We have already referred to the observations by Rothe, Marino, Neynaber, Rol, and Trujillo† of the absorption of rare-gas beams by argon, in which the pressure calibration needs to be corrected for pumping error. This could be done by comparison with the Ar–Ar results of Rothe and Neynaber‡ in which the error was eliminated. Applying this correction to the data of Rothe, Marino et al. for Ne–Ar, Kr–Ar, and Xe–Ar collisions and analysing the resultant values to obtain the van der Waals constant gives the results shown in Table 16.20. The value given for Ar–Ar is derived directly from the observations of Rothe and Neynaber. Values are also included for Kr–Kr derived from the latter observations.

TABLE 16.20

van der Waals constant for interactions of rare-gas atoms with argon atoms and for mutual interaction between krypton atoms

	van der Waals constant (erg cm$^6 \times 10^{-60}$)				
	I	II	III	IV	V
He–Ar	11·8	8·1	9·2	—	—
Ne–Ar	—	20·1	18·8	18·2	—
Ar–Ar	—	58·5	63	59·5	102·4
Kr–Ar	—	82	88·5	83	—
Xe–Ar	—	124	125	121	—
Kr–Kr	—	112	125	118	204

I, derived from analysis of observations of Düren et al. and of Beier.
II, derived from analysis of corrected observations of Rothe, Marino et al. (value for Ar–Ar is from Rothe and Neynaber).
III, calculated by Kingston§ from (212).
IV, calculated by Dalgarno and Davison‖ from (213).
V, derived from analysis of viscosity data for argon.††

For comparison, theoretical values are also included. The agreement seems very satisfactory but it must be remembered that the analysis for the He–rare-gas interactions is not yet satisfactory. Furthermore the Ar–Ar and Kr–Kr interactions derived from viscosity data (column V of Table 16.20) give van der Waals constants nearly twice as large. Once again there is clearly need for further experiments.

† loc. cit., p. 1414.   ‡ loc. cit., p. 1411.
§ loc. cit., p. 1416.   ‖ loc. cit., p. 1386.
†† HIRSCHFELDER, J. O., CURTISS, C. F., and BIRD, R. B., *Molecular theory of gases and liquids*, p. 1110 (Wiley, New York, 1954).

12. Collisions involving atoms which may interact in more than one way

12.1. *Atoms both in ^2S states—collisions between hydrogen atoms*

12.1.1. *Formulae for the cross-sections.* The classic case in which two atoms may interact in more ways than one is that of hydrogen. Two H atoms interact very differently according as the spins of the atomic electrons are parallel or antiparallel (see Chap. 13, § 1.1). In the former case, corresponding to the $^3\Sigma_u$ state of the H_2 molecule, the interaction is very similar to that between two rare-gas atoms, being predominantly repulsive except at large separations where a van der Waals attraction with asymptotic form $-C/r^6$ exists. If the spins are opposed, corresponding to the $^1\Sigma_g$ ground state of H_2, the interaction, while asymptotically of the same form as for the $^3\Sigma_u$ state, becomes strongly attractive at smaller distances before turning repulsive. Thus, whereas for the $^3\Sigma_u$ case the attractive well-depth is of the order 500κ, for $^1\Sigma_g$ it is about eighty times larger. Fig. 16.57 illustrates the actual interactions. The greatly increased well-depth arises from the existence of a chemical attraction which tends to occur when there are unpaired electrons in each atom.

For discussion of collisions between unpolarized hydrogen atoms it is strictly necessary not only to take into account the existence of the two interactions but also the symmetry between the protons. Let $f_g(\theta), f_u(\theta)$ be the amplitudes for scattering through an angle θ in the C.M. system by the $^1\Sigma_g$ and $^3\Sigma_u$ interactions respectively. If the protons were distinguishable or if one were a proton and the other a deuteron, the differential cross-section would be given by

$$\bar{I}(\theta)\,d\omega = \{\tfrac{1}{4}|f_g|^2+\tfrac{3}{4}|f_u|^2\}\,d\omega. \tag{230}$$

Allowance for the indistinguishability of the protons gives†

$$\bar{I}^{(s)}(\theta)\,d\omega = \tfrac{1}{32}\{|f_g(\theta)+f_g(\pi-\theta)|^2+3|f_g(\theta)-f_g(\pi-\theta)|^2+ \\ +3|f_u(\theta)+f_u(\pi-\theta)|^2+9|f_u(\theta)-f_u(\pi-\theta)|^2\}. \tag{231}$$

Except for very low temperature collisions $f_{g,u}(\theta)$ and $f_{g,u}(\pi-\theta)$ overlap very little so that

$$\bar{I}^{(s)}(\theta)\,d\omega \simeq \tfrac{1}{8}\{|f_g(\theta)|^2+|f_g(\pi-\theta)|^2+3|f_u(\theta)|^2+3|f_u(\pi-\theta)|^2\}$$
$$= \tfrac{1}{2}\{\bar{I}(\theta)+\bar{I}(\pi-\theta)\}. \tag{232}$$

† See, for example, BUCKINGHAM, R. A., FOX, J. W., and GAL, E., *Proc. R. Soc.* A**284** (1965) 237.

Writing f_g and f_u in terms of phase shifts in the usual way,

$$f_{g,u} = \frac{1}{2ik}\sum_l i^l(2l+1)\{\exp(2i\eta_l^{g,u})-1\}P_l(\cos\theta), \qquad (233)$$

where k is the wave number for the relative motion, we have

$$\bar{Q}_t^{(s)} = (\pi/k^2)\sum_l (2l+1)\{\omega_{l,g}\sin^2\eta_l^g + 3\omega_{l,u}\sin^2\eta_l^u\}, \qquad (234)$$

where
$$\omega_{l,g} = \tfrac{1}{2}\{2-(-1)^l\},$$
$$\omega_{l,u} = \tfrac{1}{2}\{2+(-1)^l\}. \qquad (235)$$

Ignoring the indistinguishability of the protons $\omega_{l,g} = \omega_{l,u}$ and

$$\bar{Q} = \tfrac{1}{4}\{Q_u + 3Q_g\}, \qquad (236)$$

where Q_u, Q_g are the total cross-sections for scattering by the separate interactions.

The same considerations apply to the viscosity cross-sections. Thus

$$\bar{Q}_\eta^{(s)} = (\pi/k^2)\sum \frac{(l+1)(l+2)}{2l+3}\{\omega_{l,g}\sin^2(\eta_{l+2}^g - \eta_l^g) + 3\omega_{l,u}\sin^2(\eta_{l+2}^u - \eta_l^u)\}, \qquad (237)$$

and
$$\bar{Q}_\eta = \tfrac{1}{4}\{Q_{\eta,g} + 3Q_{\eta,u}\}. \qquad (238)$$

Similarly for diffusion,
$$\bar{Q}_d = \tfrac{1}{4}\{Q_{d,g} + 3Q_{d,u}\}. \qquad (239)$$

$\bar{Q}_d$ could be measured in principle from observations of diffusion of D atoms into H, or vice versa, or from nuclear magnetic resonance experiments as for ^{3}He (see p. 1406).

12.1.2. *The H–H interactions.* The van der Waals constant for interaction between hydrogen atoms has been calculated very accurately by Pauling and Beach† and may be taken with confidence as $6\cdot25\times10^{-60}$ erg cm^6 (6·499 a.u.).

The deep attractive well of the $^1\Sigma_g$ interaction can be determined from analysis of the band spectrum of H_2. Thus in Fig. 16.57 values of the interaction derived in this way are shown. Many calculations have been carried out to reproduce the observed data. One of the most successful, particularly for separation $> 4a_0$, has been that of Dalgarno and Lynn‡ based on a perturbation method. Their results, which for large r tend to the correct van der Waals interaction, are shown in Fig. 16.57. Application of the same method to the $^3\Sigma_u$ interaction gives results which are probably also quite good. At $2a_0$, for example, the

---
† PAULING, L. and BEACH, J. Y., *Phys. Rev.* **47** (1935) 686.
‡ DALGARNO, A. and LYNN, N., *Proc. phys. Soc.* **69** (1956) 821.

agreement with the results obtained by James, Coolidge, and Present†
using an elaborate variational method, is good.

The most detailed calculations for differential, total, and transport
cross-sections have been carried out using analytical fits to the interactions given by Dalgarno and Lynn, which are in atomic units,

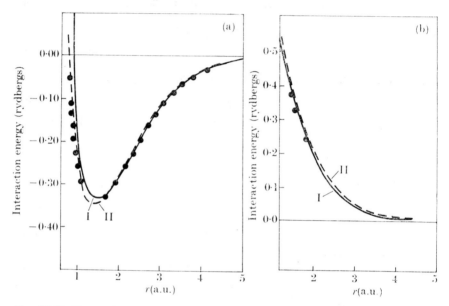

FIG. 16.57. Interactions $V(r)$ between H atoms. (a) $^1\Sigma_g^+$ state: I, Dalgarno and Lynn; II, Kolos and Wolniewicz; ● from vibrational spectrum. (b) $^3\Sigma_u^-$ state: I, Dalgarno and Lynn; II, Kolos and Wolniewicz; ● James, Coolidge, and Present.

$$\begin{align}
{}^1\Sigma_g, \quad V &= 2r^{-1}\{1-1\cdot903r+0\cdot9642r^2-0\cdot1666r^3\} \quad (r<1) \\
&= -65\cdot227e^{-1\cdot78r}+15\cdot64r^6e^{-3\cdot65r}-13r^{-6}-428r^{-8} \quad (4<r) \\
&= \{-0\cdot3388+5\cdot19\times10^{-3}r-3\cdot3275\times10^{-4}r^2-9\cdot3675 10^{-4}r^3\} + \\
&\quad +0\cdot4469\{1-e^{0\cdot86(1\cdot5-r)}\}^2 \quad (1<r<4); \tag{240 a}
\end{align}$$

$$\begin{align}
{}^3\Sigma_u, \quad V &= 2r^{-1}\{1+1\cdot167r+0\cdot039r^2+0\cdot518r^3+0\cdot864r^4\}e^{-2\cdot3r} \quad (r<2) \\
&= 5\cdot7722r^2e^{-2\cdot35r}+181r^3e^{-6\cdot034r} \quad (2<r<4) \\
&= 20\cdot533e^{-1\cdot77r}-13r^{-6}-428r^{-8} \quad (4<r). \tag{240 b}
\end{align}$$

It is noteworthy that in both cases a term in r^{-8} with a relatively large coefficient appears.

† JAMES, H. M., COOLIDGE, A. S., and PRESENT, R. D., *J. chem. Phys.* **4** (1936) 187.

Kolos and Wolniewicz† have recently calculated the interactions out to separations of $10a_0$ by a variational method. Their results are possibly more accurate than those of Dalgarno and Lynn but have not, at the time of writing, been used for the calculation of total and transport cross-sections. We show in Fig. 16.57 a comparison between the interactions which they obtain and that of Dalgarno and Lynn. The differences are not very great.

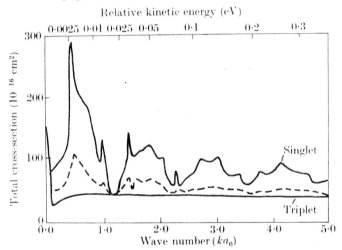

FIG. 16.58. Variation with relative wave-number k of the total elastic cross-sections for impact of H atoms in the singlet ($^1\Sigma_g$) and triplet ($^3\Sigma_u$) cases respectively as indicated. ---- Weighted mean.

12.1.3. *The total elastic cross-sections for* H–H *collisions.* Detailed calculation of the total cross-section have been carried out by Fox and Gal‡ using the interaction of Dalgarno and Lynn and calculating the phase shifts from the appropriate Schrödinger equations by electronic computation. Fig. 16.58 shows the results obtained from the separate $^1\Sigma_g$ and $^3\Sigma_u$ interactions. As would be expected, because of the strong attraction, the $^1\Sigma_g$ cross-section Q_g exhibits numerous fluctuations while the largely repulsive $^3\Sigma_u$ interaction gives rise to a nearly constant cross-section Q_u except at very low energies of relative motion. Fig. 16.58 shows the appropriate weighted mean cross-section $\bar{Q} = \tfrac{1}{4}Q_g + \tfrac{3}{4}Q_u$.

Fig. 16.59 illustrates differential cross-sections when the wave number of relative motion $= 0.5/a_0$, corresponding to a relative kinetic energy of 0·004 eV. These show not only the difference between the singlet (g) and triplet (u) cross-sections but also the effect of allowance for proton symmetry on the mean cross-section $\bar{Q}$.

† KOLOS, W. and WOLNIEWICZ, L., *J. chem. Phys.* **43** (1965) 2429.
‡ FOX, J. W. and GAL, E., *Proc. phys. Soc.* **90** (1967) 55.

To give some indication of the differences between results given by the Dalgarno–Lynn and Kolos–Wolniewicz interactions we show in Fig. 16.60 a comparison between the locations of resonance levels given by the two interactions for the singlet case.

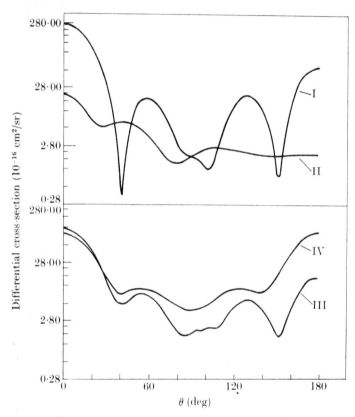

Fig. 16.59. Differential elastic cross-section for impact of H atoms with relative wave number $0.5/a_0$ (relative kinetic energy 0.004 eV). I, singlet ($^1\Sigma_g$) interaction only. II, triplet ($^3\Sigma_u$) interaction only. III, weighted mean but not allowing for proton symmetry. IV, weighted mean including proton symmetry.

When the relative angular-momentum quantum number is l we define an effective potential $V_{\text{eff}}(r, l)$ as in §§ 5.1.1 and 5.4.4 by

$$V_{\text{eff}}(r, l) = V(r) + l(l+1)\hbar^2/2Mr^2, \qquad (241)$$

where M is the reduced mass and $V(r)$ the actual interaction. $V_{\text{eff}}(r, l)$ will have an outer maximum at some value r_m of r. A plot of $V_{\text{eff}}(r_m, l)$ against $l(l+1)$ will be such that all resonance levels must fall below the curve. These levels are located by the condition that the phase η_l must change by π as the relative energy passes through a level. Usually the change is so sharp that this offers no difficulty, but for levels just below the barrier maximum the change is more gradual and the level

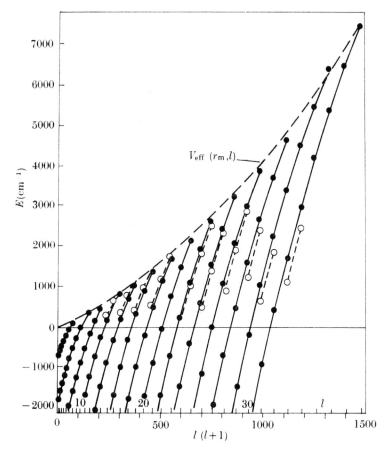

FIG. 16.60. Location of resonance and bound levels for the $^1\Sigma_g$ interaction between H atoms (see Fig. 16.13). -●-●- Calculated from the interaction of Kolos and Wolniewicz; -○-○- calculated from the interaction of Dalgarno and Lynn.

was taken as given by the inflexion point of η_l as a function of relative energy.

Fig. 16.60 shows a comparison of results obtained using the two different interactions, those for the Kolos–Wolniewicz interaction being due to Waech and Bernstein.† It will be seen that, although the results are distinguishable, they are not very different. A further point to

† WAECH, T. G. and BERNSTEIN, R. B., *J. chem. Phys.* **46** (1967) 4905.

notice is the continuity in the level distribution between bound and resonance states. This enables one to associate a vibrational quantum number v with each set of resonance states as shown.

Waech and Bernstein† have also calculated the level widths for the resonance states, defined by (see Chap. 9, § 9)

$$\Gamma = 2/(\partial \eta_l/\partial E)_{\max}. \tag{242}$$

Some of these are given in Table 16.21.

TABLE 16.21

Widths and lifetimes of resonance levels for $^1\Sigma_g$ interaction of H atoms

v	l	Γ (cm^{-1})	Lifetime (10^{-13} s)	Relative k.e. (eV)	Relative velocity (10^6 cm s^{-1})
0	38	84±0·5	0·63	0·923	1·888
0	37	6·2±0·5	8·5	0·801	1·759
1	36	150±10	0·35	0·792	1·749
2	33	21±2	2·5	0·574	1·492
3	31	25±1	2·1	0·483	1·365
4	29	19±10	3·0	0·400	1·243
5	27	23±4	2·3	0·329	1·127
6	25	29±3	1·8	0·267	1·016
7	23	32±1	1·7	0·216	0·913
8	21	46±3	1·1	0·172	0·814

12.1.4. *The viscosity of atomic hydrogen.* The first calculations of the viscosity of atomic hydrogen were carried out using the classical formulae (43) and (44 b) even though orbiting (§ 5.1.1) is important for the interaction at temperatures as high as 10 000 °K. However, it appears from comparison with the accurate quantal calculations of Buckingham, Fox, and Gal,‡ that no serious error is introduced into the calculated viscosity by these orbiting effects.

Fig. 16.61 shows the respective viscosity cross-sections Q_η^g, Q_η^u as calculated from the quantal formulae with and without allowance for the indistinguishability of the protons and employing the interactions derived by Dalgarno and Lynn.§ Values calculated by Dalgarno and Smith‖ from classical theory, using the same interactions, are given for comparison. Orbiting in the $^1\Sigma_g$ case leads to fluctuations about the classical value for $ka_0 > 1$ but these tend to average out when the viscosity is calculated.

† loc. cit.
‡ BUCKINGHAM, R. A., FOX, J. W., and GAL, E., *Proc. R. Soc.* **A284** (1965) 237.
§ loc. cit., p. 1419.
‖ DALGARNO, A. and SMITH, F. J., ibid. **267** (1962) 417.

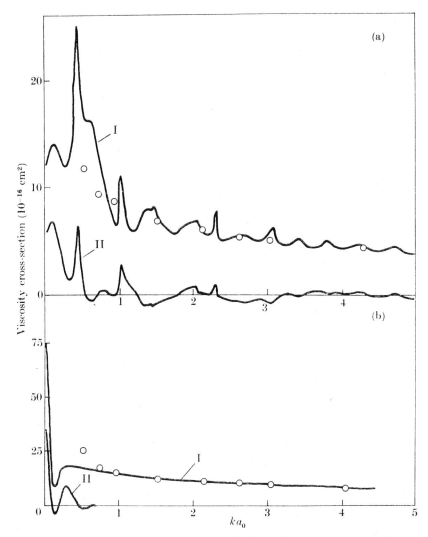

FIG. 16.61. Variation with relative wave-number k of the viscosity cross-sections for H atom impact: (a) in the singlet ($^1\Sigma_g$) state, (b) in the triplet ($^3\Sigma_u$) state. Curves I, calculated by Buckingham, Fox, and Gal without allowance for proton symmetry; ○ calculated by Dalgarno and Smith using classical theory. Curves II, the difference between the cross-sections without allowance for proton symmetry and that with such allowance.

For the $^3\Sigma_u$ case, where orbiting does not arise except at very low wave-numbers, the classical theory gives good results for $ka_0 > 1$.

Table 16.22 gives calculated values of the viscosity and thermal conductivity, showing the effect of nuclear symmetry and the validity of the classical theory at not too low temperatures.

Only meagre experimental evidence is available with which to compare these calculated values.

In 1928† Harteck determined the coefficient of viscosity of partly dissociated hydrogen containing about $2\frac{1}{2}$ per cent by volume of water

TABLE 16.22

Calculated coefficients of viscosity (η) and thermal conductivity (σ) for atomic hydrogen

Temp. (°K)	η (μP)			σ (μW cm^{-1} °K^{-1})
	Symm.	Unsymm.	Classical	Unsymm.
1	3·351	2·304	—	71·29
2	3·188	2·927	—	90·54
4	3·372	3·672	—	113·6
6	3·962	4·170	—	129·0
8	4·608	4·655	—	144·0
10	5·246	5·187	—	160·5
15	6·811	6·669	—	206·3
20	8·414	8·267	—	255·8
30	11·71	11·51	—	356·2
50	17·61	17·31	16·9	535·5
100	28·38	28·32	27·9	876·0
150	37·52	37·42	—	1158
200	45·76	45·70	45·6	1414
250	53·39	53·46	—	1654
300	60·68	60·84	60·9	1882
350	67·76	67·93	—	2101
400	74·65	74·77	75·0	2313

vapour. He used a flow method and obtained observations at −80, 0, and 100 °C. These were measured relative to the wet molecular hydrogen. Eight years later Amdur‡ analysed Harteck's data, neglecting the effect of the water vapour and assuming that both the atoms and molecules behaved as weakly attracting rigid spheres. No further experimental work was carried out until 1959 when Fox and Smith§ determined the viscosity of gaseous mixtures at 23 °C resulting from the dissociation of wet hydrogen in a radio-frequency discharge. They observed the rate of damping of oscillations of a pendulum in the gas. In this way they found that the viscosity of molecular hydrogen containing water vapour as in Harteck's experiments was 7 per cent greater than for the dry gas. Because of this Browning and Fox‖ repeated Harteck's experiments

† HARTECK, P., *Z. phys. Chem.* **139**A (1928) 98.
‡ AMDUR, I., *J. chem. Phys.* **4** (1936) 339.
§ Fox, J. W. and SMITH, A. C. H., ibid. **33** (1960) 623.
‖ BROWNING, R. and Fox, J. W., *Proc. R. Soc.* A**278** (1964) 274.

using effectively dry hydrogen. Dissociation was produced in an r.f. discharge and the partly dissociated gas allowed to diffuse through a U-tube. The total gas-pressure was measured at both ends of the tube with a Pirani gauge, the concentration of atomic hydrogen at the same sections by a Wrede gauge. Measurements were made of the viscosity of mixtures containing a fractional concentration of atomic hydrogen of about 60 per cent at three temperatures, 190, 274, and 373 °K.

A considerable amount of analysis is required to obtain the coefficient of viscosity of atomic hydrogen at the three temperatures from these measurements. According to Hirschfelder, Curtiss, and Bird,† the coefficient of viscosity of a mixture containing a fractional concentration α of atomic hydrogen is given by

$$\eta = \frac{1+Z}{X+Y}, \qquad (243)$$

where
$$X = \frac{\alpha_1^2}{\eta_1} + \frac{\alpha_1 \alpha_2}{\eta_{12}} + \frac{\alpha_2^2}{\eta_2},$$

$$Y = \tfrac{3}{5} A_{12} \left\{ \frac{1}{2} \frac{\alpha_1^2}{\eta_1} + \frac{9}{4} \frac{\eta_{12}}{\eta_1 \eta_2} \alpha_1 \alpha_2 + 2 \frac{\alpha_2^2}{\eta_2} \right\},$$

$$Z = \tfrac{3}{5} A_{12} \left[\tfrac{1}{2} \alpha_1^2 + 2 \left\{ \frac{9}{8} \left(\frac{\eta_{12}}{\eta_1} + \frac{\eta_{12}}{\eta_2} \right) - 1 \right\} \alpha_1 \alpha_2 + 2\alpha_2^2 \right]. \qquad (244)$$

Here η_1, η_2 are the coefficients of viscosity of pure H and H_2 respectively. $\alpha_2 = 1-\alpha_1$ is the fractional concentration of H_2. η_{12} is a fictitious viscosity coefficient for a pure gas of molecular weight $2M_1 M_2/(M_1+M_2)$, where M_1, M_2 are the molecular weights of H and H_2 respectively. The interaction between these molecules is to be taken the same as that between H and H_2. A_{12} is a function of the reduced temperature $\kappa T/\epsilon$ and may be calculated when the H–H_2 interaction is known.

To determine this interaction the quantal calculations of Margenau‡ for the triangular and linear configurations of H–H_2 were used. Averaged over all orientations with appropriate weighting these calculations can be fitted either to a (12, 6) interaction with $\epsilon/\kappa = 32·3$ °K and $r_m = 3·088$ Å,§ or to an (exp α, 6) interaction with $\epsilon/\kappa = 16·7$ °K, $r_m = 3·487$ Å, and $\alpha = 12·45/a_0$.∥

With these interactions η_{12} and A_{12} may be calculated and hence, through (243–4), η_1 may be derived from the observed values of η. Fig. 16.62 shows the results for η_1 obtained on the assumption of the two forms for the H–H_2 interaction. Good agreement is found between these values and those calculated directly by Buckingham, Fox, and Gal†† which are also shown in Fig. 16.62. This agreement is better when the (12, 6) form is assumed for the H–H_2 interaction than for the (exp α, 6) form.

† HIRSCHFELDER, J. O., CURTISS, C. F., and BIRD, R. B., *Molecular theory of gases and liquids*, p. 529 (Wiley, New York, 1954).
‡ MARGENAU, H., *Phys. Rev.* **64** (1943) 131 and **66** (1944) 303.
§ CLIFTON, D. G., *J. chem. Phys.* **35** (1961) 1417.
∥ WEISSMAN, S. and MASON, E. A., ibid. **36** (1962) 794.
†† loc. cit., p. 1424.

An alternative method of analysis is in terms of the diffusion coefficient D_{12} in cm² s⁻¹ at 1 atm, of H in H_2 which is related to η_{12}, in micropoise, by†

$$D_{12} = 7\cdot 327 \times 10^{-5} T A_{12} \eta_{12}. \tag{245}$$

It may also be calculated directly in terms of the H–H_2 interaction from the usual formula (7). A check for the self-consistency of the analysis is therefore to derive

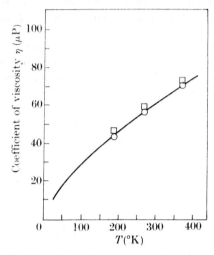

FIG. 16.62. Variation with temperature T of the viscosity of atomic hydrogen. —— calculated by Buckingham, Fox, and Gal. Derived from observations of Browning and Fox, □ on assumption of (exp α, 6) interaction between H and H_2, ○ on assumption of (12, 6) interaction between H and H_2.

η_{12} from (243) and (244) using the observed values of η and η_2 and calculated values of η_1 (see Table 16.22) and A_{12}. From this D_{12} may be obtained from (245). This gives the values shown in Fig. 16.63 for both assumed forms of H–H_2 interactions. These values are compared in the same figure with those calculated directly from the formula (7), which show no significant difference between results obtained with the (12, 6) and (exp α, 6) forms. Again there is good agreement, especially when, in the calculation of η_{12} and A_{12}, the (12, 6) form is assumed.

It is worth remembering that there still remains an uncertainty in the observed viscosities of the H–H_2 mixtures because, in the analysis of the flow, it is assumed that the fractional transfer of tangential momentum to the walls of the flow tube is the same for atoms and for molecules.

Some further evidence about the H–H_2 interaction, derived from total cross-section measurements is discussed in § 13.4.

† HIRSCHFELDER, J. O., CURTISS, C. F., and BIRD, R. B., *Molecular theory of gases and liquids*, p. 532 (Wiley, New York, 1954).

12.2. Atoms both in ^2S states—collisions between alkali-metal atoms

Just as for two hydrogen atoms, two alkali-metal atoms can interact in two distinct ways depending on whether the spins of the outer S electron in each atom are parallel or antiparallel. Also the interactions have the same qualitative form as in hydrogen. Nevertheless, provided the relative energy of collision is small compared with the

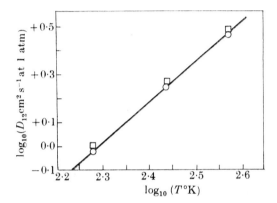

Fig. 16.63. Mutual diffusion coefficient D_{12} for H–H$_2$ mixtures. ——— calculated from (245) and the observed values of the viscosity of H$_2$ and of H–H$_2$ mixture. The results are the same for both assumed H–H$_2$ interactions. □ calculated directly assuming (exp α, 6) interaction between H and H$_2$. ○ calculated directly assuming (12, 6) interaction between H and H$_2$.

well depth for the $^3\Sigma_u$ interaction, the van der Waals constant can be derived from observation of the total collision cross-section just as for the cases discussed in § 9.3 for collisions between atoms which can interact in only one way. Considerable interest, therefore, attaches to the observation of total cross-sections for collisions between alkali-metal atoms. Pressure measurements of absorbing alkali-metal vapour can be made independent of McLeod gauge calibration. Moreover, it is possible to derive the van der Waals constant C_{AB} for interaction of two alkali-metal atoms A and B, either by studying the absorption of a beam of atoms A in a vapour composed of atoms B or vice versa, and this provides a useful check on the vapour pressure measurements. On the theoretical side the calculation of C_{AB} for alkali-metal atoms should be relatively accurate, depending as it does mainly on measurable quantities.

12.2.1. *Experimental method.* Buck and Pauly† have carried out measurements of this kind using apparatus with dimensions as indicated

† Buck, U. and Pauly, H., *Z. Phys.* **185** (1965) 155.

in Fig. 16.64. The principle of the method is the same as in corresponding measurements of cross-sections for scattering of alkali-metal atoms by rare-gas atoms, the only essential differences being the use of alkali-metal vapour as absorber.

The design of the scattering chamber is illustrated in Fig. 16.65. As the pressure of absorbing vapour is determined by the lowest temperature within the chamber it is essential that uniformity of temperature be

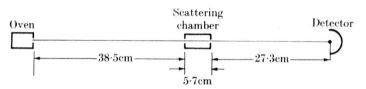

FIG. 16.64. Geometry of the apparatus used by Buck and Pauly.

maintained as closely as possible. For this reason the chamber was constructed of copper. The beam entered the chamber through a channel (10 in Fig. 16.65 (a) and (b)) 2 mm long, 2 mm high, and 0·1 mm broad, and left through a slit (4 in Fig. 16.65 (a)) 2 mm high and 0·05 mm broad. Alkali-metal vapour was introduced from a side chamber shown in section in Fig. 16.65 (b). The temperature of this chamber could be controlled as required so as to provide the desired vapour pressure in the main scattering chamber. Because of their low heat conductivity, high mechanical strength, and low expansion coefficient, quartz rods were used to support the chamber. The temperature could be measured at three points in the block by thermo-elements. At a temperature of 365 °K the maximum temperature difference between points of the scattering chamber did not exceed 0·5 °C.

The detecting wire was of tungsten 8 μm in diameter and 3 mm long giving an angular resolution of 35" according to the Kusch criterion. Discrimination against background was provided in the usual way by mechanically chopping the beam after issuing from the oven and using phase-sensitive detection with a selective amplifier.

The pressure in the scattering chamber was determined from the vapour pressure–temperature relation for the alkali-metal vapour concerned. By using a semiconductor with resistance-temperature characteristic of the form

$$R = R_0 \exp(-b/T),$$

where R_0 and b are constants, as a resistance thermometer, the pressure p could be determined conveniently from the resistance since p also varies exponentially with $1/T$.

A further check on these pressure determinations could be made by observing the flux of target particles on to the detecting wire when the incident beam was cut off. This varies as $n/T^{\frac{1}{2}}$, where n is the concentration of atoms and T the temperature in the scattering chamber. An independent check on relative concentrations is therefore available.

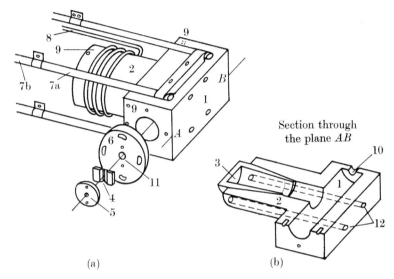

FIG. 16.65. (a) Arrangement of the scattering chamber in the experiments of Buck and Pauly. (b) Sectional drawing through the plane AB of the scattering chamber. 1, copper-walled chamber. 2, heat shield. 3, alkali-metal container. 4, collimator slit. 5, rotatable disc to which 4 is attached. 6, flange. 7, supporting rods. 8, water-cooling tubes. 9, holes for heating wires and thermoelements. 10 and 11, inlet and outlet canals. 12, holes for heating coils.

Good evidence of the reliability of the measurement of the concentration n of scattering particles is provided by the observations shown in Fig. 16.66 of $\ln(I/I_0)$ as a function of the particle density for the scattering of different alkali-metal atoms by rubidium. I_0 is the intensity of the beam when the scattering chamber is evacuated and I that when it contains n atoms/cm³. The linearity of the plots shown in Fig. 16.66 is very good, showing that the absorption law is followed well if measured values of n are used.

12.2.2. *Analysis of data.* The results were analysed as for alkali-metal–rare-gas collisions, using the method outlined in § 9.3. Because of the interchangeability of beam and target atoms all pressure measurements could be made in terms of the vapour pressure–temperature relation of one chosen alkali metal. For this purpose the observations of Langmuir and Taylor† for caesium were chosen. Referred to a selected mean relative velocity $3{\cdot}87 \times 10^4$ cm s^{-1} the observed total cross-sections for different pairs of atoms were then as given in Table 16.23. In deriving

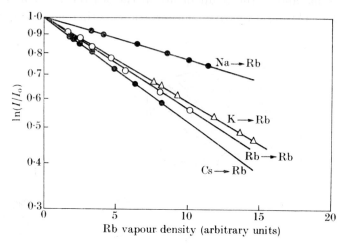

Fig. 16.66. Observed variation of $\ln(I/I_0)$ with density of the scattering vapour for alkali-metal atomic beams scattered in rubidium.

Table 16.23

Observed total cross-sections (in 10^{-16} cm^2) for collisions between pairs of alkali-metal atoms at a mean relative velocity $3{\cdot}87 \times 10^4$ cm s^{-1}, using the caesium vapour pressure curve as standard

Beam atom \ Target atom	Na	K	Rb	Cs
Na	1100	1193	1321	1340
K	1195	1369	1382	1477
Rb	1310	1384	1397	1572
Cs	1340	1477	1513	1869

these values the pressure of the alkali-metal vapour was calibrated against the vapour pressure of caesium by requiring that $Q_{\text{Cs-X}} = Q_{\text{X-Cs}}$, where X denotes an alkali-metal atom other than caesium and $Q_{\text{Cs-X}}$ denotes the cross-section measured by passing a Cs beam through the vapour X. It is then seen that there is good consistency in all the other cases, i.e. $Q_{\text{YX}} = Q_{\text{XY}}$, where X and Y are different alkali-metal atoms, other than caesium.

† Taylor, J. B. and Langmuir, I., *Phys. Rev.* **51** (1937) 753.

As the key measurement of vapour pressure was checked in so many different ways it is difficult to see how any serious error arises from this source. A further check can be made with the earlier observations of Estermann, Foner, and Stern[†] for Cs–Cs collisions using the method described in § 7.2.3. They found a cross-section of 2350×10^{-16} cm² at a most probable velocity of $2 \cdot 4 \times 10^4$ cm s⁻¹, whereas the value given in Table 16.23, when reduced to the same velocity by assuming the $v^{-2/5}$ law (101), is 2355×10^{-16} cm². The closeness of the agreement is fortuitous but is nevertheless satisfactory.

Table 16.24 gives the van der Waals constants derived from the cross-sections given in Table 16.23. Values calculated by Dalgarno and Kingston,[‡] using the formula (210), are given for comparison. The agreement is not satisfactory. Some of the discrepancy may possibly

TABLE 16.24

van der Waals constants C_{AB} (in units 10^{-60} erg cm⁶) for interaction between alkali-metal atoms

	Na obs.	Na calc.	K obs.	K calc.	Rb obs.	Rb calc.	Cs obs.	Cs calc.
Na	880	1500	1081	—	1380	—	1443	—
K			1522	3280	1561	—	1839	—
Rb					1602	5360	2053	—
Cs							3310	4880

be due to the large magnitude of C_{AB} which renders a perturbation treatment doubtful but it has been shown by Smith (F. J.)[§] that the $^3\Sigma_u$ interaction, at the relative velocities concerned, almost certainly gives rise to an oscillatory variation of the cross-section with relative velocity of the type discussed in § 6.3.1. It is therefore desirable that data be obtained at other relative velocities, but, since the discrepancies appearing in Table 16.24 are all in the same sense, it seems unlikely that all could be due to oscillatory departures from the $v^{-2/5}$ law.

12.3. *The viscosity and thermal conductivity of atomic oxygen*

At altitudes between 150 and some hundreds of kilometres in the earth's upper atmosphere atomic oxygen is the principal constituent. A knowledge of the thermal conductivity of this oxygen allotrope, at

[†] loc. cit., p. 1354. [‡] loc. cit., p. 1387.
[§] SMITH, F. J., *Molec. Phys.* **10** (1965–6) 283.

temperatures of the order of 1000–2000 °K which prevail at these altitudes, is necessary in order to interpret many upper atmosphere phenomena. Direct measurement of either the conductivity or the viscosity is very difficult so reliance must be placed on theory to provide the information. Unfortunately the problem is quite complicated. The bound state of atomic oxygen is a ^3P state and two atoms in such states can interact in no less than eighteen different ways!

At large separations r the dominant contribution arises from quadrupole–quadrupole interactions, so

$$V(r) \sim \frac{A}{r^5} - \frac{C}{r^6}, \qquad (246)$$

where A depends on the molecular state involved while the van der Waals constant C is the same for all states. When r is so large that $V(r)$ is smaller than the fine-structure separation of the ^3P levels the molecular states are distinguished by the total atomic angular momentum quantum number J and by $\Omega = |M_{J_2} + M_{J_1}|$, the magnitude of the component of total angular momentum about the molecular axis. This gives rise to fourteen separate interactions each of which is associated with a different value of A. We refer to this as region I.

At somewhat smaller but still large separations (region II) the interaction energy is large enough to decouple the electron spins from the orbital motion in each atom. The molecular states are then characterized by the atomic orbital angular-momentum quantum-number L and by $\Lambda = |M_{L_2} + M_{L_1}|$, the component of total orbital angular momentum about the molecular axis, giving rise to four states with distinct values of A.

Finally, at still smaller separations (region III) chemical forces become important and eighteen states arise. Table 16.25 lists these states and shows how they associate in four groups in region II with the corresponding four values of A. These values were calculated from a formula given by Knipp.† The absolute values are proportional to the mean square radius $\langle r^2 \rangle$ of the orbit of the outermost electron. Values given in Table 16.25 were obtained using an approximate wave function, given by Slater's rules, for which $\langle r^2 \rangle = 1 \cdot 45 a_0^2$. A check on the reliability of this value is provided from a comparison of observed and calculated polarizabilities of the atom as the latter depend mainly on $\langle r^2 \rangle$. Good agreement is found, the calculated value being $0 \cdot 83 \times 10^{-24}$ cm^3, the observed $0 \cdot 77 \times 10^{-24}$ cm^3.

† KNIPP, J. K., *Phys. Rev.* **53** (1938) 734.

The van der Waals constant may be obtained from the Slater–Kirkwood formula (218) using the observed value for the atomic polarizability. We obtain $C = 9 \cdot 45 \times 10^{-60}$ erg cm⁶.

From the known separation of the fine-structure components of the ground state we can regard region I as prevailing until the interaction

TABLE 16.25

The quadrupole–quadrupole interaction constants for various molecular states of O_2 which dissociate into ^3P atoms

States			Statistical weights	A	
				(a.u.)	erg cm⁵ × 10⁻⁵³
$^5\Pi_g$	$^3\Pi_u$	$^1\Pi_g$	2/9	−4·033	−3·62
$^5\Sigma_g^+$	$^3\Sigma_u^+$	$^1\Sigma_g^+$			
$^3\Sigma_u^-$	$^3\Sigma_g^-$	$^1\Sigma_u^-$	4/9	0·00	0·00
$^5\Pi_u$	$^3\Pi_g$	$^1\Pi_u$			
$^5\Delta_g$	$^3\Delta_u$	$^1\Delta_u$	2/9	1·008	0·908
$^5\Sigma_g^+$	$^3\Sigma_u^+$	$^1\Sigma_g^+$	1/9	6·050	4·17

reaches a value of 150κ erg so that for temperatures below about 150 °K we need only use the interaction in this region. Between 150 and 300 °K the main contribution to the viscosity comes from region II but at higher temperatures region III cannot be ignored. The determination of the eighteen separate interactions in this region presents a difficult problem.

Six of the eighteen molecular states $^3\Sigma_g^-$, $^1\Delta_g$, $^1\Sigma_g^+$, $^1\Sigma_u^-$, $^3\Delta_u$, and $^3\Sigma_u^+$ are attractive so that information is available from molecular spectra. By the standard Rydberg–Klein–Rees† method the shape and depth of the potential well can be determined with reasonable accuracy. No comparable information is available about the remaining twelve states. Vanderslice, Mason, and Maisch‡ have estimated the interactions for these cases by appealing to valence bond theory in its simplest form—the perfect pairing approximation according to which the interaction energy in a particular state is given by

$$V = \sum{}'J_{ij} - \tfrac{1}{2}\sum{}''J_{ij} - \sum{}'''J_{ij}. \qquad (247)$$

Here J_{ij} is the exchange energy for two electrons in the respective ith and jth states. $\sum'$, $\sum''$, $\sum'''$ denote summations over all orbitals with paired spins, with non-paired spins, and with parallel spins respectively.

† RYDBERG, R., *Z. Phys.* **73** (1931) 376; KLEIN, O., ibid. **76** (1932) 226; REES, A. L. G., *Proc. phys. Soc.* **59** (1947) 998.
‡ VANDERSLICE, J. T., MASON, E. A., and MAISCH, W. G., *J. chem. Phys.* **32** (1960) 515.

From the interactions derived for the bound states the exchange energies may be determined so that, if the individual molecular states may be identified in terms of the pairing or otherwise of the spins of atomic orbitals, estimates can be made of the interactions for the other states.

Konowalow, Hirschfelder, and Linder[†] estimated the viscosity and thermal conductivity in the range 5 to 350 °K using the interactions in regions I and II and a semi-empirical procedure which has proved effective in many other cases. Smith (F. J.)[‡] has elaborated this work by carrying out a thoroughgoing classical calculation of the viscosity. He used the region III interactions of Vanderslice, Mason, and Maisch and joined them smoothly to the region II values derived from (247) with the values of A given in Table 16.25. A difficult problem arises when interactions occur between the various potential-energy curves. As the atoms approach along one curve does the interaction follow the curve throughout or do transitions occur to other curves? This question is discussed in more detail in Chapter 18, § 8.4.1. It is likely that, when the velocity of relative motion is not too high, the probability of transitions occurring between the curves is small. Still, it is difficult to determine how small the velocity must be for this to be valid. Smith therefore carried out calculations on two different assumptions. In one he assumed that no transitions occur. For the second he noted that two selection rules limit the possible transitions to those in which the g, u and $+$, $-$ symmetries remain unchanged (see Chapter 12, § 2.2). This makes it possible to assign the eighteen interactions to four groups such that transitions can occur between states in a group but not with states in any other group. As an opposite extreme to the first assumption Smith averaged the interaction within such groups and calculated the viscosity cross-section for each such average. A properly weighted over-all average was then taken. Results obtained with the two assumptions differed by less than the errors arising from other uncertainties.

Table 16.26 gives Smith's calculated values of the viscosity and thermal conductivity, together with some values calculated by Yun and Mason[§] by essentially the same method.

12.4. *Observation of the anisotropy of van der Waals forces*

Suppose that a beam of atoms in a state with total angular momentum quantum number J passes through an inhomogeneous magnetic field so

---

[†] KONOWALOW, D. D., HIRSCHFELDER, J. O., and LINDER, B., *J. chem. Phys.* **31** (1959) 1575.
[‡] SMITH, F. J., Thesis, Queen's University, Belfast (1966).
[§] YUN, K. S. and MASON, E. A., *Physics Fluids* **5** (1962) 380.

that separated beams with fixed values of M_J become available. How will the total cross-section for scattering of such beams by, say, rare gas atoms depend on M_J? We cannot expect any difference between states with the same $|M_J|$ so to examine this question we must use atoms in states with $J \geqslant 1$. A convenient choice for experiments should be a substance that

TABLE 16.26

Calculated viscosity and thermal conductivity of atomic oxygen

Temp. (°K)	Thermal conductivity (10^{-4} cal °C^{-1} s^{-1} cm^{-1})	Viscosity (10^{-4} P)	
		Smith	Yun and Mason
100	0·42	0·91 (0·97)	
300	0·93	2·0 (2·1)	
1000	2·2	4·7	4·7
3000	4·9	10·5$_5$	10·5
10 000	12·3	26·5	26·7

Bracketed values of the viscosity are those given by Konowalow, Hirschfelder, and Linder (loc. cit.).

is readily detectable with precision because small effects are to be looked for. One atom fulfilling these conditions is gallium which, while being sufficiently volatile at 1600 °K, may be detected by the Langmuir–Taylor ionization technique. At 1000 °K about half the atoms issuing from the oven will be in the $^2P_{\frac{3}{2}}$ state, thereby fulfilling the condition that $J \geqslant 1$. Successful experiments using gallium beams scattered by rare-gas atoms have been carried out by Berkling, Schlier, and Toschek.†

12.4.1. *The magnitude of the effects expected.* We begin by estimating the magnitude of the effect to be expected. By following through the usual second-order perturbation theory of the van der Waals interaction (see § 9.3.1) it is found that the interaction corresponding to different values of J and M_J can be expressed in terms of two quantities γ_S and γ_D so that, for the various cases,

$$V = -C_{JM_J}/r^6,$$

where we have

$$C_{\frac{1}{2},\pm\frac{1}{2}} = \tfrac{2}{9}\gamma_S + \tfrac{4}{9}\gamma_D,$$
$$C_{\frac{3}{2},\pm\frac{1}{2}} = \tfrac{1}{6}\gamma_S + \tfrac{13}{30}\gamma_D + (\tfrac{1}{6}\gamma_S + \tfrac{1}{30}\gamma_D)\cos^2\theta,$$
$$C_{\frac{3}{2},\pm\frac{3}{2}} = \tfrac{5}{18}\gamma_S + \tfrac{41}{90}\gamma_D - (\tfrac{1}{6}\gamma_S + \tfrac{1}{30}\gamma_D)\cos^2\theta. \qquad (248)$$

† BERKLING, K., SCHLIER, CH., and TOSCHEK, P., *Z. Phys.* **168** (1962) 81; TOSCHEK, P., ibid. **187** (1965) 52.

θ is the angle between the axis of quantization, determined by that of the aligning magnetic field, and the line joining the centres of mass of the colliding atoms. The interaction averaged over all orientations is given by
$$\bar{V} = -C/r^6,$$
where
$$C = \tfrac{2}{9}\gamma_S + \tfrac{4}{9}\gamma_D. \tag{249}$$

To determine the total elastic cross-section Q_{J,M_J} for scattering of atoms in the state J, M_J we may use the approximate formula of Schiff† which is a generalization of that (104) which yields the expression (103) for scattering by an isotropic van der Waals interaction. The Schiff formula is

$$Q = \int_{-\infty}^{\infty}\int_{-\infty}^{\infty} \sin^2\left\{(1/2\hbar v) \int_{-\infty}^{\infty} V(x,y,z)\,dz\right\} dx\,dy, \tag{250}$$

where v is the velocity of relative motion. z is measured along the direction of incidence. Using (250) we find, for
$$V = -Cr^{-6}(1 + a\cos^2\theta),$$
$$Q = Q_0(1 + \alpha\cos^2\Theta),$$
where Θ is the angle between the initial direction of relative motion of the colliding atoms and the axis of quantization. α and Q_0 are given by

$$\alpha = -\frac{a}{10} + \ldots, \tag{251 a}$$

$$Q_0 = 8\cdot083(C/\hbar v)^{\frac{2}{5}}\left(1 + \frac{a}{6} + \ldots\right). \tag{251 b}$$

Applying these results to the interactions (248) we find, when $\Theta = 0$,

$$\frac{\Delta Q}{\bar{Q}} = \frac{Q_{\frac{3}{2},\frac{1}{2}} - Q_{\frac{3}{2},\frac{3}{2}}}{\tfrac{1}{2}(Q_{\frac{3}{2},\frac{1}{2}} + Q_{\frac{3}{2},\frac{3}{2}})} \simeq (40 + 20\chi)^{-\frac{2}{5}}\left[(39 + 15\chi)^{\frac{2}{5}}\left\{1 - \frac{1 + 5\chi}{195 + 75\chi}\right\} - (41 + 25\chi)^{\frac{2}{5}}\left\{1 - \frac{1 + 5\chi}{205 + 125\chi}\right\}\right], \tag{252}$$

where $\chi = \gamma_S/\gamma_D$. The analysis that leads to (248) shows that χ is nearly independent of the nature of the target atom provided it is one of the rare gases. Using the Hartree–Fock wave functions for gallium, χ comes out to be $0\cdot49$, giving
$$\Delta Q/\bar{Q} = 0\cdot014. \tag{253}$$

It is clear that to detect the effect it is necessary to observe small percentage differences between cross-sections, and this constituted the particular difficulty of the experiment.

† loc. cit., p. 1325.

12.4.2. The experimental method.

The general arrangement of the apparatus is illustrated in Fig. 16.67.

The beam issuing from the oven passed through a velocity selector and thence through an inhomogeneous magnetic field which aligned the atoms and separated them into beams corresponding to the different

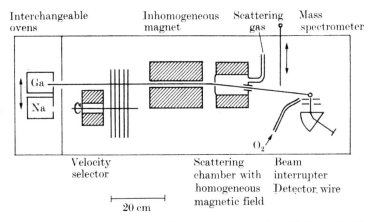

Fig. 16.67. General arrangement of the apparatus used by Berkling, Schlier, and Toschek for investigating the anisotropy of the long-range interaction between gallium and rare-gas atoms.

Zeeman states. By choosing the strength of the field either the $^2P_{\frac{3}{2},\frac{3}{2}}$ or $^2P_{\frac{3}{2},\frac{1}{2}}$ beam could be directed through the absorbing gas chamber which was pervaded by an homogeneous magnetic field to maintain the alignment. The intensity of the beam after issuing from the absorber was measured by the Langmuir–Taylor hot wire technique.

The oven was operated at 1600 °K, at which temperature a background vacuum pressure 2×10^{-5} torr could be maintained. With an oven slit of dimensions 0.13×4 mm, 3 mm deep, an ion current of order 10^{-12} A was measured from one Zeeman component.

The inhomogeneous magnetic field was of the 'two-wire' type used so extensively in atomic-beam experiments. It was necessary to use a velocity selector which was of the type designed by Bennewitz† because otherwise the thermal velocity spread would obscure the separation of the Zeeman components.

The beam left the scattering chamber through a canal with cross-sectional dimensions 3×4 mm and length 4 cm. This provided sufficient resistance to flow to prevent a serious build-up of pressure elsewhere

† BENNEWITZ, H. G., Dissertation, Bonn (1956).

along the path of the beam. The chamber pressure, usually between 1 and 5×10^{-5} torr, was measured with an ionization gauge.

The detector was a tungsten wire of 50 μm diameter. It was necessary to maintain a surface coating of high work-function to obtain a satisfactory detection efficiency. This was achieved by oxidation but, as the wire had to be heated to reduce the time-constant of the detector, for reasons explained below, the coating had to be continually renewed by exposing the wire to a stream of oxygen. Despite the presence of the gas a pressure of about 4×10^{-6} torr was maintained throughout the main vacuum chamber.

As K$^+$ ions were given off from the wire a coarse mass analyser was introduced to separate out the wanted Ga$^+$ ions from the total emitted ion current.

The angular resolution was about 45″, which was fully adequate for the measurement of the total elastic cross-section.

To detect and measure the small fractional difference between the cross-sections $Q_{\frac{3}{2},\frac{3}{2}}$ and $Q_{\frac{3}{2},\frac{1}{2}}$ it was necessary to use a special procedure for eliminating background effects. This was done by introducing a switching system which made it possible to measure, in a chosen succession, the signals from either beam and from the background over a long interval of time. Fig. 16.68 illustrates the signal sequence.

By switching the energizing current of the analysing magnet either the $P_{\frac{3}{2},\frac{3}{2}}$ or $P_{\frac{3}{2},\frac{1}{2}}$ beam could be directed through the scattering chamber to the detector. Another switch could magnetically operate a baffle that interrupted the beam. We distinguish the 4 signals as follows.

Magnetic field	Beam baffle	Number
High ($^2P_{\frac{3}{2},\frac{3}{2}}$)	Out	1
High ($^2P_{\frac{3}{2},\frac{3}{2}}$)	In	2
Low ($^2P_{\frac{3}{2},\frac{1}{2}}$)	Out	3
Low ($^2P_{\frac{3}{2},\frac{1}{2}}$)	In	4

Each operation was chosen to last only long enough to make a satisfactory measurement. The limitation came from the time-constant of the detector and the shortest practicable time required was 0·64 s. Measurements were taken in the order 42312413... and integration was carried out over 20–40 min. Short-time fluctuations were thereby averaged out and the effects of a linear drift eliminated. A comparative device, based on measurements of the magnetic field with a Hall probe, was introduced to eliminate hysteresis and thermal effects in the magnet. The ratio of the magnetizing currents which gave detector signals from the $(\tfrac{3}{2}, \tfrac{3}{2})$ and $(\tfrac{3}{2}, \tfrac{1}{2})$ beams was found to be 3·004 instead of 3·0.

There remains the question of the probability of depolarizing collisions occurring within the scattering chamber. The absence of a magnetic moment in the target atoms ensures that this probability will be small and measurements made in experiments concerned with optical pumping (see Chap. 18, § 7) have shown that it is less than 10^{-4} per collision. It may therefore be ignored in the present context.

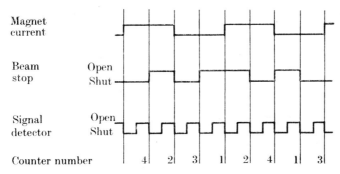

FIG. 16.68. Illustrating the signal sequence used in the experiments of Berkling, Schlier, and Toschek.

It is, of course, necessary, as usual, to analyse the observations to allow for the thermal velocity distribution of the target atoms. There is no difficulty in extending the analysis of § 7.2.1 to allow for the presence of an angular term and this was done.

Measurements were carried out alternately at zero pressure in the scattering chamber and at various pressures p. Let $(I_{M_J})_p$ be the intensity measured for the beam with a particular M_J when the chamber pressure is p. Then

$$(I_{\tfrac{1}{2}}/I_{\tfrac{3}{2}})_p = \exp\{-n(Q_{\tfrac{1}{2}}-Q_{\tfrac{3}{2}})l\}(I_{\tfrac{1}{2}}/I_{\tfrac{3}{2}})_0, \qquad (254)$$

where n is the concentration of target atoms in the chamber at pressure p and l is the path length of the beam in the chamber. If we write

$$1+\alpha = (I_{\tfrac{1}{2}}/I_{\tfrac{3}{2}})_p/(I_{\tfrac{1}{2}}/I_{\tfrac{3}{2}})_0, \qquad (255)$$

then, because $Q_{\tfrac{1}{2}}-Q_{\tfrac{3}{2}}$ is small,

$$\alpha \simeq -n(Q_{\tfrac{1}{2}}-Q_{\tfrac{3}{2}})l. \qquad (256)$$

If I_p/I_0 is the mean ratio of the intensity of the two beams then

$$\ln(I_p/I_0) = -n\bar{Q}l, \qquad (257)$$

where $\bar{Q}$ is the mean cross-section.

Hence if α defined by (255) is plotted against $\ln(I_p/I_0)$ a straight line should result. Fig. 16.69 illustrates such plots for the four target gases He, Ne, Ar, and Xe. Furthermore, when

$$\ln(I_p/I_0) = 1,$$
$$\alpha = (Q_{\frac{1}{2}} - Q_{\frac{3}{2}})/\bar{Q}, \tag{258}$$

giving the desired fractional difference between the cross-sections.

12.4.3. *The observed results.* In this way the values obtained for a beam velocity $(2\kappa T/M)^{\frac{1}{2}}$, where T is the oven temperature and M the mass of a gallium atom, are as given in Table 16.27.

TABLE 16.27

Cross-sections for collisions between gallium atoms, in different Zeeman states, and rare-gas atoms

Target gas	Xe	Kr	Ar	Ne	He
Mean cross-section $\bar{Q}$ (10^{-16} cm²)	675	517	471	176	132
$\Delta Q/\bar{Q}$	0·0095	—	0·0079	−0·040	0·0195

It will be seen that the fractional difference is much the same for argon as for xenon, in agreement with theoretical expectation that it should be nearly independent of the target rare gas. On the other hand, this does not apply to helium and neon. For the latter atom $\Delta Q/\bar{Q}$ is negative, which is only possible if a dominating contribution comes from the repulsive part of the interaction that was ignored in the theoretical analysis. This is not surprising because the relative velocity is likely to be in the high velocity region (see § 6.3.1) for both helium and neon. While $\Delta Q/\bar{Q}$ is not negative for helium, it is much larger than for argon and xenon, showing that the long-range interaction is inadequate to account for it.

The value of χ observed from the mean of the results for xenon and argon comes out to be 0·19 as compared with a somewhat larger theoretical value 0·49.

A further check on the assumptions made in the theoretical estimates follows because they also require that the van der Waals constant should depend on the nature of the rare-gas atom through its polarizability. The cross-sections $\bar{Q}$ given in Table 16.27 should therefore be approximately proportional to $\alpha^{2/5}$, where α is the polarizability of the gas atom concerned. This gives

$$Q_{\text{He}} : Q_{\text{Ne}} : Q_{\text{Ar}} : Q_{\text{Kr}} : Q_{\text{Xe}} = 0\cdot44 : 0\cdot57 : 1\cdot0 : 1\cdot18 : 1\cdot43$$

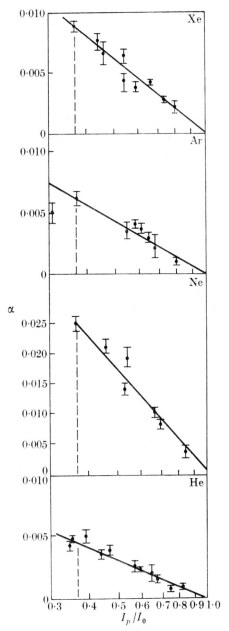

Fig. 16.69. Observed variation of I_p/I_0 (note logarithmic scale) with α, as defined by (257) and (255), respectively, for collisions of Ga atoms with He, Ne, Ar, and Xe atoms. The broken line corresponds to $\ln(I_p/I_0)_a^{-1} = 1$ (see (258)).

as compared with the experimental ratios
$$0\cdot 28:0\cdot 37:1\cdot 0:1\cdot 10:1\cdot 43.$$
The agreement is good for the three heavier gases but is poor for helium and neon. This supports the suggestion that for the light gases the collisions are in the high-velocity region.

There is little doubt of the reality of the observed differences between the cross-sections for the two Zeeman states. The successful measurement of such small differences in cross-sections opens up many possibilities of more precise measurements of the dependence of cross-sections on beam velocity, etc.

13. Collisions involving molecules

13.1. *Special features associated with such collisions*

If one or both of the colliding systems are molecules rather than atoms certain further possibilities arise.

In the first instance the interaction no longer depends only on the separation of the centres of mass but also on the molecular orientation. At large separations r the leading terms are of the form
$$-A_{\text{dd}}r^{-3}-B_{\text{qq}}r^{-5}-(C_{\text{disp}}+C_{\text{ind}})r^{-6}. \tag{259}$$
A_{dd} arises only if the molecules both possess permanent dipole moments while B_{qq} arises from interaction between the permanent quadrupole moments. C_{disp}, as for atoms, is due to interaction between induced dipoles in both systems while C_{ind} arises from that between an induced dipole in one system and a permanent dipole in the other.

For interaction between an atom and a molecule $A = B = 0$. C_{ind} will also vanish unless the molecule has a permanent dipole moment. Both C_{disp} and C_{ind} are dependent on molecular orientation but for many purposes the anisotropy is not very important and good results can be obtained by averaging over all orientations before carrying out the analysis. If α_1, α_2 are the polarizabilities of the two systems and μ_2 the permanent dipole moment of the molecule then, as in (218), the averages $\bar{C}_{\text{disp}}$, $\bar{C}_{\text{ind}}$ over all orientations are given by
$$\bar{C}_{\text{disp}} = \frac{3e\hbar}{2m^{\frac{1}{2}}} \frac{\alpha_1 \alpha_2}{(\alpha_1/n_1)^{\frac{1}{2}}+(\alpha_2/n_2)^{\frac{1}{2}}}, \tag{260}$$
n_1, n_2 being the number of electrons in the outer shell of the respective systems, and
$$\bar{C}_{\text{ind}} = \alpha_1 \mu_2^2. \tag{261}$$

If the interacting systems are two homonuclear molecules,
$$A = 0 = C_{\text{ind}}. \tag{262}$$

The quadrupole term vanishes when averaged over all orientations and for many purposes may be neglected. Under these conditions the interaction between a polar molecule and a homonuclear molecule is formally the same as between a polar molecule and an atom.

If both molecules are polar the situation is much more complicated. $\bar{C}_{\text{ind}}$ becomes $\alpha_1\mu_2^2+\alpha_2\mu_1^2$ while A_{dd} is now finite. Although $\bar{A}$, the average over all orientations, vanishes, the very long range of the interaction makes it important in many circumstances. Because it is strongly orientation-dependent, it is of particular importance in producing rotational transitions so we shall defer consideration of collisions between polar molecules until Chapter 17, § 5.

At closer distances we can expect the average interactions to be representable by the form
$$V(r) = \epsilon f(r/r_{\text{m}})$$
as for atoms (see § 6.1), with f given by the (12, 6), (12, α, 6), or (exp α, 6) form.

A further complication arises in the interpretation of data on collisions involving molecules. This is the fact that, in addition to elastic scattering, inelastic collisions, leading to excitation of rotation, will occur unless the relative kinetic energy is very low. It is only for the lightest molecules that it is at all practicable to work at such low temperatures that excitation of rotation cannot occur.

The question immediately arises as to how to derive an effective interaction from such data as the absolute value of the total cross-section, measured by the usual absorption technique described in § 7, and its variation with relative impact velocity, when rotational excitation occurs. Bernstein, Dalgarno, Massey, and Percival† have shown that, if the rotational states are very strongly coupled and the impact energy is such as to make excitation of a large number of such states possible, a statistical analysis may be applied which shows that the total cross-section Q, the sum of the elastic and all inelastic cross-sections, is determined by the interaction in effectively the same way as is the total elastic cross-section for collisions between atoms. This convenient result may be more general but this has not been established. On the other hand there is no experimental evidence that contradicts it.

The existence of excited rotational states leads to resonance phenomena in collisions between molecules which are formally exactly similar to the phenomena discussed in Chapter 9 which occur in the collisions of electrons with atoms. In particular, consider collisions between an atom and a diatomic molecule at energies

† BERNSTEIN, R. B., DALGARNO, A., MASSEY, H. S. W., and PERCIVAL, I. C., *Proc. R. Soc.* A**274** (1963) 427.

of relative motion below the threshold for rotational excitation. Nevertheless, because of the existence of the excited rotational states, contributions to the elastic cross-section will come from virtual excitation followed by de-excitation of such states. The duration of the virtual excitation will be short, except when the relative energy is close to that for which a bound state of relative motion exists for the atom in the mean field of the rotationally excited molecule. Resonance effects will appear at relative energies which will fall within a comparatively narrow band just below each rotational threshold, provided the anisotropic component of this interaction is not too large. If it is large enough, then multiple virtual processes involving more than one excited rotational state will be important and the resonance energies will not be localized closely below each threshold. In the former case the resonances will be similar to those for electron-atom collisions, in the latter to nuclear collisions.

Levine, Johnson, Muckerman, and Bernstein† have carried out sample calculations using an interaction energy between atom and molecule:

$$V(\mathbf{R}, \mathbf{r}) = V_0(r) + aP_2(\mathbf{r}.\mathbf{R})\epsilon(r_\mathrm{m}/r)^{12}. \tag{263}$$

Here $\mathbf{R}, \mathbf{r}$ are the coordinates of internal motion and of relative motion respectively and $V_0(r)$ is a (12, 6) potential with parameters ϵ, r_m. a is an adjustable asymmetry parameter. The method employed was that of the truncated eigenfunction expansion as in Chapter 9, § 3. For typical cases, for a around 2·0 they find resonance levels of width of order 0·01 to 0·05 of the threshold energy for rotational excitation. Such fine resonance effects are well beyond the reach of experimental detection at the time of writing.

There is, of course, the further possibility that a chemical reaction may occur between the colliding molecules. If this is so the interpretation of an observed total cross-section is more doubtful, although there are certainly some circumstances in which it may be regarded, as far as its relation to the interaction between the initial systems is concerned, as if it were the total elastic cross-section.

We shall discuss in some detail the excitation of inner molecular motions and certain aspects of the study of chemical reactions using crossed molecular beams in Chapter 17. Meanwhile, we shall give a short discussion of the results of total cross-section measurements in which one or both of the colliding systems are molecules on the assumption that no essential new feature is introduced by excitation of rotation and vibration or by chemical reactions. Collisions between polar molecules will not be considered at this stage because they are essentially involved with rotational excitation and are discussed in Chapter 17, § 5.6.

13.2. *Collisions of* K *and* Cs *atoms with various molecules*

Fig. 16.70 illustrates the variation of the total cross-section with relative velocity for collisions of K atoms with N_2 molecules observed

† LEVINE, R. D., JOHNSON, B. R., MUCKERMAN, J. T., and BERNSTEIN, R. B., *Chem. Phys. Lett.* (*Netherlands*) **1** (1968) 517.

by Pauly.† If the effective scattering potential varies as $-C/r^s$, the analysis of (163 b) shows that the mean free path of K atoms of velocity v in a gas of n N_2 molecules/cm³, at temperature T, is given by

$$\lambda^{-1} = nQF(s,x), \qquad x = v(M_2/2\kappa T)^{\frac{1}{2}}, \qquad (264)$$

where $F(s,x)$ is given by (164) and Q is the cross-section for a collision of a K atom with an N_2 molecule (of mass M_2) with relative velocity v.

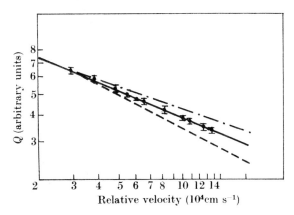

FIG. 16.70. Observed variation of the total cross-section Q for K–N_2 as a function of relative velocity. ● Exptl. points. Calculated on assumption: —— of an inverse sixth-power ($s = 6$) interaction; - - - of an inverse fifth-power ($s = 5$) interaction —·—·— of an inverse seventh-power ($s = 7$) interaction.

In Fig. 16.70 the variation of Q with v is given from observed values of λ, taking $s = 6$. The resulting variation of Q is seen to follow closely the 2/5 power law that is consistent with this assumption.

Pauly did not measure absolute cross-sections but Rothe and Bernstein‡ have made an extensive series of observations in the course of their investigation, described in § 9.3, of the van der Waals constants for interaction of K and Cs atoms with rare-gas atoms. Tables 16.28 and 16.29 extend the results of Table 16.11 to collisions with a great number of molecules.

On the whole the agreement between the calculated and observed values is remarkably good and in some cases, for collisions with polar molecules, shows the importance of including the contribution C_{ind}. The most marked discrepancies occur for collisions with H_2 and D_2.

There are very few results available from other experiments to compare with those given in Table 16.28. For K–H_2 the experiments of

† PAULY, H., Z. Naturf. **15a** (1960) 277.
‡ ROTHE, E. W. and BERNSTEIN, R. B., J. chem. Phys. **31** (1959) 1619.

TABLE 16.28

Effective van der Waals constants C for interactions of potassium atoms with various molecules

Molecules	C in 10^{-60} erg cm^6			
	Observed	Calculated		
		C_{disp}	C_{ind}	Total C
H$_2$	47 (106)†	107	—	107
D$_2$	46	105	—	105
O$_2$	290	217	—	217
N$_2$	305	238	—	238
CO	363	264	—	264
NO	376	234	1	235
HCl	506 (67)‡	344	36	380
HBr	606 (431)§ (7560)‖	460	21	481
CO$_2$	513	356	—	356
CS$_2$	1250	1057	—	1057
H$_2$O	335	199	115	314
N$_2$O	567	399	1	400
H$_2$S	624	475	35	510
SO$_2$	806	514	89	603
CH$_4$	345	342	—	342
CF$_4$	569	394	—	394
SiF$_4$	651	462	—	462
SF$_6$	809	625	—	625
CCl$_4$	1927 (2360)††	1369	—	1369
SiCl$_4$	1940	1501	—	1501
SnCl$_4$	2112	1809	—	1809
NH$_3$	358	291	73	364
H$_2$CO	602	381	181	562
(CH$_3$)$_2$CO		826	276	1102

All the observed values except those in brackets are due to Rothe and Bernstein (loc. cit.).

Lulla, Brown, and Bedersen,† using a magnetic velocity selector give a value for C in close agreement with the calculated. By studying the variation of cross-section with relative velocity v they were able to determine the low-velocity range (see § 6.3.1) over which the cross-section varied as $v^{-2/5}$. In Rothe and Bernstein's experiments measurements were made only at thermal velocities so the apparent value for C

† LULLA, K., BROWN, H. H., and BEDERSEN, B., *Phys. Rev.* **136** (1964) A 1233.
‡ ACKERMAN, M., GREENE, E. F., MOURSUND, A. L., and ROSS, J., *J. chem. Phys.* **41** (1964) 1183.
§ AIREY, J. R., GREENE, E. F., KODERA, K., RECK, G. P., and ROSS, J., ibid. **46** (1967) 3287.
‖ VON HUNDHAUSEN, E. and PAULY, H., *Z. Phys.* **187** (1965) 305.
†† HELBING, R. and PAULY, H., ibid. **179** (1964) 16.

may be in error due to the conditions falling within the high-velocity region.

There have been a number of experiments in which the interaction between K atoms and hydrogen halide molecules have been studied (see Chap. 17, § 6), directed particularly towards the study of the chemical reactions
$$\text{K} + \text{HX} \to \text{KX} + \text{H}.$$

From rainbow-scattering observations parameters ϵ_m and r for the K–HX interaction have been determined. From these C may be

TABLE 16.29

Effective van der Waals constants C for interactions of caesium atoms with various molecules

Molecules	C in 10^{-60} erg cm^6			
	Observed	Calculated		
	(Rothe and Bernstein)	C_disp	C_ind	Total C
H_2	61	120	—	120
D_2	73	118	—	118
N_2	316	267	—	267
CO_2	500	398	—	398
H_2O	456	223	142	365
CH_4	421	383	—	383
CF_4	649	440	—	440
SF_6	923	698	—	698
$SiCl_4$	1867	1684	—	1684
H_2CO	676	426	224	650
$(CH_3)_2CO$	1373	925	341	1266

obtained but it will be seen by reference to bracketed values for K–HCl and K–HBr in Table 16.28 that they vary by more than an order of magnitude. We shall discuss these experiments in more detail in Chapter 17, § 6.2. The only other available comparison of experimental results is for K–CCl_4 where the agreement is reasonable.

13.3. *Collisions between dipolar molecules and other (non-polar) molecules*

To investigate the importance and nature of the interaction at long range between molecules possessing permanent dipoles, Schumacher, Bernstein, and Rothe[†] carried out a number of total cross-section measurements, using the same technique as in the experiments of Rothe

[†] SCHUMACHER, H., BERNSTEIN, R. B., and ROTHE, E. W., *J. chem. Phys.* **33** (1960) 584.

and Bernstein described in the preceding section but with the K beam replaced by one of caesium chloride (CsCl) which can be detected by the same hot-wire ionization method. In the course of this work, the results of which will be discussed in Chapter 17, § 5.6.1, observations were also made of total cross-sections for collisions with atoms and with non-polar molecules. For these cases the long-range interaction is of the van der Waals form but the constant includes the additional term $\bar{C}_{\text{ind}}$ of (260).

TABLE 16.30

Comparison of observed and calculated cross-sections for collision of CsCl molecules with various atoms and non-polar molecules, relative to those for collisions with argon atoms

Atom or molecule	$\bar{C}_{\text{ind}}/\bar{C}_{\text{disp}}$	$Q/Q(\text{argon})$	
		Observed	Calculated
He	0·865	0·37	0·32
Ne	0·79	0·60	0·55
Kr	1·03	1·35	1·24
Xe	1·14	1·55	1·51
H_2	1·09	0·52	0·46
D_2	1·09	0·55	0·52
SF_6	0·87	1·74	1·68
$SiCl_4$	1·06	2·24	2·34
CH_4	1·04	1·01	1·04
CF_4	0·80$_5$	1·33	1·37

Table 16.30 gives a comparison between observed and calculated values of the total cross-section for collisions with various non-polar gases relative to those for argon—the magnitude of any correction due to the pumping effect in the McLeod gauge used to calibrate the argon pressure is thereby eliminated. The agreement is remarkably good and the importance of including $\bar{C}_{\text{ind}}$ is manifest.

Cross, Gislason, and Herschbach,[†] also in the course of an experimental investigation of the collisions between polar molecules, measured cross-sections for collisions of KCl and CsCl with Kr and Xe and of CsBr with Xe. The technique they employed was that of measuring the attenuation of a velocity-selected beam after passage through the gas in a scattering chamber. To avoid the difficulty of determining the gas pressure, particularly when the gas is reactive, as in the main experiment in which the gas was a polar compound, the attenuation of a second

[†] CROSS, R. J., GISLASON, E. A., and HERSCHBACH, D. R., *J. chem. Phys.* **45** (1966) 3582.

beam, of Rb atoms, normal to the main beam, was used to monitor the pressure. Cross-sections measured relative to the monitor beam were calibrated by comparison with observations made with the same target gas but with the main beam replaced by K atoms. The absolute values of the cross-sections for the latter collisions were taken to be as calculated.

TABLE 16.31

Comparison of observed and calculated cross-sections and their velocity-variation for collisions of polar molecules with atoms

Colliding system	$\bar{C}_{disp}$	$\bar{C}_{ind}$	Q_{calc}	Q_{obs}	$-\left(\dfrac{\partial \ln Q}{\partial \ln v}\right)_{calc}$	$-\left(\dfrac{\partial \ln Q}{\partial \ln v}\right)_{obs}$
			10^{-16} cm^2			
KCl–Kr	234	272	(a) 870	—	(a) 0·4	—
			(b) 780	720	(b) 0·53	0·49
KCl–Xe	313	442	(a) 1023	—	(a) 0·4	—
			(b) 900	700	(b) 0·54	0·48
CsCl–Kr	267	273	(a) 893	—	(a) 0·4	—
			(b) 700	650	(b) 0·65	0·60
CsCl–Xe	390	443	(a) 1065	—	(a) 0·4	—
			(b) 800	640	(b) 0·64	0·59
CsBr–Xe	440	460	(a) 1098	—	(a) 0·4	—
			(b) 770	800	(b) 0·67	0·60

(a) Assuming perfect resolution.   (b) Allowing for finite resolution.

The angular resolution of the apparatus, according to the Kusch criterion, was 2·9′ for the main beam and 20′ for the monitor beam. Correction for the resolution was made using the formula (176). The resolution parameter ρ for the apparatus geometry is $2 \times 10^{-4} k Q^{-\frac{1}{2}}$, where k is the wave number of the incident molecules and Q is the total cross-section, and is $\simeq 1$ for CsCl–Xe collisions. The resolution correction is important, not only for the magnitude of the cross-section, but also for its variation with velocity. Table 16.31 compares not only calculated and observed cross-sections Q but also calculated and observed values of $\partial \ln Q/\partial \ln v$ which gives the power law for the variation of Q with relative velocity. It will be seen that, when allowance is made in the calculations for the finite resolution, there is good agreement between theory and experiment, though it must be realized that the comparison is based on the assumption that the corresponding cross-sections for collisions in which K atoms replace the alkali halides are given correctly by the theory.

The effect of the finite resolution on $\partial \ln Q/\partial \ln v$ is very marked for the caesium halides and, if not allowed for, would have suggested that the effective interaction behaved like r^{-4} instead of r^{-6}. When allowance is made the results confirm the validity of the r^{-6} dependence.

13.4. The $H-H_2$ interaction

We have already referred to the $H-H_2$ interaction in § 12.1.4 in connection with the analysis of experimental data on the viscosity of partly

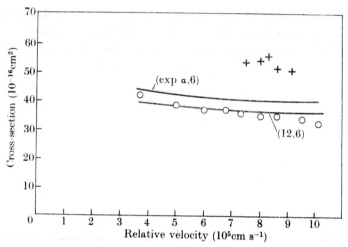

Fig. 16.71. The total cross-section for $H-H_2$ collisions. —— calculated. + observed by Harrison. ○ observed by Fluendy, Martin, Muschlitz, and Herschbach (normalized to agree with that calculated with (12, 6) potential at a relative velocity of 7×10^5 cm s^{-1}).

dissociated hydrogen. Harrison[†] has measured total cross-sections for the scattering of H atoms by H_2 molecules with the same apparatus as that used for corresponding measurements for He–He collisions (see § 10.3). His results are illustrated in Fig. 16.71 and compared with cross-sections calculated[‡] assuming the (12, 6) and (exp α, 6) interactions referred to in § 12.1.4. As for He–He collisions Harrison's data lie somewhat above the calculated values.

Fluendy, Martin, Muschlitz, and Herschbach,[§] using the apparatus described on p. 1361 (see Fig. 16.24), have measured the variation of the $H-H_2$ cross-section with H beam velocity. Conversion of their results to a relative velocity scale gives the results shown in Fig. 16.71. As no absolute measurements were made, these data are normalized to agree with that calculated using the (12, 6) potential at a relative velocity of

[†] HARRISON, H., J. chem. Phys. **37** (1962) 1164.
[‡] Fox, R. E. and SALTER, P., private communication.
[§] loc. cit., p. 1361.

7×10^5 cm s^{-1}. It will be seen that the observed variation with velocity is considerably faster than that calculated or indeed that observed by Harrison over a smaller velocity range. Clearly the position is not yet satisfactory.

13.5. *The H_2–H_2 interaction*

According to quantum theory the interaction between two hydrogen molecules can be represented as a chemical repulsion V_c and a long-range combination of van der Waals and quadrupole–quadrupole interactions V_1. If (χ_1, ψ_1), (χ_2, ψ_2) are the respective polar angles of the molecular axis referred to the vector separation $\mathbf{r}$ of the centres of mass as polar axis then

$$V_c = A(r) + B(r)(\cos^2\chi_1 + \cos^2\chi_2) + C(r)\cos^2\chi_1 \cos^2\chi_2, \qquad (265)$$

where A, B, and C decrease exponentially for large r, and

$$\begin{aligned}V_1 =\ & (a_1 r^{-6} + b_1 r^{-5}) + (a_2 r^{-6} + b_2 r^{-5})(\cos^2\chi_1 + \cos^2\chi_2) + \\ & + (a_3 r^{-6} + b_3 r^{-5})\{\sin^2\chi_1 \sin^2\chi_2 \cos^2(\psi_1 - \psi_2)\} + \\ & + (a_4 r^{-6} + b_4 r^{-5})\{\sin^2\chi_1 \sin^2\chi_2 \cos(\psi_1 - \psi_2)\} + \\ & + (a_5 r^{-6} + b_5 r^{-5})\cos^2\chi_1 \cos^2\chi_2. \end{aligned} \qquad (266)$$

a_i and b_i are constants which are such that, when V_1 is averaged over all molecular orientations to give $\bar{V}_1$, the quadrupole-quadrupole term proportional to r^{-5} vanishes. They have been calculated by Britton and Bean,† from whose results we find $\bar{V}_1 = -C/r^{-6}$, where $C = 10\cdot 5 \times 10^{-60}$ erg cm^6.

As an indication of the relative importance of the anisotropic terms, consider the collisions between two parahydrogen molecules at low temperatures for which the rotational quantum numbers J, M_J are both zero. The effective interaction energy is then the average over all orientations of $V_c + V_1$. Taking A, B, and C as calculated by Evett and Margenau,‡

$$V_{pp} = \bar{V}_c + \bar{V}_1 = (14\cdot 49 + 0\cdot 466r)e^{-1\cdot 8594r} - 10\cdot 96 r^{-6}, \qquad (267)$$

in atomic units. If we now consider a collision between a para and an ortho hydrogen molecule, both in their lowest rotational states, the average over all orientations of the axis of the para molecule gives an interaction

$$V_{op} = V_{pp} + \Delta V_{op}(r) P_2(\cos\chi), \qquad (268)$$

where (χ, ψ) are the polar orientation angles of the axis of the ortho molecule. ΔV_{op} is given by

$$\Delta V_{op} = (0\cdot 466r - 1\cdot 039)e^{-1\cdot 8594r} - 0\cdot 736 r^{-6}. \qquad (269)$$

† BRITTON, F. R. and BEAN, D. T. W., *Can. J. Phys.* **33** (1955) 668.
‡ EVETT, A. A. and MARGENAU, H., *Phys. Rev.* **90** (1953) 1021.

It is clear that the anisotropic term is relatively quite small so that for many purposes it may be ignored.

13.5.1. *The mean interaction—evidence from transport and total cross-sections.* The observed data on the viscosity of hydrogen has been analysed in the same way as the corresponding data for helium (see § 10).

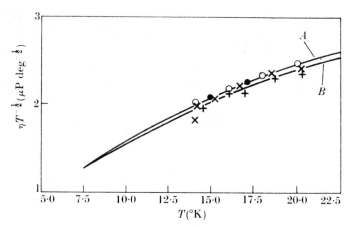

Fig. 16.72. The viscosity η of H_2 at low temperatures T °K. Observed: × van Itterbeek and Claes; ● van Itterbeek and van Paemel; ○ Keesom and Keesom; + Becker and Misenta. Calculated with (exp α, 6, 8) potential, the quantum parameter B being 17: ——— A with $C = 10\cdot5 \times 10^{-60}$ erg cm⁶; ——— B with $C = 11\cdot5 \times 10^{-60}$ erg cm⁶.

Ignoring any anisotropic interaction Buckingham, Davies (A. E.), and Davies (A. R.)† showed that a good fit to the data over a temperature range from 2 to 500 °K can be obtained assuming an interaction of the form (223) with $\alpha = 13\cdot5$, $\beta = 0\cdot2$. Taking the quantum parameter $B = 2M\epsilon r_m^2/\hbar^2$, where M is the reduced mass, as 17, the comparison of observed‡ and calculated coefficients in the low- and high-temperature range is illustrated in Figs. 16.72 and 16.73 respectively. Two sets of

---

† BUCKINGHAM, R. A., DAVIES, A. E., and DAVIES, A. R., *Joint Conference on Thermodynamic and Transport Properties of Fluids*, Instn Mech. Engrs and IUPAC, London, 1957; BUCKINGHAM, R. A., DAVIES, A. R., and GILLES, D. C., *Proc. phys. Soc.* **71** (1958) 457.

‡ *Low temperature* (Fig. 16.72): VAN ITTERBEEK, A. and CLAES, A., *Physica, 's Grav.* **5** (1938) 938; VAN ITTERBEEK, A. and VAN PAEMEL, O., ibid. **7** (1940) 265; KEESOM, W. H. and KEESOM, P. H., ibid. **29**; BECKER, E. W. and MISENTA, R., *Z. Phys.* **140** (1955) 535. *High temperature* (Fig. 16.73): VAN ITTERBEEK, A. and CLAES, A., loc. cit.; TRAUTZ, M. and BAUMANN, R. B., *Annln der Physik* **2** (1929) 733; TRAUTZ, M. and BINKELE, H. E., ibid. **5** (1930) 561; TRAUTZ, M. and ZINK, R., ibid. **7** (1930), 427; TRAUTZ, M. and HEBERLING, R., ibid. **20** (1934) 118; TRAUTZ, M. and HUSSEINI, I., ibid. **20** (1934) 121; JOHNSTON, H. L. and MCCLOSKEY, K. E., *J. phys. Chem., Ithaca* **44** (1940) 1038; WOBSER, R. and MÜLLER, F., *Kolloidbeihefte* **52** (1941) 165; BECKER, E. W. and MISENTA, K., loc. cit.

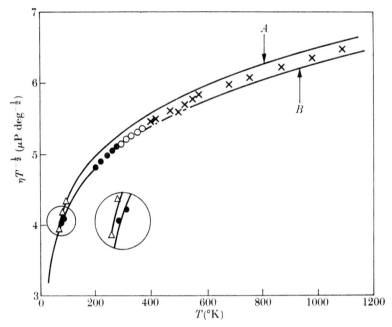

FIG. 16.73. The viscosity η of H_2 at intermediate and high temperatures T °K. Observed: × Trautz and others; ● Johnston and McCloskey; ○ Wobser and Müller; △ Becker and Misenta. Calculated with (exp α, 6, 8) potential, the quantum parameter being 17: ——— A with $C = 10\cdot 5 \times 10^{-60}$ erg cm^6; ——— B with $C = 11\cdot 5 \times 10^{-60}$ erg cm^6.

TABLE 16.32

Interaction parameters for molecular hydrogen derived from analysis of viscosity observations

Interaction	(ϵ/κ) (°K)	r_m (Å)	C (10^{-60} erg cm^6)
Buckingham, Davies, and Davies† (exp 13·5, 6, 8)	33·3	3·380	10·9
Mason and Rice‡ (exp 14, 6)	37·0	3·337	12·44
Cohen, Offerhaus, van Leeuwen, Roos, and de Boer§ (12, 6)	37·0	3·287	12·86

calculated values are shown corresponding to the van der Waals constant having the values 10·5 and 11·5 × 10^{-60} erg cm^6. A very good fit is

† loc. cit., p. 1454.
‡ MASON, E. A. and RICE, W. E., *J. chem. Phys.* **22** (1954) 522, 843.
§ COHEN, E. G. D., OFFERHAUS, M. J., VAN LEEUWEN, J. M. J., ROOS, B. W., and DE BOER, J., *Physica, 's Grav.* **22** (1956) 791.

obtained with $B^2 = 16\cdot 7$ and $C = 11\cdot 0 \times 10^{-60}$ erg cm⁶. The corresponding values of the parameters ϵ and r_m are given in Table 16.32 which also includes the parameters that give the best fit assuming interactions of (exp 14, 6) and (12, 6) type respectively. It is clear that the range within which each of these parameters must lie is quite

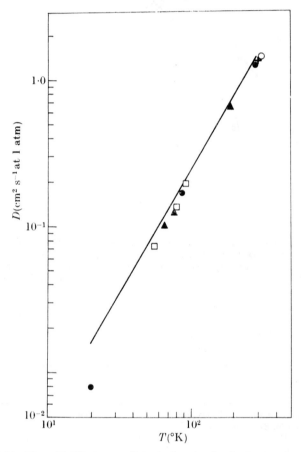

FIG. 16.74. The self-diffusion coefficient of molecular hydrogen. —— calculated with (exp 13·5, 6, 8) potential. Observed: ● Harteck and Schmidt; ○ Heath, Ibbs, and Wild; + Waldmann; ▲ Bendt; □ Lipsicas.

restricted. The agreement of the van der Waals constant with that calculated by Britton and Bean† ($10\cdot 5 \times 10^{-60}$ erg cm⁶) is quite good.

Two further checks of the derived interactions are available from the self-diffusion coefficient and the total cross-section. The former has been measured either by observing the diffusion of ortho into para

† loc. cit., p. 1453.

hydrogen† or of deuterium into hydrogen.‡ In the latter case the self-diffusion coefficient $D(\mathrm{H}_2)$ is obtained from that directly measured $D(\mathrm{H}_2\text{–}\mathrm{D}_2)$ by the relation (see (7))

$$D(\mathrm{H}_2) = [2M(\mathrm{D}_2)/\{M(\mathrm{H}_2)+M(\mathrm{D}_2)\}]^{\frac{1}{2}} D(\mathrm{H}_2\text{–}\mathrm{D}_2)$$

where $M(\mathrm{D}_2)$, $M(\mathrm{H}_2)$ are the masses of D_2 and H_2 molecules respectively. Fig. 16.74 compares $D(\mathrm{H}_2)$ obtained in this way with calculated values using the (exp 13·5, 6, 8) interaction with the parameters as in Table 16.32. The agreement is quite good except at the lowest temperature where the observed values are somewhat lower than the calculated and those obtained by the measurement of the spin-lattice nuclear magnetic relaxation time in H_2 (see Chap. 17, § 5.8).§

Fig. 16.75 compares the total elastic cross-section calculated using the same (exp 13·5, 6, 8) interaction with the total cross-section observed by Harrison‖ using the same equipment as for the observation of He–He (see § 10.3) and H–H_2 (see § 13.4) collisions. Again allowing for some uncertainty in the observed absolute values the agreement is quite good, better than might be expected from the corresponding comparisons for He–He (Fig. 16.55) and H–H_2 (Fig. 16.71). Earlier observations by Minten and Osberghaus†† give a considerably larger cross-section, $1\cdot 7 \times 10^{-14}$ cm², at thermal velocities.

13.5.2. *The anisotropic component of the interaction.* To determine the anisotropic component with any accuracy it is necessary to seek phenomena that would not occur if it did not exist. The most obvious of these is, of course, the excitation of rotation in molecular collisions. We shall, however, defer the discussion of such inelastic collisions to Chapter 17. A related process that arises from the anisotropic component is that in which a change of rotational orientation occurs during the collision without any change in the magnitude of the total angular momentum. Such collisions are elastic in that the energy of relative translation is unaltered.

Their rate may be determined from observations of the nuclear spin-lattice relaxation time in hydrogen gas at low temperatures. Although these collisions are elastic they are most appropriately dealt with in

---

† HARTECK, P. and SCHMIDT, H. W., *Z. phys. Chem.* **B21** (1933) 447; HEATH, H. R., IBBS, T. L., and WILD, N. E., *Proc. R. Soc.* **A178** (1941) 380; WALDMANN, L., *Naturwissenschaften* **32** (1944) 223, *Z. Naturf.* **1** (1946) 59.
‡ BENDT, P. J., *Phys. Rev.* **110** (1958) 85.
§ LIPSICAS, M., *J. chem. Phys.* **36** (1962) 1235.
‖ loc. cit., p. 1411.
†† MINTEN, A. and OSBERGHAUS, O., *Z. Phys.* **150** (1958) 74.

relation to rotational excitation. An account of the relevant experimental data and its interpretation in terms of the anisotropic interaction is therefore given in Chapter 17, § 5.8.

Further evidence on the anisotropic component of the interaction is provided by the observations by Becker and Stehl† of the viscosity of hydrogen as a function of the relative para-ortho concentrations over

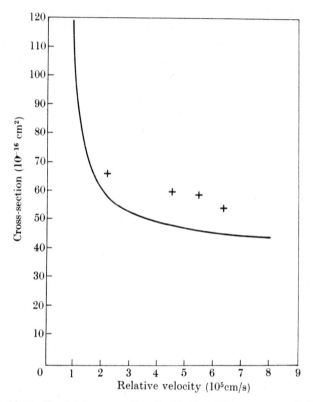

Fig. 16.75. The total cross-section for H_2–H_2 collisions. —— calculated with (exp 13·5, 6, 8) potential. + observed by Harrison.

a temperature range extending from 15 to 100 °K. From these data it is possible to derive three mean viscosity cross-sections $\bar{Q}_\eta(o\text{–}o)$, $\bar{Q}_\eta(o\text{–}p)$, $\bar{Q}_\eta(p\text{–}p)$ which refer to collisions between two ortho-molecules, an ortho- and a para-molecule, and two para-molecules respectively. We may write

$$\bar{Q}_\eta(o\text{–}p) = \bar{Q}_\eta(p\text{–}p) + \Delta_1 Q_\eta(o\text{–}p) + \Delta_2 Q_\eta(o\text{–}p),$$

where $\Delta_1 Q_\eta$ arises from the different symmetry relations that apply in

† BECKER, E. W. and STEHL, O., Z. Phys. **133** (1952) 615.

the two types of collisions and $\Delta_2 Q_\eta$ from the effect of the anisotropic interaction that does not arise in the p–p case. Similar considerations apply to $\bar{Q}_\eta(o$–$o)$.

Fig. 16.77 illustrates the magnitude of the effects as observed, giving $(\Delta_1 Q_\eta + \Delta_2 Q_\eta)/\bar{Q}_\eta(p$–$p)$ as a function of temperature. On the same diagram are given the values of $\Delta_1 Q_\eta/\bar{Q}$ calculated by Buckingham, Davies, and Gilles† and by Cohen, Offerhaus, van Leeuwen, Roos, and

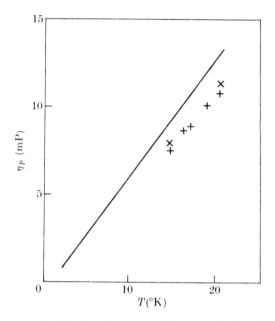

FIG. 16.76. The viscosity of para-H_2. —— calculated by Takayanagi and Niblett. × observed by Rietveld and van Itterbeek (loc. cit., p. 1409). + observed by Becker and Misenta (loc. cit., p. 1454).

de Boer‡ using central interactions of (exp α, 6) and (12, 6) types respectively (see Table 16.32) which give good values for $\bar{Q}_\eta(p$–$p)$ over the temperature range concerned. At temperatures above 15 °K the observed value of $|\Delta Q/\bar{Q}|$ is certainly greater than $|\Delta_1 Q/\bar{Q}|$ as calculated, suggesting that $|\Delta_2 Q/\bar{Q}|$ is important.

Niblett and Takayanagi§ have carried out an elaborate calculation using the full interaction (268). To determine $Q_\eta(o$–$p)$ it is then necessary to solve pairs of coupled equations and a great deal of computation is required. Fig. 16.76 shows the comparison between their calculated

† loc. cit., p. 1454. ‡ loc. cit., p. 1455.
§ NIBLETT, P. D. and TAKAYANAGI, K., *Proc. R. Soc.* A**250** (1959) 222.

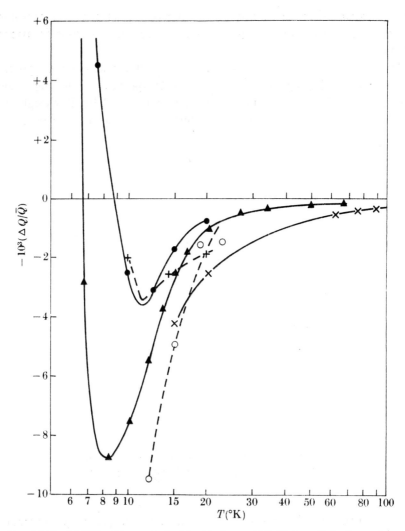

Fig. 16.77. $\Delta Q/\bar{Q}$ for H_2. × derived by Becker and Stehl from their observations. ● $\Delta_2 Q/\bar{Q}$ calculated by Takayanagi and Niblett. + $\Delta Q/\bar{Q}$ calculated by Takayanagi and Niblett. ○ $\Delta_1 Q/\bar{Q}$ calculated by Cohen, Offerhaus, van Leeuwen, Roos, and de Boer. ▲ $\Delta_1 Q/\bar{Q}$ calculated by Buckingham, Davies, and Gilles.

values and the observed values for the viscosity of para-hydrogen, while in Fig. 16.77 their calculated values of $\Delta Q/\bar{Q}$ are given. Although detailed agreement is not found their results show that $\Delta_1 Q$ becomes more important than $\Delta_2 Q$ as the temperature rises above 12 °K. It is also noteworthy that the interaction assumed is the result of quantum-theoretical calculation from first principles. In view of the complexity of the interacting systems it is gratifying that the agreement is as good as it is.

13.6. *Interaction of H_2 molecules with rare-gas atoms*

Total cross-sections for collisions between H_2 and He have been measured by Harrison[†] with the apparatus referred to in §§ 10.3, 12.1.3,

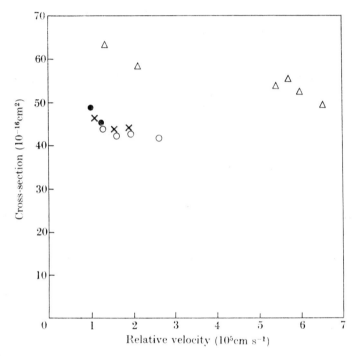

FIG. 16.78. Observed total cross-sections for H_2 and D_2 collisions with ^{3}He and ^{4}He atoms. Observed by Moore, Datz, and van der Valk: ○ H_2–^{3}He, ● D_2–^{3}He, × H_2–^{4}He. Observed by Harrison △ H_2–^{4}He.

and 13.5.1 and by Moore, Datz, and van der Valk,[‡] using the apparatus referred to in § 10.3. The latter authors compared cross-sections for H_2–^{3}He, D_2–^{3}He, and H_2–^{4}He. Results obtained are shown in Fig. 16.78. While in both experiments the variation of the cross-section with relative velocity is similar, Harrison's absolute values are considerably higher. Further evidence about the H_2–He interaction, including the anisotropic component, is available from nuclear spin relaxation measurements in mixtures of H_2 and ^{4}He (see Chap. 17, § 5.8).

[†] loc. cit., p. 1452.   [‡] loc. cit., p. 1411.

Feltgen, Gaide, Helbing, and Pauly† have observed the velocity dependence of the total cross-section for collisions of H_2 and D_2 with Ne, Ar, Kr, and Xe. They used the apparatus referred to on p. 1413. For collisions with all but Ne, 'glory' oscillations (§ 6.3.1) were found from which the products ϵr_m of the potential parameters were obtained, a (12, 6) interaction being assumed. Separate determination of ϵ and r_m was also attempted from absolute values of the cross-sections deduced by comparison with values found for the collisions with Ar, Kr, and Xe in an earlier experiment (see p. 1413). Results obtained are given in Table 16.33.

TABLE 16.33

Potential parameters for interaction of H_2 and D_2 with rare-gas atoms

	(ϵ/κ) °K	r_m (10^{-8} cm)
H_2–Ar	73 (63)	3·34 (3·60)
–Kr	77·5 (77)	3·71 (3·67)
–Xe	86 (86)	3·90 (3·93)
D_2–Ar	70	3·32
–Kr	77·5	3·71
–Xe	86	3·90

Bracketed values are derived from parameters for pure gases combined by the mixture rules.

For comparison, values obtained by applying the usual mixture rules

$$\epsilon = (\epsilon_1 \epsilon_2)^{\frac{1}{2}}, \qquad r_m = \tfrac{1}{2}(r_{m_1} + r_{m_2})$$

to parameters ϵ_1, ϵ_2, derived from analysis in terms of a (12, 6) interaction of the transport properties of the pure gases, are also given. The agreement, though possibly fortuitous, is quite good, especially for Kr and Xe.

The observed results for collisions with Ne do not show glory oscillations, presumably because, under the experimental conditions, these collisions took place mainly within the high velocity range (§ 6.3.1).

Landorf and Mueller,‡ using the apparatus described in § 7.3, measured the total cross-section for Ar–H_2 collisions at thermal energies. Assuming, as seems reasonable, that their result gives the mean cross-section in a low-velocity range in which the glory oscillations have been smeared out by the velocity distribution, their observed cross-section $1·97 \times 10^{-14}$ cm² yields a van der Waals constant $C = 65 \times 10^{-60}$ erg cm⁶. This is to be compared with $31·5 \times 10^{-60}$ erg cm⁶ from the Slater–Kirkwood formula (218) and 28×10^{-60} erg cm⁶ derived from the parameters given in Table 16.33.

† FELTGEN, R., GAIDE, W., HELBING, R., and PAULY, H., *Phys. Lett.* **20** (1966) 501.
‡ LANDORF, R. W. and MUELLER, C. R., *J. chem. Phys.* **45** (1966) 240.

17

COLLISIONS BETWEEN ATOMS AND MOLECULES UNDER GAS-KINETIC CONDITIONS—COLLISIONS INVOLVING EXCITATION OF VIBRATION AND ROTATION—REACTIVE COLLISIONS BETWEEN MOLECULAR BEAMS

WE now begin the discussion of inelastic collisions between atoms and/or molecules. These may involve electronic, vibrational, and/or rotational transitions as well as collisions in which rearrangement of atoms takes place. In this chapter we limit the discussion to collisions in which no electronic transitions occur. Nevertheless, there remains much that must be taken as outside the scope of this book. Thus, under the heading of atomic rearrangement collisions, are included all chemical reactions in the gas phase. In a somewhat arbitrary fashion we choose for detailed consideration the study of vibrational and rotational excitation and transfer between molecules which are not too complex. Even this field is quite wide so we concentrate mainly, but not exclusively, on transitions between low-lying vibrational states. Although we shall not attempt to discuss chemical reactions in general, an account will be given in the last sections of this chapter of the study of reaction rates using crossed molecular beams, partly because it is a natural extension of the work described in Chapter 16 on the non-reactive scattering of such beams and partly because it has great interest and importance as a new analytical tool for investigating chemical-reaction mechanisms in the gas phase.

1. Introductory remarks concerning the probability of vibrational and rotational transitions in gas-kinetic collisions

Under this head we must consider a number of possibilities which may be listed as follows.

(a) Excitation or deactivation of vibration, involving transfer between vibrational and translational motion.

(b) Transfer of vibration between molecules.

(c) Excitation or deactivation of rotation, involving transfer between rotational and translational motion.

(d) Transfer of rotation between molecules.

(e) Transfer of energy between vibration and rotation.

(f) Collisions in which energy transfer between all these modes occur.

Of these six possibilities by far the most is known, both from experiment and theory, about (a) and to a much lesser extent (b) and (c). This is because there are a number of different physical phenomena, such as the dispersion and absorption of ultrasonic waves, the high-speed flow of gases, and the thickness of shock-wave fronts, which depend on the collision rates concerned. The study of these phenomena yields a great deal in addition to the much less comprehensive information available from optical techniques which provide the main source of information about the remaining processes listed above.

Three-body collisions that lead to recombination of atoms to form a molecule can be regarded as coming under (a), because the process essentially involves transitions from open states of relative motion of the atoms to closed vibrational states in which there is a transfer of energy to that of the motion of the third body relative to the centre of mass of the recombining atoms. The discussion of these collisions would take us too far into the theory of chemical kinetics and we shall concentrate our attention mainly on transitions between low-lying vibrational states.

We begin by applying a semi-classical argument which enables us to determine the conditions under which the probability of vibrational excitation in a collision will be small.

Consider a collision between two molecules A and B, the first of which is vibrationally excited. If the molecules approach each other with a velocity small compared with the mean relative velocity of the vibrating atoms in A, the latter will have plenty of time to readjust themselves to the slowly changing conditions without a transition taking place—the impact will be nearly adiabatic. Classically, we may represent this by considering the amplitude of vibration that will be set up by applying a disturbing force to an oscillator of natural frequency ν. This disturbance will vary with the time t according to some function $F(t)$. To determine the effect on the oscillator we expand $F(t)$ in a Fourier integral. It is only the components of this expansion which have frequencies between ν and $\nu+d\nu$ which will produce any appreciable forced oscillation. In order that these components of $F(t)$ should be strong it is necessary that the time τ of collision should not be large compared with a natural period of the oscillator. Excitation will be weak if $\tau\nu \gg 1$. As τ is of the order a/v, where a is the range of interaction between the

atoms and v their relative velocity, the condition for weak excitation becomes
$$av/v \gg 1. \tag{1}$$
In this expression $h\nu$ is the energy of the vibrational quantum excited and a/v is the time of collision. a may be taken as of the order of gas-kinetic radii and v as the relative velocity of the colliding systems when widely separated, unless a strong attractive force exists between them as, for example, when there is a chemical affinity between atoms in the colliding systems. If such a force exists the relative motion may be so strongly accelerated during the collision that the correct value to take for v may be much larger.

The condition (1) may be recast in a slightly different form which is sometimes useful. If d is the amplitude of vibration and M the reduced mass associated with it, then
$$d \simeq (\hbar/\pi\nu M)^{\frac{1}{2}},$$
so that (1) becomes
$$\frac{a}{d}\left(\frac{\hbar\nu}{\pi M v^2}\right)^{\frac{1}{2}} \gg 1.$$
Under gas-kinetic conditions we may write
$$v^2 \simeq \kappa T/M_1,$$
where M_1 is the mass of the lighter molecule concerned in the collision and T the absolute temperature. We thus have, finally, the condition
$$\frac{a}{d}\left(\frac{\hbar\nu M_1}{\pi\kappa T M}\right)^{\frac{1}{2}} \gg 1, \tag{2}$$
as that which must be satisfied in order that the chance of excitation of a vibrational quantum $h\nu$ should be small.

For most molecules $h\nu$ will be greater than κT only for transitions between the deepest vibrational levels for which $a \gg d$. The chance of vibrational deactivation under these conditions will certainly be small. On the other hand, if $h\nu < \kappa T$, the condition may still be satisfied because $a \gg d$, but when $h\nu/\kappa T$ is sufficiently small the probability of excitation or deactivation per collision will be comparable with unity.

As we shall see in §§ 3 and 4 these features are exhibited by the experimental data. The dispersion of sound is due to the slowness of vibrational deactivation when (2) is satisfied but $h\nu < \kappa T$. On the other hand, the study of sensitized band fluorescence, for example, reveals cases in which $h\nu/\kappa T$ is so small that (2) is not satisfied and exchange of vibrational and translational energy occurs freely. A more detailed discussion of these questions will be given in § 4 after the nature of the experimental evidence has been described.

Turning to the excitation of rotational energy, we may obtain the condition for it to occur readily on collision by substituting in (1)

$$h\nu = \tfrac{1}{2}I\omega^2,$$

where the angular momentum $I\omega$ is of order $\hbar$. The condition becomes

$$\frac{a\hbar}{4\pi I v} \gg 1.$$

Substituting further, $I = MR^2$, where M is the reduced mass for the internal molecular motion and R is of the order of the interatomic separation, we obtain the condition as

$$ah/8\pi^2 R^2 M v \gg 1. \tag{3}$$

In all cases, except for the lightest molecules at very low temperatures, the wavelength h/Mv is very small compared with either a or R, which are of the same order of magnitude, so that the condition is not satisfied. Accordingly, it is to be expected that excitation or deactivation of molecular rotation will take place very readily in gas-kinetic collisions. This is generally confirmed by the experimental evidence discussed in § 5 below.

We shall first discuss the study of vibrational transitions, beginning with an account of the experimental methods that may be used for this purpose.

2. Experimental methods for studying the probability of vibrational transitions occurring on impact between molecules

2.1. *The dispersion and absorption of high-frequency sound*

Most of the information available about vibrational deactivation has come from observations of dispersion and absorption of high-frequency sound in gases.

The well-known Laplace formula for the velocity v of sound in an ideal gas at pressure p and density ρ,

$$v = (p\gamma/\rho)^{\frac{1}{2}}, \tag{4}$$

where γ is the ratio of the specific heats of the gas, includes no dependence on the frequency of the sound. If account is taken of viscosity and of heat conduction and radiation, an absorption coefficient α/λ given by

$$\alpha = \frac{4\pi^2}{v\rho\lambda}\left(\tfrac{4}{3}\eta + \frac{\gamma-1}{c_p}\sigma\right), \tag{5}$$

where η is the viscosity, σ the heat conductivity, λ the wavelength, and c_p the molar specific heat at constant pressure, is found as well as a slight dependence of velocity on frequency which is far too small to explain the observed dispersion of high-frequency sound. The observed absorption in the same frequency range is also much greater than that given by the formula (5) and depends on the frequency. It is only in monatomic gases that the classical formulae (4) and (5) are found to hold.

The explanation of these anomalies was first given by Herzfeld and Rice,† after which the theory was developed in detail by a number of authors.‡ The Laplace formula (4) supposes that the thermal changes that take place as the sound wave travels through the gas occur so slowly that the changes in the excitation of all the degrees of freedom that contribute appreciably to the specific heat can follow with very little time-lag. If this is not true, then the effective specific heat, and hence the velocity of propagation, will depend on the frequency. Furthermore, there will be a phase difference between the pressure and density fluctuations, giving rise to absorption.

We now put this in more definite terms and see how, from measurements of dispersion and absorption, the rate of vibrational excitation and deactivation by collision can be determined. We shall consider the simplest case in which the effect arises from the slowness of transitions between the ground and first excited vibrational levels.

We may still start from the formula (4) with the ratio γ written in the form $1+R/c_v$, where c_v is the molar specific heat at constant volume and R the gas constant. The only modification is that we must regard c_v as a complex function c_ω of frequency $\omega/2\pi$, the imaginary part corresponding to the phase difference between the pressure and density fluctuations. Writing the complex velocity **v** as $v(1+i\alpha/4\pi)$, say, the absorption coefficient is then given approximately by α/λ.

We write c_ω for the effective specific heat at frequency ω so that c_∞ is the limit for infinitely high frequencies. At these frequencies a vibrational degree of freedom can no longer play any part in the specific heat, whereas at a lower frequency it can do so. Hence, if $h\nu$ is the excitation energy of the first excited state, and T is the temperature,

$$c_\omega - c_\infty = h\nu \Delta n_1/\Delta T,$$

where n_1 is the number per mole of vibrationally excited molecules.

The time rate of change of n_1 will be given by

$$\frac{\partial n_1}{\partial t} = -\gamma_{10} n_1 + \gamma_{01} n_0.$$

† HERZFELD, K. F. and RICE, F. O., *Phys. Rev.* **31** (1928) 691.
‡ See, for example, BOURGIN, D. G., *Nature, Lond.* **122** (1928) 133; *Phys. Rev.* **50** (1936) 355; *J. acoust. Soc. Am.* **5** (1933) 57; KNESER, H. O., *Annln Phys.* **11** (1931) 761; RUTGERS, A. J., ibid. **16** (1933) 350.

γ_{10} is the number of deactivating collisions made by an excited molecule per second, γ_{01} the corresponding number of exciting collisions made per second by a normal molecule. They will be related by

$$\gamma_{01}/\gamma_{10} = g \exp(-h\nu/\kappa T),$$

where g is the ratio of the statistical weights of the ground and excited vibrational states.

With a periodic disturbance of frequency $\omega/2\pi$ we may replace $\partial/\partial t$ by $i\omega$, so that

$$i\omega n_1 = -\gamma_{10} n_1 + \gamma_{01} n_0$$

and, as $n_0 + n_1$ is constant,

$$\Delta n_1 = (n_0 \Delta\gamma_{01} - n_1 \Delta\gamma_{10})/(i\omega + \gamma_{01} + \gamma_{10}).$$

We then have

$$\Delta n_1/(\Delta n_1)_{\omega=0} = (\gamma_{01} + \gamma_{10})/(i\omega + \gamma_{01} + \gamma_{10}),$$

so that

$$h\nu \frac{\Delta n_1}{\Delta T} = h\nu \left(\frac{\Delta n_1}{\Delta T}\right)_{\omega=0} \frac{\gamma_{01} + \gamma_{10}}{(i\omega + \gamma_{01} + \gamma_{10})}.$$

Hence

$$c_\omega - c_\infty = \frac{\gamma_{01} + \gamma_{10}}{i\omega + \gamma_{01} + \gamma_{10}} (c_0 - c_\infty), \tag{6}$$

where c_0 is the specific heat at very low frequencies. This gives

$$v^2 = \frac{p}{\rho} \left\{ 1 + R \frac{1 + i\omega\tau}{c_0 + i\omega\tau c_\infty} \right\}, \tag{7}$$

where $\gamma_{01} + \gamma_{10} = 1/\tau$.

As in all cases of practical importance the imaginary part of v is small we obtain, to a good approximation,

$$v^2 = (v_0^2 \omega_i^2 + v_\infty^2 \omega^2)/(\omega_i^2 + \omega^2), \tag{8a}$$

$$\alpha = 2\pi(v_\infty^2 - v_0^2)\omega_i \omega/(v_0^2 \omega_i^2 + v_\infty^2 \omega^2). \tag{8b}$$

v_0 and v_∞ are the velocities of propagation in the limiting cases of very low and very high frequencies respectively and

$$\omega_i = c_0/c_\infty \tau.$$

The variation of these expressions with $\log \omega$ is illustrated in Fig. 17.1. v^2 has a point of inflexion at $\omega = \omega_i$ and α a maximum at

$$\omega = \omega_m = \omega_i v_0/v_\infty.$$

Both v^2 and α may be measured as functions of frequency. From such measurements ω_i may be obtained. Thus, from (8a),

$$\omega_i^2 = \omega^2(v_\infty^2 - v^2)/(v^2 - v_0^2), \tag{9}$$

and the velocities v_∞ and v_0 may be calculated from the known specific heats and vibrational constants of the molecules concerned.

Once τ, the relaxation time, is known the collision cross-section for deactivation, Q_v, is given from the relation

$$\tau^{-1} = \gamma_{10}+\gamma_{01}$$
$$= n_0 Q_v(2\pi\kappa T/M)^{\frac{1}{2}}\{1+g\exp(-h\nu/\kappa T)\}, \tag{10}$$

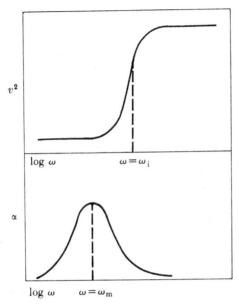

FIG. 17.1. Illustrating the variation with angular frequency ω of the velocity v and the quantity α (absorption coefficient = α/λ) for high-frequency sound in a dispersive gas.

where M is the mass of a gas molecule. In this Chapter we shall, unless otherwise stated, give relaxation times taking n_0 equal to Loschmidt's number, i.e. corresponding to 1 atmosphere pressure at 0 °C.

The formulae may easily be generalized to cases in which more than one vibrational mode contributes. In (6), c_0-c_∞ is the contribution to the equilibrium specific heat due to the particular vibrational mode concerned. When there are a number of modes A, B,..., whose contributions to the equilibrium specific heats are c_A, c_B,... respectively and whose relaxation times are τ_A, τ_B,..., then (6) is replaced by

$$c_\omega-c_\infty = c_A/(1+i\omega\tau_A). \tag{11}$$

With the values of τ all different this gives rise to a step-like dispersion curve involving a number of different dispersive regions. In many cases, however, the effective relaxation times are so nearly equal that separate regions cannot be distinguished. This is probably because excitation is transferred between the modes without difficulty (see § 3.1.3).

2.1.1. *Variation of dispersion and absorption with pressure, temperature, and impurity content.* According to the theory outlined above, the relaxation time must vary inversely as the gas pressure so that the dispersive region should shift to higher frequencies as the pressure increases. This provides a test of the essential validity of the theory. It must be remembered, however, that in some cases the rate of vibrational deactivation by collision is so slow that the possibility of radiation may not be negligible. This would include in τ a term independent of pressure. At the other extreme, at high pressures, three-body collisions may have some influence, introducing a quadratic pressure variation.

Study of the effect of temperature on the dispersion and absorption gives information about the variation of the deactivation cross-section Q_v with relative kinetic energy of the colliding molecules. If the temperature T is too low ($h\nu/\kappa T \gg 1$), the contribution to the specific heat from vibrational excitation will be small at all frequencies and no appreciable dispersion or absorption effects due to vibrational relaxation will be detectable.

If the deactivation cross-section is very sensitive to the detailed properties of the colliding molecules it is to be expected that small admixtures of impurities might have a profound influence on dispersion and absorption. This can only be so if the molecules of the impurity are much more effective in deactivating the excited molecules of the main gas than are normal molecules of the same gas. An impurity must, therefore, either shift the dispersive region to higher frequencies or have no effect. Study of the effect of controlled admixture of 'impurities' leads to valuable information about the deactivation cross-sections when the colliding molecules are unlike.

2.1.2. *Methods for measuring the velocity and absorption of high-frequency sound waves*

2.1.2.1. *The ultrasonic interferometer.* Most measurements of the velocity of high-frequency sound have been made with the interferometer technique first introduced by Pierce.†

Consider a plane source S of plane sound waves of a definite angular frequency ω. If we take the plane of S as the xOy plane and the direction of propagation of the waves as that of Oz we may write the pressure p and particle velocity u in the gas through which the wave propagates as

$$p = P\exp\{i\omega(x/v-t)\}, \qquad (12\,\mathrm{a})$$

$$u = (P/\rho v)\exp\{i\omega(x/v-t)\}. \qquad (12\,\mathrm{b})$$

† PIERCE, G. W., *Proc. Am. Acad. Arts Sci.* **60** (1925) 271.

Here v is the velocity of propagation of the sound wave and ρ is the gas density. We have, at this stage, assumed that there is negligible absorption of the sound wave. The ratio p/u is known as the *acoustic impedance* and provides a measure of the reaction of the sound wave on the radiator. The ultrasonic interferometer operates through the measurement of this radiation reaction on the source when a reflector is placed parallel to it at different distances from it.

If P_r is the pressure amplitude due to the reflected wave we now have, in place of (12),

$$p = P\exp\{i\omega(x/v-t)\}+P_r\exp\{-i\omega(x/v+t)\}, \qquad (13\text{a})$$

$$u = (1/\rho v)[P\exp\{i\omega(x/v-t)\}-P_r\exp\{-i\omega(x/v+t)\}]. \qquad (13\text{b})$$

As a result of the reflection there will, in general, be a reduction of amplitude and a phase change. We may express this by writing

$$P_r/P\exp(2i\omega l/v) = e^{-2\psi}, \qquad (14)$$

where $\psi = \gamma - i\delta$ and l is the distance of the reflector from the source. $\psi = 0$ corresponds to reflection from a perfectly rigid piston. We now have

$$p = 2P\exp(-i\omega t - \psi + i\omega l/v)\cosh\{\psi+(x-l)i\omega/v\}, \qquad (15\text{a})$$

$$u = (2P/\rho v)\exp(-i\omega t - \psi + i\omega l/v)\sinh\{\psi+(x-l)i\omega/v\}, \qquad (15\text{b})$$

giving for the acoustic impedance

$$Z = p/u = \rho v\coth\{\psi+(x-l)i\omega/v\}. \qquad (16)$$

So far we have not allowed for absorption of the sound in the gas. In terms of the absorption coefficient per wavelength α, (15a) becomes

$$p = 2P\exp(-i\omega t-\psi+i\omega l/v-\tfrac{1}{2}\alpha l/\lambda)\cosh\{\psi+(x-l)(i\omega/v-\tfrac{1}{2}\alpha/\lambda)\} \qquad (17)$$

and the acoustic impedance

$$Z = \rho v\coth\{\psi+(x-l)(i\omega/v-\tfrac{1}{2}\alpha/\lambda)\}. \qquad (18)$$

At $x = 0$ we have, writing $\psi = \gamma-i\delta$,

$$Z = \rho v\frac{\sinh(2\gamma+\alpha l/\lambda)+i\sin(2l\omega/v+2\delta)}{\cosh(2\gamma+\alpha l/\lambda)-\cos(2l\omega/v+2\delta)}. \qquad (19)$$

The impedance is purely reactive when

$$\delta+\omega l/v = n\pi, \qquad (20\text{a})$$

$$\delta+\omega l/v = (n+\tfrac{1}{2})\pi, \qquad (20\text{b})$$

$$(n = 0, 1, 2,...).$$

In the former case $\quad Z = \rho v\coth(\gamma+\tfrac{1}{2}\alpha l/\lambda)$

and the radiation resistance is a maximum, whereas in the latter

$$Z = \rho v\tanh(\gamma+\tfrac{1}{2}\alpha l/\lambda)$$

and the radiation resistance is a minimum.

Fig. 17.2 illustrates the variation of the impedance with l. It follows that, if the impedance can be monitored as a function of l, the velocity v of the sound waves is given by

$$v = (\omega/\pi)\Delta l,$$

where Δl is the interval between successive minima or successive maxima. To obtain γ and α† we write S_n as the difference between the impedance signal measured at the nth maximum and the interpolated minimum signal (see Fig. 17.2) at the same value l_n of l. Then

$$S_n = A\{\coth(\gamma+\tfrac{1}{2}\alpha l_n/\lambda) - \tanh(\gamma+\tfrac{1}{2}\alpha l_n/\lambda)\}$$
$$= 2A/\sinh(2\gamma+\alpha l_n/\lambda)$$

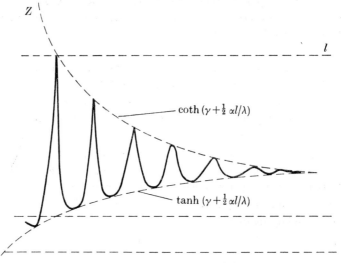

FIG. 17.2. Variation of the acoustic impedance with source-reflector separation l in an ultrasonic interferometer.

where A is an instrumental constant. We now have

$$2\gamma + \tfrac{1}{2}(n-\delta/\pi)\alpha = \operatorname{arsinh}(2A/S_n). \tag{21}$$

A is the magnitude of the signal when $n \to \infty$ and is the difference between the signal when the reflector is moved to a very great distance and that when no gas is present.

In practice the source is a quartz crystal driven at its resonance frequency by a tuned parallel resonant circuit. The acoustic reaction due to the reflection may be monitored by measuring the current in this circuit.‡ The interferometer may be made recording§ by using a piston reflector which is driven by a screw so as to vary the separation l.

To obtain results of sufficient accuracy it is necessary that the

† STEWART, J. L. and STEWART, E. S., *Phys. Rev.* **77** (1950) 143; *J. acoust. Soc. Am.* **24** (1952) 22.
‡ HUBBARD, J. C., *Phys. Rev.* **38** (1931) 1011; **41** (1932) 523; **46** (1934) 525.
§ STEWART, J. L. and STEWART, E. S., *J. acoust. Soc. Am.* **24** (1952) 22.

frequency be tuned very closely (to at least one part in 10^6), that the crystal and reflecting surfaces should be accurately parallel, and that the crystal radiates plane waves.

It is convenient to tune the frequency by adjusting the current to a minimum with the reflector effectively at infinity. A sensitive test of mistuning is the appearance of asymmetry in the current peaks recorded.

The presence of other modes of vibration can be detected by the appearance of multiple peaks in the current record when the crystal and reflector are not accurately parallel. They may also be detected by schlieren photography of the lateral part of the radiation field.

Lack of parallelism between source and reflector introduces apparent additional absorption losses. Adjustment is therefore made† by levelling the crystal relative to the reflector so that the acoustic reaction is a maximum while moving the reflector back and forth to compensate for any change in mean distance between them.

2.1.2.2. *Further methods for measurement of sound absorption.* The most direct method of measuring the absorption of sound is to measure the decrease in amplitude of the pressure oscillations in a plane sound wave with distance from the source. To avoid complications due to the production of standing waves by reflection from the receiver the sound source may be pulsed‡ so that the attenuation of the pulse on arrival may be directly measured. If, as usual, the source and receiver are enclosed it is necessary to eliminate the effects of reflections from the walls by covering the inside of the walls with absorbing material. Also, if the source does not generate plane waves an additional apparent absorption will result due to beam divergence. To eliminate this the observed absorption coefficient may be plotted as a function of the reciprocal of the gas pressure. If beam divergence is unimportant the plot should be a straight line passing through the origin.

A second method of determining the absorption coefficient is essentially similar to the determination of relaxation times by line width measurements (see Chap. 2, §§ 5, 6; Chap. 5, § 7.2; Chap. 18, §§ 3.4, 6.2.3, and 7). The apparatus is basically similar to that used in the ultrasonic interferometer except that the source is now attached to a movable piston while the reflector remains at rest. If the piston is oscillated with a velocity amplitude U then, in the notation of (17), we have

$$(2P/\rho v)\exp(-\psi - \tfrac{1}{2}\alpha l/\lambda)\sinh\{\psi - l(i\omega/v - \alpha/2\lambda\} = U, \qquad (22)$$

† ALLEMAN, R. S., *Phys. Rev.* **55** (1939) 87.
‡ PARKER, J. G., ADAMS, C. E., and STAVSETH, R. M., *J. acoust. Soc. Am.* **25** (1953) 263; TEMPEST, W. and PARBROOK, H. D., *Acoustica* **7** (1957) 354.

so that the pressure at the reflector is given by

$$P_{x=l} = \tfrac{1}{2}\rho v U \cosh\psi \operatorname{cosech}\{\psi - l(i\omega/v - \tfrac{1}{2}\alpha/\lambda).\tag{23}$$

Hence

$$|P_{x=l}| = \tfrac{1}{2}\rho v U(\cosh 2\gamma + \cos 2\delta)^{\frac{1}{2}}[\{\cosh(2\gamma + \alpha l/\lambda) - \cos(2\delta + 2l\omega/v)\}]^{-\frac{1}{2}}.\tag{24}$$

We now consider the variation of $|P_{x=l}|$ with frequency ω. We denote by ω_n the nth resonant angular frequency, for which

$$\delta + l\omega_n/v = n\pi \quad (n = 1, 2, \dots).\tag{25}$$

If ω is an angular frequency close to ω_n and $\gamma + \alpha l/\lambda \ll 1$ then

$$|P_{x=l}| = 2^{-\frac{3}{2}}l^{-1}\rho v^2 U(\cosh 2\gamma + \cos 2\delta)^{\frac{1}{2}}[(\gamma v/l + \alpha v/2\lambda)^2 + (\omega - \omega_n)^2]^{-\frac{1}{2}}.\tag{26}$$

The amplitude of the pressure at the reflector therefore varies with frequency as $\{(\omega - \omega_n)^2 + \mu^2\}^{-\frac{1}{2}}$ where

$$\mu = v(\gamma/l + \alpha/2\lambda).\tag{27}$$

Hence by observation of the form of $|P_{x=l}|$ as ω varies about ω_n, μ may be determined. By making observations at different tube lengths the terms γ/l and $\alpha/2\lambda$ in (27) may be separated. Usually, however, with suitable choice of materials γ/l can be made less than a few per cent of α/λ.

As an example, in the experiments of Parker[†] on O_2–H_2, O_2–D_2, and O_2–He mixtures, using this technique, the resonance tube was a precision bore Pyrex glass cylinder with inner diameter $2 \cdot 005 \pm 0 \cdot 002$ in, thickness $\tfrac{1}{4}$ in, and length $29\tfrac{5}{8}$ in. The sound source was a cylindrical brass piston of diameter $1\tfrac{1}{2}$ in and length 1 in, excited acoustically through a bellows by an electromagnetic shaker. The sound detector was a condenser microphone.

A further method for measurement of α is to observe the decay time of a sound field excited in a cavity. If sound is generated in a cavity of volume V and the source is then sharply cut off the intensity of the sound in the room falls off as $\exp(-avt/4V)$, where t is the time since cut-off, v is the velocity of the sound waves, and

$$a = \alpha_\mathrm{v} V + \alpha_\mathrm{s} S,$$

where α_v is the volume absorption coefficient of the sound in the gas, α_s the absorption coefficient of the inner surface materials of the cavity, and S the total surface area exposed to the sound. If measurements

[†] PARKER, J. G., J. chem. Phys. **34** (1961) 1763.

of the decay time are made for cavities with different ratios S/V, α_v and α_s may be separately determined. If the sound is of a definite wavelength then $\alpha_v = \alpha/\lambda$. Experiments of this kind were first carried out by Kneser and Knudsen.† In later experiments the diffuse sound field generated in these experiments has been replaced by a normal mode of oscillation of a cavity of simple geometrical shape. Thus Edmonds and Lamb‡ used an arrangement as illustrated in Fig. 17.3 in which the cavity

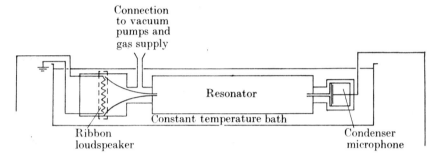

Fig. 17.3. General arrangement of the experiment of Edmonds and Lamb for measuring decay times of normal acoustical oscillations of gas in a cavity.

was a precision-bore Pyrex tube of internal diameter $9 \cdot 80 \pm 0 \cdot 01$ cm, length $60 \cdot 1$ cm, and wall thickness $6 \cdot 5$ mm. Energy was fed into the resonator from a ribbon loudspeaker through a horn and probe tube. The intensity of acoustical excitation was monitored by a condenser microphone coupled to the opposite end of the resonator.

2.1.3. *Some notes on the derivation of relaxation times from measurements of ultrasonic dispersion and absorption.* Since the vibrational relaxation is inversely proportional to the gas pressure p it follows that, if this is the only source of dispersion or absorption, the sound velocity v and the absorption coefficient per wavelength α are functions of ω/p, where ω is the angular frequency of the sound wave. It is therefore possible to simulate change of frequency by change of pressure, and this is often useful in extending the range of observations. In doing this, however, it is important to remember that we have assumed that the gas concerned is ideal. Corrections have to be made to allow for gas imperfection, starting from the formula for the sound velocity

$$v^2 = (\partial p/\partial \rho)_S,$$

† Kneser, H. O., *J. acoust. Soc. Am.* **5** (1933) 122; Kneser, H. O. and Knudsen, V. O., *Annln Phys.* **21** (1934) 682.
‡ Edmonds, P. D. and Lamb, J., *Proc. phys. Soc.* **71** (1957) 17.

where ρ is the gas density and the entropy S is constant, and an empirical equation of state. The velocity so corrected is called the idealized velocity.

The measured absorption coefficient will include the classical contribution (5) as well as, in certain circumstances, that from wall losses. Experiments in monatomic gases have established the validity of the classical contribution (5) which may be calculated and subtracted off. If wall losses are important and cannot be calculated they may be separated by making observations with different ratios of exposed surface area to gas volume.

As an indication of the magnitudes of the effects observed, in N_2O the idealized velocity at 25 °C changes from 268 m s^{-1} at low frequencies to 280 m s^{-1} at such frequencies that no contribution is made from the vibrational heat capacity. For more complex molecules the changes are somewhat greater. The dispersion range may be covered by observing over a pressure range of 0·1–2 atm at a fixed frequency of 250 kc s^{-1}.

In CO_2, for example, the maximum value of α due to vibrational relaxation is 0·230 at 20 kc s^{-1} atm^{-1}. On the other hand, the classical absorption coefficient only attains this value at 30 Mc s^{-1} atm^{-1}.

Most measurements have been made from room temperature to 600 °K which does not quite overlap, at the higher temperature end of the range, with the lowest conveniently attainable temperature 1000 °K in shock tube investigations (see § 2.2).

2.2. *Shock-wave investigations*

2.2.1. *Principles of the shock tube and its application to relaxation measurements.* The great advantage of the use of shock waves is that they provide a means for rapidly increasing the pressure and temperature of a gas.

The formation of a shock wave may be understood from a number of different points of view. Thus the velocity of a sound wave in a gas increases with pressure. It follows that, if a sound wave of finite amplitude is generated in a gas, the crests move faster than the troughs until eventually the wave form has a saw-tooth appearance with a sharp discontinuity at the front of each saw-tooth which constitutes a shock wave.

Alternatively, consider a tube of gas closed at one end and with a piston at the other. A slight acceleration of the piston will generate a pressure disturbance in the gas which propagates with the speed of sound. A second slight acceleration will generate a second disturbance which will propagate a little faster than the first because the gas will itself be in mass motion with the velocity attained by the piston in its first acceleration. The second disturbance will then eventually catch up on the first. Thus if the piston is accelerated to a finite velocity by a succession of small accelerations the successive small disturbances will catch each other up to generate down the tube a discontinuity of pressure and temperature. Behind this shock front the compressed and heated gas moves with the velocity of the piston.

In practice a shock wave is generated in a somewhat different way. The shock tube is usually cylindrical, of constant cross-section, though for the generation of especially strong shocks it may include a convergent region in which the cross-section decreases in going along the tube. By means of a diaphragm, gas at high pressure in the rear section is isolated from low pressure gas in the forward section. Generation of the shock wave is brought about by sudden puncturing of the diaphragm. Initial turbulence due to the finite bursting time of the diaphragm soon smoothes out and thereafter the shock travels uniformly down the tube. In this arrangement the high-pressure gas released by the puncturing of the diaphragm plays the part of the piston. The driver gas does not mix appreciably with the shocked gas and the composition boundary between them travels down the tube behind the shock front.

Once a steady condition has been reached we may distinguish the following four regions (see Fig. 17.4).

I. To the extreme left of the composition boundary we have the as yet unexpanded driver gas with pressure, density, and temperature p_2, ρ_2, T_2, respectively.

II. To the extreme right of the tube we have the as yet uncompressed gas with pressure, density, and temperature p_0, ρ_0, T_0.

III. The region to the left of the composition boundary consisting of expanded driver gas with pressure, density, and temperature p_2^f, ρ_2^f, T_2^f.

IV. The region to the right of the composition boundary on the left of the shock front, consisting of compressed 'shocked' gas with pressure, density, and temperature p_1, ρ_1, T_1.

Now let us consider the motion relative to the shock front. If the velocity of the shock front relative to the tube is V, then gas is entering the front from the right with velocity V. It follows from the conservation of mass that gas must be leaving the front to the left at a velocity U relative to the front, where

$$\rho_0' V = \rho_1' U. \tag{28}$$

Again, to satisfy the conservation of momentum and of energy respectively,

$$p_0 + \rho_0 V^2 = p_1 + \rho_1 U^2, \qquad (29)$$

$$E_0 + \tfrac{1}{2}V^2 + p_0/\rho_0 = E_1 + \tfrac{1}{2}U^2 + p_1/\rho_1. \qquad (30)$$

Here E_0 and E_1 are the internal energy contents per unit mass of the gas on the right- and left-hand sides of the shock front respectively.

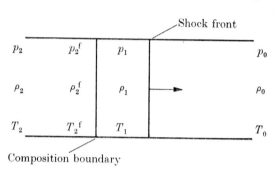

Fig. 17.4.

Also, assuming that the gas densities are low enough, we have

$$p_1/\rho_1 = \kappa T_1/m, \qquad (31)$$

where κ is Boltzmann's constant and m the mass of a gas atom.

In a monatomic gas, provided the temperature T_1 is not so large that appreciable electronic excitation occurs, we have

$$E_{1,0} = \tfrac{3}{2}\kappa T_1/m. \qquad (32)$$

It is possible with certain assumptions to calculate V in terms of the initial conditions by using a relation derived by Taylor.† He showed that if the driven gas is of the same composition as the compressed gas, the velocity of the composition boundary relative to the tube is given by

$$W = 2\{V_s/(\gamma-1)\}\{1-(p_2^f/p_2)^{(\gamma-1)/2\gamma}\}, \qquad (33\text{ a})$$

where $V_s = (\gamma p_0/\rho_0)^{\frac{1}{2}}$ is the velocity of sound in the gas and γ the ratio of the specific heats. If the shocked gas behind the shock front moves with the same velocity at all sections up to the composition boundary,

$$W = V - U. \qquad (33\text{ b})$$

Since, at this boundary, there is no discontinuity of pressure so that $p_2^f = p_1$, (33) together with (28), (29), and (30) provide four equations to determine V, U, ρ_1 and p_1 in terms of p_0, ρ_0, and p_2, use being made

† TAYLOR, G. I., quoted in PAYMAN, W. and SHEPHERD, W. C. F., *Proc. R. Soc.* A186 (1946) 293.

of (31) and (32). In practice it is usual to measure V directly because the breaking of the diaphragm which initiates the shock is usually somewhat more ragged than assumed so that the theoretical value of V may not be of sufficient accuracy when applied to the real circumstances. Furthermore, it cannot be assumed that (33 b) is valid (see Chap. 20, § 3.1.5).

It is easy to show that

$$V^2 = V_s^2\{1+(\gamma+1)(\xi-1)/2\gamma\}, \tag{34}$$

where $\xi = p_1/p_0$. For a weak shock $\xi \simeq 1$ and $V \simeq V_s$. On the other hand for $\xi \gg 1$, $V \gg V_s$ and the Mach number

$$M = V/V_s = \{1+(\gamma+1)(\xi-1)/2\gamma\}^{\frac{1}{2}}. \tag{35}$$

The temperature T_1 of the shocked gas is given by

$$T_1/T_0 = \xi\{1+(\gamma-1)(\xi-1)/2\gamma\}\{1+(\gamma+1)(\xi-1)/2\gamma\}^{-1}, \tag{36}$$

so that, for strong shocks, the temperature ratio is proportional to the shock strength and hence to the square of the Mach number.

For application to the measurement of relaxation rates the most important aspect of the shock wave to study is the transition region at the shock front. For a monatomic gas the thickness of this region is of the order of a few free paths for gas-kinetic collisions. Because of the very rapid change in the thermodynamic parameters in this region the usual gas-kinetic transport theory can only be used with confidence for weak shocks. In that case the shape and thickness of the transition region may be calculated in terms of the viscosity η and thermal conductivity σ. Let p, ρ, T, v be the pressure, density, temperature, and velocity of the gas relative to the shock front at a point distant z from the shock front, measured along the direction of propagation of the shock. Then we have[†]

$$\rho_0 V = \rho v = \rho_1 U = \text{a constant } a, \tag{37 a}$$

$$aV+p_0 = av+p+\tfrac{4}{3}\eta\,dv/dz = aU+p_1 = \text{a constant } b, \tag{37 b}$$

$$aE_0+p_0 V+\tfrac{1}{2}aV^2 = aE+bv-\tfrac{1}{2}av^2-\sigma\,dT/dz$$
$$= aE_1+p_1 U+\tfrac{1}{2}aU^2, \tag{37 c}$$

together with $\qquad p = p(\rho, T), \qquad E = E(\rho, T), \tag{37 d}$

the equation of state of the gas.

[†] BECKER, R., Z. Phys. **8** (1922) 321; MORDUCHOW, M. and LIBBY, P. A., J. aeronaut. Sci. **16** (1949) 674; GRAD, H., Communs pure appl. Math. **5** (1952) 257.

Elimination of p, ρ, and T from these equations gives a second-order differential equation for v as a function of z. In the special case of an ideal gas in which
$$10\kappa\eta/3m\sigma = 1$$
an analytical solution may be obtained. According to simple kinetic theory $10\kappa\eta/3m\sigma$ should equal $8/9$ for monatomic gases so that the special case provides a realistic representation of the transition region. Fig. 17.5 shows the form of this region calculated in this way for various

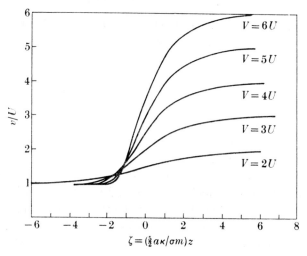

Fig. 17.5. Variation of the velocity v of the gas, relative to the shock front, with distance z along a shock tube, in the neighbourhood of the shock front, V being the velocity of the shock wave and U as defined in (28).

values of V, the shock velocity. The reduced scale of z is given in units of $2\sigma m/5a\kappa$. In terms of the simple kinetic-theory model this is of the order $(\bar{c}/V)l$, where l is the mean free path of the gas atoms and $\bar{c}$ the root mean square velocity. Since $\bar{c}$ and V are comparable it can be seen from Fig. 17.5 that the shock-front thickness is of the order of a few mean free paths only. A more elaborate analysis carried out by Mott-Smith,† which does not depend on first-order solutions of the transport equations, gives results of the same order.

We now proceed to examine how these considerations will be modified for a gas of molecules possessing internal degrees of freedom. In general, the rotational degrees of freedom will come to equilibrium very quickly but this will not be true of the vibration except for quite complex molecules. We may therefore distinguish two separate regions behind

† Mott-Smith, H. M., *Phys. Rev.* **82** (1951) 885.

the shock front. One immediately behind will have a thickness comparable with that for a monatomic gas. It will differ only in that full allowance must be made for rotational excitation. The equations (28)–(31) still apply but the energy per gramme now includes the contribution

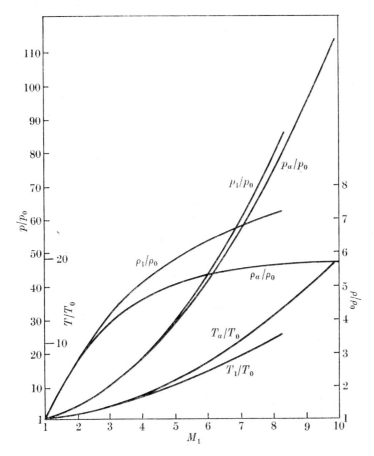

Fig. 17.6. Density (ρ), pressure (p), and temperature (T) ratios for shock waves in oxygen. Subscript 0 distinguishes initial values, a the region a behind the shock front, and 1 the final equilibrium values behind the shock front.

from the rotational specific heat. Thus for linear molecules, in (32), E is now $\frac{5}{2}\kappa T$. This region is usually referred to as region a in which the pressure, density, and temperature have the respective values p_a, ρ_a, and T_a. Fig. 17.6 illustrates the ratios p_a/p_0, ρ_a/ρ_0, T_a/T_0 of these quantities to their initial values for shock waves in oxygen, as a function of Mach number.

Extending behind the shock front for a relatively much greater distance, beyond region a, will be a region b in which the vibrational degrees of freedom are gradually becoming excited, ultimately to reach the equilibrium value. As the excitation energy for the vibration must come from translation and rotation the gas temperature will be falling in this region to reach a final equilibrium value T_1. For the calculation of the final equilibrium values p_1, ρ_1, T_1 it is only necessary to add the appropriate contribution from the vibrational specific heat, as well as from the rotational specific heat, to the energy content per unit mass in (32). At working temperatures it will, in contrast to the rotational specific heat, still be temperature dependent. Fig. 17.6 shows the ratios of the ultimate parameters p_1, ρ_1, T_1 to the initial ones, again for oxygen. It will be noted that, in attaining the ultimate equilibrium from state a, the pressure changes by a relatively small amount, even for quite strong shocks.

To calculate approximately† the extent of region b we note that, since the increase of vibrational energy E_{vib} is gained at the expense of energy of translation and rotation,

$$dE_{\text{vib}} = -c'_p \, dT, \tag{38}$$

where c'_p is the specific heat per unit mass at constant pressure arising from translation and rotation alone, c_{vib} is the contribution from vibration.

Hence
$$\int_{E_{\text{vib}}(T)}^{c_{\text{vib}} T_1} dE_{\text{vib}} = -\int_T^{T_1} c'_p \, dT, \tag{39}$$

so, since c'_p is independent of T,

$$E_{\text{vib}}(T) + c'_p T = c_{\text{vib}} T_1 + c'_p T_1$$
$$= c_p T_1. \tag{40}$$

Now, if τ is the relaxation time for vibration and $\bar{E}_{\text{vib}}(T)$ is the vibrational energy ($c_{\text{vib}} T$) in *equilibrium* at temperature T,

$$\frac{dE_{\text{vib}}}{dt} = \tau^{-1}\{\bar{E}_{\text{vib}}(T) - E_{\text{vib}}\}. \tag{41}$$

Hence, from (40),
$$-c'_p \frac{dT}{dt} = \tau^{-1}\{c_{\text{vib}} T - E_{\text{vib}}\}. \tag{42}$$

Also, from (40),
$$E_{\text{vib}} = c_p T_1 - c'_p T, \tag{43}$$

so that
$$\frac{d(T-T_1)}{dt} = -\frac{c_p}{c'_p} \frac{(T-T_1)}{\tau}. \tag{44}$$

Integrating, we have
$$\frac{T-T_1}{T_a-T_1} = \exp\left\{-\frac{c_p t}{c'_p \tau}\right\}. \tag{45}$$

† BLACKMAN, V., *J. Fluid Mech.* **1** (1956) 61.

In terms of density we have

$$\frac{\rho_1-\rho}{\rho_1-\rho_a} = \frac{\rho_1}{\rho_a}\exp\left(-\frac{c_p t}{c_p' \tau}\right)\left[1+\frac{(\rho_1-\rho_a)}{\rho_a}\exp\left(-\frac{c_p t}{c_p' \tau}\right)\right]^{-1}. \qquad (46)$$

As $(\rho_1-\rho_a)/\rho_a$ is small the density also approaches its ultimate equilibrium value exponentially with time constant $\tau c_p'/c_p$.

In terms of distance z from the shock front the equilibrium is approached as $\exp\{-c_p z/c_p' U\tau\}$. For a typical diatomic gas τ is of the order 10^{-6} s or higher and U 10^4 cm s^{-1} so that the thickness of the region b is of the order 10^2 cm!

If the temperature or density can be observed as a function of distance behind the shock front the vibration relaxation time may be determined from (45) or (46) respectively.

Hydrogen and deuterium are unique in that the relaxation time for rotation is quite long and the above analysis applies with appropriate reinterpretation of the symbols. Thus c_p becomes c_p' while c_p' must now be replaced by c_p'' the translational specific heat.

Even for other gases a small effect is observed due to the finite time required for rotational relaxation. This may be taken into account approximately† by adding to the shear viscosity $4\eta/3$ in (37) a bulk viscosity K given by

$$K = \frac{n\kappa^2 T}{c_V^2} c_i \tau_i, \qquad (47)$$

where n is the concentration of gas molecules, c_V the specific heat of the gas at constant volume, and c_i the contribution to the specific heat from the relaxing degree of freedom for which the relaxation time is τ_i.

2.2.2. *Measurement techniques.* The essential measurements to be made are those of the velocity of the shock wave and of the density (or temperature) behind the shock front. The latter measurements need to be carried out over both the very short region a and the much more extended region b.

A convenient way to measure the shock velocity is by observing the change of resistance of fine gold-covered films due to heating by the shock wave. In typical experiments‡ these films have been of the order of a few hundred atoms thick evaporated on to sound-recorder mending tape about 2 mils thick, ½ mm wide, and 7 mm long, mounted in the shock tube perpendicular to the direction of shock propagation. The resistance of a film was between 15 and 20 Ω. Through the resistance

† TISZA, L., *Phys. Rev.* **61** (1942) 531.
‡ BLACKMAN, V., loc. cit., p. 1482.

change on passage of the shock a voltage signal of around 10 mV was communicated to an oscilloscope. With devices of this sort the velocity may be measured to about 0·5 per cent.

The density 'discontinuity' across region a and its variation over the extended region b is most conveniently observed by the Mach–Zehnder interferometric technique.† Passage of the shock wave past a selected observation point triggers off an intense flash of light which is collimated and split into two parallel beams in the interferometer. One of these beams passes through the shocked gas and the other through an unshocked sample of the same gas. The beams are then allowed to interfere to produce a system of fringes. This system will change as the density of the shocked gas changes. Let δ be the fringe shift, at some point, as compared with the system obtained before the shock front passes. Then if ρ is the density in the gas at this point

$$\rho/\rho_1 = 1 + W\delta/p_1, \tag{48}$$

where
$$W = Tp_s\lambda/T_s l(\mu_s - 1), \tag{49}$$

l being the width of the tube, μ_s, T_s, and p_s the refractive index, temperature, and pressure respectively of the gas at n.t.p. and λ the wavelength of the light used.

Fig. 17.7 shows some typical photographs of fringes taken under these conditions.‡ For a monatomic gas such as argon all that can be seen is a sharp discontinuity at the shock front but, in CO_2 and N_2O, although region a appears as a discontinuity, the gradual approach to vibrational equilibrium in region b is also clearly seen. From such photographs the vibrational relaxation time may be obtained, using (46).

A 'schlieren' method of greater sensitivity§ has been developed which is capable of observing density gradients that would yield a fringe shift of less than 0·002 fringe/mm. This is specially useful for the determination of fast relaxation times at 1550 °K or higher for light gases such as D_2 which have low vibrational heat capacity and low molar refractivity.

A laser beam is partially reflected by a mirror to pass across a section of the shock tube after which it is focused on to a knife edge. The light passing the knife edge is detected by a photomultiplier, amplified, and displayed on an oscilloscope. The passage of a density gradient across the laser beam deflects it and so changes the recorded signal. The laser light, which is transmitted by the mirror and does not pass through the

---

† WINCKLER, J., *Rev. scient. Instrum.* **19** (1948) 307; CURTISS, C. W., EMRICH, R. J., and MACK, J., ibid. **25** (1954) 679.
‡ GRIFFITH, W., BRICKL, D., and BLACKMAN, V., *Phys. Rev.* **102** (1956) 1209.
§ KIEFER, J. H. and LUTZ, R. W., *J. chem. Phys.* **44** (1966) 658.

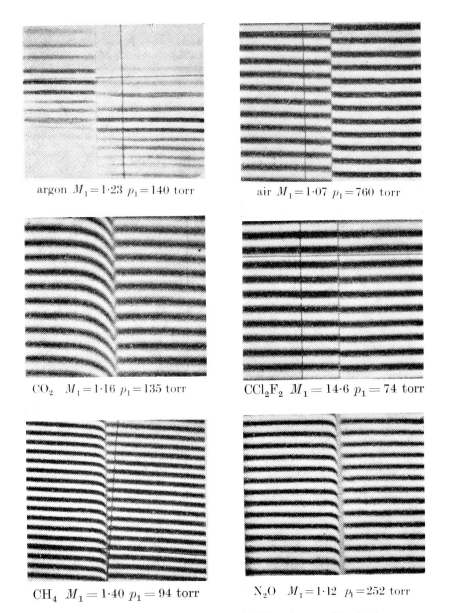

FIG. 17.7. Interferograms of shock waves in different gases. The shock wave, of Mach number M_1, is moving to the right in all cases.

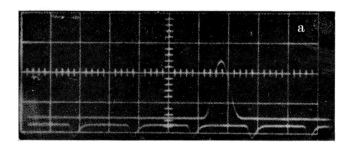

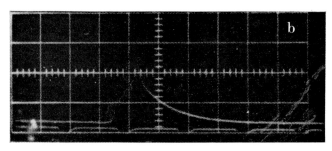

Fig. 17.8. Typical records taken by the schlieren method of the density gradient behind a shock front: (a) in pure argon, $V = 1\cdot 905 \times 10^5$ cm s^{-1}, $p_0 = 9\cdot 7$ torr; (b) in a mixture 40 per cent D_2–60 per cent Ar, $V = 1\cdot 719 \times 10^5$ cm s^{-1}, $p_0 = 70\cdot 8$ torr, $T_1 = 1720$ °K, $\tau = 0\cdot 504$ μs. The bottom sweep in each record contains 1 μs timing markers, the time increasing from left to right.

shock tube, is detected by a second photomultiplier and oscilloscope so that small variations in the laser intensity during recording may be allowed for.

The deflexion of a single ray of the laser beam is given by
$$D(x,t) = RLW\, \partial\rho(x,t)/\partial x,$$
where x is an axial coordinate with origin at the centre of the beam and increasing towards the diaphragm of the shock tube. R is the specific refractivity of the test gas, L is the distance from the focusing telescope to the knife edge, W the internal diameter of the shock tube, and $\partial\rho(x,t)/\partial x$ the density gradient along the tube at the point x. If the power distribution across the laser beam is given by
$$P(x,y) = (P_0/\pi\sigma^2)\exp\{-(x^2+y^2)/\sigma^2\},$$
where P_0 is the total beam power, σ the standard deviation of the distribution, and y is measured normal to x and to the direction of propagation, the fractional change of signal due to a deflexion D of the beam initially centred at the knife edge is given by
$$\Delta S/S_0 = -\tfrac{1}{2}\mathrm{erf}(D/\sigma_k)$$
$$\simeq -(D/\pi^{\frac{1}{2}}\sigma_k)\{1-(D^2/3\sigma_k^2)+...\},$$
where σ_k is the value of σ at the knife edge. Hence, the total fractional signal change is given by
$$\Delta S/S_0 = -(RLW/\pi^{\frac{1}{2}}\sigma_k)\int_{-\infty}^{\infty}\mathscr{P}(x)(\partial\rho(x,t)/\partial x)\,\mathrm{d}x,$$
where $\mathscr{P}(x)$ gives the power distribution across the wave front of the beam.

In the relaxation zone we have
$$\frac{\partial\rho}{\partial x} = (\Delta\rho/V\tau)\exp\{-(t+x/V)/\tau\},$$
where V is the shock velocity, $\Delta\rho$ the total change in density, and τ the apparent relaxation time in laboratory coordinates. We then have
$$\Delta S/S_0 = -(RLW\,\Delta\rho/\pi^{\frac{1}{2}}\sigma_k V\tau)e^{-t/\tau}\int_{-\infty}^{\infty}\mathscr{P}(x)e^{-x/V\tau}\,\mathrm{d}x$$
$$= -(RLW/\pi^{\frac{1}{2}}\sigma_k)\left(\frac{\partial\rho}{\partial x}\right)_{0,t-\Delta t},$$
where Δt depends on the width of $\mathscr{P}(x)$ and the values of t and V. If we take
$$\mathscr{P}(x) = \int_{-\infty}^{\infty} P(x,y)\,\mathrm{d}y,$$
$\Delta t = \sigma^2/4V^2\tau$ and in most experimental conditions is negligibly small. In any case if $\partial\rho/\partial x$ is truly exponential a time shift is of no importance.

Fig. 17.8 shows two typical records of the deflexion signal. (a) was taken with pure argon in the shock tube and shows a single spike, the width of which is probably due to curvature of the shock front near the wall. The origin for the relaxation shown in (b), observed with a mixture of D_2 in Ar, is taken as the peak of the initial spike. Results obtained for the vibrational relaxation time of D_2 in pure D_2 gas are discussed on p. 1518.

To obtain information about the thickness of the shock front in a monatomic gas or region a in a polyatomic gas it is necessary to use a different technique. The most effective one, introduced by Cowan and Hornig,† depends on measurement of the variation of the reflectivity R of the shock front with angle of incidence θ. In terms of the rate of change $d\mu/dz$ of the refractive index across the shock front

$$R = \tfrac{1}{4}(1+\tan^4\theta)|F(2\cos\theta/\lambda)|^2, \tag{50}$$

where
$$F(u) = \int_{-\infty}^{\infty} \frac{d\mu}{dz}\exp(-2\pi i z u)\, dz. \tag{51}$$

Since $d\mu/dz$ can be obtained from the density profile at the shock front it is possible to use (51) to check any assumed profile. The method is sensitive to this profile because the shock thickness is of the order of the wavelength of visible light.

In application, the light beam passes through the shock tube at a chosen angle to the tube axis so that, as the shock propagates, it reflects light from the beam. The intensity of the reflected light, which is about 10^{-6} of that incident, is recorded by a photomultiplier. By varying the direction of the light beam R may be obtained as a function of θ.

The effectiveness of this technique has been checked by application to monatomic gases.‡ A good analytical representation of the shock profile is to take

$$\frac{\mu-\mu_0}{\mu_1-\mu_0} = 1+\exp(4z/L), \tag{52}$$

where μ_0, μ_1 are the refractive indices in the gas before and after the shock, corresponding to densities ρ_0, ρ_1 respectively and z is measured from the centre of the shock wave in the direction of its propagation. L may then be taken as the shock thickness. Fig. 17.9 shows the comparison between observed and calculated reflectivities for shock waves in argon, assuming (52) and fixing L to give the best agreement. It will be seen that the fit is quite good.

Having obtained L in this way for shock waves of different Mach number, comparison may be made with theoretical values calculated as described on p. 1479 above. To obtain good agreement it is necessary to allow for the observed variation of the viscosity with temperature.

† COWAN, G. R. and HORNIG, D. F., *J. chem. Phys.* **18** (1950) 1008; GREENE, E. F., COWAN, G. R., and HORNIG, D. F., ibid. **19** (1951) 427.
‡ GREENE, E. F. and HORNIG, D. F., ibid. **21** (1953) 617.

Fig. 17.10 shows the comparison of observed and calculated thicknesses for argon and for helium. The agreement is quite good in both cases. As the results are confined to Mach numbers less than 2 the first order transport theory seems to be quite adequate for calculating the shock thickness.

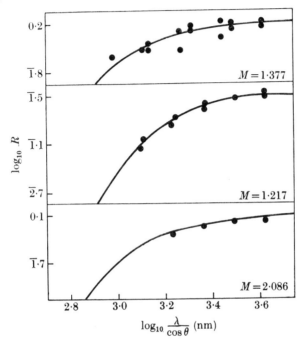

FIG. 17.9. Comparison of measured and calculated shock front reflectivities R for shock waves in argon with three different Mach numbers M, as functions of $\lambda/\cos\theta$, where λ is the wavelength of the light and θ the angle of incidence.

Applications of the technique to the study of rotational relaxation are described in § 5.1.

2.2.3. *Optical techniques for temperature measurement.* It is possible in certain cases to determine the temperature distribution directly in region b behind the shock front and hence to obtain vibrational relaxation times.

One ingenious method due to Gaydon[†] depends on the fact that sodium resonance radiation is readily quenched by the presence of a gas composed of polyatomic molecules which can take up vibrational energy. It may therefore be argued that, if the vibrational and translational temperatures of the gas are different, the excitation temperature of the

[†] CLOUSTON, J. G., GAYDON, A. G., and GLASS, I. I., *Proc. R. Soc.* A**248** (1958) 429.

sodium will be in equilibrium with the former. To apply this in practice a line-reversal technique† is used.

Light with wavelength around that of the sodium line is selected from the radiation emitted by a Pointolite source running at an adjustable

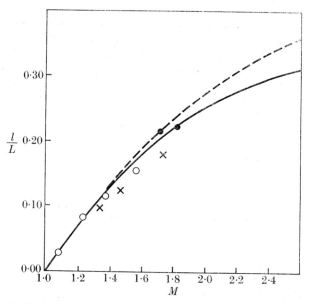

Fig. 17.10. The variation with Mach number M of the ratio l/L of mean free path to the thickness L of the shock front, for shock waves in argon and helium. Argon: —— calculated; ○ observed by Andersen and Hornig (reflectivity method); × observed by Talbot and Sherman (temperature measurement); Helium: – – – calculated; ● observed by Sherman (temperature measurement). For references see Hornig, D. F.‡

temperature in the range 2000–3000 °K, and passed through a window in the shock tube to be received on a photomultiplier whose output is recorded by a cathode-ray oscillograph responsive only to transient signals. If the shocked gas contain sodium vapour such signals will be recorded as the shock front passes across the light beam. Passage of a region which is at a vibrational temperature lower than the temperature of the Pointolite source will give rise to reduced signals due to absorption. The reverse will be the case if the vibrational temperature is higher.

Applied to shocks in nitrogen with Mach number between 5 and 7, this method showed low vibrational temperatures near the shock front. Some typical records are reproduced in Fig. 17.11. At a Pointolite

† GAYDON, A. G. and WOLFHARD, H. G., *Flames, their structure, radiation and temperature* (Chapman and Hall, London, 1933).
‡ HORNIG, D. F., *J. Phys. Chem.* **61** (1957) 856.

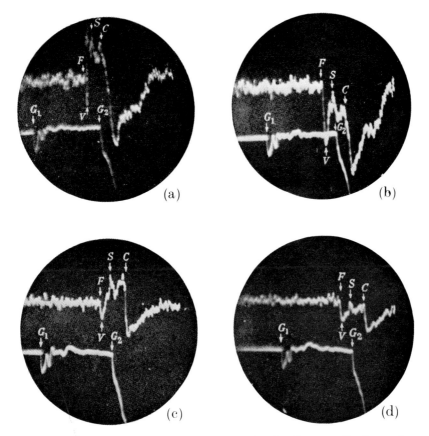

FIG. 17.11. Reproductions of oscillograms taken by Clouston, Gaydon, and Glass showing time-reversal temperatures behind shock waves in nitrogen. Lower traces are temperature records from two gold films at different points along the shock tube, from which the shock-wave velocity is determined. Upper traces are photomultiplier signals recording the intensity of light from the source after passage through the shock wave. Upward deflexion denotes emission and downward absorption. F denotes the shock front, V the absorption dip due to vibrational lag in N_2, and C the commencement of temperature fall at a contact surface.

(a) Mach number $M = 6.60$; temperature) T_B) of source $= 2400\,°K$; calculated equilibrium temperature $(T_e) = 2495\,°K$.
(b) $M = 6.42$; $T_B = 2405\,°K$; $T_e = 2390\,°K$.
(c) $M = 6.32$; $T_B = 2200\,°K$; $T_e = 2330\,°K$.
(d) $M = 6.08$; $T_B = 2200\,°K$; $T_e = 2190\,°K$.

Note how the absorption dip V broadens as the equilibrium temperature falls.

temperature of 2450 °K the half-life of the absorption signal was about 2×10^{-5} s. From such observations the relaxation time at the pressure concerned may be calculated, allowing for the difference in time scale between the flowing gas and observations from a fixed viewing point. It comes out to be 8×10^{-4} s at atmospheric pressure. At 2250 °K this was found to increase to $1\cdot 6\times 10^{-3}$ s.

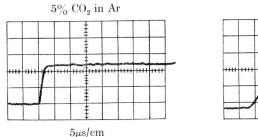
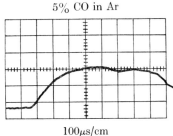

FIG. 17.12. Reproductions of records showing the infra-red emission near 2400 cm⁻¹ from a shocked mixture (a) of 5 per cent CO_2 in Ar, (b) of 5 per cent CO in Ar. Time increases to the right, emission increases vertically.

A second quite different method, introduced by Windsor, Davidson, and Taylor† is suitable for application to molecular gases that are infrared active. It consists in measuring the intensity of infra-red emission at points behind the shock front. By selecting the wavelength range it is possible to study processes involving different vibrational levels.

In experiments carried out by Hooker and Millikan‡ the observation section of the shock tube used windows of CaF_2 or CsBr. By means of a set of CaF_2 lenses, light emitted from the centre of the shock tube could be focused on to a p-type, gold-doped-germanium, infra-red detector cooled with liquid nitrogen. The spatial resolution of the optical beam was set at a width of about 2 mm by an aperture stop on the field lens. Time resolution was better than 2 μs as checked by observations of the emission from strong shocks in mixtures of 5 per cent CO_2 with argon, under conditions for which the vibrational relaxation time was less than 1 μs. This is illustrated in the record reproduced in Fig. 17.12 (a), taken with infra-red radiation with wave number near 2400 cm⁻¹. Fig. 17.12 (b) shows by contrast a slow rise of emission. In this case the wave number is close to that for the transition from the first vibrational state to the ground state, observed for a mixture of

† WINDSOR, M. W., DAVIDSON, N., and TAYLOR, R. L., *Seventh Symposium on Combustion*, p. 80 (Butterworth, London, 1959).
‡ HOOKER, W. J. and MILLIKAN, R. C., *J. chem. Phys.* **38** (1963) 214.

5 per cent CO in argon. The rise to a flat maximum is due to vibrational relaxation and the drop at later times is due to the onset of turbulence. By suitable choice of initial conditions this onset may be delayed until the upper limit to the time of observation is determined by the arrival of the composition boundary between the driver gas and the shocked gas.

Measurements were also made using radiation with wave number near 4200 cm^{-1} which includes that due to the first overtone emission, i.e. that between the second vibrational state and the ground state.

In analysing data of this kind[†] it must be realized that the spectral resolution is not good enough to exclude transitions involving the same change Δv of vibrational quantum number v between higher states. Thus all transitions with $\Delta v = 1$ will contribute to the observed fundamental emission, all with $\Delta v = 2$ to the first overtone and so on.

It has been shown by Montroll and Shuler[‡] that the process of vibrational relaxation from an initial Boltzmann distribution proceeds through a continuous series of Boltzmann distributions provided collisional excitation occurs in one-quantum jumps only. This being so we may associate a vibration temperature T with each stage of the process (see p. 1558).

Consider a diatomic molecule composed of atoms of mass M_1, M_2 respectively. Let r_e be the equilibrium separation and ν the vibration frequency. The potential energy curve for the ground electronic state can be expanded about r_e in the form

$$V(\xi) = 2\pi^2\nu^2 M r_e^2 (\xi^2 + a_1\xi^3 + ...), \qquad (53)$$

where $M = M_1 M_2/(M_1+M_2)$ and $\xi = (r-r_e)/r_e$. According to the theory of emission of radiation from molecular states the total intensity that will be observed as emission of the fundamental band when the vibration temperature is T will be proportional to[§]

$$I^{(1)} = \tfrac{1}{2}\gamma^2(\mu')^2 \sum_{v=0}^{\infty} v n_v, \qquad (54)$$

where $\gamma^2 = h/4\pi^2 M\nu r_e^2$ and $\mu' = d\mu/d\xi$, $\mu(\xi)$ being the electric dipole moment of the molecule. n_v is the number of molecules per cm^3 in the vth vibrational state. Thus

$$n_v = n\exp(-hv\nu/\kappa T)\{1-\exp(-h\nu/\kappa T)\}, \qquad (55)$$

n being the total number of molecules per cm^3. Writing $\exp(-h\nu/\kappa T) = \rho$,

$$I^{(1)} = \tfrac{1}{2}\gamma^2(\mu')^2 n\rho/(1-\rho). \qquad (56)$$

Similarly the intensity emitted in the first overtone is proportional to

$$I^{(2)} = \tfrac{1}{8}\gamma^4(\mu''+a_1\mu')^2 \sum_{v=0}^{\infty} v(v-1)n_v, \qquad (57)$$

[†] Decius, J. C., J. chem. Phys. **32** (1960) 1262.
[‡] Montroll, E. W. and Shuler, K. E., ibid. **26** (1957) 454.
[§] Crawford, B. L. and Dinsmore, H. L., ibid. **18** (1950) 983.

where $\mu'' = \mathrm{d}^2\mu/\mathrm{d}\xi^2$ and a_1 is defined by (53). Using (55) this gives
$$I^{(2)} = \tfrac{1}{8}\gamma^4(\mu''+a_1\mu')^2 n\rho^2/(1-\rho)^2. \tag{58}$$

The dependence of ρ on the time t can be expressed in terms of the relaxation equation
$$E-E_\infty = (E_0-E_\infty)\exp(-t/\tau), \tag{59}$$
in which E is the total vibrational energy at the time t when the temperature is T, E_0 that initially and E_∞ that in the final equilibrium state, and τ is the vibrational relaxation time. Thus
$$E = nh\nu\rho/(1-\rho). \tag{60}$$

From (56), (59), and (60) we have
$$I^{(1)} - I^{(1)}_\infty = (I^{(1)}_0 - I^{(1)}_\infty)\exp(-t/\tau), \tag{61}$$
and from (58), (59), and (60)
$$\{I^{(2)}\}^{\frac{1}{2}} - \{I^{(2)}_\infty\}^{\frac{1}{2}} = \{(I^{(2)}_0)^{\frac{1}{2}} - (I^{(2)}_\infty)^{\frac{1}{2}}\}\exp(-t/\tau). \tag{62}$$

In both cases I_0 denotes the initial and I_∞ the final equilibrium intensity. Hence
$$\ln\{1 - I^{(1)}/I^{(1)}_\infty\} = \alpha^{(1)} - t/\tau, \tag{63 a}$$
where
$$\alpha^{(1)} = \ln\{1 - (I^{(1)}_0/I^{(1)}_\infty)\}, \tag{63 b}$$
and
$$\ln[1 - \{I^{(2)}/I^{(2)}_\infty\}^{\frac{1}{2}}] = \alpha^{(2)} - t/\tau, \tag{64 a}$$
where
$$\alpha^{(2)} = \ln\{1 - (I^{(2)}_0/I^{(2)}_\infty)^{\frac{1}{2}}\}. \tag{64 b}$$

Thus, if for the fundamental emission $\ln\{1-I/I_\infty\}$ is plotted against t a straight line plot should result with slope τ^{-1}. A linear plot with the same slope should also result if, for the overtone emission, $\ln\{1-(I/I_\infty)^{\frac{1}{2}}\}$ is plotted against t. Fig. 17.13 illustrates results obtained by Hooker and Millikan† plotted in this way. The measurements of the fundamental and overtone emissions were made on separate shocks of nearly the same velocity and hence the same final equilibrium temperature T. Good linear plots were obtained for both emissions and the respective relaxation times, 172 and 190 μs, obtained from the plots are equal within experimental error.

Measurements of relaxation times were made for pure CO and for mixtures of CO with N_2, Ar, and H_2. The results obtained are discussed in § 3.

2.3. *The effect of persistence of vibration in gas dynamics*

In the flow of compressible fluids about obstacles, compressions and rarefactions occur which are accompanied by temperature changes. The rate of change of temperature is determined by the velocity of flow and the scale of the obstacles. If the rates of change are faster than, or comparable with, the rate of vibrational deactivation, the vibrational degrees of freedom will not come to thermodynamical equilibrium with

† loc. cit., p. 1489.

the other degrees of freedom. The transfer of energy between the vibrational and the other degrees of freedom will therefore take place by an irreversible process leading to an increase of entropy.

Such effects may give rise to additional loss in high-speed turbines at high temperatures, which may be comparable with that due to skin friction. In wind-tunnel and similar tests it is often convenient to substitute for the working gas another one which may be used more

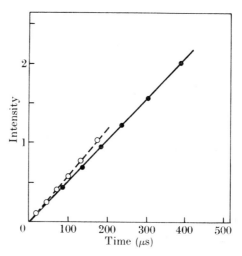

FIG. 17.13. Plots of emission intensity I against time for the fundamental and first overtone of shock-excited CO.
–O–O–O–O $\ln\{1-(I/I_\infty)\}$ for fundamental emission.
–●–●–●–● $\ln\{1-(I/I_\infty)^{\frac{1}{2}}\}$ for overtone emission.

conveniently at a given Mach or Reynolds number. Care must be taken in adopting such a procedure because the substitute gas may exhibit a substantially different vibrational heat-lag from that of the actual gas, resulting in a different behaviour under the flow conditions. A detailed investigation of these effects has been carried out by Kantrowitz.†

The experimental method introduced by Kantrowitz utilizes the arrangement illustrated in principle in Fig. 17.14. Gas enters the chamber A in which it settles at pressure p_0 and temperature T_0. It expands adiabatically to pressure p_1 and temperature T_1 through a faired orifice O of suitable design. Gas that is flowing along the axial stream line of an impact tube I is brought to rest at the nose during which its pressure rises to p_2 and temperature to T_2. If this process is also sufficiently slow to be adiabatic, the molar entropy and energy of

† KANTROWITZ, A., J. chem. Phys. **10** (1942) 145; **14** (1946) 150.

the gas which has reached equilibrium at the nose of the impact tube will be the same as the corresponding quantities in the chamber A. Hence $p_2 = p_0$ and the reading of the alcohol manometer will be zero. We now consider circumstances in which the expansion through the faired orifice O is still so slow that it may be followed by all degrees of

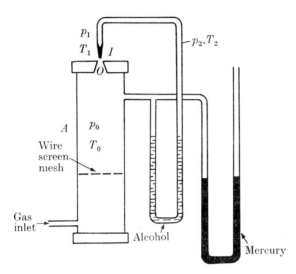

FIG. 17.14. Illustrating the principle of the apparatus used by Kantrowitz to study the lag in vibrational deactivation in a flowing gas.

freedom, but the compression at the nose of the impact tube is so fast that the vibrational degrees of freedom do not contribute to the heat changes.

Let c_p be the specific heat per unit mass of gas at constant pressure, c_p' that due to all but the vibrational degrees of freedom. Then, if u is the velocity of flow, we must have for the slow expansion

$$c_p T + \tfrac{1}{2} u^2 = \text{constant}, \tag{65}$$

so
$$c_p T_0 = c_p T_1 + \tfrac{1}{2} u_1^2,$$

where u_1 is the velocity on leaving the faired orifice. For the compression at the nose we have, on the other hand,

$$c_p' T + \tfrac{1}{2} u^2 = \text{constant}, \tag{66}$$

so
$$c_p' T_2 = c_p' T_1 + \tfrac{1}{2} u_1^2.$$

Hence
$$c_p (T_0 - T_1) = c_p' (T_2 - T_1).$$

Also, for the separate adiabatic processes we have

$$(p_2/p_1)^{R/c_p'} = T_2/T_1, \qquad (p_0/p_1)^{R/c_p} = T_0/T_1,$$

where R is the gas constant. Eliminating T_2 we find

$$p_0/p_2 = \left\{\frac{c'_p}{c_p-(c_p-c'_p)T_1/T_0}\right\}^{c'_p/R} (T_0/T_1)^{(c_p-c'_p)/R}. \tag{67}$$

This relation was checked in CO_2 by choosing the dimensions of orifice and impact tube so that the time of expansion was of order 10^{-4} s and that of compression about 100 times shorter. The orifice was a hole in

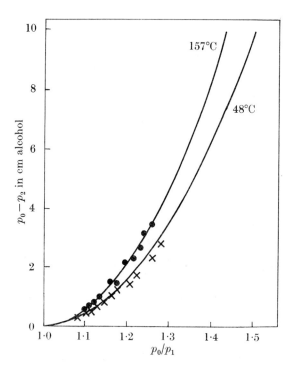

FIG. 17.15. Comparison of calculated pressures with those observed by Kantrowitz in CO_2 using the apparatus illustrated in Fig. 17.14. × experimental points (48 °C); ● experimental points (157 °C); ——— calculated.

a $\frac{1}{2}$-in plate, the glass impact tube of 0·005 in diameter, and the chamber pressure such that the gas velocity at the orifice was between 300 and 600 ft s^{-1}. As the relaxation time for CO_2 is about 6×10^{-6} s the conditions assumed in the equations (65) and (66) are both well satisfied. Fig. 17.15 illustrates the comparison between the observed values of p_0-p_2 for different values of p_0/p_1 at two temperatures T_0, 48 and 157 °C. The agreement is very satisfactory. It was also checked that no pressure difference occurred when the CO_2 was replaced by nitrogen.

To extend these considerations to determine relaxation times by choosing comparable compression times it is necessary to develop a more general theory of the entropy increase which accompanies flow.

Let $u(x, y, z)$ be the speed at some point of the fluid in a state of steady flow and ϵ the energy per unit mass in the vibrational degrees of freedom in excess of that for equilibrium partition at the translational temperature T. We wish to calculate the entropy increase ΔS in proceeding along a stream line with a given element of fluid from a point at time t_0 to another at time t.

The rate of heat flow per unit mass from the vibrational excitation will be proportional to ϵ, so that the rate of entropy change due to this flow will be given by

$$\frac{dS}{dt} = k\epsilon \left(\frac{1}{T} - \frac{1}{T_{\text{vib}}} \right), \tag{68}$$

where T_{vib}, the temperature of the vibration, is given by

$$\epsilon = c_{\text{vib}}(T_{\text{vib}} - T), \tag{69}$$

c_{vib} being the vibrational specific heat $c_p - c_p'$. We now have, eliminating T_{vib},

$$\frac{dS}{dt} = k\epsilon \left(\frac{1}{T} - \frac{1}{T+T_1} \right), \tag{70}$$

where

$$T_1 = \epsilon/c_{\text{vib}}.$$

Hence

$$\Delta S = \int_{t_0}^{t} k\epsilon \left(\frac{1}{T} - \frac{1}{T+T_1} \right) dt. \tag{71}$$

If the flow is such that p and T are nearly equal to their ambient values, k is practically constant throughout and

$$\Delta S \simeq \frac{k}{c_{\text{vib}} \overline{T}^2} \int_{t_0}^{t} \epsilon^2 \, dt, \tag{72}$$

$\overline{T}$ being the mean temperature. It remains to identify k and determine ϵ^2 in terms of the flow pattern.

Since along a stream line $c_p T + \tfrac{1}{2} u^2 + \epsilon$ will be constant we have

$$\frac{d\epsilon}{dt} + c_p \frac{dT}{dt} = -\frac{1}{2} \frac{du^2}{dt}. \tag{73}$$

To eliminate dT/dt we note that, if q_{vib} is the heat content of the vibration,

$$\frac{dq_{\text{vib}}}{dt} = -k\epsilon \quad \text{and} \quad \epsilon = q_{\text{vib}} - c_{\text{vib}} T,$$

so

$$c_{\text{vib}} \frac{dT}{dt} = -\frac{d\epsilon}{dt} - k\epsilon. \tag{74}$$

This gives for ϵ the first-order differential equation

$$\frac{c_p'}{c_{\text{vib}}} \frac{d\epsilon}{dt} + \frac{kc_p}{c_{\text{vib}}} \epsilon = \frac{1}{2} \frac{du^2}{dt}, \tag{75}$$

from which ϵ may be determined, provided we assume that the disturbance of the flow pattern by the vibrational heat lag is small so that u is given from the appropriate standard hydrodynamical solution.

k may be related to the relaxation time τ by noting that, if there were no flow, we would have, from (75),

$$\frac{d\epsilon}{dt} = -k\epsilon c_p/c_p'.$$

This shows that kc_p/c_p' is to be taken as $1/\tau$.

Kantrowitz applied this analysis to the flow through a faired orifice and to the compression at an impact tube to obtain the entropy change ΔS when the time of compression is comparable with the relaxation time. Knowing ΔS the pressure change may be calculated from the relation

$$S = R\ln(p_0/p_2).$$

From a series of measurements of p_0/p_2 in CO_2 Kantrowitz obtained a relaxation time corresponding to a probability of vibrational deactivation of about 1/30 000 per collision at a temperature of 306 °K. This agrees well with the values derived from ultrasonic dispersion and absorption measurements (see Fig. 17.33). The reality of the effect in gas dynamics was thus established and an alternative method of obtaining vibrational deactivation probabilities became available.

2.4. *Spectroscopic methods for measurement of vibrational relaxation time for molecules in ground electronic states*

The first spectroscopic experiments that yielded semi-quantitative information about the rate of collisional deactivation of low-lying vibrational levels of the ground state of a diatomic molecule were those of Dwyer.[†] He studied the band absorption of iodine in the visible region. Under normal conditions, at ordinary temperatures, the strongest transitions in these bands are from the ground vibrational states. When the iodine is excited by electron impact the proportion of molecules in the first vibrational state is correspondingly increased. This is because downward transitions from electronically excited molecules often end in this level rather than the ground level. Observations of the band absorption during the discharge revealed this effect clearly. When the discharge was cut off it was found to persist for at least one-thirtieth of a second, showing that the excited vibration was able to survive over 7000 collisions.

Another early experiment was carried out by White[‡] in which an intense electric discharge was passed through cyanogen gas. This produced CN radicals strongly populated in the first excited vibrational

[†] Dwyer, R. J., *J. chem. Phys.* **7** (1939) 40.
[‡] White, J. U., ibid. **8** (1940) 79.

state through electronic transitions from higher levels. The half-life of the excited state was measured by absorption spectroscopy as 3 ms during which time some thousands of collisions occurred.

Recently it has been possible to observe the quenching of resonance fluorescence arising from resonant excitation of a low-lying vibrational level while the decay of vibrational excitation produced by flash photolysis and by laser pulses has been studied by infra-red emission or absorption spectroscopy. We now describe some experiments of this kind which have proved very fruitful.

2.4.1. *The quenching of infra-red fluorescence.* Suppose that, by some exciting agency of constant strength, a number R of molecules A is raised per second to an excited vibrational state so that at any instant there are n'_A excited molecules present. If τ_r is the lifetime of these molecules towards emission of radiation, then the intensity of radiation emitted per second will be proportional to n'_A/τ_r.

We suppose now that the conditions are such that this radiation does not suffer absorption before emergence from the gas to an observation chamber. This requires that the molecules A are present at a sufficiently low pressure. Otherwise they will absorb much of the radiation by a resonance process and re-emit it one or more times so that imprisonment of the radiation occurs and the emergent intensity is not proportional simply to n'_A/τ_r.

If the excited atoms lose their excitation solely by radiation then, in equilibrium,
$$n'_A/\tau_r = R. \tag{76}$$

On the other hand, if a foreign gas of molecules B, which are capable of removing the vibrational excitation of a molecule A in a collision at a gas-kinetic velocity, is present, a number $n''_A Z$ of excited molecules will be deactivated per second by these collisions. Z is the number of deactivating collisions made per second by an excited molecule A and is given in terms of the effective collision cross-section Q for the deactivation by the usual gas-kinetic formula

$$Z = 2n_B Q\{2\kappa T(M_A+M_B)/\pi M_A M_B\}^{\frac{1}{2}}, \tag{77}$$

where n_B is the number of molecules B/cm³, T is the absolute temperature, and M_A, M_B are the respective masses of the atoms A and molecules B. We may write $Z = 1/\tau_c$, where τ_c is the mean lifetime towards deactivation by collision. We now have, for equilibrium,

$$R = n''_A(Z+1/\tau_r), \tag{78}$$

provided the strength of the exciting agency is unchanged, and the addition of the foreign gas does not alter its effectiveness in producing excited atoms, i.e. by producing line broadening. Comparing (78) and (76) it will be seen that the foreign gas decreases the intensity of emitted radiation in the ratio

$$n''_A/n'_A = 1/(1+\tau_r Z) = (1+\tau_r/\tau_c)^{-1}. \tag{79}$$

Measurement of this quantity therefore gives $Z\tau_r = \tau_r/\tau_c$. The radiative lifetime τ_r may be measured by suitable experiments so that Z and hence Q may be derived.

In practice the method is not easy to apply. It is, of course, necessary that the molecule be infra-red active but, for a molecule such as CO, the mean radiative lifetime of the first excited vibrational state is as long as 0·033 s as determined from optical absorption studies (the wavelength of the resonance radiation is 4·67 μm). To observe resonance fluorescence it is necessary that the lifetime towards collisional deactivation should not be much greater than for radiation. At atmospheric pressure this requires that a vibrationally excited molecule should be able to survive 10^8 collisions or so. Polyatomic molecules are very effective in producing vibrational deactivation and their fractional concentration must be kept to less than 1 part per million. Obviously the carbon monoxide needs to be very pure for the experiment to succeed. Nevertheless, Millikan† has, using a flow method, been able to observe vibrational fluorescence for this gas and has studied its quenching by addition of various gases.

Fig. 17.16 illustrates the gas handling arrangement. The carbon monoxide entering the system from a cylinder was Matheson chemically pure grade, containing 200 parts per million of CO_2 as the main polyatomic impurity. Further purification was carried out by passage through the two cold traps T_1 and T_2 shown in Fig. 17.16. Each trap was made in the form of a copper helix 10 ft long and 4 in diameter, the tubing being of ½ in diameter. In T_1 the coil was packed with ⅛-in pellets of aluminium and in T_2 with copper wool. Before operation each trap was cleaned by flame heating while being purged with nitrogen. T_1 was then immersed in a bath of dry ice and T_2 in one of liquid oxygen. The aluminium in T_1 was necessary to remove some impurity that otherwise passed through, even when the trap was cooled to liquid nitrogen temperature. It is important to avoid deactivation of vibrationally excited molecules by collisions with a containing wall so that an 'open-air' flow system was used. The purified carbon monoxide was contained in an

† MILLIKAN, R. C., J. chem. Phys. 38 (1963) 2855.

annular flow of inert gas, either nitrogen or oxygen, moving with the same velocity. For this purpose a porous plug 'burner' of the type used in flame experiments was employed (see Chap. 18, § 2.1.1). The carbon monoxide issued from the central porous disc (3·5 cm diameter) and the containing flow of inert gas from an annular porous ring 1·2 cm wide. Flow speeds were in the range 5–20 cm s^{-1}.

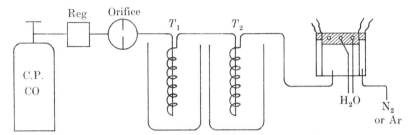

Fig. 17.16. Gas-handling arrangement in Millikan's experiments on the quenching of infra-red fluorescence of CO.

The optical arrangement in the later version of this experiment is illustrated in plane view in Fig. 17.17. Rich methane–oxygen flames were used as sources of the infra-red radiation and fluorescence was looked for in the same plane around 90° from the direction of incidence. The spectrometer was a Perkin–Elmer 12G prism-grating unit with thermocouple detector. Calibration of the detector sensitivity was carried out by turning off the light sources and heating the flowing gas. A rise of 2 °C in temperature gave a signal from pure CO of the same strength as the background. The water cooled light trap (see Fig. 17.17) was introduced to reduce background emission.

The first experiments were designed to detect fluorescence. With T_1 cooled to -80 °C with dry ice and T_2 to -154 °C with freezing isohexane, a signal equivalent to 70 °C rise in gas temperature was obtained with the exciting lights switched on. When T_2 was allowed to warm up to -120 °C the signal disappeared, to be restored again on cooling once more to -154 °C. At -120 °C the vapour pressure of CO_2 is high enough to introduce as much as 1 per cent of CO_2 into the flowing CO. There seems little doubt that the emission signal was really due to fluorescence which, with T_2 allowed to warm up, was quenched by vibrational deactivation by CO_2. It is difficult to see how impurities could have such a pronounced effect if the emission signal were due to thermal radiation or scattered light. Addition of as little as 4 p.p.m. of water vapour to pure CO reduced the fluorescence signal by half.

The radiative lifetime τ_r has been determined† from absorption experiments as $0·033\pm0·001$ s. It is possible in principle to determine the lifetime by observing the decay of fluorescence as one proceeds downstream from the exciting source. With pure CO the time constant for decay came out to be four times longer than that derived from the radiation lifetime. This discrepancy was traced to imprisonment of the

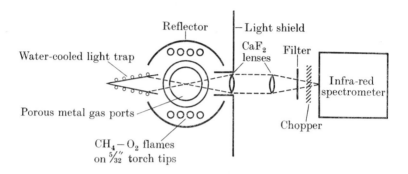

Top view

Fig. 17.17. Plane (top) view showing the optical arrangement in Millikan's experiments (see Fig. 17.16).

resonance radiation. Thus, when the CO was diluted with argon, the variation of intensity with distance downstream from the source became more nearly exponential and the decay time closer to that expected. It was not possible to continue the dilution until agreement was obtained because the signal-to-noise ratio increased with dilution. Fig. 17.18 illustrates the observed variation of the time constant for decay with the proportion of CO in a CO–Ar mixture.

With pure CO and trapped radiation it was possible to detect a signal as far as 4 cm downstream from the light source. This means that the fluorescence persists for as much as $0·2$ s, during which time a vibrationally excited molecule will have suffered as many as 10^9 collisions.

For the quenching experiments it was necessary to use very pure gases, subjected to the same purifying procedures as the CO. Provided the emitting gas is optically thin the analysis of p. 1497 may be applied. Thus if I_0 is the fluorescent intensity in the absence of quenching gas and I that when the gas is present at a partial pressure p atmospheres,

$$I_0/I = (1+\tau_r/\tau_c\,p), \tag{80}$$

† BENEDICT, W. S., HERMAN, R., MOORE, G. E., and SILVERMAN, S., *Astrophys. J.* **135** (1962) 277.

where τ_r is the radiative lifetime and τ_c the vibrational relaxation time due to the presence of the foreign gas at a pressure of 1 atmosphere. If the emitting gas is optically thin the intensity of emission from a mixed CO–Ar stream should vary linearly with the CO content. Furthermore, if I_0/I is plotted as a function of p for a particular quenching gas the slope and intercept on the axis of zero p should not depend on the CO content. Fig. 17.19 illustrates results of this kind obtained with

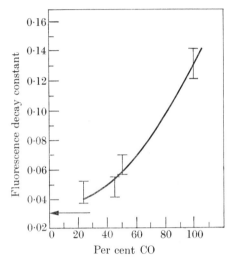

Fig. 17.18. Observed variation of the time constant for decay of the $1 \rightarrow 0$ infra-red fluorescence with the proportion of CO in a CO–Ar mixture. The radiative lifetime for the $1 \rightarrow 0$ transition is indicated by ←.

helium as quenching gas in a mixture with CO and Ar. It will be seen that the same linear plots are obtained at CO concentrations of 0·34 and 0·5 per cent but, at higher concentrations, different lines are obtained. Hence, in all the quenching measurements a mixture of 0·34 per cent CO in Ar was used as fluorescing gas. Results obtained for the times τ_c for different quenching gases are discussed in § 3.

A special experiment was carried out to compare the effectiveness of ortho- and para-hydrogen as quenching gases. The quenching by normal hydrogen was compared with that by hydrogen which was prepared by boiling from liquid H_2 and shown to be 97 ± 2 per cent para-hydrogen. Tests were carried out to show that the para-hydrogen was not converted to normal hydrogen in the apparatus. Fig. 17.20 shows the linear plots of the intensity ratios I_0/I for the two gases. A striking difference was found, the reality of which was further checked by catalytic conversion of the para-hydrogen to normal hydrogen, whereupon the same results were obtained as for cylinder hydrogen. From the data, relaxation times τ_c of 6.5×10^{-5} and 1.43×10^{-4} s were obtained for collisional deactivation by pure para- and pure ortho-hydrogen respectively (see also p. 1532).

2.4.2. *The use of flash spectroscopy.* A technique similar to flash photolysis may be used to determine the time rate of decay of intensity of a vibrationally excited species in the presence of different gases. The

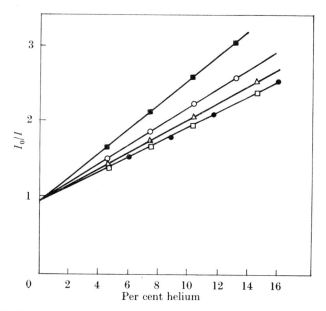

FIG. 17.19. Plots of the ratio I_0/I of the unquenched to quenched infra-red fluorescent intensity as a function of the helium percentage in a CO–Ar–He mixture. The different plots refer to the different CO percentages as follows: ■ 6 per cent, ○ 2·3 per cent, △ 1 per cent, □ 0·5 per cent, ● 0·34 per cent.

gas under investigation is subjected to an intense flash of light of some μs duration, with energy of order 1000 J or more, which produces the excitation. Observations may be made by absorption spectroscopy at various delay times of the order 10^{-5} to 10^{-3} s after the flash. By plate photometry the variation of intensity of a particular absorption line with delay time may be determined.

Fig. 17.21 illustrates the general arrangement of a typical experiment of this type. The gas under investigation is contained in a quartz tube about 50 cm long and 2 cm diameter. The flash lamp consists of a similar tube filled with krypton which is placed alongside. To produce the flash a condenser bank is discharged between two tungsten electrodes.

Photography of the absorption spectrum of the gas at a selected short interval after flashing is carried out by triggering a less energetic flash in a second flash tube. The light from this flash, after passing through the gas under study, enters a spectrograph and its spectrum is photographed.

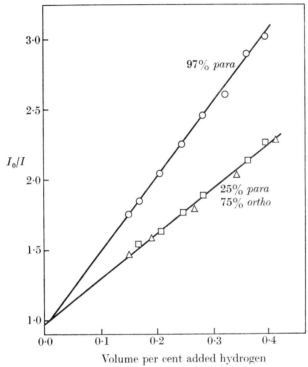

FIG. 17.20. Plots of the ratio I_0/I of the unquenched to quenched fluorescent intensity as a function of the hydrogen percentage in a CO–H_2 mixture at 288 °K. The two plots refer to hydrogen of different para–ortho composition as indicated.

This method has been applied by Callear† to investigate the deactivation of the first excited vibrational level of the ground ($X^2\Pi$) state of nitric oxide. The flash produces electronic excitation. Radiative transitions from the excited electronic states to the ground electronic state then populate particularly the first vibrational level of that state. Fig. 17.22 reproduces some photographs showing the variation of intensity of the 1–2 absorption band of NO ($X^2\Pi$) with delay time after the flash, for three different gas mixtures. Results obtained for

† CALLEAR, A. B., *Applied optics*, Supp. 2 of *Chemical lasers* (1965), p. 145.

vibrational relaxation times and corresponding probabilities of deactivation per gas kinetic collision, by photometry of plates such as are shown in Fig. 17.22, are given in Table 17.2 and discussed in § 3.1.

Lipscomb, Norrish, and Thrush,† in the course of a study of the flash photolysis of chlorine dioxide and of nitrogen dioxide, in a large excess of inert gas, observed the formation of O_2 molecules in the ground electronic state but possessing up to eight quanta of vibrational energy.

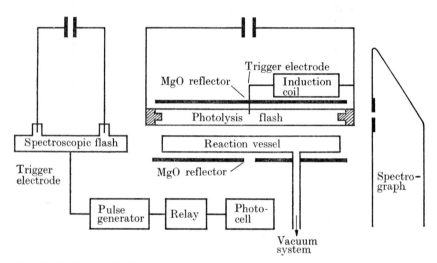

FIG. 17.21. Schematic diagram of a flash photolysis apparatus (Norrish and Thrush).

They were able to estimate the efficiencies of various molecules in deactivating O_2 from the sixth to the fifth vibrational level. These results are discussed in § 3.2.

2.5. *Spectroscopic methods for measurement of vibrational relaxation times for molecules in excited electronic states*

The observation of collisional deactivation of vibrational levels associated with molecules in excited electronic states offers considerable further scope for the study of vibrational relaxation. In particular, it is comparatively easy to produce electronically excited molecules with highly excited vibration. The spacing of the vibrational levels at these excitations will normally be very small compared with κT so that deactivation can take place in very small steps and the conditions are nearly classical. Corresponding to these circumstances it is found that

† LIPSCOMB, F. J., NORRISH, R. G. W., and THRUSH, B. A., *Proc. R. Soc.* A**233** (1956) 455.

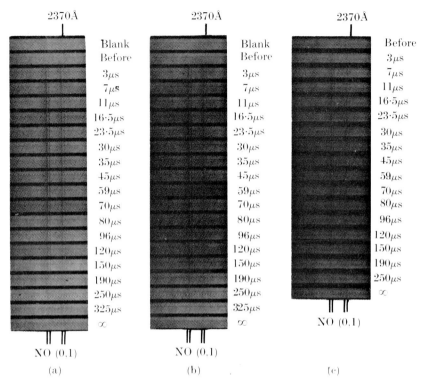

Fig. 17.22. Time-resolved absorption spectra taken after flashing three different mixtures of NO with other gases. The rise and fall of the absorption from the $v = 1$ level of the ground $X\ ^2\Pi$ state is clearly seen. (a) NO–N$_2$ partial pressures 2·2 and 430 torr. (b) NO–Kr partial pressures 2·2 and 430 torr. (c) NO–He partial pressures 2·25 and 540 torr.

vibrational deactivation takes place with ease, in contrast to the strong persistence of vibration observed, for example, from the dispersion and absorption of ultrasonic waves.

The first quantitative study was made by Roessler.† He investigated the effect of the rare gases on the sensitized band fluorescence of iodine excited by the green mercury line 5461 Å. Absorption of this line produces exclusive excitation of an electronic state in which the vibrational quantum number v is 26, so that the excited level is one in which the molecule possesses vibrational excitation equal to three-fifths of the dissociation energy of the particular electronic state.

The effect of foreign gas is not only to produce quenching, so that the total intensity of the fluorescent radiation is reduced, but is also apparent in the appearance of new bands. These are due to the collision excitation of new vibrational levels. The change of vibrational quantum number in the collision is never greater than ± 2. In addition a broadening of the lines indicated that rotation was also being excited by collision. This aspect will be discussed further in § 5.

To obtain quantitative information about the cross-sections for collisions in which the vibration is changed, Roessler measured the following intensity ratios:

(a) the ratio R_1 of the total fluorescent intensity before and after the introduction of the foreign gas at pressure p;
(b) the ratio R_2 of the intensity of the lines beginning on the level with $v = 26$ before and after introduction of the foreign gas at pressure p;
(c) the ratio R_3 of the intensity of the lines beginning on levels other than that with $v = 26$ to those beginning on that level, when foreign gas is present at pressure p.

To obtain the cross-sections for quenching and for vibrational transitions the following method was used.

Let N_0 be the number of exciting quanta absorbed per cm³ s, n_p the equilibrium concentration of molecules excited to the level with $v = 26$, and n_i the concentration of molecules in another vibrational level to which the molecule may be transferred on collision. Then, in equilibrium,

$$N_0 = n_p(p\delta + p \sum \epsilon_i + \tau_r^{-1}) - p \sum n_i \epsilon_i. \tag{81}$$

On the right-hand side n_p/τ_r represents the loss per cm³ s due to radiation, δ that due to quenching, and $pn_p \sum \epsilon_i$ that due to collision-induced vibrational changes. The term $p \sum n_i \epsilon_i$ represents the rate of repopulation per cm³ s due to collision-induced transitions back from other vibrational levels.

† ROESSLER, F., Z. Phys. 96 (1935) 251.

In the same way we may write the equilibrium relations for the gain and loss of molecules from other vibrational states:

$$pn_p \epsilon_i = n_i(p\delta + p\epsilon_i A + \tau_r^{-1}).$$

The factor A has been introduced to allow for transitions to states other than that with $v = 26$.

We now have, writing $\epsilon = \sum \epsilon_i$,

$$R_2 = \tau_r N_0/n_p = 1 + p\tau_r\delta + p\tau_r\epsilon - p^2\tau_r^2 \sum_i \epsilon_i^2/(1+p\tau_r\delta+p\tau_r A\epsilon_i) \qquad (82)$$

and
$$R_3 = n_i/n_p = p\tau_r\epsilon_i/(1+p\tau_r\delta+p\tau_r A\epsilon_i). \qquad (83)$$

If it is assumed that ϵ_i is independent of the level concerned, so that we may write

$$\epsilon = \sum \epsilon_i = z\epsilon_i,$$

where z is the number of occupied levels, we may eliminate A to give

$$R_2 - 1 - p\tau_r\delta = p\tau_r\epsilon(1-R_3). \qquad (84)$$

To correct for the variation of ϵ_i with the particular vibrational level, Roessler replaced $1-R_3$ by $1-BR_3$, where B is a correcting factor to be determined. By comparison with an exact treatment with helium as foreign gas, a value 0·85 was found for B, practically independent of the relative importance of transitions involving changes ± 1, ± 2, in the quantum number v. It was therefore taken as 0·85 for all other gases.

Having measured R_2 and R_3 and taking for the radiative lifetime $\tau_r = 10^{-8}$ s the relation (84), with the correcting factor B, gives ϵ in terms of δ. The measurement of R_1 gives δ directly (see the discussion of quenching in § 2.4.1) so ϵ may be derived.

From δ and ϵ the respective cross-sections may be obtained in the usual way. The absolute values are not likely to be very accurate, but there is no doubt that the cross-sections for collisions involving vibrational transfer are in all cases comparable with the gas-kinetic. In fact the actual values found are from 5 to 10 times larger. This is in sharp contrast to the results obtained from the dispersion and absorption of sound in which large vibrational quanta are involved and to the spectroscopic observations of Dwyer† in the deactivation of low vibrational levels of the ground electronic state of iodine (see p. 1496).

A similar investigation was carried out by Durand‡ who studied the effect of the rare gases on the fluorescence of sulphur excited by a magnesium spark line to the vibrational level $v = 8$ in the first excited electronic state. In this level of the S_2 molecule the energy of vibration is about one-quarter of the dissociation energy. Collisions leading to transitions to the level $v = 10$ produced quenching by pre-dissociation which could be separately distinguished in the spectrograms from transitions to $v = 7$ or 9 which led to new bands. The results agree qualitatively

† loc. cit., p. 1496.
‡ DURAND, E., *J. chem. Phys.* 8 (1940) 46.

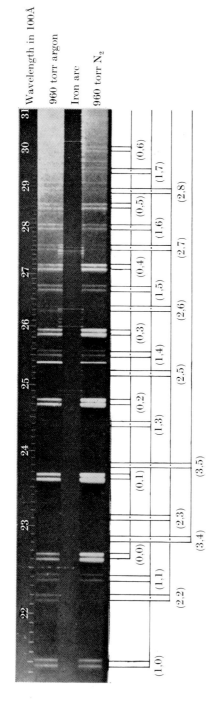

FIG. 17.23. Spectra illustrating the effect of Ar and N₂ on the fluorescence of the γ bands of NO. NO pressure 1 torr. The $v' = 0, 1, 2,$ and 3 progressions are indicated.

with those of Roessler, though the cross-sections are somewhat smaller. This may be because the vibrational quanta concerned are rather larger but the quantitative accuracy of the method may not be high enough for the difference to be significant.

Callear[†] has investigated the effect of nitrogen in producing deactivation of low-lying vibrational states associated with the excited $A^2\Sigma^+$ state of nitric oxide. This state is the upper state for the γ-bands that arise in transitions to the ground $X^2\Pi$ state. The study of the quenching of resonance fluorescence of the γ-bands is complicated by pressure broadening. To overcome these difficulties advantage was taken of the observed fact that argon and nitrogen are equally effective in producing broadening.

Fig. 17.23 illustrates the effect of N_2 on the fluorescence of NO. It is clear, by comparing spectra in N_2 and in Ar, that the N_2 suppresses the γ ($v' > 1$) progressions and enhances the γ ($v' = 0$) progression. The former effect is due to vibrational deactivation of NO ($A^2\Sigma, v' > 1$) by the N_2 but the enhancement of γ ($v' = 0$) arises not only by population of $v' = 0$ from $v' = 1$ lines by collision but also because, with the illumination employed (the source was a high-pressure xenon arc emitting a continuum down to about 1850 Å), the $C^2\Pi$ state is also excited. Collisional deactivation of this state also populates selectively the $v' = 0$ level of $A^2\Sigma^+$.

To separate these effects an ammonia filter was introduced in the exciting beam. This cuts off below 2200 Å and so limits excitation to $A^2\Sigma^+$ ($v = 0$). The enhancement of the γ ($v' = 0$) progression when the ammonia filter is removed is a measure of the contribution to the population of $A^2\Sigma^+$ ($v = 0$) by processes other than direct light absorption.

Fig. 17.24 illustrates the observed enhancements for mixtures of 0·6 torr NO with N_2 and Ar as functions of the pressure of the main gas. It will be seen that argon has only a very small effect up to pressures of 1000 torr, the probability of deactivation of $A^2\Sigma^+$ ($v = 1$) per collision with an argon atom being certainly less than 2×10^{-4}. On the other hand, it will be seen from Fig. 17.24 that N_2 has a marked enhancing effect. At low pressures the steep rise with N_2 pressure is due to quenching of $C^2\Pi$ and at higher pressures to vibrational relaxation from upper vibrational levels of $A^2\Sigma^+$.

Having established that argon does not produce appreciable vibrational relaxation, the effect of N_2 was determined by observing the intensity of the various γ-progressions in a mixture of Ar and N_2 as a

[†] CALLEAR, A. B., loc. cit., p. 1503.

function of composition, keeping the total pressure constant. The behaviour of levels with v' ranging from 0 to 3 was observed. In each case, filters were employed to prevent excitation of levels higher than the one under consideration.

As usual, the reciprocal intensity of the particular progression under study should vary linearly with the partial pressure of N_2 and from the slope of a linear plot of this kind the rate of deactivation by N_2 may be

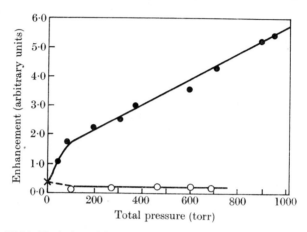

FIG. 17.24. Variation with pressure of the main gas of the enhancement of the NO γ ($v' = 0$) progression in Ar and N_2. × 0·6 torr NO, ● 0·6 torr NO+N_2, ○ 0·6 torr NO+Ar.

obtained, provided the radiative lifetime τ_r of the $A^2\Sigma^+$ state is known. From absorption measurements† of the f-value of the γ-bands τ_r is found to be $2\cdot2\times10^{-7}$ s, nearly independent of the vibrational quantum number.

Fig. 17.25 shows some typical linear plots obtained. The derived results for the rates of the transitions $v' \to v'-1$ where $v = 3, 2$, and 1 respectively, due to impact with N_2 molecules are discussed in § 3.

2.5.1. *The use of laser-induced fluorescence.* Hocker, Kovacs, Rhodes, Flynn, and Javan‡ have measured vibrational relaxation times in CO_2 by observing the rate of decay of fluorescence induced in the gas by an intense pulse of Q-switched laser radiation of wavelength 10·6 μm.

A level diagram of the lowest vibrational levels of CO_2 is given in Fig. 17.26. The normal modes (see Chap. 11, Fig. 11.4) consist of a symmetrical stretch, a bending vibration, and an asymmetrical stretch.

---

† BETHKE, G. W., *J. chem. Phys.* **31** (1959) 662; WEBER, D. and PENNER, S. S., ibid. **26** (1957) 860; CALLEAR, R. B., loc. cit., p. 1503.

‡ HOCKER, L. C., KOVACS, M. A., RHODES, C. K., FLYNN, G. W., and JAVAN, A., *Phys. Rev. Lett.* **17** (1966) 233.

A quantum state (l, m, n) is distinguished by the number of quanta of excitation in each of these modes. The laser radiation arises from transitions between the lowest asymmetrical and symmetrical stretching modes (001) and (010) (see Fig. 17.26).

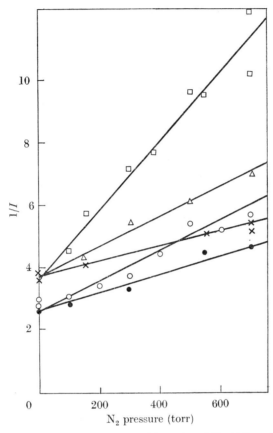

FIG. 17.25. Plots of the reciprocal intensity $(1/I)$ of the γ bands of NO as a function of the partial pressure of N_2 in NO–N_2–Ar mixtures at a total pressure of 700 torr. —●— (1, 1) band; —○— (2, 2) band, 1 torr NO; —×— (1, 1) band; —△— (2, 2) band; —□— (3, 4) band, 2 torr NO.

The CO_2, together with other gases under investigation, could be introduced into a short sample tube with infra-red transmitting windows placed inside the resonator of a Brewster-angle laser system. The tube was also provided with a similar window and was connected to a gas handling system so that chosen pressures of CO_2 and other gases could be introduced.

When the laser is Q-switched the laser beam produces sudden changes in the population of the (010) and (001) levels. The relaxation back to equilibrium of the (001) level was monitored from the intensity of the

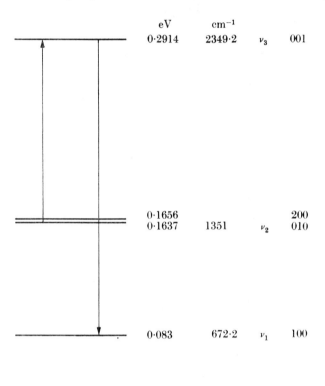

Fig. 17.26. Low-lying vibrational levels of CO_2. A level denoted by (l, m, n) involves l quanta in the bending mode, m in the symmetrical, and n in the asymmetrical stretching modes.

infra-red emission at $4 \cdot 26\,\mu$ ($2349 \cdot 2$ cm^{-1}) from the (001) → (000) transition, which could be observed through the side window of the sample tube.

Fig. 17.27 is a typical plot of the reciprocal of the decay time against the gas pressure in pure CO_2. As usual the slope gives the rate of quenching by collisions, while the departure from linearity at low pressures is due to loss by diffusion to the walls. From results such as these the quenching rate in pure CO_2 was found to be 385 s^{-1} torr^{-1}, which corresponds to a cross-section of $3 \cdot 3 \times 10^{-19}$ cm^2 and a relaxation time at 1 atm of $3 \cdot 5 \times 10^{-6}$ s.

2.6. *The spectrophone*

A further method of measuring relaxation times for conversion of vibrational into translational energy depends on the optic-acoustic effect. This was observed as long ago as 1881 by Tyndall[†] and by Röntgen.[‡]

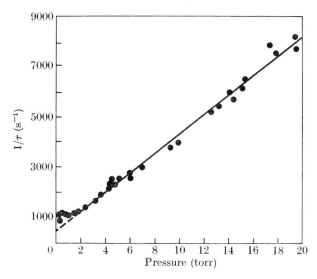

FIG. 17.27. Plot of the reciprocal of the quenching time τ for the $(001) \to (000)$ radiation of CO_2 as a function of pressure.

It occurs in the following way. Consider a gas the molecules of which can absorb infra-red radiation, thereby producing vibrational excitation. If such a gas is irradiated by infra-red light, the intensity of which is modulated at an acoustic frequency, the modulation will be transmitted to the translational motion of the gas molecules through transfer from the excited molecular vibrations. This results in the generation of acoustic oscillations in the gas, with the modulation frequency. The amplitude and phase of these oscillations may be measured by suitable acoustic techniques and it was pointed out by Gorelik[§] in 1946 that information about the rate of transfer of energy in the gas from vibration to translation could be derived from these measurements.

[†] TYNDALL, J., *Proc. R. Soc.* **A31** (1881) 307, 478.
[‡] RÖNTGEN, W. C., *Phil. Mag.* **11** (1881) 308.
[§] GORELIK, G., *Dokl. Akad. Nauk SSSR* **54** (1946) 779.

This possibility is of special interest because, by a suitable choice of infra-red radiation, excitation in the gas can be confined to a single vibrational mode so that the relaxation time for that mode, as distinct from other modes, can be obtained. Furthermore, it is possible to study vibrational relaxation in a gas at such low temperatures that, without irradiation, very few of the molecules would be in vibrationally excited states.

If ρ_i is the energy density of infra-red radiation, the time rate of change of the concentrations n_1 and n_0 of vibrationally excited and ground state molecules is given by

$$\frac{dn_1}{dt} = -n_1(\gamma_1 + B_{10}\rho_i + A_{10}) + n_0(\gamma_0 + B_{10}\rho_i)$$

$$= -\frac{dn_0}{dt}. \tag{85}$$

Here γ_1 and γ_0 are as in § 2, p. 1468 and A_{10} and B_{10} are the Einstein A and B coefficients for the radiative transition from the excited to the ground state.

Writing $n_1 + n_0 = n$, a constant, we have

$$\frac{dn_1}{dt} = -\Gamma_1 n_1 + \Gamma_0 n, \tag{86a}$$

where
$$\Gamma_1 = \gamma_1 + \gamma_0 + 2B_{10}\rho_i + A_{10}, \tag{86b}$$
$$\Gamma_0 = \gamma_0 + B_{10}\rho_i. \tag{86c}$$

We are concerned with the solution of (86 a) when

$$\rho_i = \bar{\rho}_i + \Delta\rho_i \cos\omega t, \tag{87}$$

ω being the modulation frequency. In this case we may write (86 a) in the form

$$\frac{dn_1}{dt} = -(\bar{\Gamma}_1 + 2\Delta\Gamma\cos\omega t)n_1 + (\bar{\Gamma}_0 + \Delta\Gamma\cos\omega t)n, \tag{88a}$$

where
$$\bar{\Gamma}_1 = \gamma_1 + \gamma_0 + 2B_{10}\bar{\rho}_i + A_{10}, \tag{88b}$$
$$\bar{\Gamma}_0 = \gamma_0 + B_{10}\bar{\rho}_i, \tag{88c}$$
$$\Delta\Gamma = B_{10}\Delta\rho_i. \tag{88d}$$

In practice, A_{10} and $B_{10}\bar{\rho}_i$ are small compared with γ_1 and γ_0, so

$$\bar{\Gamma}_1 \simeq \gamma_1 + \gamma_0, \qquad \bar{\Gamma}_0 \simeq \gamma_0. \tag{89}$$

Also, if $h\nu \gg \kappa T$, $\gamma_1 \gg \gamma_0$; so we take henceforward

$$\bar{\Gamma}_1 = \gamma_1, \qquad \bar{\Gamma}_0 = \gamma_0.$$

An approximate solution for such times that all transients have died out may be obtained by assuming

$$n_1 = \bar{n}_1 + \Delta n_1 \cos(\omega t - \delta). \tag{90}$$

On substitution in (88 a) we find, on equating separately to zero terms independent of t and proportional to $\cos\omega t$, $\sin\omega t$, respectively, while neglecting all terms varying with angular frequency 2ω, that

$$\bar{n}_1 = (\gamma_0/\gamma_1)n, \tag{91a}$$
$$\Delta n_1 = \gamma_1^{-1}\Delta\Gamma(1+\omega^2/\gamma_1^2)^{-\frac{1}{2}}, \tag{91b}$$
$$\delta = \arctan(\omega/\gamma_1). \tag{91c}$$

An exact solution of (88 a) shows that, provided

$$B_{10}\Delta\rho_i/\omega \ll 1, \tag{92}$$

(91 b) and (91 c) are certainly accurate enough for our purposes.

The periodic variations in n_1 will produce fluctuations in heat input through the transfer of vibrational to translational motion. Thus, if dQ/dt is the rate of change of heat input per cm^3,

$$\frac{dQ}{dt} = h\nu(\gamma_1 n_1 - \gamma_0 n_0) \simeq h\nu\gamma_1\{n_1 - n_0\gamma_0/\gamma_1\}, \tag{93}$$

$h\nu$ being the excitation energy of the vibrational mode concerned.

The oscillating term in Q arising from $\Delta n_1 \cos(\omega t - \delta)$ is therefore given by

$$\Delta Q \sin(\omega t - \delta), \tag{94}$$

where
$$\Delta Q = (h\nu/\omega)\Delta\Gamma(1+\omega^2/\gamma_1^2)^{-\frac{1}{2}}. \tag{95}$$

The temperature T is related to Q by the equation

$$\frac{\rho_g c}{\sigma}\frac{\partial T}{\partial t} = \nabla^2 T + \frac{1}{\sigma}\frac{dQ}{dt}, \tag{96}$$

where σ is the thermal conductivity, c the heat capacity, and ρ_g the density of the gas. Boundary conditions depending on the shape of the containing vessel and the temperature distribution at its surface must also be satisfied. Since we have assumed $h\nu \gg \kappa T$, c contains no appreciable contribution from vibration and is a constant.

In the simplest case of a thermally isolated system the term in $\nabla^2 T$ may be neglected so that

$$\frac{\partial T}{\partial t} = \frac{1}{\rho_g c}\frac{dQ}{dt}, \tag{97}$$

and the oscillatory term in T is given from (94) and (95) by

$$\Delta T \sin(\omega t - \delta), \tag{98a}$$

where
$$\Delta T = \frac{h\nu}{\omega\rho_g c}\frac{\Delta\Gamma}{(1+\omega^2/\gamma_1^2)}. \tag{98b}$$

A more nearly practical case is that in which the gas is enclosed in a sphere of radius a whose walls are maintained at a constant temperature. In that case the mean temperature throughout the sphere, at time t, contains a term varying with frequency ω which is nearly equal to (98a), provided†

$$\omega \gg \frac{\sigma\pi^2}{\rho_g c a^2}. \tag{99}$$

A similar result holds for a cylindrical container of radius a provided

$$\omega \gg \frac{\sigma}{\rho_g c}\alpha_1^2, \tag{100}$$

where α_1 is the largest root of the equation

$$J_0(a\alpha) = 0.$$

† DELANY, M. E., Thesis, London, 1959.

The pressure change can now be obtained from the relation

$$\frac{\Delta p}{\Delta T} = \frac{p_0}{T_0},$$

where p_0 and T_0 are the initial pressure and temperature respectively.

Thus, provided conditions such as (99) or (100) are satisfied for the appropriate containing vessel, the oscillating pressure of modulation frequency $\omega/2\pi$ is given by

$$\frac{h\nu}{\omega \rho_g c} \frac{p_0}{T_0} \frac{\Delta \Gamma}{(1+\omega^2/\gamma_1^2)} \sin(\omega t - \delta), \qquad (101\,\text{a})$$

with
$$\delta = \arctan(\omega/\gamma_1). \qquad (101\,\text{b})$$

In principle, it is possible to determine γ_1 either by measuring the phase shift δ or the variation of amplitude of the oscillating pressure with ω. The latter method is difficult to apply because it is necessary to cover a range of modulation frequency which includes one or more resonance frequencies of acoustic oscillation of the absorption chamber. Nevertheless, an attempt has been made by Woodmansee and Decius[†] to use this method by observing the resonances independently and then making due allowance for them. Their results for carbon monoxide are, however, almost certainly vitiated by the inadequate purity of the gas they used.

The modulation frequency for use in phase-shift measurements must be chosen, so as to avoid resonance effects, as less than one-quarter of the lowest resonance frequency. On the other hand, the frequency must be sufficiently high if corrections due to heat conduction to the walls of the chamber are to be unimportant.

As an illustration of the technique, and of the difficulties encountered in applying it, we describe the experiment of Cottrell, Macfarlane, Read, and Young,[‡] using a variety of polyatomic gases as working substances.

Fig. 17.28 shows the general arrangement of the apparatus. The source of infra-red radiation was an electrically heated nichrome rod about 0·3 cm diameter. This radiation was focused on the spectrophone cell by silvered mirrors. Filters to cut out either short wave (wave number > 2100 cm^{-1}) or long wave (wave number < 1700 cm^{-1}) radiation could be inserted. Modulation of the radiation at frequencies of 250 and 450 c/s was introduced through 12-in diameter chopping discs driven by a 3000 rev/min synchronous motor. The spectrophone cell was of stainless steel with polished sodium chloride windows through which the radiation passed, the absorption path being 3 cm. One side of the cell incorporated

---

[†] WOODMANSEE, W. E. and DECIUS, J. C., *J. chem. Phys.* **36** (1962) 1831.
[‡] COTTRELL, T. L., MACFARLANE, I. M., READ, A. W., and YOUNG, A. H., *Trans. Faraday Soc.* **62** (1966) 2655.

a condenser microphone with a flat audiofrequency response and low inherent noise.

The phase of the microphone signal was compared with that of a reference signal obtained from a photocell which received radiation from a lamp interrupted by the same chopping disc as the infra-red radiation.

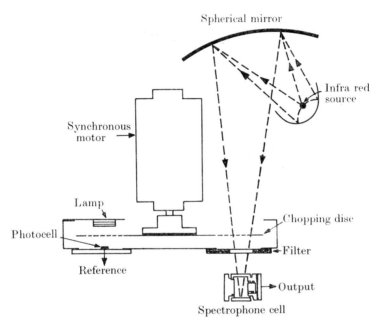

FIG. 17.28. General arrangement of the spectrophone experiments of Cottrell, MacFarlane, Read, and Young.

The first experiments with a particular spectrophone cell were made at a modulation frequency of 250 c/s and the results obtained for the variation with pressure of the phase of the signal relative to that of the reference signal are shown in Fig. 17.29 (a). According to the theory the phase should tend to a constant value as the pressure increases, but there is little sign of this in the observations. With $CClF_3$ the phase increased with pressure and was still increasing at pressures of 600 torr. On the other hand, CH_4, CO_2, and also NH_3 (apart from a preliminary decrease at very low pressures) showed phases which at first increased with pressure to a maximum and then decreased at higher pressures. The reason for this complex behaviour of $CClF_3$ and CO_2 was traced to the presence of resonances in the cell at 260 c/s and 430 c/s respectively. The effect of the resonances was clearly seen when the signal amplitude was plotted against the modulation frequency.

It was found that the resonance arose from the presence of pumping connection holes between the cell and the microphone system. When the holes were sealed off the resonance disappeared and much more satisfactory results, shown in Fig. 17.29 (b), were obtained.

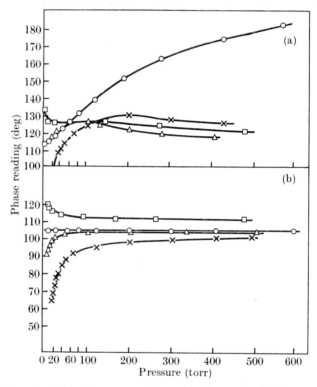

FIG. 17.29. Variation of the phase of the spectrophone signal with gas pressure observed by Cottrell et al. using (a) the unmodified spectrophone cell and (b) the modified cell. ○ $CClF_3$ (unfiltered radiation), △ CH_4 3020 cm^{-1} vibration excited, × CO_2 2349 cm^{-1} vibration excited, □ NH_3 3336 cm^{-1} vibration excited. The phase lead relative to the reference signal increases upwards.

For all the working substances the phase relative to that of the reference signal tended to a constant value at high pressures, as shown in Fig. 17.29 (b). Furthermore, at low pressures, for CO_2 there was a phase lag, relative to the constant limiting value, which increased as the pressure decreased. $CClF_3$ showed no change of phase over the entire pressure range. NH_3, on the other hand, showed a phase lead at low pressures which could be interpreted as an effect of heat conduction to the walls of the chamber, the condition corresponding to (99) not being satisfied. There is some slight evidence of an effect of this kind for CH_4

as the phase increases very slowly as the pressure decreases, but at pressures below 100 torr this effect is masked by a rapidly increasing phase lag due to vibrational relaxation.

By filtering out the low-frequency radiation, results were obtained for the relaxation of the 2349 cm^{-1} vibration of CO_2 and the 3020 cm^{-1} (ν_3) vibration of CH_4. The relaxation of the 1306 cm^{-1} (ν_4) vibration of CH_4

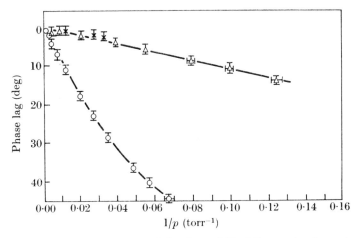

FIG. 17.30. Observed variation of the phase shift of the spectrophone signal with reciprocal gas-pressure for CO_2 and CH_4, compared with that calculated assuming respective relaxation times of 7 and 0·9 μs. ○ CO_2 2349 cm^{-1} vibration; △ CH_4 3020 cm^{-1} vibration; × CH_4 1306 cm^{-1} vibration.

was studied by filtering out the high-frequency radiation. Fig. 17.30 shows the observed phase-lags for these cases. It will be seen that, for CH_4 there is no appreciable difference between the results for the ν_3 and ν_4 vibrations. The results for each gas are seen to fit well to those expected from (101 b) with relaxation times, at 1 atm pressure, of 7 μs and 0·9 μs for CO_2 and CH_4 respectively.

Results were also obtained for N_2O and for COS. For N_2O the relaxation time for the 2223·5 cm^{-1} vibration was found to be about 50 per cent higher than for the combined 1285 and 588·1 cm^{-1} vibrations. For NH_3 and HCN heat conduction was dominant, yielding increasing phase leads at low pressures.

The observed relaxation time for the ν_3 mode (see p. 1510) of CO_2 (7 μs) may be compared with earlier measurements by the spectrophone technique, 11 μs due to Delany[†] and 12 μs to Slobodskaya and Gasilevich.[‡]

[†] loc. cit., p. 1513.
[‡] SLOBODSKAYA, P. V. and GASILEVICH, E. S. *Optika Spectrosk.* **7** (1959) 58.

A comparison of these and other data obtained by the spectrophone method with those obtained by other methods is made in § 3.1.3.

3. Results of experimental observations of vibrational relaxation times

3.1. *Relaxation times for low-lying vibrational levels*

3.1.1. *Pure diatomic gases.* The cross-sections for vibrational deactivation of diatomic molecules in collision with molecules of the same kind are very small at ordinary temperatures and consequently difficult to measure. Indeed, for hydrogen, the vibrational quanta are so large and the vibrational deactivation cross-sections so small under these conditions that no observations are possible at room temperatures. On the other hand, it is possible to observe ultrasonic dispersion and absorption due to relaxation of rotational energy. Results obtained are discussed in § 5. Some evidence of vibrational deactivation at temperatures above 2000 °K has been obtained by Gaydon and Hurle† from vibrational temperature measurements in a shock tube (see p. 1487). They find relaxation times of 4×10^{-7} s at 2280 °K and 8×10^{-7} s at 2600 °K. These correspond to probabilities of deactivation per collision of about 7.5×10^{-5} and 3.7×10^{-5} respectively which are somewhat higher than expected (see p. 1546).

It is somewhat less difficult to study the vibrational deactivation of deuterium. White‡ observed a relaxation time of $4 \pm 2 \times 10^{-6}$ s at 1400 °K using a shock tube with interferometer measurement of gas density behind the shock front (see p. 1484). However, this technique had to be pushed to the limit because of the small density changes in the relaxation zone due to the low vibrational heat capacity and low molar refractivity of D_2. Kiefer and Lutz§ have carried out more extensive measurements over the range 1100–3000 °K in which they determined the density distribution behind the shock front by the schlieren method described on p. 1485.

According to the elementary theory described below,

$$\tau = C\exp\{AT^{-\frac{1}{3}}\}, \quad (102)$$

where T is the absolute temperature and C and A are constants. Kiefer and Lutz find that their result for pure deuterium can indeed be represented in the form (102) by

$$\tau = 2.7 \times 10^{-10}\exp\{110.5/T\}^{\frac{1}{3}} \text{ s}. \quad (103)$$

† GAYDON, A. G. and HURLE, I. R., *Eighth Symposium on Combustion*, Pasadena, 1960, p. 309.
‡ WHITE, D. R., *J. chem. Phys.* **42** (1965) 447.
§ KIEFER, J. H. and LUTZ, R. W., ibid. **44** (1966) 658.

Vibrational relaxation can be observed, though with some difficulty, at ordinary temperatures in other diatomic gases. O_2, N_2, and CO are still so light that the vibrational deactivation cross-sections are very small at room temperature for collisions in the pure gas. Accurate determination of the cross-section is rendered more difficult by the fact that the presence of quite small amounts of polyatomic impurity, such as water vapour, reduces the relaxation time very much (see p. 1532). This difficulty is less serious for heavy molecules such as Cl_2 or Br_2 and also at the higher temperatures covered by the shock tube technique because deactivation cross-sections for collisions between similar molecules are much higher in these cases.

In Fig. 17.31 we summarize the results obtained for a number of diatomic gases in the form of logarithmic plots of the observed relaxation time τ, reduced to s.t.p., against $T^{-\frac{1}{3}}$, where T °K is the absolute temperature.

It will be seen that for all cases the plots, in accordance with (102), are truly linear at such temperatures that the relaxation time is $< 10^{-3}$ s. At lower temperatures the observations, while becoming less consistent with each other, begin to deviate markedly from the values expected from the linear relation established at higher temperatures. It may be that some of these deviations are due to the aforementioned difficulty of obtaining accurate results when the observations are very sensitive to the presence of impurities.

On the basis of the linear portions of the plots shown in Fig. 17.31 Millikan and White obtained the following empirical formula for the relaxation time τ,

$$\ln \tau = 1 \cdot 16 \times 10^{-3} \mu^{\frac{1}{2}} \Theta^{\frac{4}{3}} (T^{-\frac{1}{3}} - 0 \cdot 015 \mu^{\frac{1}{4}}) - 18 \cdot 42, \qquad (104)$$

where μ is the reduced mass of the colliding molecules in a.m.u. and $\Theta = h\nu/\kappa$, the characteristic temperature for the molecular vibrations concerned. This gives agreement with observed values, usually within 50 per cent and often much better.

The probability of vibrational deactivation per collision corresponding to the observed relaxation times may be obtained approximately by calculating the collision frequency using a Lennard–Jones form of interaction between the gas molecules, the parameters having been chosen to fit observations of transport and virial coefficient data (see Chap. 16, §§ 10, 11). Illustrative values obtained in this way for a number of diatomic gases at different temperatures are given in Table 17.1. As the calculated collision frequencies are only rough approximations the temperatures of the observations have been rounded off.

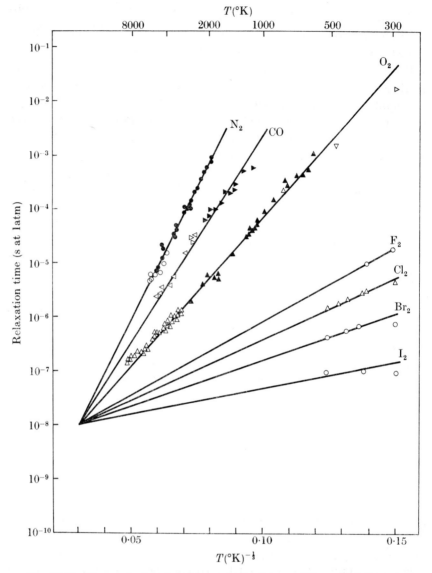

FIG. 17.31. Observed relaxation times τ (reduced to s.t.p.) in various diatomic gases, plotted as functions of $T^{-\frac{1}{3}}$, where T is the temperature.

N_2, MILLIKAN, R. C. and WHITE, D. R., J. chem. Phys. **39** (1963) 98.
BLACKMAN, V., J. Fluid Mech. **1** (1956) 61.
CO, HOOKER, W. J. and MILLIKAN, R. C., J. chem. Phys. **38** (1963) 214.
GAYDON, A. G. and HURLE, I. R., Eighth Symp. Combust., Pasadena, California, 1961, p. 309 (1962).
MATTHEWS, D. L., J. chem. Phys. **34** (1961) 639.
O_2, WHITE, D. R. and MILLIKAN, R. C., ibid. **39** (1963) 1803.
HOLMES, R., SMITH, F. A., and TEMPEST, W., Proc. phys. Soc. **81** (1963) 311.
LOSEV, S. A. and GENERALOV, N. A., Soviet Phys. Dokl. **6** (1962) 1081.
SHIELDS, F. D. and LEE, K. P., J. acoust. Soc. Am. **35** (1963) 251.
F_2 SHIELDS, F. D., ibid. **34** (1962) 271.
Cl_2,
Br_2, } SHIELDS, F. D., ibid. **32** (1960) 180.
I_2,

TABLE 17.1

Typical values of the probability of vibrational deactivation per collision between similar diatomic molecules at different temperatures

N_2 T(°K)		CO T(°K)		O_2 T(°K)		F_2 T(°K)		Cl_2 T(°K)		Br_2 T(°K)		I_2 T(°K)	
550	$(1 \cdot 6) \times 10^{-8}$ (1)	300	10^{-9} (4)	300	10^{-8} (7)	300	8×10^{-6} (11)	300	$3 \cdot 2 \times 10^{-5}$ (12)	300	$1 \cdot 8 \times 10^{-4}$ (14)	400	$2 \cdot 6 \times 10^{-3}$ (14)
700	$(5) \times 10^{-8}$ (1)	1200	8×10^{-7} (5)	500	10^{-7} (8)	375	$1 \cdot 5 \times 10^{-5}$ (11)	440	$2 \cdot 6 \times 10^{-5}$ (13)	450	$4 \cdot 6 \times 10^{-4}$ (14)	500	$4 \cdot 2 \times 10^{-3}$ (14)
800	$1 \cdot 6 \times 10^{-7}$ (2)	1500	2×10^{-6} (5)	1000	4×10^{-6} (9)			660	$2 \cdot 3 \times 10^{-5}$ (14)				
1200	$4 \cdot 7 \times 10^{-7}$ (2)	2200	$1 \cdot 6 \times 10^{-5}$ (3)	1400	3×10^{-5} (9)			660	8×10^{-5} (14)				
2200	$7 \cdot 7 \times 10^{-6}$ (3)	2550	$3 \cdot 4 \times 10^{-5}$ (3)	1900	$1 \cdot 5 \times 10^{-4}$ (9)			660	$7 \cdot 7 \times 10^{-4}$ (15)				
2500	$1 \cdot 2 \times 10^{-5}$ (3)	3800	2×10^{-4} (6)	2200	$1 \cdot 6 \times 10^{-3}$ (10)			900	$3 \cdot 2 \times 10^{-3}$ (15)				
3500	$6 \cdot 7 \times 10^{-5}$ (4)	4500	5×10^{-4} (6)	3200	5×10^{-3} (10)			1500	$1 \cdot 1 \times 10^{-2}$ (15)				
4200	$1 \cdot 5 \times 10^{-4}$ (4)			4000	2×10^{-2} (10)								
4900	$2 \cdot 7 \times 10^{-4}$ (4)			6000	4×10^{-2} (10)								
5500	$4 \cdot 5 \times 10^{-4}$ (4)			8000									

References

(1) HUBER, P. W. and KANTROWITZ, A., *J. chem. Phys.* **15** (1947) 275. Impact tube method (p. 1495). These results are uncertain as they were obtained from observations in N_2 containing 0·05 per cent H_2O by correcting for the latter admixture.
(2) LUKASIK, S. J. and YOUNG, J. E., ibid. **27** (1957) 1149. Acoustic resonance tube method (p. 1473).
(3) GAYDON, A. G. and HURLE, I. R., loc. cit., p. 1518. Shock tube, measuring vibrational temperature by sodium line reversal (p. 1487).
(4) BLACKMAN, V., *J. Fluid Mech.* **1** (1956) 61. Shock tube, measuring density by interferometry (p. 1484).
(5) HOOKER, W. J. and MILLIKAN, R. C., *J. chem. Phys.* **38** (1963) 214. Shock tube, measuring infra-red emission (p. 1489).
(6) MATTHEWS, D. L., ibid. **34** (1961) 639. Shock tube measuring density by interferometry (p. 1484).
(7) HOLMES, R., SMITH, F. A., and TEMPEST, W., *Proc. phys. Soc.* **81** (1963) 311. Acoustic resonance tube (p. 1473).
(8) SHIELDS, F. D. and LEE, K. P., *J. acoust. Soc. Am.* **35** (1963) 251. Acoustic resonance tube (p. 1473).
(9) WHITE, D. R. and MILLIKAN, R. C., *J. chem. Phys.* **39** (1963) 1803. Shock tube, measuring density by interferometry (p. 1484).
(10) LOSEV, S. A. and GENERALOV, V. A., *Dokl. Akad. Nauk SSSR* (1961); *Soviet Phys. Dokl.* **6** (1961) 1081. Shock tube, measuring density by ultra-violet absorption.
(11) SHIELDS, F. D., *J. acoust. Soc. Am.* **34** (1962) 271. Acoustic resonance tube (p. 1473).
(12) RICHARDSON, E. G., ibid. **31** (1959) 152. Acoustic interferometer (p. 1470).
(13) EUCKEN, A. and BECKER, R., *Z. phys. Chem.* **27B** (1934) 235. Acoustic interferometer (p. 1470).
(14) SHIELDS, F. D., *J. acoust. Soc. Am.* **32** (1960) 180. Acoustic resonance tube (p. 1473).
(15) SMILEY, E. F. and WINKLER, E. H., *J. chem. Phys.* **22** (1954) 2018. Shock tube, measuring density by interferometry (p. 1484).

It will be seen that the probabilities behave generally as expected. For the lighter molecules, N_2, CO, and O_2, with large vibrational quanta, the probability is of order 10^{-7} to 10^{-8} even at 500 °K. In contrast, at this temperature, it is already as high as 10^{-4} for Cl_2 and is as large as 4.2×10^{-3} for I_2. F_2, on the other hand, is an exception as a light molecule as the probability for it is nearly as large as 10^{-5} already at room temperature. An even more exceptional case is that of NO.

TABLE 17.2

Probability of vibrational deactivation of NO per collision with different molecules at room temperature

Molecule	NO	N_2	CO	H_2O	Kr
Probability of deactivation					
(Callear)†	$3.5_5 \times 10^{-4}$	4×10^{-7}	2.5×10^{-5}	7×10^{-3}	10^{-8}
(Bauer, Kneser, and Sittig)‡	3.7×10^{-4}				

Table 17.2 gives the probability of deactivation for collisions with other NO molecules measured by Callear using the flash spectroscopic technique (see p. 1502) and by Bauer, Kneser, and Sittig using an acoustic pulse absorption method. These two quite different methods give results that agree well but which are about 10^4 times greater than for either N_2 or O_2. On the other hand, the probability is very much lower for collision with N_2 or Kr (see Table 17.2 and § 3.1). An even more remarkable result is that the probability of deactivation of the first vibrationally excited level of the upper $A^1\Sigma^+$ state of NO by impact with ground state NO molecules is as large as 0.3, as determined from the spectroscopic observations.§ This is 100 times larger than the already very large value for vibrational deactivation of ground state NO and yet the vibrational constants are not very different for the two electronic states. Observations for NO have also been made at higher temperatures using shock tube techniques.‖

3.1.2. *Mixtures of diatomic molecular gases with rare gases.* We now consider the observations that have been made of the vibrational relaxation times for diatomic gases in the presence of rare gases from which it is possible to deduce the probabilities for deactivation of vibrational excitation by impact with a foreign atom. In this section we confine

---

† loc. cit., p. 1507.
‡ BAUER, H. J., KNESER, H. O., and SITTIG, E., *J. chem. Phys.* **30** (1959) 1119.
§ BROIDA, H. P. and CARRINGTON, T., loc. cit., p. 1583.
‖ ROBBEN, F., *J. chem. Phys.* **31** (1959) 420.

ourselves to mixtures with monatomic gases so that the only process that can be effective is one in which energy transfer between vibration and translation takes place.

Under these circumstances the relaxation time τ for a mixture of a molecular gas A with a rare gas B is given by

$$\frac{1}{\tau} = \frac{(1-x)}{\tau_{AA}} + \frac{x}{\tau_{AB}}, \qquad (105)$$

where τ_{AA} is the relaxation time for the pure gas A and τ_{AB} that for a mixture of A in a large excess of B. x is the molar fraction of atoms B in the mixture.

If the added gas is polyatomic, transfer of vibrational excitation may take place between molecules of the main and admixed gases. This we shall discuss in § 3.1.3.

In Fig. 17.32 observed relaxation times for mixtures of argon and helium with CO and O_2 are shown plotted against $T^{-\frac{1}{3}}$, where T °K is the absolute temperature as in Fig. 17.31 for the pure gases. For comparison the linear portions of the plots for the pure gases shown in that figure are reproduced in Fig. 17.32.

It will be noted that, whereas the relaxation time due to argon is longer than for the pure gas, that for helium is much shorter. This reflects the strong dependence on the reduced mass of the colliding systems which is shown in the empirical formula (104) and which, in fact, applies equally to the mixtures with the rare gases as to the pure gases.

3.1.3. *Pure polyatomic gases and mixtures with rare gases.* As the vibrational quanta concerned are much smaller, vibrational relaxation times for polyatomic gases are readily determined by ultrasonic methods at room temperature. An interesting question concerns the independence, or otherwise, of the relaxation of the different modes of vibration. In most cases there is little or no evidence of multiple relaxation processes presumably because the excitation energy is rapidly distributed among the different modes through collisions (see, however, § 4.4.2, p. 1557). There are some exceptions in which two relaxation times are clearly identified. These are cases in which the energy separation between the first and second excited modes is relatively high.

A great amount of attention has been paid to studies in pure CO_2 and in mixtures of CO_2 with other gases. Historically, the first observations of dispersion of sound waves through vibrational relaxation were made in 1925 by Pierce† in CO_2. Although many observations have been made by different observers the consistency of the results is not high, mainly

† loc. cit., p. 1470.

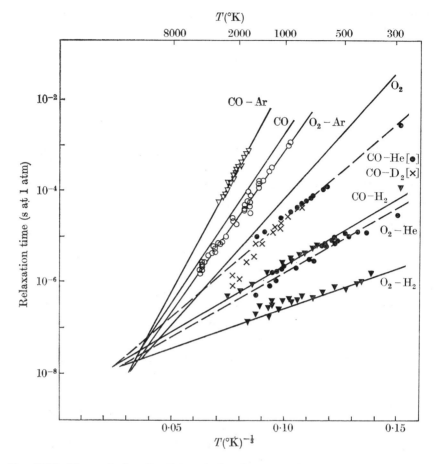

FIG. 17.32. Observed relaxation times τ (reduced to s.t.p.) for CO and for O_2 infinitely dilute in Ar, He, and H_2, and for CO in D_2 plotted as functions of $T^{-\frac{1}{3}}$, where T is the temperature. For comparison the plots giving the best fits to the data for the pure gases (see Fig. 17.31) are also shown.

CO–Ar, HOOKER, W. J. and MILLIKAN, R. C., *J. chem. Phys.* **38** (1963) 214.
O_2–Ar, WHITE, D. R. and MILLIKAN, R. C., ibid. **39** (1963) 1807.
    CAMAC, M., ibid. **34** (1961) 448.
CO–He, MILLIKAN, R. C., ibid. **38** (1963) 2855.
    MILLIKAN, R. C., ibid. **40** (1964) 2594.
O_2–He, WHITE, D. R. and MILLIKAN, R. C., ibid. **39** (1963) 1807.
    PARKER, J. G., ibid. **34** (1961) 1763.
    HOLMES, R., SMITH, F. A., and TEMPEST, W., Paper presented at the International Conference on Acoustics, Copenhagen, 1962.
CO–H_2, MILLIKAN, R. C., *J. chem. Phys.* **38** (1963) 2855.
    HOOKER, W. J. and MILLIKAN, R. C., ibid. **38** (1963) 214.
O_2–H_2, WHITE, D. R. and MILLIKAN, R. C., ibid. **39** (1963) 2107.
    PARKER, J. G., ibid. **34** (1961) 1763.
CO–D_2, WHITE, D. R., loc, cit. p. 1536.

because the results are very sensitive to the presence of impurities, particularly water vapour.

The three fundamental modes of vibration consist of a bending mode (ν_1) O—C—O (with arrows up on C and down on O's), a symmetric stretching mode (ν_2) O→C←O, and an asymmetric stretching mode (ν_3) ←O—C→←O with frequencies 2·00, 4·16, and $7·05 \times 10^{13}$ s^{-1} (see Fig. 17.26). At room temperature the respective contributions to the specific heat are 1·8, 0·14, and 0·002 cal mol^{-1} deg^{-1}, so that ultrasonic measurements will not be affected by the relaxation of the asymmetric stretching mode ν_3 and little by that of the mode ν_2. Acoustic observations give a single relaxation time which is a little over 6 μs corresponding to a deactivation probability of about 2×10^{-5} per collision. Experiments using the spectrophone (§ 2.6) in which the ν_3 mode is exclusively excited do not give significantly different results within the accuracy of the observations, which is not very high. Thus, Cottrell, MacFarlane, Read, and Young† find 7·0 μs as compared with earlier spectrophone observations of Delany‡ (11 μs) and Slobodskaya§ (12 μs) in which the relaxation time for pure CO_2 was derived, by extrapolation to zero N_2 pressure, from observations in a mixture with N_2.

Fig. 17.33 (a) shows a plot of observed relaxation time in 'pure' CO_2 as a function of $T^{-\frac{1}{3}}$, where T °K is the absolute temperature. The spread in the observed values is manifest. We shall return to a discussion of the interaction between the different modes in CO_2 in connection with theoretical calculation of the various relaxation rates in § 4.4.2.

Observations of the relaxation time for mixtures of CO_2 with He are shown in Fig. 17.33 (b). As for diatomic gases these times are shorter than for the pure gas.

Another gas that has been investigated by many observers is N_2O. On the whole the results, shown in Fig. 17.34 as a plot against $T^{-\frac{1}{3}}$, are more consistent than for CO_2, possibly because the usual impurities, such as H_2O, are less important. Results obtained for N_2O–He mixtures are also given in Fig. 17.34, showing the usual shortening of the relaxation time.

The relaxation times for H_2O are very short and vary quite slowly with the temperature T, as may be seen from Fig. 17.35. For NH_3, the relaxation time at room temperature is $1·5 \times 10^{-9}$ s, according to ultrasonic velocity measurements using an acoustic interferometer (§ 2.1.2.1),

† loc. cit., p. 1514.
‡ loc. cit., p. 1513.
§ SLOBODSKAYA, P. V., *Izv. Akad. Nauk SSSR* **12** (1948) 656.

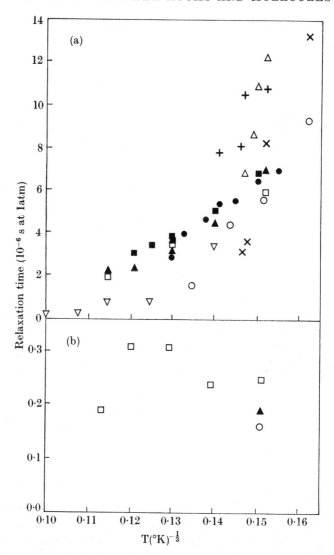

FIG. 17.33. Observed relaxation times (reduced to s.t.p.), (a) for pure CO_2 and (b) for CO_2 infinitely dilute in He, plotted as functions of $T^{-\frac{1}{3}}$, where T is the absolute temperature.

○ EUCKEN, A. and BECKER, R., Z. phys. Chem. 27B (1934) 235.
□ KÜCHLER, L., ibid. 41B (1938) 199.
▲ EUCKEN, A. and NÜMANN, E., ibid. 36B (1937) 163.
× VAN ITTERBEEK, A. and MARIENS, P., Physica, 's Grav. 5 (1938) 153.
+ VAN ITTERBEEK, A. and MARIENS, P., ibid. 7 (1940) 125.
△ VAN ITTERBEEK, A., DE BRUYN, P., and MARIENS, P., ibid. 6 (1939) 511.
● SHIELDS, F. D., J. acoust. Soc. Am. 29 (1957) 450.
■ SHIELDS, F. D., ibid. 31 (1959) 248.
▽ SMILEY, E. F. and WINKLER, E. H., J. chem. Phys. 22 (1954) 2018.

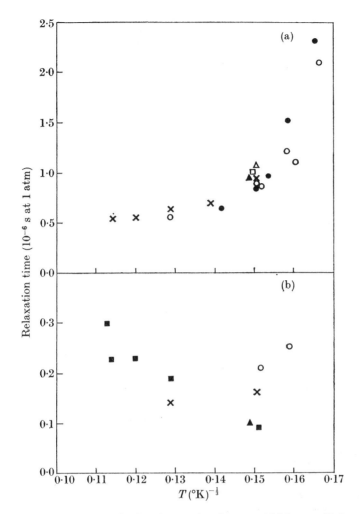

FIG. 17.34. Observed relaxation times (reduced to s.t.p.) (a) for pure N_2O and (b) for N_2O infinitely dilute in He, plotted as functions of $T^{-\frac{1}{3}}$.

○ EUCKEN, A. and JAACKS, H., Z. phys. Chem. 30B (1935) 85.
● BUSCHMANN, K. F. and SCHÄFER, K., ibid. 50B (1941) 73.
× EUCKEN, A. and NÜMANN, E., ibid. 36B (1937) 163.
△ WIGHT, H. M., J. acoust. Soc. Am. 28 (1956) 459.
▲ WALKER, R. A., ROSSING, T. D., and LEGVOLD, S., NACA Tech. Note 3210 (Washington, 1954).
□ ARNOLD, J. W., McCOUBREY, J. C., and UBBELOHDE, A. R., Proc. R. Soc. A248 (1958) 445.
■ KÜCHLER, L., Z. phys. Chem. 41B (1938) 199.

carried out by Cottrell and Matheson.† This very short time is rather surprising as the vibrational frequency involved is relatively high and it has been suggested‡ that, in some way, the inversion phenomenon in NH_3 might be responsible. The inversion frequency is such that during a collision one or two inversions may occur. Little effects of this kind are likely to arise in ND_3 for which the inversion rate is about 10 times slower, and Cottrell and Matheson find a more normal relaxation time $1 \cdot 3 \times 10^{-8}$ s.

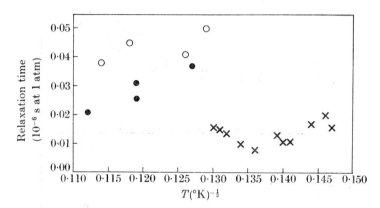

FIG. 17.35. Observed relaxation times (reduced to s.t.p.) for pure H_2O plotted as functions of $T^{-\frac{1}{3}}$.

● HUBER, P. W. and KANTROWITZ, A., J. chem. Phys. **15** (1947) 275.
○ EDEN, D. C., LINDSAY, R. B., and ZINK, H., Trans. Am. Soc. mech. Engrs **83** (1961) 137.
× FUJII, Y., LINDSAY, R. B., and URUSHIHARA, K., J. acoust. Soc. Am. **35** (1963) 961.

SO_2 is the simplest molecule which exhibits two distinct relaxation times. Fig. 17.36 reproduces results obtained by Lambert and Salter‡ in which the apparent specific heat at constant volume derived from ultrasonic velocity measurements is shown as a function of $\log(\nu/p)$, where ν/p is the ratio of the frequency ν of the sound wave to the gas pressure p (see § 2.1.3). It will be seen that a good fit with the observed results is obtained when a double relaxation process with appropriate constants is assumed, but no such fit can be obtained with a single process alone. The relaxation times for the two processes, derived from the observations of Lambert and Salter, are given in Table 17.3, together with the corresponding deactivation probabilities per collision.

For SO_2 the frequencies of the fundamental modes are 519, 1151, and 1360 cm^{-1}, which at room temperature contribute respectively 1·189,

† COTTRELL, T. L. and MATHESON, A. J., Trans. Faraday Soc. **59** (1963) 824.
‡ LAMBERT, J. D. and SALTER, R., Proc. R. Soc. **A243** (1957) 78.

0·224, and 0·112 cal mol^{-1} deg^{-1} to the specific heat. It is likely that the shorter relaxation time refers to deactivation of the lowest excited mode and the larger to that of the two higher modes combined.

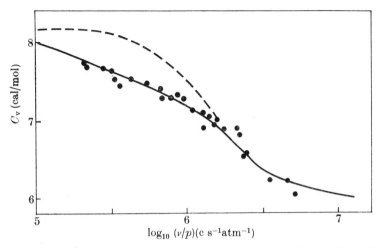

FIG. 17.36. Apparent specific heat at constant volume obtained from ultrasonic dispersion measurements in SO$_2$ as a function of log(ν/p), where ν is the frequency of the sound wave and p the pressure, at a temperature of 102 °K. ● observed results; ---- best fit assuming single relaxation process; —— best fit assuming a double relaxation process.

TABLE 17.3

Relaxation times and deactivation probabilities for collisions in pure SO$_2$

Temp. (°K)	Relaxation times (s)		Deactivation probabilities per collision	
	Process 1	Process 2	Process 1	Process 2
293	5·6×10^{-8}	5·6×10^{-7}	2·6×10^{-3}	2·8×10^{-4}
375	8·9×10^{-8}	8·9×10^{-7}	1·9×10^{-3}	2·0×10^{-4}
473	5·0×10^{-8}	5·0×10^{-7}	3·3×10^{-3}	3·9×10^{-4}

The only other gases for which two vibrational relaxation times have been observed are methylene chloride† (CH$_2$Cl$_2$), the first in which it was observed, and ethane (C$_2$H$_6$).‡ In all three cases the ratio of the frequency of the lowest excited mode to that of the next above is greater than 2 whereas in most other polyatomic molecules it is less than 2. It appears that, because it is necessary to transfer energy between one

† SETTE, D., BRESALON, A., and HUBBARD, J. C., *J. chem. Phys.* **23** (1955) 787.
‡ LAMBERT, J. D. and SALTER, R., *Proc. R. Soc.* A**253** (1959) 277.

quantum of the upper mode and two or three quanta of the lower, the over-all rate of transfer between the modes, which can only occur on collision,† is slower than that of deactivation of the lower mode. For CO_2, the frequency of the bending mode is very close to being exactly twice that of the symmetrical stretching mode so that transfer between them should be abnormally rapid through resonance effects (cf. Table 17.8). This is probably the reason why only a single relaxation time is observed in this case.

Lambert and Salter‡ have pointed out a dependence of the vibrational relaxation time, for polyatomic molecules at 300 °K, on the number of hydrogen atoms present in the molecule. This they displayed in the form of plots of observed relaxation times against the lowest fundamental frequency for different molecules as in Fig. 17.37. It will be seen that, for molecules containing no hydrogen atoms, the observed values fall approximately on one straight line and those with two or more on another with considerably smaller slope. Molecules with one hydrogen atom fall in between.

The explanation of these results is probably related to that of certain observations of the relative relaxation times for hydrides and the corresponding deuterides. Although, for the deuterides, the mean relative velocity of impact is smaller and the reduced mass larger than for the hydrides, both of which tend to increase the relaxation time, the fact that the fundamental frequency is lower for the deuterides would be expected to dominate and lead to a shorter relaxation time. Cottrell and Matheson found, however, by ultrasonic dispersion measurements with an acoustic interferometer that the relaxation times of the hydrides of carbon,§ nitrogen‖, silicon,§ and phosphorus‖ are shorter than for the corresponding deuterides (see Table 17.4). We have already discussed the case of nitrogen which is likely to be complicated by inversion transitions. It was suggested by Cottrell and Matheson§ that, in molecules containing hydrogen atoms which rotate about the main core with relatively high velocities, vibrational deactivation occurs through collisions with these atoms. Energy transfer would be expected to take place with comparative ease because of the small reduced mass and high relative velocity involved in such a collision. The effectiveness of

---

† HERZFELD, K. F., *Thermodynamics and physics of matter* (editor Rossini, F. D.), Section H, p. 616 (Oxford University Press, 1955).
‡ LAMBERT, J. D. and SALTER, R., *Proc. R. Soc.* A**253** (1959) 277.
§ COTTRELL, T. L. and MATHESON, A. J., *Proc. chem. Soc.* (1962) 17; *Trans. Faraday Soc.* **58** (1962) 2336.
‖ Ibid. **59** (1963) 824.

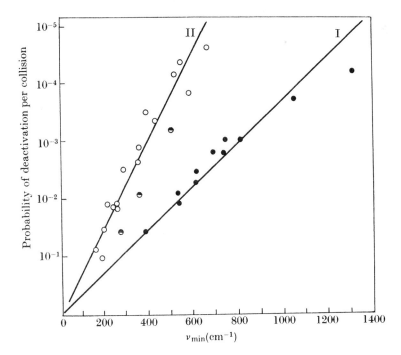

Fig. 17.37. Plots of the probability of vibrational deactivation for collision at 300° for various molecules against the fundamental vibration frequency $\nu_{\min}$, listed in order from the bottom upwards.

○ molecules containing no H atom: $C_2F_4^{(1)}$, $CF_2Br_2^{(2)}$, $BrCl^{(2)}$, $CF_2Cl_2^{(2)}$, $CFCl_3^{(2)}$, $CCl_4^{(3)}$, $CF_3Br^{(2)}$, $CF_3Cl^{(2)}$, $SF_6^{(4)}$, $CF_4^{(1)}$, $CS_2^{(5)}$, $N_2O^{(6)}$, $COS^{(7)}$, $Cl_2^{(8)}$, and $CO_2^{(6)}$.

◐ molecules containing 1 H atom: $CHCl_2F^{(2)}$, $CHCl_3^{(3)}$, $CHClF_2^{(2)}$, $CHF_3^{(1)(2)}$.

● molecules containing 2 or more H atoms: $CH_2ClF^{(2)}$, $CH_3I^{(9)}$, $CH_2F_2^{(1)}$, $CH_3Br^{(2)(9)}$, $C_2H_2^{(2)}$, $CH_3Cl^{(2)}$, $C_2H_4O^{(5)}$, $C_2H_4^{(10)}$, cyclo $C_3H_6^{(10)}$, $CH_3F^{(9)}$, $CH_4^{(11)}$.

References

1. Fogg, P. G. T. and Lambert, J. D., Unpublished results.
2. Rossing, T. D. and Legvold, S., *J. chem. Phys.* **23** (1955) 1118.
3. Sette, D., Busala, A., and Hubbard, J. C., ibid. **23** (1955) 787.
4. O'Connor, C. L., *J. acoust. Soc. Am.* **26** (1954) 361.
5. Angona, F. A., ibid. **25** (1953) 1116.
6. Eucken, A. and Nümann, E., *Z. phys. Chem.* B **36** (1937) 163.
7. Eucken, A. and Aybar, S., ibid. **46** (1940) 195.
8. Eucken, A. and Becker, R., ibid. **34** (1934) 235.
9. Hanks, P. A. and Lambert, J. D., Unpublished results.
10. Corran, P.G., Lambert, J. D., Salter, R., and Warburton, B., *Proc. R. Soc.* A**244** (1958) 212.
11. Cottrell, T. L. and Martin, P. E., *Trans. Faraday Soc.* **53** (1957) 1157.

collisions of this kind would be much less marked for deuterides because of the mass doubling.

This process, which is essentially one of energy transfer from vibration in one molecule to rotation in the other, would account for the main features of the Lambert–Salter plot shown in Fig. 17.37. Referring to Table 17.4 we see that AsD_3 has a little shorter relaxation time than AsH_3 but, as will be discussed in greater detail in § 4.5, if it were not

TABLE 17.4

Observed relaxation times at 300 °K *for hydrides and the corresponding deuterides*

Hydride	CH_4[†]	NH_3[‡]	SiH_4[†]	PH_3[‡]	AsH_3[§]	HCl[∥]
Relaxation time (s)	2×10^{-6}		1.1×10^{-7}	1.96×10^{-7}	3.1×10^{-7}	1.1×10^{-2}
Deuteride	CD_4[†]	ND_3[‡]	SiD_4[†]	PD_3[‡]	AsD_3[§]	DCl[∥]
Relaxation time (s)	3.9×10^{-6}	0.13×10^{-7}	2.0×10^{-7}	3.43×10^{-7}	2.2×10^{-7}	1×10^{-2}

for the possibility of vibration–rotation transfer in AsH_3 the relaxation time of AsH_3 would be expected to be much larger than it is. The same applies to HCl and DCl. On p. 1501 we gave an account of the experiments of Millikan and Osburg (see Fig. 17.20) which provide direct evidence that cross-sections for deactivation of CO by H_2 depend on the rotational state of the H_2, so that vibration-rotation transfer must play a significant role in these cases also.

3.1.4. *The effect of diatomic and polyatomic impurities on the vibrational relaxation times of molecular gases.* So far we have considered only the observed results for the relaxation times of mixtures of molecular with atomic gases. In these cases the effect of the admixed rare gas is determined solely by the extent to which it modifies the rate at which vibrational energy of the main gas is transferred to energy of relative translation. Molecular impurities may act, however, through partial transfer of vibrational energy between the molecules of the main gas and of the impurity.

[†] COTTRELL, T. L. and MATHESON, A. J., *Proc. chem. Soc.* (1962) 17; *Trans. Faraday Soc.* **58** (1962) 2336.
[‡] Ibid. **59** (1963) 824.
[§] COTTRELL, T. L., DOBBIE, R. C., MCLAIN, J., and READ, A. W., ibid. **60** (1964) 241.
[∥] FERGUSON, M. G. and READ, A. W., ibid. **63** (1967) 61.

Thus consider a mixture involving two kinds of molecules A, B. Denoting vibrationally excited molecules by A*, B*, we now have the possible processes:

$$A^* + A \rightarrow A + A, \qquad (106\,a)$$

$$A^* + B \rightarrow A + B, \qquad (106\,b)$$

$$A^* + B \rightarrow A + B^*, \qquad (106\,c)$$

$$B^* + B \rightarrow B + B, \qquad (106\,d)$$

$$B^* + A \rightarrow B + A. \qquad (106\,e)$$

Processes (106 a) and (106 b) are the only ones that would occur if B were monatomic. (106 c) is the vibrational-transfer process, while (106 d) and (106 e) lead to the deactivation of vibrationally excited molecules B. The effects expected depend on the rates of the different reactions.

We suppose that the relaxation time for B is much shorter than for A. Thus, if the vibration-transfer reaction (106 c) is much faster than (106 d) or (106 e) the total vibrational energy will relax at a rate determined by the faster of (106 d) and (106 e). If it is (106 d) this rate will be proportional to x^2, the square of the mole fraction of B in the mixture.

On the other hand, if (106 c) is slower than (106 d) or (106 e) though faster than (106 a) or (106 b), it will be the rate-determining process for the relaxation of A. B will relax independently so there will be two relaxation times τ_A, τ_B that, τ_A, of A being the longer. Both τ_A^{-1} and τ_B^{-1} will vary linearly with the molar fraction x of B.

In principle, it is possible, through observations of the vibrational relaxation of a gas mixture as a function of composition, together with data on the relaxation times for the pure gases, to determine not only what are the rate-determining processes in each case but also the magnitudes of the rates involved. Relatively few experiments have been carried out on these lines, although a great deal of data have been obtained about the relaxation of gases containing small admixtures of known impurities.

In 1962 Lambert, Edwards, Pemberton, and Stretton[†] made ultrasonic dispersion measurements on a number of mixtures of polyatomic gases, all of which showed double dispersion and linear dependence of both reciprocal relaxation times on molar composition. At about the same time Valley and Legvold[‡] observed dispersion for mixtures of

---

[†] LAMBERT, J. D., EDWARDS, A. J., PEMBERTON, D., and STRETTON, J. L., *Discuss. Faraday Soc.* **33** (1962) 61.
[‡] VALLEY, L. M. and LEGVOLD, S., *J. chem. Phys.* **36** (1962) 481.

C_2H_4 and C_2H_6. Pure C_2H_6 is one of the few pure gases for which double dispersion has been observed. One dispersion region is associated with the torsional oscillations which are at much the lower frequency. Mixtures of C_2H_4 and C_2H_6 also show double dispersion† in which the torsional oscillations relax independently as in pure C_2H_6 while the remaining vibrations in both molecules relax together, as when the vibrational transfer reaction (106 c) is much the fastest reaction. Lambert, Edwards, Pemberton, and Stretton later found that mixtures of SF_6 and $CHClF_2$ also show single dispersion with a non-linear dependence of reciprocal relaxation time on composition.

Table 17.5 summarizes the results of these experiments in terms of probabilities P_{AB} of vibrational energy transfer per collision. The corresponding probabilities P_{AA}, P_{BB} for deactivation in homomolecular collisions between molecules A and between molecules B respectively are also given. In deriving the transfer probabilities P_{AB} from the raw experimental data the full relaxation equations for the system were set up in terms of the rates of the reactions (106). Taking P_{AA} and P_{BB} as given, the remaining rates were determined to give the best fit to the observations.

The fact, for example, that the reciprocal of the relaxation time for SF_6–C_2F_4 mixtures varies nearly linearly with molar composition and exhibits double dispersion whereas that of SF_6–$CHClF_2$ mixtures is clearly nonlinear and shows single dispersion can be understood in terms of the ratio of the rates of (106 c) and (106 d). For the latter mixture (106 c) is faster whereas in the former the reverse is the case.

In all the cases shown there is a close approach to resonance between the fundamental frequencies of each molecule or between that of A and an overtone of B. This may well be significant in determining the vibrational transfer rates.

All the cases investigated by ultrasonic dispersion methods which are given in Table 17.5 involve quite complex molecules. Unfortunately the information available about mixtures of diatomic with diatomic, or even polyatomic, molecules is mainly semi-quantitative at best. Perhaps the most accurate results available for transfer between diatomic molecules are those obtained by spectroscopic methods, so we give in Table 17.5 the results of Callear for NO ($X^2\Pi$) using flash excitation and time-resolved spectroscopy (see § 2.4.2), of Millikan from observation of the quenching of infra-red fluorescence of CO (§ 2.4.1), and of Callear and

† LAMBERT, J. D., PARKS-SMITH, D. G., and STRETTON, J. L., *Proc. R. Soc.* A**282** (1964) 380.

Smith from observation of the quenching of the γ-bands of NO (§ 2.5). Although there is no direct evidence that the deactivation proceeds through vibrational transfer—the emission of infra-red radiation from the 'impurity' molecules has not yet been observed—it is very likely

TABLE 17.5

Probabilities per collision for intermolecular vibrational energy transfer at 300 °K

		ν_A	ν_B (cm^{-1})	$\Delta\nu$ (cm^{-1})	P_{AA}	P_{AB}	P_{BB}
Singly dispersing mixtures							
SF$_6$	CHClF$_2$†	344	369	25	10^{-3}	2×10^{-2}	0.82×10^{-2}
C$_2$H$_4$	C$_2$H$_6$†‡	810	821.5	11.5	10^{-3}	2.5×10^{-2}	1.3×10^{-2}
Doubling dispersing mixtures†§							
CCl$_2$F$_2$	CH$_3$OCH$_3$	260	250	10	1.4×10^{-2}	0.2	> 0.3
CH$_3$Cl	CH$_3$OCH$_3$	732	250	18	2.4×10^{-3}	1.4×10^{-2}	> 0.3
SF$_6$	CH$_3$OCH$_3$	344	164	16	10^{-3}	1.2×10^{-2}	> 0.3
CHF$_3$	C$_2$F$_4$	507	507	0	6×10^{-4}	2×10^{-2}	0.18
SF$_6$	C$_2$F$_4$	344	190	36	10^{-3}	1.4×10^{-2}	0.18
CF$_4$	C$_2$F$_4$	435	220	5	4.4×10^{-4}	0.9×10^{-2}	0.18
From spectroscopic observations							
NO($A^2\Sigma^+$)	N$_2$§	2341	2330	11		1.2×10^{-3}	
NO($X^2\Pi$)	CO§	1876	2143	267		10^{-4}	
NO($X^2\Pi$)	N$_2$§	1876	2330	454		2×10^{-6}	
CO	O$_2$‖	2143	1554	589		2×10^{-7}	
CO	CH$_4$††	2143	1534	609		3×10^{-5}	

that this is the process responsible. In all these cases the transfer probability is much smaller than for the polyatomic molecules, but this may be largely associated with the fact that the energy differences between the fundamental frequencies in the colliding molecules are relatively much higher.

Although the experimental data are far from precise there is no doubt that the observed vibrational relaxation times of diatomic and simple

† LAMBERT, J. D., PARKS-SMITH, D. G., and STRETTON, J. L., loc. cit., p. 1534.
‡ VALLEY, L. M. and LEGVOLD, S., loc. cit., p. 1533.
§ CALLEAR, A. B., loc. cit., p. 1503.
‖ MILLIKAN, R. C., loc. cit., p. 1498.
†† MILLIKAN, R. C., *J. chem. Phys.* **42** (1965) 1439.

polyatomic gases are very sensitive to the presence of polyatomic impurities. Interpreted in terms of the probability of deactivation in a collision with an impurity molecule, results for O_2 range from $2 \cdot 4 \times 10^{-4}$ for CCl_4† and H_2S,† $8 \cdot 4 \times 10^{-4}$ for C_2H_2,† $2 \cdot 5$†–$4 \cdot 0$‡ $\times 10^{-3}$ for NH_3 and 5×10^{-3}‡ for H_2O. These are much greater than for O_2–O_2 collisions and much smaller than for the corresponding collisions between impurity molecules. It is probable, but not certain, that in most cases vibrational transfer is responsible for the deactivation.

For CO_2, at room temperature, the presence of a small amount of water vapour certainly changes the relaxation time drastically but the results of different observers do not agree very well. Apart from a five times higher value found by Knudsen and Fricke,§ most of the results‖ indicate that deactivation by collision with a water molecule is about 1000 times as probable as by collision with a second CO_2 molecule. For deactivation of N_2O, collisions with H_2O are relatively less effective, being about 70–100 times more so than for N_2O–N_2O collisions. Whether in these cases the deactivation process is one of vibrational transfer is still uncertain (see, however, § 4.4.2, p. 1559, in which further details of the relaxation of CO_2–H_2O mixtures are discussed in terms of a detailed theory, without allowance for vibrational transfer).

In addition to these data there exists a number of measurements of the effect of diatomic impurities such as H_2, D_2, N_2, and CO on vibrational relaxation. At room temperature these impurities have much the same effect as rare gas impurities of about the same molecular weight. This similarity persists to quite high temperatures as may be seen from Fig. 17.32. However, White,†† in a study of the relaxation of a mixture of $0 \cdot 1$ D_2 with $0 \cdot 9$ CO over the range 940–2800 °K, obtained evidence that, at the higher temperatures, a new relaxation process, which may well be vibration transfer, becomes important. The technique employed was that of the shock tube. Relaxation times were determined both by gas density measurements using interferometry (see p. 1484) and by observation of the infra-red emission from CO near $2100\,\mathrm{cm^{-1}}$ (see p. 1489). From these observations, relaxation times for CO dilute in D_2 were

† KNÖTZEL, H. and KNÖTZEL, L., *Ann. Phys.* **2** (1948) 393.
‡ KNUDSEN, V. O., *J. acoust. Soc. Am.* **5** (1933) 112; **6** (1935) 199.
§ KNUDSEN, V. O. and FRICKE, E., ibid. **12** (1940) 233.
‖ EUCKEN, A. and BECKER, R., *Z. phys. Chem.* **27**B (1934) 235; EUCKEN, A. and NÜMANN, E., ibid. **36**B (1937) 163; MCCOUBREY, J. C., unpublished, quoted by COTTRELL, T. L., *Molecular energy transfer in gases* (Butterworths, 1961); GUTOWSKI, F. A., *J. acoust. Soc. Am.* **28** (1956) 478; VAN ITTERBEEK, A. and MARIENS, P., *Physica, 's Grav.* **7** (1940) 125.
†† WHITE, D. R., *J. chem. Phys.* **45** (1966) 1257.

obtained on the assumption that no vibrational transfer occurs in CO–D_2 collisions. Reference to the plot of these times against $T^{-\frac{1}{3}}$, where T is the absolute temperature, shows that for $T < 1600$ °K the plot is linear and coincident with that observed for CO dilute in He (see Fig. 17.32). At higher temperatures, however, the points deviate very markedly from linear. No such departure is found for H_2. This can be understood if vibrational transfer is involved because the smaller vibrational quanta for D_2 are more nearly in resonance with those of CO.

Millikan and White† have searched for evidence of vibrational transfer between N_2 and CO which would be expected to be relatively fast because there is a fairly close approach to resonance (2330 cm^{-1} for N_2 as compared with 2143 cm^{-1} for CO). They used a similar technique to that employed by White in the experiments on D_2–CO mixtures. At temperatures between 3200 and 4200 °K, in which both interferometric density measurements, giving the relaxation time of the main gas, and infra-red emission observations, giving that of the CO, could be made, it was found that both gave the same relaxation time within 20 per cent. At 3600 °K the relaxation time for pure N_2 is four times that for pure CO, indicating that vibrational exchange occurs fast enough in the mixture to bring both constituents to the same vibrational temperature during the relaxation.

We have already drawn attention to the experiments of Millikan and Osburg (see p. 1532), which revealed a remarkable difference in the effectiveness of ortho- and para-hydrogen in producing vibrational relaxation of CO.

In § 4.4.2 we shall discuss in some detail certain detailed features of the relaxation of CO_2–H_2 mixtures in terms of Marriott's detailed theory, which allows for transitions between the low-lying levels of the CO_2 molecule.

3.2. *Relaxation times for higher vibrational levels*

In § 1, p. 1465 it has been pointed out that, whereas the probability of energy transfer between vibration and translation in a collision at ordinary temperatures is very small if the vibrational level concerned is low-lying, this no longer applies if the level is a highly excited one.

Although less detailed quantitative evidence is available about cross-sections for deactivation of such levels there is little doubt that these conclusions are generally valid.

† MILLIKAN, R. C. and WHITE, D. R., *J. chem. Phys.* **39** (1963) 98.

As explained in § 2.5, electronically excited molecules are often produced with highly excited vibration and most of the evidence is obtained from experiments involving such molecules.

The results obtained by Roessler[†] for I_2 and by Durand[‡] for S_2, in which the vibrational excitation is to levels with vibrational quantum numbers v of 26 and 8 respectively, and in which the deactivating atoms were those of the rare gases, both indicate probabilities of deactivation of order unity. On the other hand, Lipscomb, Norrish, and Thrush,[§] in their experiments on the flash photolysis of ClO_2 and of NO_2 in a large excess of rare gas, observed the rate of the vibrational deactivation from the $v = 6$ to the $v = 5$ level associated with the ground electronic state. They obtained probabilities per collision which are very small, comparable with those for deactivation of the first vibrational state. This is not surprising, however, because the energy difference between levels with $v = 6$ and $v = 5$ is not very different from that between $v = 1$ and $v = 0$.

Callear[||] has observed the dependence of the relaxation rate due to collisions with N_2, for vibrational levels of the excited $A^2\Sigma^+$ electronic state of NO (see p. 1507), on the vibrational quantum number. He finds that the probability per collision increases from $1 \cdot 2 \times 10^{-3}$ to $2 \cdot 4 \times 10^{-3}$ to 5×10^{-3} as v increases from 1 to 3. It is likely, however, that the deactivating process in this case is one of vibrational transfer, and the results cannot be regarded as evidence of an increasing probability of transfer to translation as the vibrational quantum number increases.

3.3. *Summarizing remarks*

The experimental data available on the probability of deactivation for collisions of low-lying vibrational states of the lightest diatomic molecules in pure gases or in mixtures with rare gases is consistent with the elementary theory of § 1. Thus, referring to (2), the probability at room temperature falls rapidly as the vibrational quanta concerned increase. As the temperature rises the probability increases rapidly so that, for molecules such as N_2, while being barely measurable at room temperature it is readily observable at the temperatures available in shock tubes. For H_2, on the other hand, with the largest vibrational quanta, it is difficult to measure even at these temperatures. According to (2) the probability of deactivation by collisions with rare-gas atoms should decrease with increase of the mass of the atom, in agreement

[†] loc. cit., p. 1505.   [‡] loc. cit., p. 1506.
[§] loc. cit., p. 1504.   [||] loc. cit., p. 1503.

with the observed result that, at a given temperature, argon is less effective than helium.

In more detail, the variation with quantum energy, temperature, and reduced mass of the colliding systems follows the empirical law (104). Particularly significant is the fact that $\ln \tau$ varies as $T^{-\frac{1}{3}}$, where τ is the relaxation time and T the absolute temperature.

Observed results, obtained by spectroscopic methods, for the rate at which transitions between highly excited vibrational states of diatomic molecules occur are also in general accordance with (2). Because the quanta concerned are small the probability of deactivation is expected to be much larger than for low-lying levels and this is found to be so.

For pure polyatomic gases the probabilities of deactivation are much greater at a given temperature because the vibrational quanta are smaller. In all but a very few cases only one relaxation time is observed, even though it would be expected at first sight that each mode of vibration would relax at a different rate. The exceptions are cases in which the energy differences between the vibrational states concerned are comparatively large. Accurate values of relaxation times for pure polyatomic gases are difficult to obtain because of the profound effect of even small concentrations of polyatomic impurities.

Molecular impurities can effect vibrational relaxation through vibrational transfer as well as direct deactivation through transfer of vibrational to translational energy. Evidence about vibrational transfer rates obtained from acoustic measurements on mixtures of varying composition is still incomplete and confined to relatively complex molecules but it is being supplemented from observations made by spectroscopic methods which are applicable to transfer collisions between diatomic as well as polyatomic molecules.

Evidence for transfer between vibrational and rotational energy is forthcoming from the observations of relaxation times of molecules containing two or more hydrogen atoms. In such molecules the hydrogen atoms acquire considerable velocities through rotation and, as they have low reduced mass, they should be effective in deactivating the vibration of an impinging molecule.

We now proceed to develop a more detailed theoretical account of the subject which attempts to interpret the observed results quantitatively and to predict the rates of processes such as vibration–vibration transfer and the nature of the relaxation processes in simple polyatomic gases which are still not thoroughly investigated experimentally.

4. Theoretical discussion of vibrational relaxation

4.1. *Head-on collisions of atoms with diatomic molecules*

4.1.1. *Application of the distorted wave method with exponential interaction.* We consider first head-on collisions between an atom A and a molecule BC. It may be assumed that the only interaction that we need to take into account in these circumstances is that between the atoms A and B which approach closely. As a good approximation to the form of this interaction which is sufficiently simple for detailed calculation to be carried out we take

$$V = V_0 e^{-ar}, \qquad (107)$$

where r is the distance between A and B. This may be written

$$V(R, \rho) = V_0 \exp\{-a(R-\lambda\rho)\}, \qquad (108)$$

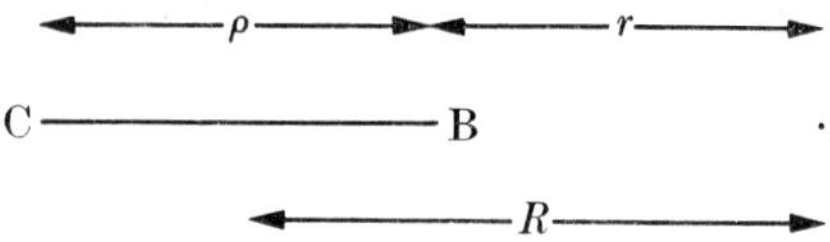

Fig. 17.38.

where R is the distance between A and the centre of mass of the molecule BC (see Fig. 17.38), ρ the nuclear separation of atoms B and C, and

$$\lambda = M_C/(M_B+M_C), \qquad (109)$$

M_A, etc. being the masses of the respective atoms. It is convenient to rewrite (108) in the form

$$V(R, x) = W_0 \exp\{-a(R-\lambda x)\}, \qquad (108\,a)$$

where $x = \rho - \rho_0$, ρ_0 being the equilibrium separation of the atoms B and C in the molecule BC and $W_0 = V_0 \exp(a\lambda\rho_0)$.

The collisions may now be treated in the same way as for collisions of electrons with atoms. Let $\chi_s(x)$ be the wave function for the sth vibrational state of BC. We expand the wave function $\Psi(R, x)$, describing the collision, in the form

$$\Psi = \sum_s F_s(R)\chi_s(x). \qquad (110)$$

The functions $F_s(R)$ must then satisfy the set of coupled differential equations corresponding to (16) of Chapter 8, namely

$$\left\{\frac{d^2}{dR^2}+k_s^2-U_{ss}(R)\right\}F_s(R) = \sum_{p \neq s} U_{sp} F_p(R). \qquad (111)$$

Here
$$k_s^2 = 2M(E-E_s)/\hbar^2, \qquad (112)$$

where E is the total energy of the system and E_s that of the sth vibrational state of the molecule. M, the reduced mass for the relative motion, is given by

$$M = M_A(M_B+M_C)/(M_A+M_B+M_C) \tag{113}$$

and
$$U_{sp} = (2MW_0/\hbar^2)Y_{sp}\,e^{-aR}, \tag{114}$$

where
$$Y_{sp} = \int e^{a\lambda x}\chi_s\chi_p\,\mathrm{d}x. \tag{115}$$

If, initially, the molecule is in the vibrational state s and we wish to determine the probability of excitation to any particular excited state we seek solutions of (111) which satisfy the boundary conditions

$$F_s \sim e^{ik_s R}+\alpha_{ss}\,e^{-ik_s R},$$
$$F_p \sim \alpha_{sp}\,e^{-ik_s R}. \tag{116}$$

The probability of excitation of the pth state from the sth during the impact is then
$$P_{sp} = (k_p/k_s)|\alpha_{sp}|^2. \tag{117}$$

Under room temperature conditions P_{sp} is very small for low-lying vibrational states, essentially due to the fact that the U_{ps} are all small for $p \neq s$. On the other hand, $Y_{sp} \ll Y_{ss}$ so $U_{ss} \gg U_{sp}$ and is not small. The distorted-wave method described in its three-dimensional form in Chapter 8, § 4.2 is then applicable. To apply this we reduce (111) to

$$\left\{\frac{\mathrm{d}^2}{\mathrm{d}R^2}+k_s^2-U_{ss}(R)\right\}F_s(R) = 0, \tag{118a}$$

$$\left\{\frac{\mathrm{d}^2}{\mathrm{d}R^2}+k_p^2-U_{pp}(R)\right\}F_p(R) = U_{s0}F_0. \tag{118b}$$

According to Chapter 8, § 4.2 (73), suitably modified for the one-dimensional case,
$$\alpha_{ps} = \int U_{sp}(R)F_s(R)\mathscr{F}_p(R)\,\mathrm{d}R, \tag{119}$$

where F_s satisfies (118a) and $\mathscr{F}_p$ the corresponding equation with U_{pp} in place of U_{ss}. They must have the asymptotic forms

$$F_s \sim \sin(k_s r+\eta), \qquad \mathscr{F}_p \sim \sin(k_p r+\mu). \tag{120}$$

The detailed application of (119) with the exponential form (107) for the interaction was first carried out by Jackson and Mott,[†] who took advantage of the fact that the equations for F_s and $\mathscr{F}_p$ are exactly soluble in analytical terms when U_{ss}, U_{pp} are given by (114) with

---

[†] JACKSON, J. M. and MOTT, N. F., *Proc. R. Soc.* **A137** (1932) 703.

$Y_{ss} = Y_{pp} \simeq 1$. Furthermore, the integral (119) may also be calculated analytically. It is then found that

$$P_{sp} = \frac{\pi^2}{4} Y_{sp}^2 \mathscr{I}_{sp}, \tag{121}$$

where
$$\mathscr{I}_{sp} = \left\{\frac{q_p^2 - q_s^2}{\cosh \pi q_p - \cosh \pi q_s}\right\}^2 \sinh \pi q_p \sinh \pi q_s, \tag{122a}$$

and
$$q_{p,s} = 2M v_{p,s}/\hbar a, \tag{122b}$$

$v_{p,s}$ being the velocities of relative motion of the atom and molecules when the latter are in the vibrational states p, s respectively.

A number of further simplifications may now be made, by restricting ourselves to transitions between low-lying vibrational states. First, since $1/a\lambda$ will be large compared with the amplitude of vibration in such states

$$Y_{sp} \simeq \int (1 + a\lambda x) \chi_s \chi_p \, dx. \tag{123}$$

If the vibrational wave functions can be well represented by Hermite polynomials we have then

$$\begin{aligned} Y_{sp} &\simeq a\lambda x_{sp} \quad (s = p \pm 1), \\ &\simeq 1 \quad (s = p), \\ &= 0 \quad \text{(otherwise)}, \end{aligned} \tag{124}$$

where
$$x_{01} = (2^{\frac{1}{2}}/4\pi)(h/M^*\nu)^{\frac{1}{2}}, \tag{125}$$

M^* being the reduced mass of the atoms B and C and ν the vibration frequency.

Again, under gas kinetic conditions, $q_{p,s}$ are both large, so that

$$\mathscr{I}_{sp} \simeq (q_p^2 - q_s^2)^2 \exp\{-\pi |q_p - q_s|\}, \tag{126}$$

provided $\pi|q_p - q_s| \gg 1$. But, since

$$\tfrac{1}{2} M(v_p^2 - v_s^2) = h\nu, \tag{127}$$

$$\begin{aligned} q_p - q_s &= (q_p^2 - q_s^2)/(q_p + q_s) \\ &= 4\pi\nu/a\bar{v}, \end{aligned} \tag{128}$$

where $\bar{v}$ is the mean relative velocity $\tfrac{1}{2}(v_p + v_s)$. Hence

$$\mathscr{I}_{sp} \simeq (256\pi^2 M^2 \nu^2/h^2 a^4) \exp\{-4\pi^2 \nu/a\bar{v}\} \tag{129}$$

and
$$P_{01} \simeq (16\pi^3 M^2 \lambda^2 \nu / M^* \hbar a^2) \exp\{-4\pi^2 \nu/a\bar{v}\}, \tag{130}$$

where
$$M^2 \lambda^2 / M^* = M_A^2 M_C (M_B + M_C) / M_B (M_A + M_B + M_C)^2. \tag{130a}$$

It is to be noted that the probability depends strongly on the slope of the exponential interaction and not on its magnitude.

4.1.2. *Temperature dependence of deactivation probability.* The temperature dependence of the vibrational relaxation time can now be predicted from (130). The probability of deactivation by impacts of relative velocity v_1 will be proportional to

$$f(v_1)\exp(-4\pi^2\nu/a\bar{v}), \tag{131}$$

where $f(v_1)$, the fraction of collisions in which the initial relative velocity is between v_1 and v_1+dv_1, is approximately equal at temperature T °K to $\exp(-\tfrac{1}{2}Mv_1^2/\kappa T)$. $\bar{v}$ is given by $\tfrac{1}{2}(v_1+v_0)$, where v_0 is the final relative velocity. The probability will be a maximum for those collisions in which

$$4\pi^2(\nu/a\bar{v})+\tfrac{1}{2}Mv_1^2/\kappa T$$

is a minimum. If we ignore the difference between $\bar{v}$ and v_1 this will occur when
$$v_1 = (4\pi^2\nu\kappa T/aM)^{\tfrac{1}{3}}. \tag{132}$$

The temperature variation of the deactivation probability, averaged over the molecular velocity distribution, should therefore be given by

$$\exp\left\{-\left(54\,\frac{\pi^4 M\nu^2}{a^2\kappa T}\right)^{\tfrac{1}{3}}\right\}. \tag{133}$$

This provides the basis for the systematic plotting of the logarithms of the relaxation times against $T^{-\tfrac{1}{3}}$ as in Figs. 17.31–17.35. Although the form (133) has been derived on many simplifying assumptions, including particularly that of head-on collisions, it nevertheless represents quite well many of the main features of vibrational relaxation for low-lying levels. Thus, according to (133), the probability decreases exponentially as $\nu^{\tfrac{2}{3}}$ as well as with $T^{-\tfrac{1}{3}}$ and, while the observed results do not determine precisely the form of the variation with ν, they do show a rapid decrease of probability as ν increases.

The great sensitivity of the magnitude of the factor (133) to the assumed values of the constants makes it very difficult to make any even semi-quantitative comparisons. However, if the vibrational quantum energy is 0·1 eV, $a = 1/a_0$, and the reduced mass M is 20 a.m.u., the factor (133) is as small as e^{-22} at room temperature. This is not inconsistent with the observations for, say, N_2 or O_2.

We may take into account the difference between $\bar{v}$ and v_1 when $h\nu/\kappa T$ is small, for then, since
$$\tfrac{1}{2}M(v_0^2-v_1^2) = h\nu,$$
$$v_0 \simeq v_1(1+h\nu/Mv_1^2),$$
and
$$4\pi^2\nu/a\bar{v} \simeq (4\pi^2\nu/av_1)(1-\tfrac{1}{2}h\nu/Mv_1^2). \tag{134}$$

This introduces an additional factor $\exp(2\pi^2 v^2 h/Mav_1^3)$. On substituting for v_1 from (132) this is simply $\exp(\tfrac{1}{2}hv/\kappa T)$. There will be conditions of intermediate temperature for a given pair of colliding systems for which this simple correction is applicable.

It is clearly necessary to elaborate in many ways the simple model we have discussed in order to attempt a closer semi-quantitative description of the phenomena. Before doing this, however, we present a semi-classical version of the quantum theory we have described above. This treatment has the advantage of simplicity and at the same time provides more insight into the nature of the process involved.

4.1.3. *Semi-classical method*

We begin by referring back to the time-dependent perturbation theory of collision processes described in Chapter 7, § 5.3.1. To apply this theory to the present problem we regard the relative motion of the atom A and molecule BC as essentially classical so that the interaction between them can be described by (108) with R now a function of time determined by the classical equation of motion

$$M\frac{d^2R}{dt^2} = -\frac{d}{dR}(W_0 e^{-aR}). \tag{135}$$

According to (108) of Chapter 7 the probability of a transition from the ground to the first vibrational state is given by $|a_1(\infty)|^2$, where

$$|a_1(\infty)|^2 = \hbar^{-2}\left|\int_{-\infty}^{\infty} V_{10}(t)\exp\{2\pi i vt\}\,dt\right|^2, \tag{136}$$

with
$$V_{10}(t) = \int V(x, R(t))\chi_1(x)\chi_0(x)\,dx$$
$$= Y_{10}W_0\exp\{-aR(t)\}. \tag{137}$$

If we take the origin of time as at the closest distance of approach

$$|a_1(\infty)|^2 = 4\hbar^{-2}\left|\int_0^{\infty} V_{10}(t)\cos(2\pi vt)\,dt\right|^2. \tag{138}$$

The solution of the classical equation (135) gives

$$t = (2/av)\operatorname{arcosh}\{e^{\tfrac{1}{2}aR}(Mv^2/2W_0)^{\tfrac{1}{2}}\}, \tag{139}$$

where v is the velocity of relative motion at infinite separation, so

$$e^{-aR} = (Mv^2/2W_0)\operatorname{sech}^2(\tfrac{1}{2}avt), \tag{140}$$

and the integral in (138) becomes

$$\tfrac{1}{2}Mv^2 Y_{10}\int_0^{\infty}\operatorname{sech}^2(\tfrac{1}{2}avt)\cos(2\pi vt)\,dt. \tag{141}$$

It may be shown by contour integration that

$$\int_0^{\infty}\operatorname{sech}^2(\tfrac{1}{2}avt)\cos(2\pi vt)\,dt = (8\pi^2 v/v^2 a^2)\operatorname{cosech}(2\pi^2 v/av). \tag{142}$$

Since $Y_{10} = (-2^{\frac{1}{2}}a\lambda/4\pi)(h/M^*v)^{\frac{1}{2}}$ we find, when $2\pi^2 v/av \gg 1$,

$$P_{10} = (16\pi^3 M^2\lambda^2 v/\hbar M^* a^2)\exp(-4\pi^2 v/av), \tag{143}$$

which agrees with (130) if v is replaced by the mean velocity $\bar{v}$. This is reasonable as we have not allowed for the change of relative velocity due to the inelastic collisions in our determination of R as a function of t.

The formula (136), which can readily be generalized to other vibrational transitions, shows clearly how the probability depends on the suddenness of the collision. Thus the integral in (136) is proportional to the Fourier component of $V_{10}(t)$ with frequency ν and the considerations of § 1 clearly apply.

The first calculations of vibrational deactivation probabilities were made by Zener† in 1931, using essentially a distorted-wave method with very simple interactions. This work was followed by that of Jackson and Mott,‡ who used the distorted-wave method consistently with the interaction (108). A little later, Zener§ developed the semi-classical method we have just described and applied it to the exponential interaction obtaining (143). Landau and Teller‖ then derived the same result under the condition $2Mv/\hbar a \gg 1$ by a purely classical treatment.

We shall apply the formula (136) in § 4.5 to discuss energy transfer on collision between vibration and rotation. Meanwhile, before discussing ways of elaborating the analysis of § 4.1.1 and to render it more realistic, we consider its extension to head-on collisions between two homonuclear diatomic molecules. This enables us to consider vibrational transfer as well as exchange of energy between vibration and translation.

4.2. *Head-on collisions between diatomic molecules with exponential interaction—vibrational transfer*

We denote the two molecules as AB and CD and suppose that B and C collide. In place of (108 a) we now have for the interaction

$$V(x_1, x_2, R) = W_0 \exp\{-a(R-\lambda_1 x_1-\lambda_2 x_2)\}. \tag{144}$$

Here R is the separation of the centres of mass of the two molecules. In terms of the separations ρ_1, ρ_2 of the atoms in the respective molecules AB, CD and their equilibrium values ρ_{10}, ρ_{20},

$$x_1 = \rho_1-\rho_{10}, \qquad x_2 = \rho_2-\rho_{20}. \tag{145}$$

---

† ZENER, C., *Phys. Rev.* **37** (1931) 556.
‡ loc. cit., p. 1541.
§ ZENER, C., *Phys. Rev.* **38** (1931) 277.
‖ LANDAU, L. D. and TELLER, E., *Phys. Z. SowjUn.* **10** (1936) 34.

λ_1, λ_2 are given in terms of the masses M_A, etc. of the respective atoms by

$$\lambda_1 = M_A/(M_A+M_B), \qquad \lambda_2 = M_D/(M_C+M_D). \tag{146}$$

Following through an exactly similar analysis to that in § 4.1.1 we find, for the probability of a collision in which the molecule AB changes its vibrational state from 1 to n and CD from 0 to m†

$$P_{10}^{nm} = \frac{\pi^2}{4}\{Y_{1,n}^{(1)}\}^2\{Y_{0,m}^{(2)}\}^2 \mathscr{J}_{1,0}^{n,m}. \tag{147}$$

If the initial excitation of AB is wholly transferred to energy of relative translation, $n = 0$ and $m = 0$ and we regain (130) except that the reduced mass M is changed to

$$M = \frac{(M_A+M_B)(M_C+M_D)}{M_A+M_B+M_C+M_D}. \tag{148}$$

It is also possible that the excitation energy of AB will be wholly or partly transferred to vibrational excitation of CD so that $m = 0$ in (147). In the special case in which $m = 1$, corresponding to (130) and (148),

$$P_{10}^{01} = \frac{M_A M_D}{M_B M_C} \frac{(M_A+M_B)(M_C+M_D)}{(M_A+M_B+M_C+M_D)^2} \frac{\Delta E^2}{\hbar^2 \nu_1 \nu_2} \exp\{-4\pi^2 \Delta E/a\hbar\bar{v}\}. \tag{149}$$

Here ν_1, ν_2 are the vibrational frequencies of AB and CD respectively and $\Delta E = h|\nu_1-\nu_2|$. This formula is valid provided

$$4\pi^2 a \Delta E/h\bar{v} \gg 1.$$

Because $h|\nu_1-\nu_2| < h\nu_1$ the exponent in (149) is much smaller than for a collision in which all of the excitation energy is transferred to that of relative translation. This is the essential reason why vibrational transfer, when it can occur, is relatively more important. This importance increases rapidly as $\Delta E \to 0$. In the case of exact resonance (149) is not valid. However, returning to the complete formula (122) for $\mathscr{J}_{sp}$ and putting $q_p = q_s$, we find

$$\mathscr{J}_{sp} \to 4q_s^2/\pi^2$$

and

$$P_{10}^{01} \to \frac{1}{4\pi^2} \frac{M_A M_D}{M_B M_C} \frac{(M_A+M_B)(M_C+M_D)}{(M_A+M_B+M_C+M_D)^2} \frac{a^2 v^2}{v^2}. \tag{150}$$

Under these conditions the probability increases only as the square of av/v instead of as $\exp(-4\pi^2 v/av)$ in the non-resonant cases.

Even in the case of exact resonance the probability of transfer may be quite small at ordinary temperatures. Thus, for vibrational energy

† TAKAYANAGI, K., *Prog. theor. Phys.*, Osaka **8** (1952) 111.

0·1 eV and similar homonuclear molecules with $M_A \simeq 20$ a.m.u. and $a = 1/a_0$,
$$P_{10}^{01} \simeq 3 \times 10^{-3}.$$

As an illustration of the dependence of the probabilities on the energy difference ΔE we show in Fig. 17.39 the variation of $\mathscr{J}_{10}^{nm}$ with ΔE for three different values of a. The reduced mass M is taken as $2 \cdot 5 \times 10^4$ electron masses and the initial wave-number of relative motion as 10 a.u.

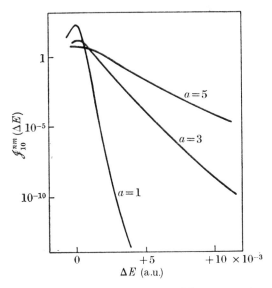

Fig. 17.39. Variation of the expression $\mathscr{J}_{10}^{nm}$ in (147) with energy change ΔE for three different values of the parameter a. M is taken as $2 \cdot 5 \times 10^4$ electron masses and the initial wave-number of relative motion as 10 a.u.

corresponding roughly to collisions between O_2 and N_2 molecules at room temperature. The sharpness of the resonance, which becomes more marked as a decreases, is apparent. As $\mathscr{J}_{10}^{nm}$ is independent of n and m, if ΔE is given the same plot applies whether or not vibration transfer occurs. The only effect of such transfer on $\mathscr{J}_{10}^{nm}$ arises through the reduction of ΔE.

4.3. *Use of a more realistic interaction*

Apart altogether from the restriction to head-on collisions, we have also concentrated attention on a single type of interaction, the exponential form (108). In practice, the basic interaction between atoms that do not attract each other chemically is of the form shown in Fig. 16.1 which, at large separations, is dominated by the van der Waals attraction.

Because of the gradual nature of the variation of the interaction with distance in this region it has little direct effect in producing vibrational deactivation. However, because of the attraction, the relative velocity of the colliding systems is increased so that they enter the region of rapidly varying interaction with greater relative velocity. An increase of the effective value of $\bar{v}$ in (130) will lead to an appreciable increase in deactivation probability.

One procedure to allow for these effects would be to represent the interaction by some such empirical form as the Lennard–Jones potential (see Chap. 16, § 6.1)

$$V(r) = \frac{\epsilon}{n-6}\left\{6\left(\frac{r_m}{r}\right)^n - n\left(\frac{r_m}{r}\right)^6\right\}, \tag{151}$$

where n is often taken as 12. Unfortunately, to carry out the calculations using this interaction, even with the semi-classical method of § 4.1.3, a great amount of computation is required and each case would have to be dealt with specially.

One approach takes account of the fact that the distorted-wave method can still be applied analytically if the interaction is of the Morse form (see Chap. 12, § 4.1)

$$V(r) = -\epsilon[2\exp\{-\tfrac{1}{2}a(r-r_m)\} - \exp\{-a(r-r_m)\}]. \tag{152}$$

Here ϵ is the depth of the potential minimum and r_m its location. Devonshire† has applied the distorted-wave method with this potential and finds, in place of (121),

$$P_{sp} = \frac{\pi^2}{4}\lambda^2 a^2 |x_{sp}|^2 (q_p^2 - q_s^2)^2 \sinh 2\pi q_s \sinh 2\pi q_p \times$$
$$\times (\cosh 2\pi q_p - \cosh 2\pi q_s)^{-2}(A_s + A_p)^2/A_s A_p, \tag{153}$$

where
$$A_{s,p} = |\Gamma(iq_{s,p} + \tfrac{1}{2} - d)|^2, \qquad d = (2M\epsilon)^{\frac{1}{2}}/a\hbar.$$

The remaining symbols are as in § 4.1.1.

If $q_{s,p} \gg 1$, as before,

$$\frac{\sinh 2\pi q_s \sinh 2\pi q_p}{(\cosh 2\pi q_p - \cosh 2\pi q_s)^2} \simeq \exp\{-2\pi|q_s - q_p|\}. \tag{154}$$

Also, if $q_{s,p} \gg d$, we may use the asymptotic expansion of the Γ-function

$$\Gamma(z) \sim (2\pi)^{\frac{1}{2}} z^{z-\frac{1}{2}} e^{-z},$$

to give
$$(A_s + A_p)^2/A_s A_p \simeq (q_p/q_s)^{-d} \exp 2\{(\pi - \theta_s)q_s - (\pi - \theta_p)q_p\}, \tag{155}$$

where
$$\theta_{s,p} = \arctan(q_{s,p}/d)$$
$$\simeq \frac{\pi}{2} - \frac{d}{q_{s,p}}. \tag{156}$$

Thus
$$(A_s + A_p)^2/A_s A_p \simeq (q_p/q_s)^{-d} \exp\{\pi|q_s - q_p|\}. \tag{157}$$

† DEVONSHIRE, A. F., *Proc. R. Soc.* A**158** (1937) 269.

Since $q_p/q_s = v_p/v_s$ and $v_p^2 = v_s^2 - 2h\nu/M$,

$$(q_p/q_s)^{-d} \simeq (1+h\nu d/Mv_s^2), \quad \text{provided } h\nu d/Mv_s^2 \ll 1.$$

On substitution in (153) we see that

$$P_{sp} \simeq \frac{256\pi^6 \nu^2 \lambda^2 M^2}{a^2 h^2} \exp\{-4\pi^2\nu/a\bar{v}\}\left\{1+\frac{2\pi\nu(2M\epsilon)^{\frac{1}{2}}}{aMv_s^2}\right\}|x_{sp}|^2. \tag{158}$$

Comparison with (130) shows that, with the assumptions we have made, the only effect of the attractive term is to introduce the factor $(q_p/q_s)^{-d}$ which, when $h\nu d/Mv_s^2 \ll 1$, can be written as $(1+h\nu d/Mv_s^2)$. This shows that the vibrational transition must take place near the closest distance of approach where its probability is determined mainly by the slope of the rapidly rising repulsion. The extra factor can be interpreted as arising from the acceleration of the relative motion by the attractive field so that the relative velocity in the reaction region near the closest distance of approach is increased and the collision made more sudden.

Herzfeld and his collaborators,† taking account of the expectation that the principal effect of the attractive part of the interaction is to accelerate the relative motion, so that the collision in the reactive region near the distance of closest approach is less nearly adiabatic, have represented the Lennard–Jones (12, 6) interaction by one of the form

$$V(R) = -\epsilon + \mu\exp(-aR), \tag{159}$$

where ϵ is the depth of the potential minimum at $r = r_m$. μ and a are then adjusted to give the best fit with the interaction for $r > r_c$, the closest distance of approach.

Two method have been used. In both, the interaction (159) is chosen to agree with the Lennard–Jones one at $r = r_c$. The second condition is taken to be either that the two curves also have the same slope at r_c (method A) or that they agree at $r = \sigma$, where σ is the separation for which the (12, 6) interaction vanishes (method B) (see Chap. 16, § 6.1).

The effect of the constant term in (159) is to change the effective relative velocity in the collision. If $\epsilon < \bar{E}$, the mean relative kinetic energy, the mean relative velocity is increased to $\bar{v}(1+\frac{1}{2}\epsilon/\bar{E})$ so that the probability of deactivation per collisions is increased by the factor $\exp(4\pi^2\nu\epsilon/aM\bar{v}^3)$. To average this factor over a Maxwellian distribution of relative velocities we need only substitute (132) for $\bar{v}$ to give simply $\exp(\epsilon/\kappa T)$.

† HERZFELD, K. F. and LITOVITZ, T. A., *Absorption and dispersion of ultrasonic waves*, p. 278 (Academic Press, New York, 1959).

It is of interest to compare the Herzfeld factor $(1+\delta_H)$ with that $(1+\delta_M)$ which arises in the calculations with the Morse potential (152). We have
$$\delta_H = 4\pi^2\nu\epsilon/aM\bar{v}^3, \qquad \delta_M = 2\pi\nu(2M\epsilon)^{\frac{1}{2}}/aMv_s^2,$$
provided both are small. Hence
$$\delta_H = \frac{2\delta_M}{\pi}\left(\frac{E}{\epsilon}\right)^{\frac{1}{2}}, \tag{160}$$
if we ignore the difference between $\bar{v}$ and v_s, E being $\tfrac{1}{2}Mv_s^2$, the initial relative kinetic energy. We would not expect the two factors to be equal as no attempt has been made to represent the Morse potential by the form (159) for $r < r_m$.

Before discussing the effectiveness of this method of attempting to correlate information from the transport and equilibrium properties, which determine the constants in the (12, 6) interaction, with that about vibrational deactivation we consider how to extend the method to deal with collisions in three dimensions.

4.4. *Three-dimensional theory of vibrational deactivation*

4.4.1. *Semi-empirical treatment.* If, instead of head-on collisions, we consider collisions in which the impact parameter is p, it is necessary to include an additional effective interaction
$$V_{\text{eff}} = Mv^2p^2/r^2. \tag{161}$$
With this term included it is no longer possible to carry out the calculations analytically with an exponential interaction either by the semi-classical or distorted-wave methods. Takayanagi† suggested that this difficulty may be circumvented by taking account of the fact that, just as with the van der Waals attraction, the interaction (161) is so gradual that its only important effect is to change the effective relative velocity of impact near the distance of closest approach r_c. He therefore proposed to allow for (161) by introducing a constant repulsion Mv^2p^2/r_c^2. The probability of a deactivating collision, in which the impact parameter is p, now becomes
$$P_{10}(p) = \frac{16\pi^3 M^2\nu\lambda^2}{\hbar M^* a^2}\exp\left\{-\frac{4\pi^2\nu}{a}\left(\frac{1}{\bar{v}}-\frac{\epsilon}{M\bar{v}^3}+\frac{p^2}{\bar{v}r_c^2}\right)\right\}, \tag{162}$$
and the cross-section for a deactivating collision is given by
$$Q_{10} = 2\pi\int_0^\infty P_{10}(p)p\,dp$$
$$= \frac{8\pi^2 M^2\nu\lambda^2}{\hbar M^* a}\exp\left\{-\frac{4\pi^2\nu}{a}\left(\frac{1}{\bar{v}}-\frac{\epsilon}{M\bar{v}^3}\right)\right\}\pi r_c^2. \tag{163}$$

† Takayanagi, K., *Prog. theoret. Phys. Japan* **8** (1952) 497. See also Herzfeld, K. F. and Litovitz, T. A., loc. cit., p. 287.

This may be averaged over the distribution of relative velocities in the usual way. The fraction of molecules which have relative velocities between v and $v+dv$ is, at temperature T,

$$f(v)\,dv = \frac{M^2 v^3}{2(\kappa T)^2}\exp\left(-\frac{Mv^2}{2\kappa T}\right)dv. \tag{164}$$

We now make the same approximation as for head-on collisions (see p. 1543) in integrating (163) over the distribution (164). The quantity $(4\pi^2\nu/av)+(Mv^2/2\kappa T)$ has a minimum (see (132)) where

$$v = v_{\mathrm{m}} = (4\pi^2\nu\kappa T/aM)^{\frac{1}{3}}, \tag{165}$$

so $\displaystyle\int Q_{10} f(v)\,dv = \frac{4\pi^2 M^4 v_{\mathrm{m}}^4}{hM^* a(\kappa T)^2}\exp\{-(54\pi^4\nu^2 M/a^2\kappa T)^{\frac{1}{3}}+$

$$+(\epsilon+\tfrac{1}{2}h\nu)/\kappa T\}\pi r_{\mathrm{c}}^2 \times \int_{-\infty}^{\infty}\exp\left\{-\frac{3}{2}\frac{M}{\kappa T}(v-v_{\mathrm{m}})^2\right\}dv$$

$$= \frac{2\pi M^4 v_{\mathrm{m}}^4}{hM^* a(\kappa T)^2}\left(\frac{2\pi\kappa T}{3M}\right)^{\frac{1}{2}}\exp\{-(54\pi^4\nu^2 M/a^2\kappa T)^{\frac{1}{3}}+(\epsilon+\tfrac{1}{2}h\nu)/\kappa T\}\pi r_{\mathrm{c}}^2. \tag{166}$$

Schwartz, Slawsky, and Herzfeld† introduce two characteristic temperatures

$$\vartheta = h\nu/\kappa, \qquad \vartheta' = 16\pi^4 M\nu^2/a^2\kappa, \tag{167}$$

in terms of which (166) may be written, using (164), as

$$\frac{M_\mathrm{A}\,M_\mathrm{C}}{M_\mathrm{B}(M_\mathrm{A}+M_\mathrm{B}+M_\mathrm{C})}\frac{\vartheta'}{\vartheta}\left(\frac{2\pi}{3}\right)^{\frac{1}{2}}\left(\frac{\vartheta'}{T}\right)^{\frac{1}{6}}\exp\left\{-\frac{3}{2}\left(\frac{\vartheta'}{T}\right)^{\frac{1}{3}}+\frac{(\tfrac{1}{2}\vartheta+\epsilon/\kappa)}{T}\right\}\pi r_{\mathrm{c}}^2. \tag{168}$$

This expression was derived by Schwartz, Slawsky, and Herzfeld,‡ who were particularly concerned with calculating the probability of deactivation per collision. To do this they divided (168) by the averaged cross-section for elastic collisions which they took as $\pi\alpha\sigma^2$, where σ is the separation at which the interaction vanishes and α is a factor that varies slowly with temperature, is of order unity, and has been tabulated.§ The mass factor given by Schwartz, Slawsky, and Herzfeld differs from (168) which refers only to collisions in which the atom A collides with the atom B of BC. Assuming that ϑ' and hence a is the same for interaction of A with B and with C, for which there is little justification unless B and C are similar, the factor $M_\mathrm{C}/M_\mathrm{B}$ in (130 a) is replaced by the mean

$$\tfrac{1}{2}\left(\frac{M_\mathrm{C}}{M_\mathrm{B}}+\frac{M_\mathrm{B}}{M_\mathrm{C}}\right) = \tfrac{1}{2}\left(\frac{M_\mathrm{C}^2+M_\mathrm{B}^2}{M_\mathrm{B}\,M_\mathrm{C}}\right),$$

---

† See HERZFELD, K. and LITOVITZ, T. A., *Absorption and dispersion of ultrasonic waves*, p. 277 (Academic Press, New York, 1959).
‡ SCHWARTZ, R. N., SLAWSKY, Z. I., and HERZFELD K. F., *J. chem. Phys.* **20** (1952) 1591.
§ loc. cit., p. 1427.

and this converts (130 a) to the factor used by Schwarz, Slawsky, and Herzfeld. In addition, they added an arbitrary factor which was supposed to take account of orientational effects. This they chose to be 1/3.

Calculations of the relaxation times and of mean probabilities of vibrational deactivation per collisions have been carried out by Herzfeld and his collaborators, using this procedure, for a number of pure gases.

TABLE 17.6

Ratio P_c/P_0 of deactivation probability per collision calculated by the Herzfeld method to that observed for different gases at different temperatures

N_2			O_2			Cl_2			CO_2†		
T (°K)	P_c/P_0		T (°K)	P_c/P_0		T (°K)	P_c/P_0		T (°K)	P_c/P_0	
	A	B		A	B		A	B		A	B
550	5×10^{-3}	0·08	1400	0·4	1·6	300	0·06	0·2	300	2	1·5
800	—	0·09	1900	0·4	1·0	1000	0·1	0·25	600	1	3
1200	—	0·15	3200	0·2	0·3	1400	0·1	0·25	1000	0·3	0·7
3500	0·55	0·65									
5400	0·5	0·6									

A and B refer to the two methods of fitting the constants in the interaction (159) (see p. 1549).

Considering the numerous simplifying assumptions which were made in deriving the formulae the agreement with observation is better than would be expected. Some examples are given in Table 17.6. In all cases of pure gases the constants a and ϵ in (159) were chosen to fit the best (12, 6) interaction derived from analysis of transport phenomena as given by Hirschfelder.‡ Results obtained by both methods of fitting are given in most cases. In all the arbitrary orientation factor of 1/3 has been included. It will be seen that the discrepancies between theory and experiment are not very different at different temperatures, showing that observed and calculated variations with temperature agree quite well. Except for N_2 at temperatures below 1000 °K the calculated results are not incorrect by more than one order of magnitude. In view of the sensitivity of the calculations to the slope of the interaction this is remarkable. It will be noted also that, in most cases, method B of fitting the form (159) to the (12, 6) interaction gives the better results.

† See p. 1523 and Fig. 17.33, which draw attention to the unreliability of the observed results for CO_2, so the ratios given are very rough.
‡ loc. cit., p. 1428.

4.4.2. *Application of the close-coupling (truncated eigenfunction) method.* Although the probability of vibrational deactivation on collision between diatomic molecules is very low at room temperature it rises rapidly with temperature and at some thousand degrees becomes comparable with unity. Under these conditions even the distorted-wave method, applied in full detail with realistic interactions, will not give good results. Furthermore, it is necessary to allow for the fact that real or virtual transitions between more than one vibrational state cannot be neglected. To calculate cross-sections for vibrational transitions it is necessary to turn to the close-coupling method discussed in Chapter 8, § 2.2.3 and applied to electron collisions with hydrogen and other atoms in that and the succeeding chapter. Marriott[†] has applied this method in detail to a number of cases and we now discuss the procedure he adopted and the results obtained.

The starting point is the three-dimensional generalization of (110), the only approximation being the neglect of nuclear rotation so that the wave functions $\chi_s(x)$ remain simply functions of the vibrational coordinates. The functions $F_s(\mathbf{R})$ now depend on the orientation of the vector $\mathbf{R}$ in space and (111) becomes

$$(\nabla^2 + k_s^2 - U_{ss})F_s(\mathbf{R}) = \sum_p U_{sp} F_p(\mathbf{R}). \tag{169}$$

Denoting the initial molecular state by χ_i, the coupled equations (169) must be solved to obtain solutions which are proper functions satisfying the boundary condition

$$F_s(R, \Theta, \Phi) \sim \delta_{is}\,\mathrm{e}^{\mathrm{i}k_s R} + R^{-1}\mathrm{e}^{\mathrm{i}k_s R} f_s(\Theta, \Phi). \tag{170}$$

The cross-section for excitation of the transition $i \to s$ is then given by

$$Q_{is} = \frac{k_s}{k_i} \int\!\!\int |f_s(\Theta, \Phi)|^2 \sin\Theta \, \mathrm{d}\Theta\mathrm{d}\Phi. \tag{171}$$

It is clearly impossible to solve the infinite set of coupled equations (169) and the close-coupling method works by truncating the expansion (110) to include $F_i(\mathbf{R})\chi_i(x)$ and only those other terms that are closely coupled to it. In this way it is possible, in suitable circumstances, to limit the number of coupled equations so solutions may be obtained using automatic computers.

To obtain the functions U_{sp} Marriott assumed that the interaction between two molecules can be written in separable form as

$$V(x_1,...,x_j; R) = V_0\,V(R)V_1(x_1)...V_j(x_j), \tag{172}$$

[†] MARRIOTT, R., *Proc. phys. Soc.* **83** (1964) 159; **84** (1964) 877; **86** (1965) 1041; **88** (1966) 83 and 617.

where $x_1,\ldots,x_j$ are the internal vibrational coordinates of the two molecules. This is essentially a generalization of (108 a). With this form

$$\hbar^2 U_{sp}(R)/2M = V_{sp}(R)$$
$$= V_0 V(R) \prod_{n=1}^{j} V_{sp}^n, \tag{173}$$

with
$$V_{sp}^n = \int \chi_s(x_n) V_n(x_n) \chi_p(x_n) \, dx_n. \tag{174}$$

The vibrational wave-functions χ were taken to be Hermite polynomials.

The function $V(R)$ was chosen to have the $(12,6)$ form appropriate to the transport data in the gas concerned. $V_n(x_n)$ was taken to have the simple exponential form

$$V_n(x_n) = \exp(-ax_n) \tag{175}$$

with a determined by the prescription (method A, p. 1549) of Herzfeld and his collaborators so as to provide a good fit with the $(12,6)$ interaction.

Finally, the constant V_0 was chosen so that

$$V_0 = 1 \Big/ \Big(\prod_{n=1}^{j} V_{00}^n \Big), \tag{176}$$

which ensured that the interaction $V_{00}(R)$ between the molecules in the ground state, averaged over the internal coordinates, was the best fit, by Herzfeld's method A, to the $(12,6)$ interaction derived from transport data.

Having chosen the interactions, the only remaining problem was the choice of the number of coupled equations to be included. Marriott began in each case by dealing first at low collision energies with the 0–1 transition, only involving two coupled states, and then including the 2 and 3 quantum states successively as the energy increased.

To solve the equations the three-dimensional functions $F_s(\mathbf{R})$ were expanded in spherical harmonics in the usual way:

$$F_s(\mathbf{R}) = R^{-1} \sum_l i^l (2l+1) G_l^s(R) P_l(\cos\Theta), \tag{177}$$

where
$$\left[\frac{d^2}{dR^2} + \left\{k_s^2 - \frac{l(l+1)}{R^2} - U_{ss}(R)\right\}\right] G_l^s = \sum_{p \neq s} U_{sp} G_l^p. \tag{178}$$

To satisfy the boundary conditions (170)

$$G_l^s \sim \delta_{is} \sin(k_s R - \tfrac{1}{2} l\pi) + \alpha_l^s e^{ik_s R},$$
$$G_l^s(0) = 0, \tag{179}$$

and
$$Q_{is} = (4\pi k_s/k_i^3) \sum_l (2l+1) |\alpha_l^s|^2. \tag{180}$$

Not only did Marriott solve the coupled equation numerically for all significant values of l with the centrifugal force term $l(l+1)R^{-2}$ fully included but in all cases except the first studied (CO) he also allowed the constant a in (175) to vary with l. This takes account of the fact that the distance of closest approach, at which the fitting to the (12, 6) potential is done, depends on l.

Table 17.7 gives some typical results[†] for vibrational excitation of CO in terms of calculated collision cross-sections at different impact energies. The importance of coupling with the higher vibrational states is apparent at the higher impact energies which, however, refer to very high temperatures. The vibrational relaxation times for pure CO calculated by Marriott[†] are compared with observed times in Fig. 17.40. It will be seen that the calculated values are somewhat low but the variation with temperature is given quite well. The temperature range covered is quite small, however, so most of the special features of Marriott's method, such as the close coupling with higher vibrational states, are unimportant.

The initial work on CO was followed by a detailed study of vibrational relaxation in pure CO_2[‡] and in mixtures of CO_2 with H_2[§] and with H_2O.[||] In Marriott's notation the vibrational states $(0, 0, 0)$, $(1, 0, 0)$, $(2, 0, 0)$, $(0, 1, 0)$, and $(0, 0, 1)$, in which the respective numbers refer to the number of quanta in the bending mode, the symmetric longitudinal mode, and the asymmetric longitudinal mode respectively (see § 3.1.3 and Fig. 17.26), are designated as the states 0, 1, 2, 3, and 4. Cross-sections for transitions between all of these states were calculated up to collision energies of 2·5 eV. Having all these cross-sections it became possible to study the effective relaxation allowing for transfer between the different modes. In particular, Marriott applied his results to calculate the time rate of decay of the vibrational energy density in CO_2 shock-excited from room temperature to 400 and 1000 °K.

The total vibrational energy density in a gas is given by

$$E_{\text{vib}} = \sum_{n=0}^{\infty} N_n E_n, \qquad (181)$$

where N_n is the population density and E_n the vibrational energy of the nth vibrational state. The N_n satisfy the coupled equations

$$\frac{dN_n}{dt} = \sum_{n=0}^{\infty} \Gamma_{mn} N_m, \qquad (182)$$

[†] MARRIOTT, R., Proc. phys. Soc. **83** (1964) 159.
[‡] MARRIOTT, R., ibid. **84** (1964) 877.
[§] ibid. **86** (1965) 1041.
[||] ibid. **88** (1966) 83.

TABLE 17.7

Cross-sections for vibrational excitation in CO as calculated by Marriott

Cross-sections (in 10^{-16} cm^2)

Relative kinetic energy (eV)	Two coupled states (0, 1)	Three coupled states (0, 1, 2)			Four coupled states (0, 1, 2, 3)		
	Q_{01}	Q_{01}	Q_{02}	Q_{12}	Q_{01}	Q_{02}	Q_{12}
0·266	0·000	0·000			0·000		
0·50	0·336×10^{-5}						
0·532			0·000	0·000		0·000	0·000
0·85		0·521×10^{-3}	0·569×10^{-10}	0·225×10^{-6}			
1·50	0·968×10^{-3}	0·147×10^{-1}	0·183×10^{-5}	0·150×10^{-2}	0·196×10^{-1}	0·55×10^{-5}	0·226×10^{-2}
2·50	0·521×10^{-1}	0·872×10^{-1}	0·165×10^{-3}	0·608×10^{-1}			

where $\Gamma_{nm} = \gamma_{nm}$ and $\Gamma_{nn} = -\sum_{m=0}^{\infty} \gamma_{nm}$ $(m \neq n)$. The γ_{nm} are generalized rate coefficients corresponding to γ_{01} and γ_{10} in the analysis of simple ultrasonic dispersion given on p. 1468. Illustrative values for the γ_{nm} calculated by Marriott,† ignoring all but the five vibrational states referred to above, are given in Table 17.8.

It will be seen that, at 400 °K, the excitation rates between the excited vibrational modes are faster than all transitions to the ground state. Furthermore, the rate of transition from the bending mode (state 1) to the ground state is much greater than for other transitions to the ground state. These are the conditions already discussed in § 3.1.3 in which the observed relaxation time is essentially that for the bending mode only. At 1000 °K this situation no longer applies to the same degree. The detailed calculations using (181) and (182) show that, at both temperatures, after about $\tfrac{1}{3}\,\mu$s the vibrational energy density falls off as

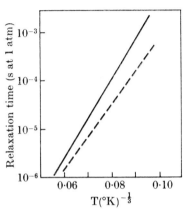

FIG. 17.40. Comparison of observed and calculated vibrational relaxation times for pure CO as functions of $T^{-\frac{1}{3}}$, where T °K is the temperature. —— best fit of observed data (see Fig. 17.31). – – – calculated by Marriott.

$\exp(-t/\tau)$, where τ is a (single) effective relaxation time. At 400 °K, $\tau = 0.13 \times 10^{-4}$ s, which is not very different from the relaxation time 0.17×10^{-4} s for the bending mode alone. This is expected from the argument just outlined. At 1000 °K, $\tau = 0.7 \times 10^{-6}$ s, which is five times smaller than that, 3.8×10^{-6} s, for the bending mode alone.

TABLE 17.8

Calculated excitation rate coefficients for different vibrational transitions in pure CO_2 at 400 and 1000 °K

Transition	01	02	03	04	12	13	14	24	32	34
Rate coefficient (cm^3 s^{-1} atom^{-1})										
400 °K	1.3×10^4	4.1	0.3	0.2×10^{-3}	4.7×10^5	4.6×10^3	1.1	1.3×10^3	9.4×10^7	140
1000 °K	8.3×10^5	5.2×10^3	630	19	4.9×10^6	1.4×10^5	4.9×10^3	4.1×10^5	9.4×10^7	4.9×10^4

† *Proc. phys. Soc.* **84** (1964) 877.

The calculated values for the effective relaxation times are about ten times larger than the most probable values derived from the observations (see Fig. 17.33) but this does not affect the interest of the analysis of the relaxation in a simple polyatomic molecule. Marriott has also used his results to check the assumption (see p. 1490) that a single vibrational

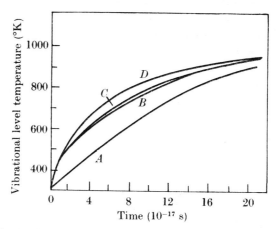

FIG. 17.41. Variation of calculated vibrational temperatures during relaxation of CO_2 from 300 to 1000 °K. Curves A, B, C, D correspond to vibrational states (100), (200), (010), and (001) respectively (see Fig. 17.26).

temperature can be specified at each stage of the relaxation process. For a particular state n the vibrational temperature T_v^n is defined by

$$N_n g_n/N_0 = \exp(-E_n/\kappa T_v^n), \qquad (183)$$

g_n being the statistical weight of the nth state. Fig. 17.41 shows the variation of T_v^n for the states 1–4 with time, during relaxation of CO_2 from 300 to 1000 °K, calculated by Marriott from (183). It will be seen that a single vibrational temperature can be specified to better than 20 per cent.

A similar elaborate analysis of vibrational relaxation was carried out by Marriott for a mixture of CO_2 with H_2† in which he introduced into the equations (182) terms involving cross-sections for vibrational excitation of CO_2 by impact with H_2. At 400 °K the rate coefficient γ_{01} for CO_2–H_2 collisions is 10^9 cm^3 s^{-1} atom^{-1}, nearly 10^4 times larger than for CO_2–CO_2 collisions. For fractional H_2 concentrations less than 10 per cent and temperatures below 400 °K the relaxation rate is essentially that of the bending modes but at higher temperatures it is more complex.

† loc. cit., p. 1555.

One interesting feature of the calculation is that it predicts that H_2 should be less effective in producing vibrational relaxation at 400 °K than at 300 °K. This result, which is in agreement with observations of Winter,† is illustrated in Fig. 17.42, which shows the calculated frequency of maximum absorption in CO_2 as a function of fractional H_2 concentration at 300, 400 and 600 °K.

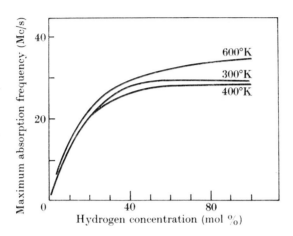

FIG. 17.42. Frequency of maximum sound absorption in CO_2–H_2 mixtures as a function of H_2 concentration at different gas temperatures, calculated by Marriott.

In a further calculation Marriott‡ investigated the effect of water vapour on relaxation in CO_2. This is well known to be very pronounced and indeed has probably contributed substantially to the unreliability of many experimental observations of relaxation times. It has been suggested that the effects are large because of chemical forces such as those that lead to the production of carbonic acid H_2CO_3. Marriott ignored such forces but included, in addition to the (12, 6) interaction, a specific additional contribution due to interaction between the permanent dipole moment of the water molecule and the induced dipole moment in CO_2. This is a non-central interaction which cannot be handled computationally, so Keesom's approximation§ was introduced. This consists in using a weighted mean of the interaction over all orientations of the dipole, the weights being Boltzmann factors which allow for the fact that the molecule spends more time in orientations for which

† WINTER, T. G., *J. chem. Phys.* **38** (1963) 2761.
‡ loc. cit., p. 1555.
§ KEESOM, W. H., *Phys. Z.* **22** (1921) 129.

the interaction is small (see also § 5.6). In this way an effective additional interaction
$$V_d = -\mu^2\alpha/R^6 \tag{184}$$
is introduced, μ being the dipole moment of the water molecule and α the polarizability of the CO_2. Although the Keesom approximation is not very good when the rotation period and time of collision are comparable, as they are under the conditions of the calculation, the errors made are probably no larger than those arising from other approximations which were made as outlined earlier.

A similar study of the vibrational relaxation of a CO_2–H_2O mixture to that for CO_2–H_2 was then carried out, account being taken of the same vibrational states of CO_2 as before. No account was taken of the inner structure of the H_2O molecule as vibrational transfer between the molecules was neglected. The results show many of the observed features. In pure CO_2 at 300 °K the absorption peak due to vibrational relaxation occurs at a frequency of about 20 kc/s. At temperatures below 600 °K, 0·1 per cent of H_2O introduces a second dominant absorption peak at a higher frequency, while with 10 per cent H_2O a third peak at a still higher frequency becomes important. This is in qualitative agreement with the observations (see Fig. 17.35). Detailed quantitative agreement is obtained if the collision cross-sections are increased by a factor of 4.

A second feature is the temperature dependence of the effect of water vapour. According to (105) the relaxation time τ for the mixture is given by
$$\tau^{-1} = (1-x)\tau_{CO_2}^{-1} + x\tau_{H_2O}^{-1},$$
where x is the molar fraction of H_2O and τ_{H_2O} is the relaxation time for CO_2 in a large excess of H_2O. Knowing τ_{CO_2} from previous calculations and calculating an effective τ for a given mixture as described above, τ_{H_2O} can be obtained. It is found that the value derived in this way depends on x, the relaxation process, involving several vibrational modes of CO_2, being a complex one. Fig. 17.43 shows the results obtained for τ_{H_2O} as a function of temperature derived from calculations carried out for different molar fractions of H_2O. An experimental curve, derived from observations with 2 per cent concentration of H_2O, is also shown. It will be seen that, as the molar fraction of H_2O assumed in the calculation increases, the theoretical curve below 700 °K resembles the observed more and more closely. If the calculated cross-sections for the CO_2–H_2O collisions were increased by a factor of 2 the observed and calculated maxima would be brought into close agreement at 350 °K. This is a very small adjustment in view of the complexity of the calculations. It is

noteworthy also that we would expect from the theory that $\tau_{H_2O}^{-1}$ should increase again with temperature beyond 700 °K.

It seems that it is not necessary to involve chemical forces to explain the most outstanding features of the relaxation of CO_2–H_2O mixtures. Allowance for the complexity of the process seems to provide an adequate interpretation.

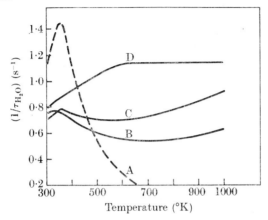

FIG. 17.43. Variation with temperature of the reciprocal relaxation time (reduced to s.t.p.) $1/\tau_{H_2O}$ for CO_2 infinitely dilute in H_2O. – – – A, observed by Eucken and Nümann; ——— B, C, D derived using the linear mixture formula (105) from theoretical calculations for mixtures containing 40, 10, and 1 per cent concentrations of H_2O respectively.

In a final calculation Marriott applied the method to H_2–H_2O mixtures allowing for coupling between the lowest five vibrational states of H_2O. The H_2–H_2O cross-sections were calculated as for CO_2–H_2O. For H_2O–H_2O collisions the interaction between the two permanent dipoles must be included. This was also statistically averaged over all dipole orientations and gives rise to a term varying as R^{-6}, the magnitude of which depends on the mean kinetic energy of the gas (see § 5.6) which was replaced in the calculations by the appropriate relative kinetic energy involved in the collision.

It was found, in agreement with observation, that the relaxation of pure H_2O is well described by a single relaxation time τ which is such that $\log \tau$ varies with temperature T as $T^{-\frac{1}{3}}$. The calculated values of τ are between 5 and 10 times too large. Detailed results were also given for mixtures with H_2.

4.5. *Elementary theory of vibration–rotation transfer*

It was pointed out in § 3.1.3 that the vibrational relaxation times for polyatomic molecules containing two or more hydrogen atoms are relatively shorter than for other molecules with approximately the same vibrational frequency. Furthermore, the relaxation times for hydrides are often shorter than for the corresponding deuterides although the latter have smaller vibration frequencies. A possible explanation of these results was suggested by Cottrell and Matheson[†] in terms of transfer of energy between vibration in one molecule and the relatively free rotation of hydrogen atoms in the other. We now describe briefly a simple theoretical model for this process introduced by Cottrell, Dobbie, McLain, and Read.[‡]

Let C_1, C_2 be the centres of mass of the two colliding molecules with $C_1 C_2 = R$. The vibration in molecule 1 is assumed to take place normal to $C_1 C_2$, the displacement of the vibrating atom A from C_1 being x. In molecule 2 the hydrogen atom H is assumed to rotate, in the plane of $C_1 C_2$ and of the vibration in molecule 1, at a distance d from C_2. When the angular displacement of C_2 H from $C_1 C_2$ is θ we have, for the distance between the H atom and the vibrating atom A,

$$\rho \simeq r\left(1 - \frac{x}{r}\frac{d}{r}\sin\theta\right),$$

where
$$r^2 = R^2 + d^2 - 2Rd\cos\theta.$$

If $V(\rho)$ is the interaction energy between A and H

$$V(\rho) \simeq V(r) - (xd\sin\theta/r)V'(r). \tag{185}$$

Treating the molecule 2 as a moving source of interaction, pursuing a rectilinear orbit so that its centre of mass makes a head-on collision with that of molecule 1, we take $\theta = \omega t$, where ω is the angular velocity of rotation of the H atom, and for simplicity write

$$R = R_0 \cosh(t/t_0),$$

where t_0 is an effective time of collision.

According to (136) the probability of a vibrational transition occurring in the collision is given by

$$P_{01} = \hbar^{-2} \left| \int_{-\infty}^{\infty} \int_{-\infty}^{\infty} \chi_0(x) V(t) \exp(2\pi i \nu t) \chi_1(x) \, dt dx \right|^2,$$

where ν is the frequency of the vibrational quantum involved and χ_0, χ_1 are the initial and final vibrational wave functions. Using the form (185) for V with the appropriate substitutions for θ and R in terms of t we have

$$P_{01} = \hbar^{-2}|x_{01}|^2 d^2 \left| \int_{-\infty}^{\infty} \exp(2\pi i \nu t) \sin\omega t \, U[\{d^2 - 2dR_0\cos\omega t\cosh(t/t_0) + R_0^2\cosh^2(t/t_0)\}^{\frac{1}{2}}] \, dt \right|^2,$$

where $x_{01} = \int x\chi_0\chi_1 \, dx$, and $U(r) = V'(r)/r$.

Cottrell, Dobbie, McLain, and Read applied this formula to compare vibrational relaxation times for the pairs CH_4, CD_4; PH_3, PD_3; and AsH_3, AsD_3. They took $V(r)$ to have the form $V_0 r^{-12}$. For each member of a pair V_0, R_0, and d are the same. Results were expressed in terms of the ratios of reduced relaxation times τ_D/τ_H for deuteride and hydride respectively where

$$\frac{\tau_D}{\tau_H} = \frac{P_{01}(H)}{P_{01}(D)} \frac{\bar{v}(H)}{\bar{v}(D)} \frac{1 - \exp(-h\nu_H/\kappa T)}{1 - \exp(-h\nu_D/\kappa T)},$$

$\bar{v}$ being the mean relative velocity of the colliding molecules so $\bar{v}(H)/\bar{v}(D)$ is the

[†] loc. cit., p. 1530.   [‡] loc. cit., p. 1532.

ratio of the collision frequency for the hydride to that for the deuteride. ν was taken as the lowest vibration frequency and ω calculated as for a classical rotator with energy taken as the mean value $\frac{1}{2}(\kappa T + h\nu)$ before and after collision and with moment of inertia that of the molecule. d was obtained from molecular structure data. R_0 was estimated from the collision diameter and t_0 taken as $R_0/\bar{v}$.

It was found that the ratio came out to be 1·7, 1·5, and 1·2 for C, P, and As respectively, agreeing much more closely with the observed ratios 1·7, 1·5, and 0·6 than do the corresponding ratios 0·6, 0·13, and 0·04 calculated on the assumption of deactivation by transfer of vibrational to translational energy as in § 4.1, the inverse twelfth power interaction being represented by an exponential form using the Herzfeld–Litovitz prescription (p. 1549).

Although this model is oversimplified it nevertheless provides considerable support for the suggestion that vibration–rotation transfer is the factor which makes the relaxation times of the deuterides longer than those of the corresponding hydrides despite the smaller vibrational quanta involved.

4.6. *Summarizing remarks on the theory of vibrational relaxation*

On the whole, the detailed theoretical description is remarkably successful. The vibrational deactivation cross-sections are very sensitive to the slope of the intermolecular interaction near the distance of closest approach so that quantitative prediction of relaxation rates, even on a semi-empirical basis using interactions derived from transport coefficient data, could hardly be expected. Nevertheless, the main features of the variation from molecule to molecule are reproduced as well as the temperature variation. At any rate, the phenomena are well understood in terms of the theory and further progress depends on more accurate data to justify more elaborate theoretical computation using more sophisticated molecular interactions.

The analysis by Marriott of the relaxation process in CO_2 and in H_2O in terms of a three-dimensional theory, which allows for rapid real or virtual transfer of vibrational energy between higher vibrational levels, is illuminating in relation to the observed existence of a single relaxation time only, even if his predicted rates are not in close agreement with the observed.

The theory provides useful estimates of the rates of vibration–vibration transfer processes between diatomic molecules which show that, for the lighter diatomic molecules, it is still not a highly probable process even in conditions of exact energy resonance. The predicted temperature variation of the transfer rate is probably reliable. The fact that Marriott was able to interpret many of the salient features of the behaviour of a CO_2–H_2O mixture without taking into account vibration–vibration transfer is of interest.

Although only a simple theory of vibration–rotation transfer has been attempted it provides support for the interpretation of the behaviour of polyatomic molecules containing two or more hydrogen atoms.

5. The excitation and deactivation of molecular rotation on impact between molecules

It was pointed out in § 1, that while under room temperature conditions the duration of a collision between molecules is long compared with the vibrational period it is either comparable with or shorter than the period of molecular rotation. As a result, whereas the probability of a vibrational transition is very small during the impact, rotational transitions can take place with high probability.

This conclusion is supported by the experimental evidence which we discuss below. In fact, the only molecules for which the chance of rotational excitation on impact at ordinary temperatures is very small are H_2 and D_2. The perturbation theory which gave satisfactory results for vibrational relaxation times for molecules at temperatures less than say 500 °K cannot therefore be applied to deal with rotational transition probabilities for any but the lightest molecules and more elaborate methods must be developed. The situation is like that for vibrational relaxation at high temperatures in which many vibrational states are closely coupled (see § 4). Because of this, and also because rotational relaxation effects are much less noticeable, comparatively little attention has been paid to the subject. However, inelastic collisions involving rotational excitation must be taken into account in the interpretation of observed data in the absorption and scattering of molecular beams. The great revival of interest in this subject (see Chap. 16 and also § 6 of this chapter) has naturally stimulated interest in this type of collision.

We shall begin by discussing both the experimental and theoretical investigations that have been carried out on the rotational relaxation of H_2 and D_2, as these are special cases in which the relaxation effects are relatively large and a perturbation treatment may be applied. This will be followed by an account of observational evidence about excitation rates for other molecules from the same types of techniques which we have described for vibrational relaxation. A brief account will then be given of the contributions made by rotational relaxation to the transport properties of polyatomic gases.

We shall next discuss the importance of inelastic collisions in which rotational energy loss occurs in molecular beam collisions. This will include an account of theories capable of allowing for the close coupling between rotational states, as well as experiments to measure specific inelastic cross-sections, and will deal in particular with collisions between polar molecules.

The section will conclude with a brief account of certain phenomena concerned with nuclear magnetic resonance in which an important role is played by collisions in which the orientation of the molecular angular momentum alone is changed, its magnitude being unaltered.

5.1. *Observed rotational relaxation in* H_2 *and* D_2

The first observations of sound absorption due to rotational relaxation in H_2 and D_2 were made by van Itterbeek and his associates† in 1937–8 using an ultrasonic interferometer at 0·6 Mc/s. Working at pressures between 0·4 and 1 atm the measured absorption coefficients exceeded the classical (see (5)) by a factor of 20 for H_2 and 10 for D_2. However, these measurements did not yield the correct variation with pressure and it was not until 1951 that they were corrected by van Itterbeek and Verhaegen.‡

Meanwhile, in 1945, Stewart (E. S.), Stewart (J. L), and Hubbard§ observed rotational dispersion for the first time using an interferometer at 3·9 and 6·4 Mc/s and pressures between 0·5 and 1 atm. A little later Rhodes‖ made dispersion observations both with normal H_2 carefully purified and with 99·8 per cent p-H_2, finding evidence of more than one relaxation process. Further evidence on this matter was obtained by Stewart and Stewart†† who made accurate absorption measurements for H_2 and D_2.

Later measurements have been made by Huber and Kantrowitz‡‡ and by Griffith,§§ using the impact tube method (§ 2.3), and by Parbrook and Tempest,‖‖ who measured the absorption coefficients of spectroscopically pure H_2 using the acoustic pulse technique.

The existence of a considerable delay time in attainment of rotational equilibrium by shocked H_2 has been established by Greene and Hornig.††† They measured the reflectivity R of the shock front in H_2 for a Mach number of 1·389, corresponding to a temperature of 373 °K, by the

† VAN ITTERBEEK, A. and MARIENS, P., *Physica, 's Grav.* **4** (1937) 609; VAN ITTERBEEK, A. and THYS, L., ibid. **5** (1938) 889.
‡ VAN ITTERBEEK, A. and VERHAEGEN, L., *Nature, Lond.* **167** (1951) 477.
§ STEWART, E. S., STEWART, J. L., and HUBBARD, J. C., *Phys. Rev.* **68** (1945) 231; STEWART, E. S., ibid. **69** (1945) 632; STEWART, J. L. and STEWART, E. S., *J. acoust. Soc. Am.* **20** (1948) 585.
‖ RHODES, J. E., *Phys. Rev.* **70** (1946) 91, 932.
†† STEWART, E. S. and STEWART, J. L., *J. acoust. Soc. Am.* **24** (1952) 194.
‡‡ HUBER, P. W. and KANTROWITZ, A., *J. chem. Phys.* **15** (1947) 275.
§§ GRIFFITH, W., *J. appl. Phys.* **21** (1950) 1319.
‖‖ PARBROOK, H. D. and TEMPEST, W., *J. acoust. Soc. Am.* **30** (1958) 985.
††† GREENE, E. F. and HORNIG, D. F., *J. chem. Phys.* **21** (1953) 617.

technique described on p. 1486. Fig. 17.44 shows their results, in which $\log R$ is plotted as a function of $\log(\lambda/\cos\theta)$, where λ is the wavelength and θ the angle of incidence of the radiation. For comparison, two theoretical curves are calculated from (37), one assuming no participation of rotational modes and the other full participation. The observations agree quite closely with the former of these showing that rotational

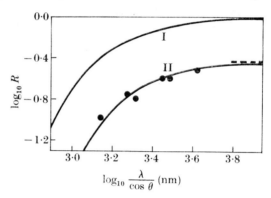

Fig. 17.44. Reflectivity of a shock front in H_2 at a Mach number 1·389. ● observed; —— I, calculated assuming full contribution from rotational modes; —— II, calculated assuming no contribution from rotational modes.

relaxation is relatively slow. Allowing for the inaccuracies of measurement it was only possible to show that rotational relaxation requires more than 150 collisions. This limit is consistent with the results of the acoustic and impact tube measurements as will be seen by reference to Table 17.9. This gives observed relaxation times on the assumption that a single relaxation process is operative. That this is certainly not correct may be seen from the differences observed between normal and para-H_2. Nevertheless, the results may be represented to a good approximation on this assumption and the values given represent some weighted means of the relaxation times involved.

It will be seen that, on the average, at the temperatures concerned rotational excitation persists through a few hundred collisions in H_2 and a comparable but smaller number in D_2. The variation with temperature is much less rapid than for typical vibrational relaxation times (see § 3.1).

5.2. *Theoretical calculation of rotational relaxation times for* H_2, D_2, *and* HD

The first theoretical estimates of cross-sections for rotational excitation of H_2 were made by Roy and Rose† using a distorted wave method.

† Roy, A. S. and Rose, M. E., *Proc. R. Soc.* A**149** (1935) 511.

Because the probability of excitation or deactivation per collision is small this method should be applicable but, unless very simple assumptions are made about the form of the interactions it is complicated to work out in practice. It is natural to try to take advantage of the simplicity of the exponential interaction (107) as far as possible.

TABLE 17.9

Observed rotational relaxation times and probabilities of deactivation per collision for H_2 and D_2

	Temp. (°K)	Relaxation time at 1 atm (10^{-8} s)	Probability of deactivation per collision × 10^3
H_2 (normal)	90		1·85†
	207	1·2	4·65‡
	273	2·3	2·85§
	285	1·1	6·25‡
	288	2·1	3·2‖
	298	1·9	3·9††
	373		< 7‡‡
H_2 (para)	90		2·8†
D_2 (normal)	273	2·0	4·8§

The interaction between two H_2 molecules has been discussed in Chapter 16, § 13.5. If we ignore the long-range attractive terms and the term $C(R)$ in (265) of that chapter, it can be written in the form

$$V(R, \chi_1, \chi_2) = V_0(R) + V_2(R)\{P_2(\cos \chi_1) + P_2(\cos \chi_2)\}, \tag{186}$$

where χ_1 and χ_2 are the angles of inclination of the internuclear axis of the two molecules to the line joining their centres which are at a distance R apart (see Fig. 17.49). $V_0(R)$ and $V_2(R)$ are short-range repulsive potentials which may be represented, at least to an initial approximation, by the exponential form (107). The simplest assumption is then to take

$$V(R, \chi_1, \chi_2) = A\mathrm{e}^{-aR}[1+\beta\{P_2(\cos \chi_1) + P_2(\cos \chi_2)\}]. \tag{187}$$

In a collision in which one of the molecules is in the ground rotational state which is spherically symmetrical, the average over all orientations of the axis of this molecule will remove the term in $P_2(\cos \chi_2)$, so we are left with

$$V(R, \chi_1) = A\mathrm{e}^{-aR}\{1+\beta P_2(\cos \chi_1)\}. \tag{188}$$

We are now able to follow a procedure very similar to that used in the calculations of cross-sections for vibrational deactivation in § 4.1.1 with $\beta P_2(\cos \chi_1)$ playing the role of $a\lambda x$ in (123).

† VAN ITTERBEEK, A. and VERHAEGEN, L., *Nature, Lond.* **167** (1951) 478.
‡ HUBER, P. W. and KANTROWITZ, A., *J. chem. Phys.* **15** (1947) 275.
§ STEWART, E. S. and STEWART, J. L., *J. acoust. Soc. Am.* **24** (1952) 194.
‖ GRIFFITH, W., *J. appl. Phys.* **21** (1950) 1319.
†† STEWART, E. S., *Phys. Rev.* **69** (1946) 632.
‡‡ GREENE, E. F. and HORNIG, D. F., loc. cit., p. 1486.

We start from a three-dimensional generalization of the eigenfunction expansion (110) in which the vibrational wave-functions $\chi_s(x)$ are replaced by the rotational wave-functions $\rho_{J,M}(\vartheta, \varphi)$ for molecule 1. The vibrational states are ignored because we are considering collisions at energies far below the threshold for vibrational excitation. We then arrive at the same equation (169) but now s, p denote the rotational states concerned, i.e. (169) may be rewritten in the form

$$(\nabla^2 + k_{JM}^2 - U_{J,M})F_{J,M}(R) = \sum_{J',M'} U_{J,M}^{J',M'} F_{J',M'}(R), \tag{189}$$

where

$$U_{J,M}^{J',M'} = \frac{2\mathcal{M}A}{\hbar^2}\beta e^{-aR} \int_0^\pi \int_0^{2\pi} P_2(\cos\chi)\rho_{J,M}(\vartheta, \varphi)\rho^*_{J',M'}(\vartheta, \varphi)\sin\vartheta\, d\vartheta d\varphi, \tag{190a}$$

and

$$U_{J,M} = \frac{2\mathcal{M}A}{\hbar^2} e^{-aR} = U_0, \text{ say.} \tag{190b}$$

$\mathcal{M}$ is the reduced mass.

For most molecules the equations (189) are closely coupled, but for H_2 and D_2 we expect the coupling to be weak so that the distorted wave approximation will be applicable. According to this approximation we consider only the functions referring to the initial and final states. For simplicity we suppose the initial state to be the ground state and the final state that with rotational quantum numbers J, M.

Since $\cos\chi = \cos\vartheta\cos\Theta + \sin\vartheta\sin\Theta\cos(\varphi - \Phi)$,

$$P_2(\cos\chi) = P_2(\cos\vartheta)P_2(\cos\Theta) + 2\sum \frac{(2-M)!}{(2+M)!} P_2^M(\cos\vartheta)P_2^M(\cos\Theta)\cos M(\varphi - \Phi), \tag{191}$$

so, representing the rotational wave-functions by normalized spherical harmonics,

$$U_{0,0}^{J,M} = (2\mathcal{M}A\beta/\hbar^2)e^{-aR}P_2^M(\cos\Theta)e^{\pm iM\Phi} \quad (M = 0, \pm 1, \pm 2) \tag{192}$$

$$= W(R)P_2^M(\cos\Theta)e^{\pm iM\Phi}, \tag{193a}$$

where

$$W(R) = (2\mathcal{M}A\beta/\hbar^2)\{(2-M)!/5(2+M)!\}^{\frac{1}{2}}e^{-aR}. \tag{193b}$$

To illustrate the method of calculation we now consider specifically the case $M = 0$. Following the distorted-wave method we approximate to (189) by the equations

$$(\nabla^2 + k_{0,0}^2 - U_0)F_{0,0} = 0, \tag{194a}$$

$$(\nabla^2 + k_{2,0}^2 - U_0)F_{2,0} = U_{2,0}^{0,0}F_{0,0}. \tag{194b}$$

Expanding $F_{0,0}$, $F_{J,M}$ as in (177) we have

$$\sum_{l'} i^{l'}\left\{\frac{d^2}{dR^2} + k_{2,0}^2 - U_0 - \frac{l'(l'+1)}{R^2}\right\}(2l'+1)G_{2,0}^{l'}(R)P_{l'}(\cos\Theta)$$

$$= W(R)\sum_l i^l(2l+1)G_{00}^l(R)P_l(\cos\Theta)P_2(\cos\Theta). \tag{195}$$

But

$$P_l(\cos\Theta)P_2(\cos\Theta) = a_l P_l + b_l P_{l+2} + c_l P_{l-2},$$

where

$$a_l = \frac{l(l+1)}{(2l+3)(2l-1)}, \quad b_l = \frac{3}{2}\frac{(l+1)(l+2)}{(2l+1)(2l+3)}, \quad c_l = \frac{3}{2}\frac{l(l-1)}{(2l+1)(2l-1)}.$$

The right-hand side of (194 b) therefore becomes

$$W(R)\sum_{l} i^l(2l+1)P_l(\cos\Theta)\Big\{a_l G_{00}^l - b_{l-2}\frac{2l-3}{2l+1}G_{00}^{l-2} - c_{l+2}\frac{2l+5}{2l+1}G_{00}^{l+2}\Big\}. \quad (196)$$

Equating the multiplying factors of $P_l(\cos\Theta)$ on both sides gives now

$$\Big\{\frac{d^2}{dR^2}+k_{2,0}^2-U_0-\frac{l(l+1)}{R^2}\Big\}G_{2,0}^l(R)$$
$$= W(R)\Big\{a_l G_{0,0}^l - b_{l-2}\frac{2l-3}{2l+1}G_{0,0}^{l-2} - c_{l+2}\frac{2l+5}{2l+1}G_{0,0}^{l+2}\Big\}. \quad (197)$$

Under almost all conditions of interest the main contribution to the total cross-section comes from large values of l. It will therefore not be a bad approximation to ignore the differences between the radial functions on the right-hand side of (197) and take l as $\gg 1$ in the expressions a_l, b_l, c_l giving

$$\Big\{\frac{d^2}{dR^2}+k_{2,0}^2-U_0-\frac{l(l+1)}{R^2}\Big\}G_{2,0}^l(R) = -\tfrac{1}{2}W(R)G_{0,0}^l(R). \quad (198)$$

If we further introduce Takayanagi's approximation of replacing the centrifugal force term $l(l+1)/R^2$ by a constant, $l(l+1)/R_c^2$, where R_c is the classical closest distance of approach for the interaction U_0, we have

$$\Big\{\frac{d^2}{dR^2}+\tilde{k}_{2,0}^2-U_0\Big\}\tilde{G}_{2,0}(R) = -\tfrac{1}{2}W(R)\tilde{G}_{0,0}(R), \quad (199)$$

where $\tilde{k}_{2,0}^2 = k_{2,0}^2-l(l+1)/R_c^2$ and $\tilde{k}_{0,0}^2$ is substituted for $k_{0,0}$ in $\tilde{G}_{0,0}$. This equation is of exactly the same form as for the corresponding equation for vibrational excitation.

We have the following correspondence.

Eqn (199)

$$\Big(\frac{d^2}{dR^2}+\tilde{k}_{2,0}^2-U_0\Big)\tilde{G}_{2,0}(R)$$
$$= -\tfrac{1}{2}W(R)\tilde{G}_{0,0}(R),$$
$$U_0 = (2\mathcal{M}/\hbar^2)A e^{-aR},$$
$$-\tfrac{1}{2}W(R) = -(\mathcal{M}A\beta/5^{\frac{1}{2}}\hbar^2)e^{-aR}$$

Eqn (111)

$$\Big(\frac{d^2}{dR^2}+k_s^2-U_{ss}\Big)F_s(R)$$
$$= U_{sp}F_p(R),$$
$$U_{ss} = (2M/\hbar^2)W_0 e^{-aR},$$
$$U_{sp} = (2MW_0/\hbar^2)Y_{ps} e^{-aR}. \quad (200)$$

It follows that the probability $P_{0,0}^{2,0}(l)$ of the rotational transition is obtained from (121) by replacing W_0 by A, $2Y_{sp}$ by $-5^{-\frac{1}{2}}\beta$, M by $\mathcal{M}$, and q_p, q_s by $\tilde{k}_{2,0}/a, \tilde{k}_{0,0}/a$ respectively. Thus

$$P_{0,0}^{2,0}(l) = (\pi^2\beta^2/5a^4)(\tilde{k}_{2,0}^2-\tilde{k}_{0,0}^2)^2\{\cosh(\pi\tilde{k}_{2,0}/a)-\cosh(\pi\tilde{k}_{0,0}/a)\}^{-2}\times$$
$$\times \sinh(\pi\tilde{k}_{2,0}/a)\sinh(\pi\tilde{k}_{0,0}/a), \quad (201)$$

and the cross-section $\quad Q_{0,0}^{2,0} = (\pi/k_{0,0}^2)\sum_l (2l+1)P_{0,0}^{2,0}(l)$

$$\simeq (2\pi/k_{0,0}^2)\int_0^\infty l P_{0,0}^{2,0}(l)\,dl. \quad (202)$$

As for vibrational deactivation

$$\tilde{k}_{2,0}^2-\tilde{k}_{0,0}^2 = 2\mathcal{M}\Delta E/\hbar^2 \quad (203)$$

and $k_{2,0}/a$, $k_{0,0}/a$ are $\gg 1$, but $4\pi^2 \Delta E/ah\bar{v}$, where $\bar{v}$ is the mean relative velocity, is no longer $\gg 1$. Thus for H_2, $\Delta E = 0.045$ eV and $a \simeq 2/a_0$, so that taking for $\bar{v}$ the mean relative velocity at 300 °K,

$$4\pi^2 \Delta E/ah\bar{v} \simeq 3.75. \qquad (204)$$

The approximation (126) is not very good but there is no difficulty in calculating $Q_{0,0}^{2,0}$ from (202). When this is done it is found that, for a relative velocity of impact 3×10^5 cm s^{-1}, which is close to the mean relative velocity at room temperature, and taking $a = 2/a_0$, $Q_{00}^{2,0}$ is $0.03\beta^2 \pi r_c^2$, where r_c is the closest distance of approach. This is of about the order of magnitude required to explain the acoustical observations if we take $\beta = 0.28$, as it then corresponds to a probability of deactivation per collision of 2.5×10^{-3}.

To obtain closer agreement with observation it is necessary to introduce more realistic interactions, including the long-range attraction. Takayanagi† allowed for this by replacing the simple exponential interaction in (188) by a Morse potential as in (152). He took

$$V(R, \chi) = D\exp\{-a(R-R_e)\} - 2D\exp\{-\tfrac{1}{2}a(R-R_e)\} + \\ + \beta D \exp\{-a(R-R_e)\}P_2(\cos\chi), \qquad (205)$$

and determined the constants to give the best fit with the H_2–H_2 interaction discussed in Chapter 16, § 13.5 (see (268) of that chapter). The values he used were, in atomic units, $D = 1.1 \times 10^{-4}$, $R_e = 6.4$, $\beta = 0.075$, $a = 1.87$. There is no difficulty in extending the method discussed above to apply to this interaction. Using again the approximation (199) Takayanagi calculated cross-sections for excitation of the $0 \to 2$, $1 \to 3$, and $2 \to 4$ transitions in H_2 and in D_2. In Fig. 17.45 (a) his results are presented in the form of plots of

$$P_J^{J'} = (2J+1)^{-1} \sum_{MM'} P_{JM}^{J'M'}$$

against the reduced incident wave number $\tilde{q}_i = \tilde{k}_i/a$. Since $\beta \simeq 0.075$ it is clear that, for the range of incident relative energies shown, $P_J^{J'} \ll 1$. As the probability of excitation is small for all incident relative angular momenta the method of distorted waves should give reliable results (see Fig. 17.46 below). Corresponding cross-sections are given in Fig. 17.45 (b).

Before discussing the comparison with the acoustic observations we must refer to the results of later calculations. Roberts‡ repeated the calculations using the same interaction as Takayanagi but not making the approximation (199), the differential equation for relative motion with different angular momenta being solved by electronic computation. His results for the cross-section, shown in Fig. 17.45 (b), agree quite well with those of Takayanagi obtained by the simple method.

† TAKAYANAGI, K., *Proc. phys. Soc.* A**70** (1957) 348; *Sci. Rept. Saitama Univ.* A**3** (1959) 65. ‡ ROBERTS, C. S., *Phys. Rev.* **131** (1963) 209.

Davison† also carried out calculations solving the differential equations numerically, but with a somewhat modified interaction in which the exponential attraction used by Takayanagi was replaced by a van der Waals term so that, averaged over all orientations of molecule 2,

$$V(R, \chi) = D\exp\{-a(R-R_e)\}\{1+\beta P_2(\cos\chi)\} - CR^{-6}\{1+\gamma P_2(\cos\chi)\}, \qquad (206)$$

with D, a, R_e, and β as before. C and γ were taken to be $11 \cdot 0$ and

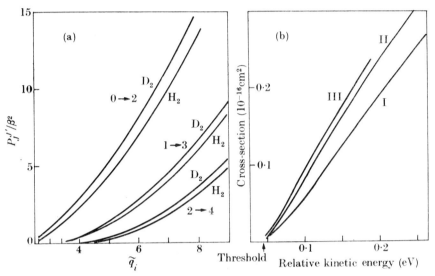

FIG. 17.45. Relating to the theoretical calculation of cross-sections for rotational excitation of H_2 in collision with a second H_2 molecule. (a) Plot of

$$P_J^{J'}/\beta^2, \text{ where } P_J^{J'} = \{1/(2J+1)\} \sum_{MM'} P_{JM}^{J'M'},$$

against the reduced initial wave-number $\tilde{q}_i = \tilde{k}_i/a$, as calculated by Takayanagi. Plots for D_2–D_2 collisions are also included. The $J \to J'$ transitions involved are indicated on each pair of curves. (b) Calculated cross-sections for the $0 \to 2$ transition taking $\beta = 0 \cdot 075$. I, Takayanagi—modified wave-number method; II, Roberts; III, Davison.

$0 \cdot 072$ a.u., which agree closely with theoretical expectation (see Chap. 16, § 13.5, eqn (268)). Davison† also made allowance for the symmetry properties of the colliding molecules (see Chap. 16, § 4.1). His results, shown in Fig. 17.45 (b), agree quite well with those of Takayanagi and of Roberts.

The cross-section Q_2^0 for the deactivation transition $2 \to 0$ is related to that Q_0^2 for the excitation by

$$Q_2^0 = \frac{1}{5} \frac{E_2}{E_0} Q_0^2, \qquad (207)$$

† DAVISON, W. D., *Proc. R. Soc.* A**280** (1964) 227.

E_0, E_2 being the relative energies of motion when the molecule 1 is in the states with $J = 0$, 2 respectively. Q_2^0, derived in this way from the calculations of Roberts and of Davison, is given in Fig. 17.46. The probability of deactivation per collision is Q_2^0/Q_d where Q_d is the total momentum-transfer cross-section for H_2–H_2 collisions. At relative

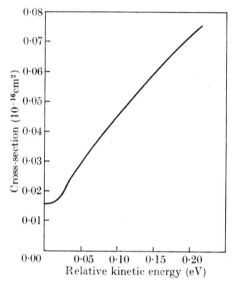

Fig. 17.46. Cross-sections for deactivation of H_2 ($J = 2$) by collision with a second H_2 ($J = 0$) molecule, calculated by Roberts.

energy E_2 corresponding to κT, where $T = 300$ °K, $Q_2^0 = 2 \times 10^{-17}$ cm². Taking Q_d as $23 \cdot 2 \times 10^{-16}$ cm² under the same conditions gives

$$Q_2^0/Q_d = 0 \cdot 86 \times 10^{-3},$$

whereas the acoustic observations (Table 17.9) give $3 \cdot 0 \times 10^{-3}$. Since Q_2^0 is closely proportional to the square of the asymmetry parameter β this comparison suggests that β should be increased, by a factor of about 1·9, to 0·14.

The same conclusion may be arrived at from an earlier application by Takayanagi† of the cross-sections which he calculated using the interaction (205) to calculate ultrasonic dispersion curves for para-H_2 and normal H_2. This he did for various values of $x = R_c^2 \beta^2$, where R_c is in Å, and found a good fit for $x \simeq 0 \cdot 1$ as shown in Fig. 17.47. Since the classical closest distance of approach at the energies concerned is near

† TAKAYANAGI, K., *J. phys. Soc. Japan* **14** (1959) 1458.

$4 \cdot 2a_0 = 2 \cdot 2$ Å this gives $\beta \simeq 0 \cdot 14_5$. In later calculations this increased value of β has been adopted.

To check the validity of the distorted wave method Allison and Dalgarno† calculated cross-sections for excitation of the 0–2 transition in H_2–H_2 collisions by the close-coupling (truncated eigenfunction) method. This involved working with the equations (189) in which all rotational states other than the ground state and those with $J = 2$ were

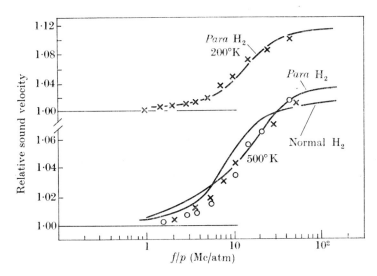

FIG. 17.47. Comparison of observed (Rhodes) and calculated (Takayanagi) ultrasonic dispersion curves for normal and para H_2 as indicated. × observed, for para H_2. ○ observed for normal H_2.

ignored but no other approximation made. In this way the problem was reduced to that of solving four coupled differential equations without having to assume that the coupling between the states with $J = 0$ and $J = 2$ was weak. Fig. 17.48 compares results obtained in this way with those calculated by the distorted-wave approximation (note that β is now taken as 0·14, as compared with 0·075 in Fig. 17.45 (b)). It will be seen that the use of this approximation is fully justified.

Although the distorted-wave method gives good results for H_2–H_2 and D_2–D_2 collisions it does not follow that it is also satisfactory for collisions between HD molecules. As pointed out by Takayanagi,‡ because the centre of mass and geometrical centre are not coincident

† ALLISON, A. C. and DALGARNO, A., Proc. phys. Soc. **90** (1967) 609.
‡ TAKAYANAGI, K., Prog. theor. Phys., Supp. 25 (1963) 1.

in these molecules, the interaction contains a term in $P_1(\cos \chi)$. Thus, referring to Fig. 17.49, we have for the interaction

$$V_0(R')+\beta V_2(R')\{P_2(\cos \chi_1')+P_2(\cos \chi_2')\}, \qquad (208)$$

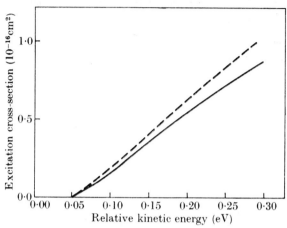

FIG. 17.48. Calculated cross-sections for excitation of the $J = 0 \to J = 2$ rotational transition in H_2. ---- using distorted wave method. —— using close coupling (truncated eigenfunction) method.

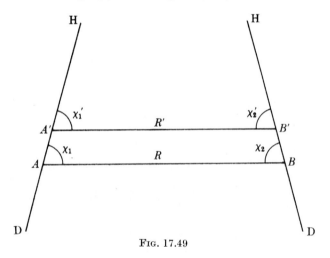

FIG. 17.49

where χ_1', χ_2' are the angles the line joining the mid-points A', B' makes with the respective nuclear axes, and R' is the length $A'B'$. This must be expressed in terms of the corresponding angles χ_1, χ_2, and $R (= AB)$. If $AA'/AB \ll 1$, and we write $R' = R(1+\delta)$, $\chi_1' = \chi_1+\mu$, $\chi_2' = \chi_2-\mu$, then

$$\delta = -\frac{\rho_0}{6R}(\cos\chi_1+\cos\chi_2), \qquad \mu = \frac{\rho_0}{6R}(\sin\chi_1-\sin\chi_2), \quad (209)$$

where ρ_0 is the equilibrium nuclear separation in either molecule. Hence (208) becomes

$$V_0(R)+V_1(R)\{P_1(\cos\chi_1)+P_1(\cos\chi_2)\}+\text{terms in }P_2\text{, etc.} \quad (210)$$

with
$$V_1(R) = \tfrac{1}{6}\rho_0\{V'_0(R)+\tfrac{2}{5}\beta V'_2(R)\}. \quad (211)$$

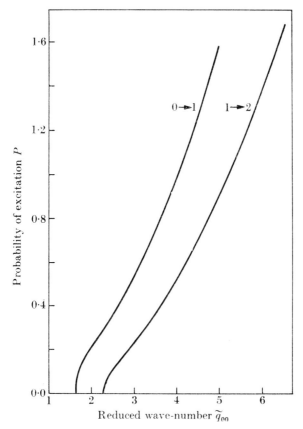

FIG. 17.50. Probability of excitation of the rotational transitions $0 \to 1$ and $1 \to 2$ (as indicated) on collision between two HD molecules as a function of the reduced wave number $\tilde{q}_{00}$.

Takayanagi has carried out detailed calculations using the Morse form (205) and the approximation (199). His results for the probability of excitation of the $0 \to 1$ and $1 \to 2$ transition as a function of the reduced wave-number $\tilde{q}_{00}$ are given in Fig. 17.50. It will be seen that, whereas for the homonuclear molecules the probability of rotational transition is very small even for $\tilde{q}_{00}$ as large as 8, it already reaches unity according to the distorted-wave method for $\tilde{q}_{00} \simeq 4$ for the heteronuclear case.

This means that, in such cases the coupling between rotational states is already strong, even at quite low relative energies of collision, and cross-sections can only be calculated accurately if this is taken into account.

5.3. *Rotational excitation of H_2 and of D_2 by collision with H and with He atoms*

The rotational excitation of H_2 by impact with H atoms as well as with H_2 molecules, is an important cooling mechanism in the interstellar medium and considerable attention has been paid to the calculation of its rate under these conditions. The most elaborate calculations for $H–H_2$ and $H–D_2$ collisions are those of Allison and Dalgarno,† using the close-coupling (truncated eigenfunction) method that they applied to $H_2–H_2$ collisions (see p. 1573).

They assumed an interaction of the form

$$V = V_0\exp(-aR) - C_0 R^{-6} + \{V_2\exp(-a_2 R) - C_2 R^{-6}\}P_2(\cos\chi), \quad (212)$$

with $\qquad V_0 = 18\cdot 93, \qquad V_2 = 12\cdot 76_5$

$\qquad\qquad C_0 = 9\cdot 233, \qquad C_2 = 1\cdot 023$ a.u.

The constants in the repulsive interactions were adjusted to give a good fit to the interaction calculated by Mason and Hirschfelder.‡ The constant C_0 was taken to be that calculated by Dalgarno and Williams§ while C_2 was estimated on the assumption that the anisotropy of the van der Waals interaction is the same as that of the electric polarizabilities.

The calculated cross-sections are given in Fig. 17.51 and compared with those calculated earlier by the distorted-wave method. As expected the latter gives good results.

Roberts,‖ as a prelude to the calculation of cross-sections for excitation of H_2 by collision with He, determined theoretically an approximate H_2–He interaction in the form

$$V = V_0 e^{-aR}\{1 + \beta P_2(\cos\chi)\}, \quad (213)$$

with $V_0 = 17\cdot 283$ a.u. (470$\cdot$10 eV), $a = 2\cdot 027$ a.u. (3$\cdot$830 Å), and $\beta = 0\cdot 375$. His cross-sections, calculated using the distorted-wave method, agree quite well with the later, more accurate results, obtained by Allison and Dalgarno using the close-coupling method (see Fig. 17.51).

† loc. cit., p. 1573.
‡ MASON, E. A. and HIRSCHFELDER, J. O., *J. chem. Phys.* **26** (1957) 756.
§ DALGARNO, A. and WILLIAMS, D. A., *Proc. phys. Soc.* **85** (1965) 685.
‖ loc. cit., p. 1570.

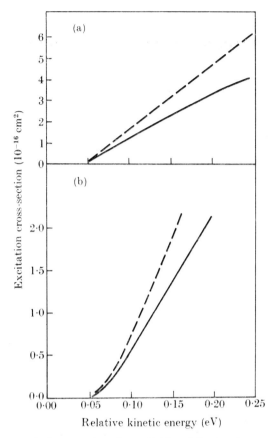

Fig. 17.51. Cross-sections for excitation of the $0 \to 2$ rotational transition in H_2: (a) by H impact, (b) by He impact. – – – calculated using distorted-wave approximation. ——— calculated by close-coupling method.

5.4. Rotational excitation of heavier molecules

5.4.1. Experimental evidence from acoustic and shock-wave observations.

Evidence of rotational relaxation in oxygen, nitrogen, and air has been forthcoming from acoustic observations at high frequencies (see Table 17.10). Although the results obtained for each gas by different observers do not agree very closely in detail they nevertheless show that rotational excitation, at room temperature, persists through ten or so collisions at most, on the average. There are indications from the data that rotational relaxation is less rapid in O_2 than in N_2.

Further information is available from shock-wave observations. Rotational relaxation is so fast that a distinct region behind the shock in

TABLE 17.10

Acoustic measurements of rotational relaxation times in O_2 and N_2 at temperatures near 300 °K

Nature of observations	Frequency range (Mc s^{-1} atm^{-1})	Relaxation time 10^{-10} s	Probability per collision
O_2			
absorption and dispersion[1]	1–60	20–48	0·08–0·03
absorption[2]	11–8000	—	0·24
absorption[3]	5–50	3·7	0·5
absorption in H_2–O_2 mixtures[4]	2–30	22	0·07
dispersion[5]	2–400	22	0·08
dispersion[6]		—	0·06
absorption[7]		5	0·25–0·5
N_2			
dispersion[7]	3–109	8·2	0·17
absorption[3]	5–50	4·7	0·3
absorption[8]	0·45–44	6	0·17–0·25
Air			
absorption and dispersion[9]		2·3–3·0	0·05–0·06
absorption and dispersion[10]		—	0·16
absorption[8]		—	0·2
absorption[2]		—	0·2

References

1 THALER, W. J., *J. acoust. Soc. Am.* **24** (1952) 15.
2 GREENSPAN, M., ibid. **31** (1959) 155.
3 PARKER, J. G., ADAMS, C. E., and STAVSETH, R. M., ibid. **25** (1953) 263.
4 PETRALIA, S., *Nuovo Cim.* **2** (1955) 241.
5 CONNOR, J. V., *J. acoust. Soc. Am.* **30** (1958) 297.
6 BOYER, R. A., ibid. **23** (1951) 176.
7 ZMUDA, A. J., ibid. **23** (1951) 472.
8 TEMPEST, W. and PARBROOK, H. D., *Acoustica* **7** (1957) 354.
9 ENER, C., GABRYSH, A. F., and HUBBARD, J. C., *J. acoust. Soc. Am.* **24** (1952) 474.
10 MEYER, E. and SESSLER, G., *Z. Phys.* **149** (1957) 15.

which this is occurring will not be observed. Instead it shows up in the fact that the observed shock-front thickness is a little greater than that calculated from (37) with η taken as the shear viscosity. This is found to be so,† not only in O_2 and N_2 but also in CO, CO_2, N_2O, CH_4, Cl_2, NH_3, and HCl.

As pointed out on p. 1483, Tisza‡ has shown that the effect of a process with relaxation time τ_r short compared with the time of measurement

† HORNIG, D. F., loc. cit., p. 1488.
‡ TISZA, L., *Phys. Rev.* **61** (1942) 531.

can be expressed, at temperature T, by a bulk viscosity

$$K = \frac{N\kappa^2 T}{c_v^2} c_r \tau_r, \tag{214}$$

where N is the number of molecules per cm³, and c_v and c_r the respective heat capacities at constant volume for all degrees of freedom and for the relaxing degree of freedom. In place of $\tfrac{4}{3}\eta$ in (37 b) we now have $\tfrac{4}{3}\eta + K$. By comparison of calculated shock front thicknesses, assuming different values of K, with those observed,† values of τ_r are found which correspond to a probability of rotational deactivation per collision of about 1/3 for O_2 and N_2 and not less than this for the other gases referred to above.

5.4.2. *Rotational relaxation and thermal conductivity of gases.* For a monatomic gas of molecular weight M the ratio σ/η of the thermal conductivity σ to the viscosity η is given by

$$\sigma M/\eta = f C_v, \tag{215}$$

where C_v is the molar heat capacity at constant volume and f is a constant very nearly equal to 5/2. In a polyatomic gas, part of the thermal energy resides in internal degrees of freedom and the transport of this fraction of the energy will affect the thermal conductivity. Eucken‡ suggested that to allow for this (215) should be generalized to

$$\sigma M/\eta = f_{tr} C_{v,tr} + f_{int} C_{v,int}, \tag{216}$$

where $C_{v,tr}$, $C_{v,int}$ are the respective contributions to the molar heat capacity from the translational and internal motion. f_{tr} was taken to be 5/2 and f_{int} as 1. Although this gives good agreement with experiment for many non-polar gases near 0 °C,§ it was found that,∥ to fit observed data taking f_{tr} as 5/2, f_{int} must be temperature-dependent. To allow for this it was then supposed that transport of internal energy occurs by diffusion, leading to

$$f_{int} = \rho D/\eta, \tag{217}$$

where ρ is the gas density and D the self-diffusion coefficient. However, it appears that this factor is almost independent of temperature and equal to 1·3 for most gases, a value which, according to observation, is approached approximately at high temperatures. To account for the discrepancies it is necessary to allow for interchange of energy between internal and translational degrees of freedom—it having been assumed

---

† HORNIG, D. F., loc. cit., p. 1488; SHERMAN, F. S., *NASA Technical Note* 3298 (1959).
‡ EUCKEN, A., *Phys. Z.* **14** (1913) 324.
§ CHAPMAN, S. and COWLING, T. G., *The mathematical theory of non-uniform gases*, 2nd edn., p. 241 (Cambridge University Press, 1952).
∥ GULLEY, E. R., *Am. J. Phys.* **20** (1952) 447.

in the above discussion that the flow of translational and internal energy proceeds independently.

Mason and Monchick† have adapted the formal transport theory of Wang Chang and Uhlenbeck‡ which takes into account inelastic collisions between gas molecules. They then find

$$f_{tr} = \tfrac{5}{2}[1-\tfrac{5}{6}\{1-\tfrac{2}{5}(\rho D/\eta)\}(C_{int}/R)(\eta/p\tau)], \quad (218\,a)$$
$$f_{int} = (\rho D/\eta)[1+\tfrac{5}{4}\{1-\tfrac{2}{5}(\rho D/\eta)\}(\eta/p\tau)], \quad (218\,b)$$

it being assumed that there is only one relaxation time τ (at 1 atm pressure) for the internal degrees of freedom. p is the pressure in atmospheres and R the gas constant.

To check the validity of (218) and obtain further information about rotational relaxation times, Mason and Monchick used a theoretical formula derived by Parker§ according to which the variation of the probability $P(T)$ of deactivation per collision with temperature (T) is given by

$$P(T) = P(\infty)\{1+\tfrac{1}{2}\pi^{\frac{3}{2}}(\epsilon/\kappa T)^{\frac{1}{2}}+(\tfrac{1}{4}\pi^2+\pi)(\epsilon/\kappa T)\}. \quad (219)$$

This formula was derived by a two-dimensional classical theory in which the interaction between the two molecules was taken to be of the form

$$V = Ae^{-aR}(1+\beta\cos 2\chi_1)(1+\beta\cos 2\chi_2)-Be^{-\tfrac{1}{2}ar}, \quad (220)$$

where R, χ_1, and χ_2 are as in Fig. 17.49 and ϵ is the depth of the minimum in the central part of the interaction (220) with $\beta = 0$.

Fig. 17.52 shows a comparison between observed values of the factor $\sigma M/\eta C_v$ as a function of temperature and that calculated from (216), (218 a), (218 b), and (219) for N_2, O_2, Cl_2, and CO_2.

For these four cases $P(\infty)$ in (219) was chosen so that $P(300\,°K) = 0.27$, 0.29, 0.18, and 0.5 respectively, values that are consistent with the data from acoustic experiments (see Table 17.10). It will be seen that the agreement is not unsatisfactory and certainly better as far as temperature variation is concerned than with the corresponding values calculated from (216) either with $f_{int} = 1$ or $= \rho D/\eta$.

H_2 is a special case because of the relatively long rotational relaxation time at 300 °K, so f_{tr}, f_{int} given by (218) reduce to $f_{tr} = 1, f_{int} = \rho D/\eta$. However, it can be seen from Fig. 17.53 that, while for $T < 400\,°K$ quite good agreement is then obtained with observed results, the observed values of $\sigma M/\eta C_v$ rise rapidly above the calculated at higher temperatures. The reason for this is not understood.

† MASON, E. A. and MONCHICK, L., *J. chem. Phys.* **36** (1962) 1622.
‡ WANG CHANG, G. K. and UHLENBECK, G. E., *Transport phenomena in polyatomic gases*, University of Michigan Engineering Series Rept. No. CM–681 (1951).
§ PARKER, J. G., *Physics Fluids* **2** (1959) 449.

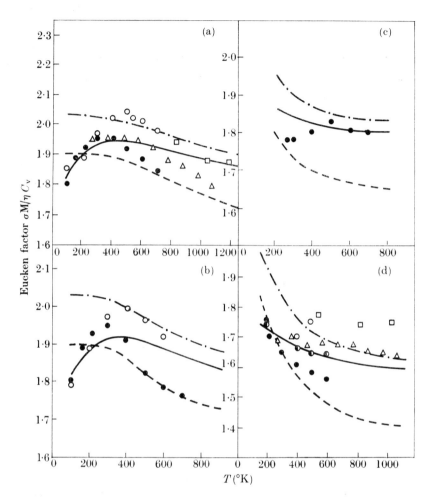

FIG. 17.52. Comparison of observed and calculated values of the Eucken factor $\sigma M/\eta C_V$ as a function of temperature for (a) N_2, (b) O_2, (c) Cl_2, and (d) CO_2.

——— calculated by Mason and Monchick.
– – – according to (216) with $f_{tr} = 5/2$, $f_{int} = 1$.
—·— according to (216) with $f_{tr} = 5/2$, $f_{int} = \rho D/\eta$.

Observed: ○ HILSENRATH, J., *Tables of thermal properties of gases*, Nat. Bur. Stand. Circular No. 564 (1955).
△ ROTHMAN, A. J. and BROMLEY, L. A., *Ind. Engng Chem. ind. (int.)* Edn **47** (1955) 899.
□ VINES, R. G., *Trans. Am. Soc. mech. Engrs (J. Heat Transfer)* **82C** (1960) 48.
● FRANCK, E. U., *Z. Elektrochem.* **55** (1951) 636.
◐ obtained from σ measurements of Franck combined with η and C_V as tabulated by Hilsenrath.

5.4.2.1. *Rotational relaxation from measurements of thermal transpiration.* If gas is maintained in two bulbs, at different temperature, connected by a capillary, a pressure difference Δp will exist between the two bulbs. The phenomenon is essentially similar to thermal diffusion (Chap. 16, p. 1304) in which the gas and the walls respectively replace the light and heavy gas.

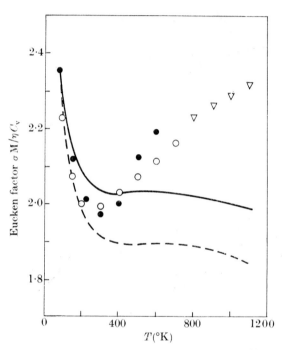

Fig. 17.53. Comparison of observed and calculated values of the Eucken factor $\sigma M/\eta C_v$ as a function of temperature for H_2.

—— calculated by Mason and Monchick which in this case reduces to (216) with $f_{tr} = 5/2$ and $f_{int} = \rho D/\eta$.
--- according to (216) with $f_{tr} = 5/2$, $f_{int} = 1$.
Observed: ○ HILSENRATH, J. (loc. cit., Fig. 17.52);
▽ BLAIS, N. C. and MANN, J. B., *J. chem. Phys.* **32** (1960) 1459;
● FRANCK, E. U. (loc. cit., Fig. 17.52).

Mason, Evans, and Watson† show that the maximum pressure difference $(\Delta p)_m$ is given in terms of the Eucken formula (216) by

$$f_{tr}\{1+(9f_{tr}/80\delta)^{\frac{1}{2}}\}^{-2} = (5d/12)(2\pi M/\kappa T_0)^{\frac{1}{2}}(T_0/\eta)(\Delta p)_m/\Delta T. \quad (221)$$

Here δ is a numerical constant $\simeq 3\pi/16$, d is the diameter of the capillaries, M the mass of a molecule, and ΔT the difference in the

† MASON, E. A., EVANS, R. B., and WATSON, G. M., *J. chem. Phys.* **38** (1963) 1808.

temperatures T_1, T_2 of the two bulbs. η is the viscosity of the gas at a certain temperature T_0 which falls between T_1 and T_2.

For rare gases $f_{\text{tr}} \simeq 2\cdot 5$, independent of T, so, for fixed ΔT, it follows from (221) that, if
$$U(T) = (\Delta p)_{\text{m}}/\{\eta(T)/(MT)^{\frac{1}{2}}\}$$
is plotted for each gas as a function of T the curves should intersect when $T = T_0$. In practice the intersection is not very clearly defined but the final results are not very sensitive to T_0. If $U_{\text{A}}(T_0)$, $U_{\text{B}}(T_0)$ are the values found for U for a rare gas A and a molecular gas B then
$$U_{\text{A}}(T_0)/U_{\text{B}}(T_0) = F_{\text{A}}/F_{\text{B}},$$
where
$$F = f_{\text{tr}}\{1+(9f_{\text{tr}}/80\delta)^{\frac{1}{2}}\}^{-2}$$
$$= 0\cdot 8744, \text{ for a rare gas.}$$
Hence
$$F_{\text{B}} = 0\cdot 8744\, U_{\text{B}}(T_0)/U_{\text{A}}(T_0).$$

By comparative measurements F_{B} may be determined and hence f_{tr}. Referring to (218a) the rotational relaxation time τ may then be obtained.

Experiments on these lines were carried out by Tip, Los, and de Vries,[†] using two circular glass tubes of 1 cm diameter as hot and cold reservoirs, connected by ten glass capillaries with diameter $0\cdot 736 \pm 0\cdot 003$ mm. Typical bulb temperatures were 300 and 550 °K and pressure differences of a few 10^{-2} torr were observed for a number of gases. Analysing these measurements as described above, probabilities of rotational excitation per collision of 0·36, 0·31, 0·67, 0·32, and 0·10 were found for N_2, O_2, CO, CO_2, and CH_4 respectively, taking T_0 as 500 °K. With $T_0 = 600$ °K the corresponding values are 0·27, 0·23, 0·53, 0·36, and 0·086. These are in reasonable agreement with results obtained by other methods (see § 5.4.1).

5.4.3. *Spectroscopic evidence.* An excellent illustration of the high probability at which rotational transitions may be excited by impact at ordinary temperatures is provided by the experiments of Broida and Carrington[‡] on the effect of foreign gases on the fluorescence of NO.

In this work, which was an extension of earlier work by Kleinberg and Terenin[§] and by Kleinberg,[||] the nitric oxide was excited by the narrow emission line of CdII at 2144 Å. This line populates only the rotational level $K = 13$ of the vibrational level $v' = 1$ and $A^2\Sigma^+$ electronic state. Collisions with other NO molecules, as well as with foreign molecules, lead to population of other states and this redistribution was

[†] TIP, A., LOS, J., and DE VRIES, A. E., *Physica, 's Grav.* **35** (1967) 489.
[‡] BROIDA, H. P. and CARRINGTON, T., *J. chem. Phys.* **38** (1963) 136.
[§] KLEINBERG, A. V. and TERENIN, A. N., *Dokl. Akad. Nauk SSSR* **101** (1955) 445, 1031.
[||] KLEINBERG, A. V., *Optika Spektrosk.* **1** (1956) 469.

studied by high resolution spectroscopic analysis of the fluorescent radiation under different pressure conditions. One complication which must be taken account in interpreting the results is that the two spin substates associated with the initially populated $K = 13$ rotational level are not equally populated by the exciting line and transitions between spin substates may occur.

In the experiments the cadmium source could be maintained at constant intensity within ± 10 per cent for many hours. The spectrograph collected the fluorescent light from a region along the axis of the tube. It consisted of an Ebert monochromator with a low-noise level photomultiplier detector (RCA type 1P28). The focal ratio of the monochromator was $f/10$ and it used a grating with 1200 lines/mm ruled over a width of 65 mm. To observe the rotational distribution observations were made in the first order with a grating blazed for 2000 Å, giving a resolution of $1{\cdot}8$ cm^{-1} near 2500 Å.

Fig. 17.54 illustrates three spectra which show some striking features of the rotational excitation process. The lowest spectrum, taken in pure NO at $0{\cdot}2$ torr pressure, shows little evidence of any redistribution of rotational energy—the eight strong lines all arise from the $K = 13$ upper state. When the NO pressure is increased to $2{\cdot}6$ torr there is more rotational redistribution but it is still not very marked. This does not mean that NO–NO collisions are relatively ineffective, for the following reason. Such collisions produce electronic deactivation quite strongly at a rate which is proportional to the NO pressure. This means that, as the pressure increases, there is proportionally less time available for rotational transitions to occur. The net result is that the spectrum in pure NO changes very little with NO pressure—that for 600 torr NO is not very different from that shown for $2{\cdot}6$ torr in Fig. 17.54. However, N_2 is relatively very ineffective in producing electronic deactivation so there is unlimited scope for rotational transitions to occur. Thus, in the third, uppermost, spectrum shown in Fig. 17.54, taken in a mixture of $0{\cdot}2$ torr of NO with $2{\cdot}5$ torr of N_2, there is very clear evidence of rotational redistribution.

Fig. 17.55 shows the rotational distributions, taking the two spin substates together, derived from analysis of spectra such as shown in Fig. 17.54, for pure NO at low pressures and for two different pressures of added argon. Again there is clear evidence of the effectiveness of Ar atoms in producing rotational transitions on impact.

Quantitative information on the rate coefficient r for rotational transfer from the initial level to any other rotational level may be

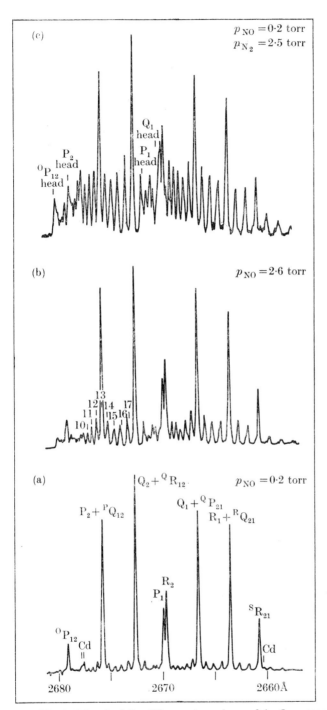

FIG. 17.54. The 1, 5 band of the NO γ system observed in fluorescence by Broida and Carrington. (a) In pure NO at a pressure of 0·2 torr. The eight strongest lines are emitted by molecules in the rotational level $K' = 13$—the branch designations are indicated. (b) In pure NO at a pressure of 2·6 torr. Neighbouring rotational levels of $K' = 13$ for the $P_2 + PQ_{12}$ branches are indicated. (c) In a mixture of NO at 0·2 torr with N_2 at 2·5 torr. Band heads are indicated.

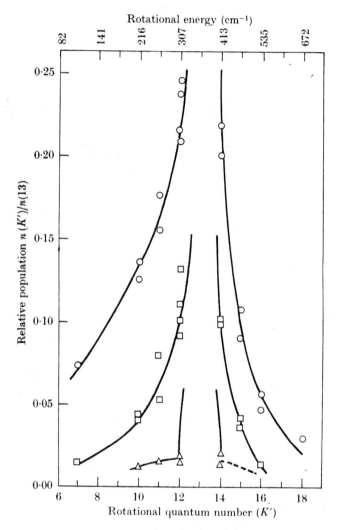

Fig. 17.55. Rotational distributions in NO and NO–Ar mixtures observed by Broida and Carrington. △ 0·055 torr NO; □ 0·046 torr NO+0·42 torr Ar; ○ 0·055 torr NO+1·3 torr Ar.

obtained as follows. If I_1 is the population in the initially excited level, $K = 13$, and I_2 the total population of all other levels,

$$r = (I_2/I_1)(1-\alpha I_2/I_1)^{-1}p^{-1}\{1+(e+v)p\},$$

where p is the pressure of the gas concerned, e and v are the corresponding coefficients for electronic and vibrational deactivation respectively, and

α allows for transitions back to the initial levels from other levels. e and v are very small for all the cases investigated, with the exception of pure NO and CO_2. Table 17.11 gives results obtained for r and the mean rotational deactivation cross-sections derived from them. It will be seen that these cross-sections are comparable with the gas-kinetic, indicating strong coupling between rotational states.

TABLE 17.11

Rate coefficient (r) and mean effective cross-sections (Q) for rotational deactivation of the $K = 13$ level of the $v = 1$, $A^2\Sigma^+$ state of NO

Deactivating gas	NO	CO_2	N_2	H_2	Ar	He
r (10^6 s^{-1} torr^{-1})	7·5	10·0	8·0	11·0	7·0	9·0
Q (10^{-16} cm^2)	13	20	14	6·6	13	7·5

Although they were not able, because of the complexity of the conditions, to analyse the observations in terms of rate coefficients for particular transitions from the $K = 13$ level, Broida and Carrington nevertheless obtained some interesting information about the relative importance of transitions involving different changes ΔJ of the rotational quantum number J. This they did by making different assumptions about the dependence on ΔJ, calculating the distributions which would result and comparing with the observed. In Fig. 17.56 the observed distribution, illustrated by the upper curve in Fig. 17.55, is compared with that calculated on three different assumptions in which the importance of higher values of $|\Delta J|$ progressively increase. It will be seen that the assumption in which transitions with $|\Delta J|$ up to five are included gives the closest approximation to the observed deactivation. On the other hand, the assumption that only transitions with $\Delta J = \pm 1$ are significant gives much too narrow a distribution. These results provide further convincing evidence of strong coupling between the different rotational levels.

The rotational transfer cross-section from the $K = 11$ rotational level of the $v' = 0$, $A^2\Sigma^+$ state of OH was measured by Carrington[†] in a mixture mainly of O_2 and H_2O and found to be 33×10^{-16} cm^2, which is of the same order as those found in the NO experiments. He also found evidence that transitions in which $|\Delta J| > 1$ are of importance.

[†] CARRINGTON, T., *J. chem. Phys.* **31** (1959) 1418; *Eighth symposium on combustion*, p. 257 (Williams and Wilkins, Baltimore, 1962).

5.5. *Rotational excitation and molecular beam experiments*

5.5.1. *Theoretical considerations.* It seems quite clear from direct observation and also from the calculations carried out for H_2, D_2, and HD that rotational excitation on impact between all but the lightest

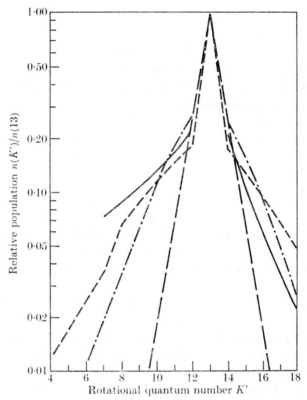

Fig. 17.56. Comparison of calculated and observed rotational distributions for a mixture of 0·055 torr NO with 1·3 torr Ar (see ○ of Fig. 17.55). Observed: ———. Calculated: — — assuming $\Delta J = \pm 1$ only; —·—· assuming $\Delta J = \pm 1, \pm 2, \pm 3$ with relative probabilities 0·5, 0·25, and 0·12 respectively; — — — assuming $\Delta J = \pm 1, \pm 2, \pm 3, \pm 4, \pm 5$ with relative probabilities 0·3, 0·2, 0·13, 0·09, and 0·06 respectively.

molecules, even under thermal conditions, is very probable. The question then arises as to the effect this will have on the interpretation of observations of cross-sections in molecular beam experiments (see Chap. 16, § 13.1).

In the extreme case of very strong coupling the rotational energy is distributed statistically among the various rotational states. Under these dominant coupling conditions, Bernstein, Dalgarno, Massey, and

Percival† showed that, for collisions between non-polar molecules, the total collision cross-section will be the same as that calculated for elastic collisions, for the same intermolecular interaction, neglecting the internal degrees of freedom. Collisions between polar molecules represent a special case because of the long range of the interaction between two dipoles. Use can be made of this to derive quite good theoretical predictions of cross-sections. This will be discussed in § 5.6 after a description in § 5.5.2 of some experimental observations of cross-sections for excitation of individual rotational states of a polar molecule TlCl in crossed-beam experiments.

If the coupling is strong, but not dominant, one approach to calculating the inelastic cross-sections is to use the close-coupling (truncated eigenfunction) method similar to that applied by Marriott to the calculation of vibrational excitation cross-sections at high temperatures (see § 4.4.2). Simple examples of such calculations are those of Allison and Dalgarno‡ for the rotational excitation of H_2 and D_2 (§ 5.3). Lester and Bernstein§ carried out a representative calculation of this kind for collisions between an atom and a homonuclear diatomic molecule in which the interaction was taken to be of the form, in the usual notation,

$$V = \epsilon\{(R_m/R)^{12} - 2(R_m/R)^6\}(1+\beta \cos \chi). \qquad (222)$$

Using this interaction, illustrative calculations were carried out for collisions in which the total angular momentum quantum number F, which remains constant throughout the collision, was taken to be 6. Close coupling between rotational states with $J = 0$ to 6 was allowed for, requiring the solution of 16 coupled equations. Results were also obtained with fewer equations—9, allowing for coupling between states with $J = 0, 2$ and 4, and 4, allowing for coupling between $J = 0$ and 2. Comparison could therefore be made between the probabilities for transition from states with $J = 0$ or 2 to $J' = 0$ or 2 obtained from solutions of 4, 9, or 16 coupled equations. Probabilities of rotational transitions are tabulated for an initial kinetic energy of relative motion $E = 1 \cdot 1\epsilon$, $(2M/\hbar^2)\epsilon R_m^2 = 1000$ and $R_m^2 M/I = 2 \cdot 351$, M being the reduced mass of the atom and molecule and I the moment of inertia of the molecule. If $\beta < 0 \cdot 15$ the distorted-wave method gives results for $0 \to 2$ rotational transitions which exceed those given by the close coupling calculation by < 20 per cent. For larger values of β the gap widens as the distorted-

---

† BERNSTEIN, R. B., DALGARNO, A., MASSEY, H. S. W., and PERCIVAL, I. C., *Proc. R. Soc.* A**274** (1963) 427.
‡ loc. cit., p. 1573.
§ LESTER, W. A. and BERNSTEIN, R. B., *Chem. Phys. Lett.* **1** (1967) 207, 347.

wave method gives results varying as β^2. Dominant coupling, distinguished by a nearly statistical distribution among rotational states, is not approached until β is much greater.

A less laborious method is applicable if the energy separation ΔE between the rotational states is such that

$$\Delta E/\hbar \ll 1/\tau_c, \tag{223}$$

where τ_c is the time of collision. Under these conditions the change in orientation of the molecular axis during the collision is very small. The collisions are the very opposite to adiabatic and a so-called 'sudden' approximation is applicable.

Ignoring all internal degrees of freedom but that of simple molecular rotation, the Schrödinger equation for the system of atom plus molecule may be written in terms of the coordinates $\mathbf{R}$ of the relative motion of the atom and the centre of mass of the molecule and the angles (ϑ, φ) of orientation of the molecular axis relative to a fixed axis, in the form

$$\left[\frac{\hbar^2}{2M}\nabla_R^2 - K(\vartheta,\varphi) + E - V(\mathbf{R};\vartheta,\varphi)\right]\Psi = 0. \tag{224}$$

$K(\vartheta, \varphi)$ is the kinetic energy, in operator form, of the molecular rotation. E, the total energy, is given by

$$E = \frac{\hbar^2}{2M}k_J^2 + \frac{\hbar^2}{8\pi^2 I}J(J+1), \tag{225}$$

where k_J is the wave number of relative motion with reduced mass M, I is the moment of inertia of the molecule, and J the initial rotational quantum number. We require solutions of (224) which have the asymptotic form for large R

$$\Psi \sim e^{i\mathbf{k}_J \cdot \mathbf{R}}\rho_J(\vartheta,\varphi) + R^{-1}\sum_{J'} e^{ik_{J'}R}\rho_{J'}(\vartheta,\varphi)f_{J'}(\Theta,\Phi), \tag{226}$$

where the $\rho_J(\vartheta, \varphi)$ are the rotational eigenfunctions of the molecule. The differential cross-section for a collision in which the molecular rotation quantum number changes from J to J', and the relative motion after the collision is in the direction (Θ, Φ) relative to the incident direction, is given by

$$I_{JJ'}(\Theta,\Phi)\,d\Omega = |f_{J'}(\Theta,\Phi)|^2\,d\Omega. \tag{227}$$

If the collision is a sudden one, it takes place under conditions in which $K(\vartheta, \varphi)$ in (224) is small compared with $(-\hbar^2/2M)\nabla_R^2$. Also, to the same approximation, we may ignore the rotational energy contribution to E. The problem then reduces to that of calculating the scattering of particles of mass M and wave vector k_J by a potential $V(\mathbf{R}; \vartheta, \varphi)$ in which ϑ, φ are constants. We now suppose this problem to be solved to give a solution Ψ_s which has the asymptotic form for large R

$$\Psi_s \sim \{e^{i\mathbf{k}_J \cdot \mathbf{R}} + R^{-1}e^{ik_J R}f_s(\Theta,\Phi;\vartheta,\varphi)\}\Lambda(\vartheta,\varphi), \tag{228}$$

where $\Lambda(\vartheta, \varphi)$ is an arbitrary function of ϑ, φ. If we take $\Lambda(\vartheta, \varphi)$ as $\rho_J(\vartheta, \varphi)$ and expand $f_s(\Theta, \Phi; \vartheta, \varphi)\rho_J(\vartheta, \varphi)$ in the form

$$\rho_J(\vartheta,\varphi)f_s(\Theta,\Phi;\vartheta,\varphi) = \sum_{J'}\langle\rho_J|f_s|\rho_{J'}\rangle\rho_{J'}(\vartheta,\varphi), \tag{229}$$

where $$\langle\rho_J|f_s|\rho_{J'}\rangle = \int_0^{2\pi}\int_0^{\pi} \rho_J(\vartheta,\varphi)f_s(\Theta,\Phi;\vartheta,\varphi)\rho_{J'}^*(\vartheta,\varphi)\sin\vartheta\,d\vartheta d\varphi, \tag{230}$$

then (228) becomes

$$\Psi_s \sim e^{i\mathbf{k}_J \cdot \mathbf{R}} \rho_J(\vartheta, \varphi) + R^{-1} \sum_{J'} e^{i k_J R} \langle \rho_J | f_s | \rho_{J'} \rangle \rho_{J'}(\vartheta, \varphi). \tag{231}$$

Comparison with (226) shows that our approximate solution is of the required form if $k_J \simeq k_{J'}$ for all J' of significance. This is consistent with the sudden approximation which therefore gives

$$I^S_{JJ'}(\Theta, \Phi) \, d\Omega = |\langle \rho_J | f_s | \rho_{J'} \rangle|^2 \, d\Omega. \tag{232}$$

Application of this result in practice is quite complicated, particularly when the collision is treated in three dimensions with the incident and final directions of relative motion and that of the molecular axis not coplanar. Kramer and Bernstein[†] have nevertheless used (232) to discuss small angle inelastic scattering. To calculate f_s he used the three-dimensional form of Schiff's semi-classical approximation[‡] (cf. (250) of Chapter 16)

$$f_s(\Theta, \Phi) = \frac{ik}{2\pi} \int_{-\infty}^{\infty} \int_{-\infty}^{\infty} \exp\{i(q_x x + q_y y)\} \left[1 - \exp\left\{ -\frac{i}{2k} \int_{-\infty}^{\infty} V(x, y, z) \, dz \right\} \right] dx \, dy, \tag{233}$$

where the scattering potential V is written in terms of cartesian coordinates x, y, z with z in the direction of incidence. q_x and q_y are given by

$$q_x = 2k \sin \tfrac{1}{2}\Theta \cos \Phi, \qquad q_y = 2k \sin \tfrac{1}{2}\Theta \sin \Phi. \tag{234}$$

A semi-classical version of the sudden approximation may be developed without difficulty.§ We suppose the relative motion to be essentially classical so that the rotator is acted on by a time-dependent potential $V(\vartheta, \varphi; p, t)$ where p is the impact parameter. The time dependence of V may be calculated from classical dynamics. We now have the time-dependent Schrödinger equation for the internal motion of the rotator

$$-i\hbar \frac{\partial \mathscr{P}}{\partial t} = \{K(\vartheta, \varphi) + V(\vartheta, \varphi; p, t)\} \mathscr{P}. \tag{235}$$

Once again, following the sudden approximation, we obtain an approximate solution by ignoring $K(\vartheta, \varphi)$. This gives

$$\mathscr{P}(t) = \exp\left\{ (i/\hbar) \int_{-\infty}^{t} V(\vartheta, \varphi; p, t) \, dt \right\} \mathscr{P}(-\infty; \vartheta, \varphi), \tag{236}$$

where $\mathscr{P}(-\infty; \vartheta, \varphi)$ is the initial form of $\mathscr{P}$. If the rotator is initially in the state J then $\mathscr{P}(-\infty; \vartheta, \varphi) = \rho_J(\vartheta, \varphi)$. We now expand $\mathscr{P}(t)$ in the form

$$\mathscr{P}(t) = \sum_{J'} a_{JJ'} \exp(iE_{J'} t/\hbar) \rho_{J'}(\vartheta, \varphi), \tag{237}$$

so

$$a_{JJ'}(p, t) = \exp(-iE_J t/\hbar) \left\langle \rho_J \middle| \exp\left\{ (i/\hbar) \int_{-\infty}^{t} V(\vartheta, \varphi; p, t) \, dt \right\} \middle| \rho_{J'} \right\rangle. \tag{238}$$

Hence, as on p. 448 of Chapter 7, the probability that the rotator is left finally in the state J' is

$$|a_{JJ'}(p, \infty)|^2 = \left| \left\langle \rho_J \middle| \exp\left\{ (i/\hbar) \int_{-\infty}^{\infty} V(\vartheta, \varphi; p, t) \, dt \right\} \middle| \rho_{J'} \right\rangle \right|^2. \tag{239}$$

[†] KRAMER, K. H., and BERNSTEIN, R. B., *J. chem. Phys.* **40** (1964) 200.
[‡] SCHIFF, L. I., *Phys. Rev.* **103** (1956) 443.
[§] See ALDER, K. and WINTHER, A., *K. danske Vidensk. Selsk. Mat. fys. Medd.* **32** (1960) no. 8.

The cross-section for the transition from J to J' is then

$$Q_{JJ'} = 2\pi \int_0^\infty |a_{JJ'}(p,\infty)|^2 p\, dp. \tag{240}$$

If, in addition to the condition (223), $(i/\hbar) \int_{-\infty}^{\infty} V(\vartheta,\varphi;p,t)\, dt \ll 1$, (239) reduces to the usual formula of time-dependent perturbation theory (see Chap. 7, (108)) but the validity of (239) does not depend on the latter condition being satisfied. It only requires that the collision be a sudden one in the sense that (223) is valid.

TABLE 17.12

Total (elastic+inelastic) (Q_t) and total inelastic cross-sections (Q_{in}) for collisions in which the interaction is given by case (b)

$E^* = E/\epsilon$	$Q_t/\pi R_m^2$	$Q_{in}/\pi R_m^2$	Q_{in}/Q_t
1	17·38	5·04	0·284
3	14·28	4·13	0·289
10	11·22	3·31	0·295
30	9·01	2·73	0·303
100	7·08	2·19	0·309

Illustrative calculations using this semi-classical form of the sudden approximation have been carried out by Bernstein and Kramer,† who chose for the interaction between atoms and diatomic molecules a potential of the form

$$V(R,\chi) = 2\epsilon(R_m/R)^{-12}\{1+b_1 P_1(\cos\chi)+b_2 P_2(\cos\chi)\}-$$
$$-\epsilon(R_m/R)^6\{1+a_2 P_2(\cos\chi)\}-6\alpha\mu q\{\tfrac{3}{5}P_1(\cos\chi)+\tfrac{2}{5}P_3(\cos\chi)\}. \tag{241}$$

ϵ and R_m are the usual parameters characterizing the central part of the potential. α is the polarizability of an atom and μ, q are respectively the permanent dipole and quadrupole moments of a molecule. If the molecule is homonuclear b_1 and c vanish. The time dependence of the interaction $V(\vartheta,\varphi;p,t)$ which appears in (239) was calculated by supposing the relative motion to be a classical trajectory for motion under the central potential $\epsilon\{(R_m/R)^{12}-2(R_m/R)^6\}$.

Fig. 17.57 illustrates the calculated variation of the total probability of an inelastic collision from the ground rotational state $J = 0$ with the square of the reduced impact parameter $p^* = p/R_m$ at reduced energy $E^* = E/\epsilon$ of 0·1. The interaction constants are such that

$$\hbar/R_m(2M\epsilon)^{\frac{1}{2}} = 0\cdot 1, \quad a_2 = 0\cdot 3, \quad \begin{cases} b_1 = b_2 = c = 0, & \text{case (a)},\\ b_1 = b_2 = 0\cdot 5,\ c = 2\cdot 0, & \text{case (b)}. \end{cases}$$

Cases (a) and (b) correspond respectively to collisions with homonuclear and heteronuclear molecules. The rapid rise of the probability to a value comparable with unity for $p^* \simeq 2$ is striking and corresponds to the onset of dominant coupling.

It was found that the total inelastic cross-section is given quite closely by $\pi R_m^2\{p^*(\tfrac{1}{2})\}^2$, where $p^*(\tfrac{1}{2})$ is the value of p^* at which the total inelastic transition probability is $\tfrac{1}{2}$ (indicated by the broken lines in Fig. 17.57).

In Table 17.12 the total (Q_t) and total inelastic (Q_{in}) cross-section for case (b) are tabulated as functions of the reduced energy E^*. Q_t has been obtained assuming

† BERNSTEIN, R. B. and KRAMER, K. H., *J. chem. Phys.* **44** (1966) 4473.

dominant coupling and so is equal to Q_{el}, the elastic cross-section for scattering of two particles interacting with the same central potential with the same reduced collision parameters. It will be seen that the ratio Q_{in}/Q_t varies much more slowly with reduced energy than does either cross-section separately.

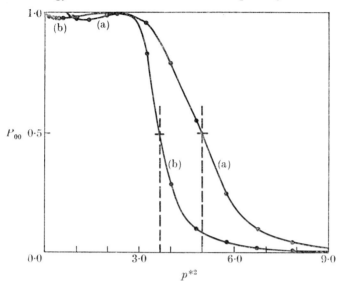

FIG. 17.57. Probability P_{00} of rotational excitation from the ground $(0, 0)$ state as a function of p^{*2}, where p^* is the reduced impact parameter p/R_m calculated by Bernstein and Kramer using the 'sudden' approximation. The reduced relative kinetic energy $E/\epsilon = 0\cdot 1$ and the interaction constants are such that

$$h/R_m(2M\epsilon)^{\frac{1}{2}} = 0\cdot 1, \quad a_2 = 0\cdot 3, \quad \begin{cases} b_1 = b_2 = c = 0, & \text{case (a),} \\ b_1 = b_2 = 0\cdot 5, c = 2\cdot 0 & \text{case (b).} \end{cases}$$

5.5.2. *Experiments on the scattering of beams of polar molecules in selected rotational states.* Just as it is possible to analyse a beam of atoms that possess finite total angular momentum into beams with selected magnetic substates by passage through an inhomogeneous magnetic field, the rotational states of a beam of polar molecules may be selected by passage through an inhomogeneous electric field.

A polar diatomic molecule in a rotational state (J, M) quantized with respect to the direction of an electric field $\mathbf{F}$, acquires an energy due to the field which is given by

$$W = (2I\mu_0^2 F^2/\hbar^2) \frac{J(J+1)-3M^2}{2J(J+1)(2J-1)(2J+3)}. \tag{242}$$

μ_0 is the dipole moment and I the moment of inertia of the molecule. If $\mathbf{F}$ is inhomogeneous the molecule is subject to a force

$$-\text{grad } W = -(2I\mu_0^2/\hbar^2) \frac{J(J+1)-3M^2}{J(J+1)(2J-1)(2J+3)} F \text{ grad } F. \tag{243}$$

As this depends on J and M the possibility of rotational state selection is clear.

A convenient analyser is one which uses an electric quadrupole field.† The analyser consists of four similar parallel cylinders the centres of the sections of which lie in the corners of a square, as in Fig. 17.58. The

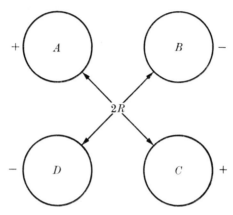

Fig. 17.58

cylinders A, C are raised to equal positive potential and B, D to equal and opposite negative potential. If U is the potential difference between neighbouring electrodes the magnitude of the field strength F varies across a section and is given by

$$F = Ur/R^2, \qquad (244)$$

where r is the distance from the axis of the quadrupole and $2R$ the shortest distance between the cylinders A, C or B, D. With this arrangement W varies as r^2 so the molecules passing through the field oscillate harmonically in transverse motion with a period depending on J and M. It follows that a beam of molecules of small divergence entering a quadrupole analyser along its axis, with uniform velocity, will be refocused at different points depending on their rotational states. This will only apply to those molecules for which $J(J+1) > 3M^2$.

Two types of experiment on the scattering of beams of polar molecules in selected rotational states have been carried out. In both the polar molecules were those of TlF. The first was concerned with observing the dependence of the total cross-section for scattering of these molecules by various atomic gases on the rotational quantum number M for fixed J, the second with the measurement of cross-sections for inelastic collisions

† BENNEWITZ, H. G., PAUL, W., and SCHLIER, C., Z. Phys. **141** (1955) 6.

with various atoms and molecules which involve rotational transitions in the TlF.

5.5.2.1. *Dependence of elastic cross-section on molecular orientation.* In many ways the first experiment carried out by Bennewitz, Kramer, Paul, and Toennies† resembles that of Berkling, Schlier, and Toschek‡ who observed the dependence of the total cross-section for scattering of magnetically oriented gallium atoms by rare gases, on the magnetic quantum number (see Chap. 16, § 12.4). In both cases the dependence arises from the presence of a non-central term in the van der Waals interaction which can be written

$$V = -\frac{C}{R^6}(1+\beta\cos^2\chi). \tag{245}$$

For the atom polar-molecule case, χ is the angle between the direction of the molecular axis and the line joining the centre of mass of the molecule to the centre of the atom—in the case of the magnetically oriented gallium atom interacting with a rare-gas atom it is the angle between the axis of quantization and the line joining the centres of mass of the interacting atoms.

As in Chapter 16, § 12.4 we may calculate the total cross-sections from Schiff's semi-classical approximation (250) of that chapter. Thus, if the initial direction of relative motion of atom and molecule is taken as the z-direction

$$Q = \int_{-\infty}^{\infty}\int_{-\infty}^{\infty} \sin^2\left\{\frac{M}{2k\hbar^2}\int_{-\infty}^{\infty} V(x,y,z)\,dz\right\} dx dy, \tag{246}$$

M being the reduced mass of atom and molecule and k the incident wave-number of relative motion. The geometry of the system is illustrated in Fig. 17.59. It is convenient to transfer the integration over x, y to one over p, ϕ as follows. Using the relations

$$R^2 = p^2+z^2, \quad \cos\theta = z/(p^2+z^2)^{\frac{1}{2}}, \quad \sin\theta = p/(p^2+z^2)^{\frac{1}{2}},$$
$$\cos\chi = \cos\theta\cos\Theta+\sin\theta\sin\Theta\cos(\phi-\Phi), \quad dxdy = p\,dpd\phi,$$

it is found that
$$Q = Q_0\,G(\Theta,\Phi), \tag{247}$$

where $Q_0 = 8\cdot08(MC/k\hbar^2)^{2/5}$ is the total cross-section for scattering by the central part of the potential (245) (see Chap. 16, (101 a)) and

$$G(\Theta,\Phi) = \frac{1}{2\pi}\int_0^{2\pi} \{1+\tfrac{1}{6}\beta\cos^2\Theta+\tfrac{5}{6}\beta\sin^2\Theta\cos^2(\Phi-\phi)\}^{2/5}\,d\phi. \tag{248}$$

We thus have the cross-section for a given orientation of the molecular axis relative to the incident direction. The rotation period of a TlF molecule in the state $J = 1$

---

† BENNEWITZ, H. G., KRAMER, K. H., PAUL, W., and TOENNIES, J. P., *Z. Phys.* **177** (1964) 84.
‡ loc. cit., p. 1437.

is about 5×10^{-11} s which is long compared with the time of collision, about 2×10^{-12} s, with, say, an argon atom at ordinary temperature. The cross-section which will be measured will therefore be

$$Q = Q_0 \iint G(\Theta, \Phi)|\rho(\Theta, \Phi)|^2 \sin\Theta \, d\Theta d\Phi, \qquad (249)$$

where $\rho(\Theta, \Phi)$ is the wave function of the molecular state concerned. The rotational quantization will normally be relative to the direction of some orienting

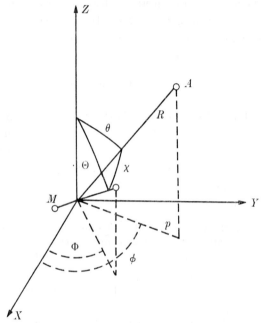

Fig. 17.59.

electric field and we consider a state with rotational quantum numbers J, M. If α is the angle the field direction makes with the incident direction we have

$$\rho(\Theta, \Phi) = \sum_{M'=-J}^{J} d^J_{M'M}(\alpha) Y_{JM'}(\Theta, \Phi), \qquad (250)$$

where $Y_{JM'}$ is a normalized spherical harmonic. We therefore have

$$Q = Q_0 \sum_{M'} \{d^J_{M'M}(\alpha)\}^2 G_{JM'}(\beta), \qquad (251)$$

where
$$G_{JM'} = \int_0^{2\pi} \int_0^\pi Y_{JM'} Y^*_{JM'} G(\Theta, \Phi) \sin\Theta \, d\Theta d\Phi. \qquad (252)$$

In particular, for $J = 1$, $M = 0$,
$$d^1_{10} = 2^{-\frac{1}{2}} \sin\alpha = -d^1_{\bar{1}0}, \qquad d^1_{00} = \cos\alpha. \qquad (253)$$

In the experiment carried out by Bennewitz et al.† the selected TlF molecules, after passage through the quadrupole polarizer, crossed an

† loc. cit., p. 1595.

atomic beam at 90° and the attenuation suffered as a result of this was measured. Comparison was made between the total cross-sections, deduced from this attenuation, for collisions of TlF (1, 0) and (1, 1) molecules, quantized with respect to the direction of relative motion in the collision, with the gas atoms concerned. To see how this may be done it is necessary to examine the kinematic as well as the geometrical relations in the system. The velocity v_m of the molecules was selected by a velocity analyser and in any case is small compared with the mean speed of the gas atoms so that the relative velocity v_r is nearly in the direction of the atomic beam. The conditions in the quadrupole polarizer were chosen so that (1, 0) molecules with the selected velocity passed through the attenuating gas beam. The direction of quantization could be changed adiabatically after passage into the scattering chamber by means of a uniform electric field across the chamber so that the (1, 0) molecules were quantized with respect to this field direction. Measurements were made with this field in two different perpendicular directions, one referred to as position (a) in which it was parallel to the atomic beam and the other (b) in which it was perpendicular to both beams. If the relative velocity were exactly along the direction of the atomic beam case (a) would provide conditions in which (1, 0) molecules, quantized with respect to the relative velocity, were scattered and case (b) those in which (1, 1) molecules quantized with respect to the same direction, were scattered. In fact, because v_r is not exactly v_a, the velocity of the attenuating atoms, the following procedure was adopted.

The effective cross-sections measured from attenuation observations in cases (a), (b) we denote by Q^a_{eff}, Q^b_{eff} respectively. Then

$$Q^a_{\text{eff}} = Q_0(v_m)\langle(v_r/v_m)^{\frac{2}{3}}\sum\{d^1_{M'0}(\alpha)\}^2 G_{1M'}\rangle, \qquad (254)$$

$$Q^b_{\text{eff}} = Q_0(v_m)\langle(v_r/v_m)^{\frac{2}{3}}\sum\{d^1_{M'0}(\tfrac{1}{2}\pi)\}^2 G_{1M'}\rangle, \qquad (255)$$

where $\langle\ \rangle$ denotes an average over the velocity distribution in the atomic beam. α differs from zero because v_r is not exactly equal to v_a and is not known a priori. In obtaining these results (see Chap. 16, (101)) it has been assumed that the cross-section varies with relative velocity v_r as $v_r^{-\frac{2}{3}}$. We now have, on substitution for the $d^1_{M'0}$ from (253),

$$Q^b_{\text{eff}}/Q^a_{\text{eff}} = G_{11}(\beta)/[\{1-S(x)\}G_{11}(\beta)+S(x)G_{10}(\beta)], \qquad (256)$$

where

$$S(x) = \langle v_r^{\frac{2}{3}}\cos^2\alpha\rangle/\langle v_r^{\frac{2}{3}}\rangle \quad \text{and} \quad x = v_m/\bar{v}_a \text{ where } \bar{v}_a = (2\kappa T/M_a)^{\frac{1}{2}}, \qquad (257)$$

T being the temperature and M_a the mass of a gas atom. In the limit $x \to 0$, $\alpha \to 0$, $S(x) \to 1$ we have

$$\lim_{x \to 0} (Q^b_{\text{eff}}/Q^a_{\text{eff}}) = G_{11}/G_{10} = Q_{11}/Q_{10}. \tag{258}$$

Hence to obtain the required ratio, $Q^b_{\text{eff}}/Q^a_{\text{eff}}$ is calculated as a function of x for different values of β giving the set of curves shown in Fig. 17.61. Observed ratios $Q^b_{\text{eff}}/Q^a_{\text{eff}}$ for a particular atom at different temperatures T are plotted on the diagram and from these extrapolation to $x = 0$ may be carried out.

Fig. 17.60 illustrates diagrammatically the arrangement of the experiment. The TlF beam from the oven passed through the quadrupole polarizer on the axis of which was a stop to intercept molecules which were moving so close to the axis as to experience little electric field. The beam then passed through the scattering chamber in the course of which the selected molecules crossed the atomic beam at right angles. An appropriate uniform electric field was applied across this chamber either parallel to the atomic beam (case (a)) or perpendicular to both beams (case (b)). The strength of this field was chosen to be large enough (> 50 V cm^{-1}) to decouple the nuclear spin from the molecular rotation (otherwise J is not a good quantum number) and small enough (< 300 V cm^{-1}) to avoid mixing rotational states with different J values. The molecular beam then passed through a velocity selector with resolution 0·14, to enter the detector. This was of the Langmuir hot-wire ionization type which is effective for detection of TlF if the surface of the hot tungsten ribbon is maintained in a state of oxidation by a continuous oxygen flow over it. The ions formed at the hot surface were mass analysed by a high transmission r.f. quadrupole analyser and the current of the selected ions measured by a secondary emission electron multiplier. Between the scattering chamber and the velocity analyser a short electric quadrupole analyser was introduced. This checked that the orienting electric field was effective in producing adiabatic rotation of the axes of the molecules so that they remained in the $(1, 0)$ state *relative to the electric field*.

Special care had to be taken in measurement of Q^a_{eff}, Q^b_{eff} because the attenuating effect of the gas diffused throughout the volume of the scattering chamber exceeded that of the atomic beam. This could be allowed for by measuring the attenuation of the molecular beam first with only diffuse gas present at a measured pressure and then with the atomic beam present as well as a measured pressure of diffuse gas. Allowance was also made for the fact that the intensity of the rotationally selected molecular beam passing through the scattering chamber in the absence of any scattering is not the same in cases (a) and (b).

The angular resolution was about 3′ which is probably not good enough to provide accurate measurements of the total cross-section for the heavier gases but should be adequate to yield good values of the ratio $Q_{\text{eff}}^{b}/Q_{\text{eff}}^{a}$.

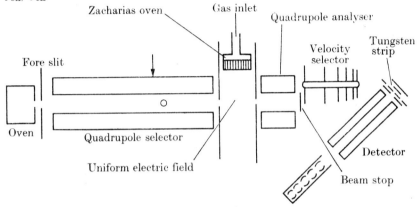

FIG. 17.60. Arrangement of apparatus in the experiments of Bennewitz et al.

Observed values of the ratio for scattering by a number of atoms at different temperatures are shown in Fig. 17.61. By extrapolation to very high temperatures ($x \to 0$ in Fig. 17.61) the ratios Q_{11}/Q_{10} given in Table 17.13 were obtained.

For argon and krypton the observed ratio, which is the same within experimental error, corresponds to an asymmetry parameter β in (245) of 0·4. The values for He and Ne are less reliable because the analysis by which they have been derived assumed that the collisions were taking place in the low velocity region (see Chap. 16, § 6.3.1) so that (247) and (248) are valid. This is a doubtful assumption for the light gases.

5.5.2.2. *Collisions involving rotational transitions.* The arrangement used in the second experiment† is illustrated in Fig. 17.62. In principle an electric quadrupole polarizer selected beams of TlF molecules in particular $(J, 0)$ states that passed through gas in a scattering chamber, across which a uniform orienting electric field was applied. As a result of the collisions in the chamber transitions to other rotational states (J', M') took place. The issuing molecular beam scattered through angles $< \tfrac{1}{2}°$ could then be rotationally selected by a second quadrupole so that molecules in particular $(J', 0)$ states alone reached the detector.

† TOENNIES, J. P., *Z. Phys.* **182** (1965) 257.

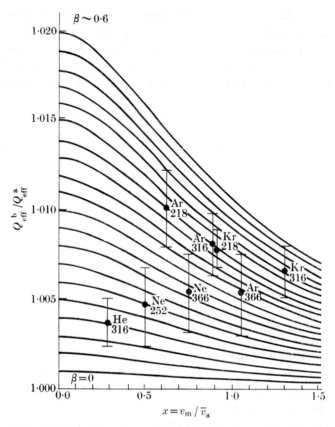

FIG. 17.61. Calculated dependence of the ratio $Q^b_{\text{eff}}/Q^a_{\text{eff}}$ on the ratio $x = v_m/\bar{v}_a$ of the velocity v_m of the TlF beam to the most probable velocity $\bar{v}_a$ of the atoms of the scattering beam, for different asymmetry parameters β. Observed values for scattering by different atoms, with probable errors, are also shown.

TABLE 17.13

Observed ratios Q_{11}/Q_{10} of total cross-sections for scattering of TlF molecules in (11) and (10) rotational states, by different gases

Gas	He	Ne	Ar	Kr
Q_{11}/Q_{10}	1·004	1·007	1·013	1·014

Fig. 17.63 shows results obtained using NH_3 at a pressure of 10^{-5} torr as the scattering gas, the path length through the gas being 5 cm. The magnitude of the polarizer potential was set at 14 kV so as to select molecules in (3, 0) states, and the detector current is shown as a function of the magnitude of the analyser potential with and without scattering

gas. It will be seen that, whereas the intensity of the (3, 0) beam at the detector was decreased by the gas there was a marked increase in the (2, 0) and (1, 0) beams. Strong evidence that these effects arise from inelastic collisions in the scattering gas was obtained from observation of the variation with gas pressure.

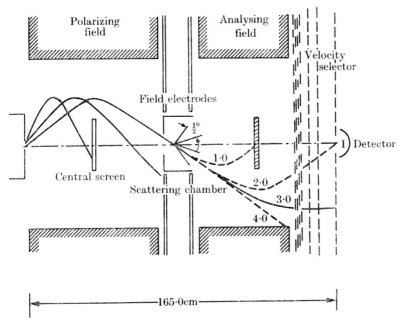

FIG. 17.62. Arrangement of apparatus in the experiments of Toennies.

Let $I(30-J0)$ be the signal when gas is present and the quadrupole voltage is set to focus molecules in $(J, 0)$ states on the detector, $I_0(30-J0)$ the corresponding signal when there is no scattering gas.

Then, if the path length through the scattering gas is x and the concentration of gas atoms is N,

$$\frac{I(30-30)}{I_0(30-30)} = \exp\{-Q_t^{30} N x\}, \tag{259}$$

where Q_t^{30} is the total cross-section for scattering of molecules initially in (3, 0) states. It must be remembered that scattering includes all inelastic collisions but only elastic collisions in which the deflexion in the laboratory system exceeds $\tfrac{1}{2}°$.

Also $$\frac{dI(30-20)}{dx} = N\{I(30-30)Q_{\text{in}}^{30,20} - I(30-20)Q_t^{20}\}, \tag{260}$$

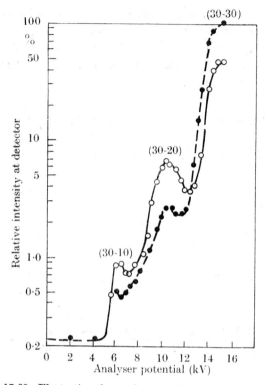

Fig. 17.63. Illustrating the production of rotational transitions in TlF due to collisions with NH_3 molecules. The points and curves show the observed variation of detector current with analyser voltage when the polarizer voltage is set to focus TlF molecules in the (3, 0) state. –●---●– when no scattering gas is present, –○–○–○– when NH_3 at 10^{-5} torr is present in the scattering chamber. The peaks arise from detection of molecules in (1, 0), (2, 0) and (3, 0) states, respectively, as the analyser voltage increases and so correspond to the transitions indicated.

where $Q_{\rm in}^{30,20}$ is the cross-section for an inelastic collision in which the transition $(3, 0) \to (2, 0)$ occurs and the deflexion is $< \frac{1}{2}°$. If we take $Q_{\rm t}^{20} \simeq Q_{\rm t}^{30}$, which is confirmed from the observations, then

$$I(30\text{–}20) = I_0(30\text{–}30) Q_{\rm in}^{30,20} Nx \exp\{-Q_{\rm t}^{30} Nx\} + \\ + I_0(30\text{–}20) \exp\{-Q_{\rm t}^{30} Nx\}. \quad (261)$$

In obtaining this result it has been assumed that no appreciable repopulation of (3, 0) states comes from re-excitation of molecules which have been deactivated to (2, 0) or (1, 0) levels, i.e. it has been assumed that
$$I(30\text{–}20) Q_{\rm in}^{20,30} \ll I(30\text{–}30) Q_{\rm t}^{30},$$
a condition well satisfied in practice.

We may write (261) in the form

$$\frac{I(30-20)}{I_0(30-20)} = \exp\{-N/N_{\mathrm{m}}\}\left\{1+\frac{Q_{\mathrm{in}}^{30,20}}{Q_{\mathrm{t}}^{30}}\frac{I_0(30-30)}{I_0(30-20)}\frac{N}{N_{\mathrm{m}}}\right\}, \quad (262)$$

where $N_{\mathrm{m}} = 1/Q_{\mathrm{t}}^{30}x$. Hence to obtain $Q_{\mathrm{in}}^{30,20}$, Q_{t}^{30} is first determined from (259) and then $I(30-20)/I_0(30-20)$ is plotted as a function of N/N_{m}. This should be of the form

$$I(30-20)/I_0(30-20) = \exp\{-N/N_{\mathrm{m}}\}\{1+aN/N_{\mathrm{m}}\}, \quad (263)$$

where $\qquad a = (Q_{\mathrm{in}}^{30,20}/Q_{\mathrm{t}}^{30})I_0(30-30)/I_0(30-20). \qquad (264)$

By suitable choice of a, a good fit to the plotted curve could be obtained and then $Q_{\mathrm{in}}^{30,20}$ may be derived from (264).

Fig. 17.64 illustrates the application of this procedure to the determination of $Q_{\mathrm{in}}^{20,30}$ for collisions of TlF with NH_3 and with CF_2Cl_2. In this case $\qquad a = (Q_{\mathrm{in}}^{20,30}/Q_{\mathrm{t}}^{20})I_0(20-20)/I_0(20-30).$

For NH_3 the observed results show that a lies between 10 and 11 while for CF_2Cl_2 the results are more definite and agree very well with that calculated for $a = 11$.

Total cross-sections were measured under the following conditions.

(a) Without rotational state selection. This permitted a relatively high angular resolution ($\simeq 1'$ of arc) but the most probable rotational state at the beam temperature has $J \simeq 33$.

(b) With the polarizer but not the analyser in operation. The angular resolution was somewhat worse ($\simeq 4'$ of arc) but the rotational state was now selected.

(c) With polarizer and analyser in operation. The angular resolution was now much worse ($\frac{1}{2}°$).

From the effective cross-sections Q_{eff} measured with a velocity selected molecular beam scattered in a gas at temperature T the cross-section $Q_t(v)$ at a relative velocity equal to the velocity v of the gas molecules may be derived in the usual way from the formula (166) of Chapter 16. Thus if the scattering potential varies as R^{-s} then

$$Q_t(v) = Q_{\mathrm{eff}}/F(s, v/v_{\mathrm{mB}}), \quad (265)$$

where $v_{\mathrm{mB}} = (2\kappa T/M_{\mathrm{B}})^{\frac{1}{2}}$, M_{B} being the mass of a molecule of the scattering gas. The function F is defined in Chapter 16 (164). This formula applies if the resolving power is sufficient to measure the full quantal cross-section. If the resolution is so bad, as it is in case (c), that a classical 'total' cross-section dependent on the resolving power is measured, this cross-section varies with relative velocity (see Chap. 16, (53 b)) as $v^{-4/s}$

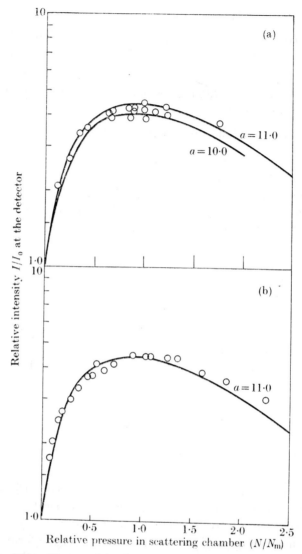

FIG. 17.64. Pressure dependence of the inelastic scattering of TlF (2, 0) to TlF (3, 0) in collisions with (a) NH_3, (b) CF_2Cl_2. ○ Observed points; ——— calculated curves for different values of the parameter a of (264) as indicated.

instead of $v^{-2/(s-1)}$ as in the quantum case. s in (265) must then be replaced by $s' = \tfrac{1}{2}s+1$.

If it is assumed that the true total cross-section is nearly independent of J, the observations made under (a) can be used to correct those made with (c) for poor resolution. This correction is difficult to make for collisions with the polar gases for which the proper value to take for s is uncertain and the resolution problems are more severe (see § 5.6). Moreover, there is likely to be a stronger dependence of the total cross-section on J, as the effective potential between two dipoles depends very much on the relation between collision time and rotational period. The inelastic cross-sections do not suffer from the small angular resolution

TABLE 17.14

Observed total and inelastic cross-sections for collisions of TlF molecules with different atoms and molecules at a relative velocity of 305 m s^{-1}

Scattering atom or molecule	Total cross-section (10^{-16} cm^2)					Inelastic cross-section (10^{-16} cm^2)
	Case (a)		Case (c)	Assumed		
	Q_{eff}	Q_t	Q_{eff}^{20}	s'	Q_t^{20}	$Q_{\text{in}}^{20,30}$
He			303			3·7
Ne	606	407	291	4	232	4·6
Ar	868	674	434	4	384	6·1
Kr	956	825	414	4	386	7·8
O$_2$	622	461	315	4	270	23·6
CH$_4$	1104	702	471	4	363	6·3
SF$_6$	1050	955	490	4	468	6·8
N$_2$O	1080	890	786	?	—	80
H$_2$O	1320	862	895	?	—	70
CF$_2$Cl$_2$	832	750	830	?	—	115
NH$_3$	1460	940	3200	?	—	580

difficulty. It is true that they do not include contributions from collisions in which the TlF molecule is deflected through more than $\frac{1}{2}°$ in the laboratory system but such collisions should not be very frequent.

Results obtained for collisions with a number of different atoms and molecules are given in Table 17.14.

It will be seen that the polar molecules behave very differently from the other scattering atoms or molecules. For the latter Q_t^{20}/Q_t is between 0·50 and 0·6, which is not unexpected because of the poorer angular resolution in case (c). In the case of the polar molecules Q_t^{20}/Q_t is closer to 1 and for NH$_3$ is over 2, even allowing for the difficulty in deriving Q_t from Q_{eff} in these cases. Equally clear are the much greater inelastic cross-sections for the polar molecules, NH$_3$ again being an extreme case.

Because of its special behaviour, other inelastic cross-sections were measured for NH$_3$ and are given in Table 17.15. It will be noted that the cross-sections for $(2, 0) \to (3, 0)$ and $(3, 0) \to (2, 0)$ transitions are approximately equal as they should be according to the principles of detailed balance.

TABLE 17.15

Observed inelastic cross-sections (in 10^{-16} cm²) for collisions of TlF with NH_3 and ND_3 molecules

State in polarizer field	State in analyser field	NH_3			ND_3
		(1, 0)	(2, 0)	(3, 0)	(2, 0)
(1, 0)		—	685	—	685
(2, 0)		—	—	580	—
(3, 0)		48	480	—	—

5.6. *Theoretical discussion of collisions between polar molecules*

In Chapter 16, § 13 we discussed the nature of the long-range forces between molecules and their effect on the collision cross-section, excluding only the case in which the molecules are both polar. This case is special in that, while the interaction is of very long range, falling off as R^{-3}, where R is the molecular separation, its anisotropic form is such that, when averaged over all orientations of the dipoles, it vanishes. Keesom† proposed that the mean effective interaction should be calculated assuming a Boltzmann distribution such that, if $V(\theta_1, \phi_1; \theta_2, \phi_2; \theta, \phi)$ is the interaction energy of two dipoles at a distance R apart when (θ_1, ϕ_1), (θ_2, ϕ_2), and (θ, ϕ) are the polar angles of the directions of the dipoles and of the line joining them, the probability of the dipole axes being within solid angle elements $d\omega_1, d\omega_2$ about these directions will be proportional to

$$\exp\{-V(\theta_1, \phi_1; \theta_2, \phi_2; \theta, \phi)/\kappa T\} d\omega_1 d\omega_2. \qquad (266)$$

This may then be used as a weight factor for the calculation of a mean value of V, treating R as fixed. In this way Keesom obtained as the leading term in this average

$$V_{d-d} = -\frac{2\mu_1^2 \mu_2^2}{3\kappa T R^6}. \qquad (267)$$

Once again an inverse sixth power interaction is obtained but we can only expect this approximation to be valid when the collision is a slow one in the sense that the collision time is long compared with the periods of rotation of the dipoles. When this condition is not satisfied we may use the 'sudden' approximation in the opposite extreme conditions. When (267) is applicable the cross-sections are comparable with those arising from van der Waals forces between atoms while the sudden approximation, which assumes energy resonance conditions, leads to

† KEESOM, W. H., *Phys. Z.* **22** (1921) 129.

very large cross-sections which, for collisions between such molecules as CH_3I and $CsCl$ exceed the van der Waals cross-section by factors of 20 or more.

A theoretical discussion of dipole–dipole collisions which is applicable under intermediate conditions has been given by Cross and Gordon.† For the calculation of the total cross-section it is convenient to use the semi-classical formulation in which the relative motion is treated by classical mechanics. At sufficiently large impact parameters even the long-range interaction will be small so the formulae (see (136)) of time-dependent perturbation theory will be applicable and the relative motion will be linear with constant velocity v.

The time-dependent interaction energy that produces rotational transitions will be $V(\theta_1,\phi_1;\theta_2,\phi_2;\theta(t),\phi(t);R(t))$, where θ, ϕ, and R are functions of the time t. Taking the polar axis along the incident direction of relative motion and the azimuthal plane as that of the collision we have

$$R = (p^2+v^2t^2)^{\frac{1}{2}}, \qquad \sin\theta = p/(p^2+v^2t^2)^{\frac{1}{2}}, \qquad \phi = 0. \tag{268}$$

V is given by
$$\frac{\mu_1\mu_2}{R^3}(\cos\epsilon - 3\cos\chi_1\cos\chi_2), \tag{269}$$

where ϵ is the angle between the axes of the dipoles and χ_1, χ_2 the angles the directions of these axes make with the line joining the dipoles. This may be expressed in terms of the angles (θ_1,ϕ_1), (θ_2,ϕ_2), (θ,ϕ) to give

$$V = \frac{\mu_1\mu_2}{R^3}\sum a_{M_1M_2}Y_{1M_1}(\theta_1,\phi_1)Y_{1M_2}(\theta_2,\phi_2)Y_{2,-M_1-M_2}(\theta,\phi), \tag{270}$$

where in terms of a Clebsch–Gordan coefficient $C(\ )$,

$$a_{M_1M_2} = -\tfrac{1}{3}(4\pi)^{\frac{3}{2}}(6/5)^{\frac{1}{2}}(-1)^{M_1+M_2}C(112;M_1M_2), \tag{271}$$

and Y_{JM} is a normalized spherical harmonic

$$Y_{JM}(\theta,\phi) = \left(\frac{2J+1}{4\pi}\right)^{\frac{1}{2}}\left\{\frac{(J-M)!}{(J+M)!}\right\}^{\frac{1}{2}}P_J^M(\cos\theta)\,e^{iM\phi}. \tag{272}$$

The probability of a transition in which the rotational states (J_1M_1), (J_2M_2), denoted together as n, of the two molecules change to $(J'_1M'_1)$, $(J'_2M'_2)$, denoted by n', is now given by (see (136))

$$P_{n,n'}(p) = \left|\frac{1}{i\hbar}\int_{-\infty}^{\infty} V(\theta_1,\phi_1;\theta_2,\phi_2;\theta(t),R(t))\rho_{J_1M_1}(\theta_1,\phi_1)\rho^*_{J'_1M'_1}(\theta_1,\phi_1)\times\right.$$

$$\left.\times \rho_{J_2M_2}(\theta_2,\phi_2)\rho^*_{J'_2M'_2}(\theta_2,\phi_2)\exp(i\Delta Et/\hbar)\,dt\right|^2, \tag{273}$$

where $\quad \Delta E = B_1\{J'_1(J'_1+1)-J_1(J_1+1)\}+B_2\{J'_2(J'_2+1)-J_2(J_2+1)\}. \tag{274}$

B_1 and B_2 are the rotational constants $\hbar^2/2I_{1,2}$, where $I_{1,2}$ are the moments of inertia of the two molecules.

Because of the form (270) for V it follows that the only possible transitions are those for which $\Delta J_1 = \pm 1$, $\Delta J_2 = \pm 1$—for large values of p we have the remarkable situation that all collisions are inelastic, $\Delta J_1 = 0$, $\Delta J_2 = 0$ being excluded.

† CROSS, R. J. and GORDON, R. G., J. chem. Phys. 45 (1966) 3571.

When J_1, J_2 are large we have
$$\Delta E = \pm 2J_1 B_1 \pm 2J_2 B_2. \tag{275}$$
We are interested in the sum $\sum_{n'} P_{nn'}(p)$ over all final states n' consistent with the allowed transitions. This has been calculated by van Kranendonk† who finds
$$\sum_{n'} P_{nn'}(p) = (4\mu_1^2 \mu_2^2 / 9\hbar^2 v^2 p^4)\{F(x^-) + F(x^+)\}, \tag{276}$$

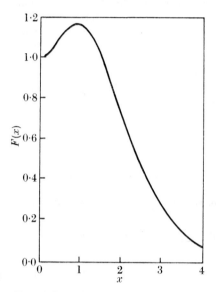

FIG. 17.65. The function $F(x)$ of (279).

where
$$x^{\pm} = |\Delta E^{\pm}| p/\hbar v, \tag{277}$$
with
$$\Delta E^+ = 2J_1 B_1 + 2J_2 B_2, \qquad \Delta E^- = 2J_1 B_1 - 2J_2 B_2. \tag{278}$$
$F(x)$ is a function which is given to a good accuracy by
$$F(x) = (1 + 2x + \tfrac{3}{4}\pi x^2 + \pi x^3) e^{-2x} \tag{279}$$
and is plotted in Fig. 17.65.

The total cross-section is given by
$$Q_t = 2\pi \int_0^\infty \sum_{nn'} P(p) p \, dp, \tag{280}$$
but we cannot use (276) for small p as the interaction can no longer be treated as a small perturbation. To evaluate (280) we proceed in the same kind of way as in the calculation of the total elastic cross-section for scattering by an inverse sth power interaction (see Chap. 16, § 5.4.1), or of the inelastic cross-section for excitation of an optically allowed transition by electron impact when the coupling is strong (see Chap. 7, § 5.3.5). We break up the range of integration at p^*, which

† VAN KRANENDONK, J., Can. J. Phys. **41** (1963) 433.

is such that for $p > p^*$ (276) applies, while for $p < p^*$ the probability may be taken as unity. This gives

$$Q_t = 2\pi \int_0^{p^*} p \, dp + 2\pi \int_{p^*}^\infty \sum P_{nn'}(p) p \, dp. \tag{281}$$

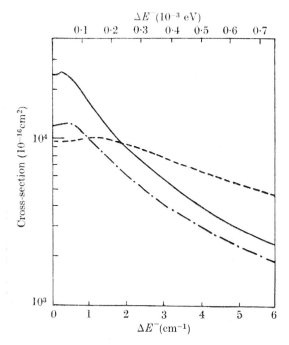

FIG. 17.66. Total cross-sections for collisions of polar molecules with CH_3I molecules $\mu_2 = 1\cdot 65\alpha$. ——— Relative velocity 400 m s^{-1}, $\mu_1 = 106\alpha$; ——— 1000 m s^{-1}, $\mu_1 = 106\alpha$; —·—· 400 m s^{-1}, $\mu_1 = 5\cdot 3\alpha$. $\alpha = 10^{-18}$ e.s.u.

We must take account also of shadow scattering which doubles the contribution from the strong interaction region. This gives

$$Q_t = 2\pi p^{*2} + (\tfrac{8}{9}\pi\mu_1^2\mu_2^2/\hbar^2 v^2) \int_{p^*}^\infty \{F(x^-) + F(x^+)\} p^{-3} \, dp. \tag{282}$$

It remains to choose p^*. This is somewhat arbitrary but little error will be made by choosing it so that

$$\sum_{n'} P_{nn'}(p^*) = 1. \tag{283}$$

The resulting cross-section is practically independent of x^+ except for very small x^+, so $F(x^+)$ may be dropped. Fig. 17.66 shows cross-sections as a function of ΔE^- for three different cases in which the dipole moments are chosen to have representative values.

The striking thing about the curves of Fig. 17.66 is their resonance character and the enormous cross-sections in the case of exact resonance where $\Delta E = 0$. The sharpness of the resonance decreases as the relative velocity increases.

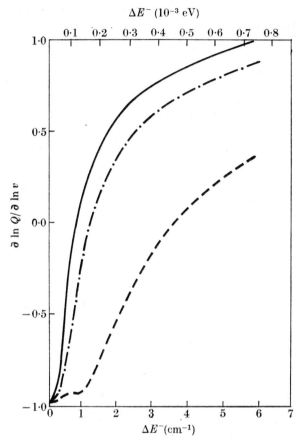

FIG. 17.67. The logarithmic derivative $\partial \ln Q/\partial \ln v$ of the total cross-section with respect to relative velocity for impact of polar molecules with CH_3I molecules ($\mu_2 = 1.65\alpha$). —— $v = 400$ m s^{-1}, $\mu_1 = 10.6\alpha$; - - - $v = 1000$ m s^{-1}, $\mu_1 = 10.6\alpha$; -·-·- $v = 400$ m s^{-1}, $\mu_1 = 5.3\alpha$. $\alpha = 10^{-18}$ e.s.u.

In Fig. 17.67 the variation of the cross-section with velocity is given for the same three cases in the form of a plot of $\partial \ln Q/\partial \ln v$ as a function of ΔE^-. For comparison it will be remembered that, for spherically symmetrical potentials varying as R^{-6} and R^{-3} respectively, $\partial \ln Q/\partial \ln v$ is equal to -0.4 and -1.0. In the exact resonance case it is indeed equal to -1 but as the resonance discrepancy ΔE increases it increases rapidly to positive values.

For applications to experimental conditions it is necessary to average the cross-sections over thermal distributions of rotational energy. If the spacings of the rotational levels are very small compared with κT we have for the mean cross-section, writing $\omega^\pm = \Delta E^\pm/\hbar$,

$$\langle Q \rangle = \int_0^\infty \int_0^\infty P_1(\omega_1)P_2(\omega_2)Q(\omega^-, \infty) \, d\omega_1 \, d\omega_2, \tag{284}$$

where
$$P_i(\omega_i) = (\omega_i/\bar{\omega}_i^2)\exp\{-\tfrac{1}{2}(\omega_i/\bar{\omega}_i)^2\} \tag{285}$$

with $\hbar\bar\omega_i = (2B_i\kappa T_i)^{\frac12}$. As ω^+ is no longer involved we drop the $^-$ and write ω for ω^-.

Cross, Gislason, and Herschbach† took into account not only the dipole–dipole cross-section (282) but also allowed for the presence of the inverse sixth power interactions by taking

$$Q(\omega) = Q_0\Gamma^2/(\omega^2+\Gamma^2) \quad (|\omega|<\Gamma'),$$
$$= Q_6 \quad (|\omega|>\Gamma'), \qquad (286)$$

where Q_6 is the background contributions from these interactions. Q_0 is the resonance cross-section $Q(0,\infty)$ which is obtained from (282) by taking

$$F(x^+) = 0, \qquad F(x^-) = 1.$$

In this case the condition (283) gives

$$p^* = p_r^* = \left(\frac{2}{3}\frac{\mu_1\mu_2}{\hbar v}\right)^{\frac12} \qquad (287)$$

and
$$Q_t = 3\pi p_r^{*2} = 2\pi\mu_1\mu_2/\hbar v = Q(0,\infty). \qquad (288)$$

Empirically it is found that $\qquad \Gamma \simeq 0\cdot 6 v/p_r^*, \qquad (289)$

while Γ' is such that $\qquad Q_0\Gamma^2/(\Gamma^2+\Gamma'^2) = Q_6$

so $\qquad\qquad \Gamma'/\Gamma \simeq (Q_0/Q_6)^{\frac12}. \qquad (290)$

If $\omega_i^2 \gg \Gamma^2$, (284) now becomes

$$\langle Q\rangle \simeq \int_0^\infty P_1(\omega_2)P_2(\omega_2)\,d\omega_2 \int_{-\Gamma'}^{\Gamma'} Q(\omega)\,d\omega + $$
$$+ \left\{1 - \int_0^\infty \int_{\omega_2-\Gamma'}^{\omega_2+\Gamma'} P_1(\omega_1)P_2(\omega_2)\,d\omega_1\,d\omega_2\right\} Q_6 \qquad (291)$$

and, on substitution of (285) and (286) for P_1 and Q the integrations may be carried out to give
$$\langle Q\rangle = A_{dd}Q_{dd} + A_6 Q_6, \qquad (292)$$

where $\qquad\qquad Q_{dd} = 1\cdot 17(Q_0/x_0)\sin 2\gamma,$

$$\gamma = \arctan(\bar\omega_2/\bar\omega_1),$$
$$x_0 = \bar\omega_{12}p_r^*/v, \qquad \bar\omega_{12} = (\bar\omega_1^2+\bar\omega_2^2)^{\frac12},$$
$$A_{dd} = (2/\pi)\arctan(\Gamma'/\Gamma), \qquad A_6 = 1 - (\tfrac12\pi)^{\frac12}\sin 2\gamma(\Gamma'/\bar\omega_{12})\exp\{-\tfrac12(\Gamma'/\bar\omega_{12}^2)\}. \qquad (293)$$

As an example, consider the collision of KCl (at 870 °K) with CH$_3$I (at 240 °K), with relative velocity 400 m s^{-1}, studied experimentally by Cross, Gislason, and Herschbach.† We have

$\bar\omega_1 = 12, \quad \bar\omega_2 = 9\cdot 3, \quad \bar\omega_{12} = 15\cdot 5, \quad \Gamma = 1\cdot 52, \quad \Gamma' = 6\cdot 80,$ all in cm^{-1},
$\gamma = 36\cdot 7°, \quad \sin 2\gamma = 0\cdot 96, \quad p_r^* = 52\cdot 3$ Å, $\quad x_0 = 6\cdot 08,$
$Q_0 = 25\,800; \quad Q_{dd} = 4770; \quad Q_6 = 1297,$ all in Å^2,
$A_{dd} = 0\cdot 860; \quad A_6 = 0\cdot 521,$
$\qquad\qquad\qquad\qquad\qquad\qquad\qquad\qquad\qquad\qquad\qquad\qquad (294)$
from which $\langle Q\rangle = 4780$ Å^2.

† Cross, R. J., Gislason, E. A., and Herschbach, D. R., *J. chem. Phys.* **45** (1966) 3582.

If the rotational levels are only closely spaced in one of the colliding molecules the problem is less simple. Cross et al. discuss the case in which only one of the transitions in one molecule appreciably overlaps those in the other.

5.6.1. *Comparison with observation.* The experiments of Cross, Gislason, and Herschbach,† concerned with the measurement of total cross-sections for collisions between polar molecules and their variation with relative velocity, have already been briefly described in Chapter 16, § 13.3 in connection with their incidental measurements of cross-sections for collisions of the polar molecules with gas atoms. Table 17.16 gives a comparison between observed and calculated results for the different pairs of polar molecules which they investigated.

Calculated values are given according to the following approximations.

(a) Taking as the interaction only $-(\bar{C}_{\text{disp}}+\bar{C}_{\text{ind}})R^{-6}$ in the notation of Chapter 16, § 13.3.
(b) Including an additional interaction $-C_{\text{K}} R^{-6}$, C_{K} being given by the Keesom approximation (267) with the temperature T replaced by $2T_1 T_2/(T_1+T_2)$ where T_1, T_2 are the temperatures of the incident and target molecules respectively.
(c) The exact resonance limit (288).
(d) $\langle Q \rangle$ given by (292).

As described in Chapter 16, § 7.2.2 it is important to modify the calculated values to allow for the imperfect angular resolution of the apparatus before comparison with observation. The form of the resolution correction is different for the different cases. For (a) and (b) it is as given by (176) of Chapter 16, for (c) it is $0\cdot0126+0\cdot4097\rho^{-\frac{2}{3}}$ where the resolution parameter (ρ) of Chapter 16, § 7.2.2, $\geqslant 1\cdot3$, and for (d) it was obtained by joining smoothly a calculated angular distribution for inelastic scattering near zero angle to the result for elastic scattering by an R^{-6} potential at larger angles.

It will be seen that approximations (a) and (c) give unsatisfactory results both for the total cross-sections and for the velocity indices. Introduction of the Keesom dipole–dipole term improves the calculated cross-section and indeed gives the best agreement with observation for collisions with HBr. For collisions with CH_3I, (d) gives the best results, particularly for the velocity index, which is small because of the strong admixture of inelastic scattering for which the cross-section increases with relative velocity.

The success of the Keesom approximation for the HBr collisions is surprising, as the theoretical justification for the approximation is

‡ loc. cit., p. 1611.

slender. It seems, however, that when Q_{dd} is small, either because the dipole moments are small or the rotation periods short, the Keesom approximation gives good results. This may be seen from the comparisons made between theory and experiment for the magnitude and temperature variation of cross-sections for collisions of CsCl with other

TABLE 17.16

Comparisons of observed total cross-sections Q and velocity variation indices $\partial \ln Q/\partial \ln v$ with those calculated by different approximations, for collisions between polar molecules

Colliding molecules	Cross-section (10^{-16} cm^2)					Velocity index ($\partial \ln Q/\partial \ln v$)				
	calculated				observed	calculated				observed
	(a)	(b)	(c)	(d)		(a)	(b)	(c)	(d)	
KCl–HBr	982	1541	17 600	1290		0·40	0·40	1·0	−0·1	
	890	1340	7000	1090	1710	0·56	0·59	1·33	0·46	0·60
CsCl–HBr	1014	1549	17 600	1290		0·40	0·40	1·0	−0·1	
	790	1130	4000	910	1270	0·70	0·73	1·33	0·54	0·72
KCl–CH$_3$I	1297	1971	36 500	4780		0·40	0·40	1·0	−0·1	
	1140	1670	11 400	3390	2050	0·56	0·59	1·33	0·23	0·25
CsCl–CH$_3$I	1346	2215	36 500	5530		0·40	0·40	1·0	−0·1	
	970	1510	6600	2810	2560	0·68	0·70	1·33	0·35	0·41
CsBr–CH$_3$I	1391	2470	37 000	5900		0·40	0·40	1·0	−0·1	
	930	1540	5800	2700	2320	0·72	0·73	1·33	0·35	0·32

polar molecules, the experiments being those of Schumacher, Bernstein, and Rothe† referred to in Chapter 16, § 13.3 in connection with the observation and analysis of cross-sections for collisions of CsCl with atoms and nonpolar molecules. In Fig. 17.68, this comparison is illustrated for collisions with NO, NH$_3$, H$_2$S, CHF$_3$, CH$_2$F$_2$, and *cis*-C$_2$H$_2$Cl$_2$. For the first three of these molecules the cross-sections are not abnormally large for an R^{-6} interaction (for NO the dipole moment is small while for the hydride molecules the rotation time is short) and assumption of the Keesom interaction gives good results. On the other hand, the cross-sections are larger for the remaining three molecules and good agreement is now obtained with approximation (d) above. The cross-sections for scattering of TlF by N$_2$O, H$_2$O, and NH$_3$ observed by Bennewitz *et al.* (Table 17.14) are also comparable with that expected for an R^{-6} interaction.

† SCHUMACHER, H., BERNSTEIN, R. B., and ROTHE, E. W., *J. chem. Phys.* **33** (1960) 584.

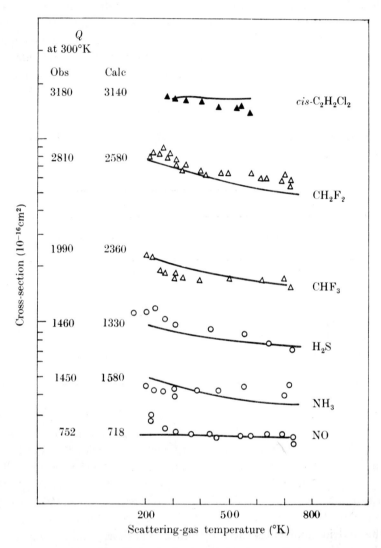

FIG. 17.68. Comparison of observed and calculated cross-sections for collisions of CsCl molecules with other polar molecules as indicated. For NO, NH_3, and H_2S the calculated cross-sections have been obtained using the Keesom averaged interaction and for CHF_3, CH_2F_2, and $C_2H_2Cl_2$ by the method of Cross and Gordon. Magnitudes of the cross-sections at 300 °K are given on the left-hand side of the diagram.

5.7. *Resonant transfer of rotational energy and the thermal conductivity of polar gases*

In § 5.4.2 we discussed the theory of the thermal conductivity of molecular gases. Two processes not present in atomic gases were taken into account. One was the 'diffusion' of rotational energy through the gas without transfer of internal energy to or from that of translation, and the other the inelastic collisions that involve such transfer. It was pointed out by Mason and Monchick[†] that, in a polar gas, the diffusion of rotational energy is substantially modified by resonant transfer of this energy between two molecules through the dipole–dipole interaction as discussed in § 5.6 above. They find that the formulae (218 a) and (218 b) for non-polar gases remain applicable for gases composed of linear polar molecules provided the diffusion coefficient D is replaced by

$$D/(1+Z'/Z_0), \tag{295}$$

where
$$Z' = (9/4)(\tfrac{1}{2}\pi)^{\frac{1}{2}} \bar{v} Q_{\text{ex}}(\bar{v}) (h^2/8\pi^2 I \kappa T)^{\frac{3}{2}},$$
$$Z_0 = \tfrac{3}{2} C_{\text{int}} T/\rho D, \tag{296}$$

in the notation of § 5.4.2. $Q_{\text{ex}}(\bar{v})$ is the exchange cross-section which is[‡] $\simeq \tfrac{2}{3} Q_t$, where Q_t is the cross-section (288), so $\bar{v} Q_{\text{ex}}(\bar{v})$ is nearly equal to $\tfrac{4}{3}\pi\mu^2/\hbar$, μ being the dipole moment of a molecule and I its moment of inertia.

More complicated expressions are obtained for Z' if the polar molecules are not linear.

The analysis was applied to the linear dipoles HCl and HBr, the nearly spherical top NH_3, the prolate top CH_3Cl, the oblate top $CHCl_3$, and the slightly asymmetric prolate tops H_2O and CH_3OH. It is found that, for HCl, HBr, and H_2O the correction to D is relatively large. Thus, at 370 °K, $Z'/Z_0 = 0.344$, 0.174, and 0.876 for the respective molecules. For CH_3Cl, $CHCl_3$, and CH_3OH, on the other hand, it is very small. The physical reason for this is that, in the three hydride molecules, the rotational quanta are much greater so the exchange process involves a considerable proportion of the energy flux. NH_3 is an intermediate case ($Z'/Z_0 = 0.092$ at 373 °K). Using the calculated values of Z' and Z_0 and treating the rotational relaxation time as an adjustable constant, very good agreement was obtained between observed and calculated values of the factor f in (215) for all the molecules. Over the temperature range concerned in each case f remains nearly constant and appreciably

[†] loc. cit., p. 1580.
[‡] $Q_{\text{ex}} = Q_t -$ the contribution from shadow scattering $= Q_t - \pi p^{*2}$ in the notation of (288).

different (usually by 10–15 per cent) from that given by either of the simpler formulae, (216) or (217). The values of the rotational relaxation time which give the good fit are reasonable, corresponding to deactivation probabilities for collision of 0·3, 0·3, 0·5, 0·6, 0·6, 0·25 for HCl, HBr, NH_3, CH_3Cl, $CHCl_3$, H_2O, and CH_3OH respectively (see § 5.4).

5.8. *Collisions involving change of molecular orientation only*

We now consider collisions of a molecule in a rotational state with quantum numbers J, M with an atom or molecule in which M but not J is changed. Such collisions do not involve any change of rotational energy but do change the orientation of the molecule. They are essentially similar to the collisions discussed in Chapter 18, § 3 in which the orientation of the angular momentum of an atom in an excited state is changed while its magnitude is unaltered.

Experimental evidence† of the rate of these collisions for H_2 molecules is forthcoming from observations of the nuclear spin-lattice relaxation times in hydrogen gas of different ortho-para composition as well as in mixtures with other gases. The relaxation time concerned is a measure of the rate at which a system of nuclear spins comes into thermodynamic equilibrium with its surroundings, usually referred to as the lattice (see Chap. 18, § 6). It is clear that this rate is determined by the fluctuating magnetic fields at each nuclear site. For a system of diatomic molecules coupling exists between the rotational angular momentum and the nuclear spin. Due to molecular collisions the rotational angular momentum will undergo transitions so altering the coupling with the nuclear spin. The fluctuating fields at the nuclei due to these transitions are the major source of relaxation. It is important to realize that it is not necessary for the *magnitude* of the rotational angular momentum to change in a transition—it is sufficient that the orientation should change.

The spin Hamiltonian H_s for two isolated hydrogen molecules in an applied magnetic field $\mathscr{H}_0$ can be written‡

$$\hbar H_s = \omega_I I_Z + \omega_J J_Z + \gamma \mathscr{H}' \mathbf{I}.\mathbf{J} + \gamma \mathscr{H}''\{\mathbf{I}_1.\mathbf{I}_2 - 3(\mathbf{I}_1.\hat{\mathbf{R}})(\mathbf{I}_2.\hat{\mathbf{R}})\}, \quad (297)$$

where $\mathbf{I}_1\hbar$, $\mathbf{I}_2\hbar$ are the total spin vectors of the two protons and $\mathbf{I} = \mathbf{I}_1 + \mathbf{I}_2$. ω_I is the proton precession frequency $-\gamma\mathscr{H}_0$ in the applied field, ω_J the corresponding precession frequency of the rotational angular

---

† BLOOM, M., *Proc. seventh int. Conf. low Temp. Phys.* Toronto, 1960, p. 61; LIPSICAS, M. and BLOOM, M., *Can. J. Phys.* **39** (1961) 881; BLOOM, M. and OPPENHEIM, I., ibid. **41** (1963) 1580; BLOOM, M., OPPENHEIM, I., LIPSICAS, M., WADE, C. G., and YARNELL, C. F., *J. chem. Phys.* **43** (1965) 1036.

‡ ABRAGAM, A., *The principles of nuclear magnetism*, p. 316 (Clarendon Press, Oxford, 1961).

momentum vector $J\hbar$. $\hat{\mathbf{R}}$ is a unit vector in the direction of the nuclear separation. $\mathscr{H}'$ and $\mathscr{H}''$ are the spin-rotation and spin–spin coupling constants which have been determined by molecular beam methods as 27 and 34 gauss respectively.†

Well below room temperature ($< 200\ °\mathrm{K}$) the energy of relative motion in a collision will usually be insufficient to produce changes of J and the gas will consist of ortho- and para-hydrogen molecules in their ground states only.

It has been shown that the spin-lattice relaxation time $\mathscr{T}$ is then given by‡

$$\frac{1}{\mathscr{T}} = \tfrac{4}{3}\gamma^2 \mathscr{H}'^2 \frac{\tau_1}{1+\omega_I^2 \tau_1^2} + \tfrac{12}{25}\gamma^2 \mathscr{H}''^2 \left\{ \frac{\tau_2}{1+\omega_I^2 \tau_2^2} + \frac{4\tau_2}{1+4\omega_I^2 \tau_2^2} \right\}. \quad (298)$$

τ_1 and τ_2 are such that, if ν_1 and ν_2 are the number of collisions per second which produce transitions $\Delta M_J = \pm 1, \pm 2$ respectively, then

$$\tau_1 = (\nu_1 + 2\nu_2)^{-1}, \qquad \tau_2 = (3\nu_1)^{-1}. \quad (299)$$

In terms of the definitions of Chapter 18, § 3.1, τ_2, τ_1 are the respective relaxation times for the rotational dipole and quadrupole polarization so that
$$\tau_1^{-1} = n Q_{\mathrm{al}} \bar{v}, \qquad \tau_2^{-1} = n Q_{\mathrm{or}} \bar{v},$$
where n is the number of molecules/cm³, $\bar{v}$ their mean relative velocity, and Q_{or}, Q_{al} are the cross-sections for disalignment and disorientation respectively.

In deriving (298) use has been made of the fact that $\omega_J \ll \omega_I$. If we assume $\tau_1 = \tau_2 = \tau$ and $\ll 1/\omega_I$ then, putting in numerical values,

$$1/\mathscr{T} = 2{\cdot}7 \times 10^{12} \tau$$
$$= 2{\cdot}7 \times 10^{12}/n Q_{dp} \bar{v}, \quad (300)$$

where Q_{dp} is the mean cross-section for a molecular collision with relative velocity v in which the rotational orientation or alignment is changed.

At first sight this result, which corresponds to most experimental conditions, seems strange, as the relaxation time increases as the reorientational collision frequency increases. However, this is because, when $\omega_I \tau \ll 1$, the nuclear spins have difficulty in following the fast fluctuations and their response falls off as the fluctuations become even faster. If $\omega_I \tau \gg 1$ then we would have $\mathscr{T} \propto \tau$ but this condition could only be satisfied at very low gas-densities.

† KELLOGG, J. M. B., RABI, I. I., RAMSEY, N. F., and ZACHARIAS, J. R., *Phys. Rev.* **56** (1939) 728.
‡ ABRAGAM, A., loc. cit., p. 1616; see also CHEN, F. M. and SNIDER, R. F., *J. chem. Phys.* **48** (1968) 3185.

(300) will be valid, when $\omega_I \tau \gg 1$, at pressures below a few hundred atmospheres, as may be seen from Fig. 17.69 which illustrates the observed variation† of $\mathscr{T}$ with gas density for normal hydrogen at 41 °K. On substitution of numerical values we find, at this temperature,

$$Q_{dp}\bar{v} = 1\cdot 3\times 10^{-11} \text{ cm}^3 \text{ s}^{-1},$$

which gives a mean cross-section

$$Q_{dp} \simeq 2\times 10^{-16} \text{ cm}^2.$$

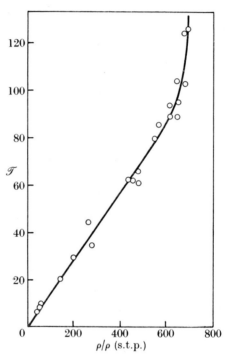

FIG. 17.69. Observed variation of spin-lattice relaxation time with gas density for normal H_2 at 41 °K. ○ Experimental points.

The observed variation of $Q_{dp}\bar{v}$ with temperature, again for normal hydrogen, is shown in Fig. 17.70.

It is possible to observe the variation of $\mathscr{T}$ with fractional parahydrogen concentration and, by extrapolation to the limiting cases when this function tends to zero and to ∞, $\mathscr{T}$ for ortho–ortho and para–ortho collisions may be obtained separately.‡ Fig. 17.71 illustrates results obtained in this way for temperatures ranging from 60 to 150 °K.

† LIPSICAS, M. and BLOOM, M., loc. cit., p. 1616.
‡ LIPSICAS, M. and HARTLAND, A., Phys. Rev. **131** (1963) 1187.

Analysis of these data in terms of intermolecular interactions has been carried out by means of a stochastic theory by Bloom and Oppenheim[†] and their associates.[‡] The reorientation cross-section is determined by the strength of the anisotropic interaction. For ortho–ortho collisions the most important term at not too high temperatures arises from the quadrupole–quadrupole interaction. It is so dominant that the orientation cross-section is effectively proportional to the fourth power of the

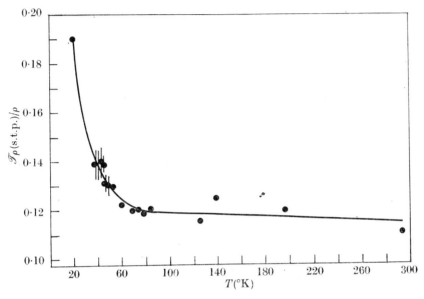

Fig. 17.70. Observed variation of $\mathscr{T}/\rho$ with temperature in normal H_2.

quadrupole moment. For ortho–para collisions no quadrupole contribution arises. The non-central component of the interaction includes a repulsive term at small separations and a van der Waals term at large (see Chap. 16, (269)). The relative importance of these terms is still not clear.

The disorientation cross-section may be calculated by the distorted-wave method in much the same way as cross-sections for rotational excitation (see §§ 5.2 and 5.3). Preliminary calculations of this kind have been carried out for ortho–para collisions using the interaction (269) of Chapter 16. It appears that the calculated cross-sections, to which $\mathscr{T}$ is proportional, are too small. This is consistent with the

[†] loc. cit., p. 1616.
[‡] BLOOM, M., OPPENHEIM, I., LIPSICAS, M., WADE, C. G., and YARNELL, C. F., J. chem. Phys. 43 (1965) 1036.

evidence from comparison of observed and calculated cross-sections for rotational excitation which requires the non-central term to be larger than that given by (269) of Chapter 16. It seems clear that very useful information about the non-central interaction can be obtained

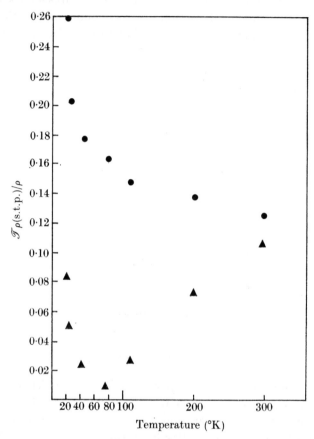

FIG. 17.71. Variation of $\mathscr{T}/\rho$ with temperature for ● pure ortho-H_2 and ▲ ortho-H_2 dilute in para-H_2.

from analysis of relaxation times in terms of distorted-wave collision theory. Further calculations using such a theory have been carried out by Riehl, Kinsey, and Waugh[†] for relaxation of ortho-H_2 infinitely dilute in He, in an attempt to determine the relative importance of the short-range repulsion and long-range van der Waals attraction in the non-central interaction between He and H_2.

† RIEHL, J. W., KINSEY, J. L., and WAUGH, J. S., *J. chem. Phys.* **46** (1967) 4546.

At temperatures above about 150 °K in H_2, as pointed out by Lalita, Bloom, and Noble,† it is necessary to take into account the effect of transitions in which the total rotational quantum number changes. Up to 750 °K, for ortho-H_2, the only transitions which need be considered are those between $J = 1$ and $J = 3$ states. In that case we find that (298) is replaced by†‡

$$\frac{1}{\mathcal{T}} = \tfrac{4}{3}\gamma^2 \mathcal{H}'^2 \tau_1' + \tfrac{12}{5}\gamma^2 \mathcal{H}''^2 \tau_2', \qquad (301)$$

where

$$n\bar{v}\tau_l' = [-\{f_1 Q_l(33)+\alpha_l f_3 Q_l(11)\}+\beta_l\{f_1 Q_l(31)+f_3 Q_l(13)\}] \times \\ \times \{Q_l(11)Q_l(33)-Q_l(13)Q_l(31)\}^{-1}. \qquad (302)$$

f_J is the fractional population of the level J, i.e.

$$f_J = (2J+1)\exp(-E_J/\kappa T) \Big/ \sum_J (2J+1)\exp(-E_J/\kappa T),$$

while $\alpha_1 = 6$, $\alpha_2 = \tfrac{2}{3}$, $\beta_1 = 6^{\frac{1}{2}}$, $\beta_2 = \tfrac{1}{3}\cdot 6^{\frac{1}{2}}$.

The cross-sections $Q(J, J')$ are defined as follows. It is assumed that the non-central component of the interaction involves only the second-order harmonics of the orientation angles and that its magnitude is small enough for the method of distorted waves to be applied. The cross-section for a collision in which a transition from state (J, M) to (J', M') occurs may then be written

$$Q(J, M; J'M') = Q(J, J')C(J'2J; M', M-M')^2, \qquad (303)$$

where $C(\ )$ is a Clebsch–Gordan coefficient. $Q_l(J, J')$ is then given by

$$\left(\frac{2J'+1}{2J+1}\right) Q_l(J, J') = (-1)^{l-J-J'} Q(J, J') \times \\ \times \left[\{(2J+1)(2J'+1)^{\frac{1}{2}}\} W(JJ'JJ'; 2l) - \delta_{JJ'} \sum_{J''} \frac{2J''+1}{2J'+1} Q(J'', J')\right], \\ (304)$$

where $W(\ )$ is a Racah coefficient.

In a full analysis of relaxation in normal or in para-H_2 allowance must also be made for contributions from collisions with para-molecules in which $J = 0$ to $J = 2$ transitions occur. At the time of writing preliminary results only have been obtained by Lalita, Bloom, and Noble, partly because of lack of complete knowledge of the cross-sections and partly because of incomplete and insufficiently accurate experimental data.

† LALITA, K., BLOOM, M., and NOBLE, J. D., preprint.
‡ CHEN, F. M. and SNIDER, R. F., loc. cit., p. 1617.

6. Reactive collisions between molecular beams

6.1. *Introduction*

Although it would take us much too far afield to discuss chemical reactions in the gas phase in general it is appropriate to conclude this chapter with a brief account of the investigation of such reactions occurring in collisions between molecules in the region of interaction between two molecular beams. The development of techniques for studying these reactions, which began with the experiments of Taylor and Datz[†] on the reaction

$$K+HBr \to KBr+H, \qquad (305)$$

is likely to prove of the greatest importance for the elucidation of the mechanism of homogeneous chemical reactions in general. This is because, for the first time, it becomes possible to observe, not only total reaction cross-sections, but also differential cross-sections and the distribution of rotational and vibrational excitation energy for particular products resulting from the reaction.

Although considerable progress has already been made in these directions there are many technical problems to be overcome, while the analysis required to obtain the desired cross-sections and energy distributions from the primary data is formidable. This is largely because the transformation from observation in the laboratory system to the centre of mass system for the interacting molecules is usually a complicated one, particularly when allowance has to be made for the initial velocity distributions of the reacting molecules.

Almost all of the experimental work which has been carried out up to the time of writing has been concerned with reactions of alkali-metal atoms with halogen-containing molecules such as

$$K+HBr \to KBr+H, \qquad (306)$$

$$K+Br_2 \to KBr+Br, \qquad (307)$$

$$K+CH_3I \to KI+CH_3. \qquad (308)$$

This is because of the relative simplicity of the methods for detecting both the elastically scattered alkali-metal atoms and the alkali-halide products. Furthermore, it is possible to obtain hot-wire detectors which distinguish between alkali metals and alkali halides. According to a prescription of Touw and Trischka[‡] a filament of platinum containing 8 per cent of tungsten can be conditioned to detect both alkali metal

---

[†] TAYLOR, E. H. and DATZ, S., *J. chem. Phys.* **23** (1955) 1711.
[‡] TOUW, T. R. and TRISCHKA, J. W., *J. appl. Phys.* **34** (1963) 3635.

and alkali halide with near 100 per cent efficiency by heating the wire in oxygen. On the other hand, by heating in methane, the wire becomes almost completely insensitive to the halide while remaining almost 100 per cent efficient for the detection of alkali-metal atoms.

A check on the performance of the detector was made by Herm, Gordon, and Herschbach[†] who deflected away the paramagnetic alkali metals in an inhomogeneous magnetic field so that the distribution of alkali-halide product could be observed directly without the need for subtraction of a large background signal. They verified the results that had been obtained by the differential background method and were able to improve the accuracy of measurement of alkali-halide angular distributions at small angles of scattering where the atom background is particularly large.

The nature of the measurements which are made may be summarized as follows:

(a) the flux of the scattered, non-reacting alkali-metal atoms as a function of laboratory angle of scattering;
(b) the flux of the product molecules as a function of laboratory angle of scattering;
(c) the velocity distribution of the product molecules at different laboratory scattering angles;
(d) the rotational energy distribution of the product molecules at different laboratory scattering angles.

Although (a) would not seem to be of much use for the study of reaction cross-sections, in fact it is often possible from such measurements, as we shall explain in § 6.2, to obtain a good deal of information about the total reaction cross-section. This may also be obtained by integrating the flux of product molecules over a sufficiently comprehensive set of measurements of type (b). It is also possible from these measurements, at least in principle, to derive the differential reaction cross-section in the centre of mass system.

The analysis of data obtained in the course of all of these measurements is greatly simplified if velocity selection is employed in the experiments. In most cases the incident alkali-metal beam has been velocity-selected while the reacting halide beam is kept as cool as possible. Few experiments have yet been carried out in which a second velocity selector has been used to make measurements of type (c). In some cases using a single velocity selector, this has been included to analyse the velocity

[†] HERM, R. R., GORDON, R., and HERSCHBACH, D. R., *J. chem. Phys.* **41** (1964) 2218.

distribution of the products, the colliding beams then possessing their thermal distributions.

The rotational distribution of the product molecules may be determined by measuring the breadth of the deflexion pattern and its variation with field strength when the molecules are passed through an inhomogeneous electric field (see p. 1640).

We shall now describe some of this work to illustrate the principles and techniques involved, without attempting a comprehensive account.

6.2. *The determination of the total reaction cross-section—relation to non-reactive scattering—complex phase shifts*†

As described in Chapter 6 the cross-section for scattering of particles of wave number k by a centre of force can be expressed in terms of real phase shifts η_l. Thus the partial cross-section for scattering of particles with angular momentum quantum number l is

$$(4\pi/k^2)(2l+1)\sin^2\eta_l. \tag{309}$$

We may analyse the scattering a little further to see the significance of the reality of the phase shifts.

The radial component of the flux of particles, with the specified angular momentum, is

$$j_r = (\hbar/2mi)\left(\psi^*\frac{\partial \psi}{\partial r} - \psi\frac{\partial \psi^*}{\partial r}\right), \tag{310}$$

where ψ is the wave function for the particles. In the scattering problem with which we are concerned (see Chap. 6, § 3)

$$\psi \sim (kr)^{-1}\{i^l(2l+1)\sin(kr-\tfrac{1}{2}l\pi)+r^{-1}c_l\,e^{ikr}\}P_l(\cos\theta) \tag{311}$$

so, for large r,

$$j_r = (\hbar/mr^2)\{-\tfrac{1}{2}i(2l+1)(c_l^*-c_l)+k|c_l|^2\}\{P_l(\cos\theta)\}^2. \tag{312}$$

For scattering of particles by a structureless centre of force the net radial flux must vanish—no particles are absorbed so the inward and outward flow must be equal. Hence,

$$|c_l|^2/(2l+1) = (c_l-c_l^*)/2ik, \tag{313}$$

and, if we put

$$c_l = (d_l-1)(2l+1)/2ik, \tag{314}$$

then

$$|d_l-1|^2 = -d_l-d_l^*+2,$$

so

$$|d_l|^2 = 1.$$

We may therefore write

$$d_l = e^{2i\eta_l}, \tag{315}$$

---

† See MOTT, N. F. and MASSEY, H. S. W., *The theory of atomic collisions*, 3rd edn, pp. 184–94 (Clarendon Press, Oxford, 1965).

where η_l is real, and we have
$$c_l = (e^{2i\eta_l}-1)(2l+1)/2ik. \tag{316}$$
Comparison with (36) of Chapter 6 shows that η_l is the same as the phase shift defined in that section.

In terms of this analysis the reality of η_l is a consequence of the vanishing of the net radial flux for each l. If the scattering centre is partially absorbing then we may still write the asymptotic form of the wave function as
$$\psi \sim (kr)^{-1}\{i^l(2l+1)\sin(kr-\tfrac{1}{2}l\pi)+(e^{2i\eta_l}-1)(2l+1)/2i\}P_l(\cos\theta) \tag{317}$$
but η_l is no longer real. Writing
$$\eta_l = \lambda_l + i\mu_l, \tag{318}$$
where λ_l and μ_l are real we see that the net outward radial flux is
$$j_r = (\hbar/4mkr^2)(2l+1)^2(e^{-4\mu_l}-1). \tag{319}$$
From this it follows that $\mu_l \geqslant 0$ since the net outward flux must be less than the inward because of the absorption.

Also, the partial cross-section Q_a^l for absorption will be given by
$$vQ_a^l = \int_0^\pi \int_0^{2\pi} j_r r^2 \sin\theta\, d\theta d\phi, \tag{320}$$
so
$$Q_a^l = (\pi/k^2)(2l+1)(1-e^{-4\mu_l}). \tag{321}$$
The partial elastic cross-section Q_{el}^l will now be given by
$$\begin{aligned}Q_{el}^l &= 4\pi|c_l|^2/(2l+1)\\ &= (2\pi/k^2)(2l+1)(\cosh 2\mu_l - \cos 2\lambda_l)e^{-2\mu_l},\end{aligned} \tag{322}$$
and the partial 'total' cross-section Q_t^l by
$$\begin{aligned}Q_t^l &= Q_a^l + Q_{el}^l \\ &= (2\pi/k^2)(2l+1)(1-e^{-2\mu_l}\cos 2\lambda_l).\end{aligned} \tag{323}$$

A complex phase shift can be considered to arise from a complex scattering potential $V(r)$ with a negative imaginary part. It is easy to show that, if we write
$$2mV(r)/\hbar^2 = U(r)-iW(r), \tag{324}$$
then the rate of absorption of particles per second per unit volume when at a distance r from the centre is $2W/\hbar$.

There is no difficulty in extending this analysis to consider inelastic scattering. We may regard the inelastic collisions as removing particles of the incident energy from the scattered beam so that in place of Q_a^l we have the total inelastic cross-section Q_{in}^l.

von Hundhausen and Pauly† investigated the effect of adding a negative imaginary potential to the typical $(n, 6)$ interaction (Chap. 16, § 6.1) between two molecules on the angular distribution of the elastic scattering at such energies that a clear rainbow effect (Chap. 16, § 6.4) would arise in the absence of the imaginary component.

The real parts λ_l of the phase shifts were taken to be the same as those η_l calculated for the real $(12, 6)$ interaction while the imaginary part was taken as

$$\mu_l \to \infty \quad (l \leqslant l_r),$$
$$= 0 \quad (l > l_r). \tag{325}$$

This corresponds to a situation in which all particles with angular momentum quantum number $\leqslant l_r$ are completely absorbed or undergo some reaction. Under the semi-classical conditions of gas-kinetic collisions this means that a reaction is assumed to take place for all collisions in which the impact parameter $p_r \leqslant l_r \hbar/Mv$, where v is the initial relative velocity of impact and M is the reduced mass of the interacting molecules, i.e. the reaction cross-section is πp_r^2.

The elastic differential scattering cross-section is then

$$I(\theta) \, d\omega = (1/4k^2) |\sum i^l (2l+1) \exp(2i\eta_l - 2\mu_l) P_l(\cos\theta)|^2 \, d\omega. \tag{326}$$

Calculations were carried out for an $(8, 6)$ potential at a relative kinetic energy E such that $E/\epsilon = 8 \cdot 4$ and $kr_m = 260$, ϵ and r_m being as in Chapter 16, § 6.1. The resulting functions $I(\theta)\sin\theta$ calculated assuming values of l_r ranging from 0 to 350 are shown in Fig. 17.72 for θ from $0°$ to $60°$. In order to avoid confusion the successive curves are displaced downwards.

A finite value of l_r leads to a sudden fall in the scattered intensity at large angles which sets in at smaller angles as l_r increases. If the rapid decrease in intensity sets in at $\theta = \theta_f$, say, then between θ_f and the rainbow angle oscillations appear which are gradually squeezed out as θ_f approaches the rainbow angle θ_r. Eventually, for sufficiently large θ_f, the primary rainbow is blurred out completely. The existence of oscillations between θ_r and θ_f is important because in the interpretation of a curve such as that for $l_r = 200$ in Fig. 17.72 the peak near $30°$ might be taken as the primary rainbow angle. To check this it is necessary to check the variation in the position of the peak with relative energy as well as its consistency with reasonable values of the potential parameters ϵ and r_m.

Using the apparatus already referred to in Chapter 16, p. 1390, in connection with their investigation of rainbow scattering in atom–atom

† VON HUNDHAUSEN, E. and PAULY, H., *Z. Phys.* **187** (1965) 305.

collisions, von Hundhausen and Pauly† observed the angular distributions of Na and K atoms scattered by reactive molecules. Three examples are shown in Fig. 17.73. For Na–HBr there is a clear single rainbow maximum, for Na–C$_2$H$_4$Br$_2$ two maxima are present while for K–SnI$_4$

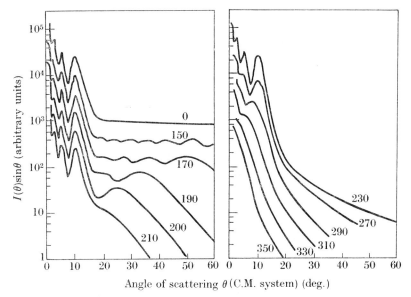

FIG. 17.72. Calculated distributions of elastic scattering per unit angle $I(\theta)\sin\theta$ for an interaction such that the real parts of the phase shifts are those for an (8, 6) interaction and the imaginary parts are such that $\mu_l \to \infty$, $l < l_r$; $= 0$, $l > l_r$. The relative mass and energy in the collision are such that $E = 8.4\epsilon$ and $2M\epsilon r_m^2/\hbar^2 = 8 \times 10^3$. Numbers on the different curves refer to the assumed value of l_r. For clarity the ordinates are displaced from curve to curve—at small angles all distributions are the same.

the rainbow maxima are absent, suggesting that the reaction cross-section is very large. A detailed analysis was carried out for Na–C$_2$H$_4$Br$_2$ using the complex potential model. The best fit obtained, which is shown in Fig. 17.73 (b), still leaves much to be desired as it gives a much more rapid fall in the scattered intensity beyond the rainbow angle than is observed. This may well be due to the assumption of a sharp boundary for the imaginary part of the potential. It is interesting to note that the theoretical curve is similar to that for $l_r = 210$ in Fig. 17.72 and the inflexion near 15° is not ascribed to a blurred primary rainbow—the primary rainbow peak is near 8°. The parameters which give the best fit are $\epsilon = 170\kappa$, $r_m = 7.5$ Å, $p_r/r_m = 0.81$, corresponding to a reaction cross-section as large as 1.2×10^{-14} cm^2.

† loc. cit.

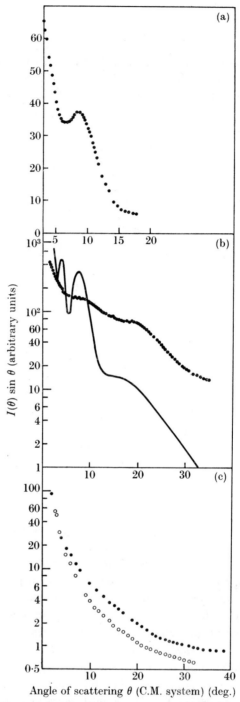

FIG. 17.73. Observed distributions of non-reactive scattering per unit angle, $I(\theta)\sin\theta$, for collisions of Na atoms with (a) HBr, at a relative velocity of 1370 m s^{-1}, (b) C$_2$H$_4$Br$_2$ at a relative velocity of 1274 m s^{-1}, and (c) K atoms with SnI$_4$ at relative velocities of 930 m s^{-1} (●) and 1090 m s^{-1} (○). For C$_2$H$_4$Br$_2$ the full line curve is one calculated for an (8, 6) interaction with $E = 11\epsilon$, $2M\epsilon r_m^2/\hbar^2 = 8\cdot55 \times 10^3$ together with an imaginary interaction which produces an infinite imaginary phase shift for $l < 250$ but has no effect for $l > 250$.

A proposal for a more definite and detailed method of deriving the reaction cross-section was made by Beck, Greene, and Ross.† In agreement with the results shown in Fig. 17.72 it is assumed that, for scattering at an angle $\theta < \theta_e$ say, the effect of reactive collisions on the elastic differential cross-section is negligible. Observed data for $\theta < \theta_e$ are then used to derive an interaction potential of (exp α, 6) or (n, 6) type (Chap. 16, § 6.1). From this interaction the elastic scattering for $\theta > \theta_e$ which would occur in the absence of reactive collisions is calculated. Let $I_{e,0}(\theta)\,d\omega$ be the differential cross-section derived in this way. Then, if $I_{\text{obs}}(\theta)\,d\omega$ is the measured cross-section, it is assumed that the probability $P(\theta)$ of a reactive collision occurring in a collision in which elastic scattering would lead to a *deviation* through θ for $\theta > \theta_e$ is

$$P(\theta) = 1 - I_{\text{obs}}(\theta)/I_{e,0}(\theta). \tag{327}$$

As the conditions are semi-classical, corresponding to each value of θ beyond the rainbow angle θ_r there is a single impact parameter p, so (327) can be converted to $P(p)$ for $p < p_e$. The reaction cross-section is then simply given by

$$Q_r = 2\pi \int_0^{p_e} P(p)\,dp. \tag{328}$$

According to the analysis in terms of complex phase shifts the cross-sections Q_{el}^l, Q_r^l, for elastic and for all inelastic collisions (including chemical reactions) are given by

$$Q_{\text{el}}^l = \frac{\pi}{k^2}(2l+1)|1-S_l|^2, \qquad Q_r^l = \frac{\pi}{k^2}(2l+1)|P_l|^2, \tag{329}$$

where

$$S_l = \exp\{2i(\lambda_l + i\mu_l)\}, \qquad P_l = 1 - \exp(-4\mu_l). \tag{330}$$

We wish to express P_l in terms of Q_{el}^l and $Q_{\text{el},0}^l$, where $Q_{\text{el},0}^l$ is the elastic cross-section that would arise if no inelastic collisions could occur, i.e. if the effective imaginary part of the scattering potential were absent. Thus

$$Q_{\text{el},0}^l = (\pi/k^2)(2l+1)|1-S_{l,0}|^2, \tag{331}$$

where $S_{l,0} = \exp(2i\eta_l)$, and it is reasonable to assume that

$$|\eta_l - \lambda_l| \ll 1. \tag{332}$$

Now,‡ writing $|1-S_l|^2 = \sigma_l$, $|1-S_{l,0}|^2 = \sigma_{l,0}$ and dropping the suffix l, we have

$$P = \sigma_0 - \sigma + 2\{1 - e^{-2\mu}\cos 2(\lambda-\eta)\}\cos 2\eta + 2e^{-2\mu}\sin 2(\lambda-\eta)\sin 2\eta$$

$$\simeq \sigma_0 - \sigma + 2(1-e^{-2\mu})\cos 2\eta, \quad \text{in view of (332).} \tag{333}$$

But $e^{-2\mu} = (1-P)^{\frac{1}{2}}$ so that

$$P^2 + P(\sigma_0^2 - 4\sigma_0 + 2\sigma) = (\sigma_0^2 - 4\sigma_0 + 2\sigma) - \sigma(\sigma-2). \tag{334}$$

† Beck, D., Greene, E. F., and Ross, J., *J. chem. Phys.* 37 (1962) 2895.
‡ Rosenfeld, J. L. J. and Ross, J., ibid. 44 (1966) 188.

If
$$P \ll \sigma_0^2 - 4\sigma_0 + 2\sigma, \tag{335}$$
then
$$P \simeq 1 - \frac{\sigma}{\sigma_0} \left\{ \frac{\sigma_0(\sigma-2)}{\sigma_0^2 - 4\sigma_0 + 2\sigma} \right\}$$
$$= 1 - \frac{\sigma}{\sigma_0} \left\{ 1 - \frac{(\sigma_0-2)(1-\sigma/\sigma_0)}{(\sigma_0-2) - 2(1-\sigma/\sigma_0)} \right\}. \tag{336}$$

Now, provided l is such that the phase shifts η_l are large, the average value of σ_0 is 2 (see, for example, Chap. 16, § 5.4.1). It would then follow that, for such values of l the average value of P_l is indeed closely given by
$$P_l = 1 - Q_{el}^l/Q_{el,0}^l. \tag{337}$$

As it is unlikely that any appreciable contribution to the reaction cross-section comes from values of l which are so large that $\eta_l < \tfrac{1}{2}\pi$ it would seem that the use of (327) would be justified. However, there are a number of reservations.

The first relates to the condition (335). To the same approximation as (337), when $\sigma_0 \simeq 2$,
$$\sigma_0^2 - 4\sigma_0 + 2\sigma \simeq 4P, \tag{338}$$
which is barely consistent with (335).

Rosenfeld and Ross† have carried out a more detailed study of the conditions for applicability of (337) and conclude that it is valid provided
$$|\partial P_r/\partial p| \ll 2k(\pi-\theta)|1-P_r|, \tag{339}$$
k being the wave number of initial relative motion. P_r is here expressed in terms of the impact parameter as in (328). This condition is well satisfied in most applications which have been made.

The second arises because P in (336) refers to the probability of all the inelastic collisions which include the possibility of internal excitation without chemical reaction as well as reactive collisions, whereas (327) is applied to reactive collisions only. Let $P_{in}(\theta)$, $P_r(\theta)$ be the separate probabilities for internal excitation and chemical reaction respectively. Then if we can write
$$I_{in}(\theta) = P_{in}(\theta) I_{el}(\theta), \tag{340}$$
where $I_{in}(\theta)\,d\omega$ is the differential cross-section for internal excitation, (337) gives
$$P_r(\theta) = 1 - \{I_{el}(\theta) + I_{in}(\theta)\}/I_{el,0}(\theta). \tag{341}$$

It would follow that, if the observed differential cross-section I_{obs} for scattering of the non-reactive atoms includes those that have produced excitation, (327) would still be valid. While it is possible that, if the sudden approximation (232) is valid, (340) is a good approximation there is no evidence that it is even roughly true under other conditions.

Finally, Kwei and Herschbach‡ have discussed the form the elastic differential cross-section will take if the interaction energy $V(r)$ possesses

† loc. cit., p. 1629.
‡ KWEI, G. H. and HERSCHBACH, D. R., *Atomic collisions* (editor McDowell, M. R. C.), p. 972 (North Holland, 1964).

not only an attractive well due to the van der Waals attraction but an inner, much deeper, well due to chemical forces. Such an interaction is already exemplified by that between H atoms with antiparallel spin (see Chap. 16, § 12.1.2). The additional strong attraction introduces, under typical conditions, a complex rainbow structure as well as features due to orbiting (Chap. 16, §§ 5.1.1 and 5.4.4). As an illustrative case, Kwei and Herschbach chose an interaction $V(r)$ of the $(12, 6)$ type defined by the usual parameters ϵ and r_{m}, for $r > r_{\mathrm{m}}$. This potential was then taken to fall monotonically for smaller r to join at $r = 0 \cdot 7 r_{\mathrm{m}}$ with an inner chemical well, again of $(12, 6)$ form, with minimum at $r'_{\mathrm{m}} = 0 \cdot 56 r_{\mathrm{m}}$ and depth $\epsilon' = 5\epsilon$. Fig. 17.74 illustrates typical angular distributions calculated from semi-classical theory, by methods described in Chapter 16, for $2M\epsilon r_{\mathrm{m}}^2/\hbar^2 = 5 \times 10^3$, for three values of E/ϵ, M being the reduced mass of the colliding systems and E the relative kinetic energy. Comparison is made with the corresponding results if the chemical attraction were absent so the interaction would be of the $(12, 6)$ type for all r with the same ϵ and r_{m}. It will be seen that, for $E = 8\epsilon$, the calculated distribution has the same features as those for $l_{\mathrm{r}} = 230$, 270, and 290 in Fig. 17.72. Thus the intensity falls more steeply at large angles and additional oscillations are introduced on the large angle side of the rainbow maximum due to the outer van der Waals attraction.

Although it is clear that the use of (327), or rather (341), to obtain reaction cross-sections cannot be accepted without reservation, particularly when there is reason to expect the presence of a chemical attraction in the inner interaction or a very high probability of excitation of inner molecular motions, it is nevertheless a useful guide, the limitations of which will become clearer as more information becomes available.

Beck, Greene, and Ross and their associates have applied (341) to obtain reaction cross-sections as a function of relative kinetic energy for collisions of K atoms with a number of molecules, including the hydrogen halides HCl, HBr, and HI as well as CH_3I, CCl_4, $SiCl_4$, $SnCl_4$, and SF_6. Fig. 17.75 illustrates the general arrangement of their apparatus.

A velocity-selected K beam intersects the molecular beam at 90° and the variation of the intensity of scattered K atoms is measured at various scattering angles Θ in the laboratory system. This may be done either by observing (a) in the plane of the intersecting beams or (b) in the perpendicular plane as shown in Fig. 17.75. Configuration (a) has the advantage that observations at all angles of scattering θ in the centre of mass (C.M.) system can be covered but faces the difficulty that for $\Theta = \tfrac{1}{2}\pi$ the detector looks directly at the molecular beam. In configuration (b) the detector always looks in a direction normal to this

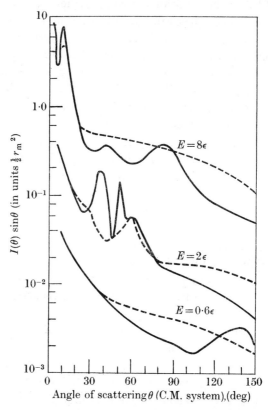

Fig. 17.74. Calculated distributions of elastic scattering per unit angle $I(\theta)\sin\theta$: ---- for a (12, 6) interaction with $2M\epsilon r_{\rm m}^2/\hbar^2 = 5\times 10^3$; —— for an interaction which is the same for $r > r_{\rm m}$ but for $r < r_{\rm m}$ joins smoothly to an inner well also of (12, 6) form with minimum at $r'_{\rm m} = 0\cdot 56$ and depth $\epsilon' = 5\epsilon$. The relative energies in the collisions are indicated on each pair of curves.

beam so there is less chance of a change in detector sensitivity due to chemical effect of the molecular beam as the observed angle of scattering changes. On the other hand, in this configuration there is always a range of angles of scattering in the C.M. system near π which cannot be observed. The width at half height of the velocity distribution transmitted by the selector was 0·084 times the central speed.

Having measured the angular distributions in the laboratory system and converted to the C.M. system, the small angle scattering up to and including the primary rainbow, if any, was fitted as well as possible by an (exp α, 6) interaction (see Chap. 16, § 6.1). From this $I_{el,0}$ in (341) was calculated and so $P_r(\theta)$ and hence $P_r(p)$ and Q_r obtained. For collisions with HBr† and with HI‡ measurements were also made of the intensity of scattered alkali halide as a function of angle of projection in the

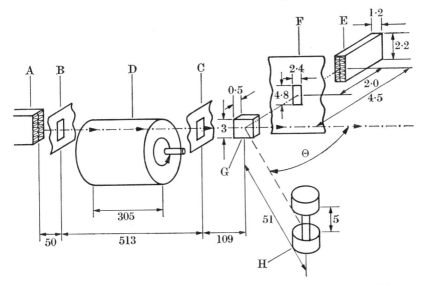

FIG. 17.75. General arrangement of apparatus used by Beck et al. for studying reactive collisions of K atoms with halogen-containing molecules. A, alkali metal oven; B, C, F, slits; D, velocity selector; E, halide source; H, detector. Dimensions are given in millimetres.

laboratory system. From these measurements Q_r could be obtained by integration over the full angular range, but the accuracy of the absolute measurements was not very high. For HI, Q_r as determined from the scattering of the K beam was between 2 and 3 times larger than from direct measurement of the KI flux. This discrepancy is still not outside the experimental uncertainty, particularly in the absolute determination of the KI flux. A similar comparison, from observations in which the K beam was not velocity-selected and the experimental accuracy was higher, gave agreement between the two values of Q_r to within 20 per cent. For HBr no comparison of absolute magnitudes could be made but the variation with initial kinetic energy of relative motion derived from the two methods was closely the same.

† AIREY, J. R., GREENE, E. F., KODERA, K., RECK, G. P., and Ross, J., J. chem. Phys. **46** (1967) 3287.
‡ ACKERMAN, M., GREENE, E. F., MOURSUND, A. L., and Ross, J., ibid. **41** (1964) 1183.

Results obtained for the total reaction cross-sections are given in Table 17.17. In some cases estimates have been made of the reaction cross-sections from the optical model calculations of von Hundhausen and Pauly (see Fig. 17.72). These are very rough as the parameters assumed in the latter calculations are not the most appropriate for the molecules concerned. However, using the curves shown in Fig. 17.72 values of l_r were derived and from these the reaction cross-sections. It will be seen from Table 17.17 that these agree well with those derived from (341).

TABLE 17.17

Total cross-sections for chemical reaction of K atoms with various halogen-containing molecules, derived from measurements of the scattered intensity of K atoms

Molecule		HCl	HBr	HI	CH_3I	CCl_4	$SnCl_4$	SF_6
Reaction cross-section (10^{-16} cm²)	(a)	3·9‡	35†	28‡	—	70§	—	70§
	(b)	—	—	—	50§	50§	100§	60§
	(c)	—	—	—	~35	—	⩾100	⩾100
Mean initial relative kinetic energy								
kcal/mol		2		2	4	5	5·7	2
eV		0·09		0·09	0·17	0·22	0·25	0·09

(a) Derived from (341). (b) Derived from optical model analysis. (c) Derived from observations of the flux of product halides.

It is encouraging that the cross-sections given in Table 17.17 are not inconsistent with those obtained in other experiments from direct measurement of the flux of product halides.

6.3. *Experimental analysis of angular, recoil, and internal energy distribution in the reaction* $K+Br_2 \rightarrow KBr+Br$

To illustrate the techniques and results obtained for the determination of the distributions, in the C.M. system, of the angle and energy of the recoiling product as well as the distribution of rotational energy, we choose the reaction which has been most thoroughly investigated in these respects—that of K with Br_2.

† AIREY, J. R., GREENE, E. F., KODERA, K., RECK, G. P., and Ross, J., loc. cit.
‡ ACKERMAN, M., GREENE, E. F., MOURSUND, A. L., and Ross, J., loc. cit.
§ AIREY, J. R., GREENE, E. F., RECK, G. P., and Ross, J., *J. chem. Phys.* **46** (1967) 3295.

Minturn, Datz, and Becker[†] were concerned particularly with measurement of the angular distribution of the products and they used a velocity-selected K beam. No attempt was made to perform an energy analysis of the KBr molecules at each projection angle. Birely and Herschbach,[‡] on the other hand, were particularly concerned with these distributions so they introduced the velocity selector into the recoil beam and worked with a thermal K beam. At about the same time, Bernstein and Grosser[§] carried out experiments on the same reaction using a velocity selector in both the incident K beam and the product KBr beam.

The kinematics of the collision, including the relation between the angles of scattering and energies of projection in the C.M. and laboratory systems, is best visualized by means of a vector diagram as shown in Fig. 17.76. From the origin O the right-angled triangle AOB is drawn with OA proportional to the velocity $v(K)$ of the K atoms and OB to that $v(Br_2)$ of the Br_2 molecules. $\vec{BA}$ then represents the initial relative velocity $\mathbf{v}_r$ and if C is a point dividing BA so that

$$BC/AC = M(K)/M(Br_2)$$

where $M(K)$, $M(Br_2)$ are the respective masses of K and Br_2, then $\vec{OC}$ represents the velocity $\mathbf{v}_{CM}$ of the centre of mass. Elastic scattering of K atoms rotates the vector $\vec{CA}$ about C, tracing out the broken circle. Thus if the velocity of the scattered K atom is in the direction $\vec{CE}$ making an angle θ with $\vec{CA}$ the velocity in the laboratory system will be $\vec{OE}$ making an angle Θ with $\vec{OA}$. θ and Θ are then respectively the angles of scattering in the C.M. and laboratory system.

In the C.M. system the total kinetic energy of the products is given by

$$E' = E + \Delta E_0 - \Delta E_{\text{int}}, \qquad (342)$$

where ΔE_0 is the excess zero-point energy possessed by the reactants over that possessed by the products, ΔE_{int} the excess internal excitation energy of the products. The magnitude $w(KBr)$ of the velocity of the KBr in the C.M. system is then given by

$$w(KBr) = \{M(Br)/(M(Br)+M(KBr))\}(2E'/\mu')^{\frac{1}{2}}, \qquad (343)$$

where
$$\mu' = M(Br)M(KBr)/\{M(Br)+M(KBr)\}. \qquad (344)$$

For the reaction
$$K + Br_2 \rightarrow KBr + Br, \qquad (345)$$

[†] Minturn, R. E., Datz, S., and Becker, R. L., *J. chem. Phys.* **44** (1966) 1149.
[‡] Birely, J. H. and Herschbach, D. R., ibid. **44** (1966) 1690.
[§] Bernstein, R. B. and Grosser, A. E., ibid. **43** (1965) 1140.

$E_0 = 45 \cdot 0 \pm 1 \cdot 5$ kcal/mol (1·94 eV). At 315 °K, 77 per cent of the Br$_2$ molecules are in the ground vibrational state with a most probable rotational energy of 0·9 kcal/mol (0·04 eV).

The velocity of the KBr in the laboratory system for a particular value of ΔE_{int}, will lie on a circle with centre C and radius proportional to w(KBr). Thus, if the laboratory angle of projection is ϑ the direction of w(KBr) will be along CP so that the angle of projection in the C.M. system is θ_p.

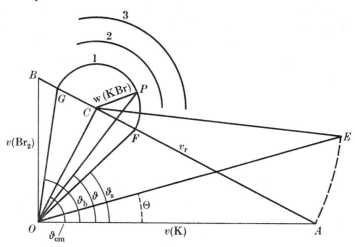

FIG. 17.76. Vector diagram illustrating the kinematics of reactive collisions between K atoms and Br$_2$ molecules.

Three extreme forms of the C.M. angular distribution can be recognized. If the distribution is isotropic so that all angles θ_p are equally probable, the most probable direction of projection in the laboratory system will be along OC, i.e. at a projection angle in that system equal to ϑ_c. On the other hand, if the direction of projection is nearly along $\mathbf{v}_r$, corresponding to zero angle of projection in the C.M. system, the most probable angle of projection in the laboratory system for a particular value of E' will be ϑ_s where $\vartheta_s < \vartheta_{\text{CM}}$—the velocity of projection in the laboratory system will be along OF in Fig. 17.76. The third case is one in which the halide molecule is projected backwards in the C.M. system, i.e. along $\mathbf{v}_r$ but in the opposite sense. Here the velocity of projection in the laboratory system will be along OG and the most probable angle of projection for a particular value of E' will be ϑ_b where $\vartheta_b > \vartheta_{\text{CM}}$.

In general, there will be a distribution of values of E' depending on the distribution of internal energy in the products, apart from any blurring due to the velocity distributions in the colliding beams. We have drawn in Fig. 17.76 a number of circles centred on C which correspond to different values of E', taking E as fixed. We may now consider the form the energy distribution of the KBr will take in the laboratory

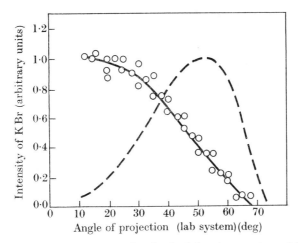

Fig. 17.77. Angular distribution in the laboratory system of KBr molecules produced in the reaction

$$K+Br_2 \to KBr+Br,$$

the most probable initial relative energy of the reactants being 2·31 kcal/mol (0·1 eV), observed by Minturn, Datz, and Becker. —— best fit of observed data; ——— distribution of the C.M. vector allowing for velocity spread in the Br_2 beams.

system. Suppose the angle of projection in the laboratory system is ϑ. Then, if E' is such that the velocity of projection relative to the C.M. is a radius vector of the circle 1 say, then the triangle of velocities is OCP with $\overrightarrow{OP}$ representing the velocity of the KBr in the laboratory system. It follows that, if E' is the most probable value of the kinetic energy of the products in the C.M. system the velocity distribution in the laboratory system will show a peak at a velocity of magnitude OP. From the position of the peak the most probable value of E' may thus be obtained without much difficulty.

Fig. 17.77 shows the angular distribution in the laboratory system observed by Minturn, Datz, and Becker[†] for a mean initial kinetic energy of 0·1 eV (2·31 kcal/mol). For comparison, the distribution corresponding to averaging the direction of the C.M. over the velocity distribution and angular divergence of the Br_2 beam is also shown. This will be

† Minturn, R. E., Datz, S., and Becker, R. L., loc. cit., p. 1635.

further broadened symmetrically about the peak if the angular distribution of the products in the C.M. system is isotropic. It will be seen that the observed peak falls at much smaller angles in the laboratory system, showing that the distribution in the C.M. system has a peak at an angle considerably less than $\frac{1}{2}\pi$.

Fig. 17.78 shows the velocity distributions of the KBr for three laboratory angles of projection observed by Birely and Herschbach† and by Bernstein and Grosser.‡

The former worked with thermal K and Br_2 beams and their observations refer to experiments in which the most probable initial relative kinetic energy E was 1·2 kcal/mol (0·052 eV). Bernstein and Grosser used a velocity-selected K beam and results are given for a velocity of 732 m s^{-1} for all three angles. This corresponds to a most probable relative kinetic energy E of 2·1 kcal/mol (0·090 eV). In addition, results are also given at an angle of 20° for K beams with velocities of 549 and 915 m s^{-1} corresponding to E equal to 1·3 and 3·3 kcal/mol (0·056 and 0·142 eV). It will be seen that there is very little variation between the different results at each angle.

Birely and Herschbach analysed their data assuming that the colliding beams were homogeneous in energy with velocity having the most probable value. Observed data in the laboratory system are shown in the contour diagram of Fig. 17.79. This includes not only the data on the velocity distribution but also earlier data, taken without a velocity selector, on the angular distribution. This is peaked at a laboratory angle of 20° which is considerably less than the C.M. angle (60°) and is in general agreement with that shown in Fig. 17.78. It is also apparent from the location of the peaks in the velocity distribution of the product KBr that the peak recoil energy increases as the angle of projection decreases.

Fig. 17.80 shows the angular distribution of the KBr product in the C.M. system derived from analysis of the data shown in Fig. 17.79. In transforming from the laboratory to the C.M. system the distributions per unit solid angle in the two systems are related by

$$I_{CM} = I_{lab} (d\Omega/d\omega), \qquad (346)$$

where
$$d\Omega/d\omega = (u/v)^2 |\cos\delta|, \qquad (347)$$

u and **v** being the velocities of the KBr in the C.M. and laboratory systems respectively and $\cos\delta = \mathbf{u}.\mathbf{v}/uv$. It will be seen that the

---

† BIRELY, J. H. and HERSCHBACH, D. R., loc. cit., p. 1635.
‡ BERNSTEIN, R. B. and GROSSER, A. E., loc. cit., p. 1635.

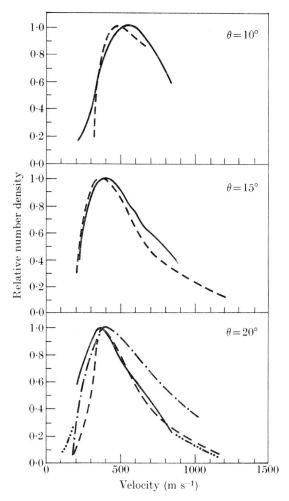

Fig. 17.78. Velocity distribution, in the laboratory system, of KBr molecules produced in the reaction

$$K + Br_2 \rightarrow KBr + Br$$

at three angles of projection in the laboratory system as indicated, observed by Birely and Herschbach using thermal K and Br_2 beams. Most probable relative energy $\bar{E}$ of reactants, 1·2 kcal/mol (0·052 eV) ———. Observed by Bernstein and Grosser with a velocity-selected K beam and thermal Br_2 beam and relative energies E', 1·3 kcal/mol (0·056 eV) —·—·—, 2·1 kcal/mol (0·090 eV) — — —, 3·3 kcal/mol (0·14 eV) —··—.

distributions are strongly peaked in the forward direction, a feature already deduced by inspection of the vector diagram in Fig. 17.79.

Fig. 17.81 shows the recoil energy distributions at various laboratory angles. It will be seen that at the most probable angle, 20°, three-quarters of the collisions result in products moving with relative kinetic

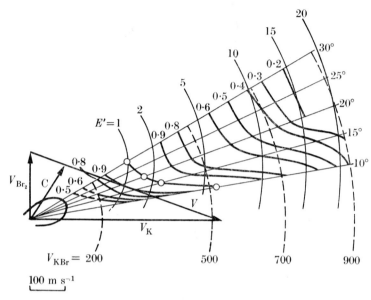

Fig. 17.79. Contour map showing the observed velocity distribution of KBr, resulting from the K–Br$_2$ reaction, in the laboratory system. The polar plot gives the observed angular distribution and the contours denote the fraction of the maximum intensity at each laboratory energy. The full-line circles are loci of recoil vectors corresponding to various values of the final relative translational energy in kilocalories per mole and the broken circles the loci of vectors corresponding to various velocities in the laboratory system in metres per second.

energy $E' < 10$ kcal/mol (0·43 eV). Since the zero point energy E_0 of the products is less than that of the reactants by 45 kcal/mol (1·93 eV) it is clear that most of the energy released in the reaction appears as internal energy of the KBr. In an earlier experiment Herm and Herschbach[†] investigated the rotational excitation of the KBr by deflexion in an inhomogeneous electric field.

The transmission through such a field depends on the ratio

$$X = \mu F/\langle E \rangle, \tag{348}$$

where F is the electric field strength, μ the dipole moment, and $\langle E \rangle$ the

[†] HERM, R. R. and HERSCHBACH, D. R., J. chem. Phys. 43 (1965) 2139.

geometric mean $(E_t E_r)^{\frac{1}{2}}$ of the translational and rotational energies of the molecules. For analysis at 30° angle of projection in the laboratory system the distribution of E_t is nearly Maxwellian about a temperature $T \simeq 755$ °K (see Fig. 17.81). It was found that the observed transmission agreed with that calculated on the assumption that the distribution of E_r is also Maxwellian with a rotational temperature of 1150 °K. This distribution is shown in Fig. 17.81.

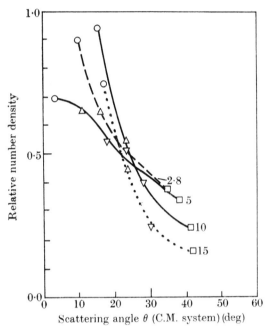

Fig. 17.80. Angular distribution of projected KBr in the C.M. system for various values of the recoil energy (indicated in kilocalories per mole) resulting from the K–Br$_2$ reaction, derived by Birely and Herschbach from their observations. Data derived from different angles in the laboratory system: ○ 10°, △ 15°, ▽ 20°, □ 30°.

It is quite clear that very little of the internal energy carried away by the products resides in rotation. Knowing the distribution of E' and of E_r the distribution of E_{vib} may be derived and is also illustrated in Fig. 17.81.

Though differing in detail, very similar results were derived by Warnock, Bernstein, and Grosser[†] from an analysis of the observations

[†] WARNOCK, T. T., BERNSTEIN, R. B., and GROSSER, A. E., *J. chem. Phys.* **46** (1967) 1685.

of energy and angular distributions in the laboratory system. They adopted an indirect procedure of successive approximations to obtain the distributions in the C.M. system. Plausible trial distributions were assumed from which the corresponding laboratory distributions were

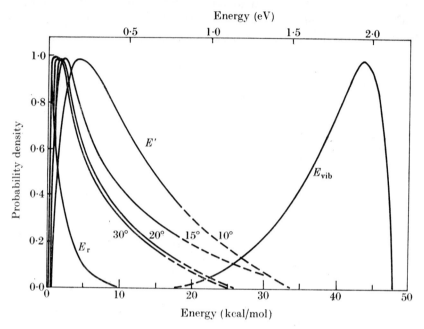

FIG. 17.81. Distribution of recoil energy E' (at various angles in the laboratory system) and of rotational and vibrational excitation E_r, E_{vib} respectively of KBr (both for $\theta = 30°$) produced from the K–Br$_2$ reaction.

calculated and compared with the observed. This suggested modifications in the assumed C.M. distributions which would improve the agreement. The procedure was repeated until agreement was obtained within the errors of observation. To render the calculations tractable it was assumed that the velocity distribution of the products is independent of the angle of projection in the C.M. system and of the incident relative kinetic energy. The velocity distribution in the Br$_2$ beam was taken to be Maxwellian and the velocity-selected K beam was assumed to be homogeneous in velocity. In the first instance the angular spread of the Br$_2$ beam was ignored but its effect was later studied separately and found to be small.

Fig. 17.82 shows the derived distributions in angle and internal energy for K atom velocities of 732 and 915 m s^{-1}. The curves giving the best fit are shown within shaded areas which indicate the limits of error.

These results are very similar to those derived from their observations by Birely and Herschbach. In particular, the angular distribution is peaked in the forward direction and most of the energy released is carried away as internal excitation of the KBr.

Estimates made of the reaction cross-section, based on the observed integrated intensity of KBr produced, divided by the total intensity of scattered K, indicate that it is of the order of 10 per cent of the total collision cross-section or $\simeq 10^{-14}$ cm^2.

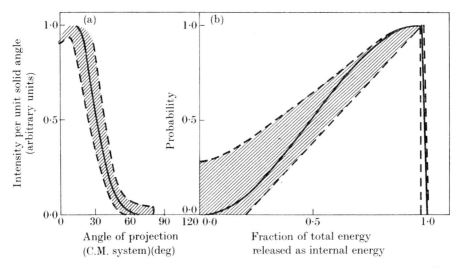

FIG. 17.82. (a) Angular distribution and (b) internal energy distribution in the C.M. system of KBr product from the K–Br$_2$ reaction, derived from observations in the laboratory system made by Bernstein and Grosser. The range of uncertainty is indicated for each distribution by the shading.

6.4. Results for reactions of alkali atoms with other halide molecules

The outstanding features of the reaction of K with Br$_2$, namely, high reaction cross-section, C.M. angular distribution peaked in the forward direction, and a large fraction of the energy released taken up in excitation of the product halide, are also exhibited by a number of other reactions. Thus in Table 17.18 we compare observed values of ϑ_s, the angle of projection in the laboratory system of maximum production of the product, with the C.M. angle ϑ_CM for some reactions of Cs atoms.† In all these cases $\vartheta_\mathrm{s} < \vartheta_\mathrm{c}$, showing that the angular distribution in the C.M. system is peaked in the forward direction. Also, although the recoil

† WILSON, K. R., KWEI, G. H., NORRIS, J. A., HERM, R. R., BIRELY, J. H., and HERSCHBACH, D. R., J. chem. Phys. **41** (1964) 1154.

energies E' derived from a vector diagram as in Figs. 17.76 or 17.79 are only nominal in that they are based on the assumption that the velocities of the interacting atoms and molecules are equal to the most probable values, there is no doubt that for all the caesium reactions E' is so much less than the total energy released (ΔE_0) that most of the latter must be carried off as internal excitation.

TABLE 17.18

Angles of projection ϑ_s in the laboratory system for maximum yield of product in various reactions of Cs atoms with halogen-containing molecules

Reaction		$Cs+Br_2$	$Cs+ICl$	$Cs+IBr$	$Cs+I_2$
Temp. of alkali metal beam (°K)		575	675	670	577
Temp. of molecular beam (°K)		325	317	317	317
ϑ_s		+8°	+6°	+5°	+13°
C.M. angle ϑ_{CM}		39°	37°	40°	45°
Nominal recoil energy	(kcal/mol)	1	5 (3)	5 (3)	1
	(eV)	0·04	0·22 (0·13)	0·22 (0·13)	0·04
ΔE_0	(kcal/mol)	60	31 (56)	39 (53)	45
	(eV)	2·6	1·35 (2·45)	1·7 (2·3)	1·95

For the reactions with ICl and IBr the unbracketed values are derived on the assumption that the product is mainly CsI, the bracketed that it is CsCl and CsBr respectively.

Finally, for all the reactions the cross-section is about 10 per cent of the total cross-section and is therefore of the order 10^{-14} cm² which is comparable with the large value for the K–Br$_2$ reaction.

Reactions of this kind are referred to as *stripping reactions* as they can be imagined to take place by the alkali atom stripping off the halogen atom as it passes by. Such reactions have been recognized in nuclear physics from which the name has arisen.

A second set of reactions has been observed to show a peaking of the angular distribution in the C.M. system in the backward direction. In these cases most of the energy released is carried off as internal energy as for stripping reactions but, unlike the latter, the reaction cross-section is quite small, being of order 1 per cent or less of the total cross-section. Up to the present all of these so-called *rebound reactions* involve collisions between alkali-metal atoms and alkyl halides, the classical example being that of K with CH_3I.† For this reaction $\Delta E_0 \sim 23$ kcal/mol (1 eV).

† HERSCHBACH, D. R., KWEI, G. H., and NORRIS, J. A., *J. chem. Phys.* **34** (1961) 1842; HERSCHBACH, D. R., *Discuss. Faraday Soc.* **33** (1962) 149.

The most probable initial kinetic energy of relative motion is 1·3 kcal/mol (0·06 eV) and the most probable rotational energy of the CH_3I 0·6 kcal/mol (0·03 eV). Accordingly, we have the vector diagram for the initial and final velocities in the plane of the beams as shown in Fig. 17.83 with the circles centred on the C.M. corresponding to different recoil energies. Fig. 17.84 illustrates typical angular distributions of the KI product in the laboratory system, observed by Herschbach, Kwei, and Norris, which show a peak at 83°. Since this is greater than the C.M.

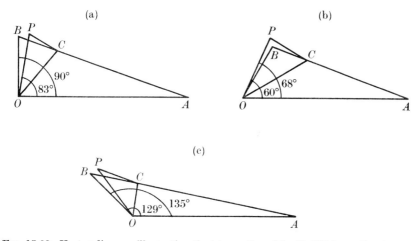

FIG. 17.83. Vector diagram illustrating the kinematics of the K–CH_3I reaction between beams intersecting (a) at 90°, (b) at 60°, and (c) at 135°.

angle ϑ_{CM} of 60° the scattering is peaked at an angle > 90° in the C.M. system as may be seen by drawing the recoil vector in the diagram of Fig. 17.83. To obtain information about E' use was made of the fact that the distribution of KI must have cylindrical symmetry about the initial direction of relative motion. Measurements of the angular deviation of the recoil vector from this direction may be made by observing the angular distribution in a plane other than that containing the beam (as on p. 1631). In this way it was found that the most probable energy of recoil in the C.M. system is small, corresponding to a nominal value of E' of 1 kcal/mol (0·04 eV). Further confirmation of this was obtained by observing the in-plane distributions when the angles of intersection of the beam were changed to 60° and to 135°, in which cases the peak angle in the laboratory system changed to 68° and 129° respectively (see Fig. 17.83). Because of the blurring by the velocity distributions of the reactants it is only possible to say that at the peak the recoil angle in the

C.M. system $\geqslant 130°$ and E' is between 1 and 8 kcal/mol (0·04–0·35 eV), but the main features are established.

Similar results have been obtained for the reactions of Na,[†] Rb,[‡] and Cs[§] with CH_3I, the peaks in the laboratory angular distribution occurring at 112°, 66°, and 52° respectively. Reactions of K, Rb, and Cs with C_2H_5I, C_3H_7I, C_5H_9I, and $C_6H_{11}I$[||] are also of this type. The evidence

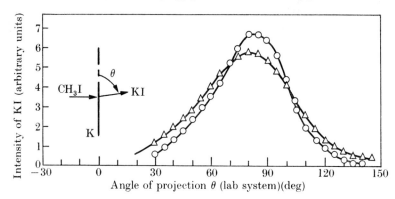

Fig. 17.84. Angular distribution in the laboratory system of KI resulting from the K–CH_3I reaction between beams intersecting at 90°. ○ and △ denote observations made in experiments at different times and refer to measurements in the plane of the two beams. The angle is measured relative to the K beam in the sense shown in the inset diagram.

indicates that, as the size of the alkyl group is increased, the fraction of the available energy released as translational motion falls rapidly.

Intermediate examples appear to be provided by reactions of alkali-metal atoms with $CHCl_3$ and CCl_4,[††] molecules for which the reaction cross-sections are of intermediate magnitude. The nominal recoil energy is again small ($\simeq 1$ kcal/mol (0·04 eV)) but the angular distributions in the laboratory system of the alkali-halide product appear to have two maxima corresponding to a distribution about a recoil angle near $\tfrac{1}{2}\pi$ in the C.M. system.

6.5. *The use of ionization detectors in the study of reactions between crossed molecular beams—the reaction of D with Br_2*

So far we have described experiments in which one of the reactants is an alkali metal and one of the products is an alkali halide so that hot

[†] WILSON, K. R., and HERSCHBACH, D. R., *Bull. Am. phys. Soc.* **7** (1962) 497; WILSON, K. R., Thesis, Berkeley, 1964.
[‡] KINSEY, J. L., KWEI, G. H., and HERSCHBACH, D. R., *Bull. Am. phys. Soc.* **6** (1961) 152.
[§] NORRIS, J. A., KWEI, G. H., KINSEY, J. L., and HERSCHBACH, D. R., ibid. **6** (1961) 339.   [||] NORRIS, J. A., Thesis, Berkeley, 1963.
[††] WILSON, K. R. and HERSCHBACH, D. R., *Bull. Am. Phys. Soc.* **9** (1964) 709.

wire ionization detectors may be used both for the elastically scattered alkali-metal atoms and for the alkali-halide product. To extend the scope of the technique to the study of reactions in which alkali-metal atoms are not involved an electron bombardment ionization detector may be used of the type described briefly in Chapter 16, p. 1345. At the time of writing very few experiments have been carried out using such detectors but we give a brief description of one of these which illustrates the technique and has provided interesting results.

In this experiment, carried out by Datz and Schmidt,† the reaction

$$D + Br_2 \rightarrow DBr + Br \qquad (349)$$

was studied. Fig. 17.85 illustrates the arrangement of the apparatus. A deuterium atom beam, produced through thermal dissociation of D_2 at 2800 °K in a tungsten oven, was modulated at 100 c/s and crossed a molecular beam of Br_2 at 90°. The detector consisted of an electron-bombardment ion source, followed by a quadrupole mass analyser and an electron multiplier. By constructing the scattering chamber from three concentric rotatable cylinders the detector could be rotated through 130° in the plane of the intersecting beams.

To reduce the background signal due to DBr formed other than through direct reaction between the beams, differential pumping was provided between the reaction and ionization regions, the latter being also surrounded by a liquid nitrogen trap. By means of the modulation of the D atom beam, phase-sensitive detectors could be employed.

Fig. 17.86 illustrates the observed angular distribution of the DBr product in the laboratory system. The kinematics of the collision can be discussed in terms of a vector diagram as on pp. 1636, 1640, account being taken of the fact that neither beam is velocity-selected. Because of the velocity spread the angular distribution of the centre of mass motion in the laboratory system is given by the broken line in Fig. 17.86. It is seen to be very similar to the observed distribution so that the kinetic energy of the product DBr in the C.M. system must be very small, despite the fact that the reaction (349) is exothermic by 1·8 eV.

Some of the surplus energy may appear in electronic excitation of the Br atom to the $^2P_{\frac{1}{2}}$ state, requiring 0·46 eV, but earlier experiments by Cashion and Polanyi,‡ in which the intensity of radiation corresponding to the $^2P_{\frac{1}{2}}$–$^2P_{\frac{3}{2}}$ transition accompanying the reaction was observed, have shown that little of the energy is taken up in this way. It therefore seems

† DATZ, S. and SCHMIDT, T. W., Fifth Int. Conf. Phys. Elect. Ionic Collisions, Leningrad, 1967, *Nauka*, Abstracts, p. 247.
‡ CASHION, J. K. and POLANYI, J. C., *Proc. R. Soc.* A**258** (1960) 570.

that most of the DBr molecules are produced with a high degree of vibrational and rotational excitation just as in other reactions involving alkali-metal atoms described earlier (see §§ 6.3, 6.4).

Datz and Schmidt carried out a further check by determining the reaction cross-section for production of the observed DBr. This was done by measuring the ratio of the integrated distribution of the product to the total number of collisions. The total collision cross-section was then

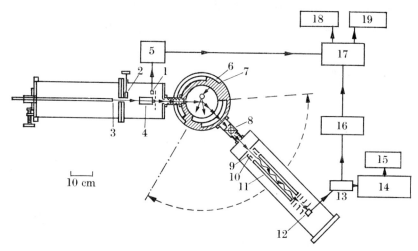

FIG. 17.85. Arrangement of the apparatus used by Datz and Schmidt to study the reaction
$$D+Br_2 \rightarrow DBr+Br.$$
1, light and photocell phase pickup; 2, beam valve; 3, primary beam source; 4, 100-cycle chopper; 5, synchronous signal amplifier; 6, cross beam source; 7, scattering chamber; 8, valve; 9, liquid N_2 trap; 10, beam ionizer; 11, quadrupole mass filter; 12, electron multiplier; 13, switch; 14, vibrating reed electrometer; 15, Brown recorder; 16, DD2 linear amplifier; 17, gate; 18, beam-on scaler; 19, beam-off scaler.

determined by observing the absorption of the D atom beam in a cell containing Br_2 and from this the reaction cross-section for production of the observed DBr obtained. This was found to be 12×10^{-16} cm² which is already 50 per cent of that derived from measurement of rate constants in experiments in the bulk gas. As some of the centre of mass motion takes place out of the detector plane due to a component of velocity of the Br_2 molecules out of the plane, 50 per cent is a lower limit and the results confirm the general conclusion that most of the excess energy appears as vibrational and rotational excitation.

6.6. *Concluding remarks*

A considerable amount of effort has been directed towards a detailed theoretical interpretation of reactions of the type we have been discussing. Since the dynamics of the collision are essentially classical the progress of the reaction may be traced out by the passage of a representative point on a hypersurface, the height of which above a chosen hyperplane is proportional to the interaction energy between the various atoms and molecules. It is possible in principle to apply a Monte Carlo method in which a statistical set of paths are followed and their consequences in terms of reaction determined. In practice, considerable simplification of the interaction hypersurface is necessary to make such

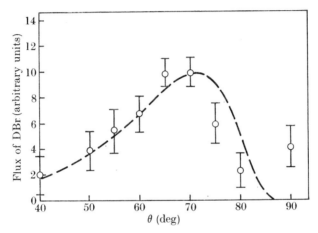

FIG. 17.86. Angular distribution (in the laboratory system) of DBr molecules produced in the reaction

$$D+Br_2 \rightarrow DBr+Br.$$

○ observed by Datz and Schmidt; ––– angular distribution of C.M. motion due to velocity distributions of D and Br_2 in the colliding beams.

a procedure practicable. Nevertheless, it has already been applied with considerable success not only to the simplest reactions such as[†]

$$H+H_2 \rightarrow H_2+H,$$

but also to the reaction[‡]

$$K+CH_3I \rightarrow KI+CH_3.$$

It would take us too far afield into the theory of chemical reactions to discuss this work here. The potentialities of the crossed-beam technique for detailed experimental analysis of chemical reactions are so great, however, that we can expect a considerable improvement in our understanding of the important factors that determine the nature and

[†] KARPLUS, M., PORTER, R. N., and SHARMA, R. D., *J. chem. Phys.* **43** (1965) 3259.
[‡] RAFF, L. M. and KARPLUS, M., *J. chem. Phys.* **41** (1964) 1267; **44** (1965) 1212.

probability of formation of different reaction products. Already it seems that many reactions proceed in such a way that the least possible energy released by the reaction appears in the form of relative translation. This applies in many other circumstances, as for instance in transfer of electronic excitation (see Chap. 18, § 4.2), but as the dynamics of the chemical reactions we have been discussing here are effectively classical it would be misleading to describe the characteristic feature of minimum change of relative translational energy as a resonance effect. It is, nevertheless, already established as an effect of considerable generality which should be capable of interpretation without need for detailed analysis in any particular case.

18

COLLISIONS BETWEEN ATOMS (AND/OR MOLECULES) UNDER GAS-KINETIC CONDITIONS—COLLISIONS INVOLVING ELECTRONIC TRANSITIONS

1. Introduction

WE now consider collisions between atoms (and/or molecules) one or both of which (before or after the collision) is in an excited electronic state. Although our limitation to gas-kinetic conditions restricts the availability of translational energy, so that inelastic collisions in which the electronic configuration in either colliding partner is changed cannot occur with significant probability, the corresponding superelastic collisions can occur and there are a number of other possibilities. These may be summarized as follows:

(a) Superelastic collisions in which an excited atom (or molecule) is deactivated by collision with another atom (or molecule). The excited state concerned may or may not be metastable.

(b) Collisions in which a transfer of excitation takes place from one atom to the other. If the atoms are of the same kind such a transfer will not involve any change ΔE in the total internal energy but, even if they are unlike, there may exist transfer processes for which ΔE is either negative or positive yet sufficiently small for the reaction to take place with a considerable probability under gas-kinetic conditions.

(c) Collisions in which there is no change in the configuration or total orbital angular momentum but changes occur in the orientation or magnitude of the total (orbital+spin) angular momentum or the magnitude of the total electronic spin. Such transitions will require changes of internal energy which are comparable with or smaller than the mean translational energy under gas kinetic conditions.

A wide variety of experimental techniques must be employed to study these reactions. In general, the procedure is different according to whether the excited states concerned are metastable or not. For that reason we shall begin by dealing with the experimental methods available for studying reactions in which the excited states are not metastable.

In such cases the progress of the reaction may be followed by observing the radiation emitted. Thus, for the collisions of type (a) above, deactivation will lead to a reduction in intensity of radiation emitted in transitions from the excited states concerned, i.e. a quenching of the emission. Again, if transfer of excitation occurs, radiative transitions will take place from the newly populated states, a phenomenon known as sensitized fluorescence. Most experimental information about the cross-sections for collisions of types (a) and (b) above in which the excited states are not metastable has come from the observation either of quenching or of sensitized fluorescence.

Experiments of this kind also yield information about processes of type (c). Thus the observation of the polarization of fluorescent radiation under different experimental conditions is a major source of information about the rates of collisions that change the direction but not the magnitude of the total electronic angular momentum in excited atoms. Reactions such as

$$\mathrm{Na}(^2P_{\frac{3}{2}}) + \mathrm{X} \to \mathrm{Na}(^2P_{\frac{1}{2}}) + \mathrm{X}$$

may be studied from the observation of sensitized fluorescence. The development of electronic and nuclear magnetic resonance techniques and that of optical pumping has greatly expanded the possibilities for the determination of other cross-sections of type (c) such as for spin-exchange or spin-reversal collisions.

We shall begin by describing the experimental study first of the quenching and polarization of fluorescent radiation and then of sensitized fluorescence. This will be followed by a discussion of the variety of methods available for investigating reactions involving metastable atoms and the results obtained thereby. After this we shall consider the application of magnetic resonance and optical pumping methods to obtain information about spin-exchange and spin-reversal collisions. A detailed theoretical discussion will follow in the concluding sections.

It is clear that many of the subjects we are concerned with here form part of photochemistry and, if we consider the full variety of reactions in which one or other colliding system is a molecule, the volume of material to discuss is far beyond the possibility of consideration in this chapter. We shall therefore restrict ourselves, as elsewhere in this book, to cases in which the colliding systems are comparatively simple. On the other hand, it would not be sensible to exclude entirely the consideration of collisions in which one of the reactants is a molecule. The role of molecular vibration as a source or sink of energy is of much interest and importance as well as being related to the subject-matter of the preceding chapter which is devoted to inelastic collisions in which no electronic transitions occur.

The search for simplifying rules that apply to the estimation of the cross-sections for collisions of the type with which we are now concerned has not proved very fruitful so far. Most effort has been devoted to investigating the dependence of the cross-section, under gas-kinetic conditions, on the change ΔE of internal energy involved, and this will be reflected both in the description of the experimental results obtained and in the theoretical discussion in § 8.

2. Quenching of radiation

We have already discussed in Chapter 17, § 2.4.1 how, from the quenching of infra-red fluorescence by foreign gases, the rate of deactivation of vibrational excitation by collisions of the excited molecules with foreign gas molecules can be determined. Exactly the same principles apply to quenching of fluorescence arising from electronic excitation. Thus, if I_0 is the intensity of fluorescent emission when no collisional deactivation of electronic excitation occurs and I is that when a concentration n of deactivating molecules is present,

$$I_0/I = (1+\tau_r Z), \qquad (1)$$

where τ_r is the lifetime of the radiating state towards radiation and Z is the number of deactivating collisions per second. Z is given by

$$2nQ(2\kappa T/\pi M)^{\frac{1}{2}}, \qquad (2)$$

where Q is the mean effective cross-section for deactivation and M is the reduced mass of the colliding molecules. Since τ_r may be measured by suitable experiments, Z and hence Q may be obtained.

The same reservations apply as in the discussion of Chapter 17, § 2.4.1. Thus it is assumed that addition of the foreign gas does not affect the rate of production of excited atoms by the incident radiation, as through line broadening. The smallness of the effective cross-section Q in some cases may impose a limitation on the method, for it may be that, to avoid broadening, the foreign gas pressure has to be so low that quenching is also negligible. Fortunately, in many cases, this is not so.

A number of sources of excited atoms have been used for quantitative experiments on quenching—absorption of resonance radiation, flash photolysis, shock-wave excitation, and optical dissociation of molecules.

2.1. *Quenching experiments with resonance radiation*

In principle, these experiments consist in exciting the atoms of a vapour such as mercury by a beam of radiation from a suitable source in which the same vapour is excited. The emission from this source will include radiation arising from transitions from the first excited state (which combines optically with the ground state) to the ground state of the atoms concerned. If radiation of this wavelength is selected to illuminate the vapour, atoms will be raised to the first excited state by resonance absorption.

A typical resonance lamp for quenching experiments has the form shown in Fig. 18.1 (a). The light trap is arranged to prevent internal reflections and so to reduce the amount of stray light present. To reduce the path, through the gas, of the emergent radiation, which is observed at 90° to the exciting beam, the entrance window has a slight projection, A, so there is no large thickness of unexcited gas between the exciting beam and the window from which the resonance radiation emerges. In this way imprisonment of the resonance radiation is avoided. Although it is possible to make allowance for imprisonment in terms of a theory such as that of Milne,[†] it renders the interpretation of the experiments very much more difficult and uncertain (see, however, § 2.1.3).

Fig. 18.1 (b) illustrates the apparatus used in the experiments of Norrish and Smith[‡] on sodium. C is the quenching cell with light trap T. The exciting radiation, emitted from the sodium lamp, L, passed through the lens A and the two glass windows M into the cell C. The resonance radiation emitted from the cell through the glass windows N was observed in the photometer P. To avoid errors due to variation in the intensity of the source L, the intensity of the resonance radiation entering the photometer was compared always with that entering from the source L via the tube R, mirror Q, iris diaphragm I, opalescent screen S, mirror V, and adjustable wedge W.

The resonance cell was enclosed in hot air at a thermostatically controlled temperature and contained specially purified sodium. Special precautions were also taken to ensure that the foreign gases introduced were very pure.

To check that imprisonment of resonance radiation was not important it was verified that, at a temperature 20° above that (130 °C) of the cell in the actual observations, the resonance radiation was emitted only from the path of the exciting beam.

It was found that foreign gases could be divided into two categories, which quench strongly and weakly respectively. For the former the

---

[†] MILNE, E. A., *J. Lond. math. Soc.* **1** (1926) 1; a further discussion has been given by HOLSTEIN, T., *Phys. Rev.* **72** (1947) 1212.

[‡] NORRISH, R. G. W. and SMITH, W. M., *Proc. R. Soc.* **A176** (1940) 295.

quenching cross-sections are greater than 10^{-15} cm² and it is only necessary to employ pressures of a few torr to make satisfactory measurements. Thus in Fig. 18.2 the observed ratio of $I_0/I = 1 + \tau_r Z$ is plotted against the foreign gas pressure of hydrogen, nitrogen, and benzene respectively. It is clear that a linear relation is found as it should be and the effects of Lorentz broadening are likely to be unimportant. This is further confirmed by reference to the measurements by Schutz† of Lorentz broadening of the sodium absorption line by helium. At the temperature

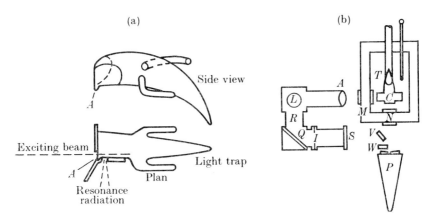

FIG. 18.1. (a) Illustrating the design of a resonance lamp. (b) Apparatus used by Norrish and Smith for studying the quenching of sodium resonance radiation.

of the quenching experiments and a pressure of 5 torr the Lorentz breadth, which does not depend markedly on the nature of the foreign gas, was found to be less than 3 per cent of the Doppler breadth.

For the strongly quenching gases measurements could therefore be made with a quenching ratio between 0·50 and 0·67 without risk of complication by Lorentz broadening. On the other hand, for the weakly quenching gases, the quenching at a pressure of 5 torr was only barely measurable and the cross-sections had to be estimated from quenching ratios of 0·97. To obtain much stronger effects it was necessary to raise the foreign gas pressure so much that Lorentz broadening became very serious. Thus it was found, for helium, that the intensity of the resonance radiation actually increased when pressures of 30–50 torr of the gas were added. The values given for the quenching cross-sections for these weakly quenching gases can therefore only be regarded as rough estimates.

† SCHUTZ, W., *Z. Phys.* **45** (1927) 30; **71** (1931) 301.

2.1.1. *Quenching experiments with flame gases.* A technique, similar in principle, has been applied by Hooymayers and Alkemade[†] and by Jenkins[‡] to the measurement of quenching cross-sections for excited metal atoms present in a hydrogen–oxygen or hydrogen–hydrocarbon flame so that the kinetic temperature is around 1500 °K or higher. Of necessity the quenching takes place in a medium containing more than one kind of quenching agent. To separate out the contributions from each of these the following procedure is adopted.

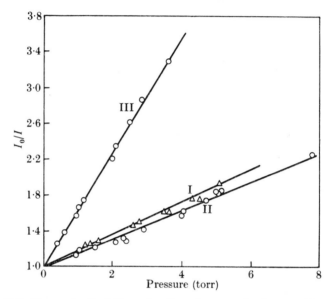

FIG. 18.2. Observed ratio I_0/I as a function of pressure for the quenching of Na resonance radiation by I, H_2; II, N_2; III, C_6H_6.

A beam of resonance radiation for the metal concerned is passed through the flame and the radiant power ΔP absorbed by the metal atoms in the flame measured. The absorbed power is used up in producing excited metal atoms M*. Some of these will return to the ground state, re-emitting the resonance radiation as fluorescence, and some will be quenched. Let P_f be the power which is emitted from the system as fluorescent radiation. The ratio $P_f/\Delta P$ is known as the fluorescent yield and will be noted by y. To relate this to the quenching rates we must allow for the fact that a fraction of the power emitted internally in fluorescent transitions will be reabsorbed before leaving the reaction

[†] HOOYMAYERS, H. P. and ALKEMADE, C. T. J., *J. quant. Spectrosc. Radiat. Transfer*, **6** (1966) 847.
[‡] JENKINS, D. R., *Proc. R. Soc.* **A293** (1966) 493.

system. The power emitted by the atoms excited directly by the beam will therefore be $P_f^0 = P_f/\alpha$, say. We then have

$$P_f^0/\Delta P = y/\alpha = (1+Z\tau_r)^{-1}, \tag{3}$$

where τ_r is the natural lifetime of the excited state and Z is the total quenching rate

$$Z = \sum Z_i, \tag{4}$$

where Z_i is the contribution from the ith constituent.

It follows that, if α and y can be measured and τ_r is known, $\sum Z_i$ is obtained. By variation of the proportions of the respective quenching constituents the separate Z_i may be derived. The main difficulty is the determination of α. This can be done by making measurements with varying concentrations of metal atoms and extrapolation to zero concentration, the procedure adopted by Hooymayers and Alkemade. However, it is very tedious, and an alternative semi-empirical method was adopted by Jenkins.

According to the theory of Kolb and Streed,† for monochromatic radiation of frequency ν

$$P_f(\nu) = P_f^0(\nu)\exp(-\tfrac{1}{2}k_\nu d)\cosh(\tfrac{1}{2}k_\nu d), \tag{5}$$

where k_ν is the absorption coefficient of the flame for the radiation and d is the flame thickness. The integral over the line contour, which is what is required, can be written in a similar form to give

$$P_f = P_f^0 \exp(-\tfrac{1}{2}\bar{k}d)\cosh(\tfrac{1}{2}\bar{k}d), \tag{6}$$

where $\bar{k}$ is the appropriate mean value of k_ν for the absorption of the resonance line concerned. If the source and absorption lines have the same shape, $\bar{k}$ is given by the fractional absorption of the primary light

$$\Delta P/P_0 = 1 - \exp(-\bar{k}d). \tag{7}$$

In practice, the lines will not have the same shape, but (7) is likely to hold to a sufficient accuracy if the source line is not much broader than the absorption line. Assuming this,

$$\alpha = P_f/P_f^0$$
$$= 1 - \tfrac{1}{2}\Delta P/P_0. \tag{8}$$

If y_0 is the fluorescent yield in the absence of self-absorption then

$$y = y_0 \alpha$$
$$= y_0(1 - \tfrac{1}{2}\Delta P/P_0), \tag{9}$$

and $\quad P_f/P_0 = \alpha P_f^0/P_0 = \alpha y_0 \Delta P/P_0 = y_0(\Delta P/P_0)(1-\tfrac{1}{2}\Delta P/P_0). \tag{10}$

† KOLB, A. C. and STREED, E. R., *J. chem. Phys.* **20** (1952) 1872.

Thus the assumptions may be checked by observing y and P_t/P_0 as functions of $\Delta P/P_0$.

In experiments with sodium in a hydrogen–oxygen flame diluted with nitrogen, Jenkins found from such observations that the results could be well fitted if instead of (8) α were given by

$$\alpha = 1 - \beta \Delta P/P_0, \tag{11}$$

with $\beta \simeq 0.59$. The departure of β from $\tfrac{1}{2}$ is presumably due to the inadequacy of the assumption (7) and to the fact that the emitting atoms are not uniformly distributed throughout the flame. β was found to vary from 0.53 to 0.69 for the variety of flames examined. To avoid determining β for each flame studied, the sodium concentration was chosen to be such that $\Delta P/P_0 < 0.15$, under which conditions differences in β amount to changes of less than 2 per cent and a mean value may be used for all flames.

Fig. 18.3 is a block diagram of the apparatus used by Jenkins. It is very similar to that used by Hooymayers and Alkemade, which was based on the earlier work of Boers, Alkemade, and Smit.† Light from the source a was focused on to the flame e. With the optical bench in this position the fluorescent light emitted through a well-defined solid angle about a direction normal to the incident beam was focused on a photomultiplier i. Alternatively, the optical bench, which was pivoted at a point beneath the centre of the burner, could be rotated so that the main beam emerging from the flame could be focused. To reduce background noise, phase-sensitive detection was employed by chopping the incident beam at d and feeding the output of the photomultiplier to a wide-band, high-input impedance, a.c. amplifier, rectified by a phase-sensitive detector drawing a reference signal from the chopper.

The source was a Pirani-type discharge lamp (Philips lamp type 93122). Long-term stability of about 1 per cent and stability throughout an absorption and fluorescence measurement of about 0.1 per cent was secured. To achieve this it was found necessary, apart from using a constant voltage supply, to seal the lamp, already double-walled, in a third, silvered, glass jacket to protect against room draughts. It was cooled by an air stream between the second and third jackets and the cooling rate could be used to control the width and intensity of the resonance line.

The burner was of a type used by Padley and Sugden,‡ consisting of a bundle of stainless steel tubes of 0.5 mm diameter. These were welded

† BOERS, A. L., ALKEMADE, C. T. J., and SMIT, J. A., *Physica, 's Grav.* **22** (1956) 358.
‡ PADLEY, P. J. and SUGDEN, T. M., *Proc. R. Soc.* **A248** (1958) 248.

along their lines of contact and sectioned into an inner pack 12 mm square, forming the experimental flame, and an outer shield flame to protect the inner from the effects of indrawn air. Metal atoms were introduced into the gas supply to the inner burner by a fine spray of an aqueous salt solution.

The flames used were in isothermal sets at temperatures of about 1800, 1600, and 1400 °K, the temperatures in each set being adjusted to be

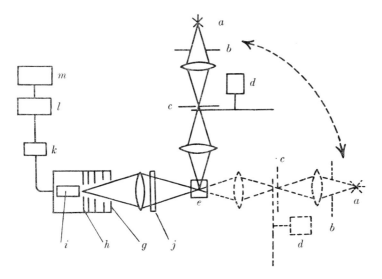

FIG. 18.3. Block diagram of the arrangement used by Jenkins for determining cross-sections for quenching of the resonance radiation of sodium by species present in flames. *a*, source; *b*, 1st aperture; *c*, 2nd aperture; *d*, chopper; *e*, flame; *g*, ω-stop; *h*, *f*-stop; *i*, photomultiplier; *j*, filter; *k*, amplifier; *l*, phase sensitive detector; *m*, recorder.

equal to within 20 °C by using a silicon-coated thermocouple. The primary beam was focused through a region in the flame about 1·5 cm above the reaction zone.

In the detector the ω-stop in the focal plane of the lens *g* restricted the solid angle of light accepted from all points within the flame while the *f*-stop *h*, in the image plane of the flame, restricted the field of view of the photomultiplier to a section of the flame a little greater than the path of the primary beam through the inner flame. The intensity of unmodulated light reaching the detector was reduced by the *f*-stop and the interference filter *j*. The ground-quartz plate, immediately ahead of the *f*-stop, scattered the light over the whole of the photomultiplier cathode. This was to eliminate the effect of differences in sensitivity due to different areas of the cathode covered by the primary beam and the fluorescent light in the plane of the f-stop.

To determine cross-sections for quenching of sodium resonance radiation by H_2, O_2, H_2O, Ar, and He, Jenkins measured the fluorescent yield in isothermal sets of fuel-rich flames of hydrogen and oxygen diluted with helium at 1800 °K. Some typical results are shown in Fig. 18.4 in the

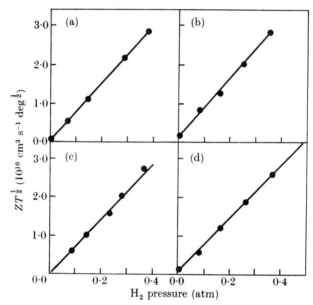

FIG. 18.4. Variation of the product $ZT^{\frac{1}{2}}$ with H_2 concentration for isothermal sets of fuel-rich hydrogen–oxygen flames diluted by rare gases. Diluted with Ar: (a) 1600 °K, (b) 1800 °K. Diluted with He: (c) 1600 °K, (d) 1800 °K.

form of plots of observed values of $ZT^{\frac{1}{2}}$, where T is the flame temperature, against the concentration of H_2. It will be seen from the fact that all the plots are linear with nearly zero intercepts on the y-axis that almost all the quenching is due to H_2. From detailed analysis it is found that the mean cross-sections $Q(X)$ for quenching by different species X are given by

$$Q(H_2) = 9 \cdot 0 \times 10^{-16}\,\text{cm}^2, \qquad Q(H_2O) = 1 \cdot 6 \pm 0 \cdot 9 \times 10^{-16}\,\text{cm}^2,$$
$$Q(\text{He}) \text{ and } Q(\text{Ar}) < 0 \cdot 3 \times 10^{-16}\,\text{cm}^2.$$

Fig. 18.5 shows a plot, for hydrogen–oxygen flames at 1800 °K diluted with helium, of $ZT^{\frac{1}{2}}$ against the concentrations of O_2 and of H_2 respectively. The two plots are linear and give closely the same intercept on the y-axis as they should. From such plots $Q(O_2)$ was determined as $38 \cdot 5 \times 10^{-16}\,\text{cm}^2$.

Results for $Q(N_2)$ were obtained by using nitrogen as a diluent in adjustable concentration in fuel-rich hydrogen–oxygen flames. $Q(CO)$ and $Q(CO_2)$ were determined by feeding carbon dioxide to hydrogen-rich flames and taking account of the fact that in such flames it is partially dissociated to CO.

In later experiments[†] the quenching of Li, Rb, K, Cs, and Tl was investigated. The results for all quenching species studied are given in

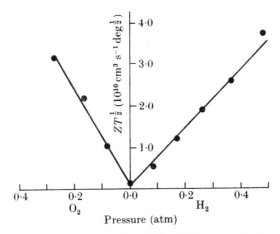

FIG. 18.5. Variation of the product $ZT^{\frac{1}{2}}$ with the concentration of H_2 and of O_2 in an isothermal set of hydrogen–oxygen flames at 1800° K diluted with helium. Left-hand side, oxygen-rich flames. Right-hand side, fuel-rich flames.

Table 18.4, p. 1693, where they may be compared with results obtained by other methods.

As mentioned earlier, Hooymayers and Alkemade[‡] used a very similar experimental arrangement. However, their burner was cylindrical, the diameter of the inner flame being about 1·8 cm and the thickness of the flame mantle about 1 cm. They obtained results for flame temperatures, which were measured by a line-reversal technique (see § 2.3), ranging from 1700 to 2200 °K. The procedure they adopted in analysing their results was also somewhat different from that used by Jenkins. It involved iteration based on results obtained for stoichiometric hydrogen–oxygen–argon flames. The burnt gases from these flames consist mainly of water and argon which were assumed to be solely responsible for the observed quenching. Working with different pressures it was then

[†] JENKINS, D. R., *Proc. R. Soc.* **A303** (1968) 453, 467; ibid. **A306** (1968) 413.
[‡] loc. cit., p. 1656.

possible to determine separately the quenching cross-sections for water and for argon, assuming that neither depends very strongly on the flame temperature. Knowing these, the corresponding cross-sections for quenching by H_2 and by O_2 could be found as functions of temperatures from observations on fuel-rich and oxygen-rich flames respectively. This permitted an improved determination of the best results for H_2O and Ar and so on. A possible criticism of this procedure is that both H_2O and Ar are very weak quenching agents, so the assumption that the influence of other stronger quenchers can be neglected, even under stoichiometric conditions, may not be justified. This is not likely to affect the determination of the larger cross-sections but may yield values for H_2O and Ar that are too large.

Reference to Table 18.4, which includes the results obtained by different methods, reveals wide inconsistencies in the results, particularly for Ar.

In addition to quenching of $Na(3\,^2P)$, Hooymayers and Alkemade also obtained cross-sections for quenching of $K(4\,^2P)$ and $K(5\,^2P)$ which are given in Table 18.4.

2.1.2. *Quenching cross-sections from lifetime measurements.* Demtröder[†] has developed a method of measuring the lifetimes of optically excited states which, while primarily designed to determine natural lifetimes, is also capable of determining quenching cross-sections. Instead of irradiating a resonance cell with a steady beam of resonance radiation the beam was modulated at an angular frequency ω, so that the incident intensity at time t is given by

$$I_{in} = I_0(1+a\sin\omega t), \tag{12}$$

where $a \leqslant 1$. If the scattered radiation is emitted from states of finite lifetime τ, the intensity of the radiation in a particular direction will be given by
$$I_{sc} = C\{1+a(1+\omega^2\tau^2)^{-\frac{1}{2}}\sin(\omega t-\phi)\}, \tag{13}$$

where $$\tan\varphi = \omega t. \tag{14}$$

Hence, by measuring the phase difference ϕ between the modulation of the scattered and incident beams, τ may be obtained. It is immaterial in this analysis whether τ is solely related to the natural lifetime or includes the quenching effect of collisions. In general,

$$\tau^{-1} = \tau_r^{-1}+Z,$$

where τ_r is the natural lifetime and Z is the quenching collision frequency.

[†] DEMTRÖDER, W., *Z. Phys.* **166** (1962) 42.

Apart from application of this technique to the determination of the natural lifetimes of a number of states of Ga, Al, Mg, Tl, and Na, Demtröder determined cross-sections for quenching of Na(^2P) by helium and nitrogen. The form of the absorption cell that he used is shown in Fig. 18.6. It was of glass in which a sodium vapour pressure between 1 and 2×10^{-4} torr was maintained by heating to 200 °C. The light passed through a neck of the cell only 8 mm thick so as to reduce imprisonment effects. The results obtained are given in Table 18.4.

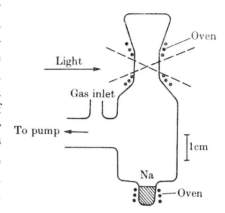

FIG. 18.6. Form of the sodium resonance absorption cell used by Demtröder.

2.1.3. *Quenching cross-sections from the rate of decay of imprisoned resonance radiation.* In the experiments that we have described so far, the imprisonment of resonance radiation has been a complication that had to be allowed for, but it is possible, given a satisfactory theory of the imprisonment process, to use the phenomenon to obtain quenching rates. Such a theory has been developed by Holstein† and checked experimentally by Alpert, McCoubrey, and Holstein.‡

If a sealed-off tube containing say, mercury vapour, is irradiated for a time with resonance radiation and the radiation then turned off, the intensity of the scattered light from the vessel will decay exponentially with a time constant T_i. According to Holstein's theory T_i is given by

$$T_i = \tfrac{5}{8} k_0 R (\pi \ln k_0 R)^{\frac{1}{2}} \tau_r, \tag{15}$$

where k_0 is the absorption coefficient at the centre of the Doppler-broadened resonance line, R the radius of the tube, and τ_r the natural lifetime of the resonance level of the mercury atom. The validity of (15) was checked for mercury by Alpert, McCoubrey and Holstein, using similar apparatus to that described below for the application to quenching.

If the resonance tube contains in addition a quenching gas, the rate of decay of the scattered light after cut-off of the incident radiation will be increased, so that the new constant T is given by

$$T^{-1} = T_i'^{-1} + Z, \tag{16}$$

† HOLSTEIN, T., *Phys. Rev.* **83** (1951) 1159.
‡ ALPERT, D., MCCOUBREY, A. O., and HOLSTEIN, T., ibid. **76** (1949) 1257.

where Z is the quenching rate and T'_i is the imprisonment time constant T_i as modified by the presence of the foreign gas. According to Holstein's theory†

$$T'^{-1}_i = T^{-1}_i + \Delta, \qquad (17)$$

where
$$\Delta = (\lambda \nu / 4\pi v)(\ln k_0 R)^{-\frac{1}{2}}.$$

Here λ is the wavelength of the radiation and v the most probable speed

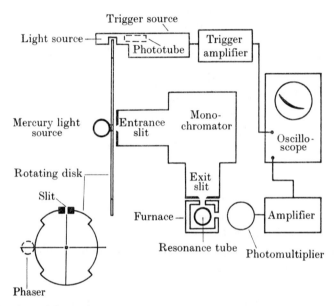

FIG. 18.7. Schematic diagram of apparatus used to determine the rate of decay of imprisoned resonance radiation.

of the quenching molecules, while ν is the collision frequency as determined from the pressure-broadening due to these molecules.

Having measured T_i and T and determined T'_i from (17), Z may be obtained.

Fig. 18.7 is a schematic diagram of the apparatus used first by Alpert, McCoubrey, and Holstein‡ to check (15) and then by Matland§ to determine the cross-section for quenching mercury resonance radiation by nitrogen. Light from a mercury discharge source was interrupted periodically, by a disc rotating at about 10 000 rev/min, before passing into a monochromator which selected resonance radiation (2537 Å) and directed it into the resonance tube. This was surrounded by an oven consisting of two independent sections, the upper being maintained at

† HOLSTEIN, T., loc. cit., p. 1663.   ‡ loc. cit., p. 1663.
§ MATLAND, C. G., *Phys. Rev.* **92** (1953) 637.

a higher temperature than the lower so that no mercury condensed on the walls at the upper part of the resonance tube.

Radiation escaping from the resonance tube was detected by a photomultiplier and amplifier, the output signal of which was recorded on the vertical axis of an oscilloscope. The horizontal trace was initiated by the trigger source as indicated in Fig. 18.7.

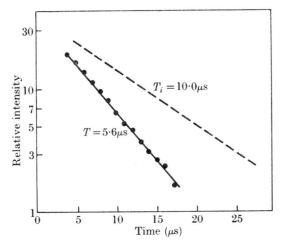

Fig. 18.8. Typical observations of the decay of the intensity of radiation escaping from a resonance tube. –●– observed with N_2 present; – – – observed in pure mercury at the same vapour density.

Great care was taken in the preparation of the resonance tube to avoid contamination by foreign gases. All measurements were carried out at the same nitrogen pressure, 0·74 torr. At much higher pressures, band fluorescence due to excitation of Hg_2 molecules became serious, while at much lower pressures the quenching effect was too small to measure accurately. Fig. 18.8 illustrates a typical set of results obtained at a particular mercury vapour density. The relative intensity of escaping radiation clearly falls exponentially with time and with a decay constant considerably greater than when no foreign gas is present.

Fig. 18.9 shows the quenching cross-section derived from such data as a function of temperature. It is consistent with the existence of an activation energy of 0·007 eV, the difference between the energy separation of the $6\,^3P_1$ and $6\,^3P_0$ levels (0·218 eV) of Hg and the energy of the first vibrational state of N_2 (0·225 eV), but it could hardly be said that the observed form of the variation with temperature positively requires that the quenching process involves the excitation of a vibrational quantum of N_2.

2.2. Use of flash photolysis and time-resolved spectroscopy

The technique of flash photolysis combined with time-resolved absorption spectroscopy can be used under certain conditions to provide a direct means of determining quenching rates. The procedure is essentially the same as that described in Chapter 17, § 2.4.2 for the study of vibrational reaction rates. Thus the excited atoms under investigation are produced from a suitable mixture of gases by means of the photoflash

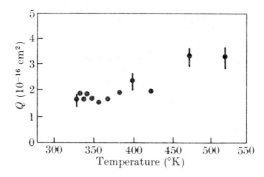

FIG. 18.9. Observed cross-section Q for quenching of the resonance radiation of mercury by N_2 as a function of the vapour temperature. ● observed values.

and the rate of decay of their concentration is followed by absorption spectroscopy. At suitable intervals after the flash the gas is irradiated by a light beam of suitable wavelength which is selectively absorbed by the excited atoms. The fractional absorption of this beam is measured by photographic photometry. If it can be assumed that, after the flash, the particular excited state is not populated by cascade transitions from higher states, or in any other way, then we have, if n^* is the concentration of the excited atoms at time t after the flash,

$$\frac{dn^*}{dt} = -\left(Z + \frac{1}{\tau_r}\right)n^*, \qquad (18)$$

where Z and τ_r have the same significance as before. Hence

$$-\frac{d\ln n^*}{dt} = Z + \frac{1}{\tau_r}. \qquad (19)$$

If the medium is optically thin so the fraction ΔI of light absorbed is proportional to n^* we have

$$-\frac{d\ln(\Delta I)}{dt} = Z + 1/\tau_r \qquad (20)$$

so that, from observation of ΔI as a function of time, $Z + 1/\tau_r$ can be

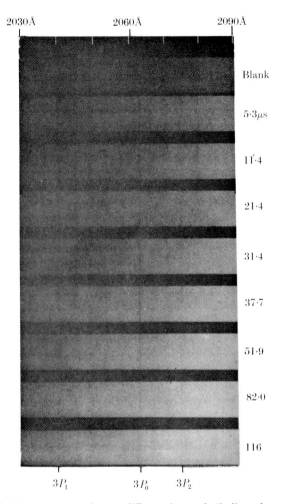

Fig. 18.10. Spectrograms taken at different intervals (indicated on the right) after flash photolysis of CSe_2, illustrating the formation and decay of $Se(4\ ^3P_0)$.

obtained. Since Z is proportional to the pressure of the quenching gas and τ_r is a constant, measurements at different quenching gas pressures make it possible to separate Z from $1/\tau_r$.

The general arrangement of the apparatus for experiments of this kind has been illustrated in Fig. 17.21. The photoflash is produced by discharging two low-inductance capacitors of 30–35 μF at 9–10 kV through the flash lamp. This produces a flash of energy 2000–3000 J which decays to half value in about 50 μs. The light pulse from the capillary spectroscopic flash lamp has a half-life of about 10 μs. The reaction vessel is of quartz, typically of 4 cm diameter and 75 cm length, mounted parallel to the photoflash lamp of 1 cm diameter and 75 cm length. Both tubes are enclosed by cylindrical reflectors of asbestos, coated with magnesium oxide or of aluminium.

We shall now describe the application of this method to the deactivation of $Se(4\,^3P_0)$. This is the highest fine-structure level associated with the ground $4\,^3P$ term so the process is one of spin-orbit relaxation either to the $4\,^3P_1$ or the ground $4\,^3P_2$ state. Callear and Tyerman† found that, if CSe_2 is flashed under the conditions described above, about 20 per cent is dissociated by the flash to produce $Se(4\,^3P_0)$ and the remainder excited to metastable states by radiation of wavelength close to 2300 Å. They‡ therefore applied absorption spectroscopy to follow the relaxation of the $Se(4\,^3P_0)$ atoms. A mixture of CSe_2 at a pressure of 0·05 torr was diluted with argon at a considerably higher pressure ($\simeq$ 50 torr) to maintain the temperature constant during a flash. Absorption spectra, with different quenching substances added, were photographed with a Hilger small quartz spectrograph. Decay of $Se(4\,^3P_0)$ was followed from the absorption of the line at 2063 Å due to the $4\,^3P_0$–$5\,^3S_1$ transition. Four exposures at each delay time were sufficient to produce measurable blackening on the Ilford Q3 photographic plate, with a slit opening of 0·05 mm.

Fig. 18.10 shows the formation and decay of $Se(4\,^3P_0)$. After about 20 μs there is little further production of atoms and relaxation is effectively complete after 100 μs. During this period removal of atoms was shown to be negligible. This was established by photometry of the 1960 Å resonance line ($4\,^3P_2$–$5\,^3S_1$) and observation of the rate of production of Se_2.

The atomic absorption at 2063 Å was found to be proportional to the square root of the optical path length and hence must vary in the

---

† CALLEAR, A. B. and TYERMAN, W. J. R., *Trans. Faraday Soc.* **61** (1965) 2395.
‡ ibid. **62** (1966) 2313.

same way with atomic concentration. Using this result, the variation of $\ln n^*$ with the time could be derived for different concentrations of quenching agents. Fig. 18.11 shows observed results for mixtures containing 25 torr pressure of Ar with added nitrogen. There is evidence that the variation of $\ln n^*$ with time is linear as expected. From slopes

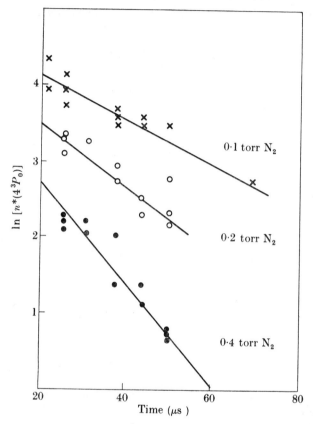

FIG. 18.11. Observed variation of $\log n^*(4\ ^3P_0)$ with time after flash photolysis of CSe_2 in Ar at 25 torr, with different partial pressures of added N_2.

of linear plots of this kind the variation of the relaxation rate with the concentration of a particular added gas could be obtained. Thus Fig. 18.12 shows the variation with argon pressure in a mixture of 0·04 torr CSe_2 with argon, Fig. 18.13 with the pressure of a gas added to a mixture of 0·04 torr CSe_2 with 25 torr of argon. The fact that the straight line in Fig. 18.12 passes very nearly through the origin checks that, in the case concerned, almost all the quenching is due to argon.

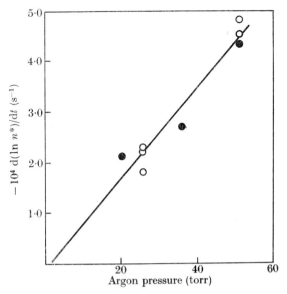

FIG. 18.12. Observed rate for deactivation of Se($4\,^3P_0$) in argon as a function of argon pressure. ○ 'cylinder' argon. ● spectroscopically pure argon.

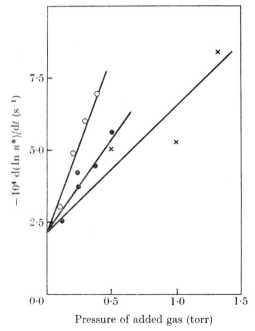

FIG. 18.13. Observed rate of deactivation of Se($4\,^3P_0$) as a function of the partial pressure of added gas in a mixture containing a partial pressure of argon of 25 torr. ○ N_2; ● O_2; × CO.

Donovan and Husain† have applied the same technique for observing spin-orbit relaxation of $Br(4\,^2P_{\frac{1}{2}})$ and $I(5\,^2P_{\frac{1}{2}})$‡ by flash photolysis of CF_3Br and CF_3I respectively. Mean cross-sections for relaxation to the ground $4\,^2P_{\frac{3}{2}}$ and $5\,^2P_{\frac{3}{2}}$ states respectively were obtained for a number of molecules.

Results obtained by this technique are given in Table 18.9 where they may be compared with results obtained for similar spin-orbit relaxation processes.

Callear and Norrish, and later Callear and Williams, have used the technique to study the nature of the processes that lead to deactivation of $Hg(6\,^3P_1)$ atoms by different molecular species. An account of this work is given in § 2.5.

2.3. Deactivation of atoms excited by shock waves

In Chapter 17, § 2.2 an account is given of shock-wave methods for raising gases to temperatures of the order of several thousands of degrees C and for investigating the reactions that occur in the shock-heated gas. Emphasis was placed in that section on vibrational relaxation. In particular, in § 2.2.3 a method for observing vibrational relaxation of N_2 behind the shock front was described which depended on the assumption that there is a high probability of excitation of sodium atoms to the resonance ^2P state by impact with N_2 molecules excited to the first vibrational state by the shock wave. In such collisions the vibrational energy of the molecule is transferred to excitation energy of the atom. If this does occur with high probability on collision then the excitation temperature of sodium present in the shocked gas should be the same as the vibrational temperature of the N_2 at each section behind the shock. As described in Chapter 17, § 2.2.3 experiments carried out by Gaydon and his collaborators§ have confirmed this picture and, from measurements of sodium reversal temperatures, the variation of vibrational temperature of N_2 behind the shock front has been determined (see Fig. 17.11).

It is possible, from comparison of the reversal temperature when full equilibrium is reached behind the shock with theoretically expected temperatures (see Chap. 17, (36)), to determine quenching rates for the particular excited metal atom concerned, under the experimental condi-

† DONOVAN, R. J. and HUSAIN, D., *Trans. Faraday Soc.* **62** (1966) 2987.
‡ ibid. 2023.
§ GAYDON, A. G. and HURLE, I. R., *Eighth int. Symp. Combust.*, Pasadena, 1960, p. 309 (Williams Wilkins, New York, 1962); HURLE, I. R., *J. chem. Phys.* **41** (1964) 3911.

tions. We suppose a trace of a metal M is present in the shocked gas. If [M], [M*] are the concentrations of ground state and excited metal atoms respectively the excitation or reversal temperature T^* is given by

$$[M^*] = (g^*/g_0)[M]e^{-E/\kappa T^*}. \tag{21}$$

E is the excitation energy and g_0, g^* the statistical weights of the ground and excited states respectively. At any section of the shock tube behind the shock we have

$$\frac{\partial[M^*]}{\partial t} = -Z_q[M^*] + Z_e[M] - \tau_r^{-1}[M^*]. \tag{22}$$

Here Z_q is the rate of deactivation by quenching collisions, Z_e that of excitation by inverse collisions, and τ_r the radiative lifetime of M*. If there is negligible self-absorption τ_r is the natural lifetime, but otherwise is an effective lifetime allowing for imprisonment. By detailed balancing

$$Z_e/Z_q = (g^*/g_0)e^{-E/\kappa T}, \tag{23}$$

where T is the temperature of the quenching species. In equilibrium we therefore have

$$[M^*] = [M](g^*/g_0)e^{-E/\kappa T} Z_q/(Z_q + 1/\tau_r). \tag{24}$$

Comparison with (21) gives

$$(1 + 1/\tau_r Z_q) = \exp\left\{(E/\kappa)\left(\frac{1}{T^*} - \frac{1}{T}\right)\right\}. \tag{25}$$

In application to shock excitation T will be the calculated temperature in the final equilibrium state behind the shock front so that, from measurement of T^* and knowledge of τ_r, Z_q may be obtained.

A detailed application of this method for studying quenching has been made by Tsuchiya[†] to sodium (^2P) deactivated by N_2, CO, and Ar. He used a shock tube, of inner diameter 10 cm, made of steel sections with inner walls plated with chromium. The driver and low-pressure sections were respectively 160 cm and 480 cm long. Hydrogen was the driver gas and the bursting diaphragms were made of Lumirror (polyethylene terephthalate). The shock velocity was measured from the time interval between signals received at upstream platinum resistance gauges 60 cm apart. From check observations made with different pairs among four such gauges it was estimated that the shock velocity could be determined to 0·5 per cent, corresponding to an uncertainty of about 30 °K in the calculated equilibrium temperature T behind the shock. Spectroscopic observations were made through two windows 410 cm downstream from the diaphragm. Argon, nitrogen, and carbon dioxide, with nominal

[†] TSUCHIYA, S., *J. chem. Soc. Japan*, Pure Chem. section **37** (1963) 828.

purities of 99·99, 99·96, and 99·5 per cent, respectively, were used without further purification. Sodium was introduced by feeding the working gas into the shock tube through a glass tube containing sodium metal heated to between 250 and 350 °C.

The principles used in determining the sodium reversal temperatures are the same as those described in Chapter 17, § 2.2.3. The two-channel monochromator was of Littrow type which could resolve the D-line doublets. In the actual experiments the $^2P_{\frac{3}{2}}$–$^2S_{\frac{1}{2}}$ line was used. It is considered that the measured reversal temperature was correct to about 50 °C at 2500 °K.

Fig. 18.14 (a), (b), (c) illustrates some of the results obtained. Fig. 18.14 (a) shows the variation of reversal temperature behind the shock front in a mixture containing 1 per cent N_2 and 99 per cent Ar. It is seen to rise to a final equilibrium value after about 200 μs which is the same as the calculated final equilibrium temperature for the shock conditions concerned. This means that, under the experimental conditions, the quenching rate Z_q is so large that the sodium reversal temperature follows the nitrogen vibrational temperature. Hence in (25) $T^* = T$ which, when vibrational relaxation is complete, will be the calculated final equilibrium temperature T_1^{calc} behind the shock (see p. 1481). A similar situation is shown in Fig. 18.14 (b) for a mixture containing 0·5 per cent of CO and 99·5 per cent Ar. On the other hand, in pure argon the situation is quite different, as shown in Fig. 18.14 (c). Thus the reversal temperature rises very rapidly but to an equilibrium value nearly 600 °C below the final calculated equilibrium temperature of the main gas. In terms of (25) this means that the quenching rate Z_q is very much smaller than under the conditions of Fig. 18.14 (a) while there is no gradual rise because the quenching atoms are soon in translational equilibrium.

Data on cross-sections for quenching by N_2 and CO were obtained using very much more dilute mixtures of N_2 (0·05 per cent) and of CO (0·03 per cent) with argon. Table 18.1 summarizes a number of observations and the quenching rates and cross-sections derived from them. It will be seen that the results from different runs are reasonably consistent and that, for both mixtures, the final equilibrium reversal temperature is about 300 °C lower than the calculated final equilibrium gas temperature (T_1^{calc}). The quenching cross-sections obtained will be discussed in relation to results from other types of experiments in § 2.7.

Considerable interest attaches to the observations in pure argon, a sample of which is given in Table 18.2. Two striking differences are seen from the data in Table 18.1. Not only is the quenching cross-section obtained nearly four orders of magnitude smaller but the results are much less consistent than for the molecular gases. It seems likely that they refer not to the true quenching cross-sections of argon but to a combined effective cross-section due to various impurities—the argon used was

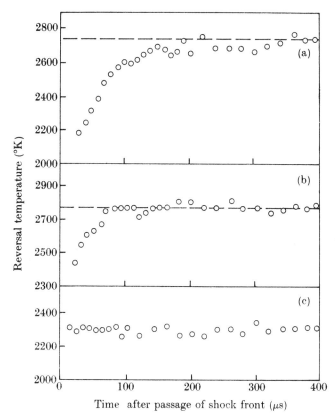

FIG. 18.14. Observed reversal temperatures after passage of a shock wave. (a) In a mixture of 1 per cent N_2 with 99 per cent Ar. (b) In a mixture of 0·5 per cent CO with 99·5 per cent Ar. (c) In pure Ar.

by no means pure to as much as one part in 10^4. Furthermore, as O_2 and H_2O, which would be among these impurities, have much shorter vibrational relaxation times than N_2 (see Chap. 17, Table 17.1) the rapid rise to the equilibrium reversal temperature shown in Fig. 18.14 (c) could still be understood. The results shown in Table 18.2 are of interest, nevertheless, in showing how small the quenching cross-section for argon collisions must be. Further discussion of this point in relation to observations by other techniques will be given in § 2.7.

TABLE 18.1

Cross-sections for deactivation of Na(^2P) by collisions with N_2 and CO, derived from shock-wave data

T_1 (°K)	N_2 (0·05% N_2–99·95% Ar) 303 °K					CO (0·03% CO–99·97% Ar) 302 °K				
v_s (10^5 cm/s)	1·657	1·649	1·700	1·690	1·675	1·648	1·657	1·675	1·680	1·680
p_1^{calc} (torr)	839	834	886	943	894	881	865	880	879	876
T_2^{calc} (°K)	2737	2713	2868	2837	2792	2709	2736	2791	2806	2806
T_r^s (°K)	2430	2395	2330	2440	2590	2405	2530	2450	2325	2270
$1/(1+1/\tau_r Z_q)$	0·32	0·31	0·14	0·25	0·51	0·32	0·48	0·29	0·17	0·13
Z_q (10^6/s)	1·40	1·46	1·43	2·35	1·39	1·43	1·65	1·36	1·68	1·33
Quenching cross-section (10^{-16} cm^2)	44	47	44	66	41	66	78·5	66	78·5	63

T_1, T_2^{calc}, and p_2^{calc} are as in Chapter 17, § 2.2.1, T_r^s is the final equilibrium reversal temperature, and v_s the shock velocity.

TABLE 18.2

Cross-sections for deactivation of Na(^2P) in 'pure' argon, derived from shock-wave data

T_1 (°K)	282	282	282	296	297	297	297	297
v_s (10^5 cm/s)	1·724	1·749	1·734	1·929	1·874	1·618	1·562	1·522
p_1^{calc} (torr)	1017	852	773	556	242	1031	1053	1074
T_1^{calc} (°K)	2921	3000	2953	3609	3422	2616	2456	2342
T_r^s (°K)	2320	2250	2170	2390	2385	2085	2020	1925
$1/(1+1/\tau_r Z_q)$	0·11	0·064	0·050	0·032	0·045	0·093	0·12	0·10
Z_q (10^6/s)	3·1	1·3	1·2	1·1	1·8	2·2	1·6	3·4
Quenching cross-section (10^{-19} cm^2)	4·1	2·2	2·3	3·5	11·3	2·9	1·7	4·1

2.4. *Deactivation of excited atoms produced by optical dissociation*

Certain molecules such as NaI dissociate on absorption of a quantum with frequency lying in a certain range, to produce an excited and a normal atom, viz.

$$\text{NaI} + h\nu \rightarrow \text{Na}(3\,^2P) + \text{I}, \qquad \lambda < 2430 \text{ Å}. \tag{26}$$

This process may be used as a source of excited atoms for quenching experiments. It has the advantage that the concentration of sodium atoms is always very small. Absorption of the emitted radiation, the D lines of sodium, therefore occurs to a negligible extent in the vapour so that there is no risk of radiation imprisonment. Foreign gases may be used even though they react chemically with sodium. It is also possible in this type of experiment to obtain some information about the variation with relative velocity of the effective cross-section for the quenching process. This is because, by varying the frequency of the dissociating radiation, the velocity of the excited atoms produced may be varied.

This method has been applied to sodium produced from NaI[†] and NaBr[‡] and to thallium produced from TlI.[§] Some of the results obtained are given in Table 18.4. An interesting study of the variation of a quenching cross-section with relative velocity of the colliding systems has been carried out by Terenin and Prileschaweja[||] for the quenching of sodium radiation by I_2. By using successively as exciting sources the resonance radiation from Fe, Tl, Sb, Ni, Cd, Zn, Mg, and Al lamps they were able to measure the quenching cross-section for velocities of the sodium atom ranging from 0.7 to 2.8×10^5 cm s^{-1}. Their results are illustrated in Fig. 18.15.

Hanson,[††] in more recent experiments, studied the variation of the cross-section with relative velocity by using a hydrogen discharge lamp as source. Used in conjunction with a reflection grating monochromator, the continuous molecular spectrum (see Chap. 13, § 1.2.4) provided a source of exciting radiation which could be varied in wavelength from 2100 to 2500 Å. By observing the yield of sodium D lines as a function of wavelength the shape of the repulsive upper potential energy curve (see Chap. 12, p. 818) for NaI was determined approximately. Knowing this the velocity with which the Na(^2P) atoms were produced on

---

[†] HANSON, H. G., *J. chem. Phys.* **23** (1955) 1391.
[‡] WINANS, J. G., *Z. Phys.* **60** (1930) 631.
[§] PRILESCHAWEJA, N., *Acta phys.-chim. URSS* **2** (1935) 647.
[||] TERENIN, A. and PRILESCHAWEJA, N., *Z. phys. Chem.* **B13** (1931) 72.
[††] HANSON, H. G., *J. chem. Phys.* **23** (1955) 1391.

dissociation could be obtained as a function of the wavelength of the dissociating radiation (cf. Fig. 12.5). The mean relative velocity in collisions between these atoms and atoms or molecules of the quenching gas was then calculated from the formula (164) of Chapter 17.

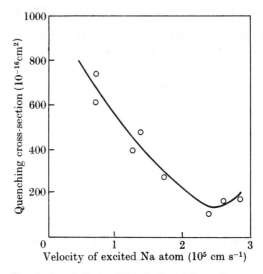

FIG. 18.15. Observed variation, with velocity of the sodium atoms, of the cross-section for quenching of excited Na atoms by I_2.

2.5. *The deactivation of* $Hg(6\,^3P_1)$ *and* $Hg(6\,^3P_0)$

In the early development of the study of quenching a great deal of attention was paid to mercury vapour as a working substance, mainly because of its experimental convenience but also partly because of its effectiveness in the photosensitization of chemical reactions.

Zemansky,[†] followed shortly after by Bates[‡] and by Evans,[§] observed the quenching of mercury resonance radiation by a variety of foreign gases. Many of their results are given in Table 18.4. It was assumed that the quenching process was of the form

$$Hg(6\,^3P_1)+M \to Hg(6\,^3P_0)+M, \tag{27}$$

and on this basis evidence was produced of a clear 'resonance' dependence of the quenching cross-section on the energy discrepancy ΔE involved in this reaction—the smaller $|\Delta E|$ the larger the cross-section.

[†] ZEMANSKY, M. W., *Phys. Rev.* **36** (1930) 919; MITCHELL, A. C. G. and ZEMANSKY, M. W., *Resonance radiation and excited atoms* (Cambridge University Press, 1934 and 1961).

[‡] BATES, J. R., *J. Am. chem. Soc.* **52** (1930) 3825; **54** (1932) 569.

[§] EVANS, M. G., *J. chem. Phys.* **2** (1934) 445.

Monatomic gases were found to be very inefficient quenchers so that in all significant cases ΔE was calculated on the assumption that the quenching molecule absorbs most of the energy difference, 0·218 eV, between the $6\,^3P_1$ and $6\,^3P_0$ levels as vibrational energy. The observations of Matland (see p. 1664 and Fig. 18.9) on quenching by N_2 are not inconsistent with this assumption but are not sufficiently definite to confirm it.

Unfortunately, later experiments have shown that, in many of the cases investigated, the quenching takes place to the ground state,

$$\text{Hg}(6\,^3P_1) + \text{M} \rightarrow \text{Hg}(6\,^1S) + \text{M}, \tag{28}$$

and no significance can be attached to the previously reported resonance behaviour. Considerable progress has, however, been made in determining the nature of the quenching process and in obtaining quenching cross-sections both for $\text{Hg}(6\,^3P_1)$ and the metastable $\text{Hg}(6\,^3P_0)$.

In 1961 Callear and Norrish† applied the technique of flash spectroscopy. The apparatus they used was essentially the same as that described in § 2.2 in connection with experiments on the flash photolysis of CSe_2. A mixture at 20 °C of mercury vapour with nitrogen was flashed with ultra-violet radiation including the mercury resonance line (2537 Å). The nitrogen pressure (almost atmospheric) was high enough to produce considerable pressure broadening of the line and under these circumstances a sufficient concentration of $\text{Hg}(6\,^3P_0)$ metastable atoms is produced to be detectable by absorption spectroscopy. Metastable atoms were also detected in flashed mixtures of mercury and argon with H_2O and with CO, but none were observed when the mercury and argon were mixed with H_2, C_2H_4, C_2H_6, NO, CO_2, O_2, or N_2O. This provided evidence that these molecules quench $\text{Hg}(6\,^3P_1)$ through the reaction (28) and not (27). Approximate cross-sections of these molecules for deactivation of $\text{Hg}(6\,^3P_0)$ were then determined by flashing a mercury–nitrogen mixture containing an admixture of the substance under study. After a delay interval of 100 μs the variation of $\text{Hg}(6\,^3P_0)$ concentration followed an exponential decay law and so, by measuring the decay constant as a function of the pressure of the deactivating substance, the cross-section could be derived in the usual way.

This work was extended by Callear and Williams.‡ They controlled the concentration of mercury vapour by continuously circulating gases through the reaction vessel and a trap containing mercury vapour, maintained at a fixed temperature in a water bath. Thorough precautions were taken to ensure a high degree of purity in the gases and vapours

† CALLEAR, A. B. and NORRISH, R. G. W., *Proc. R. Soc.* A**266** (1962) 299.
‡ CALLEAR, A. B. and WILLIAMS, G. J., *Trans. Faraday Soc.* **60** (1964) 2158.

used to produce deactivation of $Hg(6\,^3P_0)$. Thus the permanent gases, N_2, H_2, O_2, CO, and Ar were passed through a tube at $-196\,°C$ packed with glass wool so as to remove condensable impurities. The last traces of O_2 were removed from the N_2 by passage over copper turnings heated to $240\,°C$ and thence through a trap cooled to $-196\,°C$. Condensable vapours were frozen at $-196\,°C$, degassed, and fractionated.

Fig 18.16 (a) shows a typical set of time-resolved absorption spectra taken after flashing a mixture of mercury vapour with 760 torr Ar and 5 torr D_2O. Fig. 18.16 (b) shows a fluorescent spectrum of mercury in the presence of 780 torr of Ar in which the 2537-Å line is strongly present. Fig. 18.16 (c) shows a similar spectrum where the Ar is replaced by N_2. The fluorescence is quenched and there is no sign of the forbidden line at 2657 Å resulting from the transition $6\,^3P_0$–$6\,^1S$.

The concentration of $Hg(6\,^3P_0)$ was monitored from the absorption of the 2967 Å line ($6\,^3D_1 \leftarrow 6\,^3P_0$). Fig. 18.17 shows, on a logarithmic scale, the observed decay of the concentration with time for a number of mixtures of mercury and N_2 (780 torr) with various deactivating molecules. The linearity of the variation for delay times greater than 100 μs is apparent. From observations such as these the deactivation cross-sections were obtained for a number of substances. These are given in Table 18.4 and it will be seen that they are much smaller than the corresponding cross-sections for deactivation of $Hg(6\,^3P_1)$.

The deactivation of $Hg(6\,^3P_0)$ in mixtures with N_2 was also studied in some detail. This deactivation can be due both to collisions with ground state $Hg(6\,^3S)$ atoms and with N_2 molecules. Callear and Norrish, in their experiments, worked with a fixed concentration of mercury vapour. This was sufficient to mask any effect of N_2 even up to atmospheric pressure. Callear and Williams worked with lower concentrations of mercury vapour so that the contribution from N_2 became significant. The decay constant λ for the $Hg(6\,^3P_0)$ concentration can be written in the form

$$\lambda = k_{N_2} p_{N_2} + k_{Hg} p_{Hg}, \tag{29}$$

where k_{N_2}, k_{Hg} are rate constants for the respective deactivation by N_2 and by $Hg(6\,^1S)$ respectively and p_{N_2}, p_{Hg} the corresponding pressures. A plot of λ/p_{Hg} against p_{N_2}/p_{Hg} takes the form of the straight line

$$\lambda/p_{Hg} = k_{N_2}(p_{N_2}/p_{Hg}) + k_{Hg}, \tag{30}$$

from the slope of which k_{N_2} can be obtained and from the intercept at the origin k_{Hg}. Fig. 18.18 shows such a plot. From observations of this kind the deactivation cross-sections may be obtained. Reference to

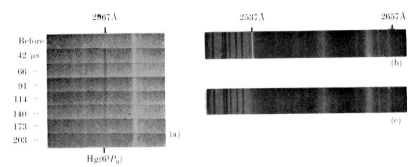

Fig. 18.16. Typical time-resolved spectra observed by Callear and Williams following flash excitation of a mixture of mercury vapour with other gases. (a) Spectra in a mixture with 5 torr $D_2O + 700$ torr Ar taken at various time intervals (indicated on the left) after the flash. The growth and decay of the $Hg(6\,^3P_0)$ line is clearly seen. (b) Spectrum showing the fluorescence of mercury in the presence of 780 torr of Ar. (c) Corresponding spectrum in the presence of 780 torr N_2. Note the absence of the 2537 Å resonance line due to quenching by N_2. The forbidden line at 2657 Å due to the $6\,^3P_0$–$6\,^1S$ transition is also absent.

18.2 UNDER GAS-KINETIC CONDITIONS

Table 18.4 shows two remarkable features of these results. The cross-section for $Hg(6\,^1S)$ is remarkably high and that for N_2 extraordinarily small.

The deactivation of $Hg(6\,^3P_1)$ and $Hg(6\,^3P_0)$ by N_2 and CO is of considerable interest because the excitation energy of both the $6\,^3P_1$ and

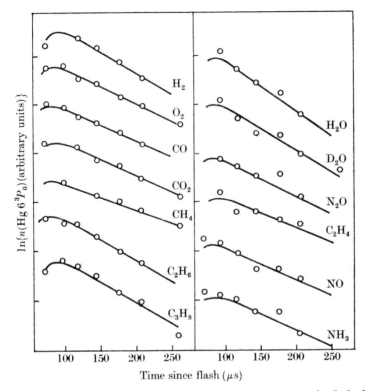

Fig. 18.17. Observed variation of $\ln n(Hg\ 6\,^3P_0)$ with time since the flash, for different gases added to a mixture of mercury vapour with a pressure of 780 torr of N_2. Partial pressures in torr in each case are as follows. H_2 0·1, O_2 0·04, CO 0·10, CO_2 1·7, H_2O 1·1, D_2O 1·1, N_2O 0·10, CH_4 1·1, C_2H_6 0·51, C_3H_8 0·50, C_2H_4 0·0014, NO 0·010, NH_3 1·0.

$6\,^3P_0$ states is insufficient to dissociate either of these molecules. Chemical effects are therefore very unlikely to be important. Already in 1932 Webb and Messenger† and Samson‡ investigated the behaviour of an excited mixture of mercury and nitrogen. Metastable mercury atoms were produced by irradiation with the 2537 Å line. After the excitation was cut off, some radiation from $6\,^3P_1$ mercury atoms persists because

---

† WEBB, H. W. and MESSENGER, H. A., *Phys. Rev.* **40** (1932) 466.
‡ SAMSON, E. W., ibid. 940.

some of the atoms return to the $6\,^3P_1$ state in superelastic collisions. The intensity of the radiation produced in this way was used as an indicator of the concentration of metastable atoms. To analyse the data, Samson assumed that both $6\,^3P_0$ and $6\,^3P_1$ atoms possess the same diffusion cross-section, allowed for the imprisonment of the 2537 Å

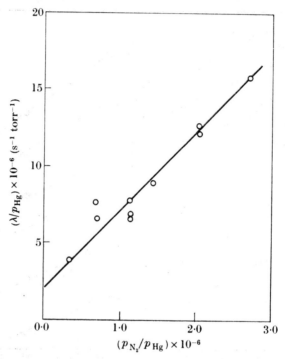

Fig. 18.18. Illustrating the deactivation of $Hg(6\,^3P_0)$ by N_2 and $Hg(6\,^1S_0)$, following (30). ○ experimental points (Callear and Williams).

radiation in the mercury vapour and took account, not only of collisions in which transitions between the $6\,^3P_0$ and $6\,^3P_1$ states occurred, but also of those in which the transition was from either to the $6\,^1S_0$ state. From the pressure variation of the decay constant for the metastable atoms separate values were found for all these cross-sections and are given in Table 18.4. The value for the quenching cross-section agrees quite well with that found by Zemansky (see Table 18.4).

A further variant of the same method was introduced by Webb.† He made use of the fact that, when a metastable atom strikes a metal plate, there is a high probability that the atom will be deactivated and an electron emitted.

† WEBB, H. W., *Phys. Rev.* **24** (1924) 113.

This technique for monitoring the concentration of metastable atoms was applied by Coulliette† to study the diffusion and destruction of metastable atoms in mercury vapour (see § 5.2.4.2). One difficulty in using this method of detection is that other highly excited species, as well as ultra-violet radiation, may also eject electrons from the metal surface. Furthermore, the nature of the surface as far as the secondary electron yield is concerned, may change with the pressure and nature of the gaseous mixture to which it is exposed.

Scheer and Fine‡ carried out some careful experiments on the quenching of $Hg(6\,^3P_1)$ by CO and N_2 in which the metal cathode used was not nickel, as in earlier experiments, but silver. They found that, while nickel was a suitable detector with N_2, it produced no ejected electron current with CO, due presumably to surface modifications by chemisorption. Silver was satisfactory with both gases. In a mixture of mercury and a foreign gas M, irradiated by 2537 Å resonance radiation, the following reactions can occur:

$$Hg(6\,^1S) + h\nu \underset{k_2}{\overset{k_1}{\rightleftharpoons}} Hg(6\,^3P_1), \qquad (31\,a)$$

$$Hg(6\,^3P_1) + M \underset{k_4}{\overset{k_3}{\rightleftharpoons}} Hg(6\,^3P_0) + M^*, \qquad (31\,b)$$

$$Hg(6\,^3P_1) + M \overset{k_6}{\longrightarrow} Hg(6\,^1S) + M', \qquad (31\,c)$$

$$Hg(6\,^3P_0) + M \overset{k_7}{\longrightarrow} Hg(6\,^1S) + M''. \qquad (31\,d)$$

In each case the appropriate rate constants associated with each process are given. Account must also be taken of loss of $Hg(6\,^3P_0)$ by diffusion to the walls of the reaction vessel.

In equilibrium we have, denoting the concentrations of $Hg(6\,^3P_0)$ and M by $[Hg_0]$, $[M]$ respectively, and writing $k_4' = k_4[M^*]/[M]$,

$$[Hg_0]/[M]^2 = I_a k_3 [k_2 k_5 + k_5(k_3 + k_6)[M] + k_2(k_4' + k_7)[M]^2 + \\ + \{k_3 k_7 + k_6(k_4' + k_7)\}[M]^3]^{-1}, \qquad (32)$$

where I_a is the number of photons absorbed/cm³ s, and the rate of loss of $Hg(6\,^3P_0)$ at the walls, being inversely proportional to $[M]$ if the foreign gas is much in excess, is written $k_5[Hg_0]/[M]$. Hence

$$\lim_{[M]\to 0} \frac{[Hg_0]}{[M]^2} = \frac{I_a k_3}{k_2 k_5}. \qquad (33)$$

---

† COULLIETTE, J. H., ibid. **32** (1928) 636.
‡ SCHEER, M. D. and FINE, J., *J. chem. Phys.* **36** (1962) 1264.

If a suitable metal surface is present in the reaction vessel the current i ejected from it will be given by

$$i = c_1[\mathrm{Hg_0}],$$

where c_1 is proportional to the rate of diffusion of metastable mercury atoms in the mixture of mercury vapour and the foreign quenching gas. If this gas is much in excess, the diffusion rate will be inversely proportional to [M] so

$$i = c_2[\mathrm{Hg_0}]/[\mathrm{M}]. \qquad (34)$$

Hence, taken in conjunction with (33),

$$\lim_{[\mathrm{M}]\to 0} \frac{i}{[\mathrm{M}]} = c_2 I_\mathrm{a} k_3/k_2 k_5. \qquad (35)$$

Both c_2 and k_5 (the rate coefficient for deactivation of $\mathrm{Hg}(6\,{}^3P_0)$ at the wall) will be proportional to the diffusion coefficient of $\mathrm{Hg}(6\,{}^3P_0)$ in the gas M. Hence, if it can be assumed that neither the chance of destruction of metastable atoms on impact with the walls nor the detection efficiency of the metal cathode vary with the nature of the quenching gas, $c_2 I_\mathrm{a}/k_2 k_5$ will also remain constant for all such gases. It follows that the initial slope of a plot of i against [M] is a relative measure of k_3, the rate constant for the quenching of $\mathrm{Hg}(6\,{}^3P_1)$ by M.

In applying this analysis in practice, Scheer and Fine used a reaction vessel of fused silica. The cathode of the electrode system was a thin silver ribbon 1 cm long and 0·5 cm wide surrounded by a cylindrical anode of nickel-wire mesh. The electrode system was shielded from the direct radiation of the resonance lamp and the connecting leads were covered with Pyrex glass. In this way the background current present in the illuminated empty vessel was reduced to about 0·1 of the operating current. The impurity content of both the gases used was checked by mass analysis and found to be less than one part in 10^3.

Fig. 18.19 shows such plots for $\mathrm{N_2}$ and CO and a silver surface. From results of this kind it is found that the quenching rates for these respective gases through the reactions (31 b) are in the ratio 1·0:1·35. On the other hand, the total quenching cross-section of $\mathrm{Hg}(6\,{}^3P_1)$ for CO is about ten times that for $\mathrm{N_2}$. This implies that, in contrast to $\mathrm{N_2}$, CO quenches largely to the ground state through the reaction (31 c).

If it is assumed that the quenching process for $\mathrm{N_2}$ is (see p. 1678 and Fig. 18.16)

$$\mathrm{Hg}(6\,{}^3P_1) + \mathrm{N_2}(v=0) + 0\cdot 07 \text{ eV} \to \mathrm{Hg}(6\,{}^3P_0) + \mathrm{N_2}(v=1), \qquad (36)$$

it follows that, at 298 °K, for which Matland observes a mean quenching cross-section of $1\cdot 3 \times 10^{-16}$ cm^2, a fraction 0·062 only of the N_2 molecules will have sufficient kinetic energy to quench in this way. The cross-section for quenching by these molecules will then be 21×10^{-16} cm^2.

For CO the mean cross-section at 298 °K for quenching to $Hg(6\,^3P_0)$, according to Scheer and Fine, is $1\cdot 7 \times 10^{-16}$ cm^2. The corresponding

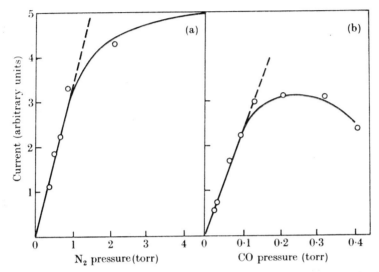

FIG. 18.19. Variation, with pressure of an added gas, of the current emitted from a silver cathode in mercury vapour at a pressure of 3×10^{-5} torr. (a) N_2. (b) CO.

activation energy is considerably smaller ($0\cdot 054_5$ eV) and about 0·16 of the molecules have sufficient kinetic energy to make the reaction possible. The cross-section for quenching by these molecules comes out to be $11\cdot 0 \times 10^{-16}$ cm^2, somewhat smaller than for N_2 despite the smaller energy discrepancy.

Karl and Polanyi† were the first to observe infra-red radiation emitted by CO molecules vibrationally excited through collisions with excited mercury atoms which had been produced by irradiating a mixture of CO and Hg vapour with 2537 Å mercury resonance radiation. Their preliminary experiments were extended by Karl, Kruus, and Polanyi‡ to obtain information about the cross-sections for excitation of different vibrational states by collisions with $Hg(6\,^3P_1)$ and $Hg(6\,^3P_0)$ atoms.

† KARL, G. and POLANYI, J. C., J. chem. Phys. **38** (1963) 271.
‡ KARL, G., KRUUS, P., and POLANYI, J. C., ibid. **46** (1967) 224.

The reaction vessel was a quartz tube of 5-cm internal diameter and 112-cm length which could be irradiated from the side by four U-shaped quartz mercury lamps. Carbon monoxide mixed with a large excess of argon or nitrogen was passed through a mercury saturator into the reaction vessel. The CO, initially Matheson CP grade (99·5 per cent pure), was further purified by passage through a 30-cm long quartz tube packed with copper turnings heated to 800 °C, then through 100 cm of tube packed with copper filings in an oven at 390 °C, and finally through three traps packed with glass wool and immersed in liquid nitrogen. The argon was 99·998 per cent (Matheson) and the N_2 99·996 per cent (Matheson prepurified) and were not further purified.

Infra-red radiation could be dispersed by a Perkin–Elmer 12G infra-red spectrometer, viewing along the quartz tube through a sodium chloride window at one end, and detected by a liquid-air cooled Infratron lead sulphide cell. Wavelength calibration was carried out by reference to mercury lines arising from fluorescence in scattering. The intensity was calibrated by replacing the reaction vessel and lamps by an oven in the form of a quartz tube, similar in shape to the reaction vessel, and held at a constant temperature of 1100 ± 10 °K along an 80-cm section. Records were taken of the CO first overtone bands when the oven was filled with CO at a pressure of 615 torr, low enough for self-absorption to be unimportant as shown by proportionality of intensity to CO pressure. A check was made at another time using HCl at 0·05 torr as emitting gas and agreement found to within 20 per cent.

The concentrations of $Hg(6\,^3P_1)$ and $Hg(6\,^3P_0)$ were measured from the absorption of 4358 and 4047 Å radiation respectively which were supplied from a secondary mercury lamp at the opposite end of the reaction tube to the infra-red detector.

Fig. 18.20 illustrates the infra-red emission observed in two typical cases, one in argon and one in nitrogen as background gas. In both cases the flow rate of the mixed gases through the reaction tube was such that a molecule was exposed to the excited mixture for about 0·18 s on the average. The band origins of the first overtone are indicated in Fig. 18.20 and it will be seen that the emission involves transitions in which the quantum number v changes by 2 for v extending up to 8 or so, but falls off rapidly for larger v. It will be seen from the energy level diagram shown in Fig. 18.21 that, if vibrational excitation were to take up almost all the available electronic excitation energy, the most likely value of v would be as high as 20. This does not seem to occur in practice. In the first instance it is important to verify that the observed infra-red

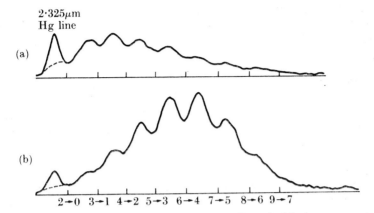

FIG. 18.20. Infra-red emission from CO vibrationally excited in mercury vapour by resonance radiation. (a) $2 \cdot 2 \times 10^{-3}$ torr CO in $5 \cdot 1$ torr Ar. (b) $1 \cdot 1 \times 10^{-2}$ torr CO in $4 \cdot 5$ torr N_2. The numbers indicate the band origins for different vibrational transitions.

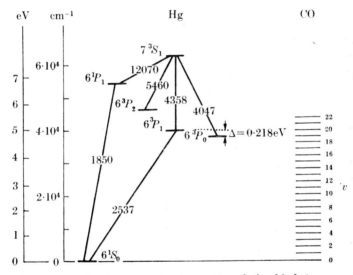

FIG. 18.21. Energy level diagram showing energy relationship between excited electronic states of Hg and vibrational levels of CO.

emission does arise from collisions of normal CO with $Hg(6\,^3P_1)$ and/or $Hg(6\,^3P_0)$ atoms.

The emission increases with Hg vapour pressure and is absent when no mercury is present. When the light intensity exciting the mercury is increased by increasing the number of lamps the intensity originating in the 6–9 vibrational levels increases proportionally. This seems to exclude the possibility that any Hg excited states, excited by step-wise absorption of photons, can be important. This leaves only $\text{Hg}(6\,^1P_1)$, $\text{Hg}(6\,^3P_1)$, and $\text{Hg}(6\,^3P_0)$. Strong evidence that $\text{Hg}(6\,^1P_1)$ is not effective is provided by the fact that the radiation required to excite it (1850 Å) is strongly absorbed by the quartz in the lamps and the walls of the reaction vessel. Furthermore, addition of 5% NaCl solution to the cooling water in the jacket surrounding the vessel, which would certainly cut out all the 1800 Å radiation, had no effect on the infra-red emission. It seems that only $\text{Hg}(6\,^3P_1)$ and $\text{Hg}(6\,^3P_0)$ can be involved.

As a further check the kinetics of the processes involved can be studied in some detail. We denote the concentrations of different constituents M, say, by [M] and, for brevity, write Hg_1^* for $\text{Hg}\,6\,^3P_1$, Hg_0^* for $\text{Hg}\,6\,^3P_0$, and Hg for ground state $\text{Hg}\,6\,^1S_0$. We then have to consider the following reactions.

The only important process leading to production of Hg_1 is absorption of radiation so
$$[\text{Hg}_1^*] \propto I[\text{Hg}]$$
where I is the light intensity, and is independent of the partial pressures of Ar, N_2, or CO.

The reactions that determine $[\text{Hg}_0^*]$ are as follows.

$\text{Hg}_1^* + \text{CO} \to \text{Hg}_0^* + \text{CO}^\dagger$,		k_1,	(37 a)
$\text{Hg}_1^* + N_2 \to \text{Hg}_0^* + N_2$,		k_2,	(37 b)
$\text{Hg}_1^* + M \to \text{Hg}_0^* + M$	(M refers to Ar, Hg, and the tube walls)	k_2',	(37 c)
$\text{Hg}_0^* \to \text{Hg} + h\nu$,		k_3,	(37 d)
$\text{Hg}_0^* + \text{CO} \to \text{Hg} + \text{CO}^\dagger$,		k_4,	(37 e)
$\text{Hg}_0^* + \text{Hg} \to \text{Hg} + \text{Hg}$,		k_5,	(37 f)
$\text{Hg}_0^* + M \to \text{Hg} + M$	(M refers to Ar, N_2 and the walls)	k_5'.	(37 g)

Here $\text{CO}^\dagger$ denotes vibrationally excited CO, k_1, k_2, k_2', etc. are the associated rate constants for the various reactions. In equilibrium

$$[\text{Hg}_0^*] = [\text{Hg}_1^*] \frac{k_1[\text{CO}] + k_2[N_2] + k_2'[M]}{k_3 + k_4[\text{CO}] + k_5[\text{Hg}] + k_5'[M]}. \qquad (38)$$

The rate of (37f) is so large (see Table 18.4) that $k_5'[M]$ can be neglected in the denominator of (38). In the presence of a large excess of N_2

$$[\text{Hg}_0^*] = k_2[\text{Hg}_1^*][N_2]/(k_3 + k_4[\text{CO}] + k_5[\text{Hg}]) \qquad (39)$$

so that, for [Hg], $[N_2]$, and the exciting light intensity I kept constant, a plot of $1/[\text{Hg}_0^*]$ against [CO] should be linear. As a first check we note that, if we can attribute the observed increase of $[\text{CO}^\dagger]$ when N_2 is substituted for Ar as due to the presence of a significant concentration of $[\text{Hg}_0^*]$ in the former case and a negligible amount in the latter, then

$$[\text{Hg}_0^*] \propto \{[\text{CO}^\dagger]_{N_2} - [\text{CO}^\dagger]_{\text{Ar}}\}/[\text{CO}], \qquad (40)$$

where $[\text{CO}^\dagger]_{N_2}$, $[\text{CO}^\dagger]_{\text{Ar}}$ denote observed concentrations of $\text{CO}^\dagger$ when the added gas is N_2 and Ar respectively. The factor [CO] in the denominator allows for the effect of varying [CO] as $[\text{CO}^\dagger]$ is approximately proportional to [CO] at low CO pressures. It is assumed that the difference in the rate of loss of $\text{CO}^\dagger$ in Ar and in

N_2 is unimportant. Fig. 18.22 shows a plot of $1/[Hg_0^*]$ against [CO] in which [Hg], $[N_2]$, and I are constant and the variation of $[Hg_0^*]$ is determined from (40). That the plot is linear is encouraging, although the assumptions made, particularly about the loss rate of $CO^\dagger$ in the different environments, have been somewhat drastic.

A second check was made as follows. In this $[Hg_0^*]$ was determined by absorption observations using the secondary lamp. When only N_2 is present in addition to the mercury we have, in a steady state,

$$k_2[N_2][Hg_1^*] - k_3[Hg_0^*]_{N_2} - k_5[Hg_0^*]_{N_2}[Hg] = 0, \qquad (41)$$

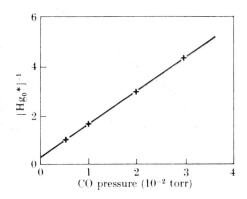

FIG. 18.22. Plot showing 'observed' variation of $1/[Hg^*(6\,^3P_0)]$ with CO pressure.

where $[Hg_0^*]_{N_2}$ denotes $[Hg_0^*]$ in the presence of N_2 only. Using $[Hg_0^*]_{CO}$ for $[Hg_0^*]$ in the presence of N_2 and CO we have, when CO is also present,

$$k_2[N_2][Hg_1^*] - k_3[Hg_0^*]_{CO} - k_5[Hg_0^*]_{CO}[Hg] - k_4[Hg_0^*]_{CO}[CO] = 0, \qquad (42)$$

so that $\qquad \{[Hg_0^*]_{N_2} - [Hg_0^*]_{CO}\}/[Hg_0^*]_{CO} = k_4[CO]/\{k_3 + k_5[Hg]\}. \qquad (43)$

Fig. 18.23 shows a plot of the left-hand side, as observed by the absorption method, against [CO]. The result is a straight line passing through the origin as it should be according to (43). From the slope, and using data for k_3 and k_5 from other experiments, k_4 is found to lie between 7.7×10^{11} and 4.8×10^{12} mol^{-1} cm^3 s^{-1}, corresponding to a mean cross-section for quenching of $Hg(6\,^3P_0)$ by vibrational excitation of CO of between 0.25 and 1.6×10^{-16} cm^2. A further estimate was obtained as described below.

If it is accepted that (39) is valid and that vibrational excitation of CO occurs only through deactivating collisions with $Hg(6\,^3P_1)$ and $Hg(6\,^3P_0)$ then, if $Q(^3P_1)$, $Q(^3P_0)$ are the corresponding cross-sections,

$$Q(^3P_1)[Hg_1^*]_{N_2} + Q(^3P_0)[Hg_0^*]_{N_2} = B[CO^\dagger]_{N_2}, \qquad (44\,a)$$

$$Q(^3P_1)[Hg_1^*]_{Ar} + Q(^3P_0)[Hg_0^*]_{Ar} = B[CO^\dagger]_{Ar}, \qquad (44\,b)$$

where B is a constant, provided the measurements are made at the same partial pressure of CO. If it is assumed that the cross-section ratio $Q(^3P_1)/Q(^3P_0)$ is the same for all vibrational states this ratio may be obtained by measurements at

a fixed infra-red frequency corresponding to one such excitation. Such observations were made at 2·51 μm which arises from the $v = 8 \to 6$ transition in CO†
and gave
$$Q(^3P_1)/Q(^3P_0) = 14\cdot 5.$$
Allowing for the uncertainties due to different rates of loss of CO† it is estimated that the ratio of the cross-sections is between 7 and 17.

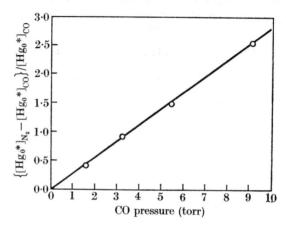

FIG. 18.23. Plot showing variation of $\{[Hg_0^*]_{N_2} - [Hg_0^*]_{CO}\}/[Hg_0^*]_{CO}$ with CO pressure.

To obtain the relative rates of excitation of different vibrational levels, spectrograms such as shown in Fig. 18.20 were first analysed to determine the distribution of initial excited vibrational levels. To do this, it was necessary to allow for the rotational structure corresponding to a rotational temperature T_R. A trial and error procedure was adopted, choosing a trial distribution N_v and rotational temperature T_R and calculating, using electronic computation, the spectral distribution from known values of the radiative transition probabilities. The trial distribution was amended after comparison of the observed and calculated spectra and the procedure repeated until good agreement was obtained.

Having obtained N_v the determination of the excitation rates for different v required consideration of the loss processes. These include, not only radiative transitions between vibrational levels but also vibrational transfer, particularly to nitrogen in the presence of that gas, as well as flow out of the reacting zone. Under conditions in which these latter losses were probably small (low pressure of N_2 and exposure times to the exciting radiation long compared with the radiative lifetime) the coupled equations may be set up for N_v, namely,

$$\frac{dN_v}{dt} = k_v[CO][Hg^*] + A_{v+2,v} N_{v+2} + A_{v+1,v} N_{v+1} - (A_{v,v-2} + A_{v,v-1})N_v. \quad (45)$$

Here $A_{v,v'}$ is the radiative transition probability between states with vibrational quantum numbers v, v' respectively. By integrating these equations along the reaction tube the relative values of N_v could be obtained. As the best fit to their observations Karl, Kruus, and Polanyi found relative values of k_v given in Table 18.3.

TABLE 18.3

Relative values of the rate coefficient k_v for excitation of different vibrational states of CO by impact with $Hg(6\,^3P_1)$ and $Hg(6\,^3P_0)$ atoms

Vibrational quantum number	v	9	8	7	6	5	4	3	2
Relative rate coefficient	k_v	1	15	35	43	48	60	70	80

Finally, the absolute value of the cross-section for vibrational excitation was obtained approximately as follows. Working with argon as admixed gas, under conditions in which loss by radiation is the only important process of vibrational deactivation, we have, say, for excitation to $v = 7$,

$$KQ_{v=7}(6\,^3P_1)[Hg_1^*][CO] + KQ_{v=7}(6\,^3P_0)[Hg_0^*][CO] + \\ + A_{8,7}[CO_{v=8}^\dagger] + A_{9,7}[CO_{v=9}^\dagger] = (A_{7,6}+A_{7,5})[CO_{v=7}^\dagger], \quad (46)$$

where KQ = the rate constant k. From measurements of $[Hg_1^*]$, $[Hg_0^*]$ by the absorption method, of [CO] from the absolute intensity of radiation and use of the corresponding distribution N_v and the ratio

$$Q(6\,^3P_1)/Q(6\,^3P_0) = 14\cdot 5,$$

a cross-section of $0\cdot 31 \times 10^{-16}$ cm² was obtained for $Q(6\,^3P_1)$. From the N_v it is then found that

$$\sum_{v\geqslant 2}^{9} Q_v(6\,^3P_1) = 3\cdot 1 \times 10^{-16} \text{ cm}^2.$$

Adding an extrapolated value for $v = 1$ gives finally

$$Q_{\text{all }v}(6\,^3P_1) = 4\cdot 1 \times 10^{-16} \text{ cm}^2,$$

which is to be compared with the cross-section for quenching of $Hg(6\,^3P_1)$ to $Hg(6\,^3S_0)$ by CO of 11×10^{-16} cm².‡

Corresponding to this

$$Q_{\text{all }v}(6\,^3P_0) = 0\cdot 28 \times 10^{-16} \text{ cm}^2,$$

compared with $0\cdot 094 \times 10^{-16}$ cm² observed by Callear and Williams.‡

‡ See Table 18.4.

The uncertainties are so large that it cannot yet be regarded as established that all deactivation of $Hg(6\,^3P_1)$ and $Hg(6\,^3P_0)$ by collisions with CO occurs through vibrational excitation but it seems very likely that this is so.

Karl, Kruus, Polanyi, and Smith‡ carried out very similar experiments with NO. Fig. 18.24 illustrates a typical spectrogram of the

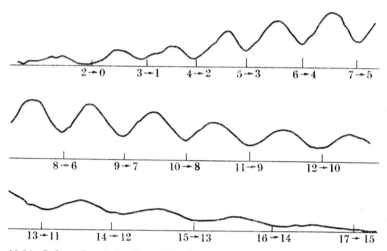

FIG. 18.24. Infra-red emission from NO vibrationally excited, in mercury vapour containing 1.8×10^{-3} torr NO and 1·8 torr N_2, by resonance radiation. The numbers indicate the band origins for different vibrational transitions.

observed infra-red emission. From the indicated band origins it will be seen that there is evidence of excitation up to levels with v as high as 17. However, the probability of excitation is already small for these highly excited levels which require only two-thirds of the total available electronic energy for their excitation. The situation seems to be similar to CO in that the most probable process is not the one with minimum transfer to translation. Similar checks to those carried out for CO indicated that the emission arose from excitation by impact with $Hg(6\,^3P_1)$ and $Hg(6\,^3P_0)$. The total cross-section for vibrational excitation could only be placed within wide limits:

$$Q_{\text{all } v}\,(6\,^3P_1) \text{ between 3 and } 46 \times 10^{-16} \text{ cm}^2,$$

$$Q_{\text{all } v}\,(6\,^3P_0) \text{ between 0·15 and } 1·1 \times 10^{-16} \text{ cm}^2.$$

The interpretation of the results is complicated by the existence of the

‡ KARL, G., KRUUS, P., POLANYI, J. C., and SMITH, I. W. M., *J. chem. Phys.* **46** (1967) 244.

excited $^4\Pi$ electronic state of NO below the $6\,^3P_1$ level of Hg. Thus the vibrational excitation could proceed by two stages:

$$\mathrm{Hg}^* + \mathrm{NO}(X\,^2\Pi) \to \mathrm{Hg} + \mathrm{NO}^*(^4\Pi),$$
$$\mathrm{NO}(^4\Pi) + \mathrm{NO}(X\,^2\Pi) \to \mathrm{NO}(X\,^2\Pi) + \mathrm{NO}^\dagger(X\,^2\Pi).$$

Finally, we refer to the experiments of Hudson and Curnutte‡ on the deactivation of $\mathrm{Hg}(6\,^3P_1)$ by thallium. These experiments, carried out with apparatus described in § 4.2.2.3, consisted in observing the quenching of mercury resonance radiation by thallium in a very similar way to the experiments on quenching of sodium resonance radiation described in § 2.1. A large mean quenching cross-section $\simeq 10^{-14}$ cm² was obtained, in contrast to the quenching by rare gases which is too small to be observed with certainty.

The present position about the deactivation of $\mathrm{Hg}(6\,^3P_1)$ and $\mathrm{Hg}(6\,^3P_0)$ is still very confusing. The rare gases are very ineffective indeed but, in contrast, ground-state mercury atoms are very effective in deactivating $\mathrm{Hg}(6\,^3P_0)$, possibly due to a chemical reaction. Many molecules quench $\mathrm{Hg}(6\,^3P_1)$ quite effectively but, whereas some do this by producing transitions to the metastable state, others quench directly to the ground state. There is evidence that the quenching process involves transfer of some or all of the electronic energy to vibrational energy, but there is no clear dependence of the cross-section on the energy discrepancy. Finally, there is the remarkable fact that the cross-sections for deactivation of $\mathrm{Hg}(6\,^3P_0)$ are all much smaller than for $\mathrm{Hg}(6\,^3P_1)$. Despite the striking difference in behaviour of these two excited states in this regard, no explanation has yet been forthcoming.

2.6. *Deactivation of* $\mathrm{Hg}(7\,^3S_1)$ *and* $\mathrm{Hg}(6\,^3D_1)$

Cojan§ has measured cross-sections for deactivation of $\mathrm{Hg}(7\,^3S_1)$ and $\mathrm{Hg}(6\,^3D_1)$ by impact with N_2 molecules. A mercury resonance cell C containing nitrogen was illuminated by a beam of radiation from a mercury arc A. The fluorescent radiation was analysed in a direction along the axis of C by a monochromator and photomultiplier. A mercury vapour lamp S was placed on the axis of C so that its radiation, after passing through C, could also be analysed by the same monochromator.

When the partial pressure of N_2 varies between 1 and 10 torr the absorption of the cell C for the green line $\lambda\,5460$ Å $(6\,^3P_2\text{--}7\,^3S_1)$ emitted by S is negligible while that of the blue line $\lambda\,4358$ Å $(6\,^3P_1\text{--}7\,^3S_1)$ is weak

‡ HUDSON, B. C. and CURNUTTE, B., *Phys. Rev.* **148** (1966) 60.
§ COJAN, J. L. and HUET, M., *C.r. hebd. Séanc. Acad. Sci., Paris* **263** (1966) 1223.

and remains constant throughout the pressure range. On the other hand, the violet line λ 4041 Å ($6\,^3P_0$–$7\,^3S_1$) is nearly completely absorbed to an extent which does not depend on the N_2 pressure. Constancy of the concentration of Hg($6\,^3P_0$) under the experimental conditions in the N_2 pressure-range concerned has been confirmed in earlier experiments by Bigeon and Cojan.‡ Their results suggest that, in the resonance cell under resonance conditions, the concentrations of $6\,^3P_2$ and $6\,^3P_1$ excited Hg atoms is negligibly small and that of $6\,^3P_0$ is relatively large and remains constant under the experimental conditions.

The intensity of the green line λ 5460 ($6\,^3P_2$–$7\,^3S_1$) re-emitted from the cell must be proportional to the concentration of Hg($7\,^3S_1$) atoms since no imprisonment can take place, the concentration of Hg($6\,^3P_2$) being negligible. Again, as the population of Hg($6\,^3P_0$) remains constant so also will the rate of population of the $7\,^3S_1$ levels, which can only occur through absorption of the violet line λ 4047 Å ($6\,^3P_0$–$7\,^3S_1$) emitted from the arc A. Under these conditions observation of the intensity I of the green line λ 5460, as a function of N_2 pressure, should follow the quenching law (see p. 1653)

$$I = \tau_r R/(1+\tau_r Z),$$

where Z is the quenching rate and τ_r the transition time. The usual plot giving $1/I$ as a function of N_2 pressure, to which Z is proportional, should therefore be linear and, from its slope, the mean quenching cross-section $\bar{Q}(7\,^3S_1)$ for quenching of Hg($7\,^3S_1$) by N_2 impact can be obtained. In this way Cojan found

$$\bar{Q}(7\,^3S_1) = 38 \times 10^{-16} \text{ cm}^2.$$

Similar experiments, based on the intensity of the line λ 3131·5 Å ($6\,^3P_1$–$6\,^3D_1$), gave $\bar{Q}(6\,^3D_1) = 472 \times 10^{-16}$ cm^2.

2.7. *Preliminary discussion of results on quenching cross-sections*

In Table 18.4 we summarize the results of measurements of cross-sections for deactivation of the excited states of various atoms made by different authors. For all but Hg($6\,^3P_1$) and Cd($5\,^3P_1$) the cross-sections given are necessarily for quenching to the ground state. In order to include only comparable transitions the results given for Hg($6\,^3P_1$) are for transitions known to be, or at least likely to be, to the ground state. Cross-sections for the spin-orbit transition Hg($6\,^3P_1$–$6\,^3P_0$) are given in Table 18.9. Unfortunately, for Cd($5\,^3P_1$) there is no knowledge of the nature of the deactivating transition so we have included all available results in Table 18.4

‡ BIGEON, M. C. and COJAN, J. L., *C.r. hebd. Séanc. Acad. Sci., Paris* **261** (1965) 353.

TABLE 18.4

Observed cross-sections (in 10^{-16} cm^2) for deactivation (to the ground state) of $Hg(6\,^3P_0)$, $Hg(6\,^3P_1)$, $Cd(5\,^3P_1)$, $Li(2\,^2P)$, $Na(3\,^2P)$, $K(4\,^2P)$, $Rb(5\,^2P)$, $Cs(6\,^2P)$, and $Tl(7\,^2S)$ by various atoms and molecules

Deactivating atom or molecule	$Hg(6\,^3P_0)$	$Hg(6\,^3P_1)$		$Cd(5\,^3P_1)$	$Li(2\,^2P)$	$Na(3\,^2P)$						$K(4\,^2P)$		$Rb(5\,^2P)$	$Cs(6\,^2P)$	$Tl(7\,^2S)$
		(a)	(b)			(a)	(b)	(c)	(d)	(e)	(f)	(a)	(b)			
He	—	—	0	—	—	—	—	0.3	—	—	—	0.2	—	0.3	1.2	0.4
Ar	—	—	0	—	—	—	—	0.3	—	—	—	0.6	0.94	0.9	3.7	0.3
Hg	23.9	—	—	—	—	—	—	—	—	—	—	—	—	—	—	—
Tl	—	~100⁵	—	—	—	—	—	—	—	—	—	—	—	—	—	—
H$_2$	0.056	18.8¹	26	2.1	1.0	23	0.34	9.0	15	9.0⁸	10⁻²	3.1	3.5	1.9	5.3	0.09
D$_2$	—	—	—	0.6	—	—	—	—	—	—	—	17.6	18.5	17	78.5	—
N$_2$	2.8×10⁻⁵	—	3	(0.07)	16	45	—	22	40	28⁹	45	39	47	17	—	20
CO	0.094	11.0¹·⁵	20.4	(0.44)	21	88	42	37	85	12⁹·¹⁰	43	48.5	31.5	78.5	—	42.5
O$_2$	0.31	44¹	63	—	39	—	—	39	64	52⁷	—	—	—	—	—	41.5
NO	1.1	77.5²	—	—	—	—	—	—	—	—	—	—	—	—	—	—
H$_2$O	0.021	3.1¹	—	—	6.0	—	—	1.5±0.9	1.0	—	—	2.8	0.42	4.0	17.3	5.5
D$_2$O	0.015	1.4³	—	—	—	—	—	—	—	—	—	—	—	—	—	—
CO$_2$	0.004	7.8¹	15.6	—	29	—	—	53	10.6	53¹⁰	—	71	80	75.5	—	102
N$_2$O	1.6	39.4⁴	—	0.13	—	—	—	—	—	—	—	—	—	—	—	—
NH$_3$	0.011	9.2¹	—	—	—	—	—	—	—	—	—	—	—	—	—	—
CH$_4$	0.02	0.2²	—	0.037	—	0.3	—	—	—	—	—	—	—	—	—	—

References

CALLEAR, A. B. and WILLIAMS, G. J., loc. cit., p. 1677.

$Hg(6\,^3P_0)$ (a) 1. ZEMANSKY, M. W., Phys. Rev. **36** (1930) 919.
$Hg(6\,^3P_1)$ 2. BATES, J. R., J. Am. chem. Soc. **52** (1930) 3825; **54** (1932) 569.
 3. EVANS, M. G., J. chem. Phys. **2** (1934) 445.
 4. ČVETANOVIC, P. J., ibid. **23** (1955) 1208.
 5. SCHEER, M. D. and FINE, J., ibid. **36** (1962) 1264.
 6. HUDSON, B. C. and CURNUTTE, B., Phys. Rev. **148** (1966) 60.
(b) BARRAT, J. P., CASALTA, D., COJAN, J. L., and HAMEL, J., J. Phys. **27** (1966) 608.

$Cd(5\,^3P_1)$ LIPSON, H. C. and MITCHELL, A. C. G., Phys. Rev. **48** (1935) 625.
$Li(2\,^2P)$ JENKINS, D. R., Proc. R. Soc. A**306** (1968) 413.
$Na(3\,^2P)$ (a) NORRISH, R. G. W. and SMITH, W. M., Proc. R. Soc. A**176** (1941) 295.

(b) DEMTRÖDER, W., Z. Phys. **166** (1962) 42.
(c) JENKINS, D. R., Proc. R. Soc. A **293** (1966) 493.
(d) HOOYMAYERS, H. P. and ALKEMADE, C. T. J., J. quant. Spectrosc. Radiat. Transfer, **6** (1966) 501.
(e) 7. KONDRATIEV, V. and SISKIN, M., Phys. Z. SowjUn. **8** (1935) 644.
 8. HANSON, H. G., J. chem. Phys. **23** (1955) 1391.
 9. KISILBACH, B., KONDRATIEV, V., and LEIPUNSKY, A., Phys. Z. SowjUn. **2** (1932) 201.
 10. WINANS, J. G., Z. Phys. **60** (1930) 631.
(f) TSUCHIYA, S., J. chem. Soc. Japan, **37** (1964) 6.

$K(4\,^2P)$ (a) JENKINS, D. R., Proc. R. Soc. A**303** (1968) 453.
$Rb(5\,^2P)$, $Cs(6\,^2P)$, $Tl(7\,^2S)$ HOOYMAYERS, H. P. and ALKEMADE, C. T. J., loc. cit. JENKINS, D. R., Proc. R. Soc. A**303** (1968) 453 (Rb, Cs); ibid. 467 (Tl).

The data given for $Hg(6\,^3P_1)$ under (a) have been obtained by direct experiments on the quenching of mercury resonance radiation and identified as representing deactivating transitions to the ground state by the experiments of Callear and Norrish‡ and of Callear and Williams§ as described on p. 1677. The result given for CO is obtained as follows. Assuming that N_2 quenches by producing $Hg(6\,^3P_0)$ alone the experiments of Scheer and Fine‖ give the corresponding cross-section for CO. This is then subtracted from the observed total quenching cross-section of CO to obtain the figure in Table 18.4.

The data under (b) for $Hg(6\,^3P_1)$ have been obtained from observation of the depolarization of resonance radiation as explained in § 3. On the whole the agreement with (a) is not too bad considering the overall lack of precision in the cross-section data.

Although N_2 is very ineffective in quenching $Hg(6\,^3P_0)$ and $Hg(6\,^3P_1)$ it readily quenches $Hg(7\,^3S_1)$ and $Hg(6\,^3D_1)$ for which the cross-sections are $3 \cdot 8 \times 10^{-15}$ and $4 \cdot 7 \times 10^{-14}$ cm^2 (see § 2.6).

It by no means follows that the process by which a given molecule quenches $Cd(5\,^3P_1)$ will be the same as for $Hg(6\,^3P_1)$ but for lack of any other information this has been assumed. The bracketed values are simply a reminder that, for $Hg(6\,^3P_1)$, N_2 and, to some extent, CO produce spin-orbit quenching.

For $Na(3\,^2P)$ there is a much greater wealth of data by different methods: (a) refers to measurement of quenching at temperatures of 400 °K while (c) and (d) are concerned with flame fluorescence at considerably higher temperatures of 1500–2500 °K; (b) refers to measurements of lifetimes at temperatures of 500 °K, but (e), which depends on optical dissociation for production of excited atoms, and (f) on shock-wave excitation, also refer to effective temperatures of 2000 °K or so. Nevertheless, it can hardly be said that any definite variation with temperature is perceptible in the results.

Assuming that no important factor enters from the temperature variation, the consistency of the results by different methods, at least as regards the relative effectiveness of different quenching molecules, is not too bad—the absolute cross-section is probably known in most cases to within a factor of 2 or so.

It will be seen that very little information is available about cross-sections for quenching by atoms. There seems little doubt that, at least for He and Ar, they are very small, but how small remains uncertain. At any rate they are so small that their quenching effect is likely to be

‡ loc. cit., p. 1677. § loc. cit., p. 1677. ‖ loc. cit., p. 1681.

masked by comparatively minor amounts of molecular impurities. The only remarkable result is for the quenching of Hg($6\,^3P_0$) by collisions with ground state atoms, which, in contrast with other cases, is by far the most effective process. Almost all molecules are relatively much more ineffective.

We have already explained on p. 1677 how it was at first thought that the cross-sections for quenching of Hg($6\,^3P_1$) by different molecules

TABLE 18.5

Minimum energy discrepancies for quenching of Na($3\,^2P$) *and* K($4\,^2P$) *by diatomic molecules*

	Quenching molecule			
	O_2	CO	N_2	H_2
Na($3\,^2P$) ΔE (eV)	+0·032	−0·066	+0·111	+0·186
K($4\,^2P$) ΔE (eV)	+0·014	−0·066	+0·062	−0·160

exhibited a resonance behaviour. This was because of the incorrect assumption that the quenching occurred through the spin-orbit transition $6\,^3P_1$–$6\,^3P_0$, whereas in fact the usual process involves transition to the ground state. For the quenching of Na($3\,^2P$) and K($4\,^2P$) no ambiguity of interpretation remains, and Hooymayers and Alkemade examined whether a resonance effect is found in these cases. They determined the minimum energy discrepancy ΔE by allowing for possible vibrational as well as electronic excitation of the quenching molecules.

The results found are given in Table 18.5. Reference to the data given in Table 18.4 shows that there is some evidence that the cross-sections decrease in the order O_2, N_2, and H_2 following the increase in $|\Delta E|$. On the other hand, for CO the cross-section, though hardly known with accuracy, is probably at least as great as for O_2, despite the fact that $|\Delta E|$ is considerably greater for CO. It is unlikely that any clear dependence on ΔE will emerge, just as for Hg($6\,^3P_1$).

We shall defer the discussion of the results obtained for deactivating collisions in which the transition involved is of spin-orbit type until § 4.1, as important data about such collisions for alkali-metal atoms are available from experiments on sensitized fluorescence.

We next describe the experiments on the polarization and depolarization of resonance radiation which have provided information about cross-sections for collisions that change the orientation and alignment of the total electron angular momentum vector without changing its magnitude. Quenching cross-sections, when appreciable, can also be obtained from these experiments and have been included in Table 18.4 (see column (*b*) for Hg($6\,^3P_1$)).

3. Polarization of resonance radiation—collisions which change the orientation and alignment of the total electronic angular momentum

3.1. *Introduction—the disorientation and disalignment cross-sections*

We have described in § 2 how observations of the variation of the intensity of emission of resonance radiation with pressure of a foreign gas can yield values for the cross-section for quenching of the resonance excitation by collisions with a gas atom. A great deal more information may be obtained if the resonance absorption cell is bathed in a magnetic field and the exciting radiation is either plane or elliptically polarized. Thus, consider the resonance excitation of Hg($6\,^3P_1$) under these circumstances. To avoid complications due to hyperfine structure we suppose that the mercury consists exclusively of the even isotope ^{202}Hg with no nuclear spin. Through the presence of the magnetic field and the polarization of the exciting radiation the Zeeman substates with $M_J = 0, \pm 1$ will not in general be equally populated. This means that the mean value of the total electronic angular momentum over the whole system will in general not vanish—there will be some degree of order in the angular momentum. At first sight it would appear that this order may be measured by the mean value $\langle J \rangle \hbar$ of the magnitude of the total angular momentum, just as in defining the polarization of a system for which $J = \tfrac{1}{2}$. However, there will be circumstances in which $\langle J \rangle = 0$ even though some degree of order is present. Thus, if the populations of the levels with $M_J = \pm 1$ are equal while that with $M_J = 0$ vanishes, $\langle J \rangle = 0$ but the angular momentum is aligned along the *z*-axis. If P_M is the population of the state M, we therefore define two quantities,

$$f = \sum P_M M / J \sum P_M, \qquad (47)$$

$$g = \sum [P_M \{M^2 - J(J+1)/3\}]/J^2 \sum P_M, \qquad (48)$$

the first of which is called the *orientation* and vanishes when $\langle J \rangle = 0$ while the second, which does not necessarily vanish under these conditions, is called the *alignment*. We shall refer to a system in which either *f*, or *g*, or both, are non-zero as *polarized*.

The orientation determines the induced magnetic moment of the system while alignment gives rise to an electric quadrupole moment which can be expressed in terms of the mean value of a second-order tensor **Q** whose five components are

$$6^{-\frac{1}{2}}(3J_z^2-J^2), \quad \pm\tfrac{1}{2}(J_zJ_\pm+J_\pm J_z), \quad \tfrac{1}{2}J_\pm^2, \qquad (49)$$

where
$$J_\pm = 2^{-\frac{1}{2}}(J_x \pm iJ_y).$$

A detailed analysis‡ of the changes that can be produced in the excited assembly through collisions either with ground state atoms of the same kind, or of foreign atoms, shows that they may be specified in terms of three cross-sections as follows:

(a) The quenching cross-section Q_q, which defines the rate of collisions which deactivate the excited atoms.
(b) A disalignment cross-section Q_{al}, which defines the rate at which collisions produce change of alignment.
(c) A disorientation cross-section Q_{or}, which defines the rate at which collisions produce change of orientation.

To see how these arise in detail it is necessary to develop the theory of the spin-density matrix. We consider an assembly of atoms all in excited states with $J = 1$. For simplicity, we suppose at first that all the atoms are in the same state, the wave function for which we may write in the form

$$\Psi = a_1\psi_1 + a_2\psi_0 + a_3\psi_{-1}, \qquad (50)$$

where $\psi_1, \psi_0, \psi_{-1}$ are the normalized wave-functions corresponding to the values $(1, 0, -1)\hbar$, respectively, of the angular momentum along the z-direction. As we are interested only in the specification of the angular momentum which corresponds to (50), we represent Ψ by the single-column matrix

$$\Psi = \begin{pmatrix} a_1 \\ a_2 \\ a_3 \end{pmatrix} \qquad (51)$$

and the Hermitian conjugate $\Psi^\dagger$ by the row matrix $\Psi^\dagger = (a_1^*, a_2^*, a_3^*)$. The matrix product:

$$\Psi\Psi^\dagger = \begin{pmatrix} a_1a_1^* & a_1a_2^* & a_1a_3^* \\ a_2a_1^* & a_2a_2^* & a_2a_3^* \\ a_3a_1^* & a_3a_2^* & a_3a_3^* \end{pmatrix} \qquad (52)$$

is called the spin density matrix ρ for the assembly.

We note that, since $|a_1|^2+|a_2|^2+|a_3|^2 = 1$,

$$\text{tr } \rho = 1. \qquad (53)$$

The mean value of any matrix **O** operating on Ψ is given by

$$\langle O \rangle = \Psi^\dagger O \Psi, \qquad (54)$$

and this may be written in terms of the density matrix ρ as

$$\text{tr}(O\rho)/\text{tr } \rho. \qquad (55)$$

‡ OMONT, A., *J. Phys., Paris* **26** (1965) 26.

The elements of the matrix ρ are bilinear in the coefficients (50) of $\psi_1, \psi_0, \psi_{-1}$. In a rotation of axes these functions transform like $-x-iy, z, x-iy$ respectively, so the a_1, a_2, a_3 transform as linear combinations of x, y, z, i.e. as vectors. The components of ρ, being quadratic in the a's, must therefore transform as those of a second-order tensor. The nine tensor components $x_i^{(1)} x_j^{(2)}$ ($i, j = 1, 2, 3$) can be analysed into sets which transform exclusively into themselves. These are known as the irreducible second-order tensors and are of three types:

(a) the scalar, δ_{ij},
(b) the axial vector $x_i^{(1)} x_j^{(2)} - x_j^{(1)} x_i^{(2)}$,
(c) the second-rank tensor $x_i^{(1)} x_j^{(2)} + x_j^{(1)} x_i^{(2)} - \tfrac{2}{3}\delta_{ij} \mathbf{x}^{(1)} \cdot \mathbf{x}^{(2)}$.

In terms of the spherical harmonics Y_q^u the three sets transform as

$$Y_0; \quad Y_1^u \ (u = 1, 0, -1); \quad Y_2^u \ (u = 2, 1, 0, -1, -2)$$

respectively. Denoting the corresponding matrix operator as $\boldsymbol{\Omega}_{qu}$ we have that

$$\mathrm{tr}(\boldsymbol{\Omega}_{qu}) = 0 \quad (q > 0). \tag{56}$$

Also the operators obey an important orthogonal property

$$\mathrm{tr}(\boldsymbol{\Omega}_{qu} \boldsymbol{\Omega}_{q'u'}^\dagger) = 0 \quad (q \neq q', u \neq u'). \tag{57}$$

This follows because $\boldsymbol{\Omega}_{qu} \boldsymbol{\Omega}_{q'u'}^\dagger$ transforms like $Y_q^u Y_{q'}^{-u'}$. It is possible to express this product of harmonics in terms of harmonics $Y_{q''}^{u-u'}$ where $|q-q'| \leq q'' \leq q+q'$. For any of the corresponding operators to have non-vanishing trace we must have $u = u'$ and $q'' = 0$. The latter condition requires $q = q'$.

The operator $\boldsymbol{\Omega}_{qu}$ may now be represented in terms of the angular momentum operators for $J = 1$,

$$\mathbf{S}_1 = 2^{-\frac{1}{2}} \begin{pmatrix} 0 & 1 & 0 \\ 1 & 0 & 1 \\ 0 & 1 & 0 \end{pmatrix}, \quad \mathbf{S}_2 = 2^{-\frac{1}{2}} \begin{pmatrix} 0 & -i & 0 \\ i & 0 & -i \\ 0 & i & 0 \end{pmatrix},$$

$$\mathbf{S}_3 = \begin{pmatrix} 1 & 0 & 0 \\ 0 & 0 & 0 \\ 0 & 0 & -1 \end{pmatrix}, \tag{58}$$

which satisfy
$$\mathbf{S}_1 \mathbf{S}_2 - \mathbf{S}_2 \mathbf{S}_1 = i\mathbf{S}_3,$$
and cyclic permutations, and

$$\mathbf{S}_1^2 + \mathbf{S}_2^2 + \mathbf{S}_3^2 = 2. \tag{59}$$

We then have

$$\boldsymbol{\Omega}_{00} = \mathbf{I}, \quad \boldsymbol{\Omega}_{10} = (3/2)^{\frac{1}{2}} \mathbf{S}_3, \quad \boldsymbol{\Omega}_{1,\pm 1} = \mp \tfrac{1}{2}\sqrt{(3)}(\mathbf{S}_1 \pm i\mathbf{S}_2),$$
$$\boldsymbol{\Omega}_{20} = \tfrac{1}{2}\sqrt{(2)}(3\mathbf{S}_3^2 - 2), \quad \boldsymbol{\Omega}_{2,\pm 1} = \pm \tfrac{1}{2}\sqrt{(3)}\{(\mathbf{S}_1 \pm i\mathbf{S}_2)\mathbf{S}_3 + \mathbf{S}_3(\mathbf{S}_1 \pm i\mathbf{S}_2)\},$$
$$\boldsymbol{\Omega}_{2,\pm 2} = \tfrac{1}{2}\sqrt{(3)}(\mathbf{S}_1 \pm i\mathbf{S}_2)^2, \tag{60}$$

the normalization being such that

$$\mathrm{tr}(\boldsymbol{\Omega}_{qu} \boldsymbol{\Omega}_{qu}^\dagger) = 3. \tag{61}$$

We may now write
$$\rho = \sum A_{qu} \boldsymbol{\Omega}_{qu}^\dagger. \tag{62}$$

In view of (55) and (56) we have

$$A_{00} = \tfrac{1}{3}, \quad A_{qu} = \tfrac{1}{3}\langle \boldsymbol{\Omega}_{qu} \rangle, \tag{63}$$

so
$$\rho = \tfrac{1}{3} \sum \langle \boldsymbol{\Omega}_{qu} \rangle \boldsymbol{\Omega}_{qu}^\dagger. \tag{64}$$

The analysis may now be extended to an assembly of atoms which have a probability distribution among the stationary states $\psi_1, \psi_0, \psi_{-1}$. All that is necessary is to average the density matrix over these probability distributions. Thus if p_1, p_2, p_3 are the probabilities of finding an atom in the respective states $\psi_1, \psi_0, \psi_{-1}$

$$\rho = \begin{pmatrix} p_1 & 0 & 0 \\ 0 & p_2 & 0 \\ 0 & 0 & p_3 \end{pmatrix}. \tag{65}$$

Instead it is possible to choose, as basis functions, any orthogonal triad obtained by linear combination of $\psi_1, \psi_0, \psi_{-1}$. Since $\psi_1, \psi_0, \psi_{-1}$ transform as $-x-iy, z, x-iy$ in a rotation of axes, we may write such a triad as the matrix product $\mathbf{R}\Psi$, where $\mathbf{R}$ is the matrix that represents a rotation of orthogonal axes in three dimensions. The general form of ρ may then be written

$$\rho = \mathbf{R} \begin{pmatrix} p_1 & 0 & 0 \\ 0 & p_2 & 0 \\ 0 & 0 & p_3 \end{pmatrix} \mathbf{R}^\dagger. \tag{66}$$

The same analysis then applies in that ρ may be expressed in the form (62). The $\langle \Omega_{1u} \rangle$ are now the average values of the spin components and the $\langle \Omega_{2u} \rangle$ of the tensor components. If both the $\langle \Omega_{1u} \rangle$ and the $\langle \Omega_{2u} \rangle$ vanish the assembly is said to be unpolarized, while if $\langle \Omega_{1u} \rangle = 0$ but $\langle \Omega_{2u} \rangle \neq 0$ the assembly still possesses a degree of spin order. The $\langle \Omega_{1u} \rangle$ determine the orientation and the $\langle \Omega_{2u} \rangle$ the alignment of the assembly. The polarization vanishes only when both orientation and alignment vanish.

In terms of the density matrix (65)

$$\langle \Omega_{10} \rangle = (\tfrac{3}{2})^{\frac{1}{2}}(p_1 - p_3), \qquad \langle \Omega_{1\pm 1} \rangle = 0,$$
$$\langle \Omega_{20} \rangle = 2^{-\frac{1}{2}}(p_1 + p_3 - 2p_2), \qquad \langle \Omega_{2\pm 1} \rangle = 0 = \langle \Omega_{2\pm 2} \rangle, \tag{67}$$

showing that, in terms of the definitions (47) and (48),

$$\langle \Omega_{10} \rangle = (\tfrac{3}{2})^{\frac{1}{2}} f, \qquad \langle \Omega_{20} \rangle = (3/\sqrt{2}) g.$$

The effect of collisions on the spin density matrix of the assembly may now be discussed. In a collision the wave function Ψ will be converted to a new function Ψ' such that

$$\Psi' = \mathbf{S}\Psi, \tag{68}$$

where $\mathbf{S}$ will be a three-by-three matrix which we refer to as the *spin scattering matrix*. It follows that the spin-density matrix is converted to

$$\rho' = \mathbf{S}\rho\mathbf{S}^\dagger. \tag{69}$$

In dealing with an assembly of a large number of atoms we must average the effect of a collision over all orientations, so we may write

$$\overline{\rho'} = \overline{\mathbf{Y}}\rho, \tag{70}$$

where $\overline{\mathbf{Y}}$ is independent of orientation. Under these circumstances we may write

$$\overline{\mathbf{Y}}\Omega_{qu} = (1 - P_q)\Omega_{qu}, \tag{71}$$

where P_q is independent of u. We may therefore define P_1, P_2 as the probabilities that, in a collision, the orientation and the alignment respectively are destroyed so that, if Q is the total collision cross-section, $P_1 Q, P_2 Q$ are the cross-sections $Q_{\text{or}}, Q_{\text{al}}$ for destruction of orientation and of alignment respectively.

In the above analysis we have been concerned only with collisions that change the polarization of the assembly and involve no change of relative kinetic energy. There is no difficulty in extending the analysis to include inelastic collisions, if required.

Methods of calculating the cross-sections $P_1 Q$, $P_2 Q$ are discussed in § 8.3. We now describe methods by which they may be measured.

3.2. Principles of the method of measurement of disorientation and disalignment cross-sections for atoms with no nuclear spin using the Hanle effect

To see how these cross-sections may be measured we consider three examples of resonance excitation of mercury ($6\,^3P_1$) by polarized light. In all three cases a constant magnetic field H is applied in the direction Oz, the light is incident in the direction Ox, and the scattered light observed along Oy (Fig. 18.25).

In case (a) of Fig. 18.25 the incident light is plane polarized with electric vector parallel to Oz. Absorption of this light will occur through π transitions ($\Delta M_J = 0$) which will produce alignment but not orientation in the excited atoms. As the ground state of mercury is $6\,^1S_0$ the excited atoms will all initially be in states with $M_J = 0$. The magnetic field will have no effect as it is parallel to the oscillating electric vector. If the mean time between collisions which can change M_J is long compared to the mean radiation lifetime of the Hg($6\,^3P_1$)$M_J = 0$ state then the emitted radiation must also be π radiation with $M_J = 0$, completely polarized parallel to Oz. However, if Z_q, Z_{al} are the rates at which quenching and disalignment collisions respectively occur and τ_r is the radiative lifetime of the excited state, the polarization is reduced to[†]

$$P = \frac{1+\tau_r Z_q}{1+\tau_r Z_q + \tfrac{2}{3}\tau_r Z_{al}}. \qquad (72)$$

Allowance must also be made for imprisonment of resonance radiation.[‡] If x is the probability that a resonance photon emitted by an atom in the resonance be absorbed by another atom before leaving the cell, (72) becomes

$$P = \frac{(1-x)+\tau_r Z_q}{1-\tfrac{4}{5}x+\tau_r(Z_q+\tfrac{2}{3}Z_{al})}. \qquad (73)$$

In case (b) of Fig. 18.25 the incident light is plane polarized with electric vector parallel to Oy. Such radiation will produce $\sigma^\pm$ transitions, corresponding to $\Delta M_J = \pm 1$, with equal probability so that the

[†] BARRAT, J. P., CASALTA, D., COJAN, J. L., and HAMEL, J., *J. Phys., Paris* **27** (1966) 608.
[‡] OMONT, A., loc. cit., p. 1697.

initial excitation will populate equally the states $M_J = \pm 1$ of $\mathrm{Hg}(6\,^3P_1)$. Once again there is alignment but not orientation. In the absence of a magnetic field the intensity emitted in the direction Oy will be very small as it is along the direction of the oscillating electrical vector. With

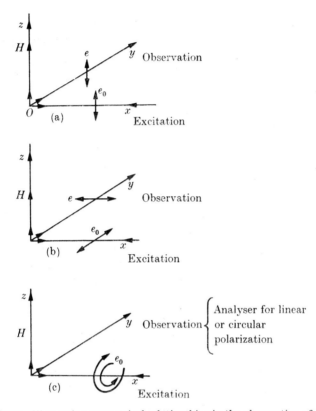

Fig. 18.25. Illustrating geometrical relationships in the observation of fluorescent radiation excited by polarized incident radiation in a constant magnetic field H along Oz. $\mathbf{e}_0$, $\mathbf{e}$ denote the electric vectors in the incident and scattered radiation respectively.

a magnetic field along Oz the precession of the active electrons about the magnetic field gives rise to circularly polarized radiation viewed along the field and to one plane polarized component when viewed in a plane perpendicular to the field, as shown in Fig. 18.25 (b). The intensity of the plane polarized radiation† observed along Oy varies as

$$\frac{4\omega^2\tau_{\mathrm{r}}^2}{1+4\omega^2\tau_{\mathrm{r}}^2}, \tag{74}$$

† Barrat, J. P., Casalta, D., Cojan, J. L., and Hamel, J., loc. cit.

where ω is the Larmor frequency corresponding to the magnetic field H and the effects of collisions and imprisonment are ignored. When these are taken into account the intensity is

$$\omega^2\tau_r^2/[\{1-\tfrac{7}{10}x+\tau_r(Z_q+Z_{al})\}^2+4\omega^2\tau_r^2]. \tag{75}$$

Finally, in case (c) shown in Fig. 18.25, we consider excitation by right- or left-handed circularly polarized radiation. In the former case only transitions with $\Delta M = +1$ will be excited leaving the $Hg(6\,^3P_1)$ atoms initially in the state with $M_J = 1$ so that both orientation and alignment will be produced. If the intensity of radiation polarized linearly along Ox is observed along Oy then it will vary with the magnetic field and the collision frequencies as in (75). However, if the intensity of circularly polarized radiation is observed along Oy it is proportional to†

$$\frac{1}{2}+\frac{\omega^2\tau_r^2}{\{1-\tfrac{7}{10}x+\tau_r(Z_q+Z_{al})\}^2+4\omega^2\tau_r^2}\pm\frac{\omega\tau_r\{1-\tfrac{1}{2}x+\tau_r(Z_q+Z_{or})\}}{\{1-\tfrac{1}{2}x+\tau_r(Z_q+Z_{or})\}^2+\omega^2\tau_r^2}, \tag{76}$$

the $\pm$ signs depending on whether right-handed or left-handed circularly polarized radiation is observed.

It is now clear how the three cross-sections can be derived. Thus, with case (a), from (73) it is possible, by taking observations at different pressures p of the quenching gas, to determine Q_q and Q_{al}. Thus we may write
$$Z_q = KQ_q p, \qquad Z_{al} = KQ_{al} p,$$
so
$$1/P = \frac{1-\tfrac{4}{5}x+\tau_r K(Q_q+\tfrac{2}{3}Q_{al})p}{1-x+\tau_r KQ_q p}. \tag{77}$$

If the plot of $1/P$ against p is linear then Q_q is negligible and Q_{al} is determined directly from the slope. On the other hand, if the plot has an appreciable curvature, the limit $p \to \infty$ gives Q_{al}/Q_q and the slope of the tangent at $p \to 0$ gives

$$-\frac{x}{5(1-x)}Q_q+\tfrac{2}{3}Q_{al}.$$

A further check may then be obtained from (75) of case (b) or (c) which gives Q_q+Q_{al}. Finally, Q_{or} is obtained from the difference in the intensities observed in case (c) with right- and left-handed circular polarization detectors. Thus, according to (76), this difference is proportional to

$$\frac{\{1-\tfrac{1}{2}x+\tau_r(Z_q+Z_{or})\}\omega\tau_r}{\{1-\tfrac{1}{2}x+\tau_r(Z_q+Z_{or})\}^2+\omega^2\tau_r^2}. \tag{78}$$

† BARRAT, J. P., CASALTA, D., COJAN, J. L., and HAMEL, J., loc. cit., p. 1700.

Once again measurements at different pressures of quenching gas enable $Q_q + Q_{or}$ to be determined and hence Q_{or} from the previously determined values of Q_q.

3.3. *Experimental measurement of cross-sections*

Observations of the polarization of the resonance radiation of mercury were made as long ago as 1922 by Rayleigh† and by Wood.‡ A little later Wood and Ellett§ were the first to observe the effect of a magnetic field on the polarization, but the first thorough investigation of these effects was made by Hanle.∥ He observed the variation of intensity with magnetic field in case (b), and this effect is often referred to as the Hanle effect. The effect of foreign gases on the polarization was first observed qualitatively by Wood‡ and the first quantitative measurements were made by von Keussler.†† These observations were made before the theoretical analysis had been sufficiently developed to allow for the contributions of three distinct cross-sections to the depolarization by collision. As an example of recent experiments in which Q_q, Q_{al}, and Q_{or} were all determined by the method discussed above we choose those of Barrat, Casalta, Cojan, and Hamel.‡‡

The general arrangement of the apparatus is illustrated in Fig. 18.26. C is the resonance cell of fused quartz in the form of a 3-cm cube, one side of which was intended to form a light trap. After evacuation of the cell to a few 10^{-6} torr, the pumps could be shut off and connection made to a drop of natural mercury which had been vacuum distilled. Under working conditions it was desirable to keep the mercury vapour pressure low so that imprisonment effects would be small. For this reason the mercury drop was maintained at -25 °C throughout an experiment. Foreign gas could be introduced through a fine leak. The cell was enclosed in a blackened box pierced with two holes to allow for entrance of the exciting radiation and exit of the scattered radiation to be observed.

The horizontal component of the earth's magnetic field was compensated by two Helmholtz coils 9 cm in diameter while the constant vertical applied magnetic field was generated through two 28-cm coils. It was

---

† RAYLEIGH, Lord, *Proc. R. Soc.* A**102** (1922) 190.
‡ WOOD, R. W., *Phil. Mag.* **44** (1922) 109.
§ WOOD, R. W. and ELLETT, A., *Proc. R. Soc.* A**103** (1923) 396; *Phys. Rev.* **24** (1924) 243.
∥ HANLE, W., *Z. Phys.* **30** (1924) 93; *Ergebn. exakt. Naturw.* **4** (1925) 214.
†† VON KEUSSLER, V., *Annln Phys.* **87** (1927) 793.
‡‡ loc. cit., p. 1700.

calibrated by using magnetic resonance in conjunction with the Hg($6\,^3P_1$) level.

Special arrangements had to be made to ensure that the exciting radiation would only excite the even isotope ^{202}Hg and not the odd isotopes ^{199}Hg and ^{201}Hg. The source was an r.f. discharge in nearly

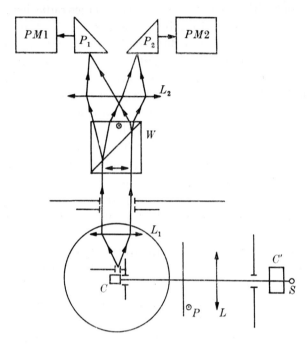

Fig. 18.26. General arrangement of apparatus used by Barrat, Casalta, Cojan, and Hamel for determining cross-sections for quenching, disalignment, and disorientation of Hg($6\,^3P_1$) atoms by impact with various foreign atoms and molecules.

pure ^{202}Hg at a pressure of a few torr. As the separations of the hyperfine components for the odd isotopes are large compared with their Doppler widths this radiation should not excite these isotopes. However, to eliminate most of the residual radiation from the source arising from the small percentage of the odd isotopes in the source, the radiation was passed through an absorption cell C', 1 cm thick, containing a little mercury enriched in the odd isotope ^{199}Hg. As a check on the absence of any significant contribution from excitation of odd isotopes the polarization in case (a) (see Fig. 18.25) was measured in the limit of vanishing mercury vapour and foreign gas pressure and found to be as high as 95 per cent.

The incident radiation could be plane polarized by passage through u.v. polarizing films P or could be circularly polarized by a quarter-wave plate.

The scattered radiation that passed out of the exit hole in the cell container was rendered parallel by a fused quartz lens L_1 and then traversed successively a collector prism W and a second fused quartz lens L_2. The prism divided the radiation into two beams, linearly polarized parallel to Ox (in the plane of the diagram) and to Oz (perpendicular to the plane of the diagram) respectively. These beams were then focused by the lens L_2 and reflected into the respective photomultipliers PM_1 and PM_2 through the totally reflecting prisms P_1 and P_2. To analyse the circular polarization of the scattered beams a quarter-wave plate could be placed before the Wollaston prism.

In carrying out the polarization measurements it was necessary to allow for instrumental asymmetries. This was done as follows. If I_x, I_z are the intensities of the scattered radiation linearly polarized parallel to Ox and Oz respectively, the currents recorded in the photomultipliers PM_1, PM_2 will be given by

$$i_1 = k_1 I_x, \qquad i_2 = k_2 I_z, \tag{79}$$

where the sensitivities k_1, k_2 will not in general be equal. To obtain I_x/I_z a half-wave plate was inserted before the Wollaston prism. This plate has the effect of changing I_x to tI_x and I_z to tI_z, where t is the transmission of the plate, but in addition exchanges the polarization, so that I_x is polarized parallel to Oz and I_z to Ox. The respective multiplier currents will now be i_1'', i_2'' where
$$i_1'' = k_1 t I_z, \qquad i_2'' = k_2 t I_x.$$
Hence $I_z^2/I_x^2 = i_2 i_1''/i_1 i_2''$.

It is also necessary, in analysing the observations, to allow for the incomplete polarization of the incident beam. Suppose that the ratio of the incident intensity polarized parallel to Oy and Oz respectively is ϵ ($\ll 1$). Then the polarization of the scattered radiation emitted in the direction Oy is modified from (73) to

$$P = \frac{(1-\tfrac{1}{2}\epsilon)(1-x+\tau_r Z_q)}{1-\tfrac{4}{5}x+\tfrac{1}{2}\epsilon(1-\tfrac{3}{5}x)+\tau_r\{(1+\tfrac{1}{2}\epsilon)Z_q+\tfrac{2}{3}(1+\epsilon)Z_{al}\}}. \tag{80}$$

x was determined from observation of the Hanle effect in case (b) (see (75)) as described below.

If P_0 is the limiting polarization when the mercury vapour pressure as well as the foreign-gas pressure tend to zero, we have

$$P_0 = \frac{1-\tfrac{1}{2}\epsilon}{1+\tfrac{1}{2}\epsilon} \tag{81}$$

and, using (77) and (80),

$$\frac{1}{P} = \frac{1}{P_0}\left(1 + \frac{3-P_0}{3}\frac{0.3x + \tau_r K Q_{al} p}{1 - x + \tau_r K Q_q p}\right) \quad (82)$$

in the presence of the foreign gas at pressure p.

Fig. 18.27 illustrates results obtained in the form of plots of $1/P$ against foreign-gas pressure for the rare gases and N_2, all of which are effectively linear. For these cases Q_q is negligible and Q_{al} is obtained

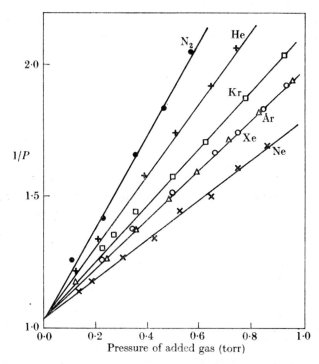

FIG. 18.27. Observed variation of reciprocal linear polarization $1/P$ of fluorescence resonance radiation observed in case (a) (see Fig. 18.25), with pressure of admixed gas as indicated.

from the slope of the line as described above. In other cases, those in which O_2, H_2, CO, and CO_2 were the foreign gases, the plots were not linear. It was then convenient to work in terms of a quantity

$$Y = \frac{3(P_0 - P)}{(3 - P_0)P}$$
$$= \frac{0.3x + \tau_r K Q_{al} p}{1 - x + \tau_r K Q_q p}. \quad (83)$$

18.3 UNDER GAS-KINETIC CONDITIONS

We then have

$$\left(Y - \frac{0\cdot 3x}{1-x}\right)^{-1} = \frac{(1-x)}{\tau_r K[Q_{al} - \{0\cdot 3x/(1-x)\}Q_q]}\frac{1}{p} + \frac{Q_q}{Q_{al} - \{0\cdot 3x/(1-x)\}Q_q}. \tag{84}$$

A plot of $\{Y - 0\cdot 3x/(1-x)\}^{-1}$ against $1/p$ should therefore be linear.

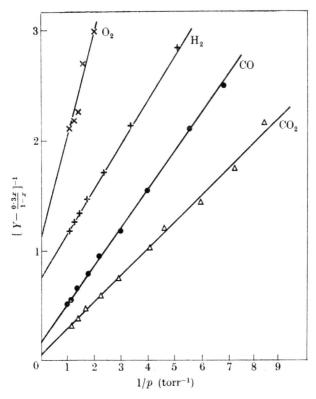

FIG. 18.28. Observed variation of the quantity $\{Y - 0\cdot 3x/(1-x)\}^{-1}$, defined in (84), with reciprocal pressure $1/p$ of added gas in case (a) (see Fig. 18.25).

From the slope, $Q_{al} - \{0\cdot 3x/(1-x)\}Q_q$, and from the intercept on the y-axis, Q_{al}/Q_q can be obtained. Fig. 18.28 illustrates typical plots of this kind.

In the determination of Q_{al} and Q_{or} from case (c) it is necessary to allow for the fact that the quarter-wave plates introduced to produce a circular polarization will not produce a phase change between the two linearly polarized components of exactly 45°. Thus, in the determination of $Q_{al} + Q_q$ from (75), a quarter-wave plate is interposed between

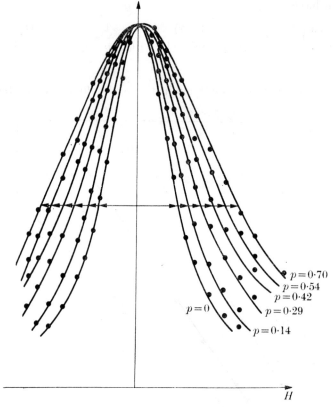

FIG. 18.29. Variation with magnetic field H of the difference signal observed in the experiments of Barrat et al. in case (c) (Fig. 18.25), using linearly polarized incident radiation. Oxygen is the added gas and curves are given for different oxygen pressures indicated in torr in each case.

the polarizing films P and the resonance cell. As a result the incident electric vector will be
$$\sin \alpha \, \mathbf{k} + i \cos \alpha \, \mathbf{j}, \tag{85}$$
where α is close to $\tfrac{1}{4}\pi$. $\mathbf{k}$ and $\mathbf{j}$ are unit vectors along Oz, Oy respectively. The scattered intensity polarized parallel to Ox, Oz respectively is then given by
$$I_x = A \frac{\omega^2 \tau_r^2}{(1+\delta)^2 + 4\omega^2 \tau_r^2}, \qquad I_z = B, \tag{86}$$
with
$$\delta = -\tfrac{7}{10}x + \tau_r(Z_q + Z_{a1}), \tag{87}$$
and A and B are constants such that $B = 2A \cot^2\alpha$.

Hence the signal on the photomultiplier which records light polarized parallel to Oz is independent of the magnetic field while that which records light polarized parallel to Ox will vary as

$$C + \frac{D}{(1+\delta)^2 + 4\omega^2 \tau_r^2}. \tag{88}$$

The sensitivities of the two photomultipliers were adjusted so that, in the limit of large magnetic field, $\omega \to \infty$, the difference between the two signals vanished. It follows that, for smaller magnetic fields the difference signal will vary as $\{(1+\delta)^2 + 4\omega^2 \tau_r^2\}^{-1}$. Fig. 18.29 shows typical

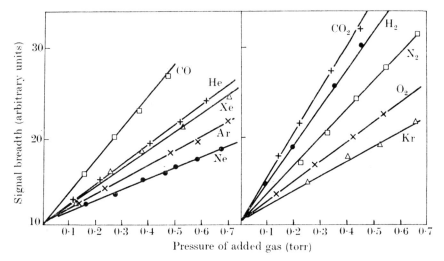

FIG. 18.30. Variation of the difference signal breadth $(1+\delta)/\tau_r$ (at half height) with pressure of added gas, as indicated in torr, corresponding to (89).

results for O_2 at different pressures p. The full width $\Delta\omega$ of the plot of the signal against ω is equal to $(1+\delta)/\tau_r$ so, using (86) and (87),

$$\Delta\omega = K(Q_q + Q_{al})p + (1 - \tfrac{7}{10}x)/\tau_r. \tag{89}$$

We see from (87) that a plot of $(1+\delta)$ against p should be linear, the intercept on the vertical giving x and the slope $Q_q + Q_{al}$. Fig. 18.30 illustrates typical plots of this kind. In this way x was found to be 0·03 at a mercury temperature of $-25\ °C$.

Finally, to make use of the formula (78) a quarter-wave plate is interposed before the Wollaston prism W. In principle the difference between the signals recorded on the two photomultipliers will be proportional to $I^+ - I^-$ and hence vary with ω, according to (78), as

$$\omega[\{1 - \tfrac{1}{2}x + \tau_r(Z_q + Z_{or})\}^2 + \omega^2 \tau_r^2]^{-1}. \tag{90}$$

If the analysing quarter-wave plate is not perfect the difference will still vary as (90) though with a different constant of proportionality. Advantage was then taken of the fact that, whereas (90) is an odd function of ω, the remaining terms in (76) are all even functions. In making the measurements the sensitivities of the photomultipliers were

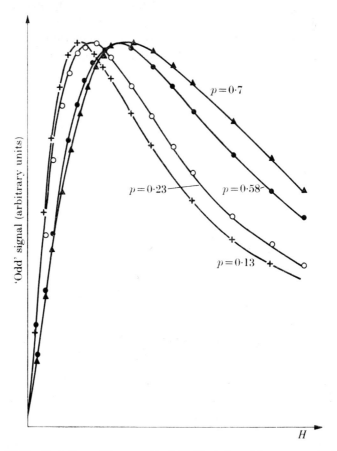

Fig. 18.31. Variation with magnetic field H of the odd component of the difference signal, in case (c) Fig. 18.25, with circularly polarized incident radiation. Argon is the added gas and curves are given for different argon pressures indicated in torr in each case.

adjusted to give a difference signal which was as nearly as possible an odd function of ω. To remove the remaining traces of the even function the difference signal for ω ranging from $-\infty$ to ∞ was analysed into odd and even parts. It was further checked that the width at half height of the former was as given by (90). Fig. 18.31 illustrates results obtained for the variation of the odd component with ω for different pressures

of argon. If the maximum of the signal occurs at ω_m then
$$\omega_m = Z_q + Z_{or} + (1 - \tfrac{1}{2}x)\tau_r. \tag{91}$$
Hence, since $Z_q = KQ_q p$, $Z_{or} = KQ_{or} p$, a plot of ω_m against the pressure p of foreign gas should be linear with slope $K(Q_q + Q_{or})$. Typical examples are given in Fig. 18.32.

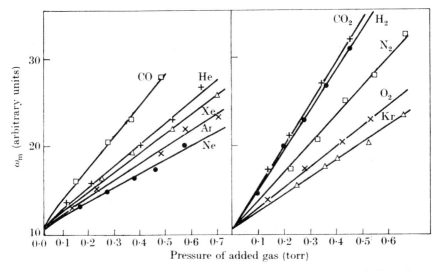

FIG. 18.32. Variation of ω_m (see (91)) with pressure of added gas as indicated.

In Table 18.6 a comparison is given between the results obtained for a number of gases in these experiments, derived from the polarization measurements of case (a) and the Hanle effect measurements of case (c). The accuracy of the measurements is such that the differences between Q_{or} and Q_{al} for the rare gases are significant. It will be seen that there is very good consistency between the values of Q_q, Q_{al}, and $Q_q + Q_{al}$.

Comparison between these results and those obtained by similar methods is given in Table 18.8, which also includes results obtained for $Cd(5\,^3P_1)$, and for $Hg(6\,^1P_1)$.

3.4. Double-resonance method

We must now draw attention to a further method of determining the cross-sections. This is the so-called 'double-resonance' method and consists in carrying out electron magnetic resonance experiments on the population of excited $^{202}Hg(6\,^3P_1)$ atoms. The effect of an r.f. magnetic field applied at right angles to the main field is to induce transitions between the sublevels of different M_J at a rate that depends on the relation between the frequency separation of these levels and the frequency of the applied r.f. field. The possible application of this technique was first suggested by Brossel and Kastler† and the first observations were made a little later by Brossel, Sagalyn, and Bitter,‡ followed by a more detailed experimental and analytical study carried out by Brossel and Bitter.§

TABLE 18.6

Cross-sections Q_q, Q_{al}, and Q_{or} observed by Barrat et al. from observations of the polarization of ^{202}Hg($6\,^3P_1$) resonance radiation in the presence of different foreign gases

Gas	Cross-sections (10^{-16} cm²) from case (a)		Cross-sections (10^{-16} cm²) from case (c)	
	Q_q	Q_{al}	Q_q+Q_{al}	Q_q+Q_{or}
He	0	40·2	39·6	45·5
Ne	0	46·5	47·1	60
Ar	0	83·3	83·1	100
Kr	0	122	124	144
Xe	0	169	173	179
H_2	26	33	56·5	61
N_2	3	135	144	144
O_2	63	53·5	112	112
CO	20·4	140	160	160
CO_2	15·7	248	270	270

They used the general arrangement shown in Fig. 18.33 in which the incident resonance radiation was polarized parallel to the constant magnetic field of a few hundred G. The intensity of resonance radiation scattered out in a direction normal to the main field and polarized parallel to it (the π component) could be measured by one photomultiplier and that of the circularly polarized σ-radiation, emitted parallel to the field, by the other. By means of a suitable resistance bridge the system could be adjusted so that no net current arose from opposing signals from the two photomultipliers when no r.f. field was applied. This eliminates errors due to fluctuation in the intensity of the light source. When the r.f. field was applied it produced transitions from the levels with $M_J = 0$ to levels with $M_J = \pm 1$ and so increased the intensity of the σ relative to the π component. This threw the bridge out of balance so a signal was recorded in the galvanometer. The experiments were carried out

† BROSSEL, J. and KASTLER, A., *C.r. hebd. Séanc. Acad. Sci., Paris* **229** (1949) 1213.
‡ BROSSEL, J., SAGALYN, P., and BITTER, F., *Phys. Rev.* **79** (1950) 196, 225.
§ BROSSEL, J. and BITTER, F., ibid. **86** (1952) 308.

under such low pressure conditions that no significant effect was produced by collisions. Fig. 18.34 shows typical results obtained by Brossel and Bitter in which the deflexion is plotted as a function of the constant magnetic field for different values of the magnitude of the r.f. field. Provided this field is not too intense the curves are of typical resonance form with a maximum at a field such that the angular frequency ω of

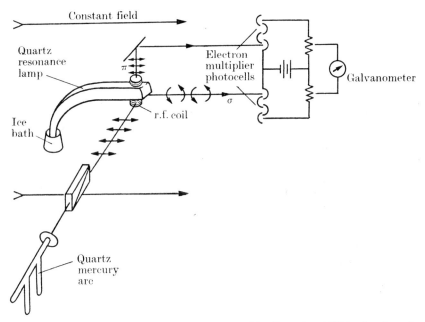

FIG. 18.33. General arrangement of apparatus used by Brossel and Bitter for observing 'double resonance'.

the r.f. field is equal to the frequency separation of the Zeeman levels with $M_J = 0, \pm 1$.

The line shape was shown to be given closely by the theoretical expression

$$\frac{\gamma^2 H_1^2}{\gamma^2 H_1^2 + (\omega-\omega_0)^2}\left\{\frac{\gamma^2 H_1^2}{4\gamma^2 H_1^2+(\omega-\omega_0)^2+1/\tau_r^2}+\frac{(\omega-\omega_0)^2}{\gamma^2 H_1^2+(\omega-\omega_0)^2+1/\tau_r^2}\right\}, \tag{92}$$

where H_1 is the magnitude of the r.f. field, γ is the gyromagnetic ratio for the $6\,^3P_1$ state, ω_0 is the Larmor frequency, γH, in the constant field, and τ_r is the radiative lifetime of the $6\,^3P_1$ state.

When $\gamma H_1 \ll 1/\tau_r$ the signal is bell-shaped but, if $\gamma H_1 \gg 1/\tau_r$, it shows two maxima (see Fig. 18.34). If the former conditions apply the width

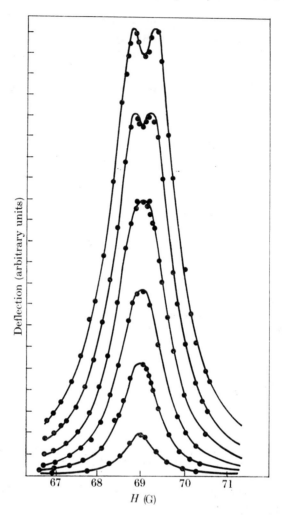

Fig. 18.34. Typical magnetic resonances observed by Brossel and Bitter in mercury vapour excited by resonance radiation. Curves with increasing peak height correspond to increasing magnitude of the r.f. field.

$\Delta\omega$ at half height is given by

$$(\Delta\omega)^2 = (4/\tau_r^2)\{1+5\cdot 8(\gamma H_1 \tau_r)^2\}. \tag{93}$$

When a foreign gas is added, so that quenching and depolarizing collisions occur, τ_r^{-1} is replaced by $\tau_r^{-1}+Z_q+Z_{al}$, where Z_q, Z_{al} are the frequencies of these respective collisions. It follows that, by measuring the line width under these conditions, Z_q+Z_{al} may be obtained.

Fig. 18.35 shows a typical plot of $\Delta\omega$ against pressure of admixed nitrogen obtained by Piketty-Rives, Grossetête, and Brossel.† Remembering that

$$Z_q + Z_{al} = K(Q_q + Q_{al})p, \tag{94}$$

where K is a constant and Q_q and Q_{al} are the quenching and disalignment cross-sections, the relation (93) suggests that, when $\gamma H_1 \ll 1/\tau_r$, the plot should be linear with a slope $2K(Q_q + Q_{al})$. From these measurements

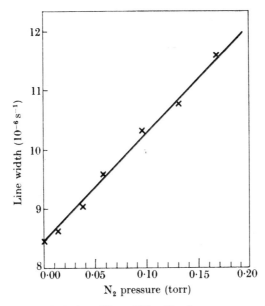

FIG. 18.35. Variation of line width with nitrogen pressure in a double resonance experiment in a mixture of mercury vapour with nitrogen (Piketty-Rives et al.).

$Q_q + Q_{al}$ was obtained as 122×10^{-16} cm² compared with 144×10^{-16} cm² measured by Barrat et al. using the Hanle effect (see Table 18.6). Further values obtained by the double resonance method are given in Table 18.8.

3.5. Depolarization of $Hg(6\,^3P_2)$

A remarkable extension of this method to determine cross-sections for depolarizing aligned $Hg(6\,^3P_2)$ metastable atoms was first introduced by Baumann‡ and further applied by Tittel.§ Since the state under

---

† PIKETTY-RIVES, C. A., GROSSETÊTE, F., and BROSSEL, J., C.r. hebd. Séanc. Acad. Sci., Paris **258** (1964) 1189.
‡ BAUMANN, M., Z. Phys. **173** (1963) 519.
§ TITTEL, K., ibid. **187** (1965) 421.

investigation is metastable it cannot be excited optically so electron excitation was used. This will produce transitions from the ground $6\,^1S_0$ state with $\Delta M_J = 0, \pm 1$, but not those with $\Delta M_J = \pm 2$. However, by application of an r.f. field under near-resonance conditions, transitions to these otherwise unoccupied levels will take place. The problem of detecting when such transitions have occurred was solved by irradiation with the green Hg line λ 5461 Å polarized parallel to the constant magnetic

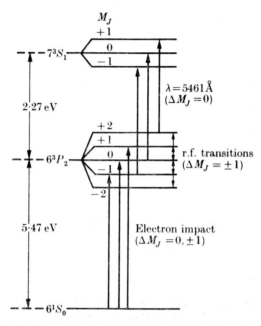

Fig. 18.36. Zeeman levels of the $6\,^3P_2$ and $7\,^3S_1$ states of mercury showing the transitions concerned in experiments on the disalignment of $6\,^3P_2$ by collisions.

field. This radiation will produce transitions, satisfying $\Delta M_J = 0$, from $6\,^3P_2$ to $7\,^3S_1$ (see Fig. 18.37). It follows from Fig. 18.36 that the only Zeeman sublevels of $6\,^3P_2$ which can contribute to the absorption are those with $M_J = 0, \pm 1$. Application of the r.f. field reduces the relative proportion of these levels and hence decreases the absorption of the polarized green line. In this way a signal of resonance shape could be obtained and the sum $Q_q + Q_{al}$ of the quenching and disalignment cross-sections derived in the usual way.

As resonance radiation is no longer involved (see § 3.8) Baumann was able to work with quite high mercury vapour pressure and thus could determine the cross-section $Q_q + Q_{al}$ for collisions with ground state mercury atoms. Tittel then extended the observations to obtain the cross-sections for depolarization by rare-gas atoms. The results are given in Table 18.8.

Similar experiments have been carried out for Cd($5\,^3P_2$) by Barrat,† the results of which are also given in Table 18.8.

3.6. *Depolarization of atoms excited by step-wise excitation*

So far we have been concerned with the depolarization of atoms excited in a single stage, either by absorption of resonance radiation or

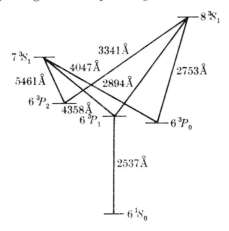

Fig. 18.37. Energy-level diagram for Hg.

by electron impact. It is possible to produce a sufficient population of atoms by step-wise excitation in certain cases to make it possible to observe depolarization of higher excited states.

The first example of such experiments was the study of the $7\,^3S_1$ level of mercury by Barrat, Cojan, and Lecluse.‡ Fig. 18.37 shows the appropriate energy level diagram. If mercury vapour is irradiated simultaneously by resonance radiation at 2537 Å and by radiation at 4358 Å, some of the atoms excited to the $6\,^3P_1$ level are raised to the $7\,^3S_1$ by further absorption of the longer-wave radiation. In practice the primary excitation was provided by two powerful lamps filled with mercury vapour in its natural isotopic state. The beams, from which the visible radiation at 4047, 4358, and 5461 Å, which could populate the $7\,^3S_1$ level by step-wise excitation, was filtered out, propagated along the direction of the uniform constant magnetic field H in which the absorption cell

---

† BARRAT, M., *C.r. hebd. Séanc. Acad. Sci., Paris* **259** (1964) 1063.
‡ BARRAT, J. P., COJAN, J. L., and LECLUSE, Y., ibid. **260** (1965) 1893; **262** (1966) 609.

was bathed. Through absorption of this unpolarized radiation the mercury atoms in the absorption cell were raised to the Zeeman levels with $M_J = \pm 1$ of the $6\,^3P_1$ states. However, at the mercury vapour pressure corresponding to saturation at 20 °C, diffusion of the resonance radiation depolarizes the Zeeman levels so there was an equal population in all three levels $M_J = \pm 1, 0$. The secondary source was a natural mercury vapour lamp providing a beam, perpendicular to the magnetic field, from which the blue line at 4358 Å was selected by suitable filters and polarized with electric vector perpendicular to the magnetic field.

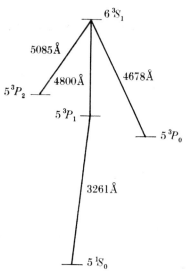

FIG. 18.38. Energy-level diagram for Cd.

Results obtained for the sum of the cross-sections for quenching and disalignment (see case (b), p. 1701, and (75)) due to collisions with different foreign gases are given in Table 18.8. Although no attempt was made in these experiments to work with even isotopes so as to avoid complications due to hyperfine structure, it seems that the separation of the hyperfine components in the two exciting lines is such that the contribution to the step-wise excitation from the odd isotopes is negligible. This was confirmed by carrying out observations with mercury in the cell, 92 per cent of which was ^{202}Hg and only 3·3 per cent consisted of odd isotopes. No difference was found from the result obtained with natural mercury.

Similar experiments[†] were carried out for the $8\,^3S_1$ level in which the secondary radiation at 2894 Å was used. Results are given in Table 18.8.

Lecluse[‡] applied the same technique to study the Hanle effect for the $6\,^3D_3$, $6\,^1D_2$, $6\,^3D_2$, and $6\,^3D_1$ levels of the $6s6d$ configuration of Hg and the results which he obtained for the sum of the quenching (Q_q) and disalignment (Q_{al}) cross-sections are also given in Table 18.8.

Laniepce[‡] has carried out similar experiments for the $6\,^3S_1$ level of cadmium (see Fig. 18.38). Because of the need to operate the discharge

[†] BARRAT, J. P., COJAN, J. L., and LECLUSE, Y., *C.r. hebd. Séanc. Acad. Sci., Paris* **262** (1966) 609.
[‡] LECLUSE, Y., *J. Phys., Paris* **28** (1967) 785; LANIEPCE, B., ibid. **29** (1968) 427.

lamps at 200 °C or so they were included within the same oven as the absorption cell. It was then impossible to use filters or polarizers between them and the cell. To produce alignment of the $6\,^3S_1$ atoms it was necessary for both the primary and secondary beams to propagate perpendicular to the magnetic field and use the polarization introduced by the field. Although the interpretation of the data was more complicated, the sum $Q_q + Q_{al}$ was determined for collisions with foreign-gas atoms. These results are given in Table 18.8.

3.7. *Effect of nuclear spin*

So far we have assumed that no complications arise from hyperfine structure. Allowance for this has been made by Omont,† who has shown that, in some cases, reasonably simple direct results can still be obtained.

We suppose that the duration of a collision is short compared with $\hbar/\Delta E$, where ΔE is the hyperfine structure energy separation. Under these circumstances, the nuclear spin can be regarded as completely decoupled from the electron spin and unaffected by the collision.

When the nuclear spin quantum number $I = \tfrac{1}{2}$, as for ^{199}Hg, there are two hyperfine structure levels with total spin quantum numbers $F = \tfrac{3}{2}, \tfrac{1}{2}$. It may be shown that,† because of the limitations imposed by the coupling between the nuclear and electron spin, an aligned as well as oriented population of atoms with $F = \tfrac{3}{2}$ may be set up but for $F = \tfrac{1}{2}$ orientation alone is possible. Hence, for atoms with $F = \tfrac{3}{2}$ we may define a disalignment cross-section $^{\frac{3}{2}}Q_{al}$ exactly as for atoms with no nuclear spin. On the other hand, when considering disorientation collisions, it is necessary to know not only $^{\frac{3}{2}}Q_{or}$ and $^{\frac{1}{2}}Q_{or}$ but also $^{\frac{1}{2},\frac{3}{2}}Q_{or} = {}^{\frac{3}{2},\frac{1}{2}}Q_{or}$, which is the cross-section for a collision in which disorientation is associated with a transfer from $F = \tfrac{1}{2}$ to $\tfrac{3}{2}$ or the inverse. Thus, if $\bar{J}(\tfrac{3}{2}), \bar{J}(\tfrac{1}{2})$ are the mean values of the total electron angular momentum associated with the hyperfine structure levels with $F = \tfrac{3}{2}, \tfrac{1}{2}$, respectively, we have

$$\left(\frac{\partial}{\partial t}\right)_{\text{coll}} \bar{J}(\tfrac{3}{2}) = -Kp\{{}^{\frac{3}{2}}Q_{or}\bar{J}(\tfrac{3}{2}) + {}^{\frac{3}{2},\frac{1}{2}}Q_{or}\bar{J}(\tfrac{1}{2})\}, \tag{95}$$

$$\left(\frac{\partial}{\partial t}\right)_{\text{coll}} \bar{J}(\tfrac{1}{2}) = -Kp\{{}^{\frac{1}{2}}Q_{or}\bar{J}(\tfrac{1}{2}) + {}^{\frac{3}{2},\frac{1}{2}}Q_{or}\bar{J}(\tfrac{3}{2})\}, \tag{96}$$

where $(\partial/\partial t)_{\text{coll}}$ is the time rate of change due to collisions and K is as in (94), so KpQ is the collision frequency.

Finally, we must introduce a further hyperfine exchange cross-section $^{\frac{1}{2},\frac{3}{2}}Q = {}^{\frac{3}{2},\frac{1}{2}}Q = Q_{he}$, which refers to collisions that change the hyperfine

† loc. cit., p. 1697.

states from $F = \frac{1}{2}$ to $\frac{3}{2}$, or the inverse, through change of the direction of the electronic angular momentum. These cross-sections are such that, if $n(\frac{3}{2})$, $n(\frac{1}{2})$ are the concentrations of atoms with $F = \frac{3}{2}, \frac{1}{2}$ respectively,

$$\left(\frac{\partial}{\partial t}\right)_{\text{coll}} n(\tfrac{3}{2}) = KpQ_{\text{he}}\left\{\frac{1}{\sqrt{2}}n(\tfrac{3}{2}) - \sqrt{2}\,n(\tfrac{1}{2})\right\}, \qquad (97)$$

$$\left(\frac{\partial}{\partial t}\right)_{\text{coll}} n(\tfrac{1}{2}) = KpQ_{\text{he}}\left\{\sqrt{2}\,n(\tfrac{1}{2}) - \frac{1}{\sqrt{2}}n(\tfrac{3}{2})\right\}. \qquad (98)$$

For the experimental determination of these cross-sections it is necessary to irradiate a resonance cell containing mercury enriched in the odd isotope ^{199}Hg (say to 99·7 per cent in typical cases) with radiation that will excite selectively states with either $F = \frac{1}{2}$, or $F = \frac{3}{2}$. A suitable source[†] for the former is a mercury lamp enriched to 98·7 per cent in the isotope ^{204}Hg. To excite selectively $F = \frac{3}{2}$ the lamp is enriched with ^{199}Hg and the radiation passed through a ^{204}Hg filter before entering the cell.

There is then no difficulty in determining $^{2}Q_{\text{al}}$ in exactly the same way as Q_{al} for atoms with no nuclear spin. To determine Q_{he} several methods may be used.

In one, due to Barrat, Cojan, and Lacroix-Desmazes,[‡] unpolarized radiation which will selectively excite states with $F = \frac{1}{2}$ is incident on the resonance cell in the direction of the uniform magnetic field and the scattered radiation observed at right angles. Under these conditions, when depolarization by radiative transitions of lifetime τ_{r} is included, (97–8) become

$$\frac{dn(\tfrac{3}{2})}{dt} = Z_{\text{he}}\left\{\frac{1}{\sqrt{2}}n(\tfrac{3}{2}) - \sqrt{2}\,n(\tfrac{1}{2})\right\} - \tau_{\text{r}}^{-1} n(\tfrac{3}{2}), \qquad (99)$$

$$\frac{dn(\tfrac{1}{2})}{dt} = R + Z_{\text{he}}\left\{\sqrt{2}\,n(\tfrac{1}{2}) - \frac{1}{\sqrt{2}}n(\tfrac{3}{2})\right\} - \tau_{\text{r}}^{-1} n(\tfrac{1}{2}), \qquad (100)$$

where R is the rate of excitation of $F = \frac{1}{2}$ by the incident radiation and $Z_{\text{he}} = KpQ_{\text{he}}$ is the rate at which hyperfine exchange collisions occur. In equilibrium

$$n(\tfrac{3}{2})/n(\tfrac{1}{2}) = \sqrt{2}\,\tau_{\text{r}} Z_{\text{he}} \Big/ \left(\frac{\tau_{\text{r}} Z_{\text{he}}}{\sqrt{2}} - 1\right). \qquad (101)$$

To measure $n(\tfrac{3}{2})/n(\tfrac{1}{2})$ it is only necessary to observe the scattered intensity at 90° with and without passage through a ^{204}Hg filter which

---

[†] SCHWEITZER, W. G., J. opt. Soc. Am. **51** (1961) 692.
[‡] BARRAT, J. P., COJAN, J. L., and LACROIX-DESMAZES, F., C.r. hebd. Séanc. Acad. Sci., Paris **261** (1965) 1627.

removes the radiation from the $F = \frac{1}{2}$ level. If I_w/I_{wo} are the ratios of these intensities we have

$$I_w = I_{\frac{3}{2}}, \qquad I_{wo} = I_{\frac{3}{2}} + I_{\frac{1}{2}},$$

where $I_{\frac{3}{2}}$, $I_{\frac{1}{2}}$ are the intensities of radiation emitted from excited atoms with $F = \frac{3}{2}, \frac{1}{2}$ respectively. Hence

$$n(\tfrac{3}{2})/n(\tfrac{1}{2}) = I_w/(I_{wo} - I_w).$$

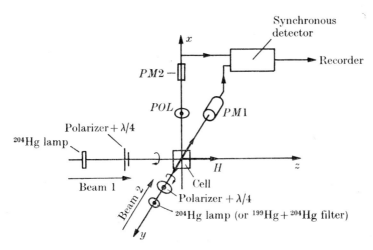

FIG. 18.39. General arrangement of apparatus used by Faroux to measure depolarization cross-sections for hyperfine states of ^{199}Hg ($6\,^3P_1$).

In an experiment of this kind it is necessary to allow for light reflected by the resonance cell by making background measurements when the cell is cooled in liquid air. Unwanted signals also arise from the presence of other isotopes in the cell or lamps. Faroux[†] has devised a phase-sensitive method of detection in which these complications do not occur. The general arrangement is as shown in Fig. 18.39.

A beam of circularly polarized radiation from a ^{204}Hg lamp is incident on a cell containing ^{199}Hg, in the direction of a constant magnetic field H. This excites mercury atoms so they are polarized in the direction of H. An r.f. field H_1 is applied at right angles at an angular frequency ω, which is in nuclear magnetic resonance for transitions between the Zeeman states of the excited atoms. This field produces a nuclear orientation which precesses around H with angular frequency ω. As a result, the absorption of a second circularly polarized beam from a ^{204}Hg lamp, incident along Oy, perpendicular to H, will be modulated at the

† FAROUX, J. P., *C.r. hebd. Séanc. Acad. Sci., Paris* **262** (1966) 1385.

same frequency. This modulation is detected by a photomultiplier and fed back to stabilize the field H at the resonance value. Because of the modulation, the emission of fluorescence radiation from each hyperfine level will also be modulated and in phase with that of the absorption. If phase-sensitive detection is used the signal obtained is free from background due to radiation scattered by the resonance cell, which is not modulated, and from the effect of other isotopes because the modulation is specific to ^{199}Hg.

To measure $^{\frac{3}{2}}Q_{\mathrm{or}}$, $^{\frac{1}{2}}Q_{\mathrm{or}}$, and $^{\frac{1}{2},\frac{3}{2}}Q_{\mathrm{or}}$ the Hanle effect may be used† as described on p. 1700. If the incident radiation excites only levels with $F = \frac{1}{2}$ the difference signal corresponding to (78), arising from levels with $F = \frac{1}{2}$, varies as

$$-\omega_{\frac{1}{2}}\tau_{\mathrm{r}}/\{(1+{}^{\frac{1}{2}}Z_{\mathrm{or}}\tau_{\mathrm{r}})^2+\omega_{\frac{1}{2}}^2\tau_{\mathrm{r}}^2\}, \tag{102}$$

while that arising from levels with $F = \frac{3}{2}$ varies as

$$^{\frac{1}{2},\frac{3}{2}}Z_{\mathrm{or}}\frac{\omega_{\frac{3}{2}}(1+{}^{\frac{1}{2}}Z_{\mathrm{or}}\tau_{\mathrm{r}})+\omega_{\frac{1}{2}}(1+{}^{\frac{3}{2}}Z_{\mathrm{or}}\tau_{\mathrm{r}})}{\{(1+{}^{\frac{1}{2}}Z_{\mathrm{or}}\tau_{\mathrm{r}})^2+\omega_{\frac{1}{2}}^2\tau_{\mathrm{r}}^2\}\{(1+{}^{\frac{3}{2}}Z_{\mathrm{or}}\tau_{\mathrm{r}})^2+\omega_{\frac{3}{2}}^2\tau_{\mathrm{r}}^2\}}, \tag{103}$$

provided $^{\frac{1}{2},\frac{3}{2}}Z_{\mathrm{or}}\tau_{\mathrm{r}}^2 \ll (1+{}^{\frac{1}{2}}Z_{\mathrm{or}}\tau_{\mathrm{r}})^2$ and $(1+{}^{\frac{3}{2}}Z_{\mathrm{or}}\tau_{\mathrm{r}})^2$, as it is in practice. Hence, by measurements with and without passage through a ^{204}Hg filter, $^{\frac{1}{2}}Z_{\mathrm{or}}$ may be obtained from (102). Similarly, if the incident radiation excites only levels with $F = \frac{3}{2}$, $^{\frac{3}{2}}Z_{\mathrm{or}}$ may be obtained from the equivalent of (102). $^{\frac{1}{2},\frac{3}{2}}Z_{\mathrm{or}}$ may then be obtained from comparison of (103) and (102).

Results obtained for these various cross-sections are given in Table 18.8.

3.8. *The measurement of cross-sections for self-depolarization*

The techniques we have been describing are applicable without special difficulty to the measurement of cross-sections Q_{al}, Q_{or} for collisions of foreign atoms with the excited atoms concerned. Considerable problems arise in some cases if the cross-section for collisions with atoms of the same kind as those excited are required. Thus consider again the case of the $6\,{}^3P_1$ state of Hg. To observe Q_{al}, Q_{or} due to collisions with ground-state Hg atoms it is necessary to work at such a high pressure of mercury vapour that resonance radiation is strongly trapped in the absorption cell. This leads to narrowing of double-resonance and Hanle-effect lines and to depolarization of the resonance fluorescence. As a result, the signals, which depend on the polarization, become weak and it is difficult to work at sufficiently high vapour pressures to observe the broadening due to collisions.

† FAROUX, J. P. and BROSSEL, J., *C.r. hebd. Séanc. Acad. Sci., Paris* **262** (1966) 41.

This difficulty is usually not serious if there are one or more intermediate levels between the excited level and the ground state to which optically allowed transitions may take place from the excited state. Photons emitted in such transitions are not trapped so that Hanle and double-resonance effects may be studied by observing the intensities and polarizations of these intermediate so-called 'cross-fluorescence' lines. An example which has been investigated experimentally by Happer and Saloman† is that of the $(6s^26p7s)\,^3P_1$ state of Pb, excited from the ground state by resonance radiation at 2833 Å. Cross-fluorescent radiation at 3639 Å, radiated in transitions to the $6s^26p^2\,^3P_1$ state, was observed. At cell temperatures below 1000 °C the concentration of atoms in this latter state due to thermal excitation was too small to produce any significant trapping. Experiments were carried out at 700 °C in a fused quartz cell containing a few milligrams of 99·75 per cent pure ^{208}Pb which possesses no nuclear spin. Results obtained are given in Table 18.8.

Other cases that present no serious problem are those in which the oscillator strength of the resonance transition is small. This does not mean that the depolarization cross-sections will also be small, as they are only proportional to the oscillator strength when this is large (see § 8.3). Under these circumstances, collision broadening may be accurately measurable at vapour pressures for which trapping is still unimportant. This is the case for the $^3P_1$ states of Cd and Zn.

For $Hg(6\,^3P_1)$ this does not apply as the oscillator strength for the resonance transition is large. However, the difficulty may be overcome by using mixtures of isotopes. Thus, if ^{200}Hg is present in a small concentration in a mixture with another isotope such as ^{202}Hg, the self-depolarization cross-sections for the $6\,^3P_1$ state of ^{200}Hg may be studied by irradiation of the mixture with resonance radiation for this state. This radiation will not be trapped in the main vapour of ^{202}Hg for which it will not be of exactly the resonance wavelength. Nevertheless, since the isotopes have the same electron structure, ^{202}Hg will depolarize ^{200}Hg on collision in exactly the same way as will ^{200}Hg atoms themselves. This is always provided the energy difference ΔE, representing the isotope shift between the states of the two isotopes, is such that $\Delta E \ll \hbar/\tau$, where τ is the time of collision, a condition usually satisfied. A useful check may be obtained by carrying out observations in different isotopic mixtures.

It is very important in experiments of this kind to ensure that the

† HAPPER, W. and SALOMAN, E. B., *Phys. Rev.* **160** (1967) 23.

incident radiation does not include any appreciable amount of resonance radiation for the concentrated isotope. To achieve this, Omont and Meunier,† in their experiments, passed the 2537 Å radiation through several filters filled with very pure mercury isotopes before allowing it

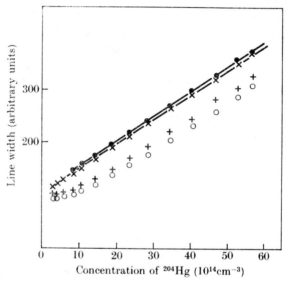

FIG. 18.40. Alignment and orientation broadening in the Hanle effect of ^{202}Hg in small concentration in ^{204}Hg. ○ alignment, + orientation, uncorrected. ● alignment, × orientation, corrected, taking fractional concentration of ^{202}Hg as 2×10^{-3}.

to enter the absorption cell. A correction for trapping was still necessary and it was carried out as follows. The line width γ may be written

$$\gamma = \tau_r^{-1}(1-\alpha x) + \nu_c, \qquad (104)$$

where τ_r is the natural lifetime of the state and ν_c the collision frequency. x increases from 0 to 1 in a known way (see § 2.1.3) as the concentration n_0 of trapping atoms goes from 0 to ∞, and α is a constant depending on the level and the process concerned. If n_0 is known, correction may be made for trapping, but the isotopic constitution of the samples used was not known accurately enough. Omont and Meunier therefore plotted γ against the total concentration N of mercury atoms, obtaining results of the type illustrated in Fig. 18.40 for the alignment and orientation broadening of the Hanle effect. To eliminate trapping effects the number n_0 was chosen so that, with the trapping term calculated from it eliminated from (104), the plot became linear. This process is shown

† OMONT, A. and MEUNIER, J., Phys. Rev. **169** (1968) 92.

in Fig. 18.40 in which it is seen that taking $n_0 = 2 \times 10^{-3} N$ gives linear plots for both the alignment and orientation broadening. This correction could not be carried out for ^{200}Hg and ^{202}Hg dilute in ^{198}Hg because the fractional concentrations were too small to show any appreciable departures from linearity in the uncorrected plots.

Table 18.7 gives results obtained by Omont and Meunier in this way for the cross-sections Q_{al}, Q_{or} for even isotopes and $^{\frac{3}{2}}Q_{al}$ for ^{199}Hg.

TABLE 18.7

Observed mean cross-sections Q_{al}, Q_{or}, $^{\frac{3}{2}}Q_{al}$ in 10^{-14} cm^2 for the $6\,^3P_1$ level in collisions between different pairs of mercury isotopes

Concentrated isotope \ Observed isotope	199	200			202		
	$^{\frac{3}{2}}Q_{al}$	Q_{or}	Q_{al}	Q_{al}/Q_{or}	Q_{or}	Q_{al}	Q_{al}/Q_{or}
198	14·2	14·0	14·9	1·06	13·0	13·7	1·05
199	—	14·9	16·1	1·08	13·6	14·0	1·03
201	—	13·6	14·6	1·07	13·3	14·0	1·04
202	14·0	13·2	14·0	1·08	—	—	—
204	12·6	13·9	14·3	1·03	13·7	14·1	1·02

It will be seen that there is quite good agreement between the results obtained for different isotopic pairs, particularly for the ratio Q_{al}/Q_{or}. In calculating average values, which are given in Table 18.8, the results for ^{200}Hg and ^{202}Hg in ^{198}Hg, which could not be corrected for radiation trapping, have been excluded, as well as for the pairs ^{202}Hg ^{199}Hg and ^{202}Hg ^{201}Hg for which the isotope shift is possibly not negligible.

3.9. *Depolarizing cross-sections for neon*

Decomps and Dumont[†] have determined disalignment cross-sections Q_{al} for collisions of neon atoms in the $2p_4$ state, with $J = 2$, with neon and with helium atoms in their ground states. They observed the change in the fluorescence light emitted by neon atoms in a cell excited by an electric discharge when illuminated with a beam of 6328 Å plane-polarized radiation from a laser in the presence of a uniform, steady magnetic field.

The energy level system of neon relevant to the interpretation of this experiment is shown in Fig. 18.41. It will be seen that the laser radiation is associated with the $2p_4$–$3s_2$ transition. The fluorescent radiation observed was usually that at 5943 Å due to the $2p_4$–$1s_4$ transition.

† DECOMPS, B. and DUMONT, M., *J. Phys., Paris* **29** (1968) 443.

Laser beams at 3·39 μm corresponding to the $3p_4$–$3s_2$ transition were also available.

Fig. 18.42 shows the geometrical arrangement. The plane-polarized laser beam was incident along the direction of the magnetic field on neon atoms in a cell excited by an electric discharge. Observations were made of the intensities of fluorescent radiation polarized along (π) and

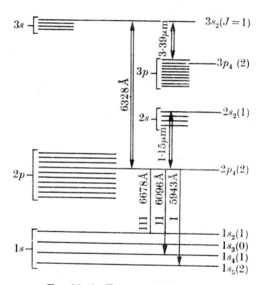

FIG. 18.41. Energy levels of neon.

perpendicular (σ) to the magnetic field, emitted in two mutually perpendicular directions perpendicular to the field, with and without the laser in operation.

Under these circumstances, which differ from those assumed in § 3.2 in which the radiation is incident normal to the magnetic field, the Hanle effect appears as follows.

Let $I_\parallel^\sigma$, $I_\perp^\sigma$ be the intensities polarized perpendicular to the field emitted in the directions Ox, Oy as in Fig. 18.42, when the laser is not operating. Then if $\Delta I_\parallel^\sigma$, $\Delta I_\perp^\sigma$ are the respective changes in these intensities produced by the laser beam, as the magnetic field varies,†

$$\Delta I_\parallel^\sigma - \Delta I_\perp^\sigma \propto \gamma/(\gamma^2 + 4\omega^2), \tag{105}$$

where $\omega = g\mu H$, g being the Landé g-factor for the level, and μ the Bohr magneton. In deriving this result it is assumed that the intensity of the laser beam is low enough for non-linear effects to be negligible.

† DECOMPS, B. and DUMONT, M., loc. cit., p. 1725.

If this is so γ may be obtained by observing $\Delta I_{\parallel}^{\sigma} - \Delta I_{\perp}^{\sigma}$ as a function of the magnetic field H. In practice, to allow for non-linear effects, γ is determined as a function of laser beam intensity and the results extrapolated to zero intensity.

If no alignment is transferred to the $2p_4$ level from $3s_2$, γ is given by

$$\gamma = \tau_r^{-1} + \nu_{a1},$$

where τ_r is the radiative lifetime and ν_{a1} the frequency of collisions

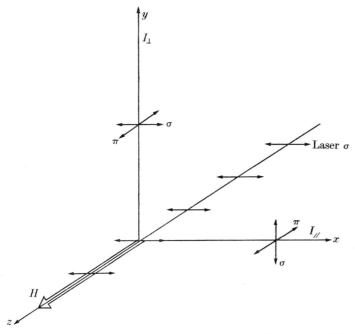

Fig. 18.42. Geometrical arrangement in the experiments of Decomps and Dumont.

producing disalignment of the $2p_4$ level. To verify that no significant contribution comes from transfer of alignment from $3s_2$, two experiments were carried out.

In the first the cell was irradiated with a laser beam at 3·39 μm ($3s_2$–$3p_4$) of such intensity as to produce the same change in population and alignment of the $3s_2$ level as that produced in the experiments with the 6328 Å radiation. As the 3·39 μm radiation has no direct effect on the $3p_4$ level, any Hanle effect observed in radiation from that level in this case is a measure of transfer from $3s_2$. No effect was observed.

Irradiation with a laser beam at 1·15 μm ($2s_2$–$2p_4$) provides the possibility of transfer of alignment from $2s_2$ instead of $3s_2$. As the spontaneous transition probability for $2s_2$ is higher than for $3s_2$ it would be expected that any transfer effect should be more marked in this case. However, no difference was found in γ for fixed pressure, composition, and discharge current from that measured with the laser beam at 6328 Å.

To separate τ_r^{-1} and ν_{al} it is necessary to make observations at different partial pressures of the gases in the cell. It is found, however, that γ

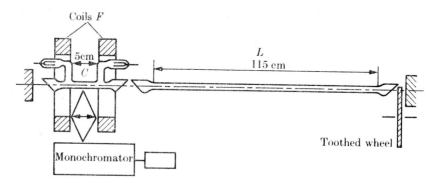

Fig. 18.43. Experimental arrangement used by Decomps and Dumont.

depends, for fixed value of the other variables, on the discharge current. Measurements were therefore made, at each set of partial pressures, for different values of the discharge current and the results extrapolated to zero current.

Fig. 18.43 shows the experimental arrangement used, corresponding to the geometry of Fig. 18.42. C, the absorption cell containing a mixture in variable proportions of helium and neon, which was excited by an electron discharge, was placed in the interior of the laser cavity. A uniform, constant magnetic field was produced throughout the cell by the field coils F. The tube L provided laser amplification for the neon radiation referred to above. To select the radiation to be amplified, mirrors, formed of multilayer dielectric materials so that their reflection coefficient only approached unity in a very narrow band of wavelength, were introduced in the laser system. Irrespective of the choice of mirrors, oscillation at 3·39 μm occurs but this may be blocked out, if desired, by introducing methane within the cavity.

The inclusion of the cell C within the laser cavity has the advantage of maximizing the energy density of radiation produced by the laser within the cell. Again, the separation of C from the amplifier tube L makes it unnecessary to adjust the discharge conditions in C to ensure laser oscillation. At the same time, the laser beam is little influenced by the magnetic field, which is localized around the cell C, or by the discharge conditions in C.

To measure directly a difference such as $\Delta I_{\perp}^{\sigma}$ or $\Delta I_{\parallel}^{\sigma}$ the laser beam was modulated into rectangular pulses of duration 4 ms. From the modulated signal in the recording photomultipliers the differences could then be directly obtained.

Disalignment cross-sections Q_{al} for collisions with neon and with helium atoms were obtained in this way and are given in Table 18.8.

3.10. *Depolarization of rubidium and caesium resonance radiation*

Gallagher[†] has used the Hanle effect to observe cross-sections for depolarization of the resonance radiation of rubidium and caesium. For the $P_{\frac{3}{2}}$ levels both alignment and orientation are involved but for $P_{\frac{1}{2}}$ only orientation. Cross-sections were obtained for disorientation and disalignment of Rb($5\,^2P_{\frac{3}{2}}$) and for the disorientation of Rb($5\,^2P_{\frac{1}{2}}$) and Cs($6\,^2P_{\frac{1}{2}}$) by various rare-gas atoms. These cross-sections, which are averages over the hyperfine components, are given in Table 18.8.

Fricke, Haas, Luscher, and Franz[‡] have used an optical pumping method to determine Q_{al} for collisions of Cs($6P_{\frac{1}{2}}$) atoms with rare-gas atoms. The principles involved in aligning ground-state alkali-metal atoms by optical pumping through absorption of circularly polarized D_2 ($P_{\frac{3}{2}} \to S_{\frac{1}{2}}$) radiation have been explained in Chapter 5, § 7.2. We consider first the case of no nuclear spin. Referring to Fig. 18.44 if the radiation is polarized right-handedly, absorption will produce transitions with $\Delta M_J = 1$, i.e. $-\frac{1}{2} \to \frac{1}{2}$ and $\frac{1}{2} \to \frac{3}{2}$ with probability in the ratio $1:3$.

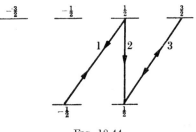

Fig. 18.44

Radiative transitions from the upper state with $M_J = \frac{1}{2}$ will take place, with relative probabilities $1:2$, to the lower levels with $M_J = -\frac{1}{2}, \frac{1}{2}$ respectively, while the upper state with $M_J = \frac{3}{2}$ can make radiative transitions only to the lower with $M_J = \frac{1}{2}$. Hence, if we denote the population of the lower levels with $M_J = \pm\frac{1}{2}$ by n_+, n_- respectively and of the upper levels with $M_J = \frac{1}{2}, \frac{3}{2}$ by $n_{\frac{1}{2}}$, $n_{\frac{3}{2}}$ respectively and if

---

[†] GALLAGHER, A., *Phys. Rev.* **157** (1967) 68.
[‡] FRICKE, J., HAAS, J., LUSCHER, E., and FRANZ, F. A., ibid. **163** (1967) 45.

there is no mixing of the upper levels by collision,

$$\frac{dn_+}{dt} = -\tfrac{3}{2}An_+ + \Gamma(\tfrac{2}{3}n_{\frac{3}{2}} + n_{\frac{1}{2}}),$$

$$\frac{dn_-}{dt} = -\tfrac{1}{2}An_- + \tfrac{1}{3}\Gamma n_{\frac{1}{2}},$$

$$\frac{dn_{\frac{1}{2}}}{dt} = \tfrac{1}{2}An_- - \Gamma n_{\frac{1}{2}},$$

$$\frac{dn_{\frac{3}{2}}}{dt} = \tfrac{3}{2}An_+ - \Gamma n_{\frac{3}{2}}.$$

Here A is the rate of absorption of the incident radiation and Γ the transition probability from the upper state.

When the excited levels are saturated

$$n_{\frac{1}{2}} = \tfrac{1}{2}An_-/\Gamma, \qquad n_{\frac{3}{2}} = \tfrac{3}{2}An_+/\Gamma,$$

so that
$$\frac{dn_+}{dt} = \tfrac{1}{3}An_-, \qquad \frac{dn_-}{dt} = -\tfrac{1}{3}An_-. \tag{106}$$

If A/Γ is small, $n_+ + n_-$ is constant throughout and the polarization P of the ground state atoms is given by

$$P = (n_+ - n_-)/(n_+ + n_-),$$

so that, from (106),
$$\frac{dP}{dt} = \tfrac{1}{3}A(1-P). \tag{107}$$

The equilibrium polarization is therefore unity.

On the other hand, if complete mixing occurs in the four upper states $M_J = \pm\tfrac{3}{2}, \pm\tfrac{1}{2}$, due to collisions, we have, if n' is the population of each such state,

$$\frac{dn_+}{dt} = -\tfrac{3}{2}An_+ + \Gamma n', \qquad \frac{dn_-}{dt} = -\tfrac{1}{2}An_- + \Gamma n',$$

so
$$\frac{dP}{dt} = A(-\tfrac{1}{2} - P), \tag{108}$$

and the equilibrium polarization is now $-\tfrac{1}{2}$.

In an intermediate case, if τ_c is the mean time between collisions and

$$\alpha = 1/\Gamma\tau_c = \tau_r/\tau_c,$$

where τ_r is the radiative lifetime of the upper state, the equilibrium polarization is given by

$$\frac{dP}{dt} = \{A/3(1+\alpha)\}(1-P) + \{A\alpha/(1+\alpha)\}(-\tfrac{1}{2} - P) = 0,$$

i.e.
$$P = \tfrac{1}{2}(2-3\alpha)/(1+3\alpha).$$

Thus the equilibrium polarization vanishes when $\alpha = \frac{2}{3}$. This provides a convenient means for determining τ_c. It is only necessary to vary the pressure of the foreign gas, to which $1/\tau_c$ is proportional, until the polarization due to the optical pumping vanishes.

We have assumed that the nuclear spin is zero but it may be shown that, for caesium with a nuclear spin quantum number $I = \frac{7}{2}$, the polarization vanishes when $\alpha = 1\cdot 82$ instead of $1\cdot 67$.

In the experiments of Fricke et al. the polarization produced by the optical pumping was monitored by a weak probing beam of D_1 radiation. By means of a special chopper this beam was circularly polarized, alternately right- and left-handedly, at 500 c/s. If the ground state atoms are polarized the absorption of these two radiations will be different so that the vanishing of P could be detected without difficulty.

The mixing cross-sections that they observed are given in Table 18.8.

3.11. *Results of measurements of depolarization cross-sections*

In Table 18.8 we summarize the results of measurements of the various depolarization cross-sections made by different authors. These results will be discussed, in relation to theory, in § 8.3.

4. Sensitized fluorescence

In preceding sections we have been concerned with the experimental study of the effect of foreign gases, or indeed of the main gas, in producing quenching or depolarization of fluorescent radiation. From such experiments measurements may be made of mean cross-sections for collisions that produce electronic transitions between an excited state of an atom or molecule and a second state. In quenching collisions this second state is often, but not always, the ground state, whereas depolarization is brought about through collisions in which the direction but not the magnitude of the total electronic angular momentum is changed.

Sensitized fluorescence occurs when the presence of a foreign gas, or an excess pressure of the main gas, leads to the fluorescent emission of radiation at wavelengths other than those emitted in the resonance fluorescence of the main gas at a low pressure. Such radiation arises through collisional excitation, either of different radiating states of the main atoms, or, through excitation transfer, of states of the foreign atoms.

We shall consider, in § 4.1, examples of the first of these possibilities in which transitions are excited between states of the main atom which differ only in the magnitude of the total electronic angular momentum quantum number J, i.e. different fine structure levels. In § 2.2 we have

TABLE 18.8

Observed depolarization cross-sections (10^{-16} cm²)

(a) Depolarization by foreign-gas atoms or molecules

Excited atom	Depolarizing atom or molecule	He	Ne	Ar	Kr	Xe	N₂	H₂	O₂
Hg(6 ³P₁)	$\bar{Q}_{al}$ (1)	38·2	47·8	82·9	119	164·5	4		
even isotopes	(2)	40·2	46·4	83·1	123	170	135	33	53·4
(no nuclear spin)	(3)	38	41	91	179	189	123	50	107
	$\bar{Q}_{or}$ (1)	42·7	57·3	101	133	178			
	(2)	45·5	60	100	144	179	144	61	112
¹⁹⁹Hg(6 ³P₁)	$^{3}_{2}\bar{Q}_{al}$ (1)	41·2	53·0	93·5	128	171			
(with nuclear spin)	$^{3}_{2}\bar{Q}_{or}$ (1)	30·9	33·4	55·0	81	108			
	$^{1}_{2}\bar{Q}_{or}$ (1)	40·4	48·2	83·5	117	157			
	$^{1}_{2},^{3}_{2}\bar{Q}_{or}$ (1)	3·7	7·5	0·13	0·16	17			
Hg(6 ¹P₁)	$\bar{Q}_{he}$ (1)	8·6	26·0	46·8	61·5	82			
	$\bar{Q}_{al}+\bar{Q}_{q}$ (4)	97·5	75·5	201	220	280	245	116	227
Hg(6 ³P₂)	$\bar{Q}_{al}$ (5)	70	80	150	200	290			
even isotopes	$\bar{Q}_{al}+\bar{Q}_{q}$ (6)	226	298	440	612	691			
(no nuclear spin)	$\bar{Q}_{al}+\bar{Q}_{q}$ (6)	239	251	408	596	660			
Hg(6 ³D₃)	$\bar{Q}_{al}+\bar{Q}_{q}$ (6)	220	261	432	585	722			
Hg(6 ¹D₂)	$\bar{Q}_{al}+\bar{Q}_{q}$ (6)	220	223	392	581	722	502		
Hg(6 ³D₂)	$\bar{Q}_{al}+\bar{Q}_{q}$ (7)	⩽ 0·15	⩽ 0·3	⩽ 0·45	⩽ 0·60	16·3	27·6		
Hg(6 ³D₁)	$\bar{Q}_{al}+\bar{Q}_{q}$ (7)	⩽ 0·45	⩽ 0·9	⩽ 1·2	4·5	94	126		
Hg(7 ³S₁)									
Hg(8 ³S₁)									
even isotopes									
(no nuclear spin)									
Cd(5 ³P₁)	$\bar{Q}_{al}$ (8)	44	53	94	132	192	157		
	$\bar{Q}_{or}$ (8)	50	60	110	148	220	157		
Cd(6 ³P₁)	$\bar{Q}_{au}+\bar{Q}_{q}$ (9)	0·6	1·2	1·2	4	1·5			
even isotope									
(no nuclear spin)									

(b) Self-depolarization

	Hg(6 $^3P_1$) even isotopes	^{199}Hg(6 $^3P_1$)	Hg(6 $^3P_2$)	Cd(5 $^3P_1$)	Cd(5 $^3P_2$)	Zn(5 $^3P_1$)	Pb($^3P_1$)	Ne($^2P_4$)
$\bar{Q}_{al}$	1460 (15)	—	210 (14)	250 (16)	97 (19)	140 (15)	7600 (17)	232 (18)
$\frac{3}{2}\bar{Q}_{al}$	—	1420 (13)	—	—	—	—	—	—
$\bar{Q}_{or}$	1400 (13)	—	—	—	—	—	—	—

Ne($^2p_4$) $\bar{Q}_{al}$ (10) — 226
Rb(5 $^2P_{3/2}$) $\bar{Q}_{al}$ (11) — 100
Rb(5 $^2P_{1/2}$) $\bar{Q}_{or}$ (11) — 57, 210
Cs(6 $^2P_{1/2}$) $\bar{Q}_{or}$ (11) — 9.0, 6.0, 9.7, 10.6, 130
Cs(6 $^2P_{3/2}$) $\bar{Q}_{or}$ (11) — 2.1, 0.8, 1.7
Cs(6 $^2P_{3/2}$) $\bar{Q}_{al}$ (12) — 100, 91, 188, 289, 356

References

1. Faroux, J. P. and Brossel, J., *C.r. hebd. Séanc. Acad. Sci., Paris* **263** (1966) 612.
2. Barrat, J. P., Casalta, D., Cojan, J. L., and Hamel, J., *J. Phys. Radium, Paris* **27** (1966) 608.
3. Piketty-Rives, C. A., Grossetête, F., and Brossel, J., *C.r. hebd. Séanc. Acad. Sci., Paris* **258** (1964) 1189.
4. Jean, P., Martin, M., and Lecler, D., *ibid.* **264** (1967) 1791.
5. Tittel, K., *Z. Phys.* **187** (1965) 421.
6. Lecluse, Y., *J. Phys., Paris* **28** (1967) 785.
7. Barrat, J. P., Cojan, J. L., and Lecluse, Y., *C.r. hebd. Séanc. Acad. Sci., Paris* **262** (1966) 609.
8. Laniepce, B. and Barrat, J. P., *ibid.* **264** (1967) 146.
9. Laniepce, B., *J. Phys., Paris* **29** (1968) 427.
10. Decomps, B. and Dumont, M., *ibid.* **29** (1968) 443.
11. Gallagher, A., *Phys. Rev.* **157** (1967) 68.
12. Fricke, J., Haas, J., Luscher, E., and Franz, F. A., *ibid.* **163** (1967) 45.
13. Omont, A. and Meunier, J., *ibid.* **169** (1968) 92.
14. Baumann, M., *Z. Phys.* **173** (1963) 519.
15. Dumont, M., Thèse, Paris, 1962.
16. Byron, F. W., McDermott, N. M., and Novick, R., *Phys. Rev.* **134** (1964) A615.
17. Happer, W. and Saloman, E. B., *ibid.* **160** (1967) 23.
18. Decomps, B. and Dumont, M., *J. Phys., Paris* **29** (1968) 443.
19. Barrat, M., *C.r. hebd. Séanc. Acad. Sci., Paris* **259** (1964) 1063.

described how, by using flash photolysis, it has been possible to measure cross-sections for such collisions between the J sub-levels of the ground term. By studying sensitized fluorescence it is possible to extend such studies to excited terms, particularly the upper levels of the D lines of the alkali-metal atoms.

This will be followed in § 4.2 by an account of experiments in which sensitized fluorescence is used to study excitation transfer. We shall first consider experiments in which the initial excitation is by resonant absorption of radiation and then discuss an important series of experiments in helium in which the initial excitation is due to electron impact.

4.1. Sensitized fluorescence and fine-structure transitions

4.1.1. *Collision-induced transitions between $^2P_{\frac{1}{2}}$ and $^2P_{\frac{3}{2}}$ states of alkali-metal atoms.* A remarkable example involving interaction of angular momenta was first observed by Wood.† He found that, if sodium vapour mixed with argon is irradiated by one of the D lines the other D line also shows up in the fluorescence. This must be ascribed to collisions with argon atoms in which a sodium atom excited to one D level undergoes a transition to the other, i.e.

$$\mathrm{Na}(3\,^2P_{\frac{3}{2}}) + \mathrm{Ar} \to \mathrm{Na}(3\,^2P_{\frac{1}{2}}) + \mathrm{Ar}. \tag{109}$$

In experiments by Lochte-Holtgreven‡ the ratio of the fluorescent intensity of the two D lines, when one only was used for excitation, was measured as a function of the argon pressure. From his results the cross-section for the process was estimated as $1\cdot 2 \times 10^{-14}$ cm² and for the inverse process about half as large.

Further experiments have since been undertaken by several investigators so that information is available, not only about the cross-sections for transitions of the type (109), but also for transitions in which the rare-gas atom is replaced by the normal metal atom concerned.

Let n_1, n_2 be the respective concentrations of $^2P_{\frac{1}{2}}$ and $^2P_{\frac{3}{2}}$ atoms respectively in the fluorescent cell. Then

$$\frac{\mathrm{d}n_1}{\mathrm{d}t} = S_1 - Z_{12}n_1 + Z_{21}n_2 - A_{10}n_1, \tag{110a}$$

$$\frac{\mathrm{d}n_2}{\mathrm{d}t} = S_2 - Z_{21}n_2 + Z_{12}n_1 - A_{20}n_2. \tag{110b}$$

In these equations S_1 and S_2 are the rates of production per cm³ of the

† Wood, R. W., *Phil. Mag.* **27** (1914) 1018; Wood, R. W. and Mohler, F., *Phys. Rev.* **11** (1918) 70.
‡ Lochte-Holtgreven, W., *Z. Phys.* **47** (1928) 362.

respective excited atoms through absorption of the incident radiation, Z_{12}, Z_{21} are the frequencies of collisions which produce the $^2P_{\frac{1}{2}} \to {}^2P_{\frac{3}{2}}$ transition and the inverse transition respectively, and A_{10}, A_{20} are the probabilities of optical transitions to the ground state from the respective excited states.

Suppose first that the cell is irradiated with the D_1 line only so $S_2 = 0$. Then, in equilibrium, from (110 b),

$$\frac{n_2}{n_1} = \frac{Z_{12}}{Z_{21}+A_{20}}. \tag{111}$$

Hence, if η_2 is the ratio of the intensity of the D_2 to the D_1 line in the fluorescent spectrum under these conditions,

$$\eta_2 = \frac{A_{20}n_2}{A_{10}n_1} = \frac{A_{20}Z_{12}}{A_{10}(Z_{21}+A_{20})}, \tag{112}$$

so
$$A_{10}\eta_2 Z_{21} - A_{20} Z_{12} = -A_{10}A_{20}\eta_2. \tag{113}$$

Similarly, if the cell is irradiated with the D_2 line only so that $S_1 = 0$,

$$\eta_1 = \frac{A_{10}n_1}{A_{20}n_2} = \frac{A_{10}Z_{21}}{A_{20}(Z_{12}+A_{10})}, \tag{114}$$

so
$$A_{20}\eta_1 Z_{12} - A_{10}Z_{21} = -A_{10}A_{20}\eta_1. \tag{115}$$

Hence, from (115) and (113),

$$Z_{12} = \frac{A_{10}(1+\eta_1)}{(-\eta_1+1/\eta_2)}, \tag{116a}$$

$$Z_{21} = \frac{A_{20}(1+\eta_2)}{(-\eta_2+1/\eta_1)}. \tag{116b}$$

From these formulae Z_{12} and Z_{21} may be obtained if η_1, η_2 are observed and the Einstein A coefficients A_{10}, A_{20} known from other experiments. If experiments are carried out using pure alkali-metal vapour Z_{12} can be written in the form
$$Z_{12} = N\bar{Q}_{12}\bar{v}, \tag{117}$$

where $\bar{v}$ is the mean relative velocity of impact between the alkali-metal atoms at the temperature of the vapour and $\bar{Q}_{12}$ is the mean cross-section for excitation of the $^2P_{\frac{1}{2}} - {}^2P_{\frac{3}{2}}$ transition. The same applies to Z_{21}.

One of the difficulties in carrying out the experiments is that trapping of the resonance radiation is hard to avoid. If it is trapped, then, in place of A_{20}, A_{10}, we must include reciprocal lifetimes $1/\tau_{1,2}$ which are determined by the diffusion time of the radiation through the vapour. To avoid this, it is necessary to work at very low vapour pressures with sensitive methods of detection.

Having determined $\bar{Q}_{12}$ and $\bar{Q}_{21}$ for impacts of the excited atoms with alkali-metal atoms, the corresponding cross-sections for impact with rare-gas atoms may be obtained by observing the change in Z_{12}, Z_{21}, produced by addition of a known partial pressure of the rare gas.

Fig. 18.45 illustrates the arrangement of apparatus used in experiments by Krause and his collaborators† in which the trapping of resonance radiation was reduced to an unimportant value. The resonance

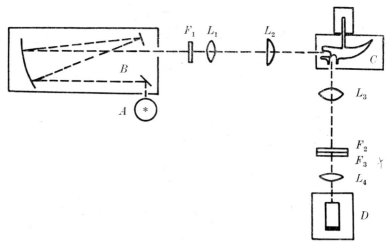

Fig. 18.45. Arrangement of apparatus used by Krause for studying sensitized fluorescence in alkali metal vapours. A, r.f. lamp; B, grating monochromator; C, oven with fluorescence cell; D, photomultiplier in a liquid-air cryostat, F_{1-3}, interference filters, L_{1-4}, lenses.

doublet was emitted from an r.f. light source. It was resolved by a grating monochromator, in series with interference filters, to a spectral purity of one part in 10^7. The emerging monochromatic beam was incident on the fluorescent cell containing the pure alkali-metal vapour or the vapour plus admixed rare gas. Fluorescent light emitted from the cell at right angles was again resolved to the incident spectral purity with interference filters and detected with photomultiplier. This was a liquid-air cooled ITT–FW—118 G tube with a very low dark current of 10^{-13} A so that very low intensities could be measured by a pulse-counting method. Additional factors that contributed to the high efficiency of the equipment in operating with very low vapour pressures

† See, for example, KRAUSE, L., *Appl. Optics*, **5** (1966) 1375; CHAPMAN, G. D., CZAJKOWSKI, M., RAE, A. G. A., and KRAUSE, L., *Abstr. IVth int. Conf. Phys. electron. and atom. collisions*, Quebec, 1965, p. 55.

were the light source, which produced sharp resonance lines of high peak intensity with no self-reversal, and the design of the fluorescence cells which is illustrated in Fig. 18.46. With the entrance and exit windows at right angles, neither the incident nor fluorescent radiation had to pass through a thick layer of unexcited alkali-metal atoms.

Before introduction of the alkali-metal sample into the side-arm of the cell the system was outgassed for several days at a temperature of 200 °C and a vacuum better than 5×10^{-7} torr.

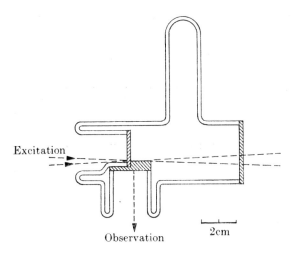

FIG. 18.46. Illustrating the design of the fluorescence cell used by Krause (see Fig. 18.45).

Fig. 18.47 illustrates some typical results showing the variation of Z_{12} with the density of rubidium atoms, in fluorescence from rubidium vapour, at three different temperatures. At densities of 4×10^{11} cm^{-3} and below the variation is linear, though Z_{12} does not vanish when extrapolated to zero density. This is presumably due to the effect of impurities. At densities above 4×10^{11} cm^{-3} there is a marked departure from linearity, which is probably due to trapping of resonance radiation.

Fig. 18.48 shows the observed variation of Z_{12} and Z_{21} with the pressure of helium atoms in caesium vapour. Over the pressure range concerned the behaviour is linear and the straight lines pass through the origin indicating that, at the caesium vapour pressure concerned, no important contribution comes from collisions with normal caesium atoms or with impurities.

Table 18.9 gives results obtained for $\bar{Q}_{12}$ and $\bar{Q}_{21}$ for a number of cases. Most of these have been obtained in the experiments described above but

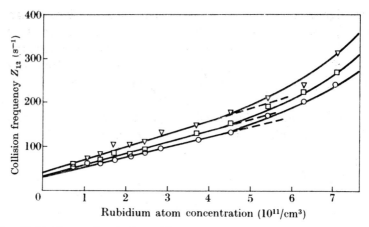

FIG. 18.47. Observed variation, with concentration of rubidium atoms, of the collision frequency for transfer of excitation from $5\,^2P_{\frac{1}{2}}$ to $5\,^2P_{\frac{3}{2}}$ levels of rubidium. ▽ 117 °C, □ 107 °C, ○ 87 °C.

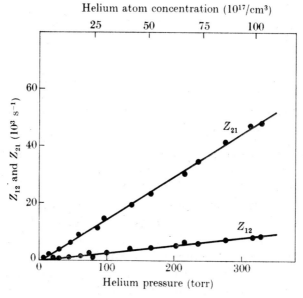

FIG. 18.48. Observed variation with helium pressure of the collision frequencies Z_{12}, Z_{21} for the transitions $6\,^2P_{\frac{1}{2}} \rightarrow 6\,^2P_{\frac{3}{2}}$ and $6\,^2P_{\frac{3}{2}} \rightarrow 6\,^2P_{\frac{1}{2}}$ induced in caesium by helium-atom impact.

some are due to Beahn, Condell, and Mandelberg† using a rather similar technique, and some to earlier investigators. The ratio $\bar{Q}_{12}/\bar{Q}_{21}$ at a temperature $T\,°\mathrm{K}$ is given, according to the principle of detailed balancing by

$$\frac{\bar{Q}_{12}}{\bar{Q}_{21}} = \frac{g_2}{g_1}\mathrm{e}^{-\Delta E/\kappa T}, \tag{118}$$

where ΔE is the energy difference between the two states and g_1, g_2 are their respective statistical weights so $g_1/g_2 = 2$. We give in Table 18.9 the temperature at which the measurements were made and a comparison between the observed and theoretical values of the ratio $\bar{Q}_{12}/\bar{Q}_{21}$. At least the trend of the observed ratios is in agreement with theory, though quantitative agreement can hardly be said to have been achieved. Agreement between results obtained by different experimenters is in about the same state.

There are many remarkable features of the results, particularly the immense range of values covered, from cross-sections considerably greater than gas kinetic for Na and K to the very low values for the rare gases in rubidium, and even more striking, in caesium. For a given alkali metal the cross-section tends to decrease in going from helium to argon or krypton and then to increase with xenon. A theoretical discussion is given in § 8.4.5.

Krause‡ has also carried out experiments in Rb–Cs mixtures and so has been able to obtain cross-sections for transfer of excitation from Rb to Cs states in which energy is released (downward transitions). His results are given in Table 18.10.

4.1.2. *The excitation of fine-structure transitions in other atoms.* We assemble in Table 18.9 (*b*) observed cross-sections for excitation of fine-structure transitions for other atoms. These have been obtained by various methods which are discussed in the sections referred to in the table. The only data available for excitation by atom impact show again the profound effect of the magnitude of the excitation energy ΔE. Thus for helium this is very small, comparable with that for sodium (see Table 18.9 (*a*)), and the cross-section is correspondingly large. On the other hand for Ne and Se ΔE is comparable with that for Rb and Cs and the cross-sections are very small.

Cross-sections for excitation by impact of diatomic molecules are much larger, even being measurable for I for which ΔE is as large as

---

† BEAHN, T. J., CONDELL, W. J., and MANDELBERG, H. I., *Phys. Rev.* **141** (1966) 83.
‡ loc. cit., p. 1736.

0·94 eV. Presumably vibrational excitation is involved but not in any very simple way.

4.1.3. *Cross-sections for 'mixing' collisions between the fine-structure levels of* $He(2\,^3P)$. The separations between the $2\,^3P_{0,1,2}$ fine-structure levels of ^{4}He are very small. Thus the lines due to transitions $2\,^3P_{0,1,2} \to 2\,^3S_1$ are close to 1·08 μm and separated by 1 and 0·08 Å respectively. Hence, whereas for the study of collisions with alkali-metal atoms in the $P_{\frac{1}{2},\frac{3}{2}}$ states it was possible to separate the D_1 and D_2 lines—and for all except sodium and lithium by use of interference filters—for the helium

TABLE 18.9

Observed mean cross-sections for excitation of fine structure transitions

(a) Excitation of transitions between $^2P_{\frac{1}{2}}$ and $^2P_{\frac{3}{2}}$ states of alkali-metal atoms

Alkali metal	ΔE (eV)	Colliding atom	T (°K)	$\bar{Q}_{12}$ (10^{-16} cm²)	$\bar{Q}_{21}$ (10^{-16} cm²)	$\dfrac{\bar{Q}_{12}}{\bar{Q}_{21}}$	$\dfrac{g_1}{g_2}e^{-\Delta E/\kappa T}$
Na	0·0021 (17·2 cm⁻¹)	He	—	—	41†		
		Ne	—	—	36†		
		Ar	—	—	65†		
				100	60‡		
K	0·0072 (58 cm⁻¹)	K	368	370	250§	1·48	1·6
		He	—	—	53†		
			368	59·5	41§	1·46	1·6
		Ne	—	—	14†		
			368	14	9·5§	1·51	1·6
		Ar	—	—	34†		
			368	37	22§	1·64	1·6
		Kr	368	61	41§	1·51	1·6
		Xe	368	104	72§	1·44	1·6
Rb	0·0295 (238 cm⁻¹)	Rb	340	53	68§	0·78	0·73
		He	—	0·10	0·12‖	—	—
			340	0·076	0·10₃§	0·74	0·73
		Ne	—	10⁻³–10⁻²	10⁻³–10⁻²‖	—	—
			340	1·7×10⁻³	2·3×10⁻³§	0·74	0·73
		Ar	—	10⁻³–10⁻²	10⁻³–10⁻²‖		
			340	1·0×10⁻³	1·6×10⁻³§	0·63	0·73
		Kr	340	6·4×10⁻⁴	1·5×10⁻³	0·43	0·73
		Xe	340	7·9×10⁻⁴	2·1×10⁻³	0·38	0·73
Cs	0·0687 (554 cm⁻¹)	Cs	311	6·4	31§	0·20	0·15
		He	311	5·7×10⁻⁵	3·9×10⁻⁴	0·15	0·15
		Ne	311	1·9×10⁻⁵	3·1×10⁻⁴§	0·06	0·15
		Ar	311	1·6×10⁻⁵	5·2×10⁻⁴	0·03	0·15
		Kr	311	8·3×10⁻⁵	18·4×10⁻⁴§	0·04	0·15
		Xe	311	7·2×10⁻⁵	27·4×10⁻⁴§	0·025	0·15

† JORDAN, J. A., Thesis, Michigan, 1964.
‡ SEIWERT, R., *Ann. Phys.* **18** (1956) 54.
§ KRAUSE, L., *Appl. Optics*, **5** (1966) 1375.
‖ BEAHN, T. J., CONDELL, W. J., and MANDELBERG, H. I., *Phys. Rev.* **141** (1966) 83.

(b) Excitation of fine structure transitions in other atoms

Atom	Transition	ΔE (eV)	Colliding atom	Cross-section 10^{-16} cm^2	Reference (section)
He	Between $2\,^3P_{0,1,2}$	9×10^{-6}, $1\cdot 2\times 10^{-4}$	He	52–68	4.1.3
Ne	$^3P_1$–$^3P_2$	0·052	Ne	$5\cdot 2\times 10^{-3}$	5.2.2.2
			He	$1\cdot 4\times 10^{-3}$	
	$^3P_0$–$^3P_1$	0·045	Ne	6×10^{-4}	
			He	6×10^{-4}	
	$^3P_0$–$^3P_2$	0·096	Ne	6×10^{-4}	
			He	6×10^{-4}	
Se	$4\,^3P_0$–$4\,^3P_1$	0·065	Ar	$4\cdot 7\times 10^{-3}$	2.2
			H$_2$	19	
			N$_2$	0·54	
			CO	0·20	
			O$_2$	0·29	
Br	$4\,^2P_{\frac{1}{2}}$–$4\,^2P_{\frac{3}{2}}$	0·434	H$_2$	$8\cdot 2\times 10^{-2}$	2.2
			D$_2$	0·14	
			N$_2$	$1\cdot 4\times 10^{-4}$	
			CO	$4\cdot 2\times 10^{-4}$	
			O$_2$	2·1	
I	$5\,^2P_{\frac{1}{2}}$–$5\,^2P_{\frac{3}{2}}$	0·942	H$_2$	2×10^{-3}	2.2
			D$_2$	3×10^{-3}	
			N$_2$	3×10^{-5}	
			CO	$7\cdot 5\times 10^{-5}$	
			O$_2$	0·59	
Hg	$6\,^3P_1$–$6\,^3P_0$	0·218	N$_2$	20·8	2.5
			CO	11·0	
NO†	$^2\Pi_{\frac{1}{2}}$–$^2\Pi_{\frac{3}{2}}$		NO	1	

lines this is not possible. However, advantage may be taken of a fortunate coincidence in that the isotope shift‡ between the levels of ^{3}He and ^{4}He is such that the $2\,^3S_1$–$2\,^3P_{1,2}$ line in ^{4}He coincides with the $2\,^3S_1$–$2\,^3P_0$ transition in ^{3}He. Hence, if ^{3}He gas containing ^{3}He atoms in the $2\,^3S$ state, produced for example by an electric discharge, is irradiated by resonance radiation from a ^{4}He lamp, then a useful proportion of these atoms will be excited selectively to the $2\,^3P_0$ level. The resonance

TABLE 18.10

Cross-sections for transfer of excitation in Rb–Cs *collisions*

Transition	$5\,^2P_{\frac{1}{2}} \to 6\,^2P_{\frac{3}{2}}$	$5\,^2P_{\frac{1}{2}} \to 6\,^2P_{\frac{1}{2}}$	$5\,^2P_{\frac{3}{2}} \to 6\,^2P_{\frac{3}{2}}$	$5\,^2P_{\frac{3}{2}} \to 6\,^2P_{\frac{1}{2}}$
Cross-section (10^{-16} cm^2)	1·5	0·5	0·9	0·3

† BAUER, H. J., KNESER, H. O., and SITTIG, E., *J. chem. Phys.* **30** (1959) 1119.
‡ FRED, M., TOMKINS, F. S., BRODY, J. K., and HAMERMESH, M., *Phys. Rev.* **82** (1951) 406.

fluorescence will include the three components $2\,{}^3P_{0,1,2} \to 2\,{}^3S$ which, by analogy with the alkali-metal case, we refer to as the D_0, D_1, and D_2 lines. In fact the separation of the D_1 and D_2 is of the same order as the Doppler width so only two components D_0 and D_3 will be observable, the latter being the sum of D_1 and D_2. By observing the variation of the intensity of these lines with cell pressure the cross-section for collisions in which $2\,{}^3P_0$ ^{3}He atoms are transferred to either $2\,{}^3P_1$ or $2\,{}^3P_2$ levels may be determined.

Experiments on these lines were carried out by Schearer.† He used as light source a capillary discharge excited with 8 W of r.f. power at 50 Mc/s modulated at 200 c/s. The gas cell was a Pyrex cylinder 3 in long and 1·2 in diameter filled with ^{3}He over a range of pressures from 0·05 to 10 torr. To provide a concentration of $2\,{}^3S_1$ metastable atoms the gas was excited by an r.f. source at 50 Mc/s. This concentration was estimated from optical pumping to be between 10^{10} and 4×10^{11} cm^{-3}. Changes in the populations of the ${}^3P_J$ levels due to the electric discharge were eliminated by detecting the change in the fluorescence at the modulation frequency. Light scattered at 90° from the cell was focused on the slits of a scanning spectrometer of resolution sufficient to separate clearly the D_0 and D_3 lines.

To avoid effects due to resonance trapping at each gas pressure the relative intensities of the D_0 and D_3 lines were measured at different total intensities $D_0 + D_3$ and the results extrapolated back to vanishing total intensity and hence vanishing concentration of $2\,{}^3S_1$ ^{3}He atoms. D_0/D_3 obtained in this way is given, for a pressure p of He, by

$$D_0/D_3 = (\alpha p^2 + \beta p + 1)/(\gamma p^2 + \delta p),$$

where

$\alpha = (Q_{10}Q_{20} + Q_{10}Q_{21} + Q_{12}Q_{20})N^2v^2\tau_r^2/p^2,$

$\beta = (Q_{20} + Q_{21} + Q_{10} + Q_{12})Nv\tau_r/p,$

$\gamma = (Q_{01}Q_{12} + Q_{21}Q_{02} + Q_{10}Q_{02} + Q_{02}Q_{12} + Q_{01}Q_{20} + Q_{01}Q_{21})N^2v^2\tau_r^2/p^2,$

$\delta = (Q_{01} + Q_{02})Nv\tau_r/p.$

τ_r is the radiative lifetime of the $2\,{}^3P$ state, N the number of ^{3}He atoms/cm^3, v the mean relative velocity of the atoms, and Q_{ij} the cross-section for excitation of the transition $2\,{}^3P_i \to 2\,{}^3P_j$. We may write

$$D_0/D_3 = b + \frac{a}{p} + \frac{c}{dp+e}, \qquad (119)$$

where $\qquad b = \alpha/\gamma, \qquad a = 1/\delta,$

† Schearer, L. D., *Phys. Rev.* **160** (1967) 76.

and c contains a factor $Q_{10}-Q_{20}$. Applying the detailed balance relationship (118) we find $b = 1/8$.

Observed results showed that D_0/D_3 varies linearly with $1/p$ and that the intercept on the axis of $1/p \to 0$ is $1/8$. This suggests that the third term in (119) is negligible, which it might well be if $Q_{10} \simeq Q_{20}$. If this is so then the slope of the plot gives a and hence $Q_{01}+Q_{02}$. In this way it was found that

$$Q_{01}+Q_{02} = 68 \pm 3 \times 10^{-16} \,\text{cm}^2 \text{ at } 300\,°\text{K}.$$

This value is consistent with the mixing cross-section determined from the line width of the excited state,[†] 53×10^{-16} cm², and from measurements similar to those described in § 3 for studying depolarization of resonance radiation which give 56×10^{-16} cm².[‡]

4.2. *Sensitized fluorescence and collisions involving transfer of excitation*

4.2.1. *Introduction.* The effect of the admission of foreign gas in quenching resonance fluorescence has been discussed in § 2.1. In certain circumstances, however, excitation may be transferred in collision to the foreign-gas molecules. Radiation from these excited molecules will then occur, giving rise to sensitized fluorescence. In many cases the transfer process will involve increase of kinetic energy so that the excited molecules will have a kinetic energy considerably in excess of the gas-kinetic value. This will appear in a Doppler effect of the emitted lines.

Observation of the intensity distribution in the spectrum of the sensitized fluorescence gives valuable information about the relative magnitudes of the cross-sections for transfer of excitation. It is difficult, however, to obtain the absolute magnitudes in this way.

The first experiments which established the existence of the effect were made by Cario and Franck,[§] who observed the sensitized fluorescence of thallium by mercury excited to the $6\,^3P_1$ level by resonance absorption of the 2537 mercury line. This level lies 4·9 eV above the ground state, but Cario and Franck found that thallium lines requiring up to 5·6 eV for excitation were excited. The additional energy presumably came from relative kinetic energy. Similar effects were observed with cadmium in place of thallium.

Sensitized fluorescence may occur due to transfer of excitation from metastable atoms. Such processes will be discussed in § 5.2 but it is of interest to point out here that evidence about the nature of the transition

[†] SCHEARER, L. D., *Phys. Rev.* **166** (1968) 30.
[‡] LANDMAN, D. A., *Bull. Am. phys. Soc.* **12** (1967) 94.
[§] CARIO, G. and FRANCK, J., *Z. Phys.* **17** (1923) 202.

responsible for the quenching of mercury resonance radiation by a particular foreign gas (see § 2.5) was obtained as long ago as 1924 by Donat† and Loria.‡ They made a study of the effect of argon, nitrogen, and hydrogen on the mercury-sensitized fluorescence of thallium. Whereas hydrogen produced a quenching effect on both the mercury and thallium emission, argon and nitrogen increased the thallium emission while decreasing that from mercury. This could be explained by supposing that the argon and nitrogen quench the mercury atoms by producing transitions to the metastable $6\,^3P_0$ level and not to the ground state. These atoms, which have a much longer lifetime towards radiation than the $6\,^3P_1$, are likely to be as effective per collision in transferring excitation to the thallium atoms. Because of their longer radiative life they would have much more chance to do so before radiating, so increasing the thallium fluorescence. Hydrogen, on the other hand, presumably quenches by some process which does not lead to the production of metastable atoms. Both these conclusions are supported by the much more recent evidence discussed in § 2.5.

4.2.2. *Experimental methods and results—energy resonance*

4.2.2.1. *Mercury-sensitized sodium fluorescence.* One of the first detailed studies of the relative effectiveness of energy transfer from a given excited atom to different states of a second atom was carried out by Beutler and Josephy.§ They studied the mercury-sensitized fluorescence of the diffuse series of sodium ($3\,^2P$–$n\,^2S$). A mixture of mercury and sodium vapour, at partial pressures of 8×10^{-3} and $0\cdot 015$ torr respectively, was irradiated with mercury resonance radiation (λ 2537). The relative intensity I of the diffuse series lines of sodium which were excited through such collision processes as

$$\mathrm{Hg}(6\,^3P_1) + \mathrm{Na}(3\,^2S) \rightarrow \mathrm{Hg}(6\,^1S) + \mathrm{Na}(9\,^2S),$$

was measured with a Zeiss microphotometer. The fraction of sodium atoms in a given excited state will be proportional to $If/\nu f_1$, where ν is the frequency of the line, f_1 the transition probability from the excited state to the final state concerned, and f that to all final states. In Beutler and Josephy's experiments f/f_1 was not determined, but the variation of I/ν with the initial state gives quite convincing results, as may be seen by reference to Fig. 18.49. This diagram includes lines originating from D as well as S states. To make these comparable it is necessary

---

† Donat, K., ibid. **29** (1924) 345.
‡ Loria, S., *Phys. Rev.* **26** (1925) 573.
§ Beutler, H. and Josephy, B., *Z. Phys.* **53** (1929) 747.

to divide I/v by the statistical weight g of the initial state, and this has been done in Fig. 18.49.

It will be seen that a sharp maximum of I/vg occurs for the $7S$ level, the excitation energy of which is only 0·020 eV higher than that of the $6\,^3P_1$ state of Hg. A weaker maximum seems also to be apparent for the $5S$ level which is closest to energy resonance with the metastable $6\,^3P_0$

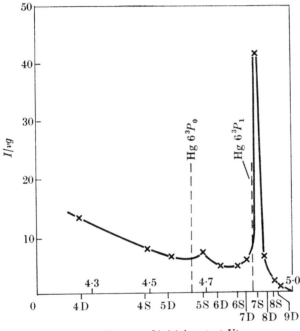

FIG. 18.49. Excitation probabilities of various sodium lines produced in collisions of the second kind with excited mercury atoms, as observed by Beutler and Josephy. The abscissae show the energies of the initial states of the Na lines. The broken lines indicate the energies of the $6\,^3P_0$ and $6\,^3P_1$ states of mercury.

state of Hg. An appreciable proportion of mercury atoms will be present in this state.

Thus the experiments indicate that the probability of transfer is greatest when the change $|\Delta E|$ in relative kinetic energy is least, i.e. when the difference in the excitation energy of initial and final states is least.

Unfortunately, Beutler and Josephy were not able to observe emission from the P states of sodium. Imprisonment prevented useful emission of radiation by transitions to the ground state while the wavelengths of $n\,^2P \to 4\,^2S$ were too long.

Over thirty years elapsed before any attempts were made to study the matter further. In 1962 Frish and Bochkova† obtained some evidence about the transfer of excitation to P states. They observed the effect on the sodium emission of adding mercury vapour to an electric discharge in a mixture of helium and sodium vapour at very low pressures. Among other lines the principal series of sodium was observed. The changes in concentration of sodium atoms in different excited states, brought about by the addition of the mercury, were determined by dividing the observed intensity of sensitized radiation in quanta/s by the transition probability for the line concerned. In general, the results are consistent with the assumption that the most likely transfer is to states for which $|\Delta E|$ is small, particularly when allowance is made for the probable presence of $Hg(6\,^3P_0)$ metastable atoms in the discharge. On the other hand, there is no evidence that the transfer to P states, which involves optically allowed transitions in both atoms, is more likely than to S or D states, as would be expected theoretically (see § 8.4.4.).

However, the presence of free electrons in these experiments complicates the interpretation and later experiments have been carried out by Kraulinya‡ and by Rautian and Khaikin§ in which the only source of excitation is optical. In the latter experiments the fluorescing mixture was confined in a spherical vessel with walls of ultra-violet-transmitting glass. The sodium and mercury were contained in side-arms whose temperatures could be separately controlled. During the experiments the sodium and mercury side-arms were maintained at 330 and 30–40 °C respectively and the temperature of the inner vessel at 350 °C, so that the concentration of sodium and of mercury was about 3×10^{14} cm^{-3}.

If I_{ki} is the intensity of radiation observed due to $k \to i$ transitions of frequency ν_{ki} in sodium, for which the transition probability is A_{ki}, then the total number of radiative transitions per second occurring from the state k to all lower states is

$$\frac{I_{ki}}{h\nu_{ki}} \frac{\sum_{i=k-1}^{0} A_{ki}}{A_{ki}}.$$

We have assumed that imprisonment of the radiation does not occur.

† FRISH, S. E. and BOCHKOVA, O. P., *Zh. éksp. teor. Fiz.* **43** (1962) 331; *Soviet Phys. JETP* **16** (1963) 237.
‡ KRAULINYA, E. K., *Optika Spektrosk.* **17** (1964) 464; *Optics Spectrosc., Wash.* **17** (1964) 250.
§ RAUTIAN, S. G. and KHAIKIN, A. S, *Optika Spektrosk.* **18** (1965) 722; *Optics Spectrosc., Wash.* **18** (1965) 406.

The rate at which the state k is populated by transfer collisions from Hg($6\,^3P_1$) is given by
$$n_0\,N_1\,Q_{0k}\,\bar{v},$$
where n_0, N_1 are the concentration of normal sodium atoms and of Hg($6\,^3P_1$) atoms. Q_{0k} is the cross-section for the transfer collision and $\bar{v}$ the mean relative velocity of impact. If the collision concerned is endothermic, an extra factor α must be included which gives the fraction of collisions in which the relative velocity is high enough for the transfer to occur.

An additional source of population of the state k is by radiative transitions from higher states. If n_m is the concentration of sodium atoms in the state m the rate of population of the state k by cascade transitions is
$$\sum_{m=k+1}^{\infty} n_m A_{mk}.$$

We must also allow for population and depopulation due to other collision processes. The population rate due to such processes we denote by
$$\sum_{s \neq k} n_s Z_{sk},$$
and that of depopulation by
$$n_k(Z_k + N_0\,Q_{k0}\,\bar{v}),$$
where $Z_k = \sum_{s \neq k} Z_{ks}$, N_0 is the concentration of normal mercury atoms and Q_{k0} the cross-section for a collision in which the excitation is transferred back to a mercury atom.

In equilibrium we then have

$$n_0 N Q_{0k}\bar{v} + \sum_{m=k+1}^{\infty} n_m A_{mk} + \sum_{s \neq k} n_s Z_{2k}$$
$$= \frac{I_{ki}}{h\nu_{ki}} \frac{\sum_{i=k-1}^{0} A_{ki}}{A_{ki}} + n_k(Z_k + N_0 Q_{k0}\bar{v}) \qquad (120)$$

so, ignoring depopulation by back transfer collisions,

$$Q_{0k} = \left\{ \frac{I_{ki}}{h\nu_{ki}} \frac{\sum_{i=k-1}^{0} A_{ki}}{A_{ki}} - \sum_{m=k+1}^{\infty} n_m A_{mk} + n_k Z_k - \sum_{s \neq k} n_s Z_{sk} \right\} \bigg/ n_0 N \bar{v}. \qquad (121)$$

Table 18.11 gives some results obtained by Kraulinya and by Rautian and Khaikin for Q_{0k} for a number of states, the concentrations N and n_0 being determined by absorption of the 4358 Å and 3302 Å lines respectively. In both sets of results the terms on the right-hand side of (121) which represent the effect of collisions in redistributing the excited state population have been ignored, while the contribution from

TABLE 18.11

Apparent mean cross-section Q_{ok} for transfer collisions between $Hg(6\,^3P_1)$ and normal sodium atoms

Excited sodium state	ΔE (eV)	Q_{ok} (10^{-16} cm^2)		
		(a1)	(a2)	(b)
10S	+0·07	0·2	—	—
9S	+0·02	2·7	—	4·5
8S	−0·05	0·9	0·9	—
7S	−0·17	0·7	0·8	—
6S	−0·37	0·5	0·7	—
5S	−0·76	0·3	0·8	—
9D	+0·09	0·5	—	—
8D	+0·04	3·0	—	—
7D	−0·03	2·4	2·4	18
6D	−0·13	2·0	2·2	—
5D	−0·29	1·8	2·8	—
4D	−0·60	1·7	4·6	—
12P	+0·16	—	—	—
11P	0·13	—	—	—
10P	0·09	—	—	—
9P	0·06	—	—	—
8P	−0·01	—	0·2	⩽ 0·008
7P	−0·10	—	—	—
6P	−0·25	—	—	—
5P	−0·52	—	—	—
4P	−1·10	—	—	—

(a1) From Kraulinya, allowing for population by cascade.
(a2) From Kraulinya, ignoring population by cascade.
(b) From Rautian and Khaikin, ignoring population by cascade.

cascade transitions is also ignored in the results of Rautian and Khaikin as well as in one set of results obtained by Kraulinya. The transition probabilities A_{ki} were taken from earlier experiments,† supplemented by calculations using the method of Bates and Damgaard.‡

The results are still consistent with the assumption that the cross-section tends to be larger the smaller the value of $|\Delta E|$ while the cross-sections for excitation of the P levels are again small. How small is difficult to judge because of the large differences between the results

† ANDERSON, E. M., BUSKO, Z. A., GRINBERG, R. O., and SAULGOZKA, A. K., *Vest. leningr. gos. Univ.* 4 (1956) 27.
‡ BATES, D. R. and DAMGAARD, A., *Phil Trans. R. Soc.* A**242** (1949) 101.

obtained in the two experiments. It must be remembered that the cross-sections given in Table 18.11 are only apparent cross-sections because redistribution of excited state population through collisions has been ignored.

Rautian and Khaikin have pointed out that, if deactivation of states such as Na($8P$) seriously reduces the apparent cross-section below a value at least comparable with, say, that for excitation of $9S$ or $7D$ then, since the small value given for $8P$ under (b) in Table 18.11 does not allow for population by cascade, we must have, in (120)

$$Z_k + N_0 Q_{k0} \bar{v} > \sum n_m A_{mk}/n_k,$$

where k refers to the $8P$ level. From their experiments they find

$$(Z_k + N_0 Q_{k0} \bar{v}) > 6 \times 10^7 \text{ s}^{-1}.$$

Since $N_0 \simeq 10^{14}$, $\bar{v} \simeq 10^5$ cm s^{-1}, $Q_{k0} \simeq Q_{0k}$, $N_0 Q_{k0} \bar{v} \simeq 10^3$. This means that $Z_k > 6 \times 10^7$ s^{-1} so that, if the depopulating process involves collisions either with normal sodium or mercury atoms, the effective cross-section would need to be of order 10^{-12} cm^2. Before ruling this out as much too large it is important to remember that we are dealing with highly excited atoms for which reactions such as (see Chap. 20, § 2)

$$\text{Na}' + \text{Na} \rightarrow \text{Na}_2^+ + \text{e},$$

are energetically possible and may have large cross-sections. The situation remains confused and we do not yet know how the true transfer cross-sections depend on $|\Delta E|$ and on the character of the transition.

4.2.2.2. *Mercury-sensitized indium fluorescence.* Donat's[†] observations of the mercury-sensitized fluorescence of indium are consistent with the assumption that the strongest transfer transitions are those for which $|\Delta E|$ is small. The excited levels of indium which lie closest to the $6\,^3P_1$ level of mercury are the $7P$ and $6D$ terms for which the energy discrepancies are 0·11 and 0·09 eV respectively. With two exceptions, all the lines observed strongly in the fluorescent spectrum could arise by cascade transitions from these terms. The exceptions are two transitions from the $7D$ and $8S$ terms but, at 900 °C, the working temperature necessary to generate a sufficient vapour pressure of indium, there will exist an appreciable number of metastable ($5\,^2P_{\frac{3}{2}}$) indium atoms with an excitation energy of 0·27 eV. ΔE for transfer from the metastable state is smallest for the $7D$ and $8S$ terms. There is again no evidence from these experiments to support the theoretical expectation, that, for

[†] DONAT, K., loc. cit., p. 1744.

given small $|\Delta E|$ the most probable transfer processes will be those in which each atom undergoes an optically allowed transition.

4.2.2.3. *Mercury-sensitized thallium fluorescence.* The mercury-sensitized fluorescence of thallium was also investigated at an early stage† but the results were more difficult to interpret. Referring to Fig. 18.50, the process with the least $|\Delta E|$ is that which excites the $8S$

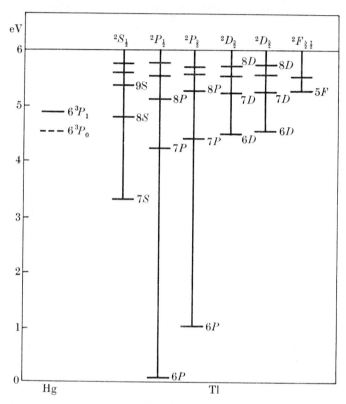

Fig. 18.50. Energy levels of thallium in relation to the $6\,^3P_0$ and $6\,^3P_1$ levels of mercury.

term ($\Delta E = 0\cdot 12$ eV). While lines originating in this level were observed, as well as quite strong lines from the $7S$ levels which can be reached via the cascade $8S$–$7P$–$7S$, the strongest line observed was one arising in a transition from the $6D_{\frac{3}{2}}$ level. For direct excitation of this level from Hg($6\,^3P_1$) $\Delta E = 0\cdot443$ V. It is difficult to see how the $6D$ terms could be populated so strongly from the $8S$ or $7P$ levels without strong lines corresponding to the populating transitions being observed.

† CARIO, G. and FRANCK, J., loc. cit., p. 1743.

However, as pointed out by Pringsheim,† the interpretation is complicated by the likely presence of metastable $Hg(6\,^3P_0)$ atoms which could excite $Tl(6D)$ with a considerably smaller energy discrepancy 0·22 eV (see Fig. 18.50), as well as Hg_2 and $HgTl$ molecules.

Again, after a long time-interval, further more quantitative experiments have been carried out. Kraulinya, Lezdin, and Silin‡ made absolute intensity measurements of a number of thallium lines emitted

TABLE 18.12

Relative intensities of mercury-sensitized fluorescent thallium lines

Transition	ΔE (eV)		Intensity/Intensity $(6\,^2D_{\frac{5}{2}}-6\,^2P_{\frac{3}{2}})$		
	$6\,^3P_1$	$6\,^3P_0$	(a)	(b)	(c)
$9\,^2S_{\frac{1}{2}} \to 6\,^2P_{\frac{3}{2}}$	0·47	0·68	0·0₃31	—	—
$8\,^2S_{\frac{1}{2}} \to 6\,^2P_{\frac{3}{2}}$	−0·08	+0·14	0·19	0·04	0·10
$8\,^2S_{\frac{1}{2}} \to 6\,^2P_{\frac{1}{2}}$			0·003	—	—
$7\,^2S_{\frac{1}{2}} \to 6\,^2P_{\frac{3}{2}}$	−1·60	−1·38	3·3	0·7	1·0
$7\,^2S_{\frac{1}{2}} \to 6\,^2P_{\frac{1}{2}}$			0·002	—	—
$7\,^2D_{\frac{5}{2},\frac{3}{2}} \to 6\,^2P_{\frac{3}{2}}$	0·33	0·55	0·004	0·3	—
$6\,^2D_{\frac{5}{2}} \to 6\,^2P_{\frac{3}{2}}$	−0·40	−0·18	1·00	1·00	1·00
$6\,^2D_{\frac{3}{2}} \to 6^2 P_{\frac{1}{2}}$			0·28	0·08	—

(a) Anderson and McFarland; Hg temp. 120 °C, cell temp. 703 °C.
(b) Kraulinya, Lezdin, and Silin; Hg temp. 50 °C, Tl temp. 710 °C, cell temp. 900 °C.
(c) Hudson and Curnutte; Hg temp. 0–30 °C, cell temp. 1070 °C.

from a thallium-mercury mixture maintained at 900 °C, supplied from side-arms with mercury maintained at temperatures of 20 and 50 °C and thallium maintained at temperatures from 400 to 800 °C. They found that the relative intensities varied little with the mercury and thallium temperatures in the range studied. Some values which they found are given in Table 18.12. It will be seen that they agree qualitatively with the much earlier observations.

Anderson and McFarland,§ a few years earlier, in the course of a study of the effect of added gases on the mercury-sensitized fluorescence of thallium, observed the relative intensities of thallium lines when no added gases were present. Some of their results, referring to fluorescence cell temperatures of 703 °C with the mercury source temperature 120 °C, are given in Table 18.12.

† PRINGSHEIM, P., *Fluorescence and phosphorescence*, p. 128 (Interscience, New York, 1949).
‡ KRAULINYA, E. K., LEZDIN, A. E., and SILIN, Y. A., *Optika Spectrosk.* **19** (1965) 154; *Optics Spectrosc., Wash.* **19** (1965) 84.
§ ANDERSON, R. A. and MCFARLAND, R. H., *Phys. Rev.* **119** (1960) 693.

It will be seen that, once again, the $6\,^2D$–$6\,^2P$, $7\,^2S$–$6\,^2P$, and $8\,^2S$–$6\,^2P$ lines are the most strongly excited but now the $7\,^2S$ transition is stronger than $6\,^2D$. Later experiments by Hudson and Curnutte,† directed towards the determination of absolute transfer cross-sections, which we shall now describe, gave rather similar results (see Table 18.13).

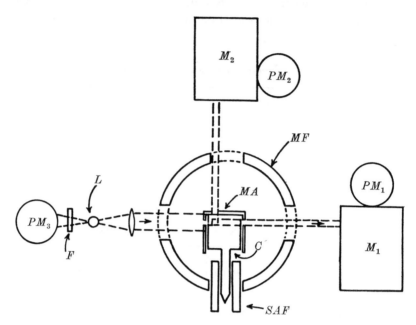

FIG. 18.51. Experimental arrangement used by Hudson and Curnutte in their study of the mercury-sensitized fluorescence of thallium.

In these experiments, which have already been referred to on p. 1691 in connection with the determination of cross-sections for quenching of $Hg(6\,^3P_1)$ by impact with thallium atoms, the arrangement was as illustrated in Fig. 18.51. Resonance radiation from a temperature-controlled Penray low-pressure mercury lamp L, collimated by a quartz lens, passed into the fluorescence cell C through a window cut in a ceramic mask MA. The intensity of the transmitted radiation was measured by a 1 P 28 photomultiplier PM_1 after passage through a Bausch and Lomb 500-mm grating monochromator M_1. Fluorescent radiation emitted from the cell in a direction normal to that of the incident light was similarly measured with a monochromator M_2 and photomultiplier PM_2. Finally, the output of the lamp L was monitored

† HUDSON, B. C. and CURNUTTE, B., *Phys. Rev.* **148** (1966) 60.

by a third photomultiplier PM_3 after passage through an interference filter with 250-Å bandwidth at half peak.

The fluorescence cell was a cube of 25 mm side with a side-arm reservoir 75 mm long. After outgassing at pressures less than 5×10^{-8} torr and temperatures above 1000 °C, the required amounts of mercury and thallium were vacuum distilled into the cell, which was then sealed off. In operation the cell was maintained at a temperature of 700 °C, or above, by a furnace MF. The side-arm temperature, which controlled the concentration of thallium atoms in the cell, was maintained about 50 °C lower than that of the main cell using an auxiliary furnace SAF. This eliminated thallium condensation in the cell windows.

The thallium concentration in the cell was determined from the known vapour pressure–temperature relation, allowance being made for the 50 °C difference between cell and side-arm temperatures. A more elaborate procedure was adopted for the measurement of the mercury concentration. At the operating temperature T, all of the mercury in the cell should be in the vapour form so the absorption of the resonance line should not vary with T. In all cases it was found to increase with T, a phenomenon considered to arise from the release of mercury atoms adsorbed on thallium. The mercury concentration in the cell was therefore determined by comparison of the line absorption in the sample cell with that in a secondary cell containing pure mercury vapour at a known pressure. The side-arm of this cell contained mercury that could be maintained at liquid nitrogen temperature or at temperatures between 0 and 30 °C. In this way the vapour pressure was fixed, and hence the pressure in the secondary cell, which was maintained at 850 °C.

The sensitivity of a 1 P 28 photomultiplier varies considerably with wavelength so that it was necessary to calibrate the multiplier PM_2, measuring the intensities of the sensitized fluorescence lines by comparison with a non-selective detector in the form of a Reeder vacuum thermopile equipped with a quartz window for use in the ultraviolet.

The radiative transition probabilities were calculated by the method of Bates and Damgaard.† Applying equation (121) with neglect of the Z_{sk}, gave the apparent cross-sections for transfer of excitation to the $8\,^2S$, $6\,^2D$, and $7\,^2P$ states, listed in Table 18.13 for three different cell temperatures, 800, 850 and 900 °C. The largest apparent cross-section is for excitation of $7\,^2P$, which requires an optically forbidden transition from the ground state. It must again be remembered that metastable

† loc. cit., p. 1748.

$\mathrm{Hg}(6\,^3P_0)$ atoms were certainly present, so the interpretation of the results is not clear cut.

Absolute transfer cross-sections have also been obtained by Kraulinya and Lezdin,† but, as they found great variability with the experimental conditions, it is difficult to include them for comparison.

TABLE 18.13

Cross-sections for transfer of excitation from mercury $6\,^3P_0$ and $6\,^3P_1$ atoms to thallium atoms

Temperature (°C)	Cross-sections in 10^{-16} cm²				
	$8\,^2S$	$6\,^2D/8\,^2S$	$7\,^2P/8\,^2S$	$7\,^2D/8\,^2S$	$9\,^2S/8\,^2S$
800	2·2	4·5	8·00	—	—
850	1·6	7·0	13	—	—
900	1·3	6·5	18	—	—
700	—	4·0	—	0·0025	0·003

In their earlier experiments, Anderson and McFarland obtained relative apparent cross-sections for excitation of some of the states, and their results are also given in Table 18.13 for a cell temperature of 700 °C. The ratio of $6\,^2D$ to $8\,^2S$ excitation cross-sections agrees well with Hudson and Curnutte's ratio—the absolute values of Table 18.13 are not accurate to much better than 30 per cent.

It seems from the experiments on sensitized fluorescence that the excitation-transfer cross-section, for given reacting species, does depend quite strongly on the internal energy discrepancy ΔE and in general is larger, the smaller is $|\Delta E|$. However, there are exceptions and, in the present semi-qualitative state of the experimental investigations, it is difficult to discuss any further systematic influences. There is no experimental evidence in support of the firm theoretical prediction (see § 8.4.4) that, provided ΔE is small, the cross-section for a transfer in which the transitions between both atoms are optically allowed will be much larger than gas-kinetic, exceeding considerably that for other transfer collisions in which one or both of the transitions are optically forbidden. The complicating influence of collision processes which can redistribute the excitation may be significant so that the interpretation of the data available at present may be more complex than assumed.

† KRAULINYA, E. K., LEZDIN, A. E., and SILIN, Y. A., *Optika Spectrosk.* **19** (1965) 154 (*Optics Spectrosc., Wash.* **19** (1965) 84).

4.2.2.4. *The enhancement of spark lines.* The enhancement of spark lines excited in a discharge by admixture of a suitable foreign gas was also studied at an early stage and provided evidence that the transfer reactions involved are most rapid when the change of internal energy involved is small.

A discharge through neon, in which metal vapour such as silver, gold, aluminium, copper, or lead is present, provides a strong source of the spark lines of the particular metal atom. The process responsible involves transfer of charge as well as excitation:

$$Ne^+ + Pb \to Ne + Pb^{+\prime}. \tag{122}$$

The most detailed study of these processes has been made by Duffendack and Gran.† They studied the enhancement of the spark lines of lead by neon, the enhancement of any level of Pb^+ being defined as the ratio of the intensity of a line originating in that level when excited mainly by neon ions to that when excited mainly by electron impact. The discharge tube was a low-voltage arc in neon in the presence of lead vapour. The lead was vaporized in a quartz crucible enclosed in a nickel cylinder. A special filament placed near the mouth of the crucible was the hot cathode, while a nickel hood, completely enclosing the filament and crucible, served as anode. The discharge was observed through a side tube in the anode.

Measurements were made of the intensities of various lead spark lines at a number of neon pressures from 5·5 to 1·6 torr, the intensity for a given arc current being independent of the pressure, as would be expected. These were compared with the corresponding intensities when the excitation was by electron impact only. This required a discharge potential of 450 V. There was little doubt that, at the neon pressures employed, the spark-line excitation was due to processes of the type (122). With the neon present, the lines were prominent even when the discharge was operated at a potential as low as 24 V, but when the pressure was reduced to 0·05 torr they did not appear even at 150 V. All intensities were measured in terms of those of certain standard lines that were assumed to be unaffected by the neon. These were chosen among lines whose initial levels were 1·6 eV below the ground level of Ne^+.

Fig. 18.52 illustrates the observed enhancements for the s, p, d, and f series of Pb^+ levels as a function of the energy discrepancy ΔE, the energy difference between the Pb^+ level and that of the ground level of Ne^+. The resonance character of the results for each series is clear, although, for the s and f series at least, the maximum enhancement appears at a small positive energy discrepancy. It is also clear that the enhancement is not determined wholly by the value of ΔE, for it is markedly different for the different series at the same ΔE.

Similar results were obtained in less extensive experiments by Duffendack and Thomson‡ on the enhancement of the copper and aluminium spark lines by neon.

An attempt was made by Duffendack and Gran† to obtain an estimate of the absolute value of the transfer cross-section, but, as this depends on various uncertain assumptions concerning the discharge conditions, the very large value they find must be regarded as very doubtful.

While there is some evidence of a 'resonance' character in excitation-transfer collisions, forthcoming from the earlier work described above,

---

† DUFFENDACK, O. S. and GRAN, W. H., *Phys. Rev.* **51** (1937) 804.
‡ DUFFENDACK, O. S. and THOMSON, K., ibid. **43** (1933) 106.

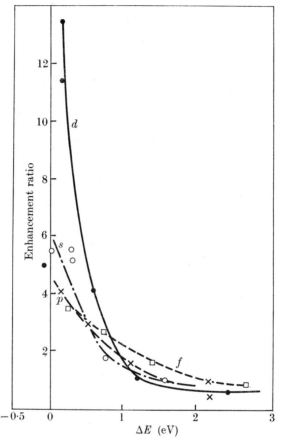

Fig. 18.52. Illustrating the enhancement of the spark lines of Pb in the presence of Ne. The observed enhancement ratio is shown as a function of ΔE, the difference between the excitation energy of the Pb II line concerned and the ionization energy of Ne. The curves refer to the s, p, d, and f series lines $-\cdot-\bigcirc\cdot$ s; $-\times--\times$ p; $-\bullet-$ d; $-\square--\square-$ f.

it is far from definite. Further evidence of a somewhat more quantitative character will be described in later sections, but first we turn to the consideration of another important general feature—Wigner's spin conservation rule. Of particular interest in this connection is the experimental information now available about collisions involving excitation transfer in helium.

4.2.3. *Wigner's spin-conservation rule.* Considerable attention has been devoted to the study of the effect of electron spin changes on the probability of transfer collisions. Owing to the very small coupling between the spin and electron orbital motion it would be expected that no change of total spin would occur in the collision. If s_1 and s_2 are initial spin quantum numbers of the electronic states of the colliding systems, the resultant spin quantum number S of the two systems taken together will have one of the values

$$s_1+s_2\ ,...,\ |s_1-s_2|. \tag{123}$$

Then we should expect that the only transfer reactions that would be important would be such that, if s_3 and s_4 are the spin quantum numbers of the final states, one of the numbers $s_3+s_4,..., |s_3-s_4|$, must be included in the set $s_1+s_2,..., |s_1-s_2|$. This rule was first enunciated by Wigner† and is known as the conservation of total spin.

As an example, suppose that $s_1 = 1$, $s_2 = 0$ so that one atom is in a triplet, the other in a singlet state. The total spin quantum number must be 1 so that, according to Wigner's rule, either $s_3 = 1$, $s_4 = 0$ or $s_3 = 0$, $s_4 = 1$. The chance of transfer when both the final states are singlets, giving $S = 0$, should be very much smaller.

Beutler and Eisenschimmel‡ tested this rule for the reactions

$$\text{Kr}(^3P_0)+\text{Hg}(6\,^1S_0) \rightarrow \text{Kr}(^1S_0)+\text{Hg}(8\,^1D_2), \tag{124a}$$

$$\text{Kr}(^3P_0)+\text{Hg}(6\,^1S_0) \rightarrow \text{Kr}(^1S_0)+\text{Hg}(8\,^3D_2), \tag{124b}$$

the first of which disobeys Wigner's rule while the second does not. The energy discrepancy in both cases is very small. The effect of adding krypton to a discharge in a mixture of mercury and helium was found to enhance the lines emanating from the triplet level much more than those from the singlet state, in accordance with the rule.

4.2.3.1. *Transfer collisions in helium—apparent breakdown of Wigner's rule.* On the other hand, strong evidence was brought forward by Skinner and Lees§ that the cross-sections for such reactions as

$$\text{He}(n\,^1P)+\text{He}(1\,^1S) \rightarrow \text{He}(1\,^1S)+\text{He}(n\,^3D) \tag{125}$$

are quite comparable with the gas-kinetic. In their experiments on the excitation of helium lines by an electron beam (see Chap. 4, § 1.3.1) they observed that, while most of the lines were emitted from the region of the beam, lines of the series $2\,^1S$–$n\,^1P$ and $2\,^3P$–$n\,^3D$ arose from a wider region. It was verified that, while the intensity of the 'spread' lines remained proportional to the beam current, it was not proportional to the gas pressure. The possibility that it arose from recombination to ions drifting out of the beam was excluded by showing that a powerful

---

† WIGNER, E., *Gött. Nachr.* (1927) 375.
‡ BEUTLER, H. and EISENSCHIMMEL, W., *Z. phys. Chem.* B**10** (1930) 89.
§ SKINNER, H. W. B. and LEES, J. H., *Proc. R. Soc.* A**137** (1932) 186.

electrostatic field applied at right angles to the beam had no effect on the relative intensity of the spreading.

There is no difficulty in providing an explanation of the spreading of the $2\,^1S-n\,^1P$ lines consistent with these results. The upper levels must be populated by resonance absorption of $1\,^1S-n\,^1P$ radiation emitted from the beam, a conclusion confirmed by an analysis of the intensity

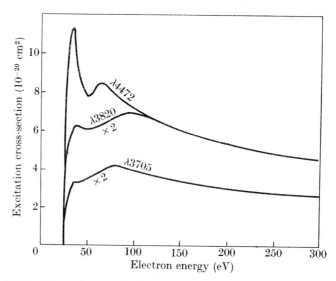

Fig. 18.53. Observed excitation functions for the 'spread' $\lambda\,4472$ ($2\,^3P-4\,^3D$), $\lambda\,3820$ ($2\,^3P-6\,^3D$) and $\lambda\,3705$ ($2\,^3P-7\,^3D$) lines of He excited by electron impact.

expected and its variation with pressure. No such possibility exists for the spreading of the $2\,^3P-n\,^3D$ lines. The similarity in the intensity–pressure relation for these and for the $2\,^1S-n\,^1P$ lines suggests that they arise from the population of $n\,^1P$ atoms outside the beam by a collision process such as (125). The strongest evidence in favour of this, which was provided by Skinner and Lees, was the form of the excitation functions observed for these lines. These are illustrated in Fig. 18.53. While the sharp maximum at low electron energies is expected for excitation to a triplet state, the second flat maximum is entirely uncharacteristic and closely resembles the excitation functions for lines originating from ^1P levels (see Chap. 4, Fig. 4.19). For the explanation to be valid, however, the process must occur with high probability in a gas-kinetic collision.

4.2.3.2. *Transfer collisions in helium—importance of transfer through F states.* Strong experimental evidence that, at least for states with

$n = 3$, the transfer reaction (125) is unimportant has recently been provided by Teter and Robertson.† They irradiated the positive column of a helium discharge with 5016 Å radiation which is strongly absorbed by $2\,^1S$ metastable helium atoms in producing the $2\,^1S$–$3\,^1P$ transition. With the strong source used, an increase in population of the $3\,^1P$ states of about 14 per cent was achieved. Consequent changes in the population of other states were looked for and, while that of $3\,^1D$ increased by about 6 per cent, that of $3\,^3P$ changed by less than 0·1 per cent. A 1·2 per cent increase in population of $3\,^3D$ was observed but could all be ascribed to small transmission of 5876 Å radiation through the filter which isolated the 5016 Å line. This radiation populated the $3\,^3D$ state through the $2\,^3P$–$3\,^3D$ transition.

The apparatus consisted of a cylindrical Pyrex discharge tube 25 cm and 1·04 cm internal diameter, around which was wrapped a jacket to contain liquid radiation filters through which the minimum path length of the radiation was 2 cm. A capillary discharge tube containing helium was wrapped in a spiral round the jacket and could be operated as a high intensity discharge (a few hundred milliamperes). The inner discharge tube contained helium at a pressure of 4 torr through which a current of 25 mA was passed.

The population of a particular state in the positive column of the inner discharge was determined from observation, along the axis of the tube, of the intensity of radiation emitted from that state in making a transition to a particular lower state. As there is a high concentration of metastable atoms in the discharge which leads to some imprisonment of radiation emitted in transitions to these states, and as such transitions were employed in the population measurements, a check was made by absorption measurements that the population of the metastable atoms themselves did not change to a significant extent as a result of irradiation by the 5016 Å radiation.

These experiments show that the cross-section for the transfer process
$$\mathrm{He}(3\,^1P)+\mathrm{He}(1\,^1S) \rightarrow \mathrm{He}(3\,^3P)+\mathrm{He}(1\,^1S) \tag{126}$$
must be less than 4×10^{-17} cm².

Meanwhile, the explanation of the apparent contradiction of the Wigner rule was given by St. John and Fowler.‡ Using the apparatus (Chap. 4, p. 191) which was primarily designed to measure optical excitation functions for electrons in helium, they confirmed the earlier observations and noted some other important features. Thus they pointed

† TETER, M. P. and ROBERTSON, W. W., *J. chem. Phys.* **45** (1966) 2167.
‡ ST. JOHN, R. M. and FOWLER, R. G., *Phys. Rev.* **122** (1961) 1813.

out that neither the observed excitation functions for the $4\,^3S$ nor the $3\,^3P$ levels show any evidence of population by transfer. At a pressure of 5×10^{-3} torr the observed excitation function for $3\,^3D\text{--}2\,^3P$ is of the type expected for excitation of a triplet state (see Chap. 4, Fig. 4.21) but at 5×10^{-2} torr it is of the form similar to that for $4\,^3D\text{--}2\,^3P$ shown in Fig. 18.53. For contrast, over this pressure range the observed excitation functions for $4\,^3S\text{--}2\,^3P$ and $3\,^3P\text{--}2\,^3S$ remain practically unaltered and of the expected form.

An $n\,^3D$ level can be populated by transfer from $n\,^1P$ directly or indirectly through transfer from $n\,^1P$ to $n\,^3P$ or to $n\,^3F$. If the first two of these possibilities were to apply we would expect them to lead to indirect population of $3\,^3P$, which does not occur. Again, thy should lead also to indirect population of $4\,^3S$ which is not observed. This leaves the possibility that the transfer reactions responsible are primarily

$$\mathrm{He}(n\,^1P)+\mathrm{He}(^1S) \to \mathrm{He}(n\,^3F)+\mathrm{He}(^1S). \qquad (127)$$

Support for this interpretation comes from theoretical calculation of the strength of the angular momentum coupling for the F states. It was shown that, for these states, there is a marked breakdown of LS coupling so the $^1F_3$ and $^3F_3$ levels are strongly mixed. Thus Lin and St. John† found a theoretical mixing ratio $1\cdot 19$ of $5\,^1F$ to $5\,^3F$. Under these conditions the transfer can be regarded as occurring to the ^1F component without violating the Wigner rule.

Detailed analysis of the observed data was then carried out by St. John and his associates in order to obtain information about the magnitude of the transfer cross-sections and their variation with n. In the first analysis‡ they used as input data the absolute apparent excitation cross-sections over a pressure range from $5\cdot 8$ to 130×10^{-3} torr as measured by St. John, Miller, and Lin.§

The production equations for the ^1P, $^{1,3}F$, ^3P, $3\,^1D$, $4\,^1D$, $3\,^3D$, and $4\,^3D$ states were written down, allowing for transfer reactions (127), for imprisonment of ^1P radiation and cascade effects. Radiative transition probabilities as given by Gabriel and Heddle‖ (see Chap. 4, Table 4.2) were used. The cross-section for excitation of the nF levels was taken to be the sum of those calculated for $n\,^1F$ and $n\,^3F$ by Massey and Mohr†† but the results are not very sensitive to this. Experimental values were

---

† LIN, C. C. and ST. JOHN, R. M., *Phys. Rev.* **128** (1962) 1749.
‡ ST. JOHN, R. M. and NEE, TSU-WEI, *J. opt. Soc. Am.* **55** (1965) 426.
§ ST. JOHN, R. M., MILLER, F. L., and LIN, C. C., *Phys. Rev.* **134** (1964) A888.
‖ GABRIEL, A. H. and HEDDLE, D. W. O., *Proc. R. Soc.* **A258** (1960) 124.
†† MASSEY, H. S. W. and MOHR, C. B. O., ibid. **140** (1933) 613.

assumed for the other apparent excitation cross-sections that occur (see Chap. 4, § 1.4.2.1). There remained the assumptions to be made about the transfer cross-sections $Q_{tr}(n)$.

$Q_{tr}(n)$ was assumed to have the form

$$Q_{tr}(n) = n^x q(x), \qquad (128)$$

where x could be taken as 1, 2, 3, 4, or 5. The $^1F_3$ and $^3F_3$ states were assumed to mix equally. In one set of calculations, referred to as model A, the $^3F_2$ and $^3F_4$ states were supposed to be inactive whereas in the other set, model B, they were considered to contribute through rapid transfer of excitation involving the $^3F_3$ state.

The best choice among the various possibilities offered in this way was made by deriving the apparent excitation functions for the $3\,^3D$, $4\,^3D$, and $4\,^1D$ levels over the observed pressure range and comparing with the shapes of those observed. By this procedure it was found that a low value of x and model A for the F states gave the best fit, the numerical values being

$$Q_{tr}(n\,^1P \to nF) = 3 \cdot 4 \times 10^{-15} n \text{ cm}^2, \text{ or } 7 \cdot 5 \times 10^{-16} n^2 \text{ cm}^2.$$

States with $n \geqslant 15$ make little contribution to the transfer process.

4.2.3.3. *Transfer collisions in helium—application of time-resolved spectroscopy.* This analysis has been considerably refined by Kay and Hughes† who applied the technique of time-resolved spectroscopy to study the processes involved. Excitation of the helium was carried out by impact of a gated electron beam, of rectangular cross-section (3 mm × 15 mm), whose energy could be varied. The beam was gated on until equilibrium was established and then gated off. The rate of decay of an excited state of interest was then measured by monitoring the intensity of a suitably chosen spectrum line using a d.c.-operated Amperex 56 AVP photomultiplier.

As it was expected that a number of different processes with different exponential decay rates would contribute to the over-all decay rate of a given excited state, the apparatus was designed to give an electronic logarithmic readout.

In this way the decay of the population of the $3\,^1P$, $4\,^1P$, $5\,^1P$, $3\,^3D$, $3\,^1D$, $4\,^1D$, and $3\,^3P$ levels was each studied at a number of different gas pressures and analysed in detail.

For $3\,^1P$, observed through the $3\,^1P$–$2\,^1S$ (λ 5015 Å) line, the decay was a simple exponential one. Fig. 18.54 shows the variation of the observed

† KAY, R. B. and HUGHES, R. H., *Phys. Rev.* **154** (1967) 61.

decay constant τ with gas pressure. Comparison with the theory of imprisonment shows good agreement if the imprisonment radius is taken to be 1·31 cm, a value used thenceforward in analysis of the results for the higher ^1P states. The decay of $3\,^3P$ measured from the $3\,^3P$–$2\,^3S$ (λ 3889 Å) transition is also purely exponential with a lifetime of 135 ns.

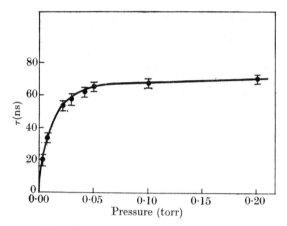

FIG. 18.54. Observed variation of the lifetime τ of the $3\,^1P$ level of helium as a function of gas pressure.

It is to be expected from the interpretation suggested by St. John and Fowler that, for the higher ^1P states, the decay will include additional components arising from transfer to F states. Thus, if N_p, N_f are the respective concentrations of atoms in ^1P and F levels of given total quantum number n, at time t after gating-off the exciting electron beam,

$$\frac{\mathrm{d}N_p}{\mathrm{d}t} = -(A_p + NvQ_p)N_p + NvQ_f N_f, \qquad (129\,\text{a})$$

$$\frac{\mathrm{d}N_f}{\mathrm{d}t} = -(A_f + NvQ_f)N_f + NvQ_p N_p. \qquad (129\,\text{b})$$

Here N is the concentration of normal atoms, A_p the radiative transition probability (allowing for imprisonment) of the ^1P level, A_f that for the F levels and Q_p, Q_f the respective cross-sections for transfer of excitation from ^1P to F and F to ^1P respectively.

Before discussing the solution of these equations we note that, if ^1Q_p, ^3Q_p are the separate cross-sections for $^1P \to {}^1F$ and $^1P \to {}^3F$ transfer and ^1Q_f, ^3Q_f those for the respective inverse processes, it follows from the principle of detailed balancing that

$$3\,^1Q_p = 7\,^1Q_f, \qquad 3\,^3Q_p = 21\,^3Q_f. \qquad (130)$$

If there is complete breakdown of LS coupling for the F states,

$$^3Q_{\mathrm{f}} = {}^1Q_{\mathrm{f}} = \tfrac{1}{2}Q_{\mathrm{f}} \quad \text{and} \quad {}^3Q_{\mathrm{p}} = 3\,{}^1Q_{\mathrm{p}},$$

so that, using (130),

$$Q_{\mathrm{f}} = (6/7)({}^1Q_{\mathrm{p}} + 3\,{}^3Q_{\mathrm{p}})$$
$$= (3/14)Q_{\mathrm{p}}. \tag{131}$$

If, however, ${}^3Q_{\mathrm{p}}/{}^1Q_{\mathrm{p}} = B$, where $B \neq 3$, we have

$$^1Q_{\mathrm{f}} = \frac{3Q_{\mathrm{p}}}{7(1+B)}, \qquad {}^3Q_{\mathrm{f}} = \frac{B}{7(1+B)} Q_{\mathrm{p}}. \tag{132}$$

In fact, it was found, from analysis of D state decay rates, that $B \simeq 3$ for $n > 4$ and in that case (131) applies. Returning to the solution of the equations (129) we have

$$N_{\mathrm{p}}(t) = C_1 e^{m_1 t} + C_2 e^{m_2 t}, \tag{133a}$$
$$N_{\mathrm{f}}(t) = C_3 e^{m_1 t} + C_4 e^{m_2 t}, \tag{133b}$$

where

$$m_{1,2} = \frac{1}{2}\Bigg[-\{(A_{\mathrm{p}}+A_{\mathrm{f}})+Nv(Q_{\mathrm{p}}+Q_{\mathrm{f}})\} \pm$$
$$\pm \{(A_{\mathrm{p}}-A_{\mathrm{f}})+Nv(Q_{\mathrm{p}}-Q_{\mathrm{f}})\}\bigg\{1+\frac{4N^2v^2Q_{\mathrm{p}}Q_{\mathrm{f}}}{(A_{\mathrm{p}}-A_{\mathrm{f}}+Nv\overline{Q_{\mathrm{p}}-Q_{\mathrm{f}}})^2}\bigg\}^{\frac{1}{2}}\Bigg]. \tag{134}$$

Since, for $n > 4$, $Q_{\mathrm{p}} \simeq 10 Q_{\mathrm{f}}$ we have approximately, when

$$NvQ_{\mathrm{p}} \ll A_{\mathrm{p}} - A_{\mathrm{f}},$$
$$m_1 = -(A_{\mathrm{f}} + NvQ_{\mathrm{f}}), \tag{135a}$$
$$m_2 = -(A_{\mathrm{p}} + NvQ_{\mathrm{p}}). \tag{135b}$$

We would therefore expect two components, one, usually the faster, essentially associated with P state decay and one with the F state decay.

Fig. 18.55 shows a typical observation of the decay of $5\,{}^1P$ at 0·1 torr which includes a long-lived component with decay time 125 ± 15 ns and a shorter-lived one with decay time 20 ± 4 ns. Using (135b), Q_{p} may be obtained from the variation of the short decay time with gas pressure. A_{p} is calculated from the imprisonment radius of the experimental tube determined from the $3\,{}^1P$ decay times and it is then found that a good fit (see Fig. 18.56) with the observed variation of m_2 with gas pressure is obtained if $Q_{\mathrm{p}} = 6\cdot 4 \times 10^{-14}$ cm². To derive Q_{f} from the long-lived component it is more convenient to proceed as follows.

When equilibrium is reached with the exciting electron beam on,

$$(A_{\mathrm{f}} + NvQ_{\mathrm{f}})N_{\mathrm{f}} = NvQ_{\mathrm{p}}N_{\mathrm{p}}, \tag{136}$$

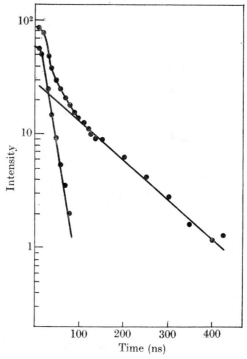

FIG. 18.55. Typical analysis of the observed decay of the population of $5\,^1P$ helium atoms at a pressure of 0·1 torr, with a fast-decaying and a long-lived component of decay times 20 ± 4 and 125 ± 15 ns respectively.

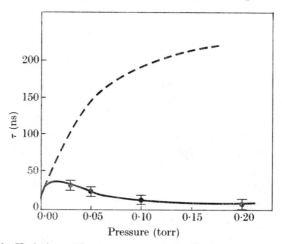

FIG. 18.56. Variation with gas pressure of the lifetime τ of the short-lived component of the observed decay of the population of $5\,^1P$ helium atoms (see Fig. 18.55). ⊕ derived from data as in Fig. 18.55; —— calculated from (135 b); --- calculated taking collisional transfer to be negligible.

provided there is no other means of populating the F levels than transfer from ^1P. If the equations (133) are solved with the initial conditions such that (136) is satisfied then, under the experimental conditions,

$$C_2/(C_1+C_2) \simeq N^2v^2Q_f\,Q_p/(A_f+NvQ_f)(m_1-m_2). \tag{137}$$

$C_2/(C_1+C_2)$ is the fraction of the intercept at $t = 0$ of the decay curve which is taken up by the long-lived component. Hence, from observation of this fraction, Q_p, m_1 and m_2 having been previously derived and A_f known, Q_f may be obtained and is found to be $0{\cdot}6\times 10^{-14}$ cm^2. This is consistent with (130).

Similar results were found for $6\,^1P$ which gave $Q_p = 13\times 10^{-14}$ cm^2, $Q_f = 0{\cdot}95\times 10^{-14}$ cm^2, again consistent with (130) when allowance is made for the probable errors. For $4\,^1P$ the long-lived component represented only about 15 per cent of the total, even at a pressure of $0{\cdot}2$ torr, so Q_f could not be obtained with accuracy. Q_p was found to be 2×10^{-14} cm^2.

The decay schemes found for the D levels were much more complicated. Fig. 18.57 (a) illustrates results for $3\,^3D$ (obtained from monitoring the $3\,^3D$–$2\,^3P$ line) at a pressure of $0{\cdot}034$ torr after excitation by 38-eV electrons, analysed into separate processes with decay lifetimes of 15, 70, 144, 260, and 700 ns. Of these, the first three are close to the theoretical lifetimes of the $3\,^3D$, $4\,^3F$, $5\,^3F$, and $6\,^3F$ levels respectively, while the longest-lived component probably includes contributions from higher ^3F levels.

Similar results were found for $3\,^1D$ (observed from monitoring the $3\,^1D$–$2\,^1P$ (λ 6678 Å line). Thus in Fig. 18.57 (b), taken at the same pressure as for $3\,^3D$ in Fig. 18.57 (a), but after excitation by 50-eV electrons, the 15, 61, 125, and 270 ns lifetimes correspond closely to those of the $3\,^1D$, $4\,^1F$, $5\,^1F$, and $6\,^1F$ levels. The long-lived 700 ns component could not be detected in this case because of the higher noise background. The shortest-lived (15 ns) component is more prominent than for ^3D, which is to be expected because of the larger cross-section for direct impact excitation of the singlet level. However, the $4\,^1F$ decay component is relatively stronger for ^1D than the $4\,^3F$ for ^3D which indicates that $4\,^1P \to 4\,^1F$ transfer is more probable than $4\,^1P \to 4\,^3F$. A more detailed analysis of the intercepts, allowing for cascade effects, gave the branching ratio B (see (132)) as $0{\cdot}5\pm 1{\cdot}0$ and $3{\cdot}0\pm 1{\cdot}6$ for $4F$, $5F$, and $6F$ respectively, suggesting that, as n increases, the breakdown of LS coupling in the F states becomes more marked and is practically complete when $n > 6$.

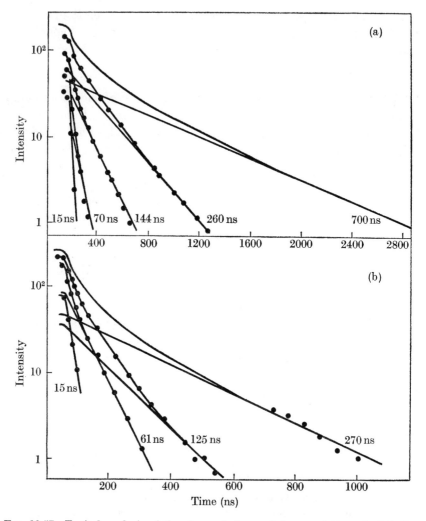

FIG. 18.57. Typical analysis of the observed decay of the population of 3D helium atoms at a pressure of 0·034 torr. (a) $3\,^3D$ excited by 38-eV electrons. (b) $3\,^1D$ excited by 50-eV electrons.

The apparent cross-section $Q'(^iD)$ for excitation of an ^iD state where $i = 1$ or 3 is now obtained as follows. It is assumed that the breakdown of LS coupling for the F states is complete. If J is the exciting electron flux, the rate of production of ^iD atoms is given by

$$JNQ(^iD) + \sum_n A(nF \to {}^iD)N_{\mathrm{f}}(n), \qquad (138)$$

where $N_{\mathrm{f}}(n)$ is the concentration of atoms in nF states, $A(nF \to {}^iD)$ is the probability for the radiative transition $nF \to {}^iD$, and $Q({}^iD)$ is the true excitation function for the ${}^iD$ state.

In equilibrium for any total quantum number n

$$-(A_\mathrm{p}+NvQ_\mathrm{p})N_\mathrm{p}+NvQ_\mathrm{f}N_\mathrm{f}+JQ({}^1P)N = 0, \qquad (139\,\mathrm{a})$$

$$-(A_\mathrm{f}+NvQ_\mathrm{f})N_\mathrm{f}+NvQ_\mathrm{p}N_\mathrm{p} = 0, \qquad (139\,\mathrm{b})$$

giving
$$N_\mathrm{f}(n) = \frac{JQ({}^1P)N^2vQ_\mathrm{p}}{A_\mathrm{p}(A_\mathrm{f}+NvQ_\mathrm{f})+NvQ_\mathrm{p}A_\mathrm{f}}. \qquad (140)$$

Hence (138) becomes
$$JNQ'({}^iD), \qquad (141)$$

where
$$Q'({}^iD) = Q({}^iD)\{1+F^{\,i}(N)\} \qquad (142)$$

with

$$F^{\,i}(N) = ({}^i\alpha)\sum_n \{Q(n\,{}^1P)/Q({}^iD)\}NvQ_\mathrm{p}A(nF \to {}^iD)\times$$
$$\times[A_\mathrm{p}(A_\mathrm{f}+NvQ_\mathrm{f})+NvQ_\mathrm{p}A_\mathrm{f}]^{-1}. \qquad (143)$$

${}^i\alpha = \tfrac{1}{4}$ or $\tfrac{3}{4}$ according as $i = 1$ or 3 and the A_p, A_f, Q_f, Q_p are those for the appropriate value of n.

$Q'({}^iD)$ in (141) is the apparent excitation cross-section and (143) shows how it should vary with gas pressure. $Q(n\,{}^1P)$ is the cross-section for excitation of the $n\,{}^1P$ state. Knowing these cross-sections as well as $Q({}^iD)$ and the transfer cross-sections, $F^{\,i}$ can be obtained. Extrapolation to large values of n is necessary. The cross-sections $Q(n\,{}^1P)$ were extrapolated on the assumption that they vary as n^{-3} (see Chap. 4, Fig. 4.22) and the transfer cross-sections assuming variation as n^{-4}, as indicated from the data for $n = 4$, 5, and 6. Fig. 18.58 (a) and (b) shows results obtained for $3\,{}^3D$ excitation by 38-eV electrons and $3\,{}^1D$ excitation by 50-eV electrons respectively. Comparison with the pressure variation of Q' observed by St. John and Nee[†] shows good agreement.

Although too much weight should not be given to the detailed numerical results, there seems little doubt that the proposed transfer mechanism suggested by St. John and Fowler[‡] is substantially correct and that the Wigner rule is not violated.

5. Collisions involving metastable atoms

We have already had occasion to refer to collisions in which metastable atoms are involved. Thus the part played by metastable mercury atoms in certain processes of sensitized fluorescence has been described in § 4.2.

[†] ST. JOHN, R. M. and NEE, T. W., *J. opt. Soc. Am.* **55** (1965) 426.
[‡] loc. cit., p. 1759.

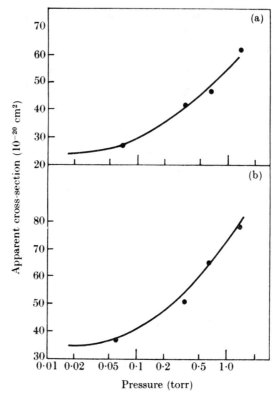

Fig. 18.58. Comparison of the pressure dependence of the apparent cross-sections for excitation of the $3D$ level of helium calculated (———) from the data obtained by Kay and Hughes with that observed (●) by St. John and Nee. (a) $3\,^3D$ excited by 38-eV electrons. (b) $3\,^1D$ excited by 50-eV electrons.

We now discuss experiments that yield more definite quantitative information about the rates of collision processes involving metastable atoms. The processes that can be studied in one way or another can be classified in three main categories as follows.

(a) Total and differential cross-sections for collisions of metastable atoms with other atoms and molecules.

(b) Processes leading to the destruction of metastable atoms, both in the gas and through diffusion to the walls of the containing vessel.

(c) Resonant processes of excitation transfer on collision with ground state atoms of the same kind.

In particular cases the data can often be analysed further so as to derive information about elastic, differential, and total cross-sections, diffusion coefficients, and cross-sections for specific inelastic collisions.

5.1. Total and differential cross-sections for collisions of metastable atoms with other atoms and molecules

5.1.1. *Measurement of total cross-sections.* The total cross-sections for collisions of metastable rare-gas atoms with other gas atoms may be measured by atomic-beam techniques in essentially the same way as for ground-state atoms (see Chap. 16, § 7). Fig. 18.59 shows schematically

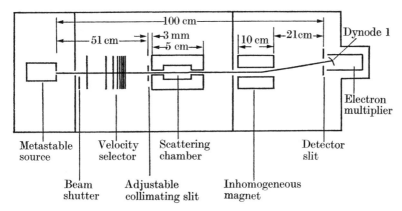

FIG. 18.59. Arrangement of apparatus used by Rothe, Neynaber, and Trujillo for measurement of total cross-sections for collisions of He metastable atoms with rare-gas atoms.

the apparatus used by Rothe, Neynaber, and Trujillo† for observing the velocity dependence of the total cross-sections for scattering of metastable $2\,^3S$ helium atoms in helium, argon, and krypton.

In place of the oven source for beams of ground-state atoms the source of metastable atoms was a low-voltage discharge between two copper plates in a Pyrex tube containing helium at about 0·07 torr pressure. The source slit, 9·5 mm high and 0·03 mm wide, was cut in the anode. The detector was an electron multiplier operated by the electron emission due to impact of the metastable atoms on the surface. The detector slit was similar to the source slit.

The beam issuing from the source slit included charged particles, photons, ground-state helium atoms, and helium atoms in non-metastable states as well as $2\,^1S$ and $2\,^3S$ metastable atoms. Charged particles were removed by electrical methods while photons could not pass the velocity selector. Excited atoms that decay in less than 3×10^{-4} s could not reach the detector. It remained to distinguish between $2\,^3S$ and $2\,^1S$ metastable

† ROTHE, E. W., NEYNABER, R. H., and TRUJILLO, S. M., *J. chem. Phys.* **42** (1965) 3310.

atoms. This was done by passage of the mixed beam through an inhomogeneous magnetic field which split the beam into three—an undeflected beam containing $2\,^3S$ atoms with the quantum number $M_J = 0$ as well as $2\,^1S$ atoms, and two beams deflected through equal and opposite angles containing $2\,^3S$ atoms with $M_J = \pm 1$. It was found that the intensity of the beam was reduced to about one-third of its intensity when the magnet was energized, providing a strong indication that the metastable atoms were very largely in $2\,^3S$ states. Furthermore, no difference was found in the total cross-section as measured for either the deflected beam, the undeflected beam with the magnet energized or with the magnet off. After these preliminary investigations the magnet was replaced by a 5-cm diameter metal tube coaxial with the beam. A positive electrical potential was applied to this tube so that any ions produced by such reactions as (see § 5.3)

$$\mathrm{He^*} + \mathrm{A} \to \mathrm{He} + \mathrm{A^+} + e, \tag{144}$$

where He* denotes a metastable helium atom, could not reach the detector. No evidence of the production of such ions was found.

The angular resolution judged by the Kusch criterion (Chap. 16, § 7.2.2) was 1' of arc in the first experiments. After a preliminary check to ensure that such high resolution was unnecessary, the collimator and detector slits were widened to give a resolution of 2'.

Absolute values of the total cross-section, not only for He $2\,^3S$ but also for $\mathrm{Ne}(2\,^3P_2)$ metastable atoms, were measured with the same apparatus by Rothe and Neynaber.† Particular attention was paid to the accuracy of the pressure measurements. These were made by means of an ion gauge calibrated by an expansion technique. The calibration was checked using a McLeod gauge operated at 0 °C so as to avoid any pumping effect (see Chap. 16, p. 1383).

5.1.2. *Discussion of results.* It is likely that, in most of these measurements, by far the largest contribution to the total cross-section comes from elastic scattering, an assumption that receives adequate justification when account is taken of the results of experiments to be discussed in § 5.4. For He* in He, collisions leading to transfer of excitation may be making a contribution of order 10 per cent at the highest relative velocities studied. In Ar and Kr the observed cross-sections for such processes as (144) (see § 5.3) are less than 3 per cent of the total.

We may therefore analyse the data, ignoring everything but the elastic scattering. This was done by the method of Chapter 16, § 7.2.1

† ROTHE, E. W. and NEYNABER, R. H., *J. chem. Phys.* **42** (1965) 3306.

using $s = 6$ for the He*–Ar and He*–Kr collisions but $s = 12$ for He*–He. Figs. 18.60 and 18.61 show the observed cross-sections as derived in this way for the three respective cases. In all cases comparison is made with corresponding results for collisions in which the He* is replaced by an Li atom that should have a very similar internal electron structure.

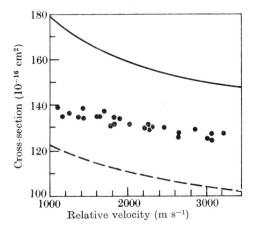

FIG. 18.60. Variation with relative velocity of the total cross-sections for collision of metastable He($2\,^3S$) and of ^{6}Li atoms with normal He atoms. ● He($2\,^3S$)–He observed by Rothe, Neynaber, and Trujillo; —— He($2\,^3S$)–He calculated by Buckingham and Dalgarno; – – – ^{6}Li–He observed by Rothe, Neynaber, and Trujillo.

As anticipated for He* in He the collisions are in the high-velocity range (Chap. 16, § 6.3.1). Indeed, the rate of fall of the cross-section with relative velocity is even slower than would be expected for $s = 12$. This may be due to a boost at higher velocity from excitation-transfer collisions, the cross-section of which increases with relative velocity in this range (see Fig. 18.96). It will be noted that the Li–He cross-section falls rather faster.

The total elastic cross-section has been calculated for He*($2\,^3S$)–He collisions by Buckingham and Dalgarno,† based on interactions‡ which they derived from quantum theory and which are discussed in more detail in connection with the inelastic transfer cross-sections (§ 8.1.3). Their results are given in Fig. 18.60. They show a more rapid increase

† BUCKINGHAM, R. A. and DALGARNO, A., Proc. R. Soc. A213 (1952) 327.
‡ DALGARNO, A. and KINGSTON, A. E., Proc. phys. Soc. 72 (1958) 1053.

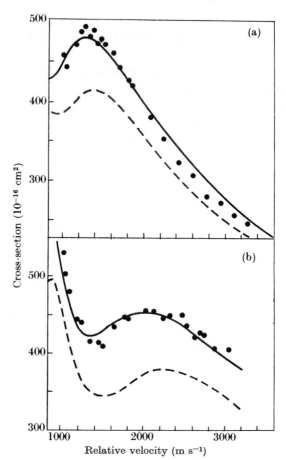

Fig. 18.61. Variation with relative velocity of the total cross-sections for collisions of metastable He($2\,^3S$) and of Li atoms with (a) Ar and (b) Kr atoms. ● He($2\,^3S$) observed by Rothe, Neynaber, and Trujillo; ⎯⎯⎯ He($2\,^3S$) calculated assuming a (12, 6) interaction; ⎯ ⎯ ⎯ Li observed by Rothe, Neynaber, and Trujillo.

as the relative velocity decreases than is observed. The significance of this in relation to the assumed interaction will be considered in § 8.1.3.

In contrast to the helium case, collisions of He* with Ar and Kr atoms are in the low-velocity region (Chap. 16, § 6.3.1) in the range of observation. The presence of undulations in the observed cross-sections as functions of velocity is apparent, in both cases, from Fig. 18.61 (a) and (b). In both cases there is a strong similarity in shape of the cross-section–velocity curve to that for the corresponding collisions in which He* is replaced by Li atoms but, as for impacts with He, the cross-section is somewhat larger for He*. From the location and width of the extrema the parameters ϵ and r_m (see Chap. 16, § 6.6) may be determined on the assumption of a (12, 6) form of interaction. The values derived are given in Table 18.14. Using these values the full cross-section velocity curve has been calculated and compared with the observed results in Fig. 18.61 (a) and (b). Comparison may be made between the values of the van der Waals constant $C\ (=2\epsilon r_m^6)$ derived from these data and those calculated from the Slater–Kirkwood formula (Chap. 16, (218)) using the polarizability of He($2\,^3S$) calculated by Dalgarno and Kingston.† The agreement is good.

TABLE 18.14

Potential parameters for interaction of He($2\,^3S$) *metastable atoms with argon and krypton atoms*

Atom	$\epsilon\ (10^{-14}\ \text{erg})$	r_m (Å)	van der Waals constant (10^{-58} erg cm^6)	
			$C\ (2\epsilon r_m^6)$	C (calc.)
Argon	0·77	5·07	2·6	2·6
Krypton	1·20	5·02	3·8	3·9

For He* in Ne the behaviour should be characteristic of the high-velocity region. At a relative velocity of $2\cdot095\times10^5$ cm s^{-1} Rothe and Neynaber obtain a cross-section of $1\cdot20\times10^{-14}$ cm^2. Furthermore, if $Q(\text{He*–A}, v)$, $Q(\text{Li–A}, v)$ denote the cross-sections for collisions of He* and of Li respectively with atom A, the relative velocity being v, then approximately, for collisions in the low velocity region (Chap. 16, § 6.3.1),

$$\frac{Q(\text{He*–A}, v)}{Q(\text{Li–A}, v)} \simeq \left\{\frac{C(\text{He*–A})}{C(\text{Li–A})}\right\}^{\frac{2}{5}},$$

where C denotes the appropriate van der Waals constant. According to the Slater–Kirkwood formula (Chap. 16, (218)), for these cases

$$\frac{C(\text{He*–A})}{C(\text{Li–A})} = \left\{\frac{\alpha(\text{He*})}{\alpha(\text{Li})}\right\}^{\frac{1}{2}},$$

where $\alpha(\text{He*})$, $\alpha(\text{Li})$ are the respective polarizabilities of He* and Li. Taking the observed value $2\cdot2\times10^{-23}$ cm^3 for Li‡ and the calculated

† The interaction between a metastable atom and a ground-state atom of the same kind is not unique (see § 8.1.3).
‡ ZORN, J. C. and CHAMBERLAIN, G. E., *Phys. Rev.* **129** (1963) 677; SALOP, A., POLLACK, E., and BEDERSON, B., ibid. **124** (1961) 1431.

value† for He* we find

$$\frac{Q(\text{He*–A}, v)}{Q(\text{Li–A}, v)} \simeq 1{\cdot}19,$$

independent of the nature of atom A, provided only that the collision is in the low-velocity region. The observed ratio is about 1·17 for argon and 1·20 for krypton, agreeing very well with these predictions.

Less detailed results are available for Ne*($^3P_2$) collisions: Rothe and Neynaber give cross-sections of 1·23, 1·43, and $3{\cdot}98 \times 10^{-14}$ cm² for collisions of Ne* of relative velocity 1·605, 1·173, and $1{\cdot}115 \times 10^5$ cm s⁻¹ with He, Ne, and Ar atoms. The first two cases fall in the high-velocity region, the third in the low-velocity region. An approximate deduction of the van der Waals constant using the Slater–Kirkwood formula (Chap. 16, (218)), and a polarizability for Ne* due to Robinson, Levine, and Bederson,‡ gives $C = 2{\cdot}0 \times 10^{-58}$ erg cm⁶ for the Ne*–Ar interaction which agrees quite well with the calculated value.

5.1.3. *Measurement of differential cross-sections.* Some information about differential cross-sections for scattering of helium metastable atoms has been forthcoming from the observations of Richards and Muschlitz.§ Essentially the method consisted in determining the variation of the total cross-section with the angular resolution.

Fig. 18.62 illustrates the general arrangement of the experiment. The atomic beam was produced by diffusion of helium through the hole H_1 in the source chamber. Some of the atoms were excited to metastable states, before passing through H_1, by electron impact, the electron beam being confined by a magnetic field of about 200 G between the poles D. The issuing atomic beam passed between two plates E, across which a potential was applied to remove charged particles, before emerging through the second defining hole H_2. The scattering chamber was contained within a cylindrical can G. The beam entered the chamber through a hole H_3 in G and a larger hole in the front SL of the chamber. It passed from the chamber through a hole H_4 in the bottom SB of the chamber to impinge on a target T. SL, SB, and the walls SC were insulated from each other and from T. When gas was present in the chamber the 'undeflected' beam produced secondary emission from the target T. To measure this, a positive potential of 16 V was applied to SB and SC with respect to SL and T which were grounded. The currents i_T^+ and i_{SL}^+ flowing to T and SL respectively were then measured. This procedure was repeated with the voltage reversed, giving reversed currents i_T^-, i_{SL}^- to T and SL. The intensity of the 'undeflected' beam was determined by i_T^+, of the scattered beam by $i_{SL}^+ + i_T^- + i_{SL}^-$ and of the total beam entering the chamber by $i_T^+ + i_{SL}^+ + i_T^- + i_{SL}^-$, provided the metastable atoms produce no ionization in the scattering gas. This confines the choice of the gas to helium and neon (see § 5.3).

---

† DALGARNO, A. and KINGSTON, A. E., loc. cit., p. 1771.
‡ ROBINSON, E., LEVINE, J., and BEDERSON, B., *Bull. Am. phys. Soc.* **9** (1964) 90.
§ RICHARDS, H. L. and MUSCHLITZ, E. E., *J. chem. Phys.* **41** (1964) 559.

18.5 UNDER GAS-KINETIC CONDITIONS

If a is the radius of the hole in SB and l is the path length in the scattering chamber then the mean angular resolution is given by

$$\Theta = l^{-1} \int_0^l \arctan(a/x)\, dx, \qquad (145)$$

ignoring the effect of finite beam width (0·05 cm in the actual experiments). a could be adjusted to values between 0·052 and 0·342 cm so that, since $l = 2\cdot 19$ cm, measurements could be made for values of Θ ranging from 6° 32′ to 25° 38′.

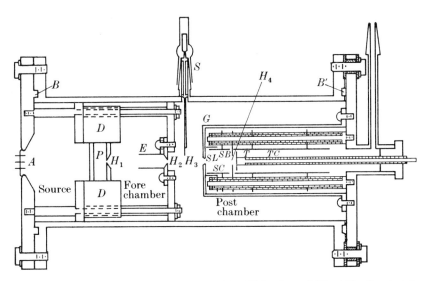

Fig. 18.62. Arrangement of apparatus used by Richards and Muschlitz for studying the scattering of metastable helium atoms in helium and neon.

It is necessary to allow for the presence of both $2\,^3S$ and $2\,^1S$ metastable atoms in the beam, including the possibility that the yield of secondary electrons for a given incident flux is different for atoms in the two states. Hence, if I_1, I_3 refer to the fluxes of $2\,^1S$ and $2\,^3S$ atoms respectively and γ_1 and γ_2 are the respective secondary emission coefficients, we have

$$i_0 = i_T^+ + i_{SL}^+ + i_T^- + i_{SL}^- = \gamma_1 I_{01} + \gamma_3 I_{03}, \qquad (146\,\text{a})$$

while
$$i_T^+ = \gamma_1 I_1 + \gamma_3 I_3. \qquad (146\,\text{b})$$

If $\bar{Q}_1$, $\bar{Q}_3$ are the mean cross-sections for collisions that deflect the atoms so that they are not collected on the target, and the chance of such collisions is small,

$$\ln(i_0/i_T^+) = nl(\bar{Q}_1 + \bar{Q}_3 R)/(1+R), \qquad (147)$$

where
$$R = \gamma_3 I_{03}/\gamma_1 I_{01} \qquad (148)$$

and n is the number of gas atoms per centimetre³ in the scattering chamber. From measurement of i_0 and i_T^+ at different gas pressures, keeping the source conditions constant, we obtain

$$\bar{Q}_3 + (\bar{Q}_1 - \bar{Q}_3)/(1+R) = \bar{Q}, \text{ say.} \qquad (149)$$

R may be varied by varying the energy of the bombarding electrons and/or the source pressure and it was determined by passing the beam through an inhomogeneous magnetic field. By plotting $\bar{Q}$ as a function of $1/(1+R)$, $\bar{Q}_3$ and $\bar{Q}_1-\bar{Q}_3$ may be obtained directly.

Fig. 18.63 illustrates $\bar{Q}_1$ and $\bar{Q}_3$, as observed for scattering in helium and in neon respectively, as functions of the mean angle of resolution Θ. These apply to conditions in which the source and scattering gas temperatures are 66 and 27 °C respectively. Conversion of these results to give differential cross-sections in the centre of mass system presents very complex problems (see Chap. 16, § 8) and has not been attempted. The results are not inconsistent with the high-resolution cross-sections for collisions with $He(2\,^3S)$ observed by Rothe and Neynaber (see § 5.1.2).

5.2. Measurement of the rates of processes which destroy metastable atoms

5.2.1. Introduction. We are concerned here with the determination of cross-sections for processes that lead to destruction of metastable atoms. As mentioned in § 5.1.1, these are, in general, much smaller than the total elastic cross-section. They include the following processes.

(a) Impact excitation to states from which optical transitions to the ground state are allowed.

(b) Collisions of the second kind in which the excitation energy is transferred to that of relative translation.

(c) Ionizing collisions of the type (144).

(d) Collisions between two metastable atoms in which the total excitation energy is concentrated in one and is sufficient to produce ionization
$$A^*+A^* \to A+A^++e. \tag{150}$$

(e) Three-body collisions leading to formation of excited molecules
$$A^*+2A \to A_2'+A. \tag{151}$$

(f) Diffusion to the containing walls.

The importance of impact excitation to higher states may vary very much from case to case. Thus in neon the first excited, $1s^2\,2s^2\,2p^5\,3s$, configuration gives rise to $^1P_1$, $^3P_0$, $^3P_1$, and $^3P_2$ terms. Of these the $^1P_1$ is strongly coupled optically to the ground state. The $^3P_1$ is also, weakly, coupled through departure from Russell–Saunders coupling but the $^3P_1$ and $^3P_2$ are metastable. Fig. 18.68 illustrates the separations of the different levels, showing that to excite the $^3P_2$ level to $^3P_1$ requires only 0·0517 eV. On the other hand, for the $2\,^3S$ and $2\,^1S$ metastable levels of helium, the nearest short-lived states $2\,^3P$ and $2\,^1P$ lie 1·1 and 0·6 eV respectively above. Under such conditions thermal destruction of metastable helium atoms by further excitation is most improbable but it may be very important for neon and, indeed, for other metastable rare-gas atoms.

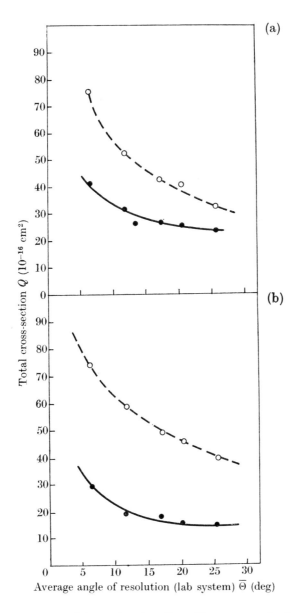

FIG. 18.63. Variation of the total cross-section with average angle of resolution Θ (in the laboratory system) for scattering of He($2\,^1S$) and He($2\,^3S$) metastable atoms in (a) helium, (b) neon. $-\bullet-\bullet-$ He($2\,^1S$). $-\bigcirc-\bigcirc-\bigcirc-\bigcirc$ He($2\,^3S$).

Collisions of the second kind to the ground state will usually involve very considerable energy transfer to translational motion, and this will render them very improbable. However, a collision of the second kind in which a $^3P_0$ neon atom makes a transition to the $^3P_1$ state liberates only 0·0446 eV and could well be important. This applies also to the heavier rare-gas atoms. Again, transitions between metastable states may occur as, for instance, the $2\,^1S$ and $2\,^2S$ states of helium, through collisions of this kind. Their importance is considerably enhanced if slow electrons are present (Chap. 4, § 3.1.1).

The importance of ionizing collisions of the type (144) was first pointed out by Penning and is often referred to as Penning ionization. Thus Kruithof and Druyvestyn† obtained information concerning the effectiveness of such reactions as

$$\text{Ne}^* + \text{A} \rightarrow \text{Ne} + \text{A}^+ + e, \tag{152}$$

in which Ne* represents a metastable neon atom, in the following way. The ionization coefficient α_i, for a swarm of electrons drifting through a gas at pressure p under a uniform electric field F, has been defined in Chapter 5, § 3.1. It is a function of F/p. For a pure atomic gas such as neon it is determined by the ionization cross-section of the atoms and the velocity distribution function for the electrons. Fig. 18.64 illustrates the observed‡ variation of α_i with F/p for highly purified neon. The effect of a small admixture of argon is very marked, particularly for small F/p, as may be seen from the observed results of Fig. 18.64. This may be interpreted in terms of the reaction (152).

Because of this, energy that would be used in producing excitation and emission of radiation is made available for ionization. From a study of the variation of α_i with argon admixture, it is possible to obtain values for the chance q that an excited neon atom passes into a metastable state and the ratio of the probability of destruction of a metastable Ne atom by collision with a Ne and an Ar atom respectively. The latter ratio comes out to be $2\cdot 4 \times 10^{-5}$. Since the probability of destruction in a Ne collision is of the order 10^{-5} (see Table 18.16), it follows that the chance of the reaction (152) occurring is of the order unity per collision. It is a reaction that may be regarded as one of exact resonance as far as the relative translational energy of the heavy particles is concerned, the electron taking up the energy excess (16·53 eV excitation

---
† KRUITHOF, A. A. and DRUYVESTYN, M. J., *Physica, 's Grav.* **4** (1937), 450; KRUITHOF, A. A. and PENNING, F. M., ibid. 430.

‡ KRUITHOF, A. A. and PENNING, F. M., loc. cit.; GLOTOV, I. I., *Phys. Z. SowjetUn.* **12** (1937) 256.

energy of Ne($^3P_2$)—15·76 eV ionization energy of Ar). It is not surprising that the probability should be high.

Reactions of the type (d) are essentially similar to those of type (c) in that no transfer of energy from internal motion to relative translation is necessary. They will only be possible under near-thermal conditions if $2E_{ex} > E_i$, where E_{ex} is the excitation energy of a metastable atom and E_i the energy required to ionize a ground state atom. This condition is satisfied for all rare-gas atoms.

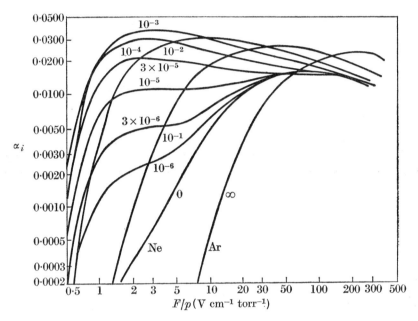

FIG. 18.64. Observed ionization coefficient α_i as a function of F/p for various mixtures of Ne and Ar. The numbers on each curve give the ratio of the argon pressure to the total pressure of the mixture.

If the rate of loss to the walls by diffusion can be separately determined the diffusion coefficient D of the metastable atoms in the gas concerned can be obtained. This may be used in conjunction with any information available about the total elastic cross-section to obtain knowledge about the interaction, or interactions, between a metastable atom and the gas atom.

An obvious method for obtaining information about the rates of processes leading to loss of metastable atoms is to observe the rate at which such atoms, present in a container of simple geometrical boundaries, disappear when the exciting source which produces them is cut off.

The rate may be measured by observing the variation in the absorption of radiation of suitable wavelength which is selectively absorbed by the metastable atoms, as described in Chapter 4, § 3.1.1. By observing the dependence of the rate of loss on the gas pressure, the metastable-atom concentration, and the electron concentration, it is possible to sort out the contributions from the different processes listed above. The method is simple in principle but suffers from the difficulty of ensuring that no atoms enter metastable states after the excitation is cut off. The usual source, a glow discharge, produces atoms and molecules in various excited states so that, after the discharge is cut off, transitions from higher states to the metastable state may occur either by radiation or in collision.

The rates of ionizing reactions such as (144) have been determined by measuring the ionization produced when an atomic beam of metastable atoms passes through the gas in question. It is also possible in principle to determine the rate of such reactions as (150) by crossing two metastable atom beams and observing the yield of ions but, in practice, there will be difficulties because of the low intensity of the obtainable current. Other reactions, which do not yield ions, are very difficult to distinguish from loss by elastic scattering.

5.2.2. *Experimental analysis of loss processes for metastable atoms in discharge afterglows.* The first experiments of this kind were carried out by Meissner and Graffunder† as long ago as 1928. They studied the loss of metastable neon from the gas in a tube after excitation by a glow discharge. The metastable neon concentration was determined by passing a beam of light, capable of selective absorption by Ne*, through the tube and measuring the absorption photographically. By operating the emitting lamp and the electrical excitation with a.c. of the same frequency but adjustable phase difference, the interval between cessation of excitation and emission of the absorbable beam could be varied at will. In many ways the method was a remarkable precursor of the microwave probing methods that have been introduced in recent years.

The technique employed by Meissner and Graffunder was applied to metastable helium and argon by Ebbinghaus‡ and by Anderson§ respectively.

5.2.2.1. *Helium.* The first observations using the microwave afterglow technique were made by Phelps and Molnar∥ in argon, neon, and

† Meissner, K. W. and Graffunder, W., *Annln Phys.* **84** (1928) 1009.
‡ Ebbinghaus, E., ibid. **7** (1930) 267.
§ Anderson, J. M., *Can. J. Res.* **2** (1930) 13.
∥ Phelps, A. V. and Molnar, J. P., *Phys. Rev.* **89** (1953) 1202.

helium. Shortly afterwards, Phelps and Pack† introduced a greatly improved method of measuring small absorption of light under pulsed conditions. The details of this technique and its application to the determination of the rate of deactivation of $2\,^1S$ metastable atoms in a helium afterglow by electron impact have been described in Chapter 4, § 3.1.1. We now describe the application‡ of this technique to determine other reaction rates, as well as diffusion coefficients, for $2\,^1S$ and $2\,^3S$ atoms in helium.

If n, n_1, n_3 are the respective concentrations of ground state, metastable singlet, and metastable triplet atoms in a pure helium afterglow at time t and n_e is the electron concentration, we have

$$\frac{\partial n_1}{\partial t} = D_1 \nabla^2 n_1 - \mu n n_1 - \beta n_e n_1, \qquad (153\,\text{a})$$

$$\frac{\partial n_3}{\partial t} = D_3 \nabla^2 n_3 - A n^2 n_3 + \beta n_e n_1, \qquad (153\,\text{b})$$

where D_1, D_3 are the diffusion coefficients of $2\,^1S$ and $2\,^3S$ He$^+$ atoms in pure helium, μn is the frequency of collisions of $2\,^1S$ atoms with ground state helium atoms which lead to destruction of the metastability, β is the rate coefficient for collisions of the second kind with electrons leading to $2\,^1S$–$2\,^3S$ transitions, and A that for three-body collisions in which the reaction is probably

$$\text{He} + \text{He} + \text{He}(2\,^3S) \rightarrow \text{He}_2' + \text{He}. \qquad (154)$$

If the $2\,^1S$ and $2\,^3S$ concentrations are both distributed in the first fundamental diffusion mode of characteristic length Λ (see Chap. 19, § 2.2.5), then, according to (153),

$$n_1 = n_1^{(0)} e^{-\nu_1 t} \qquad (155)$$

where
$$\nu_1 = \Lambda^{-2} D_1 + \mu n + \beta n_e, \qquad (156)$$

and
$$n_3 = n_3^{(3)} e^{-\nu_3 t} - n_3^{(1)} e^{-\nu_1 t} \qquad (157)$$

where
$$\nu_3 = \Lambda^{-2} D_3 + A n^2, \qquad (158)$$

$$n_3^{(3)} - n_3^{(1)} = n_3^{(0)}, \qquad (159)$$

$$n_3^{(1)} = \beta n_e n_1^{(0)} / (\nu_1 - \nu_3). \qquad (160)$$

$n_1^{(0)}, n_3^{(0)}$ are the initial concentrations of $2\,^1S$ and $2\,^3S$ atoms respectively.

The determination of β has been described in Chapter 4, § 3.1.1. To obtain D_1 and μ it is necessary to study the variation of the $2\,^1S$ concentration in an afterglow with a low electron concentration. Fig. 18.65 shows the observed variation of ν_1 with gas pressure when the electron

† PHELPS, A. V. and PACK, J. L., *Rev. scient. Instrum.* **26** (1955) 45.
‡ PHELPS, A. V., *Phys. Rev.* **99** (1955) 1307.

concentration is less than 10^8 cm^{-3} and $\Lambda^2 = 0{\cdot}053$ cm^2. This curve may be fitted well if $D_1 = 440 \pm 50$ cm^2 s^{-1} at 1 torr and μ is independent of the pressure and given by

$$\mu = \bar{Q}_{\mathrm{dm}}\,\bar{v}, \tag{161}$$

where $\bar{v}$ is the mean relative velocity of impact and $\bar{Q}_{\mathrm{dm}} = 3 \times 10^{-20}$ cm^2. Under these conditions the $2\,^3S$ metastable atom concentration did not show a build-up at early times, indicating that the $2\,^1S$ destruction process involved does not lead to production of $2\,^3S$ atoms.

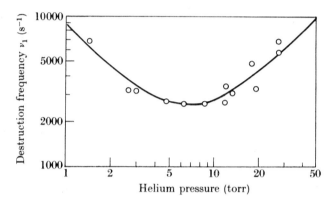

FIG. 18.65. Observed frequency ν_1 for destruction of metastable He($2\,^1S$) atoms in a helium afterglow, as a function of helium pressure at 300 °K, under conditions of low electron concentration ($< 10^8$ cm^{-3}) and with $\Lambda^2 = 0{\cdot}053$ cm^2.

Turning now to the determination of A and D_3, Fig. 18.66 illustrates the observed variation of ν_3 with pressure p for two different absorption cells and later times, for which the second term of (157) is negligible. These data are very well fitted by taking

$$D_3 = 470 \pm 25 \text{ cm}^2 \text{ s}^{-1} \text{ at 1 torr,}$$
$$A = 0{\cdot}26 \pm 0{\cdot}03 \text{ s}^{-1} \text{ torr}^{-2}$$
$$= 2{\cdot}5 \times 10^{-34} \text{ cm}^6 \text{ s}^{-1}.$$

Further experiments were carried out to determine the nature of the three-body destruction process. If (154) is correct, the He$_2'$ molecule will be in a metastable $2\,^3\Sigma_u$ state and its concentration can be monitored by measuring the absorption of the 4650 Å band of the $3\,^3\Pi_g - 2\,^3\Sigma_u$ system. Fig. 18.67 shows the results of such observations. The behaviour of the concentration n_m of the metastable molecules as a function of time is very similar to that of n_3 (cf. Chap. 4, Fig. 4.67). The initial rise is due to production from $2\,^3S$ atoms, while the linear fall, on a semi-logarithmic plot, at later times, is due to diffusion and volume destruction of the molecules. The difference between n_m observed and n'_m extrapolated linearly to initial times from late times, should vary as n_3. That this is indeed so is seen from Fig. 18.67 (cf. Fig. 4.67) and this provides support for the interpretation in terms of (154).

From observations of the rate ν_m of destruction of metastable molecules at different pressures, their diffusion coefficient is found to be 310 ± 50 cm² s⁻¹ at 1 torr. Observations at pressures up to 97 torr in

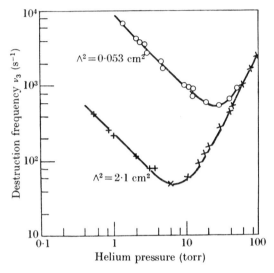

FIG. 18.66. Observed frequency ν_3 for destruction of metastable He(2 ³S) atoms in a helium afterglow as a function of helium pressure at 300 °K for two different absorption cells with $\Lambda^2 = 0\cdot053$ and $2\cdot1$ cm² respectively.

a cell with $\Lambda^2 = 2\cdot1$ cm² were made to show up any volume destruction processes. The rate of such processes was found to be less than 10^{-2} of that for the $2\,^3S$ atoms. The data also show that the natural lifetime of the $2\,^3\Sigma_u$ molecules is $> 0\cdot05$ s.

5.2.2.2. *Neon.* The study of processes involving neon metastable atoms† in a neon afterglow is considerably more complicated by the need to take into separate account the variation of the concentrations of atoms in $^1P_1$, $^3P_0$, $^3P_1$, and $^3P_2$ states ($1s_2$–$1s_5$, respectively, in the Paschen notation) (see Fig. 18.68), the second and last of which alone are metastable. Moreover, because the $^3P_1$‡ and $^1P_1$ states make optical

---

† PHELPS, A. V., *Phys. Rev.* **114** (1959) 1011.
‡ Transitions from this level to the ground state take place through partial breakdown of Russell–Saunders (*LS*) coupling.

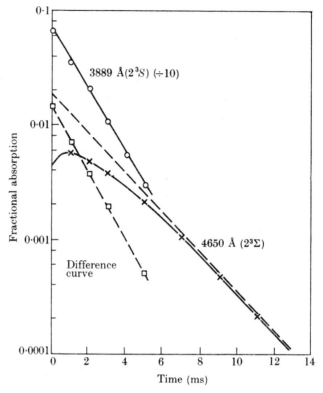

Fig. 18.67. Observed variation with time of the absorption of the 3889 Å line and 4650 Å band by He($2\,^3S$) atoms and He$_2$($2\,^3\Sigma$) molecules respectively.

transitions to the ground state, allowance must be made for resonance trapping of this radiation in the gas.

Let n_0, n_1, and n_2 denote the concentrations of $^3P_0$, $^3P_1$, and $^3P_2$ atoms, n_s that of $^1P_1$, and N that of ground-state atoms. We not only assume, as usual, that the spatial distribution of the excited atoms is close to that of the fundamental diffusion mode but also that, for $^3P_1$ and $^1P_1$, it is close to that of the fundamental mode for diffusion of the resonance radiation. Under these conditions we have

$$\nu_2 = -\frac{1}{n_2}\frac{\partial n_2}{\partial t} = \Lambda^{-2}D_2 + \gamma_2 N^2 + g_{12}a_{12}N(1-n_1/g_{12}n_2) + $$
$$+ g_{02}a_{02}N(1-n_0/g_{02}n_2) + g_{s2}a_{s2}N(1-n_s/g_{s2}n_2), \qquad (162)$$

$$\nu_1 = -\frac{1}{n_1}\frac{\partial n_1}{\partial t} = \nu_r + a_{12}N(1-g_{12}n_2/n_1) + g_{01}a_{01}N(1-n_0/g_{01}n_1) + $$
$$+ g_{s1}a_{s1}N(1-n_s/g_{s1}n_1), \qquad (163)$$

$$\nu_0 = -\frac{1}{n_0}\frac{\partial n_0}{\partial t} = \Lambda^{-2}D_0 + \gamma_0 N^2 + a_{02}N(1-g_{02}n_2/n_0) + $$
$$+ a_{01}N(1-g_{01}n_1/n_0) + g_{s0}a_{s0}N(1-n_s/g_{s0}n_0). \qquad (164)$$

In these equations D_0 and D_2 are the respective diffusion coefficients of $^3P_2$ and $^3P_0$ metastable atoms in pure neon and γ_2 and γ_0 are corresponding three-body destruction rate coefficients. ν_r is the fundamental frequency for decay of $^3P_1$ resonance radiation. According to Holstein's theory† it is given by

$$\nu_r = 0 \cdot 105 (\lambda/\tau^2 R)^{\frac{1}{2}}, \tag{165}$$

where τ is the natural lifetime of the $^3P_1$ state, λ is the wavelength of the resonant radiation, and R the tube radius. a_{ij} is the rate coefficient for collisions of the

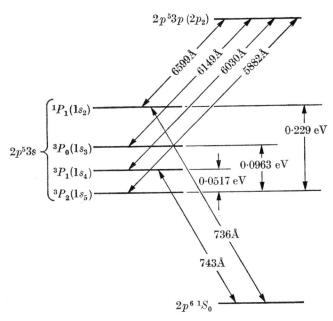

FIG. 18.68. Energy-level diagram for neon showing the ground state, the four ($1s_2$–$1s_5$) levels of the $1s^2\,2s^2\,2p^5\,3s$ configuration and one level ($2p_2$) of the $2p^5\,3p$ configuration, together with the wavelengths of optical transitions between the levels (see also Fig. 18.77).

second kind between ground-state atoms and atoms in the ^3P_i state which cause transitions to a lower ^3P_j state, while a_{sj} is the corresponding coefficient in which the initial state is the $^1P_1$ state. $g_{ij}\,a_{ij}$ is the corresponding rate coefficient for the inverse process of excitation from the ^3P_j to the ^3P_i state so that, at a temperature T,

$$g_{ij} = \frac{2j+1}{2i+1} e^{-\Delta E_{ij}/\kappa T}, \tag{166}$$

where ΔE_{ij} is the energy excess of the ^3P_i over the ^3P_j state. From the known values of ΔE_{ij} (see Table 18.15 and Fig. 18.68) the g_{ij} may be calculated and are given in Table 18.15.

As in the equations (162)–(164) the only coefficients known are Λ, ν_r, and the g_{ij}, the problem of determining the remainder is a formidable one. Nevertheless,

† HOLSTEIN, T., *Phys. Rev.* **72** (1947) 1212; **83** (1951) 1159.

advantage may be taken of the observed data to introduce simplifications sufficient to derive useful information.

The relative concentrations n_2/n, n_3/n, n_0/n of the excited atoms, where n is some convenient reference number, were determined by Phelps from measurements of the relative absorption of radiation of wavelengths 5882 Å, 6096 Å, and 6266 Å respectively. Neglecting pressure broadening effects the relative absorption coefficients for these three lines are as 1: 5·3: 14·1, according to observations of Krebs† and Garbuny.‡

TABLE 18.15

Values of g_{ij} for P states of neon at 300 °K

Transition	$^3P_1$–$^3P_2$	$^3P_0$–$^3P_2$	$^1P_1$–$^3P_2$	$^3P_0$–$^3P_1$	$^1P_1$–$^3P_1$	$^1P_1$–$^3P_0$
ΔE (eV)	0·0517	0·0963	0·229	0·0446	0·177	0·133
g_{ij}	$8·08 \times 10^{-2}$	$4·76 \times 10^{-3}$	$8·40 \times 10^{-5}$	$5·91 \times 10^{-2}$	$1·042 \times 10^{-3}$	$1·755 \times 10^{-2}$

The neon used in Phelps' experiments was reagent grade and was further purified by operating a cataphoresis discharge during the experiments when the pressure was between 10 and 100 torr.

Fig. 18.69 shows a typical set of observations of the relative concentrations of $^3P_2$, $^3P_1$, and $^3P_0$ atoms, at times up to 4 ms from the beginning of the afterglow, when the neon concentration was 4×10^{17} atoms/cm³ and $\Lambda^2 = 0·050$ cm². It will be seen that, in the early stage of the afterglow, n_0 is relatively much larger than at later times. The same features are observed in the decay rate. It is clear from Fig. 18.69 and Table 18.15 that, at early times, $g_{02} n_2/n_0$ and $g_{01} n_1/n_0$ are both very small compared with unity. Because of the relatively large value of the characteristic diffusion frequency ν_r for $^1P_1$ resonance radiation it is safe to assume also that $n_s/g_{0s} n_0$ is $\ll 1$. This leaves

$$\nu_0 = \Lambda^{-2} D_0 + \gamma_0 N^2 + (a_{02} + a_{01} + g_{s0} a_{s0}) N. \tag{167}$$

At sufficiently small values of N the three-body term $\gamma_0 N^2$ is negligible. Under these conditions we may determine D_0 and $a_{02}+a_{01}+g_{s0} a_{s0}$ by plotting $\nu_0 N \Lambda^2$ against $N^2 \Lambda^2$ as illustrated in Fig. 18.70.§ We then obtain, at 300 °K,

$$D_0 = 163 \text{ cm}^2 \text{ s}^{-1} \text{ at 1 torr}, \quad a_{02} + a_{01} + g_{s0} a_{s0} = 8 \times 10^{-15} \text{ cm}^3 \text{ s}^{-1}.$$

In general, we expect (see p. 1868) that the rate of a particular reaction will decrease as the energy discrepancy ΔE increases. We would therefore expect that $a_{s0} < a_{02}$. Since $g_{s0} = 0·017$ it is reasonable to assume that

$$a_{02} + a_{01} = 8 \times 10^{-15} \text{ cm}^3 \text{ s}^{-1}. \tag{168}$$

We now turn to the analysis of the equation for n_2. It may be seen from Fig. 18.69 that, at late times, the decay rates ν_1 and ν_2 are equal. Fig. 18.71 shows the ratio

† KREBS, K., Z. Phys. **101** (1936) 604.
‡ GARBUNY, M., ibid. **107** (1937) 362.
§ As indicated in the caption, data obtained by DIXON, J. R. and GRANT, F. A., Phys. Rev. **107** (1957) 118 are also included.

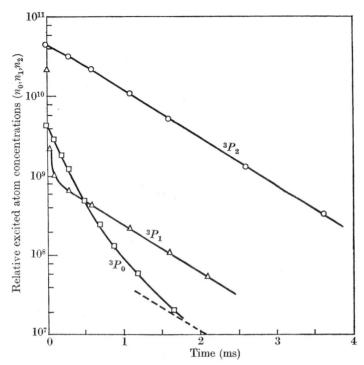

FIG. 18.69. Variation with time of the relative concentrations n_2, n_1, n_0 respectively of $^3P_2$, $^3P_1$, and $^3P_0$ neon atoms in a neon afterglow with a neon concentration of 4×10^{17} cm^{-3} and a normal diffusion area $\Lambda^2 = 0.050$ cm^2.

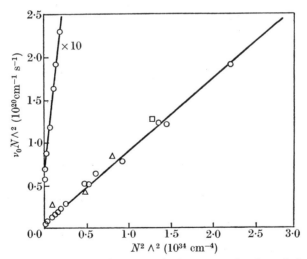

FIG. 18.70. Variation of $\nu_0 N\Lambda^2$ with $N^2\Lambda^2$ for a neon afterglow, derived from data such as those illustrated in Fig. 18.69. ν_0 is the destruction frequency for $^3P_0$ neon atoms. Observed by Phelps $\triangle$ for $\Lambda^2 = 0.050$ cm^2, $\square$ for $\Lambda^2 = 2.1$ cm^2. Observed by Dixon and Grant $\bigcirc$ for $\Lambda^2 = 2.48$ cm^2.

of the relative concentrations n_1/n_2, n_0/n_2, also as a function of N at a temperature of 300 °K. Using the fact derived above that

$$a_{02} < 8 \times 10^{-15} \text{ cm}^3 \text{ s}^{-1}$$

together with the data of Figs. 18.69 and 18.70, we see that the maximum value of

$$g_{02} a_{02} N(1 - n_0/g_{02} n_2) \ll \nu_2,$$

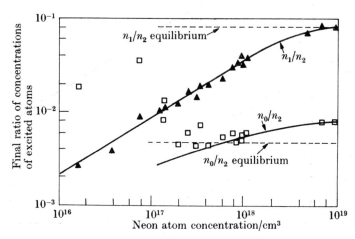

FIG. 18.71. Observed variation with neon concentration n of the ratios n_1/n_2, n_0/n_2 of the concentrations of $^3P_1$ and $^3P_0$ neon atoms to $^3P_2$ atoms during the final exponential decay of these concentrations in a neon afterglow.

so this term may be dropped from (162). It is also reasonable to neglect the term $g_{s2} a_{s2} N(1 - n_s/g_{s2} n_2)$ for the following reasons. We would expect that $a_{s2} < a_{12}$ because $\Delta E_{s2} > \Delta E_{12}$ so

$$g_{s2} a_{s2} < g_{s2} a_{12}$$
$$= 0 \cdot 001 g_{12} a_{12}, \quad \text{from Table 18.15.}$$

Also we expect, again because of the high rate of diffusion of $^1P_1$ resonance radiation, that $n_s/g_{s2} n_2 \ll 1$. The term concerned is therefore $< 10^{-3} g_{s2} a_{s2} N$ and may be neglected in comparison with the third term on the right-hand side of (162). We are now left with

$$\nu_2 = \Lambda^{-2} D_2 + \gamma_2 N^2 + g_{12} a_{12} N(1 - n_1/g_{12} n_2). \tag{169}$$

Again, for observations at sufficient low gas densities the three-body term can be ignored. Hence, by plotting measured values of $\Lambda^2 \nu_2$ against measured values of $g_{12} N(1 - n_1/g_{12} n_2)$, D_2, and a_{12} may be obtained. Fig. 18.72 shows the derived values as functions of gas temperature.† Although some of the data from earlier experiments, in which n_1/n_2 was not measured, have been used in obtaining these results, care was taken, in these cases, to use data obtained at such low neon concentrations (see Fig. 18.71) that $n_1/g_{12} n_2$ is negligible compared to unity.

† As noted in the figure caption, data obtained from the experiments of GRANT, F. A. and KRUMBEIN, A. D., *Phys. Rev.* **90** (1953) 59 and of PHELPS, A. V. and MOLNAR, J. P., ibid. **89** (1953) 1202 are also included.

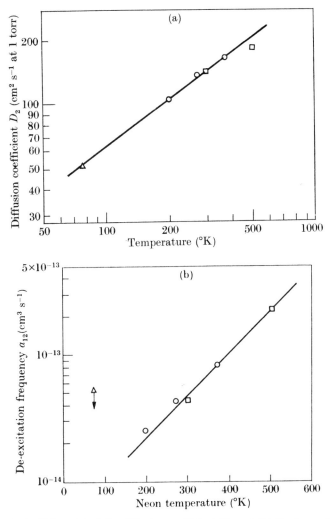

FIG. 18.72. Values (a) of the diffusion coefficient D_2 and (b) the de-excitation frequency a_{12}, for Ne($^3P_2$) atoms in neon afterglow. □ from observations of Phelps. ○ from observations of Grant and Krumbein. △ from observations of Phelps and Molnar.

From observations at high values of N, γ_2 may be determined, D_2 being known from the analysis at low N. It is then found that at 300 °K

$$\gamma_2 = 5 \cdot 0 \times 10^{-34} \text{ cm}^6 \text{ s}^{-1}.$$

The relative importance of the three terms in (169) at different neon concentrations is shown in Fig. 18.73. At low temperatures the third term in (169) is unimportant and γ_2 may be determined directly from the decay rate at large N, without requiring knowledge of n_1/n_2. In this way it is found that, at 77 °K,

$$\gamma_2 = 5 \cdot 0 \times 10^{-35} \text{ cm}^6 \text{ s}^{-1}.$$

So far we have not obtained the coefficients a_{02}, a_{01} separately but merely their sum. Some evidence about their separate values may be obtained by analysis of $\partial n_0/\partial t$ at later times. Neglecting the last term in (164) for the reasons already given, and working at neon concentrations small enough for $\gamma_0 N^2$ to be negligible, if γ_0 is of the same order as γ_2, we have

$$N^{-1}(\nu_0 - \Lambda^{-2}D_0)(1 - g_{01}n_1/n_0)^{-1} = a_{01} + a_{02}(1 - g_{02}n_2/n_0)(1 - g_{01}n_1/n_0)^{-1}. \quad (170)$$

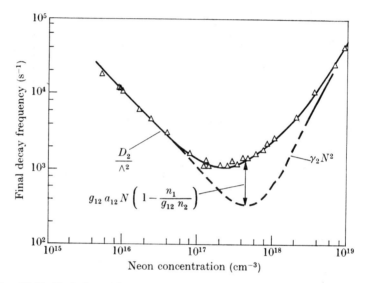

Fig. 18.73. Variation with neon concentration of the final decay frequency for the $^3P_2$ and $^3P_1$ states (see Fig. 18.69) in a neon afterglow at 300 °K in a cell with $\Lambda^2 = 0{\cdot}050$ cm². Analysis of the decay frequency into separate contributions (see (169)) is shown.

Using data as shown in Fig. 18.69 for ν_0, n_1/n_0 and n_2/n_0, together with D_0 as determined above, we may plot the left-hand side of (170) as a function of $(1 - g_{02}n_2/n_0)(1 - g_{01}n_1/n_0)^{-1}$ as in Fig. 18.74. This gives a linear relation from which we obtain $a_{01} = a_{02} = 5 \times 10^{-15}$ cm³ s⁻¹. This is consistent within experimental error with (168).

A useful check on the analysis is obtained by analysing equation (163) for ν_1. From analysis of the equations for ν_0 and ν_2 we have $a_{12}/g_{01}a_{01} \geqslant 80$. Using this result in connection with the data of Fig. 18.71 we have

$$a_{12}(1 - g_{12}n_2/n_1)/g_{01}a_{01}(1 - n_0/g_{01}n_1) \geqslant 20, \quad (171)$$

for $N < 5 \times 10^{18}$/cm³. Also, as $\Delta E_{s1} > \Delta E_{12}$, we expect $a_{s1} < a_{12}$ so

$$g_{s1}a_{s1} < 10^{-3}a_{12}.$$

We may therefore neglect the last two terms on the right-hand side of (163), giving

$$\nu_r = \nu_1 - a_{12}N(1 - g_{12}n_2/n_1). \quad (172)$$

On substitution of observed values we find $\nu_r = 5{\cdot}3 \times 10^4$ s⁻¹, independent of N and the gas temperature T, for $N > 2 \times 10^{17}$/cm³. This is to be compared with $4{\cdot}7 \times 10^4$ s⁻¹ calculated from (165) using the natural lifetime of the $^3P_1$ state,

$1\cdot 3 \times 10^{-8}$ s. This was derived from the observed† natural lifetime $1\cdot 2 \pm 0\cdot 6 \times 10^{-9}$ s for the $^1P_1$ state and the calculated ratio $13\cdot 2$ of the $^3P_1$ to the $^1P_1$ lifetime.

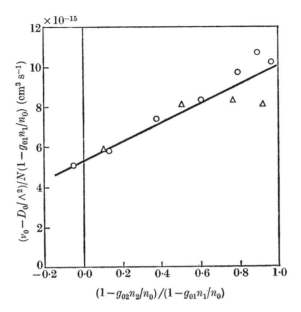

FIG. 18.74. Illustrating the determination of a_{02} and a_{01} from (170).

Table 18.16 summarizes the main results of the analysis. The small values of the destruction cross-sections, despite the small energy discrepancies, are noteworthy.

Dixon and Grant‡ have measured the decay frequencies for $^3P_2$ and $^3P_0$ metastable atoms in a mixture of helium and neon. Their results may be analysed as for the case of pure neon but the reaction rates must be regarded as appropriate to the mixture. If there are N_1 neon atoms and N_2 helium atoms present/cm³, any two-body destruction frequency a_{ij}^{12}, as measured for the mixture, is related to the frequencies a_{ij}^1, a_{ij}^2 for the pure gases neon and helium respectively by

$$a_{ij}^{12} = (a_{ij}^1 N_1 + a_{ij}^2 N_2)/(N_1 + N_2). \qquad (173)$$

Similarly, it is a good approximation for the diffusion coefficient of a metastable atom in the mixture to write

$$1/D_i^{12} = (x/D_i^1) + (1-x)/D_i^2, \qquad (174)$$

† SCHUTZ, W., *Ann. Phys.* **18** (1933) 705; SCHILLBACH, H., ibid. **18** (1933) 721; SHORTLEY, G. H., *Phys. Rev.* **47** (1935) 295.
‡ DIXON, J. R. and GRANT, F. A., *Phys. Rev.* **107** (1957) 118.

where D_i^1, D_i^2 are the corresponding diffusion coefficients in the respective pure gases and $x = N_1/N_2$.

Using these relations and data obtained from analysis of Dixon and Grant's results for a fractional helium concentration of 0·65, values for the diffusion coefficient and deactivation rates of ^3P neon atoms in pure helium given in Table 18.16 were derived by Phelps.

5.2.2.3. *Argon.* The metastable-atom decay rates in an argon afterglow have not been studied in such detail as in neon or helium. In their original experiments, Phelps and Molnar† obtained for the diffusion coefficient of $^3P_2$ metastable argon atoms in argon 54 cm² s⁻¹ at 1 torr. As far as rates of destruction are concerned, they obtained evidence of the existence both of two-body and three-body processes. For the former the mean cross-section at 300 °K is 2×10^{-20} cm² while for the latter it is $1 \cdot 2 \times 10^{-37}$ cm²/cm³ corresponding to a rate coefficient of $3 \cdot 6 \times 10^{-33}$ cm⁶ s⁻¹.

5.2.2.4. *Effect of impurities.* Benton, Ferguson, Matsen, and Robertson‡ studied the effect of addition of a controlled amount of a heavier rare-gas impurity on the time dependence of the concentrations of $2\,^3S$ and $2\,^1S$ metastable atoms of helium, using an apparatus very similar to that of Phelps. The equations (153) now take the form

$$\frac{\partial n_1}{\partial t} = -\nu_1^0 n_1 - \lambda_1 n_i n_1 - \beta(\Delta n_e)n_1, \tag{175 a}$$

$$\frac{\partial n_3}{\partial t} = -\nu_3^0 n_3 - \lambda_3 n_i n_3 + \beta(\Delta n_e)n_1, \tag{175 b}$$

where ν_1^0, ν_3^0 are as given in (156) and (158), n_i is the concentration of impurity atoms and λ_1, λ_3 are the rate coefficients for destruction of the respective metastable atoms by collision with these atoms, almost certainly through the ionizing reactions (144). Δn_e denotes the excess concentration of free electrons due to the presence of the impurity.

In equation (175 b) the contribution from the excess electrons is small because n_1 decays very rapidly compared with n_3. It then follows that

$$\ln(n_3/n_0) = -(\nu_3^0 + \lambda_3 n_i)t \tag{176}$$

so that, from the change in slope of a plot of $\ln(n_3/n_0)$ against t as n_i is varied, λ_3 may be obtained.

† loc. cit., p. 1780.
‡ BENTON, E. E., FERGUSON, E. E., MATSEN, F. A., and ROBERTSON, W. W., *Phys. Rev.* **128** (1962) 206.

The problem of determining λ_1 is more difficult because the contribution from deactivation by the excess electrons is not negligible and reliable results could not be obtained.

The major experimental difficulty in this type of experiment is to control and measure accurately the concentration of the impurity. The system used for filling the absorption cell is illustrated diagrammatically in Fig. 18.75. That portion enclosed in the oven was baked at 400 °C

TABLE 18.16

Data for reaction rates of ^3P neon atoms in neon and helium at 300 °K

	Diffusion coefficient (cm² s⁻¹ at 1 torr)		Three-body destruction coefficient
State	Neon	Helium	Neon
$^3P_2$	163	564	5.0×10^{-34} cm⁶ s⁻¹
$^3P_0$	163	564	—

Rates of destruction by collisions of second kind with normal atoms

Transition	ΔE, energy discrepancy (eV)	Rate (cm³ s⁻¹)		Mean cross-section (cm²)	
		Neon	Helium	Neon	Helium
$^3P_1$–$^3P_2$	0.0517	4.2×10^{-14}	1.9×10^{-14}	5.2×10^{-19}	1.4×10^{-19}
$^3P_0$–$^3P_1$	0.0446	5×10^{-15} $\}$	8×10^{-15}	6×10^{-20} $\}$	6×10^{-20}
$^3P_0$–$^3P_2$	0.0963	5×10^{-15}		6×10^{-20}	

for two 12-hour periods between which the internal cell electrodes and barium getter wires were outgassed with an induction heater. After the second bake-out, the barium getters were flashed, leaving an ultimate background vacuum of 10^{-8} torr. Helium containing a fractional concentration of 3×10^{-5} of neon was leaked through the liquid-nitrogen cooled charcoal trap into a cataphoresis tube. Operation of the latter at a current of 50 mA concentrated impurities which were removed by the charcoal trap 2. The helium, already very pure, was then leaked into a second cataphoresis tube 2 and the cell chamber. A further cataphoresis was then carried out by running a 50 mA current through tube 2. A 10 mA current was also run from one electrode of the cell to the anode of tube 2. By following this procedure, measurements of destruction frequencies in pure helium were reproducible to about 5 per cent.

A complication that arises when an impurity is deliberately introduced is that even rare gases are adsorbed on clean glass surfaces. The impurity pressure must therefore be measured after it is introduced so that adsorption has already taken place. The impurity was leaked through trap 3 to a pressure of a few torr as indicated by an oil manometer. It was then leaked through the variable leak into both the cell and impurity chambers until the desired pressure was attained. This was read by an ion gauge operated at 1-mA grid current to reduce pumping action

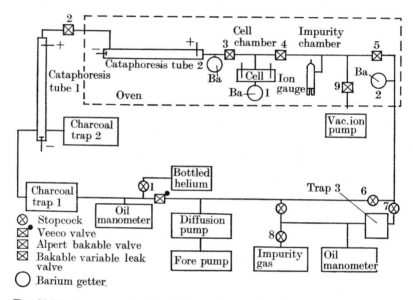

Fig. 18.75. Arrangement of the filling system used in the experiments of Benton, Ferguson, Matsen, and Robertson.

within it. For the same reason the gauge was turned off as soon as possible. After allowing time for the impurity to come to equilibrium with the surfaces the cell and impurity chambers were pumped down to pressures of 10^{-2} torr or less. The latter was then refilled to a specific pressure close to the original one. Pure helium was then admitted at the working pressure to the chamber and then, by operation of valve 4, the impurity and helium were allowed to diffuse together to give equilibrium partial pressures determined by the relative chamber volumes.

The ionization gauge was calibrated against a McLeod gauge using nitrogen. Relative gauge sensitivities for other gases were taken to be those given by Schulz† (He, Ne, Ar, H_2, and N_2) and by Dushman and Young‡ (Kr and Xe).

† Schulz, G. J., *J. appl. Phys.* **28** (1957) 1149.
‡ Dushman, S. and Young, A. H., *Phys. Rev.* **68** (1945) 278.

The barium getter was removed for observations with Xe, H_2, and N_2 and, for the last two gases, charcoal trap 3 also. Measurements with H_2 and N_2 were very difficult because they were rapidly cleaned up when the pulsed discharge was operated. Only absorption measurements for a single delay time could be made for each filling of these gases. For O_2 and CO_2 the effect was so marked that no data could be obtained. As an illustration of the magnitude of the additional loss rate for $2\,^3S$ helium atoms when an impurity is added Fig. 18.76 compares the observed

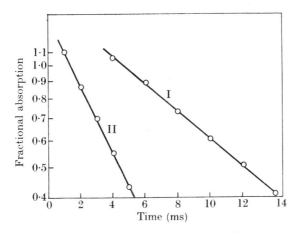

FIG. 18.76. Observed fractional absorption of the 3889 Å line as a function of time in an afterglow: I, in pure helium at 10 torr pressure; II, in the same helium to which 10^{12} atoms/cm³ of krypton have been added.

fractional absorption of the 3889-Å line as a function of time in pure helium at 10 torr pressure and in the same helium to which 10^{12} atoms/cm³ of krypton have been added. This represents a fractional impurity concentration of only 3×10^{-6}.

The three-body coefficient A was found to be $3 \cdot 0 \pm 0 \cdot 3 \times 10^{-34}$ cm⁶ s⁻¹ which agrees well with that, $2 \cdot 5 \pm 0 \cdot 3 \times 10^{-34}$ cm⁶ s⁻¹, found by Phelps. On the other hand, the diffusion coefficient D_3 was found to be 560 ± 50 cm² s⁻¹ at 1 torr which is appreciably larger than that, 470 ± 25 cm² s⁻¹, found by Phelps. This may be due to the admixture of the second diffusion mode as the electrode system favoured the corresponding configuration.

The values found for the mean cross-sections $\bar{Q}_{\mathrm{md}}$ for destruction of $2\,^3S$ metastable atoms in different gases, with the exception of Ne, are given in Table 18.20. These have been obtained from

$$\bar{Q}_{\mathrm{md}} = \lambda_3/\bar{v}, \tag{177}$$

where $\bar{v}$ is the mean relative velocity of collision at the temperature of observation (300 °C). Comparison will be made later with results obtained by other methods.

5.2.2.5. *Excitation transfer between metastable helium and normal neon atoms.* The destruction of He($2\,^3S$) atoms in neon is of special interest because the process concerned is one of excitation transfer to neon atoms.

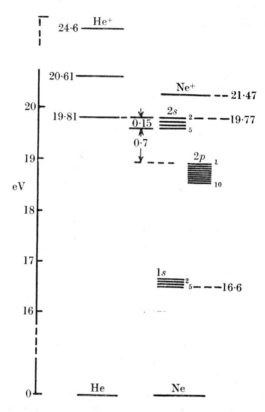

FIG. 18.77. Energy-level diagram for He and Ne atoms showing the energy relationships between the He metastable levels and the $2p$ and $2s$ excited levels of Ne. (The Paschen notation is used for the neon levels.†)

Fig. 18.77 illustrates the energy relations involved. It will be seen that the $2\,^3S$ level of He lies close to the critical $2s$† levels of neon but considerably above the excited $2p$† levels. The transfer takes place mainly

---

† We are using here the Paschen notation for the neon levels according to which the $2s$ arise from the configuration, in the usual notation, $2p^5\,4s$ and the $2p$ from $2p^5\,3p$. The subscripts in the Paschen notation denote the order of the sublevels as shown in Fig. 18.77 (see also Fig. 18.68).

to the 2s levels and it is possible in this way to maintain, in a suitable activated He–Ne mixture, a concentration of Ne(2s) in excess of Ne(2p). This inversion of population was used by Javan, Bennett, and Herriott[†] to produce a gaseous laser operating at five different wavelengths in the near infra-red.

Javan et al. determined the cross-sections for the transfer process by observing the rate of decay of the concentrations of $He(2\,^3S)$, $Ne(2s)$, and $Ne(2p)$ excited atoms in the afterglow of a pulsed r.f. discharge in a helium–neon mixture. For the excited neon atoms, the decay rates were monitored from the intensity of suitable radiative transitions from the excited levels, while that for $He(2\,^3S)$ was determined from the absorption of $2\,^3S$–$2\,^3P$ radiation. It was verified that the concentrations of all three species decayed at the same rate. From observations of the relative intensities at a fixed time in the afterglow it was established that the transfer from $He(2\,^3S)$ to $Ne(2p)$ was not occurring to a significant extent. From the decay rates the mean cross-section for the transfer process

$$He(2\,^3S)+Ne(1s) \to He(1\,^1S)+Ne(2s) \tag{178}$$

was found to be $3 \cdot 7 \pm 0 \cdot 5 \times 10^{-17}$ cm². This is somewhat larger, but not inconsistent with, the total destruction cross-section for $He(2\,^3S)$ in neon of $2 \cdot 8 \times 10^{-17}$ cm², found by Benton, Ferguson, Matsen, and Robertson in the course of their experiments described above. Javan et al. were also able to determine the diffusion coefficient for $He(2\,^3S)$ in He and obtained a value closely agreeing with that found by Phelps (see p. 1782) of 470 cm² s⁻¹ at 1 torr.

Beterov and Chebotaev[‡] further analysed the transfer cross-section from a study of the characteristics of a He–Ne laser operating at two different frequencies, one of these (11 523 Å) corresponding to the $2s_2$–$2p_4$[§] and the other (11 614 Å) to the $2s_3$–$2p_5$[§] transition.

The relative power gain per unit length for a laser operating at a wavelength λ between levels 1 and 2 is given by

$$\gamma = \frac{(\pi \ln 2)^{\frac{1}{2}}\lambda^2}{4\pi^2 \tau \Delta \nu}(n_2 - n_1 g_2/g_1).$$

Here n_2, n_1 are the concentrations of atoms in the levels 2, 1 respectively, g_2, g_1 the respective statistical weights of the levels, $\Delta \nu$ the Doppler width

---

[†] JAVAN, A., BENNETT, W. R., and HERRIOTT, D. R., Phys. Rev. Lett. 6 (1961) 106.
[‡] BETEROV, I. M. and CHEBOTAEV, V. P., Optika Spectrosk. 20 (1966) 1078 (Optics Spectrosc., Wash. 20 (1966) 597).
[§] See footnote † opposite.

of the line radiated in transitions between the levels, and τ the natural radiative lifetime.

The change $\Delta\gamma$ in γ due to a change Δn_2 in the concentration of atoms in the upper level 2 is therefore given by

$$\Delta\gamma = a\,\Delta n_2,$$

where $$a = (\pi \ln 2)^{\frac{1}{2}}\lambda^2/4\pi^2\tau\,\Delta\nu. \tag{179}$$

Beterov and Chebotaev applied the relation (179) to a helium–neon laser, producing changes in the concentrations of neon atoms in the respective upper levels $2s_2$ and $2s_3$ by irradiating the laser with light from a helium discharge lamp. This changed the concentration of $2\,^3S$ metastable helium atoms in the laser and hence that of the appropriate excited neon atoms through excitation transfer. Thus if Δn_m is the increase in concentration of $2\,^3S$ atoms

$$\Delta n_2 = n_{\text{Ne}}\,\bar{v}\,\bar{Q}_{\text{tr}}\,\tau_2\,\Delta n_m,$$

where $\bar{v}$ is the mean relative velocity of the atoms and n_{Ne} the concentration of neutral neon atoms in the laser, $\bar{Q}_{\text{tr}}$ is the mean excitation transfer cross-section, and τ_2 the effective lifetime of the excited neon level concerned. Δn_m may be measured by optical absorption and $\Delta\gamma$ from observation of the threshold for operation of the laser under the different conditions. In the experiment of Beterov and Chebotaev the operating threshold was determined by using a calibrated attenuator in the form of a plane glass plate placed within the laser resonator. When the helium lamp was switched on the threshold was achieved at a smaller angle of rotation of the plate. The difference in power involved could then be calculated from the Fresnel formula.

In this way they found transfer cross-sections to the $2s_2$ and $2s_3$ levels of 2·4 and $1\cdot 4 \times 10^{-17}$ cm² respectively, which are not inconsistent with Javan's total cross-section for decay of $\text{He}(2\,^3S)$ atoms in neon ($3\cdot 7 \times 10^{-17}$ cm²).

Transfer of excitation from $\text{He}(2\,^1S)$ to neon presents a wider variety of possibilities than for $\text{He}(2\,^3S)$. Thus the $3s$, $4s$, $5s$, $3d$, $4d$, and $5d$ levels of neon all lie within 10^{-2} eV of $\text{He}(2\,^1S)$. The afterglow method is, however, not suitable for studying the rates of these processes because the lifetime of $\text{He}(2\,^1S)$ in such an afterglow is shorter than the time (10–20 s) taken for the electrons to come to thermal equilibrium with the main gas. Colombo, Marković, Pavlović, and Peršin[†] have obtained

---

[†] COLOMBO, L., MARKOVIĆ, B., PAVLOVIĆ, Z., and PERŠIN, A., *J. opt. Soc. Am.* **56** (1966) 890.

information about some of the reaction rates using comparatively simple equipment. They observed the effect of addition of helium on the intensity of emission of different neon lines from an electrodeless discharge in neon. Enhanced intensities were observed for those lines for which the upper levels were close to the $2\,^1S$ and $2\,^3P_0$ levels of helium. Assuming that the enhancement is all due to the appropriate transfer collisions the data may be analysed in the same way as for the sensitized fluorescence experiments discussed in § 4. The cross-sections obtained are given in Table 18.17. These were derived from observations with a weak exciting discharge so that electronic excitation and cascade population from highly excited upper states were relatively unimportant.

TABLE 18.17

Cross-sections for transfer of excitation in thermal collisions between He $(2\,^1S$ *or* $2\,^3P_0)$ *and normal neon atoms derived from the observations of Colombo et al.*

He state\Ne state	Cross-section (10^{-16} cm^2)			
$2\,^1S$	$3s_5$	$3s_2$	$4d_4$	$4s_1$
	4·8	2·8	0·93	1·5
$2\,^3P_0$	$5d_5$	$4d_3$	$4s_5$	
	1·5	0·32	0·85	

5.2.3. *Metastable atoms in flowing afterglows.* A detailed account of the technique of the flowing afterglow as applied to the measurement of the rates of ionic reactions is given in Chapter 19, § 3.4.5. The use of the technique for investigating the rates of reactions involving metastable atoms is more difficult. The main problem is that of determining the concentration of these atoms at any particular section of the flow tube. The usual methods, based on optical absorption or on electron ejection from a metallic surface, are impracticable, the former from the small diameter of the flow tube and the latter because of geometrical difficulties. Interpretation of the results, when obtained, is also more complicated than in a static afterglow because of the parabolic velocity distribution across a section of the flow tube.

Nevertheless, Huggins and Cahn† have been able to make measurements on the variation of the concentration of He ($2\,^3S$) atoms along a flow tube and interpret the results in terms of volume and diffusion loss processes. Referring to Fig. 18.78, let $v(r)$ be the flow velocity along the tube at a point distant r from the axis. Then, for the triplet metastable concentration n_3 we have, in pure helium,

$$D_3\nabla^2 n_3 - v(r)\frac{\partial n_3}{\partial z} + Ap^2 n_3 = 0, \qquad (180\,\mathrm{a})$$

where D_3 is the diffusion coefficient of the $2\,^3S$ atoms and $Ap^2 n_3$ is the rate of their

† HUGGINS, R. W. and CAHN, J. H., *J. appl. Phys.* **38** (1967) 180.

destruction by the three-body process (154), when the gas pressure is p, the axis of the tube being taken as the z-direction. The problem is to solve (180 a) taking for $v(r)$ the parabolic distribution for viscous flow†

$$v(r) = (|\nabla p|/4\eta)(1-r^2/a^2),$$

where ∇p is the pressure gradient, η the coefficient of viscosity of the flowing gas,

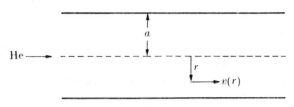

FIG. 18.78

and a the radius of the tube. An exact solution of (180 a) is not possible but, treating the middle term as small so that an approximate solution has the form

$$n_3 = R_i(r)\exp(-k_i^2 z),$$

where the k_i are eigenvalues, a variational approximation to the lowest eigenvalue k_0 may be found. Taking as a trial function

$$R(r) = \alpha(1-r^2/a^2)+\beta(1-r^2/a^2)^2,$$

Huggins and Cahn find

$$k_0^2 v_0 p = 7{\cdot}342(D_3 p/a^2)+1{\cdot}258Ap^3, \tag{180 b}$$

where $v_0 = |\nabla p|/4\eta$ is the velocity on the axis of the flow tube.

To determine k_0 it is only necessary to determine the relative variation of triplet metastable atom concentration with distance along the flow tube.

This they did in a flow tube 1 cm diameter and 120 cm long, through which the helium flowed with a velocity of about 5×10^3 cm s^{-1}, by introducing oxygen into the flow system and observing the intensity of the O_2^+ second negative bands (see p. 988) excited in collisions with the triplet metastable helium atoms. In earlier experiments in a static afterglow on the concentration of triplet metastable atoms decaying in the presence of O_2, Collins and Robertson‡ showed that the intensity of the bands varies linearly with the metastable atom concentration as monitored from optical absorption measurements, even when appreciable concentrations of He^+ and He_2^+ are also present. Excitation of the bands could occur through collisions with metastable He_2 ($a\,^3\Sigma$) molecules in high vibrational states but, at the pressures concerned (5–40 torr), such molecules would rapidly be deactivated to the ground vibrational state which is about 1 eV below the threshold for excitation.

An exponential decay with z was observed, yielding the decay constant k_0^2. A plot of $v_0 pk_0^2$ against p^3 proved to be linear as predicted from (180 b). The intercept then gave D_3 and the slope the three-body deactivation coefficient. In this way D_3 was found to be 490 ± 7 cm^2 s^{-1} at 1 torr and $A = 2{\cdot}4 \times 10^{-34}$ cm^6 s^{-1}, in good agreement with the results of static afterglow observations (§ 5.2.2.1).

† LAMB, H., *Hydrodynamics*, p. 585 (Cambridge University Press, 1932).
‡ COLLINS, C. B. and ROBERTSON, W. W., *J. chem. Phys.* **40** (1964) 701.

Huggins and Cahn also made absolute measurements of triplet metastable atom concentration by injecting argon into the flowing afterglow. The electron concentration produced through Penning ionization was then measured by a resonance cavity method as in static afterglow measurements. Complications are introduced by the distortion of the electron cloud by the axial flow and its radial variation.

5.2.4. *Data about metastable atoms from analysis of electron decay rates in discharge afterglows*

5.2.4.1. *Helium and neon.* The reaction between two metastable helium atoms,
$$He^* + He^* \rightarrow He + He^+ + e, \tag{181}$$
is energetically possible without transfer of energy to or from relative translation of the helium atoms. Already in 1937–9 Schade† and Buttner‡ had drawn attention to its possible importance as a source of ionization in excess of that produced by avalanche electrons in a Townsend discharge. In 1952 Biondi§ was able to interpret certain aspects of the variation of electron concentration in a helium afterglow in terms of the reaction and was able to obtain information about its rate.

The technique involved was that of measuring electron concentration as a function of time in the afterglow by the methods described in Chapter 2, § 5.1. The gas to be studied was contained in a cylindrical quartz bottle (0·88 in radius, 1·5 in high) surrounded by a cavity resonant at about 3000 Mc/s. By means of a pulse from a magnetron lasting about 300 μs the gas was ionized and the variation of electron concentration in the afterglow after the pulse determined by measuring the resonant frequency.

The vacuum system handling the gas was of glass and metal throughout. Metal valves, which could be baked out at high temperatures, were used for the gas metering and the complete system was baked at 420 °C for 12 hours or more before each run. This reduced the background pressure to 10^{-9} torr. Impurities in the gas samples were present to less than 1 part in 10^5.

Fig. 18.79 illustrates the observed variation of electron concentration n_e with time t in the first 7–8 ms after cessation of the exciting pulse, in pure helium and neon respectively. In both gases the same behaviour is observed. After 3–4 ms $\ln n_e$ falls off linearly with t but, if this is extrapolated linearly back to the initial time, it yields values of n_e considerably greater than n_e as observed. The difference, $\ln(\Delta n_e)$, again

---
† SCHADE, R., *Z. Phys.* **105** (1937) 595.
‡ BUTTNER, H., ibid. **111** (1939) 570.
§ BIONDI, M. A., *Phys. Rev.* **88** (1952) 660.

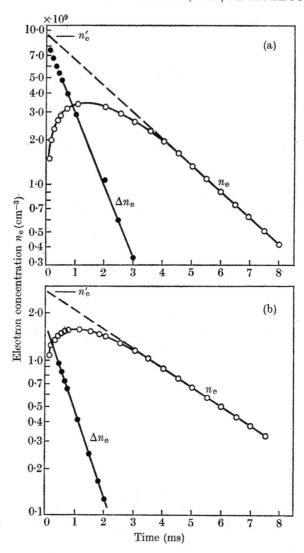

Fig. 18.79. Observed variation of electron concentration n_e with time in (a) a helium afterglow at a pressure of 3·1 torr and (b) a neon afterglow at 1·56 torr. -O-O-O-O observed points. -●—●- difference between observed points and the linear extrapolation of ln n_e to zero time.

falls linearly with t. This shows that the variation of n_e with t can be written in the form
$$n_e = n_e^{(1)} e^{-\nu_1 t} - n_e^{(2)} e^{-\nu_2 t}, \qquad (182)$$
where $\nu_2 > \nu_1$ as in the analysis of the time variation of the concentration of He($2\,^3S$) atoms and $2\,^3\Sigma$ He$_2$ molecules discussed above (see p. 1782 and Fig. 18.67). The origin of the second term can therefore be ascribed to

an electron-production process with a decay rate ν_2. It was found that the importance of the second term was reduced if the afterglow was irradiated with the 20 580 Å line which excites $2\,^1S$ helium atoms to the $2\,^1P$ state from which they may return to the ground state through allowed emission of radiation. This indicates that at least atoms in one of the metastable states are concerned. The only process involving metastable atoms which could yield electrons is (181) and it is reasonable to consider that this is in fact responsible.

Under conditions in which volume recombination of electrons may be ignored, the equations, for the concentrations n_e, n_m of electrons and of metastable atoms respectively, take the form

$$\frac{\partial n_m}{\partial t} = D_m \nabla^2 n_m - \mu n n_m - \delta n_m^2, \qquad (183\,a)$$

$$\frac{\partial n_e}{\partial t} = D_a \nabla^2 n_e + \delta n_m^2. \qquad (183\,b)$$

Here D_m is the diffusion coefficient and μ the rate coefficient for volume destruction of the metastable atoms, no distinction being made between $2\,^1S$ and $2\,^3S$ states. δ is the rate coefficient for (181). D_a is the ambipolar diffusion coefficient (see Chap. 19, § 2.2.5) for electrons and ions in the afterglow. With the usual assumption about diffusion occurring in the fundamental mode we have, from (183 a), if the term δn_m^2 is small,

$$n_m \simeq n_m^0 e^{-\nu_m t}, \qquad (184\,a)$$

where
$$\nu_m = \Lambda^{-2} D_m + \mu, \qquad (184\,b)$$

Λ being the fundamental characteristic diffusion length of the container (see Chap. 19, § 2.2.5). On substitution in (183 b) we find

$$n_e = n_e^{(1)} e^{-\nu_e t} - n_e^{(2)} e^{-2\nu_m t}, \qquad (185\,a)$$

where
$$\nu_e = \Lambda^{-2} D_a, \quad n_e^{(2)} = \delta (n_m^0)^2 / (2\nu_m - \nu_e). \qquad (185\,b)$$

From observations of the kind shown in Fig. 18.79, ν_e and ν_m may be determined. The variation of ν_m with pressure gives D_m and μ. Comparison is made in Table 18.18 with the values so obtained and those derived by Phelps† from the afterglow absorption experiments described in § 5.2.2.1 above. The agreement is remarkably good as far as the diffusion coefficients are concerned. The two-body destruction coefficients for helium are also consistent, when allowance is made for the fact that $2\,^3S$ atoms are not destroyed in this way and Biondi could not separate effects arising from the two kinds of metastable atoms. For neon, Biondi's destruction rate appears to be a little low although,

† loc. cit., pp. 1781, 1783.

in this case, the complexities are greater, as is clear from § 5.2.2.2 above, and Biondi's low value may be due to the presence of an appreciable fraction of $^3P_0$ metastable atoms (see Table 18.16). In any case there seems little doubt that the observed phenomena have been correctly interpreted.

TABLE 18.18

Comparison of diffusion coefficients and two-body destruction coefficients for metastable helium and neon in their parent gases as derived from absorption and electron concentration afterglow experiments

	Diffusion coefficient cm² s⁻¹ at 1 torr		Two-body destruction coefficient cm³ s⁻¹	
	(a)	(b)	(a)	(b)
He–He	440±50 ($2\,^1S$) 470±25 ($2\,^3S$)	520	6×10^{-15} ($2\,^1S$)	$1 \cdot 9 \times 10^{-15}$
Ne–Ne	163	200	$4 \cdot 2 \times 10^{-14}$	8×10^{-15}

(a) From absorption measurements (Phelps).
(b) From electron concentration measurements (Biondi).

$n_e^{(2)}$ of (185 a) may be determined from the difference at the initial time between the value $n_e^{(1)}$ to which the later time linear plot in Fig. 18.79 extrapolates and the observed initial value of n_e. In this way $\delta(n_m^0)^2$ in pure helium at 2·5 torr was found to be about 5×10^{-10} cm³ s⁻¹. No simultaneous observations of n_m^0 by absorption were made but, from the experiments of Phelps, it is likely that n_m^0 is of the order 10^{10} cm⁻³ giving δ of order 10^{-9} cm² s⁻¹. The corresponding average cross-section for the process (181) at room temperature would then be of order 10^{-14} cm².

A little later Phelps and Molnar,† in their earlier experiments with the absorption technique (see p. 1780), obtained evidence on the rate of the reaction (181) between He($2\,^3S$) metastable atoms. Under afterglow conditions in which the He($2\,^3S$) concentration was high they observed a linear variation of the reciprocal of this concentration with time. This is illustrated in Fig. 18.80 for an afterglow in helium at a pressure of 10 torr. A variation of this kind is to be expected if the metastable atoms are largely being destroyed by the process (178). In that case

$$\frac{dn_m}{dt} = -\delta n_m^2,$$

so

$$n_m^{-1} = (n_m^0)^{-1} + \delta t. \tag{186}$$

† loc. cit., p. 1780.

δ may then be determined from the slope of the linear plot provided n_m is known absolutely. This involved reliance on a theoretical absorption coefficient for the 3889 Å line and gave a value for the mean cross-section for the process at 300 °K of 10^{-14} cm^2.

Further experiments, by a different method, were carried out by Hurt.† He observed the variation of He($2\,^3S$) concentration near the

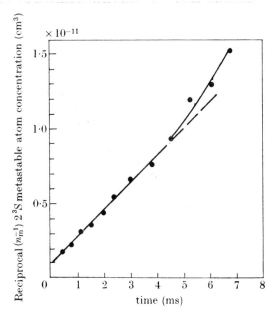

FIG. 18.80. Observed variation with time of the reciprocal (n_m^{-1}) of the $2\,^3S$ metastable atom concentration in a helium afterglow at 300 °K and 10 torr pressure.

cathode of a d.c. discharge, with perpendicular distance x from the cathode, using the absorption technique. Working under conditions of high concentration of He($2\,^3S$) and small loss by diffusion in directions parallel to the cathode, we have

$$D_m \frac{d^2 n_m}{dx^2} = \delta n_m^2. \tag{187}$$

If dn_m/dx is small at large values of x this gives, approximately,

$$(n_m)^{-\frac{1}{2}} = (\delta/6D_m)^{\frac{1}{2}} x + (n_m^0)^{-\frac{1}{2}}, \tag{188}$$

where n_m^0 is the concentration of He($2\,^3S$) at the cathode. Fig. 18.81 shows a plot of $n_m^{-\frac{1}{2}}$ against x, as observed by Hurt, verifying that it is

† HURT, W. B., J. chem. Phys. **45** (1966) 2713.

indeed close to linear. If D_m is known δ may be obtained from the slope of the plot. Unfortunately, the temperature in the neighbourhood of the cathode was as high as 520 °K while the value of D_m obtained by Phelps (§ 5.2.2.1) is for 300 °K. Assuming it to be independent of temperature T, we obtain a mean cross-section for the process (181) of

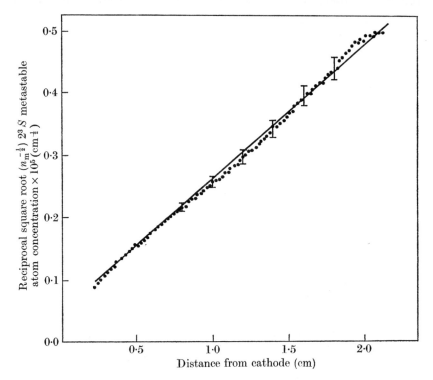

Fig. 18.81. Observed variation with distance from the cathode of the reciprocal square root $(n_m^{-\frac{1}{2}})$ of the $2\,^3S$ metastable atom concentration in a d.c. discharge in helium. ● average observed values. I probable error in observed values.

$1 \cdot 2 \times 10^{-14}$ cm^2, while on the assumption of a variation as $T^{\frac{3}{2}}$ it becomes 9×10^{-15} cm^2. Similar results were obtained from observations at helium pressures of 14, 19, and 26 torr.

It seems likely that, in Biondi's experiments, the metastable atoms concerned were in the $2\,^1S$ rather than the $2\,^3S$ state. Nevertheless, the cross-section for the process (181) appears to be much the same for both.

5.2.4.2. *Mercury.* Biondi† has also observed the reaction corresponding to (181) in afterglows in pure mercury vapour. The difficulty

† Biondi, M. A., *Phys. Rev.* **90** (1953) 730.

here is that, owing to the large masses of the mercury atoms, the electrons come to thermal equilibrium very slowly. From the known momentum-transfer cross-sections of electrons in mercury vapour (Chap. 1, p. 29; Chap 2, p. 47; Chap. 5, p. 336) the relaxation time is about $10^{-4}/p$ s, where p is the mercury pressure in torr. At pressures less than 1 torr the electrons will therefore not be in thermal equilibrium within the measuring period of milliseconds. The equation (185 a) for the electron concentration assumes that ν_e is not changing with time over the measuring period but, if the mean electron energy is changing, the ambipolar diffusion coefficient D_a, which largely determines ν_e, will also be changing. However, there is evidence, from experiments carried out by Mierdel,† using Langmuir probes to determine the electron temperature, that the electron temperature falls rapidly during the first few hundred microseconds to about 2000 °K and then changes very slowly. Electrons produced by the reaction

$$\mathrm{Hg}(6\,^3P_2) + \mathrm{Hg}(6\,^3P_2) \rightarrow \mathrm{Hg}^+ + \mathrm{Hg} + e, \qquad (189)$$

have a kinetic energy of 0·5 eV and it is considered that this is the supply of energetic electrons which maintains the temperature near 2000 °K against the loss by diffusion of the more energetic electrons to the walls (diffusion cooling; see also Chap. 19, § 2.2.5). During the time in which the electron temperature is constant (185 a) is applicable and, by measurement at different vapour pressures, the diffusion coefficient of Hg* in Hg and its two-body destruction rate may be obtained. The values so derived, which presumably apply to $6\,^3P_2$ metastable atoms, may be compared with results from other methods which refer to $6\,^3P_0$ metastable atoms. In § 2.5 we have discussed the deactivation of $\mathrm{Hg}(6\,^3P_0)$ metastable atoms by impact with normal Hg atoms. The cross-section observed by Callear and Williams for this process is $2·4 \times 10^{-15}$ cm² while Biondi finds 8×10^{-17} cm² for the corresponding process for $\mathrm{Hg}(6\,^3P_2)$.

Of the earlier methods for studying the diffusion and destruction of metastable mercury atoms in mercury vapour that of Coulliette,‡ carried out as long ago as 1928, is of considerable interest and was remarkably advanced for the time.

His apparatus (illustrated in Fig. 18.82) contained mercury vapour, the pressure of which could be varied by changing the temperature of the glass walls. Electrons from the cathode K were accelerated to the grid by a potential difference of 4·9 eV,

---

† MIERDEL, G., *Z. Phys.* **121** (1943) 574.
‡ COULLIETTE, J. H., *Phys. Rev.* **32** (1928) 636.

just too small to produce metastable mercury atoms. At intervals of about 1 ms rectangular voltage pulses applied an additional potential difference of 0·7 V between G_1 and K and metastable mercury atoms were produced near G_1. These diffused in the space between G_1 and the concentric hemispherical collecting electrode C. Their arrival at C was detected by the ejection of electrons which were collected by the concentric hemispherical grid G_2. Normally the potential difference between G_2 and C was such that there was no flow of current. At regular intervals after the application of the pulse between G_1 and K, however, another rectangular pulse could be applied between C and G_2 to attract secondary electrons to G_2 and thus to give a measure of the rate of arrival of metastable

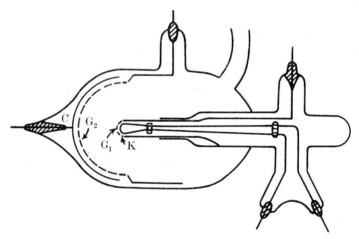

Fig. 18.82. Arrangement of apparatus used by Coulliette for studying the diffusion of metastable mercury atoms in mercury vapour.

atoms at that instant. The experiment was carried out with variable time-intervals t (up to about 2 ms) between the $G_1 K$ pulse and the $G_2 C$ pulse. In this way the variation with t of the rate of arrival of metastable atoms at C could be determined. For very small values of t (less than 30 μs) effects could be observed due to direct photo-electric emission from C, but these had become entirely negligible by the time the metastable atoms started to arrive.

Owing to the spherical symmetry of the arrangement the diffusion equation could be solved explicitly, giving for the rate of arrival of metastable atoms at C at a time t after their production at G_1,

$$i = A \sum_{n=0}^{\infty} \frac{(-1)^n}{n} \left\{ \sin \frac{n\pi a}{b} - \frac{n\pi a}{b} \cos \frac{n\pi a}{b} \right\} \exp\{-n^2 \pi D t / b\}, \qquad (190)$$

where a, b are the radii of the hemispherical grid G_1 and collector C respectively. In Coulliette's apparatus these radii were respectively 0·63 cm and 3·80 cm. D is the diffusion coefficient for the mercury metastable atoms in mercury vapour.

Fig. 18.83 shows a typical curve obtained by Coulliette for a temperature of the tube of 65 °C corresponding to a concentration of mercury atoms of $1·05 \times 10^{15}$ atoms cm^{-3}. The figure also shows the curve computed from (190) for a value of D of 57 cm² s⁻¹ at 1 torr which gives the best fit with the experimental data. This is to be compared with 42 cm² s⁻¹ at 1 torr observed at 350 °K for Hg($6\,^3P_2$) by Phelps.

McCoubrey† has also determined the diffusion coefficient of $6\,^3P$ metastable mercury atoms in mercury by studying the decay of the persistent band fluorescence of mercury vapour produced by irradiation with 2537 Å radiation. The interpretation of the phenomenon is that the radiation produces $6\,^3P_1$ atoms which are converted to $6\,^3P_0$ metastable atoms in collisions. These metastable

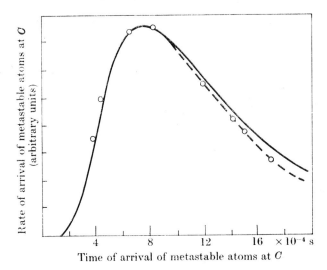

Fig. 18.83. Rate of arrival of metastable Hg atoms at the collector in Coulliette's apparatus at various time-intervals after their production has ceased. –O–O– observed rate of arrival; —— curve calculated assuming a diffusion coefficient of 57 cm² s⁻¹ at 1 torr.

atoms then combine with normal mercury atoms to form the $^3O_u^-$ state of Hg₂. The contribution to the decay rate due to loss of $6\,^3P_0$ atoms by diffusion to the walls of the containing vessel may be separated out by observing decay rates for containers of two different geometries. In this way a diffusion coefficient of 55 cm² s⁻¹ at 1 torr was found.

5.2.4.3. *Effect of impurities.* A very similar technique was used by Biondi‡ to derive the rate coefficients for Penning ionization by metastable atoms of helium and neon. If an ionizable impurity is present and the conditions are adjusted so that the reaction (189) makes no important contribution, the only modification required in the equations (183) is the replacement of δn_m^2 by γn_m where $\gamma = n_i \bar{Q}_{pi} \bar{v}$, n_i being the concentration of the impurity, $\bar{Q}_{pi}$ the mean cross-section for the Penning ionization process (144), and $\bar{v}$ the mean relative velocity of

† McCoubrey, A. O., *Phys. Rev.* **84** (1951) 1073. ‡ loc. cit., p. 1801.

collision. Observations were carried out for admixtures of argon and of mercury in helium and of argon in neon. The values obtained for $\bar{Q}_{\rm pi}$ are given in Table 18.20 and are discussed, in relation to other values obtained in other types of experiments, in § 5.3.4.

5.3. *Further study of Penning ionization by metastable atoms*

We have already (see §§ 5.2.2.4 and 5.2.4.3) described methods for measurement of cross-sections for destruction of metastable atoms of He and Ne in afterglows by addition of controlled amounts of an added impurity gas. The destructive process is almost certainly Penning ionization (see p. 1778). We now describe other methods for studying this process and discuss the results obtained by these methods as well as by those described above.

5.3.1. *Enhancement by metastable atoms of ionization produced by alpha particles.* The ionization produced by alpha particles in helium is very sensitive to the presence of impurities, which can increase it substantially. This effect is almost certainly due to ionization of the impurities by metastable helium atoms through the Penning ionization process

$$\mathrm{He^*+X \to He+X^++}e. \tag{191}$$

Information about the rates of these processes may be obtained by introducing a controlled amount of impurity into pure helium and observing the increase in ionization with impurity concentration.

We assume that the helium pressure is so high that the loss of metastable atoms by diffusion to the walls is negligible. If n, $n_{\rm i}$, $n_{\rm m}$, $n_{\rm e}$ are the concentrations of helium atoms, impurity atoms, helium metastable atoms, and electrons respectively then, in equilibrium,

$$q = \mu n_{\rm m} n + \lambda n_{\rm m} n_{\rm i}, \tag{192}$$

where q is the rate of metastable atom production per cm³ s by the alpha particles, $\mu n_{\rm m} n$ is the rate of loss per cm³ s by destruction processes in the gas through collisions with helium atoms which do not lead to ionization, and $\lambda n_{\rm m} n_{\rm i}$ is the rate of destruction through the Penning process (191) which produces ion pairs. The total rate of production of ion pairs per cm³ s is then given by

$$N = N_0 + \lambda n_{\rm m} n_{\rm i}, \tag{193}$$

where N_0 is the rate of production when no impurity atoms are present. Since, from (192),

$$n_{\rm m} = q/(\mu n + \lambda n_{\rm i}), \tag{194}$$

we have

$$(N-N_0)^{-1} = q^{-1} + \lambda/\mu q C, \tag{195}$$

where $C = n_i/n$ is the fractional concentration of impurity. Hence, by plotting $(N-N_0)^{-1}$ as a function of $1/C$, a linear relation should be obtained from the slope of which $\lambda/\mu q$ may be derived and $1/q$ from the intercept on the axis of $(N-N_0)^{-1}$. In practice, the determination of $1/q$ in this way is not very accurate (see Fig. 18.86) and this also affects the accuracy with which the ratio λ/μ can be derived.

An alternative, rather simpler, procedure may also be used. When the two processes of destruction of metastable atoms occur at the same rate
$$\mu/\lambda = n_i/n = C_0, \text{ say,} \qquad (196)$$
where C_0 is the fractional concentration of impurity under these conditions. The variation of the rate of ionization with impurity concentration has the general form shown in Fig. 18.85. At the concentration at which saturation occurs all the metastable atoms are lost through collisions with impurities which produce ionization. The differences in the ordinates of the curve at zero concentration and at the concentration at which saturation begins is proportional then to the total number of metastable atoms produced per alpha particle. The ordinate representing half this number can then be obtained and, from Fig. 18.85, the corresponding concentration C_0.

Both these methods were applied by Jesse and Sadauskis† to analyse observations which they made of ionization produced by alpha particles in helium, neon, and argon in the presence of controlled impurity concentration. Fig. 18.84 shows a schematic diagram of the ionization chamber used, with the major dimensions indicated. Ions produced by single alpha particles from polonium were collected and fed into a vibrating reed electrometer connected to a chart recorder.

As usual in experiments of this kind it is essential to work with very pure gases, including the 'impurities'. The chamber was constructed with quartz insulators so that it could be baked and pumped for 12 hours at 200 °C. The gases used were taken from breaker-flasks and were already of very high purity. Nevertheless, it was necessary further to purify the helium and neon by continuous circulation through a coconut charcoal tube cooled in liquid nitrogen. Thus, when the chamber was filled with apparently pure helium and the circulation started, a marked decrease of ionization per alpha particle was observed and continued to decrease until a steady value was reached. Further circulation then produced no change. It was assumed that once this state was reached no appreciable contribution to the ionization was coming from impurities.

† JESSE, W. P. and SADAUSKIS, J., *Phys. Rev.* **100** (1955) 1755.

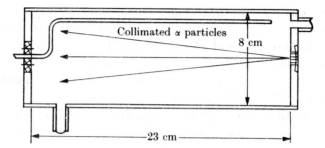

Fig. 18.84. Schematic diagram of the ionization chamber used by Jesse and Sadauskis.

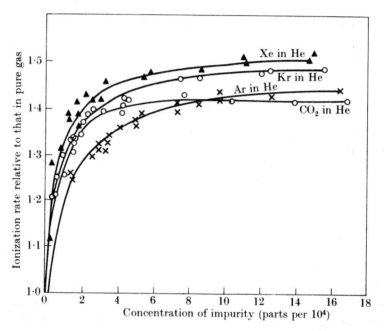

Fig. 18.85. Variation of the ionization rate for alpha particles in helium with concentration of added impurity.

The pressure of the main gas was about 400 torr and the controlled impurity concentrations were varied up to values of order 1 part in 1000.

Fig. 18.85 illustrates typical results obtained for the variation of the relative ionization per alpha particle with concentration of Ar, Kr, Xe, and CO_2 in helium, while Fig. 18.86 shows a number of linear plots following the relation (195), all taken at a pressure of 875 torr of helium. From results such as these λ/μ may be obtained as explained above.

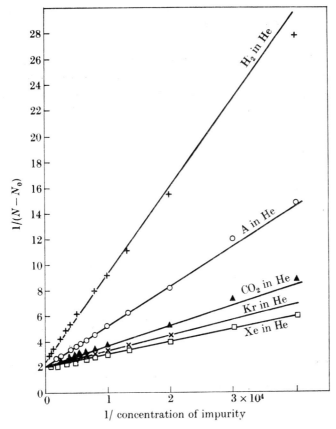

Fig. 18.86. Linear plots of observed values of $1/(N-N_0)$ against $1/C$ illustrating the validity of (195) for the effect of impurities on the rate of ionization by alpha particles in helium.

Before discussing the results of such an analysis there remains a further important effect, namely the variation of μ/λ with the pressure of pure helium. Fig. 18.87 illustrates this variation for impurities of H_2, Ar, and N_2. Its linear nature shows that

$$\mu = a+bp, \qquad (197)$$

where a and b are constant and p is the helium pressure. This is consistent with the results of the afterglow investigations which suggest that, whereas $2\,^1S$ metastable atoms in pure helium are destroyed by a two-body process, the $2\,^3S$ atoms are mainly destroyed in three-body collisions. Also, since the pressure dependence is independent of the nature of the impurity we would expect that, when adjusted to have the same intercepts on the p-axis, the pressure variation should be the same for all impurities. This is the case for H_2, Ar, and N_2 within experimental error but for other impurities, Kr, Xe, CO_2, and C_2H_4, no dependence on helium pressure is found.

A possible explanation of this result depends on the fact that the three-body destruction process for $2\,^3S$ metastable atoms leads to production of metastable $2\,^3\Sigma$ He_2 molecules. These molecules are also capable of ionizing impurity molecules but the internal energy available

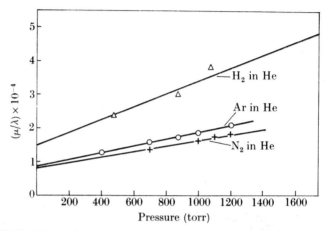

FIG. 18.87. Observed variation with helium pressure of the ratio μ/λ of the rate constant for destruction of metastable helium atoms by collision with normal helium atoms (μ) and with impurity atoms (λ).

is less, being the difference in energy of the $2\,^3\Sigma$ ground state at its equilibrium nuclear separation and that of the unstable $1\,^1\Sigma$ ground state at the same nuclear separation. This is about 15 eV and it is probably significant that the pressure dependence is only observed for impurities with ionization energy greater than 15 eV. For these alone the three-body destruction process does not lead to further ionization.

It is difficult to test this explanation in quantitative terms. Using the data obtained by Phelps† for the two- and three-body destruction rates for $2\,^1S$ and $2\,^3S$ in pure helium (§ 5.2.2.1), the relative proportion of the respective metastable atoms produced per alpha particle to give the pressure variation shown in Fig. 18.87 comes out to be about 1·5:1. This is somewhat greater than expected on theoretical arguments. The cross-section for production of $2\,^3S$ atoms by impact of He^{++} is very small indeed as the process could only occur through direct reversal of electron spin. Impact with an He^+ ion would be much more effective

† loc. cit., p. 1781.

as the excitation could then occur through exchange of electrons of opposite spin between the atom and ion. At the experimental pressures, only a fraction of the alpha particles will have captured an electron and Platzman† estimates that the ratio of $2\,^1S$ to $2\,^3S$ production in the experimental conditions should be about 10:1. The matter can only be resolved by obtaining more precise data about destruction rates.

Table 18.19 gives the results obtained for the ratio λ/μ for a number of cases studied in neon as well as helium. If the ratio depends on the pressure of the main gas, two values are given, one obtained at a pressure of 875 torr and the other extrapolated back to zero pressure.

TABLE 18.19

Ratios of the rate coefficients λ for destruction of metastable helium and neon atoms through ionization by an impurity to those, μ, for destruction by collision with the parent gas atoms. Q_{pi}, Q_{dm} are the corresponding mean effective cross-sections

Impurity		$10^{-3}\,\lambda/\mu$		$10^{-3}\,Q_{\mathrm{pi}}/Q_{\mathrm{dm}}$
		875 torr (Main gas pressure)	Zero (Main gas pressure)	Zero (Main gas pressure)
Ar		6·9	11·2	4·9
H_2		3·0	6·9	1·3
N_2		6·5	12·8	4·6
Kr	in		16·9	23·3
Xe	He		20·5	28·5
Hg			670	950
CO_2			15·1	20·4
C_2H_4			14·0	18·5
Ar			4·2	4·8
Xe	in Ne		23·0	30·4
H_2			4·6	1·9

Since the internal energy available from metastable argon atoms is only 11·6 eV there are relatively few possibilities to investigate when argon is the main gas. Observations were, however, made of increased ionization in the gas when either C_6H_6, C_2H_2 or C_2H_4 were added. These have respective ionization energies 9·2, 11·4, and 10·5 eV, all less than 11·6 eV. No values of λ/μ were derived for these cases.

5.3.2. *Measurement of cross-sections using metastable atom beams.* The method used by Richards and Muschlitz‡ to observe the differential

---

† PLATZMAN, R. L., *J. Phys., Paris*, **21** (1960) 853.
‡ RICHARDS, H. L. and MUSCHLITZ, E. E., *J. chem. Phys.* **41** (1964) 559.

cross-section for scattering of helium metastable atoms by various atoms has been described in § 5.1.3. Sholette and Muschlitz† have used essentially the same equipment to measure cross-sections for ionization of various gases by $2\,^3S$ and $2\,^1S$ helium atoms.

Referring to Fig. 18.62, we now suppose that the gas in the scattering chamber can be ionized by the metastable atom beam entering through the hole H_3. The undeviated atoms were detected at the target T, the elastically scattered atoms at the walls and bottom of the cylinder SC, while the ions produced were drawn by a suitable potential to the lid SL and detected there. Thus, if SC and SB are made negative, electron currents are read on T and SL. The sum i_s of these currents is then given by

$$i_s = i_{sc} + i^-, \qquad (198)$$

where i_{sc} is the current due to electrons ejected from SC and SB by elastically scattered metastable atoms and i^- that due to electrons produced in ionizing collisions with gas atoms. If SC and SB are made positive, a positive current is recorded on SL due to the positive ions produced,

$$i_{SL}^+ = i^+, \qquad (199)$$

and one i_T^+ on T due to electrons ejected from T by undeviated metastable atoms.

If $\bar{\gamma}$ is the mean number of electrons ejected from the detector per metastable atom per impact, the total number of metastable atoms entering the scattering chamber per second is proportional to

$$i_0 = (i_s - i_{SL}^+ + i_T^+)/\bar{\gamma}, \qquad (200)$$

all of the current being taken as positive.

The number of ionizing collisions produced per second by these atoms is proportional to

$$i_{\text{ion}} = i_T^+$$
$$= i_0\{1 - \exp(-N\bar{Q}_{\text{pi}}l)\}, \qquad (201)$$

where N is the number of ionizable atoms or molecules per centimetre^3 in the scattering cylinder and l the path length of a metastable atom in the cylinder. $\bar{Q}_{\text{pi}}$ is the mean cross-section for Penning ionization.

Using the value 0·29 of $\bar{\gamma}$ observed by Stebbings‡ $\bar{Q}_{\text{pi}}$ may be obtained. There remains the problem of determining the respective contributions of $2\,^1S$ and $2\,^3S$ metastable atoms. It was found that, for all but one of the gases investigated, change in slope of the plot of $\ln(i_T^+/i_0)$ against N (or rather, gas pressure) resulted when the energy of the electrons

---

† SHOLETTE, W. P. and MUSCHLITZ, E. E., *J. chem. Phys.* **36** (1962) 3368.
‡ STEBBINGS, R. F., *Proc. R. Soc.* **A241** (1957) 270.

producing the metastable atoms by impact excitation of the normal helium atoms leaving the source chamber was changed from 30 to 35 and thence to 45 eV. This suggests that $\bar{Q}_{\mathrm{pi}}$ does not change with the proportion of each metastable atom present (see Chap. 4, § 2.2.2). If it can also be assumed that the respective electron ejection yields, γ_1 and γ_3, are equal then it would follow that the ionizing cross-sections $\bar{Q}_{\mathrm{pi}}^{(1)}$ and $\bar{Q}_{\mathrm{pi}}^{(3)}$ for $2\,^1S$ and $2\,^3S$ atoms respectively are also equal.

A further test was made using an inhomogeneous magnetic field to deflect the $2\,^3S$ metastable atoms. In this way the ratio R of $2\,^1S$ to $2\,^3S$ metastable atoms could be obtained. These were compared with the values to be expected on the assumption of the relative electron excitation cross-sections of the respective states derived by Frost and Phelps† (Chap. 5, § 3.3.1) and equality of γ_1 and γ_3. Good agreement was obtained at all three electron energies, 30, 35, and 45 eV, studied.

Finally, measurements were made in nitrogen with and without the magnetic field. Assuming $\gamma_1 = \gamma_3$ these give $\bar{Q}_{\mathrm{pi}}^{(3)}$ directly and the values obtained agreed with $\bar{Q}_{\mathrm{pi}}$.

For H_2 a significant difference between $\bar{Q}_{\mathrm{pi}}^{(1)}$ and $\bar{Q}_{\mathrm{pi}}^{(3)}$ was found and in this case the separate cross-sections were obtained using values of R derived as above.

In this type of experiment the effect of impurities is not quite so serious but the helium used was introduced through a liquid-air cooled charcoal trap and was estimated to contain less than 1 part in 10^3 of impurities. The scattering gas samples, taken from flasks of reagent grade gas, contained even less impurity, no greater than 1 part in 10^4.

Fig. 18.88 illustrates typical plots of $\ln(i_\mathrm{T}^+/i_0)$ against gas pressure obtained for argon, at three electron energies.

5.3.3. *The energy distribution of the electrons produced by Penning ionization.* In Chapter 4, § 2.1.2.2 we describe the apparatus (see Fig. 4.45) used by Čermák for determining the relative cross-sections for excitation of helium $2\,^1S$ and $2\,^3S$ metastable atoms by electron impact. An atomic beam of helium atoms was crossed by the exciting electron beam. Charged particles produced by electron impact were then removed by an appropriate electric field so that the beam, containing metastable $2\,^1S$ and $2\,^3S$ atoms, entered a collision chamber containing argon in which they were detected from the Penning ionization they produced. To discriminate between $2\,^1S$ and $2\,^3S$ atoms a retarding potential analysis was made of the energies of the secondary electrons. These will be different for the two excited states, being the

† FROST, L. S. and PHELPS, A. V., *Westinghouse Res. Rep.* 6–94439–6–R3 (1957).

difference between the excitation energy of the state and the ionization energy (15·75 eV) of argon, i.e. 4·86 and 4·06 eV for $2\,^1S$ and $2\,^3S$ atoms. The analysis was carried out by extracting the electrons between narrow channels perpendicular to the atomic beam so that only electrons moving in this direction entered the retarding system. Fig. 4.46 shows a typical result in which the electron energy distribution, obtained by differentiation of the retarding potential–current characteristics, exhibits two clear maxima at energies 0·80 eV apart, close to the expected values 4·06 and 4·86 eV respectively.

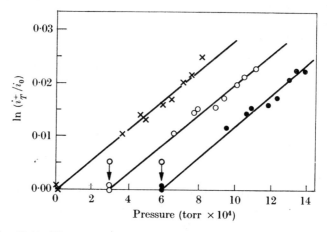

FIG. 18.88. Plots of $\ln(i_T^+/i_0)$ against pressure for Penning ionization in argon at three electron energies: ● 45 eV, ○ 35 eV, × 30 eV.

In a later investigation Čermak used the same apparatus to investigate the energy distribution of the secondary electrons produced by Penning ionization of a number of molecules. In these cases the molecular ion can be produced in different vibrational states and it by no means follows that the distribution among these states will be given correctly by the Franck–Condon principle—the collision is one between heavy particles. Nevertheless, since it is essentially a resonance process in that the kinetic energy of relative motion between the heavy systems is unchanged by the collision, there is a strong presumption that the Franck–Condon principle will still be useful in determining the vibrational distribution among the product ions.

Čermak's results for N_2 and CO shown in Figs. 18.89 and 18.90 respectively are consistent with this. In these figures the retarding potential–current characteristic and its derivative di/dV are shown, the latter giving the electron energy distribution. For N_2, peaks are found

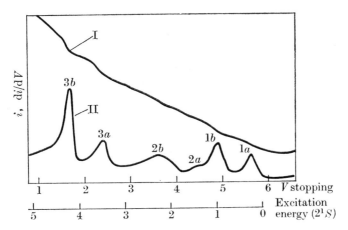

FIG. 18.89. Energy distributions of secondary electrons due to Penning ionization of N_2. I, observed retarding potential–current characteristics. II, differential characteristics di/dV proportional to the electron energy distribution, as indicated. For ease of interpretation, an energy scale is included which is based on the difference between the excitation energy of $He(2\ {}^1S)$ and the first ionization energy of N_2 as zero. Peaks due to ionization by $2\ {}^1S$ and by $2\ {}^3S$ are distinguished as a and b respectively while 1, 2, 3 distinguish the electronic state of the N_2^+ ion as $X\ {}^2\Sigma_g^+$, $A\ {}^2\Pi_u$, and $B\ {}^2\Sigma_g^+$ respectively.

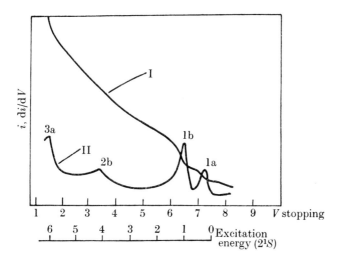

FIG. 18.90. Energy distribution of secondary electrons due to Penning ionization of CO. I, observed retarding potential–current characteristics; II, differential characteristics di/dV, proportional to the electron energy distribution, as indicated. For ease of interpretation, an energy scale is included based on the difference between the excitation energy of $He(2\ {}^1S)$ and the first ionization energy of CO as zero. Peaks due to ionization by $2\ {}^1S$ and by $2\ {}^3S$ respectively are distinguished as a and b while 1, 2, 3 distinguish the electronic state of the CO^+ ion as $X\ {}^2\Sigma$, $A\ {}^2\Pi$, and $B\ {}^2\Sigma$ respectively.

corresponding to the production of N_2^+ in the ground $X\,^2\Sigma_g^+$ and the two excited $A\,^2\Pi_u$ and $B\,^2\Pi_g^+$ states by $2\,^1S$ and by $2\,^3S$ atoms. The energy separations check well with the values determined by the energy difference (0·80 eV) between the $2\,^1S$ and the $2\,^3S$ helium states and those, 1·1 and 3·3 eV, between the respective excited states and the ground state of N_2^+. The peaks for the ground state and the $B\,^2\Pi_g^+$ state are

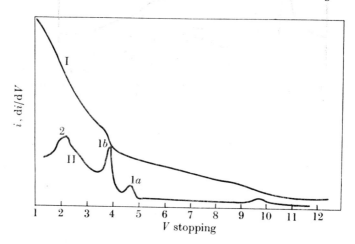

FIG. 18.91. Energy distribution of secondary electrons due to Penning ionization of NO. I, observed retarding potential current characteristic; II, differential characteristic di/dV, proportional to the electron energy distribution as indicated.

narrower than for the $A\,^2\Pi_u$ as would be expected from the Franck–Condon principle and the potential energy curves of the two levels (see Fig. 13.46). Thus, whereas the internuclear separation in the $X\,^2\Sigma_g^+$ and $B\,^2\Pi_g^+$ states (1·116 and 1·075 Å respectively) is close to that of the ground ($X\,^1\Sigma_g^+$) state of the neutral molecule, that for the $A\,^2\Pi_u$ state (1·190 Å) is appreciably larger. The same feature is seen in the photoelectron spectra for N_2 shown in Fig. 14.51.

For CO the situation is similar but exaggerated. The distribution due to production of $A\,^2\Pi$ CO$^+$ ions is even broader than for N_2^+. In fact, as seen from Fig. 18.90, the peak due to production of this state from $2\,^1S$ metastable atoms is no longer visible. This is in qualitative agreement with the photoelectron spectrum (Fig. 14.51).

The situation is less clear for NO, results for which are shown in Fig. 18.91. The first and second ionization potentials of NO are at 9·24 and 14·2 eV respectively so that electron energy peaks would be expected at 10·36 and 6·4 eV, for production from $2\,^1S$, and at 0·8 eV lower energy for production from $2\,^3S$. There are only faint signs of a peak near 10 eV and none at all near 6·4 eV. Two peaks separated by 0·8 eV are found at lower energies which could be interpreted as arising from production of NO+ in the $^3\Delta$ state for which 16·55 eV is required. It may be that, in this case, the Franck–Condon principle is not operative. However, reference to the photoelectron spectrum of NO (Fig. 14.60) does show similar features in that there is little evidence for production of NO+ at an ionization energy of 14·2 eV whereas there is a considerable rise near 16·5 eV due to production of $NO^+(^3\Delta)$.

5.3.4. *Discussion of results.* Results obtained for the cross-sections for ionization by $2\,^1S$ and $2\,^3S$ helium atoms are given in Table 18.20. For comparison, results obtained by other methods are also given.

TABLE 18.20

Observed cross-sections $Q_{pi}^{(1)}$, $Q_{pi}^{(3)}$ for ionization of various atoms and molecules by impact at thermal energies with $2\,^1S$ and $2\,^3S$ metastable atoms respectively

Atom or molecule	$Q_{pi}^{(3)}$ (10^{-16} cm²)				$Q_{pi}^{(1)}$ (10^{-16} cm²)		$\bar{Q}_{pi}$ (10^{-16} cm²)	
	(a)	(b)	(c)	(e)	(a)	(b)	(e1)	(e2)
Ar	7·6±0·5	6·6	3·8	0·9	7·6±0·5	—	(7·0)	1·5
Kr	9±2	10·3	—	—	9±2	—	33	7
Xe	12±3	13·9	—	—	12±3	—	40	8·5
Hg	—	—	—	—	—	—	1370	285
H₂	2·6±0·5	6·0	—	—	1·7±0·5	—	1·9	0·4
O₂	14±1	—	—	—	14±1	—	—	—
N₂	7±1	6·4	—	—	7±1	—	6·6	1·4
CO	7±1	—	—	—	7±1	—	—	—
CO₂	—	—	—	—	—	—	29	6·1
C₂H₄	—	—	—	—	—	—	26	5·5

(a) Sholette, W. P. and Muschlitz, E. E., loc. cit., p. 1816.
(b) Benton, E. E., Ferguson, E. E., Matsen, F. A., and Robertson, W. W., loc. cit., p. 1792.
(c) Phelps, A. V. and Molnar, J. P., loc. cit., p. 1780.
(e) Biondi, M. A., loc. cit., p. 1801.
(e1) Jesse, W. P. and Sadauskis, J., loc. cit., p. 1811, normalized to $7·0 \times 10^{-16}$ cm² for argon.
(e2) Ibid., taking for the two-body destruction cross-section for $2\,^1S$ helium atoms in helium the value 3×10^{-20} cm² obtained by Phelps, A. V. (see p. 1782).

For the triplet cross-section $Q_{pi}^{(3)}$ the agreement with the results obtained by Benton, Ferguson, Matsen, and Robertson,† using the afterglow absorption technique (§ 5.2.2.4), is remarkably good except for hydrogen. For argon, comparison may also be made with the results

† loc. cit., p. 1792.

of Phelps and Molnar† using the same technique and of Biondi using the afterglow method described in § 5.2.4.3. The agreement here, particularly with the latter, is poor.

No other results are available with which to compare the singlet cross-sections $Q_{\text{pi}}^{(1)}$. However, the results of Jesse and Sadauskis,‡ described in § 5.3.1, provide some mean, $\bar{Q}_{\text{pi}}$, of $Q_{\text{pi}}^{(1)}$ and $Q_{\text{pi}}^{(3)}$, possibly giving greater weight to $Q_{\text{pi}}^{(1)}$. Since from these experiments the ratio of $\bar{Q}_{\text{pi}}$ to the two-body destruction cross-sections Q_{dm} for $2\,^1S$ atoms in pure helium is alone obtained, it is necessary to normalize the data in some way to obtain absolute values for comparison. In Table 18.20 this is done in two different ways. The first merely takes the value of $\bar{Q}_{\text{pi}}$ for argon as $7\cdot 0\times 10^{-16}$ cm², which agrees well with the other data for this gas. This gives the values in column (e1) which, for H_2 and N_2, agree fairly well with those obtained by the other methods in columns (a) and (b) for $Q_{\text{pi}}^{(1)}$ and $Q_{\text{pi}}^{(3)}$. For Kr and Xe the agreement is poor. The alternative values given in column (e2) are obtained by assuming for Q_{dm} the value 3×10^{-20} cm² found by Phelps. These are all considerably smaller than the other sets of values, suggesting that Q_{dm} is actually about four times larger. However, this conclusion must be regarded as very tentative because the detailed interpretation of the observations made by Jesse and Sadauskis still leaves much to be desired.

5.4. *Measurement of the cross-sections for transfer of excitation between metastable and normal helium atoms*

5.4.1. *Lower limits from beam experiments.* The cross-section for transfer of excitation between $2\,^3S$ metastable and normal helium atoms and its variation with relative velocity is of considerable interest theoretically as the atoms are sufficiently simple for a detailed calculation to be carried out. Furthermore, the theory suggests certain special features that may be present in the interaction between the two atoms which would have the effect of causing the transfer cross-section to rise quite rapidly with relative energy, at least over the whole thermal region (see § 8.1.3). Direct measurement of the cross-section is very difficult, though Richards and Muschlitz§ were able to place a lower limit on its magnitude using the atomic-beam apparatus described in § 5.1.3 and illustrated diagrammatically in Fig. 18.62.

If a beam of metastable helium atoms is passed through helium gas at low pressure no backward elastic scattering of the atoms from the beam can take place except for a very small amount due to the velocities

† loc. cit., p. 1780.   ‡ loc. cit., p. 1811.   § loc. cit., p. 1774.

of the gas atoms. However, a gas atom may often move off in the backward direction after collision with one in the beam. If, during such an impact, transfer of excitation takes place the atom will be observed as a metastable atom apparently scattered backwards. Thus, referring to Fig. 18.62, any positive current to SL will be due to metastable atoms formed in this way, the apparent angle of scattering in the laboratory system being 135°. From the observed current, Richards and Muschlitz placed a lower limit of 2×10^{-16} cm^2 to the transfer cross-section Q_{tr} at 320 °K.

5.4.2. *Optical-pumping method.* The most complete data about Q_{tr} have been obtained in some remarkable experiments by Colegrove, Schearer, and Walters† who used the transfer process to produce, through optical pumping, a 40 per cent polarization of ^{3}He atoms in a gas.

The effectiveness of optical pumping in aligning $2\,^3S$ ^{4}He atoms in ^{4}He gas was first established by Colegrove and Franken.‡ These experiments were extended by Schearer.§ Metastable atoms were produced in a Pyrex bulb containing ^{4}He gas at low pressure. Optical pumping was carried out by applying a weak magnetic field to the sample while irradiating it along the direction of the field with right-hand circularly polarized $2\,^3S_1$–$2\,^3P_0$ radiation from a helium lamp. This radiation produces transitions between the Zeeman sub-levels of the $2\,^3S_1$ and $2\,^3P_0$ states which satisfy the selection rule $\Delta M_J = +1$, M_J being the magnetic quantum number. Referring to the energy level diagram shown in Fig. 18.92 this means that the only transition involved is from the $2\,^3S_1$ state with $M_J = -1$. Reverse transitions

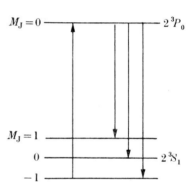

FIG. 18.92. Zeeman sub-levels of the $2\,^3S_1$ and $2\,^3P_0$ states of helium, illustrating optical pumping by absorption of right-hand circularly polarized $2\,^3S_1$–$2\,^3P_0$ radiation.

will, however, take place with equal probability from all three sub-levels and hence that with $M_J = -1$ will be relatively depopulated. As this process proceeds the absorption coefficient of the gas for the pumping light will therefore fall until an equilibrium is reached.

---

† COLEGROVE, F. D., SCHEARER, L. D., and WALTERS, G. K., *Phys. Rev.* **132** (1963) 2561.
‡ COLEGROVE, F. D. and FRANKEN, P. A., ibid. **119** (1960) 680.
§ SCHEARER, L. D., *Advances in quantum electronics*, pp. 23 (Columbia University Press, New York, 1961).

To establish that alignment has resulted, an oscillatory magnetic field at radio-frequency is applied transversely to the direction of the pumping light. The steady field is then slowly varied while the intensity of the transmitted light is monitored. A sharp reduction in transmission is observed when the frequency of the oscillatory field is in resonance with that required to produce a transition from an upper Zeeman sub-level to the $M_J = -1$ level. The width of the resonance peak is given by

$$\Delta \nu = \frac{1}{\pi}\{\tau^{-2}+\gamma^2 H_1^2\}^{\frac{1}{2}}, \qquad (202)$$

where H_1 is the amplitude of the oscillating field, $\gamma = e\hbar/2m$, and τ is the mean time between transitions which change the spin-alignment in the absence of the oscillatory field.

In ^{4}He the only processes which contribute to τ are those due to absorption of photons from the pumping light and to collisions of $2\,^3S$ metastable atoms with ions and electrons present in the discharge. Transfer of excitation in collisions between $2\,^3S$ metastable atoms and ground-state atoms can produce no change in spin alignment because both the total angular momentum and its component along the field must remain unchanged during the collision. Since a helium atom in the ground state has zero total angular momentum it is clear that transfer of excitation necessarily involves also transfer, unchanged, of the component of total angular momentum along the field. The situation is quite different when ^{4}He is replaced by ^{3}He because in this case the ^{3}He nucleus possesses a total nuclear spin quantum number $I = \frac{1}{2}$. The ground state, in a weak magnetic field, is a doublet corresponding to $M_F = \pm\frac{1}{2}$. In a transfer collision between ^{3}He atoms in the ground and $2\,^3S$ states there is now the possibility of a change ± 1 in the magnetic quantum number of the $2\,^3S$ state associated with a change ± 1 in that of the ground state.

At first sight it would seem that, because of these transfer collisions, the $2\,^3S$ metastable ^{3}He atoms would be reoriented so rapidly that negligible polarization would result from optical pumping. In fact the angular momentum is transferred to the ground state ^{3}He atoms so that, if other processes for disorientation of these atoms are sufficiently slow, the transfer process will build up polarization of the ground state atoms. The metastable and normal atoms are thus tightly coupled and the net result of the optical pumping is that both are polarized. In this way it has been possible to build up a 40 per cent alignment of normal ^{3}He atoms.

If the usual magnetic-resonance technique is now applied, as described above for $2\,^3S$ ⁴He atoms in ⁴He, three resonance lines will be observed, two corresponding to transitions between Zeeman sub-levels of the $2\,^3S$ state and one, at a much lower frequency, to a transition between the Zeeman sub-levels of the ground state. The widths of the $2\,^3S_1$ resonance lines are given by (202) but now the relaxation time includes in addition a major contribution from the excitation transfer collisions. In fact

$$\tau^{-1} = \tau_{\rm tr}^{-1} + \tau_{\rm r}^{-1} + \tau_{\rm p}^{-1}, \qquad (203)$$

where $\tau_{\rm tr}$, $\tau_{\rm r}$, $\tau_{\rm p}$ are relaxation times due to excitation transfer, to disorienting collisions with electrons and ions, and to optical pumping respectively. $\tau_{\rm r}$ and $\tau_{\rm p}$ will be of the same order as for ⁴He and are both $\gg \tau_{\rm tr}$. It follows that $\tau_{\rm tr}$, and hence the mean cross-section for excitation transfer under the thermal conditions concerned, may be obtained.

The actual transitions involved are indicated on the energy level diagram for a ³He atom in a weak external magnetic field shown in Fig. 18.93. The total nuclear spin quantum number $I = \tfrac{1}{2}$ and so the ground state is split into two Zeeman levels, the frequency separation being 3·2 Mc s⁻¹ G⁻¹. Associated with the $2\,^3S_1$ state there will be two hyperfine structure levels with total angular momentum quantum number $F = \tfrac{3}{2}, \tfrac{1}{2}$ respectively. The former splits into four Zeeman sub-levels with frequency separation 1·9 Mc s⁻¹ G⁻¹ and the latter into two with twice as great a separation. Optical pumping takes place via transitions to the $2\,^3P_0$ state. This may be done without appreciable excitation of the $2\,^3P_1$ and $2\,^3P_2$ states by taking advantage of the fact that radiation from the $2\,^3S_1$–$2\,^3P_1$ and $2\,^3S_1$–$2\,^3P_2$ transitions in ⁴He is of the correct wavelength to excite the $2\,^3S_1$–$2\,^3P_0$ transition in ³He but not that to $2\,^3P_1$ or $2\,^3P_2$ (see Fig. 18.92 and § 4.1.3).

Colegrove, Schearer, and Walters† measured $\tau_{\rm tr}$ over a wide range of temperature by this method. Fig. 18.94 shows a schematic illustration of the apparatus they used. The lamps were Pyrex tubes filled with pure ⁴He at about 4-torr pressure. Light from the lamps was circularly polarized by passage first through a Polaroid type HR linear polarizer for the infra-red and then a quarter-wave plate of cellophane 0·001 inch thick with axes oriented at 45° to the axis of the linear polarizer.

The experimental bulb was a Pyrex vessel of 5-cm diameter filled with ³He over a pressure range from 0·1 to 5 torr at room temperature.

† COLEGROVE, F. D., SCHEARER, L. D., and WALTERS, G. K., *Phys. Rev.* **135** (1964) A353.

This must be so pure that the contribution to the relaxation time due to collisions with impurity atoms should be negligible compared with that from τ_r which is of the order 10^{-4} s. The effective cross-sections Q_{or}^{im} for disorientation collisions with impurities could be of the order

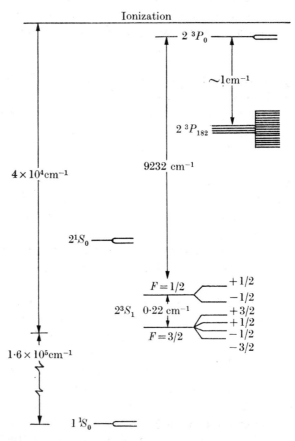

Fig. 18.93. Energy levels of ^{3}He atoms in a small external magnetic field.

10^{-14} cm^2 or so at room temperature. Hence, if the frequency of these collisions, given by $n_{im}\bar{v}Q_{or}^{im}$, where n_{im} is the impurity concentration and $\bar{v}$ the mean relative velocity of impact, is to be $\ll 10^4$, n_{im} must be $\ll 10^{13}$ cm^{-3} corresponding to a partial pressure $\ll 10^{-4}$ torr. The bulb was first heated under vacuum and then the walls were cleaned up by letting in helium and exciting a bright discharge. The impure gas was then pumped out and replaced by ^{3}He, purified by passage through a liquid helium trap, and the bulb was then sealed. To make possible measurements over a wide temperature range it could be placed in a Dewar flask or in an oven.

Metastable atoms were produced in the bulb by means of an electrodeless discharge. The pumping light was monitored by a lead sulphide detector. A pair of Helmholtz coils provided the external magnetic field.

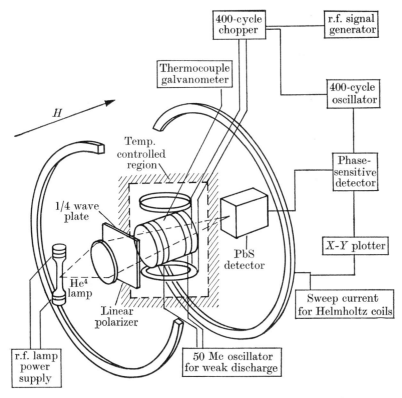

FIG. 18.94. Arrangement used by Colegrove, Schearer, and Walters for determining the cross-section for transfer collisions between 2 ^3S and normal helium atoms as a function of temperature.

The oscillatory field, which destroys the polarization under resonance conditions, was provided through a coil with axis perpendicular to that of the Helmholtz coils. To observe a resonance line the oscillating field was set at a fixed frequency between 1 and 10 Mc/s, chopped at 400 c/s to make possible phase-sensitive detection, and the external field slowly swept through resonance.

It was verified that the width of the resonance line was independent of the intensity of the pumping light so that τ_p^{-1} was negligible. Comparison with line widths observed with ^{4}He showed that τ_r^{-1} was also negligible. Fig. 18.95 shows a typical plot of the square of the line width (to half maximum) as a function of H_1^2. The intercept on the vertical axis of this plot gives $1/\pi^2\tau_{\text{tr}}^2$. From the values of τ_{tr} so derived the mean transfer cross-section Q_{tr} obtained is shown as a function of gas temperature in Fig. 18.96. The value at room temperature is compatible with the limits

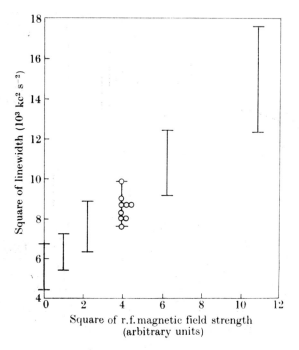

FIG. 18.95. Observed variation of the square of the width of the $2\,^3S_1$ helium magnetic resonance line with the mean square amplitude of the applied r.f. magnetic field.

placed upon it from the experiments of Richards and Muschlitz ($Q_{\text{tr}} \gg 10^{-16}$ cm^2). The fact that the cross-section rises monotonically with temperature from a low value at the lowest temperature observed is of considerable interest in connection with the form of the interaction between He($2\,^3S$) and normal He atoms and will be discussed in § 8.1.3.

5.5. Reaction rates involving metastable oxygen atoms

5.5.1. *Introduction.* So far we have confined the discussion of the passage of metastable atoms through gases to metastable rare gas and mercury atoms. These are comparatively easy to work with because they are formed from gases or vapours which are monatomic. Considerable interest attaches to the study of the behaviour of other metastable atoms, particularly those of oxygen and nitrogen because of their importance in the interpretation of upper atmospheric phenomena.

As discussed in Chapter 8, § 9.1 the ground configuration of atomic oxygen gives rise to three terms, ^3P, ^1D, and ^1S with energy increasing in that order. As with all terms arising from the same configuration, transitions between them are optically forbidden so that the ^1D and ^1S states are metastable. However, optical transitions between them

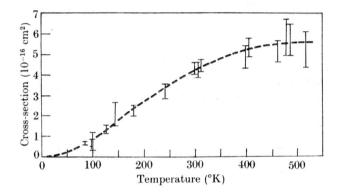

Fig. 18.96. Mean cross-section for transfer of excitation in collisions between 2 ^3S and normal helium atoms as a function of gas temperature, derived from magnetic resonance experiments.

may be observed under certain conditions. Thus the ^1S–^1D transition gives the famous green line at 5577 Å with a transition probability of 1·3 s^{-1} while ^1D–^3P gives the red doublet at 6300 and 6364 Å, the respective transition probabilities being 6·9 and 2·2×10^{-3} s^{-1}. Because these lines may be observed under laboratory conditions, the rates of reactions leading to deactivation of the upper metastable states may be determined, at least in principle, by quenching techniques. The analysis of § 2.1 may be used but with some modification of the symbols involved.

We consider first the intensity of emission of a forbidden line in a gas containing, in addition to normal oxygen atoms, in concentration n_0, other atoms or molecules, in concentrations n_i. If R is the rate of production of the metastable O atoms concerned, then the concentration n'_0 of the metastable atoms will be given by

$$n'_0/\tau = R, \tag{204}$$

where the lifetime τ is determined not only by loss through radiation but by deactivation collisions and by loss to the container walls. Thus

$$\tau^{-1} = \tau_{\rm r}^{-1} + \tau_{\rm dm}^{-1} + \tau_{\rm w}^{-1}, \tag{205}$$

where τ_r, τ_{dm}, and τ_w refer respectively to the three modes of loss. This extension of the significance of τ to include τ_w is necessary because τ_r is now so long. τ_{dm} will be given by

$$\tau_{dm}^{-1} = \mu_0 n_0 + \sum \mu_i n_i, \qquad (206)$$

where μ_0 is the rate of deactivation by collisions with normal O atoms, μ_i that by collisions with atoms or molecules i.

If now a further quenching gas is added in concentration n_q without changing the rate of production of metastable atoms, the concentration of these latter will be changed to n_0'' where

$$n_0''(\tau^{-1} + \tau_q^{-1}) = R. \qquad (207)$$

Here
$$\tau_q^{-1} = \mu_q n_q + \tau_{wq}^{-1} - \tau_w^{-1}. \qquad (208)$$

τ_q is the rate of quenching by the added gas and τ_{wq}^{-1} is the rate of loss to the walls when this gas is present. In general τ_{wq} will not be equal to τ_w.

Under suitable pressure conditions, for a given geometry, τ_w and τ_{wq} may be made large compared with τ_{dm} and τ_r. Under these circumstances

$$n_0'/n_0'' = 1 + \tau_q^{-1}/(\tau_r^{-1} + \tau_{dm}^{-1}) \qquad (209)$$
$$= 1 + \mu_q n_q/(A + \mu_0 n_0 + \sum \mu_i n_i), \qquad (210)$$

where $A\ (=1/\tau_r)$ is the A coefficient for the forbidden radiative transition. Hence, if the ratio n_0'/n_0'' of the forbidden line intensities is measured as a function of the concentration n_q of the added quenching gas a linear plot should result, the slope being

$$\mu_q/(A + \mu_0 n_0 + \sum \mu_i n_i). \qquad (211)$$

If the reciprocal of the slope is now plotted as a function of n_0 a further linear plot is obtained, the intercept at zero n_0 being $(A + \sum \mu_i n_i)/\mu_q$ and the slope $\mu_0 n_0/\mu_q$. To obtain definite results the initial conditions of emission of the line should be such that $\sum \mu_i n_i$ is negligible. If this can be done then, knowing A, μ_q is determined from the intercept and μ_0 from the slope of this second graph. Linearity in both plots verifies the assumption that wall loss is negligible because the rate of such loss varies inversely as the pressure instead of being proportional to the pressure. It also verifies the assumption that the rate of production of metastable atoms is independent of the partial pressures of the different gases present.

It may not always be possible to select experimental conditions to meet the requirements of the above analysis, particularly the smallness of wall losses. Usually, under these circumstances, the contribution from wall loss may be separated out by making observations as a function of total pressure and of container dimensions (see § 5.5.2). Unfortunately it is assumed in (207) that the rate R of production of metastable atoms is unchanged if the total pressure and container dimensions are varied. In quenching experiments concerned with deactivation of excited states from non-metastable states there is no difficulty in satisfying this assumption by using optical excitation. This is not possible for metastable atoms which are produced normally in an electric discharge at rates that depend on the electron temperature and concentration, both of which may vary with total gas pressure and container dimensions. If wall loss is important there is no way of checking, from the linearity of a plot, that the rate of production of metastable atoms has remained effectively constant during the observations at different pressures and with different geometries.

5.5.2. *Experimental methods and results for quenching of* $O(^1S)$. The first experiments, those of Kvifte and Vegard,[†] suffered from this difficulty. They excited the green line by an electrical discharge through oxygen or oxygen–neon mixtures in a cylindrical tube. The variation of the intensity of the green line emission with discharge-tube current, pressure, fractional neon concentration, and tube radius was studied. To obtain some check on the variation of the production rate of ^1S atoms the intensity of the green line was measured relative to that of a line at 5555 Å which arises through a transition from the $1s^2 2s^2 2p^3(^4S)7s\,^3S_0$ state. Unfortunately, as the threshold excitation energy for this state is 13·16 eV, the probability of excitation is likely to be very sensitive to changes in the electron temperature and energy distribution so that it is a very unreliable standard. However, even if the production rate of metastable atoms is supposed to remain constant, it is only possible to determine the ratio of the collision loss to the wall loss. This is because, under the experimental conditions, τ_r is long compared with τ_w or τ_{dm}. From observations made with neon containing a small percentage of admixed oxygen, by varying the oxygen partial pressure, either keeping the neon pressure or the composition fixed, evidence was obtained of collision loss due to deactivation by O_2. As a rough estimate, at a neon pressure of 30 torr in a tube of radius 0·6 cm, the collision loss is comparable with the wall loss at an oxygen partial pressure of 4 torr. The diffusion coefficient of $O(^1S)$ atoms in neon is probably not very different from that of neon in neon, about 350 cm^2 s^{-1} at 1 torr. Assuming this value, the wall loss may be calculated and in this way a value of about 10^{-15} cm^3 s^{-1} was found for $\mu(O_2)$. There was no evidence of any

[†] KVIFTE, G. and VEGARD, L., *Geofys. Publ.* **17** (1947) No. 1, p. 3.

dependence on neon pressure, from which it follows that the deactivation collision frequency for neon must be less than 10^{-17} cm^3 s^{-1}.

These estimates are subject to large uncertainties. Later experiments,† using a flow technique and excitation of the O atom to the metastable state through the three-body reactions,

$$O+O+O \to O_2+O(^1S), \tag{212}$$

$$N+N+O \to N_2+O(^1S), \tag{213}$$

$$N+O+O \to NO+O(^1S), \tag{214}$$

have provided more definite results. The principle of the experiments is as follows.

Pure nitrogen gas flows through a region to which a breakdown potential is applied. The discharge in this region produces N atoms. Further downstream, nitric oxide is injected at a known rate so that the reaction

$$NO+N \to N_2+O \tag{215}$$

proceeds quantitatively. This provides a concentration of atomic oxygen which can be determined from the measured rate of injection of NO. A known partial pressure of some quenching gas may then be added further downstream after which the mixture passes through a large spherical bulb. Reactions (212–14) occurring in this bulb lead to emission of the green line whose intensity may be measured as a function of the concentration of O, N, N_2 and of the quenching gas. Experiments may also be carried out in which the incoming nitrogen is mixed with a buffer gas such as helium or neon which has effectively no quenching properties.

In this type of experiment the wall loss rate in the spherical bulb can be reduced to negligible proportions by using a bulb of large radius and a high pressure of buffer gas. The analysis sketched out in § 5.5.1 above may therefore be applied.

Fig. 18.97 illustrates schematically the arrangement used by Young and Black. Chemically pure nitrogen entered the system through a needle-valve from a set of four high-pressure cylinders and, after passage through 40 ft of coiled copper tubing immersed in liquid nitrogen, entered a quartz tube of 1-cm internal diameter. A discharge was excited in this tube by an 800-W Raytheon diathermy unit. Excited and ionized species, other than ground-state nitrogen atoms, formed in the discharge were removed by requiring the issuing gas to pass through a glass-wool plug before expanding in steps into the Pyrex glass flow

† YOUNG, R. A. and SHARPLESS, R. L., *J. chem. Phys.* **39** (1963) 1071; YOUNG, R. A. and BLACK, G., *Planet Space Sci.* **14** (1966) 113; *J. chem. Phys.* **44** (1966) 3741.

tube of 4 in internal diameter. This tube made two 180° bends before entering the 72-litre observation bulb. By means of a large Roots pump a linear flow velocity of as much as 8 m s^{-1} could be maintained.

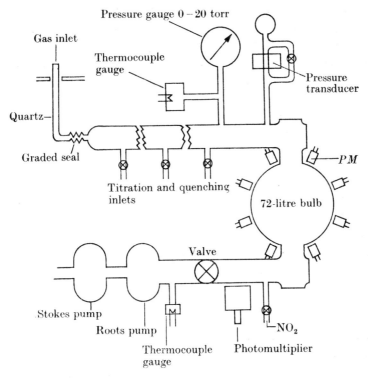

FIG. 18.97. Arrangement used by Young and Black in their experiments on the quenching of O(^1S) and on the chemiluminescence of dissociated oxygen.

Nitric oxide was added 515 cm upstream from the entrance flange of the observation bulb, the quenching gas 400 cm upstream. 60 cm downstream from the exit flange of the bulb, nitrogen dioxide was added as a titrant for atomic oxygen through the reaction

$$NO_2 + O \rightarrow NO + O_2. \tag{216}$$

These gases were introduced through inlets of the perforated-ball type extending into the centre of the stream. The rate at which the titrants were added was determined from the rate of decrease of pressure in calibrated reservoirs of the titrant gas, this being measured with differential pressure diaphragm transducers. Similar transducers were used to determine the partial pressure of added quenching gases by comparison with the flow system pressure—a volume of 100 cm^3 of gas at this pressure was trapped on one side of the transducer by shutting off a bypassing stopcock before adding the quenching gas. The quenching transducer could also be used to determine the relation between the flow rate of a titrant and the partial pressure it would produce in the main flow system if no reaction occurred. This was done by measuring the titrant flow rate simultaneously with the rise in pressure of the main flow system when no atomic nitrogen was present, as determined by the quenching transducer. For adequate sensitivity these measurements were made with high titrant flow rates.

Radiation emitted in the course of titration was observed with pairs of photomultipliers located at points 90 cm and 305 cm upstream and 110 cm downstream from the respective entrance and exit flanges of the observation bulb.

Emission from the bulb could be observed by several photomultipliers each looking along a diameter. The bulb was housed in a heavy plywood box whose interior was painted white. Blackened honeycomb collimators limited the field of view of each photomultiplier to an angle of about 5°. Interference filters of various types were used to limit the wavelength response as desired.

It was found that the intensity of the green line of atomic oxygen was greatest when [O], the concentration of atomic oxygen, is approximately equal to one-half that [N] of atomic nitrogen. Complication in interpretation of the results arose from the fact that the interference filter transmitted nearby radiation from the first positive bands of N_2 and the β-bands of NO. However, N_2O proved a very effective relative quencher of the green line and the background signal arising from the molecular bands could be determined by addition of N_2O in sufficient concentration to suppress the green line without appreciably affecting the bands.

Fig. 18.98 illustrates a typical set of results, showing the variation of the photomultiplier signal, for a fixed N_2 pressure, as a function of the partial pressure of added NO, and hence of the partial pressure of $O(^3P)$ atoms in the bulb. The relative strength of the background, determined as described, is shown. Fig. 18.99 shows the effect of addition of a quenching gas on the intensity of emission of the green line. This was obtained by adjusting the NO flow rate to give maximum intensity in this line before adding the quenching gas and then maintaining this flow rate throughout. It will be seen that the ratio I_0/I of the intensity before and after addition of the quenching gas varies linearly with the partial pressure of this gas

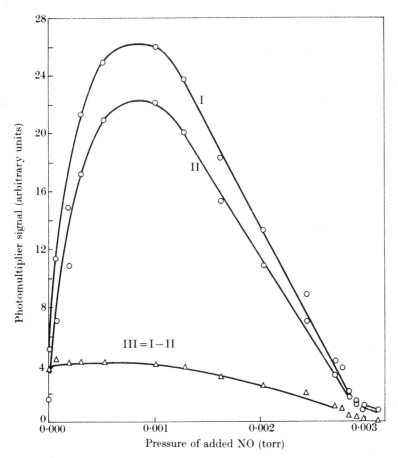

Fig. 18.98. Variation of the photomultiplier current, recording the emission passing the 5577 Å filter, with pressure of NO added to nitrogen at a pressure of 1 torr. I, without N_2O present ($O(^1S)$+background). II, with N_2O present in such concentration as to quench all $O(^1S)$ (background only). III, difference signal I–II taken to be due to $O(^1S)$ emission.

as expected from the analysis of § 5.5.1. According to the analysis the slope of this line is proportional to

$$\mu_q/(A+\mu_0 n_0), \tag{217}$$

where μ_q is the quenching rate per added quenching molecule. μ_0 is the rate of deactivation of $O(^1S)$ molecules by collisions with the main gas of concentration n_0 and A is the A coefficient for the green line transition. It being assumed that wall loss is negligible, observations with different pressures of N_2 or rare gases show that they act solely as buffer gases, so μ_0 arises from collision with $O(^3P)$ atoms only.

Fig. 18.100 now shows a plot of the reciprocal slopes of lines such as that in Fig. 18.99, as a function of the partial pressure of $O(^3P)$. The latter is determined both from the flow rate of NO, provided it is not in excess of that required to transform all N into O, and from the NO_2 titration. As would be expected the plot is again linear. Knowing A (136 s^{-1}), μ_q is obtained from the intercept at zero pressure of atomic oxygen and μ_0 from the slope.

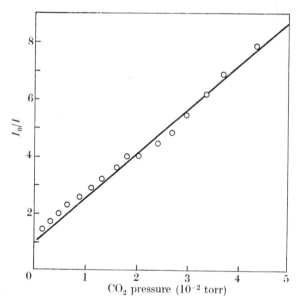

Fig. 18.99. The observed ratio I_0/I of the intensity of the green line of atomic oxygen before and after addition of CO_2, as a function of the CO_2 partial pressure. The atomic oxygen pressure is 0.0_370 torr in a pressure of 1·5 torr of N_2.

Table 18.21 gives values obtained in this way for the quenching rates for different quenching molecules.

It will be noted that the quenching rate by O_2 is about 100 times faster than found by Kvifte and Vegard. The nature of the quenching reaction with O is either

$$O(^1S)+O(^3P) \to O(^1D)+O(^1D)+0\cdot3 \text{ eV}, \qquad (218\,\text{a})$$
$$O(^1D)+O(^3P)+2\cdot2 \text{ eV}, \qquad (218\,\text{b})$$
or
$$O(^3P)+O(^3P)+4\cdot2 \text{ eV}. \qquad (218\,\text{c})$$

The large energy discrepancies in (218 b) and (218 c) indicate very small cross-sections under thermal conditions but the remaining reaction violates the Wigner spin conservation rule (see § 4.2.3).

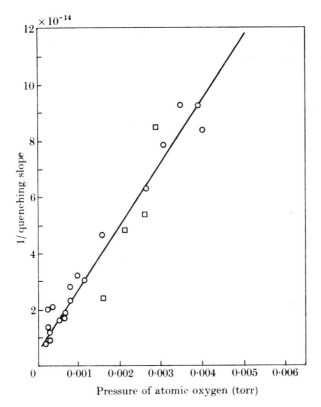

FIG. 18.100. Plot of the inverse slope of a quenching curve, as in Fig. 18.99, against the atomic oxygen pressure. The observations were made in N_2 with admixed CO_2. The points ○ were obtained with $O(^1S)$ produced from

and □ from
$$N+N+O \rightarrow N_2+O(^1S)$$
$$O+O+O \rightarrow O_2+O(^1S).$$

TABLE 18.21

Observed reaction rates and mean cross-sections for quenching of $O(^1S)$ atoms by impact with various molecules

Atom or molecule	Quenching rate ($cm^3\ s^{-1}$)	Mean quenching cross-section (cm^2)
N_2O	5.9×10^{-15}	5.8×10^{-18}
O	1.8×10^{-13}	1.4×10^{-18}
O_2	1.0×10^{-13}	0.8×10^{-18}
CO_2	2.5×10^{-14}	2.5×10^{-19}
H_2	3.0×10^{-15}	1.1×10^{-20}
He, A, Kr	$< 10^{-17}$	$< 10^{-22}$

5.5.3. *Excitation of* $O(^1S)$ *in three-body recombination.* Young and Black also investigated the rates of the three-body reactions (212–4) which produce the green line. When no additional quenching gas is present the equilibrium relation between rates of production and loss of $O(^1S)$ atoms is

$$\mu_{00}[O]^3 + \mu_{01}[O]^2[N] + \mu_{02}[O][N]^2 = \{\mu_0[O] + A\}[O(^1S)], \quad (219)$$

[M] denoting the concentration of the species M. $\mu_{00}, \mu_{01}, \mu_{02}$ are the rate coefficients of the respective reactions (212), (213), and (214). It is assumed, as before, that

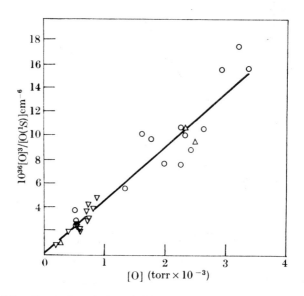

FIG. 18.101. Observed variation of $[O]^3/[O(^1S)]$ with [O] in a reaction bulb in which [N] = 0. ○ Data in N_2. △ Data in 1 per cent N_2/Ar mixture. ▽ Data in 1 per cent N_2/He mixture.

wall loss of $O\,(^1S)$ atoms is negligible. In particular, under conditions in which [N] vanishes,
$$[O(^1S)] = \mu_{00}[O]^3/\{\mu_0[O] + A\}. \quad (220)$$

Fig. 18.101 illustrates the observed variation of $[O]^3/[O\,(^1S)]$ as a function of [O] under these conditions. The relation is again a linear one and it will be noted that the results do not depend on the nature of the buffer gas. From the slope μ_0/μ_{00} is determined and hence, since μ_0 is obtained from the quenching experiments, μ_{00}. Because A is very small compared with $\mu_0[O]$ over most of the range the intensity of the green line varies as the square of the concentration of atomic oxygen. It is found that $\mu_{00} = 1 \cdot 5 \times 10^{-36}$ cm⁶ s⁻¹. This is close to the rate required for the production of the green line in the night airglow.

Approximate values were also found for μ_{01} and μ_{02} by least squares fitting of the curve giving the observed relation between $\{\mu_0[O]+A\}I(O\,^1S)$, where $I(O\,^1S)$ is the intensity of the green line, and [O] with [N] allowed to vary. The results obtained are given in Table 18.22.

TABLE 18.22

Rates of three-body reactions giving rise to $O(^1S)$ atoms

Reaction	Buffer gas	Reaction rate (cm^6 s^{-1})		
		N_2	Ar	He
$O+O+O \rightarrow O_2+O(^1S)$		1.5×10^{-34}	1.5×10^{-34}	1.5×10^{-34}
$N+N+O \rightarrow N_2+O(^1S)$		$10^{-33}+5 \times 10^{-33}p$	2×10^{-32}	4×10^{-32}
$N+O+O \rightarrow NO+O(^1S)$		3×10^{-33}	3×10^{-33}	$3 \times 10^{-33}+1 \times 10^{-33}p$

$p =$ pressure of buffer gas in torr.

5.5.4. *Quenching of metastable $O(^1D)$.* Young, Black, and Ung[†] have observed the formation of $O(^1D)$ in the photodissociation of O_2 by radiation at 1470 Å from a microwave discharge in xenon. The O_2, which was continuously pumped through the observation bulb, contained about 10-torr pressure of helium to prevent diffusion of $O(^1D)$ atoms to the walls. Fig. 18.102 shows the observed variation of intensity of emission of the red line at 6300 Å as a function of O_2 pressure. The interpretation of this variation is obscure. Quenching of the red line by addition of foreign gases, such as N_2, CO, N_2O, CO_2, and H_2, was observed but, because of the apparent complexity of the reactions involved, no rate coefficients could be obtained.

More definite information[‡] is available from observations of the day airglow in the upper atmosphere.

Using a rocket-borne photometer, the height profile of the volume intensity of red line emission has been measured[§] with a considerable degree of reliability. Denoting this by $\rho(z)$, where z is the altitude above ground, we have

$$\rho(z) = An(z), \qquad (221)$$

where $n(z)$ is the concentration of $O(^1D)$ atoms at the height z and A is the A-coefficient for emission of the red line. If $q(z)$ is the rate of production and Z the quenching rate for $O(^1D)$, assumed to be large compared with A, we also have, in equilibrium,

$$Zn(z) = q(z), \qquad (222)$$

so
$$Z = Aq(z)/\rho(z). \qquad (223)$$

It seems very likely that the only important process leading to production of $O(^1D)$ in the height range studied is through the photo-dissociation

$$O_2+h\nu \rightarrow O(^3P)+O(^1D), \qquad (224)$$

[†] YOUNG, R. A., BLACK, G., and UNG, A. Y.-M., *J. chem. Phys.* **45** (1966) 2702.
[‡] HUNTEN, D. M. and MCELROY, M. B., *Rev. Geophys.* **4** (1966) 303.
[§] WALLACE, L. and MCELROY, M. B., quoted by HUNTEN, D. M. and MCELROY, M. B., loc. cit.

which constitutes the Schumann–Runge continuum. The cross-section for this process is well known and the intensity of the effective solar radiation (wavelength < 1750 Å) has been measured with rocket-borne equipment. Less reliable data are available about the height variation of the O_2 concentration, particularly above 100 km. Fortunately, observations were made near the time of the rocket flight that provided the airglow data.

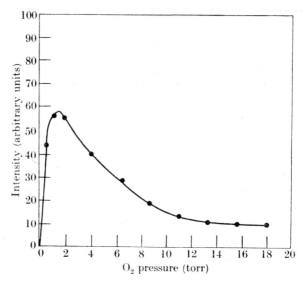

FIG. 18.102. Observed variation of the intensity of emission of the red 6300 Å OI line with O_2 pressure in a mixture with helium, at a pressure of 10 torr, acting as a diffusion buffer.

Using these results, Z is found to vary with height in a similar way to a major constituent, N_2 or O_2. From the height profile of N_2, given, for example, from a standard atmosphere compilation, it is found that $\mu(N_2)$, the rate coefficient for quenching by N_2, would need to be about 5×10^{-11} cm³ s⁻¹ to provide the quenching rate derived from the rocket experiments. If O_2 were to be effective the rate coefficient $\mu(O_2)$ would need to be 4–5 times greater.

Kvifte and Vegard in their experiments (see p. 1831) measured the ratio of red line to green line intensity in a discharge in pure oxygen, using a number of discharge tubes with radii R varying from 0·39 to 3·11 cm. The intensity ratio was not found to change much over this sixty-fold change of R^2. Assuming that wall loss is dominant for the narrower tubes and quenching for the wider tubes the constancy of the ratio requires that the rate coefficients for quenching of ^1S and ^1D atoms are in the same ratio as the wall loss rates. The latter are not likely to differ very much because they are determined by the diffusion coefficients of the respective atoms in O_2. This suggests that $\mu(O_2)$ is much the same for $O(^1D)$ as for $O(^1S)$. From the results of Young and Black,[†] $\mu(O_2)$ is only 1.0×10^{-13} cm^3 s^{-1} which is far too small to account for the quenching of the red line in the day airglow. Support is therefore given to the assignment of N_2 as the principal quenching agent in this case. This is consistent with the fact that, whereas the green line is emitted strongly from concentrations of atomic oxygen mixed with N_2, the red line is never observed under such circumstances though it is emitted quite strongly in the presence of O_2.

DeMore and Raper have[‡] found a strong quenching of $O(^1D)$ in liquid nitrogen which they ascribe to a two-body process which could be effective in the gas phase. The reaction is supposed to proceed in two stages, the first leading to formation of a vibrationally excited N_2O molecule

$$O(^1D)+N_2 \to N_2O'. \tag{225}$$

This excited complex has an appreciable lifetime, as it takes time for the excess energy to reconcentrate on the O–N_2 bond. During this period a transition may take place to a repulsive triplet state of N_2O leading to the dissociation

$$N_2O' \to O(^3P)+N_2(^1\Sigma).$$

If the liquid nitrogen data are to be interpreted in this way the rate coefficient $\mu(N_2)$ would be of the order required by the upper atmospheric data. However, DeMore and Raper suggest that a similar two-body reaction could occur for O_2 but in that case $\mu(O_2)$ would be much larger than estimated above. The position is still somewhat uncertain at the time of writing.

6. Collisions involving exchange of electron spin

6.1. *Introduction*

We now consider collisions in which a transfer of electron spin takes place between two atoms or molecules without accompanying transfer of excitation. Collisions of this kind are of interest and importance for various reasons. Thus we have discussed above an example of the vital role played by resonance charge transfer in transferring alignment or polarization from one set of atoms, $2\,^3S$ metastable ^{3}He, which can be

---

[†] loc. cit., p. 1832.
[‡] DeMore, W. and Raper, O. F., *J. chem. Phys.* **37** (1962) 2048.

polarized by optical pumping, to a second set, normal ^{3}He, which cannot be polarized directly. The simplest example of such a transfer by spin exchange alone is the polarization of free electrons through exchange with optically aligned sodium atoms, a process first demonstrated by Dehmelt[†] and described in Chapter 5, § 7.2. Shortly after, Novick and Peters[‡] produced exchange polarization in a rubidium–sodium mixture and Franken, Sands, and Hobart[§] were able to polarize potassium atoms by exchange collisions with sodium atoms and free electrons. Dehmelt also suggested extending this method to polarize hydrogen atoms, and this was soon demonstrated by Anderson, Pipkin, and Baird.[||] Since then there have been many experiments of this kind.

The width of a nuclear magnetic-resonance line in a pure gas or vapour such as that of an alkali metal is largely determined by the frequency of spin-exchange collisions. This is essentially because the exchange changes the correlation of electron and nuclear spin in either atom.

Thus if I is the total nuclear spin quantum number and M_I its projection on the direction of the uniform and constant magnetic field, collisions in which exchange of electrons of opposite spin occur will change this correlation, provided M_I is not the same for both colliding atoms. Furthermore it is only necessary for half the spins in an assembly of polarized atoms to be reversed in order to reduce the magnetic moment to zero. Hence, taking both factors into account, if Q_{se} is the cross-section for spin exchange the cross-section which contributes to the line width will be $2\{2I/(2I+1)\}Q_{\text{se}}$.

Spin exchange produces an equilibrium distribution between the hyperfine structure levels in a mixture of atoms partially oriented in a weak, uniform, and constant magnetic field H. Thus, consider an assembly of N_A atoms A and N_B atoms B which are partially oriented so that the total angular momentum of the assembly along the field is $\mathcal{M}_z$. Let the possible z-components of the total electron+nuclear spin of the hyperfine structure levels be, in units $\hbar$,

$$M_1, ..., M_n, ... \text{ for atoms A,}$$

$$M'_1, ..., M'_s, ... \text{ for atoms B.}$$

Then, in equilibrium, the fraction of atoms A with z-component of total spin $M_n \hbar$ is given by

$$e^{\mu M_n} / \sum_q e^{\mu M_q}, \tag{226}$$

where

$$N_A \frac{\sum M_q e^{\mu M_q}}{\sum e^{\mu M_q}} + N_B \frac{\sum M'_p e^{\mu M_{p'}}}{\sum e^{\mu M_{p'}}} = \mathcal{M}. \tag{227}$$

[†] DEHMELT, H. G., *Phys. Rev.* **109** (1958) 381.
[‡] NOVICK, R. and PETERS, H. E., *Phys. Rev. Lett.* **1** (1958) 54.
[§] FRANKEN, P., SANDS, R., and HOBART, J., ibid. **1** (1958) 118.
[||] ANDERSON, L. W., PIPKIN, F. M., and BAIRD, J. C., ibid. **1** (1958) 229; **4** (1960) 69.

$-\mu^{-1}$ thus behaves as a spin temperature but it is important to remember that it may be < 0.

Thus, consider a mixture of ^{14}N ground-state atoms in spin exchange equilibrium with ^{23}Na $^2P_{\frac{3}{2}}$ excited atoms. The ground state of ^{14}N is $^4S_{\frac{3}{2}}$ and the nuclear spin quantum number $I = 1$ so the total spin quantum number $F = \frac{5}{2}, \frac{3}{2},$ or $\frac{1}{2}$. For ^{23}Na $^2P_{\frac{3}{2}}$, $I = \frac{1}{2}$, $F = 2$ or 1. The relative density of population of the levels with different M in the two atoms is then as shown in Fig. 18.103 with $a = e^\mu$.

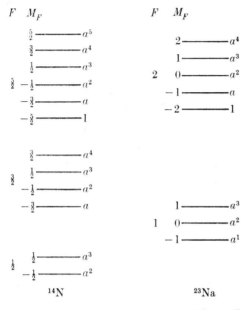

FIG. 18.103. The relative population of the different Zeeman levels in ^{14}N($^4S_{\frac{3}{2}}$), with nuclear spin quantum number $I = 1$, and ^{23}Na($^2P_{\frac{3}{2}}$) with nuclear spin quantum number $I = \frac{1}{2}$, in spin-exchange equilibrium.

While spin exchange affects the lifetime of a particular hyperfine structure state it does not produce relaxation of an oriented system once it has attained spin equilibrium. However, the rate at which such an equilibrium is set up or destroyed is determined by the rate at which spin-exchange collisions occur. Thus the relaxation time for alignment of a set of atoms through spin exchange from a second set is clearly determined by the rate at which the spin exchange occurs between the two sets. The application to the determination of cross-sections for spin-exchange collisions between electrons and alkali metal atoms has already been discussed in Chapter 5, § 7.2.

The importance in astrophysics of spin-exchange collisions has already been pointed out in Chapter 8, § 3. This is particularly in connection with the estimation of the rate of population of the upper hyperfine structure level of H from which the 21·1-cm emission line arises. A further contribution to this rate comes from spin-exchange collisions between hydrogen atoms (see §§ 6.2.5 and 8.2.1) so information about the cross-section for this process is of astrophysical interest. Apart from such special interests, the process of spin exchange is of such importance in magnetic resonance as to justify a thorough investigation. We shall now proceed to discuss the methods that have been used to study spin exchange experimentally and the results obtained. A theoretical discussion is given in § 8.2.

6.2. Methods for measuring spin-exchange cross-sections

6.2.1. *From the relaxation time of a hyperfine population.* In principle one of the most direct ways of measuring a spin-exchange cross-section for collisions between identical atoms (with the same non-zero nuclear spin) is to produce a population inversion among the hyperfine levels associated with the ground state and then observe the relaxation time for return to the normal distribution. The inversion may be produced by hyperfine optical pumping and the decay of the inverted population monitored by observing the variation with time of the absorption of a suitable optical probing signal. Provided the inversion of the hyperfine level population is not associated with any electron spin polarization of the system the relaxation will be exponential and the relaxation time τ given by

$$\tau^{-1} = \tau_w^{-1} + \tau_{se}^{-1}. \tag{228}$$

τ_{se}^{-1} is the spin-exchange collision frequency which is proportional to the gas pressure while τ_w^{-1} arises from relaxation due to collision with the walls of the container, or to strong magnetic field inhomogeneities, and is independent of gas pressure. τ_{se} may therefore be separated from τ_w in the usual way by making observations at different gas pressures.

If the hyperfine population relaxes in the presence of relaxation of net electron-spin polarization, the interaction between the nuclear and electron spin couples the two relaxation processes so that the decay of either no longer follows a single exponential law but is a combination of two exponentials. It is then more difficult to sort out the contribution from spin exchange.

Having obtained τ_{se}, and hence the spin-exchange collision frequency, it still remains to determine the mean spin-exchange cross-section Q_{se} from the relation

$$1/\tau_{se} = n\bar{v}Q_{se}, \tag{229}$$

where n is the number of atoms per centimetre^3 and $\bar{v}$ is their mean relative velocity. This requires measurement of n, which is a matter of some difficulty because the working substance is usually an alkali metal vapour. In experiments using this method to determine τ_{se}, n has been measured by optical absorption methods.

Although we have been considering spin exchange between identical atoms there is no difficulty in extending the technique to include exchange between different isotopes or even different atoms.

Different techniques may be used in monitoring the relaxation of the hyperfine population by optical probing. In the method introduced by Franzen[†] the procedure is as follows. The system is optically pumped, during which time the degree of population inversion is monitored from the absorption of the pumping light. As inversion develops the gas becomes more transparent. Soon after the inversion reaches a steady value the pumping light is cut off and the system allowed to relax in the dark. At a selected time t after cut-off the degree of inversion remaining is monitored by observing the absorption of a pulse of the pumping radiation. This is repeated many times with different probing times t so the time decay of the inversion can be built up.

An alternative procedure introduced by Bouchiat and Grossetête[‡] is to use, not a probing pulse but a continuously operating light beam of such low intensity that it does not seriously perturb the hyperfine population. In this way the relaxation can be followed continuously instead of having to build up from a series of observations at different times after cut-off of the pumping light. This method has the advantage that it eliminates effects due to variability of the pumping radiation and it allows flexibility in choice of the probing light. On the other hand, the magnitude of the probing signal is much reduced in order that the probing light should not disturb the hyperfine population.

As an example of an elaborate experiment using Franzen's method we describe the apparatus and techniques used by Gibbs and Hull[§] for determining spin exchange cross-sections for Rb–Rb and Rb–Cs collisions. Fig. 18.104 illustrates the general arrangement of their apparatus.

^{87}Rb has a nuclear spin quantum number $I = 3/2$ so that there are two hyperfine levels with $F = 2, 1$ associated with the ground $^2S_{\frac{1}{2}}$ state. The transition $5\,^2S_{\frac{1}{2}} \to 5\,^2P_{\frac{3}{2}}$ is produced by absorption of 7800 Å radiation from a natural rubidium lamp. However, such radiation will be

[†] Franzen, W., *Phys. Rev.* **115** (1959) 850.
[‡] Bouchiat, M. A. and Grossetête, F., *J. Phys. Paris*, **27** (1966) 353.
[§] Gibbs, H. M. and Hull, R. J., *Phys. Rev.* **153** (1966) 132.

absorbed by both hyperfine levels. To remove the components which are absorbed from the $F = 2$ level the radiation is passed through a filter cell containing ^{85}Rb buffered with a pressure of argon of around 50 torr. It so happens that the hyperfine levels of this isotope (for which $I = 5/2$) are so placed as to absorb the unwanted radiation with remarkable effectiveness. This is illustrated in Fig. 18.105. In a similar way 7947 Å radiation from the $5\,^2S_{\frac{1}{2}} \to 5\,^2P_{\frac{3}{2}}$ transition could be used.

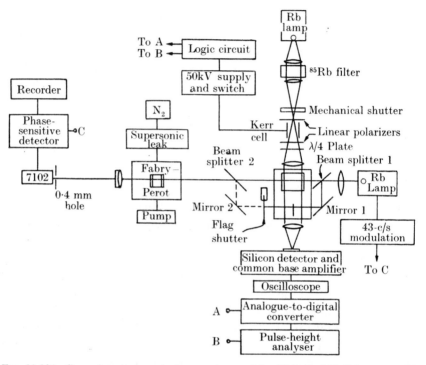

FIG. 18.104. General arrangement of apparatus used by Gibbs and Hull for measuring spin exchange cross-sections for Rb–Rb and Rb–Cs collisions.

The intensity of the pumping radiation was monitored by a silicon photovoltaic cell with peak sensitivity at 8000 Å and amplifier connected to an oscilloscope.

It is important to be able to cut the pumping radiation on and off sharply. The optical pumping signal was usually only about 1 per cent of the total light signal and the relaxation time could be as short as 10 ms, so that the light intensity should reach 99·9 per cent of its final value in less than 1 ms. It must be possible to vary the off-interval of the shutter from 1 ms to seconds while the on-interval should be longer than 100 ms to allow time for attainment of equilibrium of the hyperfine pumping in each cycle. To achieve this performance a Kerr cell was used as the shutter. The linear polarizers of the cell were arranged to be normally transmitting as a fast rise time was more important than a fast decay time. With the cell employed, the rise time to 99·9 per cent of the final level was about 0·3 ms while the time for the light signal when cut off to fall to 10 per cent of the initial value was about 0·5 ms. It was found

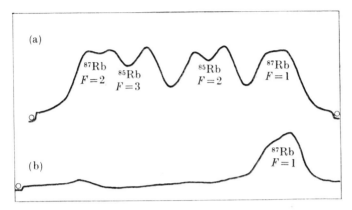

FIG. 18.105. Illustrating the effect of an ^{85}Rb filter on the profile of the 7800 Å line emitted from a natural rubidium lamp. (a) Profile without filter. The origin of each peak is indicated in terms of the isotope responsible and the final total spin quantum number F. (b) Profile with filter leaving a single peak due to transitions ending on the ^{87}Rb $^2S_{\frac{1}{2}}$ level with $F = 1$.

that, when operating in the cut-off mode, the cell did not extinguish the light completely, the residual intensity being about 3 per cent of the initial. This was serious for observations at long 'off' intervals so that, for such observations, an electrically operated mechanical shutter, synchronized with the Kerr cell, was used.

As explained earlier, it is important to avoid introduction of net electron spin polarization by the pumping radiation. It was difficult to destroy the polarization introduced by the Kerr shutter so that, in fact, the radiation issuing from the cell was circularly polarized. This produced longitudinal polarization in the vapour but this could be completely destroyed by application of a resonant r.f. magnetic field.

The atomic concentration in the vapour was determined by absorption measurements using a scanning Fabry–Perot interferometer so that the profiles of emission and absorption lines could be traced out. The scanning was effected by evacuating the interferometer chamber and then allowing dry nitrogen to enter through a supersonic leak in the form of a fine capillary about 1 cm long. As constructive interference occurs for normal incidence when

$$2\mu t = m\lambda, \qquad (230)$$

where μ is the refractive index, t the plate thickness, and λ the wavelength, we see that the change Δm in the order m due to a change $\Delta\mu$ in μ is given by

$$\Delta m = 2t\,\Delta\mu/\lambda. \qquad (231)$$

In going from a good vacuum to dry nitrogen at 1 atm, $\Delta\mu$ is about 3×10^{-4} so $\Delta m = 7.5$ orders for $t = 1$ cm and λ 8000 Å.

Reference to Fig. 18.104 shows the general arrangement employed. The appropriate radiation from a resonance lamp was interrupted at 86 c/s, so as to permit phase-sensitive detection, and then split into two beams by a mirror system. One beam traversed the vapour cell and the other bypassed it, but both were brought back to pass parallel to each other through the interferometer. By means of a flag shutter operating at 0.15 c/s the beams were switched off alternately so that each could be separately observed after passage through the interferometer. The detector was a cooled (RCA 7102) photomultiplier, operated at 1200 V, and a lock-in amplifier.

For light absorbed through a transition between states with quantum numbers J, F, M and J', F', M' respectively the absorption coefficient $k(\nu)$ integrated over the line width is given by†

$$\int k(\nu)\,\mathrm{d}\nu = (\lambda^2/8\pi\tau)n(2J'+1)\times$$

$$\times \sum_q \begin{pmatrix} F' & 1 & F \\ -M' & q & M \end{pmatrix}^2 (2F+1)(2F'+1)\begin{Bmatrix} J' & F' & 1 \\ F & J & 1 \end{Bmatrix}^2, \qquad (232)$$

where λ is the wavelength of the radiation, τ the total radiative lifetime for the $J' \to J$ transitions, n the number of ground-state atoms per cm³ and the bracketed symbols are $3j$ and $6j$ symbols. The problem is then to determine the integrated absorption coefficient from the interferometer observations.

Let $f(\nu)\,\mathrm{d}\nu$ be the intensity of radiation of frequency between ν and $\nu+\mathrm{d}\nu$ incident from the resonance lamp. The observed intensity $I_0(\nu)\,\mathrm{d}\nu$ after passage through the interferometer will differ from $f(\nu)\,\mathrm{d}\nu$ because of instrumental factors which can be represented by a slit function $g(\nu'-\nu)$ so that

$$I_0(\nu) = \int f(\nu')g(\nu-\nu')\,\mathrm{d}\nu'. \qquad (233)$$

† GIBBS, H. M., Thesis, University of California, 1965.

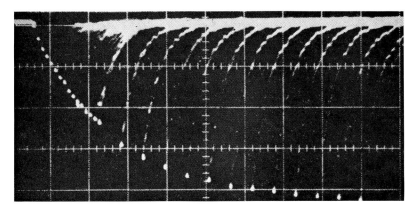

Fig. 18.106. Typical oscillogram showing the relaxation of an inverted hyperfine population in Rb vapour. The vertical scale is proportional to the intensity of the transmitted light. The transients result from reopening the shutter after different time intervals since closure. They show the return to the optical pumping equilibrium. The bright spots from which each begins trace out the decay of the inverted population.

Again, if $k(\nu)$ is the true absorption coefficient of the vapour in the cell for radiation of frequency ν, the intensity $I(\nu)$, recorded after passage through a length l of the vapour cell as well as the interferometer, will be given by
$$I(\nu) = \int f(\nu')\exp\{-k(\nu')l\}g(\nu-\nu')\,\mathrm{d}\nu'. \tag{234}$$

The analysis of observations of $I_0(\nu)$ and $I(\nu)$ to determine $k(\nu)$ is simplified when the absorption coefficient is known to be of gaussian shape, which will be the case at low vapour densities with no buffer gas present so Doppler broadening predominates. Under these circumstances, at a temperature T °K,†
$$\int k(\nu)\,\mathrm{d}\nu = \pi^{\frac{1}{2}}k(\nu_0)\,\Delta\nu/2(\ln 2)^{\frac{1}{2}}, \tag{235}$$
where
$$\Delta\nu = 2(2\kappa T\ln 2/M)^{\frac{1}{2}}\nu_0/c, \tag{236}$$

M being the mass of a gas atom and ν_0 the frequency at the peak. If this formula is applicable the shape of $k(\nu)$ is known and it only remains to determine the scale factor $k(\nu_0)$. To do this and at the same time check the assumptions, the slit function $g(\nu-\nu')$ was first determined from observations on a narrow line of argon. Knowing g, the true emission line profile $f(\nu)$ was determined by inversion of (233). The scale factor $k(\nu_0)$ in (235) was then adjusted so as to yield an absorption profile $I(\nu)$ in best possible agreement with that observed. Very good agreement could be obtained in this way, showing that the shape of the true absorption function $k(\nu)$ was indeed close to gaussian with width given by (236). Having determined $k(\nu_0)$, $\int k(\nu)\,\mathrm{d}\nu$ follows from (235) and thence n from (232), using previously determined values of the lifetime τ.

In determining $I(\nu)/I_0(\nu)$, the separated beams were first brought to the same intensity by adjustment of the optical system when no absorbing vapour was present. Using the flag switch, allowance could be made for any fluctuations in the light source during the subsequent absorption measurements.

The absorption cell, roughly cubical of edge 5 cm, was coated with Paraflint to reduce the rate of relaxation at the wall. Fig. 18.106 illustrates a typical oscillogram showing the relaxation in ^{87}Rb. The vertical axis has been amplified twenty-five times so that the relaxation of the excited hyperfine population can be traced from the bright spots on the record which results from superposition of the traces for many 'off' intervals.

† MITCHELL, A. C. G. and ZEMANSKY, M. W., *Resonance radiation and excited atoms*, p. 96 (Cambridge University Press, 1961).

In obtaining the relaxation time from records such as these, correction must be made for the finite optical thickness of the vapour which results in the relation between the hyperfine population difference and the optical absorption being not exactly linear.

Fig. 18.107 shows typical results obtained for the variation of $1/\tau$ with the concentration n of ^{87}Rb atoms. The relation is closely linear, with slope proportional to Q_{se}. The intercept on the axis $n = 0$ gives $1/\tau$.

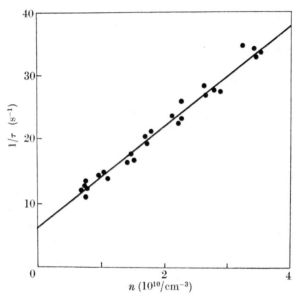

FIG. 18.107. Variation of the inverse relaxation time, $1/\tau$, for an inverted hyperfine population in ^{87}Rb, with atomic concentration n.

Results obtained for ^{87}Rb–^{87}Rb and ^{87}Rb–^{133}Cs collisions are given in Table 18.24 in comparison with results obtained by other methods and for other colliding systems.

Bouchiat and Grossetête,† using the continuous probing method with a probing beam 50–500 times weaker than the pumping beam, have compared the spin-exchange cross-sections for ^{87}Rb–Cs, ^{85}Rb–Cs, and Cs–Cs collisions. This they were able to do without recourse to vapour pressure–temperature relations for either Rb or Cs. Three cells were used, containing vapours composed as follows.

Cell I: pure Rb (N_{I}/cm^3);
Cell II: mixed Rb ($N_{\text{II}}/\text{cm}^3$) and Cs($n_{\text{II}}$)/cm^3;
Cell III: pure Cs ($n_{\text{III}}/\text{cm}^3$).

† loc. cit., p. 1845.

The relaxation times for the inverted hyperfine population of Rb and Cs atoms in the three cells due to spin exchange are then given by

I $[\tau(\text{Rb})]^{-1} = Q_{\text{se}}(\text{Rb–Rb})N_{\text{I}}\,\bar{v}(\text{Rb–Rb})$;

II $[\tau(\text{Rb})]^{-1} = Q_{\text{se}}(\text{Rb–Rb})N_{\text{II}}\,\bar{v}(\text{Rb–Rb}) + Q_{\text{se}}(\text{Rb–Cs})n_{\text{II}}\,\bar{v}(\text{Rb–Cs})$,

$[\tau(\text{Cs})]^{-1} = Q_{\text{se}}(\text{Rb–Cs})N_{\text{II}}\,\bar{v}(\text{Rb–Cs}) + Q_{\text{se}}(\text{Cs–Cs})n_{\text{II}}\,\bar{v}(\text{Cs–Cs})$;

III $[\tau(\text{Cs})]^{-1} = Q_{\text{se}}(\text{Cs–Cs})n_{\text{III}}\,\bar{v}(\text{Cs–Cs})$.

Here $Q_{\text{se}}(\text{Rb–Rb})$, $Q_{\text{se}}(\text{Rb–Cs})$, $Q_{\text{se}}(\text{Cs–Cs})$ are spin-exchange cross-sections for Rb–Rb, Rb–Cs, and Cs–Cs collisions respectively. $\bar{v}(\text{Rb–Rb})$, etc., are mean relative velocities at which the respective impact occurs. The ratios $N_{\text{II}}/N_{\text{I}}$, $n_{\text{II}}/n_{\text{III}}$ were determined by comparison of the optical absorption of the appropriate resonance radiation. It follows that, from the observed relaxation times, the ratios of the cross-sections could be obtained and they were found to be equal within the limit of accuracy of the experiments. Taking the absolute value of n_{III} as given by the vapour pressure temperature data for Cs, which are much more consistent than for Rb, the absolute value of the spin-exchange cross-section came out to be $2 \cdot 4 \times 10^{-14}$ cm^2.

6.2.2. *Use of optical pumping to Zeeman levels.* Although spin exchange has no effect on the relaxation of an electron spin-polarized assembly of similar atoms it can play an important role in transmitting polarization from one set of atoms to another. The simplest case, in principle, that in which one of the species involved is a free electron, has already been discussed in Chapter 5, § 7.2 and it was shown how the cross-section for spin-exchange collisions between electrons and alkali-metal atoms could be determined. Essentially, if we have two species A and B present between which spin-exchange collisions can occur, and the spin polarization of B is maintained zero, then spin-exchange collisions between atoms A and B make an important contribution to the relaxation of the spin polarization of A.

As an example we describe briefly the experiments of Jarrett[†] on the spin-exchange cross-section for collisions between ^{85}Rb and ^{87}Rb atoms.

In these experiments ^{87}Rb atoms were aligned by optical pumping and the polarization was then extended to ^{85}Rb through spin exchange. The theory of the method neglects the nuclear spins as the d.c. magnetic field used was insufficient to resolve the separate Zeeman components of the hyperfine structure.

The pumping radiation was circularly polarized so as to excite ^{87}Rb atoms from the $^2S_{\frac{1}{2}}$ level with $M_J = -\frac{1}{2}$. If we denote the respective

[†] JARRETT, S. M., *Phys. Rev.* **133** (1964) A111.

number of ^{87}Rb atoms/cm^3 with $M_J = \pm\tfrac{1}{2}$ by $N_\pm$, respectively, then the intensity I of pumping radiation decreases with the length x penetrated in the pumping cell according to the equation

$$\frac{dI}{dx} = -N_- Q_a I, \qquad (237)$$

where Q_a is the absorption cross-section of ^{87}Rb ($M_J = -\tfrac{1}{2}$) atoms for the radiation. The polarization P of the Rb atoms is given by

$$P = \frac{N_+ - N_-}{N}, \qquad (238)$$

where $N = N_+ + N_-$. We may therefore rewrite (237) in the form

$$\frac{dI}{dx} = -\tfrac{1}{2} N Q_a (1-P) I. \qquad (239)$$

Hence, if I_0 is the incident intensity, I that after passage through the cell of length l,

$$I = I_0 e^{-\alpha(1-\bar{P})}, \qquad (240)$$

where $\qquad \alpha = \tfrac{1}{2} N Q_a l$

and $\bar{P}$ is the mean value of the polarization along the path transversed.

If an oscillating magnetic field is applied normal to the d.c. field, with frequency equal to that necessary to produce transitions between the $M_J = -\tfrac{1}{2}$ and $+\tfrac{1}{2}$ levels of ^{87}Rb, then $\bar{P}$ is reduced to zero. Hence, if S is the difference between the light intensity transmitted when the r.f. field is applied and when it is off,

$$S = -\alpha I_0 e^{-\alpha} \bar{P}, \qquad (241)$$

provided $\bar{P} < 0.1$.

If instead of applying the r.f. field at the resonance frequency for $M_J = -\tfrac{1}{2}$ to $+\tfrac{1}{2}$ transitions in ^{87}Rb we apply it at the resonance frequency for the corresponding transition for ^{85}Rb then the mean polarization of the ^{87}Rb atoms will be changed to $\bar{P}_1$. Hence, if S_1 is the difference between the intensity of light transmitted when this r.f. field is present and when it is off

$$S_1 = -\alpha I_0 e^{-\alpha} (\bar{P} - \bar{P}_1). \qquad (242)$$

We must now calculate $\bar{P}$ and $\bar{P}_1$. If P, p are the polarizations of the ^{87}Rb and ^{85}Rb atoms, $\bar{Q}_{se}$ the mean spin exchange cross-section and $\bar{v}$ the mean relative velocity of the atoms, we have, as in (49) and (50) of Chapter 5,

$$\frac{dP}{dt} = \tfrac{1}{2} Q_a I - \{\tfrac{1}{2} Q_a I + 2R\} P - aN\lambda P + aN\lambda p, \qquad (243\,\mathrm{a})$$

$$\frac{dp}{dt} = -2Rp - N\lambda p + N\lambda P, \qquad (243\,\mathrm{b})$$

where $\lambda = \bar{v}\bar{Q}_{\text{se}}$, R is the relaxation rate from each Zeeman sub-level, and $a = n/N$, where n is the number of ^{85}Rb atoms/cm^3.

When both polarizations are in a steady state so that $\dfrac{\mathrm{d}P}{\mathrm{d}t} = \dfrac{\mathrm{d}p}{\mathrm{d}t} = 0$

$$P = \left\{1 + \frac{4R}{IQ_a}\left(1 + \frac{aN\lambda}{2R+N\lambda}\right)\right\}^{-1}. \tag{244}$$

Again, when $p = 0$, as will be the case when the r.f. field is applied at the ^{85}Rb Zeeman resonance frequency, the steady state polarization of ^{87}Rb is given by

$$P_1 = \left\{1 + \frac{4R+2N\lambda}{IQ_a}\right\}^{-1}. \tag{245}$$

The intensity I depends on the polarization but to a good approximation it may be taken, for small polarization, as equal to that in the absence of polarization viz.:

$$I = I_0 e^{-\frac{1}{2}NQ_a x}, \tag{246}$$

so that
$$\bar{P} = \frac{1}{l}\int_0^l P\,\mathrm{d}x, \qquad \bar{P}_1 = \frac{1}{l}\int_0^l P_1\,\mathrm{d}x$$

may both be calculated. We then find that, using (241) and (242),

$$\frac{N\bar{v}Q_{\text{se}}}{R} = \frac{f(1+a)}{a(1-f)}\left[1 + \left\{1 + \frac{4a}{(1+a)^2}\frac{(1-f)}{f}\right\}^{\frac{1}{2}}\right], \tag{247}$$

where $f = S_1/S$.

Hence, from observation of f and determination of the vapour density, Q_{se}/R may be obtained. It remains then to devise a means of determining R. To do this we note that, if the cell temperature is low enough, the spin exchange rate, which is proportional to the rubidium concentration, will be negligible. It follows from (243 a) that if $P = 0$ at $t = 0$ and the ^{87}Rb resonance r.f. field is switched off at this time, then

$$P = P_0(1-e^{-t/T}),$$

where
$$T^{-1} = \tfrac{1}{2}Q_a I + 2R,$$
$$P_0 = \tfrac{1}{2}Q_a I/(\tfrac{1}{2}Q_a I + 2R).$$

By observing the value of T with incident light intensity I_0 and $\tfrac{1}{2}I_0$, R may then be obtained.

In the experimental arrangement, the light from a Varian spectral lamp, using rubidium enriched in ^{87}Rb, passed through an interference filter to isolate the D_1 line. To remove any light from the lamp capable of being absorbed by ^{85}Rb, the light passed through a filter cell containing 99 per cent pure ^{85}Rb. It was then circularly polarized by passage through a polaroid linear polarizer and a quarter-wave plate. The intensity of the light transmitted through the pumping cell containing natural rubidium vapour was monitored by an RCA phototube. The resonances were produced by using pulsed r.f., and phase-sensitive detection was employed.

As in the experiments of Gibbs and Hull the rubidium concentration was determined by optical absorption using a scanning Fabry–Perot interferometer.

Measurements at 90 °C at which the concentration of ^{87}Rb atoms was found to be $3\cdot33\pm0\cdot37\times10^{11}$ atoms/cm^3 gave $S_1/S = 0\cdot107\pm0\cdot001$. Taken in conjunction with the observed value of R (413 ± 21 s^{-1}) these results give $Q_{se} = 1\cdot70\pm0\cdot21\times10^{-14}$ cm^2.

6.2.3. *Nuclear magnetic-resonance method.* We have pointed out in § 6.1 that spin-exchange collisions contribute to the width of nuclear magnetic resonance lines. As an illustration of the experimental determination of spin-exchange cross-sections from the measurement of such line widths, we describe the experiments of Moos and Sands[†] in ^{87}Rb and ^{85}Rb. In their experiments the magnitude of the nuclear magnetic-resonance absorption signal, calibrated by comparison with the corresponding signal from a weighed sample of copper sulphate (CuSO$_4$,5H$_2$O) crystals, was used to determine the rubidium concentration. As usual, the signal refers to observations at a fixed angular frequency ω and variable steady magnetic field H. At the frequency used, 9000 Mc/s, the resonance values of H ranged from 2000 to 4000 G, corresponding to high field conditions.

The absorption can be expressed in terms of an imaginary component of susceptibility χ_i where

$$\chi_i = \frac{N\mu_{ij}^2}{(2I+1)(2S+1)\kappa T}\frac{\omega_{ij}\tau}{1+(\omega_{ij}-\omega)^2\tau^2}. \tag{248}$$

Here N is the number of atoms per cm^3, I and S are the total nuclear and electronic spin quantum numbers of the states concerned, $\omega_{ij}(H)$ is the angular frequency of the resonance transition, ω that of the incident radiation, τ is the relaxation time, T is the absolute temperature, and μ_{ij} is the dipole moment associated with the transition. It is given by

$$\mu_{ij} = g\mu_\text{B}\langle i|\mathbf{S}\cdot\mathbf{H}_1|j\rangle/H_1, \tag{249}$$

where H_1 is the amplitude of the applied r.f. field $\mathbf{H}_1$, $\mathbf{S}$ is the total electron spin, g the Landé factor for the levels concerned, and μ_B the Bohr magneton.

[†] Moos, H. W. and Sands, R. H., *Phys. Rev.* **135** (1964) A591.

Fig. 18.108 is a block diagram of the microwave spectrometer used. The magnetic field was modulated at 100 kc/s to permit selective amplification and phase-sensitive detection of the wanted signal. With the crystal detector operated at constant bias current, the signal displayed on the recorder is proportional to the square root of the incident

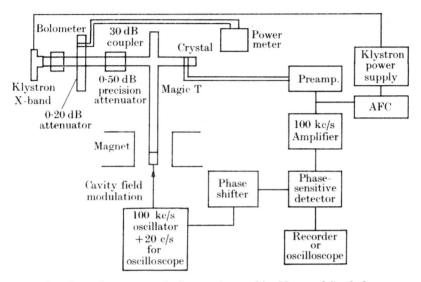

FIG. 18.108. General arrangement of apparatus used by Moos and Sands for measurement of the width of nuclear magnetic resonance lines in ^{87}Rb and ^{85}Rb.

power and hence is a voltage recorder. The signal S is proportional to

$$W_0 \, q_\mathrm{L} \, \eta (\partial \chi_\mathrm{i} / \partial H) H_\mathrm{M}^\mathrm{eff}, \qquad (250)$$

where W_0 is the square root of the incident power and q_L is the loaded quality factor of the cavity. η is given by

$$\int_\mathrm{sample} H_1^2 \, dV \Big/ \int_\mathrm{cavity} H_1^2 \, dV \qquad (251)$$

and the effective modulation of the 'steady' magnetic field $H_\mathrm{M}^\mathrm{eff}$ is given by

$$H_\mathrm{M}^\mathrm{eff} = \int_\mathrm{sample} H_\mathrm{M} H_1^2 \, dV \Big/ \int_\mathrm{sample} H_1^2 \, dV. \qquad (252)$$

With this equipment it is not the line shape which is displayed but its derivative with respect to the magnetic field H, which has the shape shown in Fig. 18.109. The maximum and minimum values occur at a separation which is given by

$$\Delta H = 2/\sqrt{3}\tau \, (\partial \omega_{ij}/\partial H). \qquad (253)$$

The recorded voltage difference ΔV between the two peaks (see Fig. 18.109) is given by

$$\Delta V = \frac{AN\mu_{ij}^2 \eta H_{\mathrm{M}}^{\mathrm{eff}}}{\kappa T(2I+1)(2S+1)} \frac{1}{(\partial \omega_{ij}/\partial H)\Delta H^2}, \qquad (254)$$

where A is a constant dependent on the incident power, cavity quality factor, etc., which does not vary in replacing the alkali metal by the

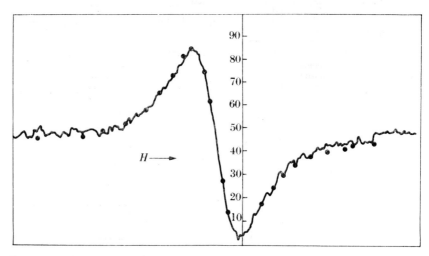

Fig. 18.109. Typical form of the derivative of the nuclear magnetic resonance line shape with respect to the steady magnetic field H.

copper sulphate. It follows that, if the relative values of η, μ_{ij}, $H_{\mathrm{M}}^{\mathrm{eff}}$, and $\partial \omega_{ij}/\partial H$ for both samples can be determined, the ratio of the concentration of alkali-metal atoms to that of the Cu^{++} ions in the copper sulphate can be obtained. As the latter concentration can be determined without difficulty that of the alkali-metal atoms may be derived.

The alkali-metal sample and the copper sulphate crystals were placed in a two-wavelength cavity made of Lava and coated on the inside with silver glaze. The quartz bulb containing the alkali metal was heated by passing warm nitrogen over it. To ensure that the r.f. fields would be the same at both samples a piece of quartz similar to that of the bulb was placed in the copper sulphate Dewar.

Fig. 18.110 illustrates some typical results obtained with ^{85}Rb in which the observed value of ΔH, given by (253) is plotted against the concentration N of ^{85}Rb atoms. Since τ is proportional to N^{-1} if spin exchange collisions are alone responsible for the line width, the plots under these conditions should be linear and pass through the origin.

It will be seen that they are indeed linear but do not all pass through the origin, presumably because of the presence of some source of line broadening independent of concentration.

Measurements were made by this technique not only for ^{85}Rb and ^{87}Rb but also for ^{133}Cs. The spin-exchange cross-sections derived from the observed relaxation times τ by the relation (see § 6.1)

$$\tau^{-1} = N\bar{v}\,2\{2I/(2I+1)\}Q_{se}$$

are given in Table 18.24 and will be seen to agree well with those measured by other techniques.

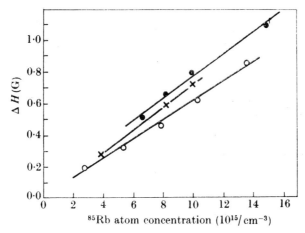

Fig. 18.110. Typical plots of ΔH (defined as in (253)) against the ^{85}Rb atomic concentration. ● × ○ experimental points.

Hildebrandt, Booth, and Barth[†] have carried out essentially similar experiments for atomic hydrogen. Their results, which have been analysed by Mazo,[‡] are given in Table 18.24.

In the course of measurements of the hyperfine splitting in ^{14}N and ^{15}N by magnetic-resonance techniques, in which the N atoms were polarized by spin exchange from optically polarized ^{23}Na atoms, Anderson, Pipkin, and Baird[§] noted that the line widths of the N hyperfine transitions were much greater than could be ascribed to spin exchange. They suggested that the process responsible involved exchange of N atoms in collisions between N atoms and N_2 molecules. For this explanation to be tenable the mean cross-section for such a collision under the experimental conditions must be $3\cdot 6\times 10^{-19}$ cm^2.

[†] Hildebrandt, A. F., Booth, F. B., and Barth, C. A., *J. chem. Phys.* **31** (1959) 273.
[‡] Mazo, R. M., ibid. **34** (1961) 169.
[§] Anderson, L. W., Pipkin, F. M., and Baird, J. C., *Phys. Rev.* **116** (1959) 87.

6.2.4. Stimulated emission method.

This method has the advantage that the spin-exchange cross-section is obtained without it being necessary to determine the atom concentration. It makes use of the phenomenon of *molecular ringing*.

Consider, for simplicity, an assembly of similar molecules each of which possesses only two stationary states. When an external electromagnetic field is applied the phases of the states, initially at random, are ordered and stimulated emission occurs. If the field is now cut off this emission does not immediately cease and the molecules are driven by their own radiation to produce the so-called ringing. Under certain circumstances the intensity of this ringing will reach a delayed maximum at some time t_m after the driving field has been cut off. If the two states concerned are essentially hyperfine structure levels, t_m depends on the spin-exchange cross-section and this forms the basis of the method.

To show how these results arise and to determine t_m we follow the simplified analysis due to Bloom.† With two states only, of energies ϵ_2, ϵ_1 such that

$$\epsilon_2 - \epsilon_1 = \hbar\omega_0, \tag{255}$$

and wave functions ϕ_1, ϕ_2, the wave function describing the system in the presence of the perturbing field can be written

$$\Psi = a_1\phi_1\exp(-i\epsilon_1 t/\hbar) + a_2\phi_2\exp(-i\epsilon_2 t/\hbar) \tag{256}$$

where, as in Chapter 7, (109),

$$i\hbar\dot{a}_1 = V_{11}a_1 + V_{12}a_2\exp(-i\omega_0 t), \tag{257a}$$

$$i\hbar\dot{a}_2 = V_{12}a_1\exp(i\omega_0 t) + V_{22}a_2. \tag{257b}$$

V_{11}, V_{12}, V_{22} are the matrix elements

$$\langle\phi_a|V|\phi_b\rangle \quad (a, b = 1, 2),$$

of the perturbing field. Transforming a_1, a_2 to c_1, c_2 where

$$a_\lambda = c_\lambda\exp\left\{-\left(i/\hbar\int_0^t V_{\lambda\lambda}\,dt\right)\right\}, \tag{258}$$

we have

$$\dot{c}_1 = fc_2, \tag{259a}$$

$$\dot{c}_2 = -f^*c_1, \tag{259b}$$

where

$$f(t) = -(i/\hbar)V_{12}\exp\left\{-i\omega_0 t - i\int_0^t (V_{22}-V_{11})\,dt/\hbar\right\}. \tag{260}$$

For our purposes, in which the perturbation is due to the action of an alternating electric field of angular frequency ω and amplitude E on the induced dipole moment of the molecular system,

$$V_{12} = -\mathbf{E}\cdot|\boldsymbol{\mu}_{12}|\cos\omega t,$$

$$V_{22} = V_{11},$$

so

$$f(t) = ip\cos\omega t\exp(-i\omega_0 t), \tag{261}$$

where $p = \mathbf{E}\cdot|\boldsymbol{\mu}_{12}|/\hbar$. $\boldsymbol{\mu}_{12}$ is now the dipole moment associated with the $1 \to 2$

† BLOOM, S., *J. appl. Phys.* **27** (1956) 785.

transition. We are particularly interested here in the resonance case in which $\omega = \omega_0$. For this case we may write

$$f(t) = \tfrac{1}{2}\mathrm{i}p, \tag{262}$$

ignoring the contribution from the oscillatory term.

On substitution in (259) we see that c_1 and c_2 both must satisfy the equation

$$\ddot{z} - (\dot{p}/p)\dot{z} + (p/2)^2 z = 0. \tag{263}$$

The substitution of

$$\theta = \int_0^t p(t)\,\mathrm{d}t, \tag{264}$$

for t gives

$$\frac{\mathrm{d}^2 z}{\mathrm{d}\theta^2} + \tfrac{1}{4}z = 0, \tag{265}$$

so

$$z = A\cos\tfrac{1}{2}\theta + B\sin\tfrac{1}{2}\theta. \tag{266}$$

Hence we have

$$c_1 = A_1 \cos\tfrac{1}{2}\theta + B_1 \sin\tfrac{1}{2}\theta, \tag{267 a}$$

$$c_2 = A_2 \cos\tfrac{1}{2}\theta + B_2 \sin\tfrac{1}{2}\theta. \tag{267 b}$$

The constants A_1, B_1, A_2, and B_2 must be such as to satisfy (259 a) and (259 b) with $f = \tfrac{1}{2}\mathrm{i}p$ and we then find that (267 a) and (267 b) may be written

$$c_1(t) = c_1(0)\cos\tfrac{1}{2}\theta + \mathrm{i}c_2(0)\sin\tfrac{1}{2}\theta, \tag{268 a}$$

$$c_2(t) = c_2(0)\cos\tfrac{1}{2}\theta + \mathrm{i}c_1(0)\sin\tfrac{1}{2}\theta. \tag{268 b}$$

The difference between the occupation numbers of the two states is

$$(|c_1|^2 - |c_2|^2) = D(t)$$
$$= \{|c_1(0)|^2 - |c_2(0)|^2\}\cos\theta -$$
$$- \mathrm{i}\{c_1(0)c_2^*(0) - c_2(0)c_1^*(0)\}\sin\theta. \tag{269}$$

When considering molecules in random phase states initially, the average value of the second term vanishes and we are left with the simple result

$$D(t) = D(0)\cos\theta. \tag{270}$$

After the driving field is cut off the system continues to radiate power at the resonance frequency, at a rate proportional to the instantaneous rate of change of the population excess. This power P is given by

$$P = \tfrac{1}{2}\hbar\omega_0 \dot{D}n, \tag{271}$$

where n is the total number of molecules per centimetre^3. P may be related to its own electric field E_r filling a cavity of volume V and loaded quality factor q_L by

$$P = \omega_0 V E_\mathrm{r}^2 / 8\pi q_\mathrm{L}. \tag{272}$$

The corresponding value for p is then given by

$$p = \mathbf{E}_\mathrm{r} \cdot |\boldsymbol{\mu}_{12}|/\hbar, \tag{273}$$

so

$$p^2 = kD, \tag{274}$$

where

$$k = 4\pi q_\mathrm{L} |\boldsymbol{\mu}_{12}|^2 n / 3\hbar V. \tag{275}$$

After cut-off we have, in the presence of state-randomizing collisions of frequency $1/\tau$,

$$D(T) = D(0)\,e^{-T/\tau}\cos\!\left(\theta + \int_0^T p\,\mathrm{d}t\right) + D(0)(1-e^{-T/\tau}), \tag{276}$$

where T is now the time since cut-off. Hence

$$\dot{D} = -pD(0)\,e^{-T/\tau}\sin\!\left(\theta + \int_0^T p\,\mathrm{d}t\right) + \{D(0)-D\}/\tau. \tag{277}$$

Of these terms, the second refers to the change in D due to collisions alone, while the first arises from ringing of those molecules which have not yet collided. In calculating p from (274) we therefore take

$$p(T) = -kD(0)\sin\!\left(\theta + \int_0^T p\,\mathrm{d}t\right)e^{-T/\tau}. \tag{278}$$

To solve this equation for p we introduce the substitution

$$\phi(T) = \int_0^T p\,\mathrm{d}t, \tag{279}$$

giving
$$\frac{\mathrm{d}\phi}{\mathrm{d}T} = -kD(0)\,e^{-T/\tau}\sin(\theta+\phi), \tag{280}$$

from which
$$\ln\{\tan\tfrac{1}{2}\theta\cot\tfrac{1}{2}(\theta+\phi)\} = kD(0)\tau(1-e^{-T/\tau}) \tag{281}$$

and hence
$$p(T) = -kD(0)\,e^{-T/\tau}\,\mathrm{sech}[kD(0)\{\tau(1-e^{-T/\tau}) - T_0\}], \tag{282}$$

where
$$\exp\{2kD(0)T_0\} = (1-\cos\theta)/(1+\cos\theta).$$

The ringing power at time T after cut-off is then given by

$$P(T) = P_\mathrm{m}\,e^{-2T/\tau}\,\mathrm{sech}^2[kD(0)\{\tau(1-e^{-T/\tau}) - T_0\}]. \tag{283}$$

It is a maximum for $T = T_\mathrm{m}$ where

$$kD(0)\tau\,e^{-T_\mathrm{m}/\tau}\tanh[kD(0)\{T_0 - \tau(1-e^{-T_\mathrm{m}/\tau})\}] = -1. \tag{284}$$

This maximum occurs at cut-off, i.e. $T_\mathrm{m} = 0$, when

$$kD(0)\tau\cos\theta = 1. \tag{285}$$

If the dominant source of relaxation is through spin-exchange collisions then $\tau = 1/nQ_\mathrm{se}\bar{v}$, where Q_se is the mean spin-exchange cross-section and $\bar{v}$ the mean relative velocity of impact. Substitution for τ and for k from (285) and (275) gives

$$Q_\mathrm{se} = 4\pi q_\mathrm{L}|\mathbf{\mu}_{12}|^2 D(0)\cos\theta/3\hbar\bar{v}V, \tag{286}$$

independent of the total atom concentration.

A more elaborate theory, which does not assume that the relaxation rates τ_1, τ_2 associated with the decay of the population inversion and the decay of the oscillatory moment are equal, shows that Bloom's semi-classical treatment gives the correct results when $\tau_1 = \tau_2$.

Vanier[†] has made the formula (286) the basis of a method for determining Q_se which he has demonstrated for rubidium. He used a rubidium maser in which the d.c. magnetic field was parallel to the z-component of the r.f. magnetic field in the cavity. The transition concerned was

[†] Vanier, J., *Phys. Rev. Lett.* **18** (1967) 333.

that of frequency 6835 Mc/s between the levels $F = 2$, $M_F = 0$ and $F = 1$, $M_F = 0$ of the ground $5\,^2S_{\frac{1}{2}}$ state. In this case $|\mathbf{\mu}_{12}|$ is equal to the Bohr magneton.

Population inversion was achieved by hyperfine pumping which produces selective population of the $F = 2$ level as described in § 6.2.1. If the light pulse is prolonged until equilibrium is set up then it may be shown by an analysis similar to that of § 3.10 that

$$D(0) = \Gamma\tau_1\{5\Gamma\tau_1+8\},$$

where Γ is the rate at which photons are absorbed per second by ^{87}Rb atoms in the $5\,^2S_{\frac{1}{2}}$ state with $F = 1$. Γ was determined from the rate at which the equilibrium under the action of the pumping radiation was set up and τ_1 by observing the variation of the signal amplitude, immediately after the microwave pulse, with the time interval since the end of the light pulse. Typical pulse lengths were 30 ms for the light and 50 μs for the microwaves, giving $\theta = \frac{1}{2}\pi$.

It remained to choose conditions so that $\tau_1 = \tau_2$ in order that (286) be applicable. The relaxation rate $1/\tau_2$ was measured under conditions in which the r.f. field produced by the atoms in the cavity had little effect on the decay of the emission itself so that the rate of decay of the field was the same as that of the emission. It was found that $\tau_1 \simeq \tau_2$ when the cavity is maintained at a temperature near 70 °C. Under these conditions Q_{se} was found to be $2\cdot0\pm0\cdot2\times10^{-14}$ cm^2 which agrees quite well with values found by other methods as will be seen from Table 18.24.

6.2.5. *Use of the hydrogen maser.* The conditions for natural oscillation of the hydrogen maser depend on the various relaxation times involved, including that due to spin exchange collisions. This may be used to measure the H–H spin-exchange cross-section as has been done by Hellwig.[†]

The general arrangement of a hydrogen maser[‡] is shown in Fig. 18.111. By means of the inhomogeneous magnetic field of a 6-pole magnet, hydrogen atoms from a r.f. discharge source in the $F = 1$, $M = 0$ and $F = 1$, $M = 1$ hyperfine states are selectively focused into an aperture in a Teflon-coated quartz bulb. This bulb is in the centre of a cylindrical r.f. cavity operating in the TE$_{011}$ mode, tuned to the frequency 1420·405 Mc/s of the $F = 1$, $M = 0$ to $F = 0$, $M = 0$ hyperfine transition. Atoms are retained in the bulb for more than one second, during which time they make 10^5 or so collisions with the wall. Nevertheless, because

---

[†] HELLWIG, H., *Phys. Rev.* **166** (1968) 4.
[‡] KLEPPNER, D., GOLDENBERG, M., and RAMSEY, N. F., ibid. **126** (1962) 603.

of their small interaction with the surface, the atoms are not seriously perturbed by the collisions. Under suitable conditions self-excited oscillations at the resonance frequency may occur.†‡

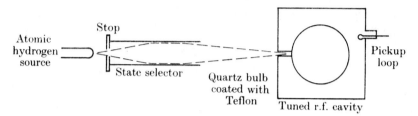

FIG. 18.111. General arrangement of a hydrogen maser.

Consider the effect, on the atoms in the $F = 1$, $M = 0$ state within the bulb, of an oscillating magnetic field
$$H_z \cos \omega t \tag{287}$$
in the same direction as the static field. At time t after application of the field the wave function of an atom will be of the form
$$\psi = a_1(t)\psi(0,0) + a_2(t)\psi(1,0), \tag{288}$$
where, at $t = 0$, $a_1 = 0$, $|a_2| = 1$, and $\psi(F, M)$ is the wave function for the hyperfine level (F, M).

It may be shown‡ that
$$a_1(t) = -\exp\{\tfrac{1}{2}i(\omega-\omega_0)t\}\sin[\tfrac{1}{2}\{(\omega-\omega_0)^2+x^2\}^{\frac{1}{2}}t]x\{(\omega-\omega_0)^2+x^2\}^{-\frac{1}{2}}, \tag{289}$$
where $x = -\mu_0 H_z/\hbar$, μ_0 being the Bohr magneton, and $\omega_0/2\pi$ is the resonance frequency. The average power radiated by a beam of N atoms/s initially is the $(1, 0)$ state is therefore
$$P = N\hbar\omega \langle |a_1|^2 \rangle_\mathrm{av}, \tag{290}$$
the average being over the time spent by an atom in its radiating state in the cavity. An atom is lost from this point of view, not only when it leaves the bulb, but also when it undergoes a radiationless transition from the upper state. These effects may all be described by a single relaxation time $1/\gamma$ so that the average is obtained by multiplying (290) by $f(t)$, where
$$f(t) = \gamma \exp(-\gamma t), \tag{291}$$
and integrating over t from 0 to ∞. This gives
$$P = \tfrac{1}{2} N\hbar\omega x^2 \{\gamma^2 + x^2 + (\omega-\omega_0)^2\}^{-1}. \tag{292}$$

As the atoms make multiple traversals of the bulb before being lost we must replace x^2 by $\langle x \rangle_\mathrm{b}^2$ where $\langle x \rangle_\mathrm{b}$ is the average of x throughout the volume of the bulb. We may write
$$\langle x \rangle_\mathrm{b}^2 = (\mu_0/\hbar)^2 \langle H_z \rangle_\mathrm{b}^2$$
$$= (\mu_0/\hbar)^2 (8\pi W/V)\eta, \tag{293}$$
where
$$W = (1/8\pi) \int_V H^2 \, dV \tag{294}$$

† KLEPPNER, D., GOLDENBERG, M., and RAMSEY, N. F., loc. cit.
‡ KLEPPNER, D., BERG, H. C., CRAMPTON, S. B., RAMSEY, N. F., VESSOT, R. F. C., PETERS, H. E., and VANIER, J., *Phys. Rev.* **138** (1965) A972.

is the energy stored throughout the volume V of the cavity, H being the peak value of the oscillating magnetic field. η is given by

$$\eta = \langle H_z \rangle_b^2 / \langle H^2 \rangle_V \tag{295}$$

and may be calculated as a function of the ratio of bulb radius to cavity diameter (in the TE_{011} mode the length and diameter of the cavity are equal).

We may now write

$$P = \tfrac{1}{2} N \hbar \omega \theta^2 / \{1 + \theta^2 + \gamma^{-2}(\omega - \omega_0)^2\}, \tag{296}$$

where
$$\theta^2 = \langle x \rangle_b^2 / \gamma^2 = W/W_c,$$

with
$$W_c = (\hbar/\mu_0)^2 V \gamma^2 / 8\pi\eta. \tag{297}$$

The condition for self-oscillation is that the power delivered to the cavity by the beam must be equal to that dissipated in the cavity. Hence

$$P(\omega = \omega_0) = \omega_0 W/q, \tag{298}$$

where q is the loaded quality factor of the cavity. At the threshold of oscillation $\theta \ll 1$ so the condition becomes

$$N_{th} = 4\pi W_c / qh$$
$$= c_1 \gamma^2, \tag{299}$$

where
$$c_1 = hV/8\pi^2 \mu_0^2 q \eta, \tag{300}$$

and can be measured without difficulty. To proceed further we must analyse the relaxation constant γ in more detail.

We first distinguish between relaxation of the population difference between the two maser levels and that of the oscillating moments. If γ_1, γ_2 are the respective relaxation constants then

$$\gamma^2 = \gamma_1 \gamma_2. \tag{301}$$

The contributions to relaxation through spin exchange, $\gamma_{1,se}$ and $\gamma_{2,se}$ respectively, can be shown to be related so that $\gamma_{1,se} = 2\gamma_{2,se}$. Assuming all other relaxation processes can be described by $\gamma_{1t} = \gamma_{2t} = \gamma_t$ we have

$$\gamma^2 = \tfrac{1}{2}\gamma_{1,se}^2 + \tfrac{3}{2}\gamma_{1,se}\gamma_t + \gamma_t^2. \tag{302}$$

Now
$$\gamma_{1,se} = n\bar{v}Q_{se}, \tag{303}$$

where n is the concentration of atoms in the bulb, $\bar{v}$ the mean relative velocity, and Q_{se} the spin-exchange cross-section. Since

$$n = N/\gamma_b V_b, \tag{304}$$

where V_b is the volume of the bulb and $1/\gamma_b$ is the mean time an atom spends within the bulb, we have
$$\gamma_{1,se} = 2c_2 N/\gamma_b, \tag{305}$$

where $c_2 = Q_{se} \bar{v}/V_b$, the factor 2 arising because, in addition to the atoms in the $F = 1, M = 0$ state there will be an equal number in the $F = 1, M = 1$ state.

From (299), (302), and (305) we have

$$N_{th} = (2c_1 c_2^2/\gamma_b^2)N^2 + (3c_1 c_2 \gamma_t/\gamma_b)N + c_1 \gamma_t^2. \tag{306}$$

If N_1, N_2 are the two roots of the equation $N_{th} = N$, the maser will oscillate when $N_1 \leqslant N \leqslant N_2$. For $N < N_1$ insufficient energy is transported by the beam to maintain oscillations, while for $N > N_2$ the rate at which spin exchange collisions occur is too great.

We have
$$N_{1,2} = (\gamma_b^2/4c_1 c_2^2)[1 - 3(\gamma_t c_1 c_2/\gamma_b) \mp \{1 - 6(\gamma_t c_1 c_2/\gamma_b) + (\gamma_t c_1 c_2/\gamma_b)^2\}^{\frac{1}{2}}]. \quad (307)$$
eglecting the quadratic term under the square root we have, if $N_2/N_1 = r^2$,
$$6\gamma_t c_1 c_2/\gamma_b = 4r/(1+r)^2,$$
so, on substitution for c_1 and c_2,
$$Q_{\text{se}} = \{4r/(1+r)^2\}(4\pi^2\mu_0^2/3hv)(\gamma_b/\gamma_t)qV_b \, \eta/V. \quad (308)$$

r may be measured directly, γ_b calculated, γ_t determined from measurements with microwave pulses applied to the maser operating at a power level well below threshold, and η calculated (see (295)).

Hellwig carried out observations with two different masers, the characteristics of which are given in Table 18.23. The values obtained for Q_{se} agree well and are estimated to be correct to 15 per cent. Comparison with values obtained by other methods is given in Table 18.24.

TABLE 18.23

	r^2	v (cm s^{-1})	γ_b/γ_t	q	V_b (cm^3)	V (cm^3)	η	Q_{se} (10^{-14} cm^2)
Maser I	17·4	3·58×10^5	0·70	3·40×10^4	1·94×10^3	1·43×10^4	2·8	0·266
Maser II	1·20	3·58×10^5	0·70	4·0×10^4	4·88×10^2	1·47×10^4	6·0	0·264

6.2.6. *Results of observations of spin-exchange cross-sections.* Table 18.24 gives the results obtained for spin-exchange cross-sections as measured by different observers using a variety of experimental techniques. It will be seen that, for Rb–Rb collisions, which have been investigated by many authors, the agreement between the different results is good. The two results for Rb–Cs also agree well.

Theoretical discussion of the results is given in § 8.2.

7. Spin-reversal collisions

We now consider collisions in which a change of orientation of electron spin occurs. Such collisions can only take place due to spin-spin and spin-orbit interaction between the colliding systems during the impact. It has already been pointed out that spin-exchange collisions between identical atoms have no effect in producing electron spin depolarization. On the other hand, it is true that, if other atoms are present with which spin exchange can occur and the mean polarizations of these atoms is held zero, relaxation of the polarized atoms will occur due to this spin exchange. If the other atoms are those of a rare gas spin-exchange with these atoms cannot occur. Relaxation of spin-polarized alkali-metal atoms in the presence of rare-gas atoms can only be due either to wall collisions or to collisions with rare-gas atoms in which spin flip occurs.

TABLE 18.24

Observed spin-exchange cross-sections

Atoms concerned	Spin-exchange cross-section (10^{-14} cm²)						
	(a)	(b)	(c)	(d)	(e)	(f)	(g)
Rb–Rb	1·5–2·6	2·6±0·4	1·85±0·2	1·70±0·20	2·2	1·9±0·2	2·02±0·20
	(h)	(f)	(i)	(j)	(k)		
Rb–Cs	2·4±0·4	2·3±0·2	Na–Na 1–3	Na–K 5	Na–Rb 2		
	(l)	(m)					
H–H	0·18	0·265					

(a) CARVER, T., *Proc. Ann Arbor Conference on Optical Pumping*, 1959, p. 29.
(b) MOOS, H. W. and SANDS, R. H., loc. cit., p. 1854.
(c) JARRETT, S. M., loc. cit., p. 1851.
(d) DAVIDOVITS, P. and KNABLE, N., *Bull. Am. phys. Soc.* **8** (1963) 352.
(e) BOUCHIAT, M. and BROSSEL, J., *C.r. hebd. Séanc. Acad. Sci., Paris*, **257** (1963) 2825; *Phys. Rev.* **147** (1966) 41.
(f) GIBBS, H. M. and HULL, R. J., loc. cit., p. 1845.
(g) VANIER, J., loc. cit., p. 1860.
(h) BOUCHIAT, M. and GROSSETÊTE, F., loc. cit., p. 1845.
(i) ANDERSON, L. W. and RAMSEY, A. T., *Phys. Rev.* **132** (1963) 712.
(j) FRANKEN, P., SANDS, R., and HOBART, J., loc. cit., p. 1842.
(k) NOVICK, R. and PETERS, H. E., *Phys. Rev. Lett.* **1** (1958) 54.
(l) HILDEBRANDT, A. F., BOOTH, F. B., and BARTH, C. A., loc. cit., p. 1857.
(m) HELLWIG, H., loc. cit., p. 1861.

To separate these effects it is only necessary to measure the spin relaxation rate as a function of the pressure of the rare-gas atoms.

Essentially the same procedure may be used to measure the relaxation rate as in the experiments described in § 6.2.1 for measurement of the corresponding rate for relaxation of an inverted hyperfine population. The only major difference is that Zeeman instead of hyperfine pumping is used to produce the spin-polarized atoms. Thus, with alkali-metal vapour the pumping radiation is the circularly polarized D_1 line which produces longitudinal spin polarization of ground state atoms. As this polarization refers to the spin alone any change requires a spin flip. After equilibrium is reached the system is allowed to relax in the dark and the state of polarization after a definite interval determined from the absorption of a probing signal of the same radiation. The complete relaxation curve is then built up by combining observations at different time-intervals after cut-off of the pumping light (cf. Fig. 18.106).

Fig. 18.112 illustrates some results obtained by Franzen† showing the variation of relaxation time with buffer-gas pressure for Xe, Kr, Ar, and Ne in rubidium vapour. In all cases the results follow the expected shape and may be analysed in an exactly similar way to corresponding decay-time plots of electron or metastable-atom concentrations in discharge afterglows discussed in § 5.2. Assuming that the decay follows the lowest diffusion mode, which gives the slowest

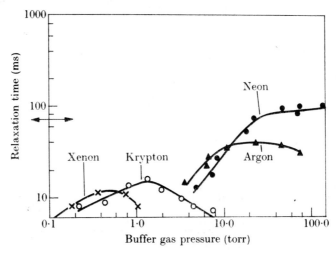

FIG. 18.112. Observed spin relaxation times for Rb vapour as a function of buffer gas pressure. The gas concerned is indicated on each curve drawn to fit best the experimental points shown.

diffusion decay time, the diffusion coefficient of rubidium in the buffer gas may be determined as well as the cross-section for spin relaxation in collisions between rubidium and rare gas atoms. The results obtained are given in Table 18.25.

A number of other combinations of alkali-metal vapour and buffer gas have been investigated, yielding results which are also given in Table 18.25. In all cases the spin disorientation cross-sections are very small, particularly for light atoms and molecules.

For collisions with rare-gas atoms the only likely mechanism for producing spin-flip is the spin-orbit interaction between the aligned electrons in the alkali-metal atom and the nucleus and electrons in the rare-gas atom. In other cases further possibilities arise through interaction of the magnetic moment of the polarized electron with either the nuclear or rotational magnetic moments of molecules such as H_2. For H_2 both

† loc. cit., p. 1845.

TABLE 18.25

Diffusion coefficients D of alkali-metal atoms in various gases and cross-sections for spin disorientation collisions between alkali-metal and rare-gas atoms

Alkali-metal atom	Gas atom	T (°K)	Diffusion coefficient D (cm² s⁻¹ at 1 atm)	Disorientation cross-section (cm²)
Na	He		$1\cdot0\pm0\cdot3$†	$3\pm4\times10^{-26}$†
	Ne		$0\cdot50\pm0\cdot17$†	$1\cdot8\pm0\cdot6\times10^{-24}$†
Rb	He		$0\cdot54$‡	$6\cdot2\times10^{-25}$‡
	Ne		$0\cdot31$§	$5\cdot2\times10^{-23}$§
	Ar		$0\cdot24$§	$3\cdot7\times10^{-22}$§
	Kr		—	$5\cdot9\times10^{-21}$§
	Xe		—	$1\cdot3\times10^{-20}$§
	H₂	343	$1\cdot34$∥	3×10^{-24}∥
			—	$2\cdot2\times10^{-24}$††
	D₂		—	$4\cdot3\times10^{-24}$††
	N₂		$0\cdot33$∥	$5\cdot7\times10^{-23}$∥
	CH₄	333	$0\cdot5$∥	8×10^{-24}∥
	C₂H₆	333	$0\cdot3$∥	$3\cdot8\times10^{-23}$∥
	C₂H₄	333	$0\cdot24$∥	$1\cdot3\times10^{-22}$∥
	C₆H₁₂	323	$0\cdot10$∥	$4\cdot5\times10^{-22}$∥

these interactions are considerably larger than for D₂ and yet the latter is observed to be more effective in producing spin-flip (see Table 18.25). This indicates that the spin-orbit effect is dominant in these cases also.

Further evidence†† of the relative ineffectiveness of electron-nuclear dipole-dipole interaction is provided by the experiments of Bouchiat, Carver, and Varnum‡‡ who succeeded in producing 0·01 per cent polarization of ³He nuclei in ³He at 2·8 atm pressure containing 10⁻³ torr of Rb atoms which were 10 per cent polarized by optical pumping. The ³He nuclei could only be polarized through interaction of the polarized Rb electron with their magnetic moments. The equilibrium polarization $P(^3\text{He})$ of the nuclei is related to that $P(\text{Rb})$ of the Rb electrons by

$$P(^3\text{He})/\tau = n(\text{Rb})\bar{v}Q_{\text{do}}\,P(\text{Rb}),$$

where τ is the relaxation time of the ³He nuclear spin, $n(\text{Rb})$ is the concentration of Rb atoms, $\bar{v}$ their mean relative velocity with respect

---

† ANDERSON, L. W. and RAMSEY, A. T., *Phys. Rev.* **132** (1963) 712.
‡ BERNHEIM, R. A., *J. chem. Phys.* **36** (1962) 135.
§ FRANZEN, W., loc. cit., p. 1845.
∥ McNEAL, R. J., *J. chem. Phys.* **37** (1962) 2726.
†† BREWER, R. G., ibid. **37** (1962) 2504.
‡‡ BOUCHIAT, M. A., CARVER, T. R., and VARNUM, C. M., *Phys. Rev. Lett.* **5** (1960) 373.

to the He atoms and Q_{do} the cross-section for disorientation of the Rb electron in collision with a ^{3}He nucleus through the dipole-dipole interaction. Since $\tau \sim 2 \times 10^3$ s and $n(\text{Rb}) = 7 \times 10^{13}$ atoms/cm^3 we find $Q_{do} = 4 \times 10^{-26}$ cm^2, which is nearly ten times smaller than the observed cross-section for spin-flip collisions between Rb and ^{4}He atoms (see Table 18.25).

A further theoretical discussion of spin-flip collisions is given in § 8.5.

8. Collisions involving excited atoms—discussion and theoretical interpretation of results

The importance of the magnitude of the so-called energy discrepancy ΔE in determining the probability of vibrational transitions occurring on impact between two molecules has been fully discussed in Chapter 17 (see especially §§ 1 and 4). ΔE is the amount by which the energy of relative translation of the colliding systems is changed by the impact. For the excitation of low-lying vibrational states at ordinary temperatures $\Delta E \tau / \hbar \gg 1$, where τ is the time of collision, and the collisions occur under nearly adiabatic conditions. This is in sharp contrast to the excitation of rotation for which the opposite condition applies, and the collisions are impulsive or sudden. Under nearly adiabatic conditions the probability of an inelastic collision falls off very rapidly as ΔE increases. We must then expect that, for thermal collisions in which electronic transitions occur, apart from transitions between fine structure levels, ΔE will often be so large that the probability of a superelastic collision occurring will be very small indeed. That this is true in general is clear from the experiments on quenching of excited atoms (see § 2.7). From such experiments no reliable evidence of quenching, under gas-kinetic conditions, by rare-gas atoms has been obtained although molecules are often quite effective quenchers.

In collisions involving transfer of electronic excitation ΔE is in general smaller and may indeed, with suitable choice of reactants, be comparable with gas-kinetic energies. The question arises as to the form of the dependence on ΔE in these cases and the magnitude of the transfer cross-section when $\Delta E \to 0$.

A simple criterion that we might be tempted to use is to suggest that the cross-section for a collision involving an electronic transition in which the energy discrepancy is ΔE will be small compared with gas-kinetic if
$$a\,\Delta E/\hbar v \gg 1. \qquad (309)$$
Here v is the initial relative velocity of impact and a is a length of order

of atomic dimensions. However, such a criterion must be used with caution. The range of interaction a will depend very much on the nature of the process and, in many cases, may be much larger than gas-kinetic radii. Also, if a is not large, allowance must be made for the variation of ΔE with separation of the colliding systems. The transition may take place at a separation R for which $\Delta E(R) \leqslant \Delta E(\infty)$ and (309) should be replaced by

$$R\,\Delta E(R)/\hbar v(R) \gg 1, \qquad (310)$$

where $v(R)$ is the relative velocity of the systems when at a separation R. In fact, the transition is most likely to take place at a value of R for which the left-hand side of (310) is a minimum and under certain circumstances (see § 8.4) this may be quite a small region.

We begin by discussing the calculation of cross-sections for collisions in which $\Delta E = 0$. Of these we first consider collisions involving excitation transfer among which we make an important distinction between the so-called symmetrical resonance cases in which $\Delta E = 0$ because the colliding systems are similar, and cases of accidental resonance in which the systems are unlike. The symmetrical resonance cases may be dealt with by a simple extension of the theory of elastic scattering and provide a convenient starting-point. In some conditions the results are applicable also to accidental resonance but this is not generally so and we defer formal discussion of these cases until after dealing with the extension of the theory to non-resonant collisions.

Before proceeding to this we discuss a number of other types of collision in which ΔE may be taken as zero, namely spin-exchange, disorientation, and disalignment collisions.

8.1. *Collisions in which exact resonance applies* ($\Delta E = 0$)—*excitation transfer*

8.1.1. *Symmetrical and accidental resonance.* We have already had occasion to discuss cases of excitation transfer in which exact resonance applies, namely in the transfer of excitation and diffusion of metastable He atoms in He. As stated above, we shall distinguish between cases of this type in which the excitation occurs between exactly similar systems, so the reaction can be written

$$A + A' \to A' + A, \qquad (311)$$

from others in which the resonance is 'accidental', viz.

$$A' + B \to A + B'. \qquad (312)$$

Penning ionization by metastable atoms (§ 5.3) is essentially of this latter kind in which, for example, A' is a metastable helium atom and B a normal argon atom. B' is then a singly ionized continuum state but the process is still one of exact energy resonance—no transfer of energy from or to kinetic energy of relative translation of the atoms is involved. For many, but not all cases such as these (see § 8.4.4), a more elaborate theory, which is essentially the limiting case of a theory in which the internal energy change $\Delta E \neq 0$, is necessary. The reason for this difference is that reactions of type (311), which we refer to as of symmetrical resonance type, proceed without any electronic transition taking place. The initial state of internal motion is described by a linear combination of two wave-functions which are respectively symmetrical and antisymmetrical in the two nuclei. In the final state the linear combination is changed and it is this change that determines the transfer cross-section. For cases of accidental resonance, as in (312), actual electronic transitions occur between two different states of the molecule AB.

8.1.2. *The case of symmetrical resonance.* In these cases a collision in which excitation transfer occurs is not experimentally distinguishable from a direct elastic collision but it is possible to make the distinction theoretically. Thus it is found in a thorough theoretical analysis that the relative angular distribution is of the general shape indicated in Fig. 18.113. It has a large maximum at zero angle and falls rapidly to a very small value. This is maintained up to an angle near 180°, after which the distribution increases to a second maximum at 180°. The first maximum can be regarded as due to direct elastic scattering, the second to collisions in which resonance excitation transfer has occurred. The two processes are only probable at angles near 0° and 180° respectively, so that they interfere very little and a classical separation is possible (for experimental applications see Chapter 24). In this way a separate cross-section for the transfer collision may be assigned, although in calculations of diffusion or mobility the effective cross-section to be taken must include both direct elastic scattering and resonance transfer (see § 8.1.3 of this chapter, § 3.3 of Chapter 19, and § 1 of Chapter 22).

As explained in Chapter 16, at large separations a normal and an excited atom of the same kind can interact in two distinct ways† so that the interaction energy is either

$$V_0(R) \pm U(R). \tag{313}$$

The existence of the term $U(R)$ is essentially due to the fact that resonance transfer of excitation may occur, i.e. the energy of the system $A + A'$

† It is assumed as usual that the normal atom possesses a complete outer shell of electrons.

is the same as that of A'+A. It may be shown that the cross-section for transfer, defined as above, is given by

$$Q_{\text{tr}} = (\pi/k^2) \sum (2l+1)\sin^2(\eta_l^{\text{g}} - \eta_l^{\text{u}}), \tag{314}$$

where $k\hbar$ is the momentum of relative motion. η_l^{g}, η_l^{u} are the phase shifts for motion with relative angular momentum $\{l(l+1)\}^{\frac{1}{2}}\hbar$ produced by the respective interactions $V_0 \pm U$. According to this formula the chance of transfer vanishes with U, as it should.

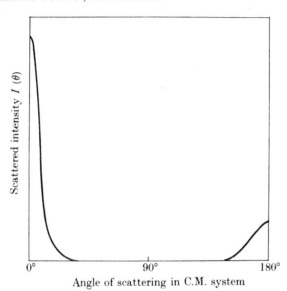

FIG. 18.113. Angular distribution in the C.M. system of the scattering of excited atoms by normal atoms of the same kind, showing the effect of excitation transfer.

An outline proof of (314) may be sketched as follows. We consider a collision between two identical atoms A, B in which the excitation initially in A is transferred to B. If the atoms are at a great distance R apart then the system is degenerate because the excitation energy may reside in either atom. When the atoms come closer and begin to interact the degeneracy is raised and the single level at infinite separation splits into two. For one of these the wave function is symmetrical, and for the other antisymmetrical, towards interchange of the two atoms. Thus, if the atoms were held fixed at a distance R apart the wavefunctions for the electron motion would be $\chi_{\text{g,u}}(R, \mathbf{r}_{\text{a}}, \mathbf{r}_{\text{b}})$ where $\mathbf{r}_{\text{a}}, \mathbf{r}_{\text{b}}$ are the coordinates of the electrons in the two atoms A, B relative to their respective nuclei. In the limit $R \to \infty$

$$\chi_{\text{g,u}}(R, \mathbf{r}_{\text{a}}, \mathbf{r}_{\text{b}}) \to 2^{-\frac{1}{2}}\{\psi_0(r_{\text{a}})\psi_1(r_{\text{b}}) \pm \psi_1(r_{\text{a}})\psi_0(r_{\text{b}})\}, \tag{315}$$

where ψ_0, ψ_1 are wave-functions for the ground and excited states concerned. $\chi_{\text{g,u}}$ satisfy the Schrödinger equation

$$\{H_{\text{a}}(\mathbf{r}_{\text{a}}) + H_{\text{b}}(\mathbf{r}_{\text{b}}) + \mathscr{V}(\mathbf{r}_{\text{a}}, \mathbf{r}_{\text{b}}, R)\}\chi_{\text{g,u}} = \{E_0 + E_1 + V_{\text{g,u}}(R)\}\chi_{\text{g,u}}, \tag{316}$$

where H_a, H_b are the Hamiltonian operators for the electrons relative to their respective nuclei and $\mathscr{V}(\mathbf{r}_a, \mathbf{r}_b, R)$ is the interaction energy between the atoms. Here E_0, E_1 are the energies of the ground and excited states respectively of the two atoms and, in the limit $R \to \infty$, $V_g(R) \to V_u(R) \to 0$.

Also, at large R,

$$V_{g,u}(R) \simeq \tfrac{1}{2} \iint \mathscr{V}(\mathbf{r}_a, \mathbf{r}_b, R) |\psi_0(r_a)\psi_1(r_b) \pm \psi_1(r_a)\psi_0(r_b)|^2 \, d\mathbf{r}_a \, d\mathbf{r}_b$$
$$= V_0 \pm U, \tag{317}$$

where
$$U = \iint \mathscr{V}(\mathbf{r}_a, \mathbf{r}_b, R) \psi_0(r_a)\psi_1(r_b)\psi_1(r_a)\psi_0(r_b) \, d\mathbf{r}_a \, d\mathbf{r}_b. \tag{318}$$

Ignoring the contributions from molecular states which do not, for large R, give rise to a normal atom and one in the particular excited state concerned, we write the wave-function describing the collision in the form

$$\Psi = F_g(\mathbf{R}) \chi_g(R, \mathbf{r}_a, \mathbf{r}_b) + F_u(\mathbf{R}) \chi_u(R, \mathbf{r}_a, \mathbf{r}_b). \tag{319}$$

If $F_g(\mathbf{R})$, $F_u(\mathbf{R})$ have the asymptotic forms

$$F_{g,u}(\mathbf{R}) \sim 2^{-\frac{1}{2}}\{e^{ikZ} + R^{-1} f_{g,u}(\Theta, \Phi) e^{ikR}\}, \tag{320}$$

where k is the wave number of relative motion, then, for large R,

$$\Psi \sim e^{ikZ} \psi_0(r_a)\psi_1(r_b) + R^{-1} e^{ikR} \{\tfrac{1}{2}\psi_0(r_a)\psi_1(r_b)(f_g+f_u) + \tfrac{1}{2}\psi_1(r_a)\psi_0(r_b)(f_g-f_u)\}. \tag{321}$$

The cross-section for a collision in which no excitation transfer occurs is then

$$Q_{\text{el}} = \tfrac{1}{4} \iint |f_g + f_u|^2 \sin\Theta \, d\Theta d\Phi, \tag{322}$$

and for one in which it does occur,

$$Q_{\text{tr}} = \tfrac{1}{4} \iint |f_g - f_u|^2 \sin\Theta \, d\Theta d\Phi. \tag{323}$$

To obtain f_g, f_u we substitute (319) into the full Schrödinger equation for the system,

$$\left\{-\frac{\hbar^2}{2M}\nabla_R^2 + H_a(\mathbf{r}_a) + H_b(\mathbf{r}_b) + \mathscr{V}(\mathbf{r}_a, \mathbf{r}_b, R) - E_0 - E_1 - E_t\right\} \Psi = 0, \tag{324}$$

$2M$ being the mass of an atom and E_t ($= k^2 \hbar^2/2M$) the energy of relative translation. Using (316), we have

$$\frac{\hbar^2}{2M} \nabla_R^2(\chi_g F_g + \chi_u F_u) - V_g \chi_g F_g - V_u \chi_u F_u + E_t(\chi_g F_g + \chi_u F_u) = 0. \tag{325}$$

We now multiply by χ_g and integrate over $\mathbf{r}_a$, $\mathbf{r}_b$. Since χ_g, χ_u are separate exact normalized eigenfunctions of (316), for each R,

$$\iint \chi_g \chi_u \, d\mathbf{r}_a \, d\mathbf{r}_b = 0, \quad \iint \chi_{g,u} \nabla_R^2 \chi_{g,u} \, d\mathbf{r}_a \, d\mathbf{r}_b = 0, \text{ and } \iint \chi_{g,u} \nabla_R \chi_{g,u} \, d\mathbf{r}_a \, d\mathbf{r}_b = 0, \tag{326}$$

so we are left with
$$\nabla^2 F_g + (k^2 - 2MV_g/\hbar^2) F_g = 0. \tag{327}$$

Similarly
$$\nabla^2 F_u + (k^2 - 2MV_u/\hbar^2) F_u = 0. \tag{328}$$

It follows then, as in Chapter 6, § 3.3,

$$f_g = (1/2ik) \sum (2l+1)\{\exp(2i\eta_l^g) - 1\} P_l(\cos\Theta), \tag{329}$$

$$f_u = (1/2ik) \sum (2l+1)\{\exp(2i\eta_l^u) - 1\} P_l(\cos\Theta), \tag{330}$$

giving, from (323), the expression (314) for Q_{tr}.

In deriving these results we have neglected the effect of electron exchange between the two atoms, but this is usually justified because the importance of these effects falls off more rapidly with atomic separation than that of the effects we have been considering—excitation transfer normally takes place at separations at which exchange is unimportant. We have also ignored the fact that the nuclei will obey either the Bose–Einstein or Fermi–Dirac statistics, but these are refinements that only affect the details of the angular distribution of the atoms after the collision.

A further complication which we have not taken into account, arises when the excited state concerned is not a ^1S state in which case the question of the reference axis for space quantization arises. For the molecular wave-functions χ the natural choice is the molecular axis but for the satisfaction of initial and final conditions reference must be made to an axis fixed in space. Further discussion of this problem is given in § 8.3 and in Chapter 23, § 6. Meanwhile, for order of magnitude estimates one can ignore it.

It is now possible to determine the conditions under which Q_{tr} is likely to be much larger than the usual gas-kinetic cross-section.

For this purpose we must consider the effectiveness of U in (313) as a scattering potential. If its range is much greater than that of V_0, Q_{tr} is likely to exceed substantially the gas-kinetic value as this is given by a scattering potential very similar to V_0. It has been pointed out in Chapter 16, § 2 that, if the excited state concerned in the transfer is one from which a transition to the ground state is associated with an electric moment of order p, then U falls off for large separations R as $R^{-(2p+1)}$. Thus, if the excited state is one which combines optically with the ground state so that p is 1, U decreases as slowly as R^{-3}. This is so slow that the differential cross-section $I(\Theta)\,d\Omega$ for scattering by such a field becomes logarithmically infinite as $\Theta \to 0$. The total cross-section, although finite, is likely therefore to be large unless the interaction, while of long range, is very weak. Actually, in these cases,

$$U(R) \sim -\mu^2 R^{-3}, \tag{331}$$

where μ is the dipole moment associated with the optical transition which occurs in either atom.

A semi-classical approximation for Q_{tr} may be obtained in very much the same way as in Chapter 16, § 5.4.1 for the total cross-section for scattering by a centre of force. With the usual relation

$$l+\tfrac{1}{2} \simeq kp, \tag{332}$$

where p is the impact parameter, (314) becomes

$$Q_{tr} = 2\pi \int_0^\infty p \sin^2(\eta_l^g - \eta_l^u) \, dp. \tag{333}$$

We now use the semi-classical expression (see Chap. 16 (104))

$$\eta_l^{g,u} = \frac{1}{\hbar v} \int V_{g,u}\{(z^2+p^2)^{\frac{1}{2}}\} \, dz, \tag{334}$$

valid when $Mv^2/V_{g,u}(p) \gg 1$. If this is satisfied then

$$\eta_l^g - \eta_l^u = \frac{1}{\hbar v} \int \Delta E\{(z^2+p^2)^{\frac{1}{2}}\} \, dz, \tag{335}$$

where

$$\Delta E(R) = V_g(R) - V_u(R). \tag{336}$$

We may now write

$$Q_{tr}^{sc} = 2\pi \int_0^\infty p \sin^2\{\phi(p)\} \, dp, \tag{337}$$

where

$$\phi(p) = \frac{1}{\hbar v} \int \Delta E\{(z^2+p^2)^{\frac{1}{2}}\} \, dz, \tag{338}$$

(334) being assumed to hold for all p. No serious error is introduced in this way, however, because for small p, $|\phi(p)|$ will be $\gg 1$. Thus we may, as usual in the derivation of a semi-classical approximation, evaluate (333) by writing

$$\int_0^\infty p \sin^2\phi(p) \, dp = \int_0^P p \sin^2\phi(p) \, dp + \int_P^\infty p \sin^2\phi(p) \, dp, \tag{339}$$

where P is such that, for $p > P$, $\phi(p) \ll \frac{1}{4}\pi$ and $\phi(P) = \frac{1}{4}\pi$. In the first integral on the right-hand side of (339) we replace $\sin^2\phi(p)$ by a mean value of $\frac{1}{2}$ and in the second by $\{\phi(p)\}^2$ so that

$$2\pi \int_0^\infty p \sin^2\phi(p) \, dp \simeq \frac{1}{2}\pi P^2 + 2\pi \int_P^\infty p\{\phi(p)\}^2 \, dp. \tag{340}$$

Detailed knowledge of $\phi(p)$ is not required when it is large so that (334) is invalid and to this extent we may regard $\sin^2\phi(p)$ as the probability that excitation transfer will occur for a collision in which the impact parameter is p.

In applying this method it must be remembered that (335) is valid only when $Mv^2/V_{g,u}(p) \gg 1$ for both potentials V_g and V_u. (337) therefore is only valid when this condition is satisfied for all significant values of p. In most instances no serious error is made by assuming (335) to be valid when $\phi(p) \ll 1$. Examples where a serious error could be made by use of (335) in a case of transfer of excitation are discussed in §§ 8.1.3 and 8.1.4 (see also Chap. 23, § 5.1.1).

If ΔE is of much longer range than V_0 so that, for $p = P$, the contribution of V_0 to $\phi(p)$ is negligible, then Q_{tr}, according to (337), is $\frac{1}{4}$ of the elastic cross-section for scattering by an interaction $\Delta E = 2U$. Hence, referring to (103) of Chapter 16, when $U \simeq -\mu^2/R^3$

$$Q_{\mathrm{tr}} = 3\pi\mu^2/v\hbar, \qquad (341)$$

where v is the velocity of relative motion. Taking μ to be of the order 10^{-18} e.s.u. we find that Q_{tr} has the large value $2\cdot 5 \times 10^{-14}$ cm². Although this is a very rough estimate it shows that, under the conditions we have assumed of exact resonance and optically allowed transitions in each atom, the transfer cross-section may greatly exceed the gas-kinetic. It must be emphasized, however, that this does not arise because of the wave aspect of the collision but because the possibility of the resonance transfer introduces a long-range interaction that would otherwise not occur. Apart from this the collision is essentially classical in character, the wavelength of relative motion being very short compared with the range of the strong interaction.

If the transfer involves a transition in each atom associated with a quadrupole moment, then $U(R) \sim (\nu^2/e^2)R^{-5}$ where ν is of the same order as μ in (341). We then find, according to the semi-classical approximation, that

$$Q_{\mathrm{tr}} = \frac{7\pi}{12}(8\nu^2/3\hbar v e^2)^{\frac{1}{2}}, \qquad (342)$$

giving a transfer cross-section of order 10^{-15} cm², not much greater than the gas-kinetic. For transitions involving higher-order moments the cross-section can be expected to exceed the gas-kinetic to an even smaller extent. In particular, if the transfer process is one involving charge exchange, $U(R)$ falls off exponentially and cross-sections greatly exceeding the gas-kinetic are not to be expected. On the other hand, even in this case $U(R)$ falls off somewhat less rapidly than the repulsive part of V_0 in (313), so that some increase above the gas-kinetic value still occurs (see Chap. 19, § 3.3).

As far as the variation of the cross-section with relative velocity is concerned, the transfer cross-section, in the case of exact resonance, falls off monotonically as the velocity increases, approximately as $v^{-2/(s-1)}$ when $U(R) \sim CR^{-s}$.

The simplest detailed application of the theory which can be made is to the transfer of excitation between a metastable and normal helium atom. Transfer collisions between hydrogen atoms are complicated by the possibility of spin exchange because the atom has a net electronic spin, and also by the existence of orbital degeneracy. Accordingly, we shall first discuss the application to the helium problem before describing the results so far obtained, with certain simplifying assumptions, for transfer collisions between H atoms.

8.1.3. *Application to collisions of metastable helium atoms in helium.* The interactions between metastable and normal helium atoms were first calculated by Buckingham and Dalgarno[†] who then proceeded to apply their results[‡] to calculate total elastic cross-sections, diffusion coefficients, and excitation-transfer cross-sections for He*–He collisions.

As usual there are two possible interactions $V_{g,u}$ for a normal helium atom with a $2\,^3S$ or $2\,^1S$ helium atom. V_g has the usual characteristic of a repulsive curve, exhibiting a minimum only at very large distances due to the van der Waals interaction. While V_u exhibits a chemical attraction, possessing a minimum about 1·6 eV below the energy at infinite separation, Buckingham and Dalgarno found that it passes over, at somewhat greater distances, to a repulsion which reaches a maximum of about 0·3 eV at a separation of about $4a_0$. At still larger distances the repulsion falls to merge gradually with the same attractive van der Waals force as for the interaction V_g. Fig. 18.114 illustrates this behaviour for both the $2\,^3S$ and $2\,^1S$ cases.

Because of the presence of this repulsive barrier in V_u as well as the strong repulsion in V_g, the interaction at distances less than $4a_0$ is of little importance in determining any of the cross-sections for encounter between metastable and normal helium atoms at thermal energies. In particular, the semi-classical formula (337) for calculating the transfer cross-section cannot be used—the repulsive barrier prevents penetration to such distances that $V_g - V_u$ is large. It is no longer correct to assume that $|\eta_l^g - \eta_l^u|$ will be large when $\phi(p)$, given by (338), with $l + \tfrac{1}{2} = kp$, is large.

To carry out detailed calculations of cross-sections for the $2\,^3S$ case Buckingham and Dalgarno[‡] assumed two different values, 10·6 and $21·2 \times 10^{-60}$ erg cm^6 for the van der Waals constant as this was only roughly estimated. Fig. 18.115 illustrates the transfer cross-sections Q_{tr}, as functions of relative wave number, which they obtained with these two different assumptions. The effect of the medium-range repulsion in reducing the cross-section rapidly towards zero as the energy of relative motion falls below 0·3 eV is clearly seen. The variation of Q_{tr},

[†] BUCKINGHAM, R. A. and DALGARNO, A., *Proc. R. Soc.* A**213** (1952) 327.
[‡] ibid. 506.

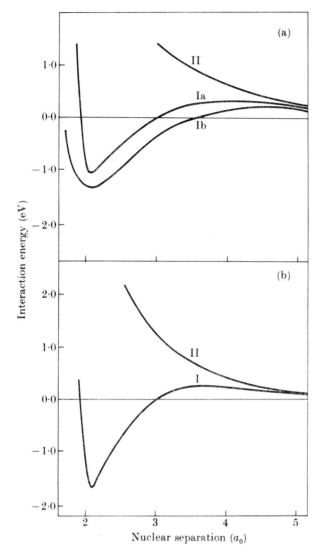

FIG. 18.114. (a) Interactions between $1S$ and $2\ ^3S$ helium atoms: I, $^3\Sigma_u$ state; I a, calculated by Buckingham and Dalgarno; I b, calculated by Poshusta and Matsen; II, $^3\Sigma_g$ state calculated by Buckingham and Dalgarno. (b) Interactions between $1S$ and $2\ ^1S$ helium atoms, calculated by Buckingham and Dalgarno: I, $^1\Sigma_u$; II, $^1\Sigma_g$.

averaged over the Maxwellian velocity distribution, as observed by Colegrove, Schearer, and Walters† as a function of mean energy of relative motion is also shown in Fig. 18.115. There is no doubt that the general form of the theoretical variation with relative energy is correct but the calculated cross-section falls off too quickly as the relative energy falls. Since the first work of Buckingham and Dalgarno, who used simple analytical approximations for the helium atomic wave functions, more

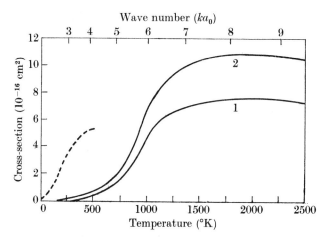

FIG. 18.115. Cross-sections for transfer of excitation between He ($2\,^3S$) and He ($1S$) atoms on impact. —— calculated by Buckingham and Dalgarno; (1) and (2) assuming van der Waals interaction energies $-10\cdot 6R^{-6}$ and $-21\cdot 2R^{-6}$ in a.u. respectively. – – – observed by Colegrove, Schearer, and Walters (see Fig. 18.96).

refined calculations have been carried out. These have confirmed the presence of the medium-range repulsive maximum in V_u but have reduced its magnitude. The most elaborate calculation to date, that of Poshusta and Matsen (see Fig. 18.114 (a))‡ gives 0·14 eV as the peak, occurring at a separation of $4\cdot 5a_0$. It seems that this reduced value will give results agreeing better with the observations.

The total cross-section, calculated by Buckingham and Dalgarno using the smaller assumed value $10\cdot 6 \times 10^{-60}$ erg cm⁶ for the van der Waals constant, has been compared with observed data in § 5.1. Since the medium-range repulsion occurring in the V_u interaction falls off quite gradually with increasing nuclear separation we should expect that the cross-section would decrease quite rapidly with increasing relative velocity, in contrast with the very slow change characteristic of a 'hard'

† loc. cit., p. 1823.
‡ POSHUSTA, R. D. and MATSEN, F. A., *Phys. Rev.* **132** (1963) 307.

repulsion (see Chap. 16, § 5.4.1). The observed variation is much more gradual than expected from the theory.

The diffusion coefficient at 1 torr as calculated by Buckingham and Dalgarno, again assuming the lower value $10 \cdot 6 \times 10^{-60}$ erg cm^6 for the van der Waals constant, varies almost linearly with temperature from 160 cm^2 s^{-1} at 100 °K to 1200 cm^2 s^{-1} at 1000 °K, the value at 300 °K being 370 cm^2 s^{-1}. This is to be compared with 470 ± 25 cm^2 s^{-1} determined by Phelps using the method discussed in § 5.2.2.1.

8.1.3.1. *The deactivation of* He($2\,^1S$) *in collisions with normal* He *atoms.* The existence of the potential maximum in the V_u interaction between normal and metastable He($2\,^1S$) atoms is of importance in the determination of the rate coefficient μ of two-body deactivation of $2\,^1S$ atoms in helium—4×10^{-15} cm^{-3} s^{-1} as measured by Phelps (§ 5.2.2.1, p. 1782). The only two-body process that can be effective is the radiative one

$$\text{He}(1\,^1S) + \text{He}(2\,^1S) \rightarrow 2\text{He}(1\,^1S) + h\nu.$$

Although in the absence of any interacting system a radiative transition from He($2\,^1S$) to He($1\,^1S$) is strongly forbidden and only proceeds through emission of two quanta,† allowed transitions may occur between the $A\,^1\Sigma_u$ state of He$_2$, which dissociates at large separations to He($1\,^1S$) and He($2\,^1S$), and the unstable $X\,^1\Sigma_g$ ground state—the dipole moment for the transition vanishes at infinite separation R between the nuclei but is finite at smaller distances.

The existence of the potential hump for the $2\,^1S$ case provides an explanation of the fact that the forbidden helium line at 601·418 Å due to the $2\,^1S$–$1\,^1S$ transition appears only when a band is also observed with the edge at 600·019 Å shading to the red.‡ This band is almost certainly due to the $A\,^1\Sigma_u \rightarrow X\,^1\Sigma_g$ transition. Since the ground $^1\Sigma_g$ state is repulsive the fact that the band edge occurs at a shorter wavelength than that of the forbidden line can only be understood if there is a potential maximum in the $A\,^1\Sigma_u$ state.§ We now discuss the importance of the maximum in determining the deactivation rate for He($2\,^1S$).

Consider a collision between two atoms interacting in the $^1\Sigma_u$ state. The relative motion is effectively classical so, if p is the impact parameter,

† DALGARNO, A., *Mon. Not. R. astr. Soc.* **131** (1966) 311.
‡ LYMAN, T., *Astrophys. J.* **60** (1924) 1; SOMMER, L. A., *Proc. natn. Acad. Sci. U.S.A.* **13** (1927) 213; KRUGER, P. G., *Phys. Rev.* **36** (1930) 855; NICKERSON, J. L., *Phys. Rev.* **47** (1935) 707; TANAKA, Y., *Sci. Pap. Inst. phys. chem. Res., Tokyo* **39** (1942) 465; TANAKA, Y. and YOSHINO, K., *J. chem. Phys.* **39** (1963) 3081.
§ NICKERSON, J. L., TANAKA, Y., and YOSHINO, K., loc. cit.

v the initial relative velocity, we have

$$\frac{dR}{dt} = \{v^2 - p^2 v^2 R^{-2} - 2V_u(R)/M\}^{\frac{1}{2}}, \qquad (343)$$

where M is the reduced mass. The probability of deactivation in such a collision, regarded as small, is then given by

$$P(p) = \int_{-\infty}^{\infty} A(R) \, dt,$$

where $A(R)$ is the Einstein A coefficient for the $^1\Sigma_u - ^1\Sigma_g$ transition when the nuclear separation is R. Hence

$$P(p) = 2\int_{R_0}^{\infty} A(R) \frac{dt}{dR} \, dR,$$

where R_0 is the closest distance of approach given by the outermost zero of the right-hand expression in (343).

The total cross-section for the process is then

$$Q(v) = 2\pi \int P(p) p \, dp$$

and the rate coefficient μ is $\overline{vQ(v)}$, the average of $vQ(v)$ over the Maxwellian distribution of relative velocities at the temperature concerned.

The calculated value of μ is very sensitive to the height of the potential maximum. It was first calculated by Burhop and Marriott[†] using the interaction derived by Buckingham and Dalgarno, shown in Fig. 18.114, in which the barrier height is 0·26 eV. They obtained a value of μ of $1·3 \times 10^{-20}$ cm^{-3} s^{-1}, too small by a factor of 10^5 or so. However, Allison, Browne, and Dalgarno[‡] repeated and extended these calculations using an interaction calculated with a much more elaborate approximate molecular wave-function including terms involving p orbitals. In this interaction the potential maximum was reduced to 0·084 eV at a separation of about $6a_0$.

They evaluated $A(R)$ using both dipole length and dipole velocity matrix elements (see Chap. 7, § 5.2.3) and their results are given in Table 18.26 as a function of gas temperature. It will be seen that the two different matrix elements give results which agree reasonably well, suggesting that the wave functions used are satisfactory. Moreover, for 250 °K the calculated values now agree very well with that observed by Phelps. In view of the sensitivity of the calculated values to the

[†] BURHOP, E. H. S. and MARRIOTT, R., *Proc. phys. Soc.* A**69** (1956) 271.
[‡] ALLISON, D. C., BROWNE, J. C., and DALGARNO, A., ibid. **89** (1966) 41.

barrier height in V_u it seems that the interaction used by Allison, Browne, and Dalgarno is more satisfactory in this respect than the more elaborate one of Scott, Greenawalt, Browne, and Matsen,† which gives a somewhat higher maximum, 0·149 eV, at a separation of about $5a_0$.

TABLE 18.26

Calculated rate coefficients μ for deactivation of He $(2\,^1S)$ *in two-body radiative collisions with* He $(1\,^1S)$ *atoms as a function of gas temperature* $(T\,°K)$

T (°K) μ (10^{-14} cm^{-3} s^{-1})	250	500	1000	2000	4000	8000	16 000	32 000
Using length matrix element	0·14	0·56	1·27	1·99	2·54	2·90	3·10	3·20
Using velocity matrix element	0·13	0·49	1·07	1·63	2·03	2·26	2·39	2·45

8.1.4. *Application to transfer of excitation on impact between hydrogen atoms.* Nakamura and Matsuzawa‡ have applied the formula (314) to the transfer collisions

$$H(2p)+H(1s) \rightarrow H(1s)+H(2p),\ddagger \qquad (344\,\text{a})$$

$$H(2s)+H(1s) \rightarrow H(1s)+H(2s).\S \qquad (344\,\text{b})$$

They ignored orbital degeneracy so that the two types of collision were treated independently.

Even under these conditions H atoms in 2s and 1s states can interact in four different ways instead of just two for He atoms in a metastable and a normal state. This is because the electron spins may be either parallel or antiparallel. Corresponding to each spin state there will be two states of opposite nuclear symmetry. Thus, when the spins are antiparallel we have interactions ^1V_g, ^1V_u corresponding to $^1\Sigma_g$ and $^1\Sigma_u$ molecular states. The corresponding transfer cross-section will be

$$^1Q_{\text{tr}} = (\pi/k^2) \sum (2l+1)\sin^2(^1\eta_l^g - {}^1\eta_l^u). \qquad (345)$$

Similarly, if the spins are parallel, giving rise to $^3\Sigma_g$ and $^3\Sigma_u$ molecular states, we have

$$^3Q_{\text{tr}} = (\pi/k^2) \sum (2l+1)\sin^2(^3\eta_l^g - {}^3\eta_l^u). \qquad (346)$$

† SCOTT, D. R., GREENAWALT, E. M., BROWNE, J. C., and MATSEN, F. A., *J. chem. Phys.* **44** (1966) 2981.
‡ NAKAMURA, H. and MATSUZAWA, M., *J. phys. Soc. Japan*, **22** (1967) 248.
§ MATSUZAWA, M. and NAKAMURA, H., ibid. **22** (1967) 392.

The mean transfer cross-section, averaged over electron spin orientations, is then
$$\bar{Q}_{tr} = \tfrac{1}{4}\{3\,{}^3Q_{tr} + {}^1Q_{tr}\}. \tag{347}$$

Nakamura and Matsuzawa calculated the interaction energies ${}^1V_g$, ${}^1V_u$, ${}^3V_g$, ${}^3V_u$ by the Heitler–London method, obtaining the results shown in

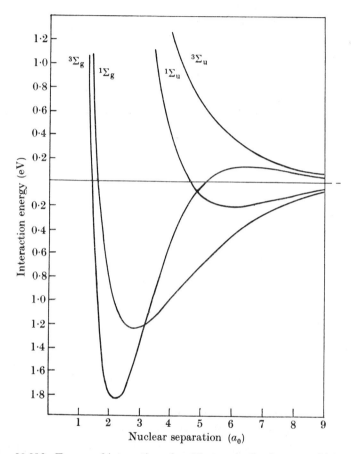

FIG. 18.116. Energy of interaction of an H atom in the 2s state with one in the ground state in the four molecular states ${}^1\Sigma_{u,g}$, ${}^3\Sigma_{u,g}$, calculated by Matsuzawa and Nakamura using the Heitler–London approximation.

Fig. 18.116. It will be seen that ${}^3V_g$ exhibits the same type of medium range repulsion as does the interaction V_u between a metastable and normal helium atom. The van der Waals interaction constant C was calculated from a simple approximate formula due to Buckingham† as

† BUCKINGHAM, R. A., *Proc. R. Soc.* A**160** (1937) 94.

−156·8 a.u. An attraction $-C/R^6$ was added arbitrarily to all interactions for $R > 6a_0$. Using these interactions $^3Q_{tr}$ and $^1Q_{tr}$ were calculated using the Langer approximation to evaluate the phase shifts involved. The results obtained are given in Table 18.27.

It will be seen that $^3Q_{tr}$ varies with relative kinetic energy in much the same way as Q_{tr} for collisions between metastable and normal helium atoms. This is because of the repulsive barrier at medium separations. $^1Q_{tr}$, on the other hand, increases monotonically as the relative kinetic energy increases and is quite large, at thermal energies, compared with the gas-kinetic cross-section for H–H collisions.

TABLE 18.27

Cross-sections, in πa_0^2, for transfer of excitation between H(2s) and H(1s) atoms

Relative kinetic energy (eV)	$^1Q_{tr}$	$^3Q_{tr}$	$\bar{Q}_{tr} = \frac{1}{4}{}^1Q_{tr}+\frac{3}{4}{}^3Q_{tr}$
0·03	101·1	10·5	33·1
0·1	69·3	22·8	34·5
0·3	53·5	34·3	39·1
1·0	51·3	36·4	40·1
3·0	45·6	36·4	38·7
5·0	41·3	31·4	33·9
10·0	32·9	29·4	30·3

Nakamura and Matsuzawa avoided difficulties about space quantization in considering the reaction (344 b), by working in terms of molecular wave functions and averaging over all p states with different orientations by averaging the resultant cross-sections. Thus H(2p) and H(1s) atoms give rise to four Σ states $^3\Sigma_g$, $^3\Sigma_u$, $^1\Sigma_g$, $^1\Sigma_u$ and four Π states $^3\Pi_g$, $^3\Pi_u$, $^1\Pi_g$, $^1\Pi_u$. From each g, u pair a transfer cross-section may be calculated if the corresponding interactions are known. We thus have $^1Q_{tr}^\Sigma$, $^3Q_{tr}^\Sigma$, $^1Q_{tr}^\Pi$, $^3Q_{tr}^\Pi$ and the over-all weighted mean is

$$\bar{Q}_{tr} = \tfrac{1}{3}\{(\tfrac{1}{4}{}^1Q_{tr}^\Sigma+\tfrac{3}{4}{}^3Q_{tr}^\Sigma)+2(\tfrac{1}{4}{}^1Q_{tr}^\Pi+\tfrac{3}{4}{}^3Q_{tr}^\Pi)\}. \quad (348)$$

The interactions were calculated by the Heitler–London method but it was found that for $^3\Sigma_u$ an admixture of an ionic state had to be included to give sensible results. In these cases the semi-classical approximation (337) could be used for the calculation of all Q_{tr} as none of the interactions include a medium-range repulsion. The results obtained for $\bar{Q}_{tr}$ are given in Table 18.28. It will be seen that, at low energies, the cross-section is very large, as would be expected from (341).

Thus μ^2 in (341) is given in units $e^2a_0^2$ as 0·555 and 1·11 for Π and Σ states respectively. On the other hand, the cross-section falls faster with increasing relative kinetic energy than for H(2s)–H(1s) transfer collisions which is again in agreement with expectation according to the discussion of p. 1875—the faster the rate of decrease of U with R for large R, the slower the rate of decrease of the cross-section with relative kinetic energy.

TABLE 18.28

Mean cross-sections, $\bar{Q}_{tr}$ in πa_0^2, for transfer of excitation between H(2p) and H(1s) atoms

Relative kinetic energy (eV)	0·1	1·0	3·0	5·0	10·0
$\bar{Q}_{tr}$	821	260	143	109	79·3

8.2. *Collisions in which* $\Delta E = 0$—*spin-exchange collisions*

In Chapter 5, § 7 we have described methods for determining cross-sections for collisions in which electron spin-exchange occurs. These methods all depend on the fact that, if the nuclei of the atoms possess spin, exchange of electron spin can lead to a change in the hyperfine structure state of the atoms. Thus, during the collision the electron and nuclear spins in each atom are uncoupled. After the collision they are coupled again but if in the meantime the electron spin in an atom has been changed due to a spin exchange, the distribution of the final states among the hyperfine structure components will also be changed.

Electron spin-exchange collisions between two atoms with single electrons in their outer shells are brought about through electron exchange, just as in the case of the corresponding collisions between electrons and such atoms discussed in Chapter 8, § 3. By following a similar analysis we obtain a total cross-section for spin exchange (cf. (64) of Chap. 8),

$$Q_{se} = (\pi/k^2) \sum (2l+1)\sin^2(\eta_l^{(1)} - \eta_l^{(3)}), \qquad (349)$$

where $\eta_l^{(1)}$, $\eta_l^{(3)}$ are the phase shifts for collisions between atoms of relative wave-number k interacting in the $^1\Sigma_g$ and $^3\Sigma_u$ states respectively.

The cross-section for excitation of a hyperfine structure transition in which the total (nuclear+electronic) spin quantum number of an atom changes from F to F' is now given by

$$Q(F, F') = \frac{2F'+1}{2(2I+1)} Q_{se}, \qquad (350)$$

where I is the nuclear spin quantum number.

If the atoms are identical, allowance must also be made for nuclear symmetry. We shall illustrate this by discussing spin-exchange collisions between hydrogen atoms, which are not only of interest in affecting the performance of the hydrogen maser (see § 6.2.5) but in providing a source of excitation of the 21-cm emission line (see pp. 521–2) through the transition $(1, 0)$ between the hyperfine structure components of the ground state.

8.2.1. *Spin-exchange collisions between* H *atoms.* According to (350) the cross-section $Q(0, 1)$ for excitation of the 21-cm emission line of hydrogen will be given by
$$Q(0, 1) = \tfrac{3}{4} Q_{se}.$$

For the calculation of the rate at which excitation of the line occurs in an assembly of hydrogen atoms in different hyperfine states, account must be taken of the fact that, in some collisions, while one of the atoms is excited to the state $F = 1$ the other is deactivated to $F = 0$. Such collisions do not lead to net excitation. Furthermore, no allowance has been made for the symmetry properties of the protons.

The latter effect introduces two spin exchange cross-sections Q_{se}^+, Q_{se}^- which are defined by (cf. (234) of Chap. 16)

$$Q_{se}^{\pm} = (\pi/k^2) \sum (2l+1) \omega_l^{\pm} \sin^2(\eta_l^{(1)} - \eta_l^{(3)}), \tag{351}$$

where
$$\omega_l^{\pm} = 1 \pm (-1)^l. \tag{352}$$

Except for very small values of k

$$Q_{se}^+ \simeq Q_{se}^- \simeq Q_{se}. \tag{353}$$

Consider now an assembly of H atoms in which n_{11}, n_{10}, n_{00} denote the number of atoms per unit volume with $F = 1$, $M = \pm 1$; $F = 1$, $M = 0$, and $F = 0$, $M = 0$ respectively. We are interested in the rate at which atoms are changed from one hyperfine state to another. Let $\nu(F, M; F', M')$ be the rate at which atoms in the hyperfine state (F, M) are transformed to those in the state (F', M') in collisions in which the relative velocity of the atoms is v_r. Then it may be shown† that

$$\nu(0, 0; 1, 0) = \tfrac{1}{4} v_r n_{00} (4 n_{11} Q_{se}^- + n_{00} Q_{se}^+),$$
$$\nu(0, 0; 1, \pm 1) = \tfrac{1}{4} v_r n_{00} (n_{10} Q_{se}^- + n_{00} Q_{se}^+),$$
$$\nu(1, \pm 1; 0, 0) = \tfrac{1}{4} v_r n_{11} (n_{11} Q_{se}^- + n_{11} Q_{se}^+),$$
$$\nu(1, \pm 1; 1, 0) = \tfrac{1}{4} v_r n_{11} (n_{11} Q_{se}^- + n_{11} Q_{se}^+),$$
$$\nu(1, 0; 0, 0) = \tfrac{1}{4} v_r n_{10} (4 n_{11} Q_{se}^- + n_{10} Q_{se}^+),$$
$$\nu(1, 0; 1, \pm 1) = \tfrac{1}{4} v_r n_{10} (n_{00} Q_{se}^- + n_{10} Q_{se}^+). \tag{354}$$

In an assembly in which the atoms have a distribution of velocities, the product of v_r with a cross-section must be averaged over this distribution.

† SMITH, F. J., *Planet. Space Sci.* **14** (1966) 929.

In equilibrium $n_{00} = n_{10} = \tfrac{1}{2}n_{11} = \tfrac{1}{4}n$ and the rates $\nu(F, F')$ at which $F \to F'$ transitions are produced by collision are given by

$$\nu(0,1) = \nu(1,0) = \tfrac{3}{32}n^2 v_{\rm r}(Q_{\rm se}^- + \tfrac{1}{2}Q_{\rm se}^+). \tag{355}$$

If (353) applies this gives

$$\nu(0,1) = \nu(1,0) = \tfrac{9}{64}n^2 v_{\rm r} Q_{\rm se}. \tag{356}$$

According to (350) we have

$$\nu(1,0) = \tfrac{3}{16}n^2 v_{\rm r} Q_{\rm se} \tag{357}$$

which is 4/3 times larger. This is because (357) includes transitions in which one atom makes a $(1,0)$ transition and the other a $(0,1)$, leading to no net change of the hyperfine state populations.

In the hydrogen maser $n_{00} = 0$, $n_{10} = n_{11}$, and

$$\nu(1,0;0,0) + 2\nu(1,0;1,1) = \tfrac{1}{4}v_{\rm r} n_{10}(2n_{10}Q_{\rm se}^- + 2n_{10}Q_{\rm se}^+)$$
$$\simeq v_{\rm r} n_{10}^2 Q_{\rm se} \tag{358}$$

so that the full spin-exchange cross-section determines the effective spin relaxation time (see § 6.2.5).

Using the phase shifts calculated in the course of their work on the total and transport cross-section for collisions between hydrogen atoms (see Chap. 16, § 12.1.3) Fox and Gal† obtained the spin-exchange cross-sections $Q_{\rm se}^-$ and $Q_{\rm se}^+$ defined in (351) and (352). These are illustrated in Fig. 18.117 (a) as functions of the relative wave number k. Fig. 18.117 (b) shows the same cross-sections, averaged over a Maxwellian distribution of velocities of the colliding atoms, as functions of the temperature T °K. It will be seen that there is an appreciable difference between the two averaged cross-sections at room temperature but it is only below 10 °K that the two behave in markedly different ways. The spin exchange cross-section measured using the hydrogen maser by Hellwig at 300 °K (see § 6.2.5 and Table 18.24) agrees quite well with the mean of the two calculated cross-sections as may be seen from Fig. 18.117 (b).

Fig. 18.118 shows the rate coefficient

$$R = \tfrac{3}{8}\bar{v}_{\rm r}(\bar{Q}_{\rm se}^- + \tfrac{1}{2}\bar{Q}_{\rm se}^+) \tag{359}$$

for transitions from the $F = 1$ to $F = 0$ states in hydrogen gas as a function of temperature T °K. On the same figure, for comparison, the corresponding rate coefficients for excitation of the transition by electron impact, using the theory and results discussed in Chapter 8, § 3, are also given. Although these latter coefficients are considerably larger than those for atom impact, the population of H atoms in interstellar space relative to electrons is probably such that the contribution from the electrons is insignificant.

† Fox, J. W. and Gal, E., unpublished.

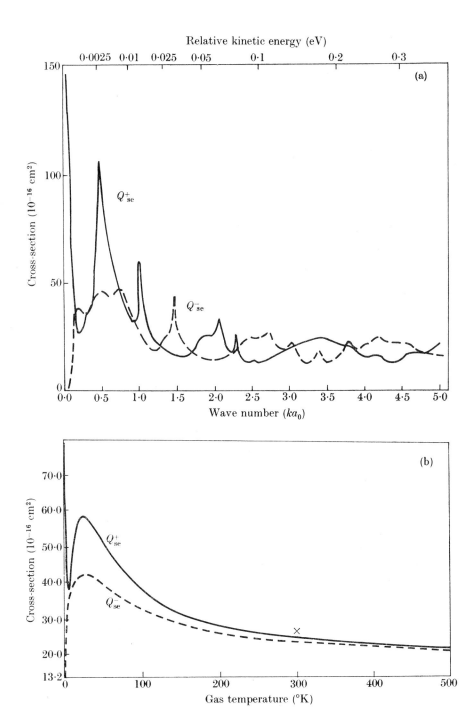

FIG. 18.117. Spin exchange cross-section Q_{se}^+, Q_{se}^- for collisions between two hydrogen atoms, calculated by Fox and Gal: (a) as functions of the relative wave number of the colliding atoms; (b) as functions of temperature after averaging over the Maxwellian velocity distribution of the colliding atoms. × observed by Hellwig.

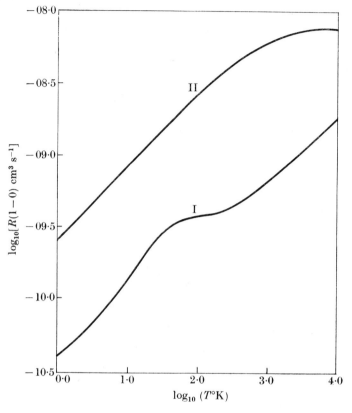

Fig. 18.118. Rate coefficients for the reactions: I, H+H ($F = 1$)→H+H ($F = 0$) calculated by Fox and Gal; II, e+H ($F = 1$)→e+H ($F = 0$) calculated using the cross-section tabulated in Chapter 8, § 3.

8.2.2. *Spin-exchange cross-sections for collisions between alkali-metal atoms.* According to the semi-classical approximation (337) the spin exchange cross-section for a collision between two one-electron atoms is given by

$$Q_{\text{se}} = 2\pi \int_0^\infty p \sin^2\phi(p) \, dp, \qquad (360)$$

where

$$\phi(p) = (1/\hbar v) \int_{-\infty}^\infty \Delta E\{(z^2+p^2)^{\frac{1}{2}}\} \, dz. \qquad (361)$$

Here

$$\Delta E(R) = V^{(1)}(R) - V^{(3)}(R), \qquad (362)$$

where $V^{(1)}$, $V^{(3)}$ are now the respective interactions of the atoms in the

$^1\Sigma_g$ and $^3\Sigma_u$ states of the pair. As usual (360) is evaluated approximately, as in (339), giving

$$Q_{se} = \tfrac{1}{2}\pi P^2 + 2\pi \int_P^\infty p\{\phi(p)\}^2 \, dp, \qquad (363)$$

where P is such that $\phi(P) = \tfrac{1}{4}\pi$.

In terms of this approximation it is only necessary to know $V^{(1)}$ and $V^{(3)}$ at large atomic separations R. Dalgarno and Rudge† obtained a suitable approximation for $\Delta E(R)$ by taking for the wave-function of an outer atomic electron at large distances from the nucleus the form (see Chap. 14, § 4.2)

$$\psi(r) = N(r/a_0)^{\gamma-1}\exp(-\alpha r/a_0), \qquad (364)$$

where $\alpha = (2E_i a_0/e^2)^{\frac{1}{2}} = \gamma^{-1}$, E_i being the ionization energy of the atom. The normalizing factor N is given by

$$\frac{(2\alpha)^\gamma}{\Gamma(\gamma+1)}\left\{\frac{\alpha}{4\pi\xi(\gamma)}\right\}^{\frac{1}{2}}, \qquad (365)$$

where ξ is not far from unity. Using this function, Dalgarno and Rudge find that, for the interaction of two atoms distinguished as 1 and 2,

$$\Delta E(R) \sim N_{12} R^\delta \exp(-\lambda R), \qquad (366)$$

where $\quad N_{12} = 2^{\gamma_1+\gamma_2}\alpha_1^{2\gamma_1+1}\alpha_2^{2\gamma_2+1}/(\alpha_1+\alpha_2)\Gamma(\gamma_1+1)\Gamma(\gamma_2+1)\Gamma(\gamma_1+\gamma_2+2),$

$$\delta = 2\gamma_1+2\gamma_2-1, \quad \lambda = \alpha_1+\alpha_2. \qquad (367)$$

We then have

$$\phi(p) \simeq (N_{12}/\hbar v)\int_p^\infty R^{\delta+1} e^{-\lambda R}(R^2-p^2)^{-\frac{1}{2}} \, dR$$

$$= (N_{12}/\hbar v)p^{\delta+1}\int_0^\infty \cosh^{\delta+1}\!\psi \exp(-\lambda p\cosh\psi) \, d\psi. \qquad (368)$$

When $\lambda p \gg 1$

$$\int_0^\infty \cosh^{\delta+1}\!\psi \exp(-\lambda p\cosh\psi) \, d\psi \simeq \int_0^\infty \exp(-\lambda p\cosh\psi) \, d\psi$$

$$= \frac{\pi}{2} K_0(\lambda p)$$

$$\sim (\pi/2\lambda p)^{\frac{1}{2}} e^{-\lambda p}, \qquad (369)$$

so $\qquad \phi(p) \simeq (N_{12}/\hbar v)(\pi/2\lambda)^{\frac{1}{2}} p^{\delta+\frac{1}{2}} e^{-\lambda p}, \qquad (370)$

and hence P is given by

$$P^{\delta+\frac{1}{2}} e^{-\lambda P}(N_{12}/\hbar v)(\pi/2\lambda)^{\frac{1}{2}} = \tfrac{1}{4}\pi. \qquad (371)$$

Also $\qquad \int_P^\infty p\{\phi(p)\}^2 \, dp \simeq (N_{12}/\hbar v)^2 (\pi/4\lambda^2) P^{2\delta+2} e^{-2\lambda P}$

$$= \pi^2 P/32\lambda, \qquad (372)$$

so, substituting in (363), $\qquad Q_{se} = \tfrac{1}{2}\pi P^2 + \pi^3 P/16\lambda. \qquad (373)$

† DALGARNO, A. and RUDGE, M. R. H., *Proc. R. Soc.* A286 (1965) 519.

From (371) we may write, in terms of a slowly varying quantity $\tilde{P}$,
$$P \simeq - (1/2\lambda)\ln\{\pi\lambda\hbar^2 v^2/8N_{12}^2 \tilde{P}^{2\delta+1}\} \qquad (374)$$
giving, from (373),
$$Q_{se} = (a - b\ln E)^2, \qquad (375)$$
where a and b are constants determined by α_1 and α_2 and E is the energy of relative motion. This formula will be valid for E small, but not so small that the semi-classical approximation is invalid.

Dalgarno and Rudge calculated a and b for collisions between one-electron atoms including hydrogen. Comparison with accurate calculations for H–H collisions showed that the simple formula underestimates the cross-section by about 30 per cent. The calculated values for collisions in an assembly at 540 °K are given in Table 18.29.

TABLE 18.29

Calculated spin-exchange cross-sections at 540 °K, *in* 10^{-14} cm^2

Atoms	Li	Na	K	Rb	Cs
H	0·5	0·5	0·6	0·6	0·6
Li	0·9	1·0	1·1	1·1	1·2
Na	—	1·1	1·2	1·3	1·4
K	—	—	1·5	1·5	1·6
Rb	—	—	—	1·6	1·8
Cs	—	—	—	—	1·9

Observed values for Rb–Rb, Rb–Cs, Na–Na, Na–K, and Na–Rb collisions are given in Table 18.24 on p. 1865. For Rb–Rb collisions, which have been the most thoroughly investigated, the observed values range from 1·5 to 2·6 × 10^{-14} cm^2 though most are close to 2 × 10^{-14} cm^2, which agrees very well with the calculated value. The two observed values for Rb–Cs, 2·3 and 2·4 × 10^{-14} cm^2, are also quite close to that calculated. Observed values for Na collisions are less reliable and are somewhat larger than the calculated, particularly for Na–K. The fact that good agreement is obtained for collisions involving the two most complex atoms, for which the theory is least accurate, suggests that further experimental investigations for the lighter atoms will improve the agreement with theory in these cases also.

8.2.3. *Spin-exchange collisions between* H *and* O *atoms.* The ground term of an O atom is a ^3P term. Transitions between the fine-structure levels ^3P_J can be excited through spin-exchange collisions with H atoms. Such collisions can be important in cooling interstellar gas.

If spin-orbit coupling is ignored an O and H atom in their ground states can interact in four different ways corresponding to $^2\Sigma^-$, $^4\Sigma^-$,

$^2\Pi$, and $^4\Pi$ states of the OH molecule (see Chap. 12, p. 812). Smith†
has shown that the cross-sections $Q(J, J')$ for excitation of the transition $^3P_J \to {}^3P_{J'}$ by H atom impact, assumed to be elastic, are given by

$$Q(J \to J') = (\pi/k^2) \sum (2l+1)\left\{\sum_{i,j} a_{ij}(J \to J')\sin^2(\eta_l^i - \eta_l^j)\right\}, \quad (376)$$

where i, j refer to the respective $^2\Sigma$, $^4\Sigma$, $^2\Pi$, and $^4\Pi$ states. The $a_{ij}(J \to J')$ are given in Table 18.30.

TABLE 18.30

The coefficients $a_{ij}(J \to J')$ in the formula (376) for the cross-section $Q(J \to J')$

i \ j	$^4\Pi$			$^2\Pi$			$^4\Sigma$			$^2\Sigma$		
	2,1	2,0	1,0	2,1	2,0	1,0	2,1	2,0	1,0	2,1	2,0	1,0
$^4\Pi$	—	—	—	4/45	−4/405	4/27	28/135	44/405	−4/3	8/135	4/405	4/81
$^2\Pi$	—	—	—	—	—	—	8/135	4/405	4/3	2/27	4/81	−4/3
$^4\Sigma$	—	—	—	—	—	—	—	—	—	2/135	−4/405	4/3

$Q(1, 2) = \tfrac{5}{3}Q(2, 1)$; $Q(0, 2) = 5Q(2, 0)$; $Q(0, 1) = 3Q(1, 0)$.

Little is known about the O–H interactions. The $^2\Pi$ interaction gives rise to the ground state of OH and it is known quite well, from spectroscopic observations, about the minimum between $1\cdot 3$ and $3\cdot 3a_0$. All the other interactions are repulsive. For estimating the spin-exchange cross-section Smith first calculated the van der Waals constant C from the Slater–Kirkwood formula (Chap. 16, (218)) as $11\cdot 5$ a.u. and, after verifying that the results were not very sensitive to the value of C, used this value throughout to give the asymptotic form of each interaction. For $^2\Pi$ he assumed this to be valid for separations $> 5a_0$ and for smaller separations extrapolated by a polynomial expansion to join with the spectroscopically determined potential for $R < 3\cdot 5a_0$. Estimates of the other interactions between $3\cdot 5a_0$ and $4\cdot 5a_0$ have been made by Mulliken.‡ These were fitted by simple exponentials and assumed to be valid for $R < 5a_0$. For $R > 5a_0$ the potential was taken, in each case, in the form

$$V(R) = -CR^{-6} + Be^{-\beta r/a_0}. \quad (377)$$

The two parameters could be adjusted to give continuity of $V(R)$ at $R = 5a_0$ and any chosen depth of the weak potential minimum due to the van der Waals attraction. Fig. 18.119 illustrates results for $Q(2, 1)$ obtained for two sets of assumptions, while Fig. 18.120 gives results

† Smith, F. J., *Planet. Space Sci.* **14** (1966) 937.
‡ Mulliken, R. S., *Rev. mod. Phys.* **4** (1932) 1.

for all cross-sections using the assumption $\beta = 2$ for all the repulsive potentials, which gives the deepest minima for the repulsive states. The results in Fig. 18.120 are also partly corrected for inelastic effects by multiplying the cross-sections given by (376) by k'/k, where k, k' are the initial and final wave numbers of relative motion, respectively.

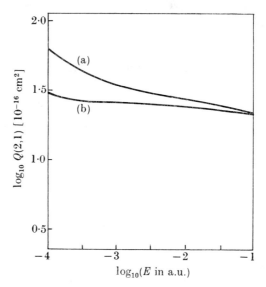

Fig. 18.119. Cross-sections for the spin change $J = 2 \to 1$ in collisions of O with H atoms calculated by Smith (F. J.) assuming an interaction of the form (377) in all repulsive states. (a) $\beta = 2\cdot 0$ for all three states. (b) $\beta = 1\cdot 5$, $2\cdot 0$, and $3\cdot 0$ for $^2\Sigma^-$, $^4\Sigma^-$, and $^4\Pi$ respectively.

8.3. Collisions in which $\Delta E = 0$—collisions producing depolarization

We now consider depolarizing collisions which produce changes in the orientation and alignment of an assembly of atoms in states with total electron-spin angular-momentum quantum number $J = 1$.† The cross-sections Q_{or}, Q_{al} for disorientation and disalignment have been defined in § 3.1 in terms of the spin-density matrix for the assembly.

Two types of collision must be considered. In both cases these will involve a ground-state atom which may or may not be of the same kind as the target excited atom. If it is of a different kind we may ignore the chance of deactivation of the target atom and consider only the change of the spin density matrix of that atom. On the other hand, when the atoms are of the same kind, excitation transfer may occur as

† We follow largely the analysis of OMONT, A., J. Phys., Paris, **26** (1965) 26.

a resonant process. Collisions in which such transfer occurs will in general lead to a change in orientation and alignment of the assembly of excited atoms and this must be taken into account. Furthermore, when the colliding atoms are of the same kind, the mean effective interaction is of much larger range and this affects the magnitude of the cross-sections and their variation with velocity.

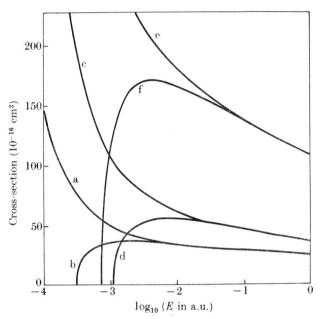

FIG. 18.120. Cross-sections for spin exchange collisions between O and H atoms calculated by Smith (F. J.). a, $Q(0, 1)$; b, $3Q(1, 0)$; c, $Q(0, 2)$; d, $5Q(2, 0)$; e, $3Q(1, 2)$; f, $5Q(2, 1)$.

As the relative motion of the two atoms will be nearly classical and rectilinear under the usual experimental conditions we may write, for the cross-sections for disorientation and disalignment respectively,

$$Q_{\text{or,al}} = 2\pi \int p P_{1,2}(p) \, dp, \qquad (378)$$

where $P_{1,2}(p)$ is the probability factor defined on p. 1699. To calculate $P_{1,2}(p)$ we may, as usual, suppose the motion of the incident atom relative to the target atom to occur with constant velocity v in a straight line passing the target at a distance p. If the interaction energy between the atoms when at separation $\mathbf{R}$ is $V(\mathbf{R})$ we may write for the time variation of the interaction between them during a collision in which t changes from $-\infty$ to ∞

$$V(t) = V(\mathbf{p} + \mathbf{v}t). \qquad (379)$$

8.3.1. *Collisions between atoms of the same electronic structure.* We consider first collisions in which the two atoms have the same electronic structure. They will still be distinguishable experimentally if they are isotopes of the same element. We shall assume that they are distinguishable but there is no difficulty in applying the analysis to the case when they are identical. It will also be assumed at first that the atoms possess no nuclear spin.

Ignoring all states except those in which one atom is in the particular excited state with $J = 1$ and the other in the ground state, we may write the wave function in the form

$$\sum_{j=1}^{3} \{a_j(t)\phi_j^{(1)}\phi_0^{(2)} + b_j(t)\phi_j^{(2)}\phi_0^{(1)}\}\exp\{-i(E_1+E_0)t/\hbar\}, \quad (380)$$

where E_0, E_1 are the energies of the ground and excited states respectively. The usual time-dependent perturbation treatment (Chap. 7, § 5.3.1) gives

$$\dot{a}_j(t) = -(i/\hbar)\sum_k \{a_k V_{jk}^{11} + b_k V_{jk}^{12}\}, \quad (381\,\text{a})$$

$$\dot{b}_j(t) = -(i/\hbar)\sum_k \{a_k V_{jk}^{21} + b_k V_{jk}^{22}\}, \quad (381\,\text{b})$$

where

$$V_{jk}^{11} = \int \phi_j^{(1)} V(t)\phi_k^{(1)*}\, d\tau = V_{jk}^{22}, \quad (382\,\text{a})$$

$$V_{jk}^{12} = \int \phi_j^{(1)} V(t)\phi_k^{(2)*}\, d\tau = V_{jk}^{21*}, \quad (382\,\text{b})$$

the integration being over the internal coordinates of the two atoms. The initial conditions are such that

$$b_j(-\infty) = 0, \qquad a_j(-\infty) = a_{j0}. \quad (383)$$

For $V(\mathbf{R})$ we take the long-range induced dipole-dipole interaction

$$V(\mathbf{R}) = -e^2 R^{-3}\{3\mathbf{d}_1.\hat{\mathbf{R}}\mathbf{d}_2.\hat{\mathbf{R}} - \mathbf{d}_1.\mathbf{d}_2\}, \quad (384)$$

where $\mathbf{d}_1$, $\mathbf{d}_2$ are given in terms of the coordinates of the respective atomic electrons by

$$\mathbf{d}_1 = -\sum_i \mathbf{r}_{1i}, \qquad \mathbf{d}_2 = -\sum_k \mathbf{r}_{2k}. \quad (385)$$

With this form for $V(\mathbf{R})$, $V_{jk}^{11} = V_{jk}^{22} = 0$ whether the atoms are like or unlike. V_{jk}^{12} and V_{jk}^{21} also do not arise if they are unlike but remain finite if they are like, for which case (381) may be written conveniently in matrix form as

$$i\hbar\dot{\Psi}_1 = \mathscr{V}(t)\Psi_2, \quad (386\,\text{a})$$

$$i\hbar\dot{\Psi}_2 = \mathscr{V}(t)\Psi_1, \quad (386\,\text{b})$$

where Ψ_1, Ψ_2 are the column matrices formed from (a_1, a_2, a_3), (b_1, b_2, b_3) respectively. After some calculation $\mathscr{V}(t)$ is found to be given by

$$\mathscr{V} = (9\lambda^3 A_{10} h/64\pi^4)(J_R^2 - \tfrac{1}{3}\mathbf{J}^2)R^{-3}, \quad (387)$$

where A_{10} is the radiative transition probability from the excited to the ground state, λ the wavelength of the radiation emitted in such a transition, and $\hbar \mathbf{J}_R$ is the operator representing the component of the spin resolved along the direction of the interatomic separation $\mathbf{R}$.

The solution of the matrix equations (386) can be written in the form

$$\mathbf{\Psi}_1(\infty) = \mathbf{S}_1 \mathbf{\Psi}_1(-\infty), \qquad (388\,\text{a})$$

$$\mathbf{\Psi}_2(\infty) = \mathbf{S}_2 \mathbf{\Psi}_1(-\infty), \qquad (388\,\text{b})$$

since we take $\mathbf{\Psi}_2(-\infty) = 0$. The density matrix after the collision is now given by the sum of the two density matrices (see (52))

$$\mathbf{\Psi}_1(\infty)\mathbf{\Psi}_1^\dagger(\infty) + \mathbf{\Psi}_2(\infty)\mathbf{\Psi}_2^\dagger(\infty)$$

$$= \mathbf{S}_1 \boldsymbol{\rho}_1(-\infty)\mathbf{S}_1^\dagger + \mathbf{S}_2 \boldsymbol{\rho}_1(-\infty)\mathbf{S}_2^\dagger \qquad (389)$$

$$= \mathbf{Y}_1 \boldsymbol{\rho}_1 + \mathbf{Y}_2 \boldsymbol{\rho}_2, \qquad (390)$$

say, where $\boldsymbol{\rho}_i = \mathbf{\Psi}_i \mathbf{\Psi}_i^\dagger$. Then, as in (71), the probabilities P_1 and P_2 of destruction of orientation and of alignment respectively are given by

$$\overline{\mathbf{Y}} \boldsymbol{\Omega}_{qu} = (1 - P_q)\boldsymbol{\Omega}_{qu}, \qquad (391)$$

where $\overline{\mathbf{Y}}$ is the average of $\mathbf{Y}$ over all directions of relative velocity and all orientations of the plane of collision.

As in (390) we may write

$$\overline{\mathbf{Y}}_1 \boldsymbol{\Omega}_{qu} = (1 - P_q^{(1)})\boldsymbol{\Omega}_{qu}, \qquad (392\,\text{a})$$

$$\overline{\mathbf{Y}}_2 \boldsymbol{\Omega}_{qu} = -P_q^{(2)}\boldsymbol{\Omega}_{qu}, \qquad (392\,\text{b})$$

where $\overline{\mathbf{Y}}_1, \overline{\mathbf{Y}}_2$ are again averages over all directions of relative velocity and all orientations of the plane of collision.

$P_1^{(1)}$ and $P_2^{(1)}$ are, as on p. 1699, the probabilities of destruction of orientation and of alignment respectively of the struck atoms. $-P_1^{(2)}$ and $-P_2^{(2)}$ are the probabilities that the orientation and alignment respectively are transferred to the colliding atoms. Thus $P_1^{(1)}+P_1^{(2)}$, $P_2^{(1)}+P_2^{(2)}$, are the respective net probabilities of destruction if the atoms 1 and 2 are indistinguishable. If they are different isotopes $P_i^{(1)}$ and $P_i^{(2)}$ can, in principle, be observed separately.

$P_0^{(1)}$ is the probability of loss of excitation by the first atom through transfer to the second, while $-P_0^{(2)}$ is the probability that it is gained by that atom. Thus, ignoring any chance of quenching to the ground state we must have

$$P_0^{(1)} + P_0^{(2)} = 0. \qquad (393)$$

To calculate the P_q we may employ a very similar technique to that used in Chapter 16, § 5.4.1, Chapter 17, § 5.6, and §§ 8.1.2 and 8.2.2 of this chapter. We first solve the equations (386) approximately when the impact parameter is so large that the P_q are small. Suppose that, as a result, we obtain solutions $P_q^\infty(p)$. As p decreases $P_q^\infty(p)$ will increase until when $p = p_0$, $P_q^\infty(p_0) = 1$. For $p < p_0$ the interaction is so strong that we can suppose, as a good approximation, that all orientation and alignment is destroyed. We therefore take

$$P_q = P_q^\infty(p) \quad (p > p_0),$$
$$= 1 \quad (p < p_0), \tag{394}$$

giving
$$Q_q = 2\pi \int_{p_0}^{\infty} P_q^\infty(p) p \, dp + \pi p_0^2. \tag{395}$$

For transfer of excitation, on the other hand, we must expect that, under conditions of strong interaction, there is an equal chance of the excitation ending up on either atom after the collision. Hence we take

$$P_0 = P_0^{(1)\infty}(p) \quad (p > p_1),$$
$$= \tfrac{1}{2} \quad (p < p_1), \tag{396}$$

giving
$$Q_{\mathrm{tr}} = 2\pi \int_{p_1}^{\infty} P_0^{(1)\infty}(p) \, dp + \tfrac{1}{2}\pi p_1^2. \tag{397}$$

p_1 may be chosen so that $P_0^{(1)\infty}(p_1) = \tfrac{1}{2}$.

Introducing the dimensionless variables

$$x = vt/p, \qquad r^2 = 1 + x^2, \qquad s = 9\lambda^3 A_{10}/32\pi^3 p^2 v, \tag{398}$$

(386) become

$$\frac{d\mathbf{\Psi}_1}{dx} = -is(1+x^2)^{-\frac{3}{2}}(\mathbf{J}_R^2 - \tfrac{1}{3}\mathbf{J}^2)\mathbf{\Psi}_2, \tag{399 a}$$

$$\frac{d\mathbf{\Psi}_2}{dx} = -is(1+x^2)^{-\frac{3}{2}}(\mathbf{J}_R^2 - \tfrac{1}{3}\mathbf{J}^2)\mathbf{\Psi}_1. \tag{399 b}$$

Choosing a reference system of axes $Ox'y'z'$ as shown in Fig. 18.121

$$\mathbf{J}_R^2 = r^{-2}(\mathbf{J}_{x'} + x\mathbf{J}_{y'})^2, \tag{400}$$

so that we may write (399) in the form

$$\frac{d\mathbf{\Psi}_1}{dx} = -is(1+x^2)^{-\frac{3}{2}}\mathbf{N}\mathbf{\Psi}_2, \tag{401 a}$$

$$\frac{d\mathbf{\Psi}_2}{dx} = -is(1+x^2)^{-\frac{3}{2}}\mathbf{N}\mathbf{\Psi}_1, \tag{401 b}$$

where
$$\mathbf{N} = \begin{pmatrix} -\tfrac{1}{6} & 0 & \tfrac{1}{2}r^{-2}(1-ix)^2 \\ 0 & \tfrac{1}{3} & 0 \\ \tfrac{1}{2}r^{-2}(1+ix)^2 & 0 & -\tfrac{1}{6} \end{pmatrix}. \tag{402}$$

For small s we have, expanding the solutions in powers of s

$$\mathbf{S}_2^\infty = -is \int_{-\infty}^{\infty} (1+x^2)^{-3/2} \mathbf{N}(x)\, dx$$
$$= -\tfrac{2}{3} is (\mathbf{J}_{x'}^2 - \mathbf{J}_{z'}^2), \tag{403 a}$$

$$\mathbf{S}_1^\infty = \mathbf{I} - s^2 \int_{-\infty}^{\infty} (1+x^2)^{-3/2} \mathbf{N}(x) \int_{-\infty}^{x} (1+x'^2)^{-3/2} \mathbf{N}(x')\, dx'\, dx$$
$$= \mathbf{I} - s^2 \left\{ \tfrac{2}{3} \mathbf{J}_{y'}^2 - \frac{\pi}{12} i \mathbf{J}_{z'} \right\}. \tag{403 b}$$

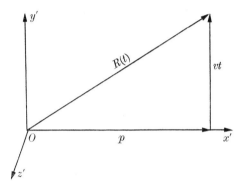

Fig. 18.121

We must now average over all orientations of the axis $Ox'y'z'$. This may be carried out without difficulty for $\mathbf{S}_1^\infty \rho \mathbf{S}_1^\infty$ since

$$\mathbf{S}_1^\infty \rho \mathbf{S}_1^\infty = (1 - \Delta \mathbf{S}_1^\infty) \rho (1 - \Delta \mathbf{S}_1^{\infty \dagger}),$$

where
$$\Delta \mathbf{S}_1^\infty = s^2 \left(\tfrac{2}{9} \mathbf{J}_{y'}^2 - \frac{\pi}{12} i \mathbf{J}_{z'} \right).$$

Hence, to first order in s^2,

$$\overline{\mathbf{S}_1^\infty \rho \mathbf{S}_1^\infty} = \rho - \rho \overline{\Delta \mathbf{S}_1^\infty} - \overline{\Delta \mathbf{S}_1^\infty} \rho$$
$$= \rho - \tfrac{8}{27} s^2 \rho, \tag{404}$$

so
$$P_0^{(1)\infty} = P_1^{(1)\infty} = P_2^{(1)\infty} = \tfrac{8}{27} s^2. \tag{405}$$

Considerably more calculation is involved in obtaining the mean of $\mathbf{S}_2^\infty \rho \mathbf{S}_2^\infty$. The most convenient method is to use the algebra of irreducible tensorial sets according to which

$$P_q^{(i)} = \delta_{i1} - \sum_{q'u'} (-1)^{q+q'} \begin{Bmatrix} 1 & 1 & q \\ 1 & 1 & q' \end{Bmatrix} |\mathrm{Tr}\{\mathbf{S}_i^\dagger \boldsymbol{\Omega}_{q'u'}\}|^2, \tag{406}$$

where $\{\}$ is a Wigner $6j$ symbol. This gives

$$P_1^{(2)\infty} = -5 P_2^{(2)\infty} = \tfrac{4}{27} s^2. \tag{407}$$

Using the simple prescription (396) we then have

$$Q_{\mathrm{or}} = 1\cdot 29 Q_{\mathrm{al}} = 1\cdot 73 Q_{\mathrm{tr}} = 0\cdot 038 A_{10} \lambda^3/v. \tag{408}$$

It will be noted that the value obtained for Q_{tr}, which allows for the dependence of the interaction on orientation, is not very different from that obtained in § 8.1.2, p. 1875, neglecting this dependence. Thus according to (341) $Q_{\text{tr}} = 6\pi^2\mu^2/hv$ where μ, the dipole moment associated with the allowed optical transition is given by $3A_{10}\lambda^3/64\pi^4$. Hence Q_{tr} neglecting orientation is $(9/32\pi^2)A_{10}\lambda^3/v = 0{\cdot}028A_{10}\lambda^3/v$ as compared with $0{\cdot}022A_{10}\lambda^3/v$ given by (408).

Hitherto we have used approximations of the type (396) for determining to a good accuracy the magnitude of the cross-section. We cannot be sure that it will give the ratio $Q_{\text{or}}/Q_{\text{al}}$ to the accuracy attainable in observation. A good deal of attention has therefore been paid to examining the accuracy of approximate methods of determining the ratio $Q_{\text{or}}/Q_{\text{al}}$.

In terms of the quantity s defined in (398) the cross-sections are given by

$$Q_{\text{or,al}} = (9\lambda^3 A_{10}/32\pi^3 v)\int_0^\infty (P_{1,2}^{(1)}+P_{1,2}^{(2)})s^{-2}\,ds, \qquad (409)$$

$$Q_{\text{tr}} = (9\lambda^3 A_{10}/32\pi^3 v)\int_0^\infty P_0^{(1)} s^{-1}\,ds. \qquad (410)$$

If $P_{0,1,2}(s)$ are known as functions of s, $Q_{\text{or,al}}$ and Q_{tr} may be calculated for any value of λ, A_{10}, and v without difficulty. The $P(s)$ may be obtained by numerical solutions of equations (401), which involve only the one parameter s, provided the appropriate averaging over all orientations of the collision axis is carried out.

$P_{1,2}(s)$ have been calculated by D'yakonov and Perel[†] and by Omont and Meunier[‡] in this way. The latter authors, as well as Watanabe,[§] have also calculated $P_0(s)$. Small differences between their results are introduced by the way in which contributions from large values of s (close collisions) are included. Numerical integration of equations (401) becomes prohibitively difficult for $s > 10$ or so. Assuming that, for such values of s, orientation and alignment are completely destroyed, Omont and Meunier obtain

$$Q_{\text{or}} = 1{\cdot}16 Q_{\text{al}} = 1{\cdot}58 Q_{\text{tr}} = 0{\cdot}036 A_{10}\lambda^3/v. \qquad (411)$$

D'yakonov and Perel give

$$Q_{\text{or}} = 1{\cdot}25 Q_{\text{al}} = 0{\cdot}038 A_{10}\lambda^3/v. \qquad (412)$$

Watanabe's value for Q_{tr} agrees with that of Omont and Meunier.

Comparison with (408) shows that the results given by the simple

---

[†] D'YAKONOV, M. L. and PEREL, V. I., *Zh. eksp. teor. Fiz.* **48** (1965) 345; *Soviet Phys. JETP*, **21** (1965) 227.
[‡] OMONT, A. and MEUNIER, J., *Phys. Rev.* **169** (1968) 92.
[§] WATANABE, T., ibid. **138** (1965) A1573; **140** (1965) AB5.

approximation are remarkably good. Figs. 18.122 illustrate the variation of the $P_i^{(1)}$ and $P_i^{(2)}$ with s^{-1}, calculated by Omont and Meunier by accurate numerical solution of (401). In the same figures the forms assumed in the simple approximation are given for comparison. It will be clear from this comparison why the simple approximation gives such good results.

A different approximation has also been used extensively and, although it gives less accurate results than the simple approximation, it is worth briefly discussing here because of its applicability in other circumstances (see §§ 8.4.2 and 8.4.5). Returning to the matrix equations (386) we note that, if the matrix functions Ψ_1, Ψ_2, $\mathscr{V}$ were scalars, the solution would simply be

$$\Psi_1(t) = \cos\left\{\int_{-\infty}^{t} \mathscr{V}(t')\,dt'\right\}\Psi_1(-\infty), \tag{413}$$

$$\Psi_2(t) = -i\sin\left\{\int_{-\infty}^{t} \mathscr{V}(t')\,dt'\right\}\Psi_1(-\infty). \tag{414}$$

It may be shown that if the matrices $\mathscr{V}(t')$, $\mathscr{V}(t'')$ at different times t', t'' commute with each other (413) and (414) are indeed the solutions of the matrix equations. In our case this condition is not satisfied but it is assumed nevertheless that (413) and (414) give good approximations to the solution.

Certain problems of interpretation still remain. However, in general, if **U** is a matrix that can be diagonalized so that

$$\mathbf{D}^{-1}\mathbf{U}\mathbf{D} = \mathbf{W}, \tag{415}$$

where **W** is a diagonal matrix, then

$$\exp\mathbf{U} = \exp(\mathbf{D}\mathbf{W}\mathbf{D}^{-1})$$
$$= \mathbf{1} + \mathbf{D}\mathbf{W}\mathbf{D}^{-1} + \tfrac{1}{2}\mathbf{D}\mathbf{W}\mathbf{D}^{-1}\mathbf{D}\mathbf{W}\mathbf{D}^{-1} + \dots$$
$$= \mathbf{D}(\exp\mathbf{W})\mathbf{D}^{-1}, \tag{416}$$

where $\exp\mathbf{W}$ is the diagonal matrix with elements $\exp(W_{ii})$.

This approximation agrees with (405) and (407) for large p (small s) and has the advantage for small p of always giving results in agreement with conservation requirements. Using it for all s in (409) and (410) one obtains

$$Q_{\text{or}} = 1 \cdot 66 Q_{\text{al}} = 2 Q_{\text{tr}} = 0 \cdot 043 A_{10} \lambda^3/v. \tag{417}$$

Comparison with (411) or (412) shows that, while (417) gives quite good results for Q_{tr}, it is considerably less satisfactory than the simple approximation (408) for Q_{or}, Q_{al} and the ratio $Q_{\text{or}}/Q_{\text{al}}$.

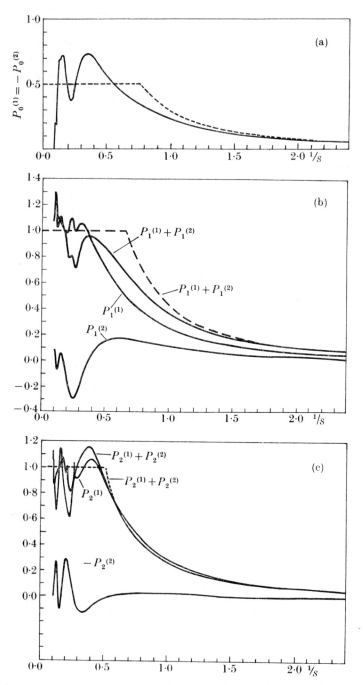

Fig. 18.122. Probabilities P_0, P_1, P_2 for excitation transfer, disorientation, and disalignment for collisions between atoms of similar electronic structure as functions of the parameter s^{-1} (see (398)). —— calculated by numerical solution of the coupled equations (399). --- according to the simple approximation (397). (a) $P_0^{(1)} = -P_0^{(2)}$; (b) $P_1^{(1)}$, $P_1^{(2)}$, $P_1^{(1)}+P_1^{(2)}$; (c) $P_2^{(1)}$, $P_2^{(2)}$, $P_2^{(1)}+P_2^{(2)}$.

8.3.2. *Collisions between atoms with different electronic structure.* If the atoms 1 and 2 have different electronic structure the allowed energy values $E_0^{(1)}, E_1^{(1)},...$ for atom 1 will differ from those $E_0^{(2)}, E_1^{(2)},...$ for atom 2 so that, corresponding to the wave function (380) for collisions of atoms 2 in the ground state with atoms 1 in one of the excited states with energy $E_1^{(1)}$, we would have just

$$\sum_{j=1}^{3} \{a_j(t)\phi_j^{(1)}\phi_0^{(2)}\}\exp\{-i(E_1^{(1)}+E_0^{(2)})t/\hbar\}. \quad (418)$$

As, however, in the notation of (382) $V_{jk}^{11} = 0$ with the assumed form (384) for V, we would merely obtain a vanishing transition probability for a depolarizing collision. To obtain a finite result† we must include terms in (418) which correspond to states in which the total energy of the two atoms differs from $E_1^{(1)}+E_0^{(2)}$. Virtual transitions between the initial state and these other states during the collision will produce an additional effective interaction. In general, the energy $\mathscr{E}$ of these states will differ from $E_1^{(1)}+E_0^{(2)}$ by an amount ΔE such that

$$\Delta E/\hbar \gg 1/\tau, \quad (419)$$

where τ is the duration of a collision. This means that many virtual transitions take place during the collision so that the effective interaction at time t due to these virtual transitions can be calculated as if the atoms were at rest, at that time, at the appropriate separation. The matrix element V_{jk} of this interaction is then given, to second order in V, by perturbation theory as

$$V_{jk} = \sum_{p,q} \langle\phi_j^{(1)}\phi_0^{(2)}|V|\phi_p^{(1)}\phi_q^{(2)}\rangle\langle\phi_p^{(1)}\phi_q^{(2)}|V|\phi_k^{(1)}\phi_0^{(2)}\rangle \times$$
$$\times (E_p^{(1)}+E_q^{(2)}-E_1^{(1)}-E_0^{(2)})^{-1}, \quad (420)$$

where $\phi_p^{(1)}, \phi_q^{(2)}$ are excited states of atoms 1 and 2 of energies E_p, E_q respectively. If we replace the denominators $(E_p^{(1)}+E_q^{(2)}-E_1^{(1)}-E_0^{(2)})$ by a mean value $\overline{\Delta E}$ then

$$V_{jk} = \langle\phi_j^{(1)}\phi_0^{(2)}|V^2|\phi_k^{(1)}\phi_0^{(2)}\rangle/\overline{\Delta E}. \quad (421)$$

In place of (386) we now have the single matrix equation

$$i\hbar\dot{\Psi}_1 = \mathscr{W}(t)\Psi_1, \quad (422)$$

where
$$W_{jk} = (V^2)_{jk}/\overline{\Delta E}. \quad (423)$$

After somewhat tedious analysis it is found that

$$\mathscr{W} = (4e^4 a_0^4/\overline{\Delta E})(\beta_1 \mathbf{I}+\beta_2 \mathbf{J}_R^2)R^{-6}, \quad (424)$$

where β_1 and β_2 involve averages of $\left(\sum_i \mathbf{r}_{1i}\right)^2$ (in the notation of (385)).

† BYRON, F. W. and FOLEY, H. M., *Phys. Rev.* **134** (1964) A625.

We then find in place of (399)†

$$\frac{d\Psi_1}{dx} = -is'(1+x^2)^{-3}(\beta_1 \mathbf{I} + \beta_2 \mathbf{J}_R^2)\Psi_1, \qquad (425)$$

where
$$s' = 4e^4 a_0^4/\hbar v \Delta E p^5. \qquad (426)$$

By carrying through an analysis similar to that for the like atom case on pp. 1896–8 it is found that, for large p,

$$P_1^\infty = \tfrac{5}{3} P_2^\infty = 7\delta^2/p^{10}, \qquad (427)$$

where
$$\delta = \frac{\pi}{4} \frac{\beta_2 e^4 a_0^4}{\hbar v \overline{\Delta E}}. \qquad (428)$$

With the simple approximation (396) this gives

$$Q_{\text{or}} = (\tfrac{5}{3})^{\frac{1}{5}} Q_{\text{al}} = 1\cdot 11 Q_{\text{al}} = \frac{5\pi}{4}(7\delta^2)^{\frac{1}{5}} = 3\cdot 0(4\beta_2 e^4 a_0^4/\hbar v \overline{\Delta E})^{\frac{2}{5}}. \qquad (429)$$

It is to be remembered that we have assumed that the effective basic interaction is the long range dipole–dipole form (384). For collisions with the lighter rare gas atoms, particularly helium and neon, an important, even dominating, contribution may come from the short range repulsion. While bearing this in mind we shall continue to assume that (384) is alone important.

8.3.3. *Numerical values and comparison with observation*

8.3.3.1. *Self-depolarization of $^3P_1$ levels.* We consider first depolarization due to collisions with atoms of the same atomic structure. For such atoms, contributions to the depolarization cross-section will come both from the first- and second-order interaction times. In Table 18.31 we give some numerical values for the disalignment cross-sections calculated by the simple approximation for the $^3P_1$ levels of Hg, Cd, Zn, and Pb calculated on the separate assumptions that the first- and second-order interactions are alone important. The values assumed for $\overline{\Delta E}$ in the calculation of cross-sections due to the second-order interaction have been taken to be the same, $\simeq 10$ eV, for all three cases. They are as estimated by Byron and Foley‡ from examination of the level diagrams for the states of each atom, taking into account the oscillator strengths of the main transitions.

It will be seen that, for Hg and Pb, for which the transition probability from the excited state is relatively high (the f-value for absorption

---

† OMONT, A., loc. cit., p. 1892. ‡ loc. cit., p. 1901.

TABLE 18.31

Comparison of observed and calculated disorientation and disalignment cross-sections (in 10^{-14} cm^2) for self-depolarization of $^3P_1$ levels of various atoms

Atom	Temp. (°K)	Lifetime of state (10^{-7} s)	Q_{al}			Q_{al}/Q_{or}		
			Resonance	Non-resonance	Observed	Resonance	Non-resonance	Observed
Hg	350	1·18	15·8	4·0	—	1·29	1·11	—
Cd		23·9	0·9	2·1	2·5	1·29	1·11	—
Zn		3200	0·044	1·3	1·4	1·29	1·11	—
Pb		21	62	—	76	1·29	—	1·21

is as high as 0·17 for Pb as compared with 0·002 for Cd), the contribution from the resonant interaction dominates. Zinc is at the opposite extreme while cadmium is intermediate. Observed cross-sections for Pb and Zn (see pp. 1722–3) agree rather better with the calculated value for the dominant contribution than would be expected in view of the simplicity of the assumed wave functions and of the value of $\overline{\Delta E}$.

The agreement is also surprisingly good for the ratio Q_{al}/Q_{or} for Pb. No observations of this ratio are available for Zn or Cd.

TABLE 18.32

Comparison of observed and calculated values of relaxation times $\bar{v}Q_{or}$, $\bar{v}Q_{al}$ in 10^{-9} cm^3 s^{-1} for an even isotope of Hg in low concentration in a large excess of a second even isotope

	Calculated			Observed
	(a)	(b)	(c)	
$\bar{v}Q_{or}$	4·3	4·2	3·96	3·77
$\bar{v}Q_{al}$	4·3	3·0	4·13	3·98
Q_{or}/Q_{al}	1·00	1·36	0·96	0·95

(a) Simple approximation (396). (b) From approximation (413). (c) From numerical solution of coupled equations (Omont and Meunier, loc. cit., p. 1724).

For Hg, much more detailed comparison between theory and experiment may be made from the observations of Omont and Meunier.† As described on p. 1724 they measured the cross-sections Q_{or}, Q_{al} for depolarization of excited atoms of one mercury isotope present in a low concentration in another isotope. When both isotopes are of even mass number and so possess no nuclear spin these observations refer to depolarization of the struck atom only and include no contribution from transfer to the colliding atom. Table 18.32 compares observed and calculated values of the relaxation times $\bar{v}Q_{or}$, $\bar{v}Q_{al}$ which are independent of temperature. It will be seen that good agreement is found for the ratio Q_{or}/Q_{al} which is very close to unity as compared with a 20 per cent higher value when the atoms are not distinguishable (see Table 18.31). The observed absolute cross-sections are also quite close to the calculated and it is noteworthy that the more accurate calculations give appreciably better agreement than do the results obtained from the simple approximation. Even this gives quite good results, considerably better than those obtained using the approximate solution (413).

† loc. cit., p. 1898.

Although the evidence is still not very extensive it strongly suggests that the theory is very satisfactory as regards depolarization in collisions with like atoms.

8.3.3.2. *Depolarization of $^3P_1$ levels in collisions with foreign atoms.* Table 18.33 gives a comparison of observed and calculated disalignment cross-sections Q_{al} for collisions of Hg and Cd atoms with rare-gas atoms. For Hg the measurements refer to an even isotope so no nuclear spin is involved but for Cd they represent an average over the natural isotopic content.

TABLE 18.33

Comparison of observed and calculated disalignment cross-sections for collisions of $\mathrm{Hg}(6\,^3P_1)$ and $\mathrm{Cd}(5\,^3P_1)$ atoms with rare-gas atoms

$\mathrm{Hg}(6\,^3P_1)\ n_1^* = 4\cdot 3,\ Z_1^* = 4\cdot 35$				$Q_{al}\ (10^{-16}\ \mathrm{cm}^2)$	
Colliding atom	n_2^*	Z_2^*	$\overline{\Delta E}$ (eV)	Calc.	Obs.†
He	1·0	1·70	28·5	27·5	39
Ne	2·0	5·85	24·8	37	47
Ar	3·0	6·75	19·8	72	83
Kr	3·7	8·25	17·3	99	121
Xe	4·0	8·25	16·1	121	168
$\mathrm{Cd}(5\,^3P_1)\ n^* = 4\cdot 0,\ Z_1^* = 4\cdot 35$					
Colliding atom					
He	1·0	1·70	28·5	24·5	44‡
Ne	2·0	5·85	24·8	32·5	53
Ar	3·0	6·75	19·8	63	94
Kr	3·7	8·25	17·3	85	132
Xe	4·0	8·25	16·1	102	192

The calculated values have been obtained, using Slater wave functions to calculate the averages involved in β_1 and β_2, and $\overline{\Delta E}$ as given in Table 18.33. For collisions with rare gases the estimation of $\overline{\Delta E}$ is not difficult because transitions to triplet states are forbidden and all attainable singlet states are close to the ionization limit (see p. 1907).

The agreement between the observed and calculated values is again better than would be expected in view of the difficulty of determining $\overline{\Delta E}$ precisely.

† These are the means between the observations of Faroux and Brossel and of Barrat *et al.* (see Table 18.8).
‡ Barrat *et al.* (see Table 18.8).

According to the simple theory the ratio Q_{or}/Q_{al} should be 1·11 in all cases given in Table 18.34. It will be seen that again the agreement with the calculated value is remarkably good.

TABLE 18.34

Observed ratios Q_{or}/Q_{al} for collisions of $Hg(6\,^3P_1)$ *and* $Cd(5\,^3P_1)$ *atoms with rare-gas atoms*

	He	Ne	A	Kr	Xe
$Hg(6\,^3P_1)$	1·17	1·25	1·20	1·14	1·06
$Cd(5\,^3P_1)$	1·14	1·13	1·17	1·11	1·16

There seems little doubt that the theory is at least as satisfactory for depolarizing collisions of excited atoms in $^3P_1$ states with foreign atoms as with atoms of the same kind, at least when nuclear spin is not involved. Before discussing the extension of the theory to cases in which the excited atoms possess nuclear spin we consider the comparison of theoretical and experimental results for depolarization of atoms in states with integral J other than $^3P_1$.

8.3.4. *Depolarization of states of integral J other than $^3P_1$.* The probability of disalignment occurring in a collision with a foreign gas, in which the impact parameter is p, may be written in the form†

$$P_2^\infty = F(J)\{\beta(J)\}^2 e^8 a_0^8/\hbar^2 v^2 \overline{\Delta E}^2 p^{10}, \tag{430}$$

when p is so large that $P_2^\infty \ll 1$.

In this expression $F(J)$ depends only on the J value of the excited state in question whereas $\beta(J)$ depends not only on J but on the configuration both of the perturbing atom and of the target atom. $\overline{\Delta E}$, the mean excitation energy which determines the effective interaction, may depend on the multiplicity of the excited level. v is as usual the velocity of relative motion in the collision.

Proceeding in the usual way, the simple approximation (396) gives for the disalignment cross-section

$$Q_{al} = \frac{5\pi}{4}[\{\beta(J)\}^2 F(J)/\hbar^2 v^2 \overline{\Delta E}^2]^{\frac{1}{4}}. \tag{431}$$

$\beta(J)$ is proportional to

$$\langle \Sigma_i r_i^2 \rangle \frac{\langle z^2 - \tfrac{1}{3}r^2 \rangle}{\langle L_z^2 - \tfrac{1}{3}L^2 \rangle} \frac{\langle L_z^2 - \tfrac{1}{3}L^2 \rangle}{\langle J_z^2 - \tfrac{1}{3}J^2 \rangle}, \tag{432}$$

† LECLUSE, Y., *J. Phys., Paris*, **28** (1967) 785.

where L^2, L_z^2 are the operators respectively representing the square of the magnitude of the total electronic orbital angular momentum and of its z-component. The first factor is the mean value of Σr_i^2 in the ground state of the perturbing atom. The other factors depend only on mean values over the excited state of the target atom. Of these, the second factor depends only on the configuration of this state and not on J.

$F(J)$ may be calculated in terms of $6j$ symbols to give

$$\frac{F(1)}{1} = \frac{F(2)}{17} = \frac{F(3)}{41}. \tag{433}$$

We consider the application to different excited states of the mercury and cadmium atom.

8.3.4.1. Hg *and* Cd($6\,^3S_1$). For these levels $Q_{\text{al}} = 0$ because $\langle z^2 - \frac{1}{3}r^2 \rangle$ vanishes. The same result follows by elementary arguments since to change the orientation of the J vector in this case requires change in the direction of electron spin which can only be effected through very weak spin-spin and spin-orbit interactions. The experiments of Barrat, Cojan, and Lecluse† on the $7\,^3S_1$ and $8\,^3S_1$ levels of Hg failed to observe any disalignment due to collisions with He, Ne, and Kr atoms, but with Xe quite considerable cross-sections were found (see Table 18.8). The reason for this is not yet clear.

For the $6\,^3S_1$ level of Cd Laniepce‡ was unable to observe disalignment due to collisions with any rare-gas atoms including Xe.

8.3.4.2. Hg($6\,^1P_1$). One difference between cross-sections for disalignment of ^1P and ^3P atoms must come from a dependence of $\overline{\Delta E}$ on the multiplicity of the level. For collisions of mercury atoms with rare-gas atoms such a dependence would be expected. Whereas all strong virtual transitions from the $6\,^3P_1$ level must be to higher states the strongest transition from $6\,^1P_1$ is to the $6\,^1S_0$ level which is 6·2 eV lower. In Table 18.33 we estimated $\overline{\Delta E}$ for $6\,^3P_1$ as the sum of the ionization energy of the rare gas and the separation, $\simeq 3\cdot 3$ eV, between the $6\,^3P_1$ level and the neighbouring levels above it to which strong virtual transitions occur. For $6\,^1P_1$, because the virtual level is 6·2 eV below instead of 3·8 eV above, $\overline{\Delta E}$ will be expected to be about 10·5 eV lower.

In addition the ratio $\langle L_z^2 - \frac{1}{3}L^2 \rangle / \langle J_z^2 - \frac{1}{3}J^2 \rangle$ is 1 for $^1P_1$ and $-\frac{1}{2}$ for $^3P_1$, giving values for the ratio

$$Q_{\text{al}}(6\,^1P_1)/Q_{\text{al}}(6\,^3P_1) = \{2\overline{\Delta E}(6\,^1P_1)/\overline{\Delta E}(6\,^3P_1)\}^{\frac{2}{3}},$$

for collisions with different rare-gas atoms, which are compared with

---
† loc. cit., p. 1717. ‡ loc. cit., p. 1718.

observed values (see Table 18.8) in Table 18.35. It will be seen that, while both sets of ratios are greater than unity in all cases, the observed ones are somewhat larger. Part of the discrepancy may be due to the very rough estimation of $\overline{\Delta E}$ in the two cases and part to uncertainty in the experimental values.

TABLE 18.35

Comparison of observed and calculated ratios $Q_{\text{al}}(6\,^1P_1)/Q_{\text{al}}(6\,^3P_1)$ for disalignment of $\text{Hg}(6\,^1P_1)$ and $\text{Hg}(6\,^3P_1)$ in collisions with rare-gas atoms

Gas	$\overline{\Delta E}(6\,^3P_1)$ (eV)	$\overline{\Delta E}(6\,^1P_1)$ (eV)	$Q_{\text{al}}(6\,^1P_1)/Q_{\text{al}}(6\,^3P_1)$	
			Calc.	Obs.
He	28·5	18	1·20	2·50
Ne	24·8	14·3	1·24	1·63
Ar	19·8	9·3	1·35	2·42
Kr	17·3	6·8	1·46	1·82
Xe	16·1	5·6	1·53	1·66

8.3.4.3. $\text{Hg}(6\,^3P_2)$ *and* $\text{Cd}(5\,^3P_2)$. For $^3P_2$ states the formula (428) applies with β_2 replaced by $\beta_2' = -\tfrac{1}{3}\beta_2$. With the appropriate averaging over all orientations of the collision frame of reference it is then found that†

$$P_2^\infty = (119/15p^{10})(\pi\beta_2 e^4 a_0^4/16\hbar v\,\overline{\Delta E})^2. \tag{434}$$

The simple approximation (395) then gives for the disalignment cross-section

$$Q_{\text{al}}(^3P_2) = (595/315)^{\frac{1}{5}} Q_{\text{al}}(^3P_1)$$
$$= 1\cdot 14 Q_{\text{al}}(^3P_1). \tag{435}$$

For depolarization of $\text{Hg}(6\,^3P_2)$ by rare-gas atoms the observed ratios $Q_{\text{al}}(6\,^3P_2)/Q_{\text{al}}(6\,^3P_1)$ are 1·80, 1·70, 1·80, 1·67, and 1·69 for He, Ne, Ar, Kr, and Xe. While greater than unity, as predicted, they are all considerably greater than the theoretical value, which is not surprising in view of the rough approximations involved.

For self-depolarization of $^3P_2$ the first-order terms vanish so that (435) should apply to these cases also. However, referring to Table 18.31, (435) predicts Q_{al} for $\text{Hg}(6\,^3P_2)$ and $\text{Cd}(5\,^3P_2)$ respectively to be 4·56 and $2\cdot 53\times 10^{-14}$ cm² respectively, more than twice the observed values 2·1 and $0\cdot 98\times 10^{-14}$ cm².

8.3.4.4. $\text{Hg}(6\,^3D)$ *and* $(\text{Hg}6\,^1D)$. Arising from the $6s6d$ configuration of Hg there are the D levels $6\,^3D_1$, $6\,^3D_2$, $6\,^3D_3$, and $6\,^1D_2$. According to (432) the ratios of the disalignment cross-sections for these levels will

† OMONT, A., loc. cit., p. 1892.

be determined solely by the ratio $\langle L_z^2 - \tfrac{1}{3}L^2\rangle / \langle J_z^2 - \tfrac{1}{3}J^2\rangle$ and the relations (433). The calculation of the mean value ratio must take into account the coupling between the $^3D_2$ and $^1D_2$ levels. It is then found that the $\beta(J)$ (see (432)) for the $6\,^3D_1$, $6\,^3D_2$, $6\,^1D_2$, and $6\,^3D_3$ levels are in the ratio $5\cdot2:2:1\cdot7:1$. Combining this with (431) it is found that

$$Q_{\text{al}}(^3D_3) : Q_{\text{al}}(^3D_2) : Q_{\text{al}}(^1D_2) : Q_{\text{al}}(^3D_1) = 1 : 1\cdot04 : 1\cdot12 : 0\cdot92, \quad (436)$$

independent of the perturbing foreign-gas atom. The observed ratios for different rare gases† are as given in Table 18.36.

TABLE 18.36

Observed ratios of cross-sections for disalignment of 6D mercury atoms by collisions with rare-gas atoms

Gas \ Level	$6\,^3D_3$	$6\,^3D_2$	$6\,^1D_2$	$6\,^3D_1$
He	1	1·05	0·98	0·99
Ne	1	0·86	0·90	0·75
Ar	1	0·94	1	0·90
Kr	1	0·97	0·95	0·95
Xe	1	0·92	1·04	1·04
Theoretical	1	1·04	1·12	0·92

It will be seen that, for all gases, the ratios are very close to unity. More detailed agreement between theory and experiment would hardly be expected as neither is sufficiently precise.

The ratio of the disalignment cross-section for $6\,^3D_1$ to that for $6\,^3P_1$, for different rare gases, may be derived from the observations.† According to the theory they should be independent of the perturbing gas. The observed ratios $Q_{\text{al}}(6\,^3D_1)/Q_{\text{al}}(6\,^3P_1)$ are 5·38, 4·73, 4·80, 4·70, and 4·26 for He, Ne, Ar, Kr, and Xe, which can be regarded as close enough to constancy to provide further support for the essential correctness of the theory.

8.3.5. *Inclusion of nuclear spin.* We now take account of the existence of nuclear spin of total quantum number I in the excited atoms.

In the stationary states of the system the nuclear spin and total angular momentum vectors **I**, **J** respectively are coupled to produce a resultant **F** such that

$$F = |J - I|, \ldots, J + I. \quad (437)$$

The hyperfine structure levels are therefore characterized by the quantum numbers F, M_F.

† LECLUSE, Y., loc. cit., p. 1906.

If the energy separation ΔE between hyperfine levels with different F is so small that
$$\Delta E/\hbar \ll 1/\tau, \qquad (438)$$
where τ is the time of collision, the nuclear spin vector $\mathbf{I}$ will not have time to precess appreciably during the collision. It will therefore be uncoupled from $\mathbf{J}$, whose orientation may be drastically changed as the collision proceeds. After the collision, however, $\mathbf{I}$ and $\mathbf{J}$ will be recoupled but, because of the reorientation of $\mathbf{J}$, this will in general produce a different hyperfine level to the initial one.

The situation is very similar to that arising due to exchange of electron spin between two atoms. Thus the nuclear spins play no direct part in the collision so the problem is one of expressing the initial distribution among the hyperfine levels in terms of a distribution among states characterized by the quantum numbers I, J, M_I, M_J, calculating the effect of the collision on the M_J distribution and then re-expressing the new distribution among the I, J states in terms of one among the F, M_F hyperfine structure levels. This may be done by use of the usual rules for combination of angular momentum eigenfunctions.

Thus we may express an angular momentum eigenfunction $|F, M_F\rangle$ for a hyperfine level with quantum numbers F, M_F in terms of those $|I, J, M_I, M_J\rangle$ for levels with quantum numbers, I, J, M_I, M_J, by
$$|F, M_F\rangle = \sum_{M_I, M_J = M_F - M_I} \langle I, J, M_I, M_J | F, M_F\rangle |I, J, M_I, M_J\rangle, \qquad (439)$$
where $\langle I, J, M_I, M_J | F, M_F\rangle$ is a vector coupling coefficient.

The effect of a collision may be expressed in terms of a mixing operator A where
$$A |I, J, M_I, M_J\rangle = \sum_{M_J'} \frac{\alpha(M_I, M_J')}{(2J+1)^{\frac{1}{2}}} |I, J, M_I, M_J'\rangle, \qquad (440)$$
$\alpha(M_I, M_J')$ being a phase factor. In terms of A the effect of the collision is to change $|F, M_F\rangle$ to
$$A|F, M_F\rangle = \sum_{M_I, M_J} \langle I, J, M_I, M_J | F, M_F\rangle \left\{ \sum_{M_J'} \frac{\alpha(M_I, M_J')}{(2J+1)^{\frac{1}{2}}} |I, J, M_I, M_J'\rangle \right\}. \qquad (441)$$

The probability that, after the collision, an atom initially in the hyperfine state F, M_F will be found in the state F', M_F' is then given by
$$|\langle F', M_F' | A | F, M_F\rangle|^2. \qquad (442)$$
This may be calculated by re-expressing (441) in terms of the $|F', M_F'\rangle$ by inversion of the coupling relation (439).

The whole procedure may be carried through in terms of the density matrix formulation as it is possible to use coupling formulae such as (439) to relate a density matrix expressed in the F, M_F representation to one expressed in the I, J, M_I, M_J representation. When this is done the results may be expressed, for a particular impact parameter p, in terms of probabilities $^{F}P_1$, $^{FF'}P_1$ which refer respectively to disorientation of the hyperfine states with quantum number F, and to transfer of orientation from the state F to F'. Similarly, we have $^{F}P_2$, $^{FF'}P_2$ which refer to alignment and $^{F}P_0$, $^{FF'}P_0$ to transfer of excitation. Omont† has shown that, when the electronic angular momentum quantum number is 1 and the depolarization is due to distinguishable atoms,

$$^{FF'}P_q = 1 - \sum_{q',u'} (-1)^{q+q'+F+F'}(2F+1)(2F'+1) \begin{Bmatrix} F & F' & q' \\ 1 & 1 & I \end{Bmatrix}^2 \times$$

$$\times \begin{Bmatrix} F & F & q \\ F' & F' & q' \end{Bmatrix} \bigg/ \{\mathrm{Tr}(S^\dagger \Omega_{q'u'})\}, \quad (443)$$

the notation being as in (406) with {} denoting a Wigner $6j$ symbol.

Using (443) the values obtained for the various probabilities for large p are given in Table 18.37, for $I = \tfrac{1}{2}$, together with the corresponding cross-sections calculated by the simple approximation (395). The ratios $Q(I = \tfrac{1}{2})/Q_{\mathrm{or}}(I = 0)$ of the cross-sections to the disorientation cross-section for an isotope with zero nuclear spin are also given.

TABLE 18.37

	$q = 0$			$q = 1$			$q = 2$		
F	$\tfrac{3}{2}$	$\tfrac{1}{2}$	$\tfrac{3}{2}$	$\tfrac{3}{2}$	$\tfrac{1}{2}$	$\tfrac{3}{2}$	$\tfrac{3}{2}$	$\tfrac{1}{2}$	$\tfrac{3}{2}$
F'	$\tfrac{3}{2}$	$\tfrac{1}{2}$	$\tfrac{1}{2}$	$\tfrac{3}{2}$	$\tfrac{1}{2}$	$\tfrac{1}{2}$	$\tfrac{3}{2}$	$\tfrac{1}{2}$	$\tfrac{1}{2}$
$(p^{10}/7\delta^2)P^\infty$	$\dfrac{1}{3}$	$\dfrac{2}{3}$	$-\dfrac{\sqrt{2}}{3}$	$\dfrac{18}{30}$	$\dfrac{20}{30}$	$\dfrac{2\sqrt{10}}{20}$	$\dfrac{26}{30}$	0	0
$(\hbar v \overline{\Delta E}/4e^4 a_0^4 \beta_2) Q$	1·01	2·03	−1·43	1·89	2·74	0·39	2·98	0	0
$Q(I = \tfrac{1}{2})/Q_{\mathrm{or}}(I = 0)$	0·34	0·68	−0·48	0·63	0·91	0·13	0·99	0	0

δ, v, $\overline{\Delta E}$, and β_2 are as in (428).

A comparison may be made between these ratios and the observed ratios for the $6\,^3P_1$ states of the Hg isotopes ^{199}Hg with nuclear spin $I = \tfrac{1}{2}$ and ^{202}Hg without nuclear spin, depolarized in collision with rare-gas atoms. Referring to the experimental results given in Table 18.8 we obtain the comparison given in Table 18.38. It will be seen that, while there are differences in detail, the agreement is remarkably good.

Finally, for self-depolarization, including collisions between an odd and even isotope, Omont‡ has derived the probabilities given in Table 18.39. s^2 is as defined in (398).

† OMONT, A., loc. cit., p. 1892.   ‡ OMONT, A., Thesis, Paris, 1967.

TABLE 18.38

Comparison of observed and calculated ratios of cross-sections for depolarization of the $6\,^3P_1$ state of ^{199}Hg and ^{202}Hg by collisions with rare-gas atoms

Colliding atom	Observed					Calculated
	He	Ne	Ar	Kr	Xe	
†$Q_{he}/Q_{or}(I=0)$	0·20	0·44	0·46	0·44	0·46	0·34
‡$Q_{or}/Q_{or}(I=0)$	0·70	0·57	0·59	0·59	0·60	0·63
‡$Q_{or}/Q_{or}(I=0)$	0·91	0·82₅	0·83	0·85	0·88	0·91
‡$Q_{or}/Q_{or}(I=0)$	0·09	0·13	0·13	0·12	0·09₅	0·13
‡$Q_{al}/Q_{or}(I=0)$	0·94	0·91	0·93	0·93	0·96	0·99

TABLE 18.39

	$q=0$			$q=1$			$q=2$		
F	$\tfrac{3}{2}$	$\tfrac{1}{2}$	$\tfrac{3}{2}$	$\tfrac{3}{2}$	$\tfrac{1}{2}$	$\tfrac{3}{2}$	$\tfrac{3}{2}$	$\tfrac{1}{2}$	$\tfrac{3}{2}$
F'	$\tfrac{3}{2}$	$\tfrac{1}{2}$	$\tfrac{1}{2}$	$\tfrac{3}{2}$	$\tfrac{1}{2}$	$\tfrac{3}{2}$	$\tfrac{3}{2}$	$\tfrac{1}{2}$	$\tfrac{1}{2}$
$(27/8s^2)P_q^{(1)\infty}$	0·80	0·88	$-0·08\sqrt{2}$	0·87	0·98	0·007$\sqrt{10}$	0·98	0	
$(27/8s^2)P_q^{(2)\infty}$	$-0·48$	$-0·24$	$-0·24\sqrt{2}$	0·11	0·04	0·022$\sqrt{10}$	$-0·01$		

For disalignment of the hyperfine state with $F=\tfrac{3}{2}$ the relevant probability $^{\ddagger}P_2^{(1)\infty}$ is seen to be $0·98\times 8s^2/27$ which is $0·98P_2^{(1)\infty}$, where $P_2^{(1)\infty}$ is the disalignment probability for a collision between even isotopes. According to the simple approximation the cross-sections are related by

$$^{\ddagger}Q_{al} = (0·98)^{\ddagger}Q_{al}, \tag{444}$$

where Q_{al} refers to collisions between even isotopes. For resonance collisions, according to (408), $Q_{al} = 15·8\times 10^{-14}$ cm², so

$$^{\ddagger}Q_{al} = 15·6\times 10^{-14}\ \text{cm}^2. \tag{445}$$

This is to be compared with 15×10^{-14} cm² observed by Omont and Meunier‡ for the disalignment of ^{199}Hg $(6\,^3P_1)$ atoms in collisions with the even isotopes ^{198}Hg and ^{202}Hg.

The more sophisticated theory in which the coupled equations (401) are solved by numerical integration for $s<9$ and the disalignment probability is assumed to be unity for larger s gives

$$^{\ddagger}Q_{al} = 17·4\times 10^{-14}\ \text{cm}^2. \tag{446}$$

8.3.6. *Impact disorientation and disalignment of atoms in excited states with $J=\tfrac{1}{2}$ or $\tfrac{3}{2}$.* So far we have applied the theory of depolarizing collisions to excited atoms in states with integral J. In § 3.10 experiments

---

† For definition of Q_{he} see p. 1720.   ‡ loc. cit., p. 1898.

were described in which cross-sections for disorientation and disalignment of alkali metals in P states with $J = \frac{1}{2}$ and $\frac{3}{2}$ were measured.

The interpretation of these experiments† depends on the relation between the fine structure separation ΔE between the $J = \frac{1}{2}$ and $J = \frac{3}{2}$ states and the time τ of collision. If $\Delta E \ll \hbar/\tau$ then, during the collision, the electron spin vector $\mathbf{S}\hbar$ remains effectively fixed in space while the orientation of the orbital angular momentum vector $\mathbf{L}\hbar$ is changed. In this case the situation is the same as in the theory of § 8.3.5 in which the total nuclear spin vector $\mathbf{F}\hbar$ remains unchanged while the orientation of the total electron angular momentum vector $\mathbf{J}\hbar$ is changed. We would therefore expect the cross-sections for disalignment and disorientation, as well as for excitation of the $J = \frac{1}{2} \to J = \frac{3}{2}$ transition, to be of order 10^{-14} cm².

On the other hand, if $\Delta E \gg \hbar/\tau$, $\mathbf{S}$ and $\mathbf{L}$ remain coupled during the collision and depolarization results from reorientation of $\mathbf{J}$. Transitions between $J = \frac{1}{2}$ and $J = \frac{3}{2}$ states will be very unlikely for, in the limit $\Delta E\tau/\hbar \to \infty$, they can only occur through reorientation of $\mathbf{S}$. The appropriate cross-section will therefore be $\ll 10^{-14}$ cm². On the other hand, for the reorientation collisions for a fixed J the theory of § 8.3.2 will apply and, at first sight, it would appear that cross-sections of order 10^{-14} cm² would be expected. This is not so, however, for $J = \frac{1}{2}$ states because in this case reorientation can only occur through

$$M_J = \tfrac{1}{2} \to M_J = -\tfrac{1}{2}$$

transitions. It has been shown by Gallagher† that, if the effective interaction has the form (423) given by the adiabatic approximation, then $M_J \leftrightarrow -M_J$ transitions are forbidden to all orders of approximation when J is half-integral. Depolarization of $J = \frac{1}{2}$ levels can therefore arise only from non-adiabatic effects, which will be small. Hence we expect the cross-section for such depolarization to be considerably less than 10^{-14} cm². On the other hand, for $J = \frac{3}{2}$, depolarization can occur through other M_J transitions and we can expect cross-sections of the same order (10^{-14} cm²) as for depolarization of excited states with integral J as in § 8.3.3.2.

The experimental evidence supports these conclusions. On substitution of the appropriate numbers we find that $\Delta E \ll \hbar/\tau$ for Li, Na and probably also K, and $\gg \hbar/\tau$ for Rb and Cs. The observed cross-sections for $J = \frac{1}{2} \to J = \frac{3}{2}$ transitions due to impact with rare-gas atoms (see Table 18.9) are of order 10^{-14} cm² for Na and K but many orders of

† GALLAGHER, A., *Phys. Rev.* **157** (1967) 68.

magnitude less for Rb and Cs, just as expected from the above considerations.

Turning now to the observed depolarization cross-sections given in Table 18.8 we see that cross-sections for both disorientation and disalignment of Rb $5P_{\frac{3}{2}}$ levels by collisions with rare-gas atoms are of order 10^{-14} cm^2 while those for disorientation of Rb $5P_{\frac{1}{2}}$ are about an order of magnitude smaller. Furthermore, for Cs $6P_{\frac{1}{2}}$, for which the ratio $\Delta E\tau/\hbar$ is greater than for Rb, the corresponding disorientation cross-sections are even smaller.

The qualitative arguments obtained above seem therefore to be supported by the observations. We shall discuss the calculation of cross-sections for $J = \frac{1}{2} \to J = \frac{3}{2}$ transitions due to collisions of excited alkali-metal atoms with rare-gas atoms a little further in § 8.4.5, account being taken of the finite value of ΔE involved. While still not much better than semi-quantitative they show the same features as expected from the qualitative analysis above.

8.4. *Collisions in which the resonance is imperfect* ($\Delta E \neq 0$)

We now consider transfer collisions in which the resonance is imperfect so that ΔE is finite. It is, nevertheless, considered to be small enough for the contributions from bound states of the colliding systems, other than those involved in the transfer, to be negligible. In that case the problem is essentially one of finding solutions to coupled equations for the relative motion of the two systems, of reduced mass M, of the form

$$\nabla^2 F_i + \{k_i^2 - (2MV_i/\hbar^2)\}F_i = (2MU/\hbar^2)F_f, \qquad (447\text{ a})$$

$$\nabla^2 F_f + \{k_f^2 - (2MV_f/\hbar^2)\}F_f = (2MU/\hbar^2)F_i, \qquad (447\text{ b})$$

which have the asymptotic form

$$F_i \sim e^{i\mathbf{k}_i \cdot \mathbf{r}} + r^{-1} e^{ik_i r} f_{ii}(\theta, \phi), \qquad (448\text{ a})$$

$$F_f \sim r^{-1} e^{ik_f r} f_{if}(\theta, \phi). \qquad (448\text{ b})$$

The transfer cross-section is then given by

$$Q_{\text{tr}} = (k_f/k_i) \iint |f_{if}|^2 \sin\theta \, d\theta d\phi. \qquad (449)$$

We suppose for simplicity that the interactions V_i, V_f, and U are all dependent on r only, in which case we can carry out the usual angular momentum analysis to give

$$Q_{\text{tr}} = (\pi/k_i^2) \sum (2l+1)|\beta_l|^2, \qquad (450)$$

where β_l is defined in terms of coupled radial equations

$$\frac{d^2 G_{i,l}}{dr^2}+\{k_i^2-(2MV_i/\hbar^2)-l(l+1)r^{-2}\}G_{i,l} = (2MU/\hbar^2)G_{f,l}, \quad (451\text{ a})$$

$$\frac{d^2 G_{f,l}}{dr^2}+\{k_f^2-(2MV_f/\hbar^2)-l(l+1)r^{-2}\}G_{f,l} = (2MU/\hbar^2)G_{i,l}. \quad (451\text{ b})$$

Thus β_l is such that the asymptotic forms of the solution of (451 a) and (451 b) which satisfy $G_{i,l}(0) = G_{f,l}(0) = 0$ are

$$G_{i,l} \sim \sin k_e r + \alpha_l e^{ik_i r}, \quad (452\text{ a})$$

$$G_{f,l} \sim (k_f/k_i)^{\frac{1}{2}}\beta_l e^{ik_f r}. \quad (452\text{ b})$$

The probability of transfer vanishes with U, which in some respects plays a similar role to $U(r)$ in (313) in the exact resonance case. However, the interaction between the two atoms in their initial states may no longer be described by a function of the form (313). Whereas when $\Delta E = 0$, $U(r)$ acts fully as a scattering potential (see p. 1874) which determines the transfer cross-section, when $\Delta E \neq 0$ its effect is reduced by the interference between the waves, representing the initial and final states of relative motion, which are no longer of the same wavelength. This is one way in which the decrease of transfer cross-section with ΔE may be understood.

Under the conditions in which we are interested the relative motion is effectively classical and we may trace out the course of events leading to a transition in the following way. V_i and V_f are the respective interactions between the atoms in their initial and final states, ignoring any modifications due to the possibility of transfer. Fig. 18.123 shows the corresponding potential energy curves $\mathscr{V}_i$ and $\mathscr{V}_f$ including the internal energy of excitation. If, at some separation R, $|V_i-V_f| = \Delta E$ the curves $\mathscr{V}_i$ and $\mathscr{V}_f$ of Fig. 18.123 cross. Two cases can thus be distinguished according to whether or not a crossing point occurs.

8.4.1. *The crossing-point case.* Crossing only occurs because the possibility of transfer has been ignored. When this is included the two interactions are replaced by

$$\tfrac{1}{2}[\mathscr{V}_i + \mathscr{V}_f \pm \{(\mathscr{V}_i - \mathscr{V}_f)^2 + \tfrac{1}{4}U^2\}^{\frac{1}{2}}] \quad (453)$$

and may be represented by the curves I and II in Fig. 18.123. The curves no longer cross and the point C, where $r = R$, is one in which the separation is a minimum and of order $|U(R)|$. If the collision is now allowed to take place infinitely slowly the interaction will be of the form of curve I, Fig. 18.123 (b), throughout and the atoms will approach and separate again in the same state as initially. This is no longer the only possibility if the collision takes place at a finite rate. As the atoms approach along curve I of Fig. 18.123 (b) there is a chance that a jump will take place to curve II. This chance will have a strong maximum at the point C and may be regarded as only finite there. After passing C the chance that the interaction will still be following curve I is $1-P$, there being a chance P that it will follow curve II instead. Eventually

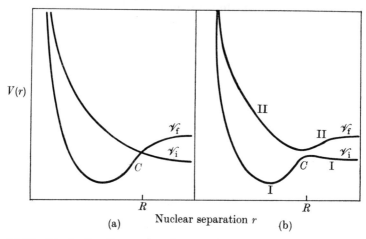

Fig. 18.123. Illustrating the crossing of potential energy curves. In the absence of interaction between the curves they follow $\mathscr{V}_i$, $\mathscr{V}_f$ with a crossing point at C. Interaction separates them to the curves I and II which approach closely but do not cross at C.

the atoms will reach the distance of closest approach and then begin to separate again. There will again be a probability P of a transition at the crossing-point C. The chance that, after separating finally beyond C, the atoms will be found to be following interaction II, will therefore be $(1-P)P+P(1-P)$. The first arises from the chance of a jump at C occurring on the way back along curve I, the second from the chance of no jump at C on the way back following curve II. The chance of a transition is therefore $2P(1-P)$.

These considerations apply for collisions with any impact parameter p as (453) is independent of p. However, the probability P will depend on p so that the 'classical' transfer cross-section will be given by

$$Q_{\text{tr}} = 2\pi \int_0^\infty 2P(1-P)p \, dp. \qquad (454)$$

P was first calculated approximately by Landau† and by Zener‡ by time-dependent perturbation theory in which the transition was assumed

† Landau, L. D., *Phys. Z. SowjetUn.* **2** (1932) 46.
‡ Zener, C., *Proc. R. Soc.* **A137** (1932) 696.

to be strongly localized near the crossing point over which region simple linear approximations were made to the form of V_i and V_f. They found

$$P = e^{-2\delta(p)}, \tag{455}$$

where
$$\delta = (\pi/\hbar v_p)U^2(R)/\{V'_i(R)-V'_f(R)\}. \tag{456}$$

v_p is the relative velocity at the crossing point, given by

$$v_p^2 = v^2(1-p^2/R^2), \tag{457}$$

where v is the initial value, and $V'_{i,f} = dV_{i,f}/dr$. (455) is a good approximation provided v_p is real. If p is such that $v_p^2 < 0$, P can be taken as zero.

A little later Stueckelberg† carried out a semi-classical analysis of the solutions of the coupled equations (451) in terms of an extension of Jeffreys' approximation (Chap. 6, § 3.7). He then found

$$2P(1-P) = 4e^{-2\delta}(1-e^{-2\delta})\sin^2\eta, \tag{458}$$

where δ is as in (456) and η is given by

$$\eta = \int_{g_i=0}^{R} g_i^{\frac{1}{2}}\,dr - \int_{g_f=0}^{R} g_f^{\frac{1}{2}}\,dr, \tag{459}$$

where
$$g_{i,f} = \tfrac{1}{2}(f_i+f_f) \pm \tfrac{1}{2}\{(f_i-f_f)^2 + \hbar^4 U^2/M^2\}^{\frac{1}{2}}, \tag{460}$$

$$f_{i,f} = k_{i,f}^2(1-p^2/r^2) + 2MV_{i,f}/\hbar^2. \tag{461}$$

In general, η is $\gg \pi$ so that we can, in (458), replace $\sin^2\eta$ by its mean value of $\tfrac{1}{2}$ so that Stueckelberg's result (458) yields the same transfer cross-sections as (455).

The significance of (458) is interesting. The probability of the transition for a fixed p is small either when δ is large or when it is small. In the former case we have nearly adiabatic conditions in the crossing-point region. Such conditions are characterized by the relation

$$\Delta E(R)\tau/\hbar \gg 1, \tag{462}$$

where τ is the time during which significant interaction occurs in the region. $\Delta E(R) \simeq U(R)$, while τ may be taken as the time required for the systems to separate a distance ΔR from the crossing point such that the energy separation increases by a factor a of order unity. Thus

$$\Delta R\{V'_i - V'_f\} = aU(R),$$

so
$$\tau = \Delta R/v_p(R) \simeq aU(R)/(V'_i-V'_f)v_p(R).$$

† STUECKELBERG, E. C. G., *Helv. phys. Acta* **5** (1932) 370.

Hence, for P to be small,

$$aU^2(R)/\hbar(V_i'-V_f')v_p(R) = \frac{1}{\pi}a\delta \gg 1.$$

The opposite extreme of δ small is the case of weak interaction for which the probability is proportional to $\{U(R)\}^2$.

The phase shift η may be analysed in terms of the phase shift produced in the semi-classical motion between the crossing point and the distance of closest approach for the two interactions. However, we shall defer discussion of this until Chapters 22 and 23 as it is related to charge-transfer and elastic-scattering processes at relatively high energies.

It is not apparent from the formulae for the crossing-point case that the transfer cross-section will depend in any simple way on the energy discrepancy ΔE at infinite separation and it certainly does not follow that, other things being equal, the cross-section will be a maximum when $\Delta E = 0$. Indeed there are many circumstances in which this is not so (see, for example, Chap. 20, § 5). Nevertheless, when $|\Delta E|$ is large compared with the incident energy of relative motion, as in the quenching of the excitation of resonance levels of alkali-metal atoms by thermal collisions with rare-gas atoms, the cross-section is determined by the nature and location of a crossing point at a quite small nuclear separation where the interaction energy is of order $|\Delta E|$. At the time of writing little further can be said about collisions of this kind which have a small cross-section, the order of magnitude of which depends on the relative behaviour of potential energy curves, as in Fig. 18.123, at small separations. Other applications of the crossing-point formulae will be discussed in greater detail in Chapter 20, § 5 and Chapter 23, §§ 5 and 6. The validity of the approximations made in deriving these formulae will be discussed in Chapter 23.

8.4.2. *The case of no crossing point.* In general the interaction U in (447) will have the asymptotic form

$$U = \alpha/r^n, \qquad (463)$$

where $n = p+q+1$, p and q being the polarity of the transitions involved in the respective atoms. Unless α is very small, transitions will take place at such large values of r that the interactions V_i, V_f in (447) will be negligible. It is convenient to work in terms of time-dependent perturbation theory assuming the usual rectilinear trajectory for the relative motion. The alternative approach through analysis of the coupled equations (451) when no crossing point occurs is less fruitful in yielding useful approximations except for the case $n = 3$, which will be discussed below.

If a_1, a_2 are the amplitudes of the initial and final states of the combined system, we have, as in Chapter 7, § 5.3.1, for collisions with impact parameter p,

$$\frac{da_1}{dt} = (-i/\hbar)a_2\, U(p,t)e^{-i\omega t}, \qquad (464\,\text{a})$$

$$\frac{da_2}{dt} = (-i/\hbar)a_1\, U(p,t)e^{i\omega t}. \qquad (464\,\text{b})$$

Here $\qquad U(p,t) = \alpha/(p^2+v^2t^2)^{\frac{1}{2}n}, \qquad \omega = \Delta E/\hbar,$

v being the constant velocity of relative motion along Oz. If solutions of (464) are found satisfying

$$|a_1(-\infty)| = 1, \qquad |a_2(-\infty)| = 0, \qquad (465)$$

then the probability of a transition is

$$P(p) = |a_2(\infty)|^2. \qquad (466)$$

In the resonance case, $\Delta E = 0$, the methods we have discussed in § 8.1.2 give

$$P(p) = \sin^2\Omega_0, \qquad (467)$$

where

$$\Omega_0 = \hbar^{-1}\int_{-\infty}^{\infty} U\{(p^2+v^2t^2)^{\frac{1}{2}}\}\,dt \qquad (468)$$

$$= (\alpha/\hbar v p^{n-1})\pi^{\frac{1}{2}}\Gamma\!\left(\frac{n-1}{2}\right)\!\Big/\Gamma\!\left(\frac{n}{2}\right). \qquad (469)$$

When $\Delta E \neq 0$ analytical solution of the equations (464) is no longer possible but three approximate methods have been applied to obtain solutions when ΔE is no longer small.

Vainshtein, Presnyakov, and Sobel'man† obtained an asymptotic solution valid in the limit $v \to 0$ which gives

$$P = \hbar^{-2}\left|\int_{-\infty}^{\infty} U(p,t)\cos\!\left\{\int_0^t \{\Delta E^2 + 4U^2(t')\}^{\frac{1}{2}}\,dt'/\hbar\right\}dt\right|^2. \qquad (470)$$

When $U \ll \omega$ this gives the usual formula of perturbation theory. For the explicit form (463),

$$P = \sin^2\Omega_0(p)\exp[-2(\Delta E/\hbar v)\{2(\alpha/\Delta E)^{2/n}\sin^2\pi/n + p^2\}^{\frac{1}{2}}], \qquad (471)$$

which tends to (467) in the limit $\Delta E \to 0$. There is no difficulty in extending the analysis to the case where

$$U = \alpha/(a^2+r^2)^{n/2}. \qquad (472)$$

This simply has the effect of replacing p^2 in the exponent by p^2+a^2.

† VAINSHTEIN, L., PRESNYAKOV, L., and SOBEL'MAN, I., Zh. éksp. teor. Fiz. **43** (1962) 518; Soviet Phys. JETP **16** (1963) 370; DE BREYN, N. G., Asymptotic methods in analysis (Russian) 11 L (1961).

Bates† obtained an approximate solution by noting that, in the two cases (i) $\Delta E = 0$ all U, and (ii)‡ $\Delta E \neq 0$,

$$U = \{C/(p^2+c^2)\}\exp[\gamma\{1-(1+p^2/c^2)^{\frac{1}{2}}\}]\operatorname{sech}\{\gamma vtc/(c^2+p^2)^{\frac{1}{2}}\}, \quad (473)$$

exact solution of (464) gives

$$P = (\Lambda^2/\Omega_0^2)\sin^2\Omega_0, \quad (474)$$

where Ω_0 is as given in (468) and

$$\Lambda(p) = \hbar^{-1} \int_{-\infty}^{\infty} U\{(p^2+v^2t^2)^{\frac{1}{2}}\}\exp(i\Delta E t/\hbar) \, dt. \quad (475)$$

Assuming that (474) also gives a good solution for the case (463) Bates obtains

$$P = \{F(p)\}^2 \sin^2\Omega_0, \quad (476)$$

where $\quad F(p) = \left\{2/\Gamma\left(\dfrac{n-1}{2}\right)\right\}(p\Delta E/2\hbar v)^{(n-1)/2} K_{\frac{1}{2}(n-1)}(\Delta E p/\hbar v), \quad (477)$

$K_{\frac{1}{2}(n-1)}$ being the usual Bessel function. For large $p\Delta E/\hbar v$

$$F(p) \sim \{\tfrac{1}{2}\pi(p\Delta E/2\hbar v)^{(n-2)}\}^{\frac{1}{2}}\exp(-p\Delta E/\hbar v)/\Gamma\{\tfrac{1}{2}(n-1)\}. \quad (478)$$

A further discussion by Skinner§ provides evidence that (476) is a good approximation in most circumstances provided $\Delta E/\hbar v$ is not so large that $|\Lambda| \ll \Omega_0$.

The third approximation, due to Callaway and Bauer,‖ uses the method described in § 8.3, p. 1899. Applied to the case of the interaction (463) they find

$$P = \sin^2\{F(p)\Omega_0\} = \sin^2\Lambda. \quad (479)$$

All these approximations tend to the correct limit when $\Delta E \to 0$ and agree with the perturbation result when α is small. They all exhibit a strong resonance effect, the probability for fixed p, a, n, and v falling off rapidly as $|\Delta E|$ increases.

Having obtained $P(p)$, the transfer cross-section can be calculated in the usual way by the same method as that applied on p. 1874 or by numerical integration. Fig. 18.124 (a) and (b) shows the variation of Q_{tr} (calculated from each approximation) with ΔE for collisions between two atoms of reduced mass 10 a.m.u., with mean relative velocity 8.5×10^4 cm s^{-1} corresponding to room temperature. In Fig. 18.124 (a), $n = 3$ and α is taken as $e^2 a_0^2$, while in Fig. 18.124 (b) $n = 5$ and $\alpha = e^2 a_0^4$.

While the results obtained from (479) and (476) are in moderate agreement they both differ very considerably in quantitative terms

† BATES, D. R., *Discuss. Faraday Soc.* **33** (1962) 7.
‡ ROSEN, N. and ZENER, C., *Phys. Rev.* **40** (1932) 502.
§ SKINNER, B. G., *Proc. phys. Soc.* **77** (1961) 551.
‖ CALLAWAY, J. and BAUER, E., *Phys. Rev.* **140** (1965) A1072.

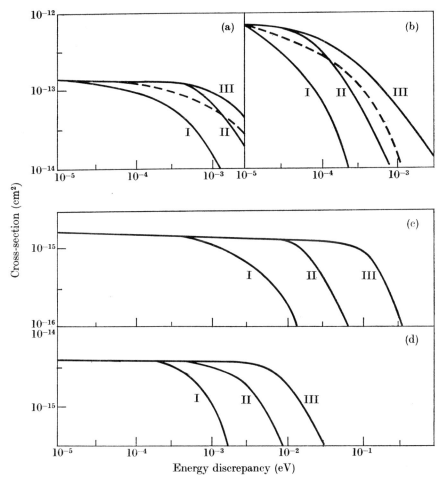

Fig. 18.124. Calculated cross-sections for transfer of excitation between two atoms as a function of the resonance defect ΔE. In all cases the reduced mass of the atoms is taken as 10 a.m.u. I, II, III are calculated using approximations (471), (474), and (479) respectively, ---- calculated using (480). (a) For $U = ea_0^2/R^3$, relative velocity of collision 2×10^{-3} a.u. (1·0 eV). (b) For $U = ea_0^2/R^3$, relative velocity of collision 4×10^{-4} a.u. (0·04 eV). (c) For $U = e^2 a_0^4/R^5$, relative velocity of collision 2×10^{-3} a.u. (1·0 eV). (d) For $U = e^2 a^4/R^5$, relative velocity of collision 4×10^{-4} a.u. (0·04 eV).

from those given from (471). Nevertheless, all approximations give qualitatively similar results. There is a marked 'resonance' effect which is sharper for the $n = 3$ case.

Stueckelberg† derived formulae for Q_{tr} through analysis of the

† loc. cit., p. 1917.

coupled equations (451) in the case of no crossing point, under semi-classical conditions. He obtained

$$Q_{\rm tr} = \pi(\alpha/\Delta E)^{2/n} f(\Delta E^{(n-1)/n} \alpha^{1/n}/\hbar v), \qquad (480)$$

where $f(x)$ has the form shown in Fig. 18.125. For $n = 3$ this gives cross-sections as functions of ΔE, which are shown in Fig. 18.124 (a) and (b) and which agree very well with those calculated from (479). On the other hand, as $f(x)$ behaves as x for small x, (480) vanishes when $\Delta E \to 0$ for $n > 3$, which is clearly incorrect.

For $n = 3$, a case for which it is almost certainly valid, (480) gives a cross-section which, considered as a function of v for a fixed ΔE has a maximum when

$$\Delta E^{\frac{2}{3}} \alpha^{\frac{1}{3}}/\hbar v = 1. \qquad (481)$$

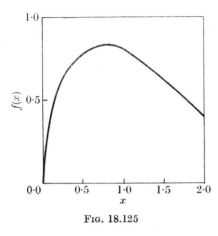

Fig. 18.125

This may be written in the form

$$a \Delta E/\hbar v = 1, \qquad (482)$$

where $a = (\alpha/\Delta E)^{\frac{1}{3}}$, i.e. the separation at which the interaction energy (463) with $n = 3$ is equal to the energy discrepancy. We thus have an example of application of the near-adiabatic criterion (309) where a definite value a/v can be ascribed to the effective time of collision.

For v such that $a\Delta E/\hbar v \gg 1$ we have nearly adiabatic conditions. Since $f(x)$ falls off as $x^2 e^{-2x}$ for large x, under these conditions

$$Q_{\rm tr} \simeq C v^2 e^{-2\lambda/v}, \qquad (483)$$

where
$$\lambda = \Delta E^{\frac{2}{3}} \alpha^{\frac{1}{3}}/\hbar = a\Delta E/\hbar.$$

Fig. 18.126 illustrates the variation of $Q_{\rm tr}$ with relative kinetic energy for three values of ΔE calculated both from (474) and from (480). For $n > 3$ the variation will be similar, the maximum occurring when

$$\Delta E^{(n-1)/n} \alpha^{1/n}/\hbar v \simeq 1. \qquad (484)$$

In many important cases, such as charge exchange or excitation transfer, involving s–s transitions, $U(r)$ will fall off exponentially with r. There is no difficulty in carrying out similar calculations to represent such cases but it is to be expected that the approximations that have been made will be much less satisfactory. Thus, because of the much shorter range of U, there is much less justification for the neglect of V_i and V_f in

(464). We have already encountered one striking example of the importance of these interactions—the transfer of excitation on impact between metastable and normal helium atoms. In cases of imperfect resonance the effects are likely to be much more generally important. Nevertheless, Rapp and Francis† have found the approximation useful for analysing

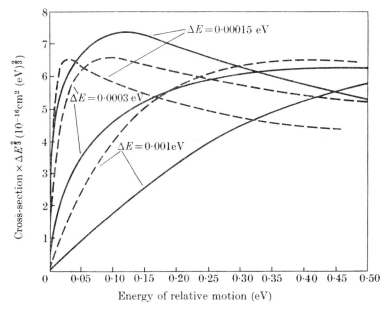

FIG. 18.126. Variation of cross-section for transfer of excitation between two atoms as a function of the energy of relative motion for different resonance defects ΔE, the reduced mass being 10 a.m.u. —— calculated using approximation (474). – – – calculated using approximation (480).

experimental data on charge-transfer cross-sections when resonances is imperfect. We shall defer discussion of their analysis until Chapter 23 as the data concerned refer to considerably higher impact energies and have been obtained using positive ion beams.

8.4.3. *Unsymmetrical or accidental resonance.* It remains to consider any special features associated with reactions in which there is a close approach to energy resonance though the colliding atoms are not similar.

If the reaction involved is one of excitation transfer in which the transitions are associated either with dipole or quadrupole electric moments the behaviour is essentially the same whether the resonance is symmetrical or merely accidental. Because of the long range of the interaction the cross-section is independent of the behaviour at close

† RAPP, D. and FRANCIS, W. E., *J. chem. Phys.* **37** (1962) 2631.

distances and depends only on the asymptotic form of U, the energy discrepancy and the relative velocity of impact. In exact resonance, therefore, the cross-section is given by (467)–(469) no matter whether the colliding systems are similar or not.

It has been emphasized by Bates and Lynn† that this does not apply to transfer reactions of the charge transfer or s–s type for which U vanishes exponentially at large r. Because of the short range of U advantage cannot be taken of the resonance to produce a transition at a large separation. By the time the separation has been reduced to R_0, say, so that U is significant, $V_i(R_0)$ and $V_f(R_0)$ will be appreciable and $|V_i(R_0)-V_f(R_0)| = \Delta E(R_0)$ will be finite. The cross-section then behaves rather as if the energy discrepancy were $\Delta E(R_0)$ instead of zero. Thus it will be very small at low relative velocities v of impact until $R_0 \Delta E/\hbar v$ is of order unity or less, so the shape of the cross-section velocity curve will resemble that for imperfect rather than exact resonance. The maximum cross-section will also be considerably smaller than for symmetrical resonance, to which the same considerations do not apply because, as mentioned in § 8.1.1, no electronic transition is involved. These conclusions are confirmed by a detailed analysis carried out by Bates and Lynn.†

8.4.4. *Summarizing remarks.* The following general conclusions emerge.

(*a*) (i) Cross-sections for excitation transfer involving optically allowed transitions in both atoms will normally be much larger (100 times or so) than gas-kinetic when energy resonance is close.

(ii) This applies whether or not the resonance is symmetrical or accidental.

(iii) The cross-section, for exact resonance, falls off as v^{-1}, where v is the relative velocity of impact.

(*b*) Similar conclusions apply to excitation transfer in which the transitions in each atom are associated with electric quadrupole moments but the resonant cross-section will be smaller, of order 10 times the gas-kinetic. The cross-section for exact resonance will fall off as $v^{-\frac{1}{2}}$.

(*c*) (i) Less regular behaviour can be expected of cross-sections for excitation transfer involving s–s transitions or for charge transfer.

(ii) For such cases the cross-section for the symmetrical resonance case will in general fall gradually with increasing velocity and will be larger than the gas-kinetic. There are exceptions to this as witnessed by transfer collisions between normal and metastable helium atoms.

† BATES, D. R. and LYNN, N., *Proc. R. Soc.* A**253** (1959) 141.

(iii) For cases in which the resonance is accidental, the cross-section will be small at low velocities, rising to a maximum, not much greater or even smaller than the gas-kinetic, before declining gradually as the velocity further increases.

(d) In *all* cases of imperfect resonance the cross-section behaves as a function of velocity in the same way as in (c) (iii). In general, for similar reactions the maximum occurs at higher velocities the greater the energy discrepancy ΔE and its magnitude is also smaller the larger is ΔE.

It cannot fairly be said at the time of writing that these conclusions have all been verified by experiment. The experimental study of excitation transfer (§ 4.2) provides general but not detailed support for the belief that the cross-section increases as ΔE decreases, other factors being equivalent. On the other hand, probably because of experimental difficulties, there is no observational evidence available which indicates that near-resonance cross-sections for transfer involving optically allowed transitions are in general larger than for other cases. Nevertheless, observed results on cross-sections for disorientation and disalignment of excited atoms by collisions with atoms of the same kind are in semi-quantitative agreement with the theory (§ 8.3.3.1), which is essentially just that for cases in which the interaction U falls off as r^{-3} and exact resonance prevails. It seems likely that, although the above conclusions lack experimental confirmation, they are nevertheless correct and provide a basis for interpretation of observed results.

8.4.5. *Collisions in which transitions occur between fine-structure levels.* We now consider collisions that lead to transitions between fine structure levels such as the $P_{\frac{1}{2}}$–$P_{\frac{3}{2}}$ levels of alkali-metal atoms. Some general theoretical features of these collisions have already been pointed out in § 8.3.6 in connection with the interpretation of the magnitudes of the cross-sections for disorientation of atoms in such excited states by collision with other like or unlike atoms.

If the energy separation ΔE between the levels is such that

$$\Delta E \ll \hbar/\tau, \qquad (485)$$

where τ is the mean duration of a collision, the electron spin vector **S** remains unchanged in the collision while the orbital angular momentum vector **L** is reoriented. After the collision **L** and **S** are recoupled giving comparable probabilities of leaving the system in a state with any value of the total electron angular momentum quantum number between $|L-S|$ and $L+S$. Since the cross-section for reorientation of **L** will be of the same order as the disorientation cross-section Q_{or} we have discussed, the cross-section for excitation of a fine structure transition will also be comparable with Q_{or}, i.e. would be expected to be of order 10^{-15}–10^{-14} cm^2.

In the opposite extreme, where $\Delta E \gg \hbar/\tau$, **S** and **L** remain coupled during the collision and change of **J** involves change of **S**. Since spin-orbit interaction is so weak, as witnessed by the cross-sections tabulated in Table 18.25, the cross-section for a collision in which **J** changes should be very small.

These general considerations are borne out very well by the observed results for the $P_{\frac{1}{2}} \rightleftharpoons P_{\frac{3}{2}}$ transitions in the alkali-metal atoms.

In proceeding along the sequence Na, K, Rb, Cs, ΔE increases steadily, the respective values being 0·0021, 0·007, 0·03, and 0·07 eV. For excitation of the transition $P_{\frac{1}{2}} \rightarrow P_{\frac{3}{2}}$ by Ar at temperatures between 300 and 400 °K the corresponding cross-sections (see Table 18.11) are, in units 10^{-16} cm^2, 60, 28, 10^{-3}, and 10^{-5}. This is just what would be expected if $\Delta E \ll \hbar/\tau$ for Na and K, $\gg \hbar/\tau$ for Rb and Cs, and increases steadily in going from Na to Cs. If we write $\tau = a/v$, where v is the mean relative velocity of collision, the results are consistent with $a = 2\cdot 5 \times 10^{-8}$ cm, a reasonable value.

It will be noticed from Table 18.9 that, in contrast, fine structure excitation cross-sections for collisions with like atoms show no marked change in passing from K to Cs (no observations are available for Na) and are of order 10^{-15} cm^2 or larger. In these cases resonant spin exchange may occur between the colliding atoms, so that the value of S for the excited atom may be changed without requiring spin-orbit interaction. We would therefore expect the cross-sections to be of the same order as for disorientation for all the atoms.

Mandelberg† has carried out detailed numerical calculations for collisions between alkali-metal atoms and rare-gas atoms using the coupled equations (Chap. 7, § 5.3.1) of time-dependent perturbation theory. All six states of the alkali-metal atom with $J = \frac{3}{2}$, $\pm M_J = \frac{3}{2}, \frac{1}{2}$ and $J = \frac{1}{2}$, $M_J = \pm\frac{1}{2}$ were included but no other states. The basic interaction between the atoms was taken to be of the van der Waals form

$$V(\mathbf{R}) = -\frac{1}{2}\frac{\alpha e^2}{R^6} r^2(3\cos^2\Theta + 1), \tag{486}$$

where α is the polarizability of the rare-gas atom and Θ is the angle

---

† MANDELBERG, H. I., *Conference on Heavy Particle Collisions*, Queens Univ., Belfast, 1968; see also CALLAWAY, J. and BAUER, E., *Phys. Rev.* **140** (1965) A 1072.

between the relative position vector **R** of the two atoms and that **r** of the alkali-metal electron relative to its nucleus.

The coupled equations were solved approximately by using an extension of the method described on p. 1899. Thus the equations take the form

$$\frac{da_n}{dt} = -(i/\hbar) \sum_m V_{nm} a_m(t) \exp(-i\omega_{mn} t), \quad (487)$$

where V_{nm} is the matrix element of V between the states n and m of energies E_n, E_m and
$$\omega_{mn} = \hbar^{-1}(E_m - E_n).$$

The suffixes distinguish the six separate states over which the sum on the right-hand side is carried out. The set (487) may be written in matrix form

$$\frac{d\mathbf{a}}{dt} = -i\mathbf{Q}(t)\mathbf{a}(t), \quad (488)$$

where
$$Q_{nm} = \hbar^{-1}V_{nm} \exp(-i\omega_{mn} t). \quad (489)$$

Corresponding to (413), (414), we take as approximate solutions,

$$\mathbf{a}(t) = \exp\left\{-i \int_{-\infty}^{t} \mathbf{Q}(t') \, dt'\right\} \mathbf{a}(-\infty), \quad (490)$$

which is exactly correct if $\mathbf{Q}(t)$, $\mathbf{Q}(t')$ commute.

A considerable amount of numerical work is still necessary even with this approximation (see p. 1899). Mandelberg, using an electronic computer, obtained the cross-sections for excitation of the fine-structure transitions by collisions with rare-gas atoms which are given in Table 18.40. Comparison with observed values, also given in the table, shows good quantitative agreement, but the calculated cross-sections exceed the observed by factors that tend to increase in going from K to Cs. This is not surprising as the approximation (490) could hardly be expected to hold well over more than four orders of magnitude in the derived cross-section.

Mandelberg's calculations illustrate clearly a further point discussed in § 8.3.6. For Rb and Cs for which $\Delta E \gg \hbar/\tau$ the cross-sections for transitions in which $\Delta J = 0$, $\Delta M_J = \pm 2$ are very much greater than for $\Delta J = 0$, $\Delta M_J = \pm 1$.

Thus for collisions of Ar with Na, K, Rb, Cs Mandelberg finds for $\Delta J = 0$, $\Delta M_J = \pm 2$, cross-sections, in 10^{-16} cm^2, of 55, 102, 123, and 135 respectively, whereas for $\Delta M_J = \pm 1$, $J = \frac{1}{2} \to J = \frac{1}{2}$, the corresponding values are 34, 16, 2, and 0·6. This means that the disorientation cross-section Q_{or}, while of order of magnitude 10^{-15} cm^2 for all alkali-metal atoms when $J = \frac{3}{2}$, falls when $J = \frac{1}{2}$ from this order of magnitude for Na to much smaller values for Rb and Cs. This is exactly as predicted from the qualitative discussion of § 8.3.6. Mandelberg obtains similar results for collisions with other rare-gas atoms.

TABLE 18.40

Cross-sections (in 10^{-16} cm^2) for excitation of $^2P_{\frac{1}{2}}$–$^2P_{\frac{3}{2}}$ transitions in alkali-metal atoms by collisions with rare-gas atoms, at 373 °K

Rare gas	Alkali metal	Na 3P		K 4P		Rb 5P		Cs 6P	
		Calc.†	Expt.‡	Calc.	Expt.	Calc.	Expt.	Calc.	Expt.
He		20	41	147	41	1·9	0·12	0·36	0·0004
Ne		27	36	9	94$_5$	0·43	0·0023	0·079	0·0003
Ar		37	65	7	22	0·25$_5$	0·0016	0·046	0·0005
Kr		39·5	—	5	41	0·16	0·0015	0·027	0·0018
Xe		43	—	4	72	0·13$_3$	0·0021	0·021	0·0027

† MANDELBERG, loc. cit. ‡ See Table 18.9.

8.5. Collisions in which electron spin-flip occurs

Experimental methods of measurement of cross-sections for collisions of rare-gas atoms and of various molecules with alkali-metal atoms which produce spin flip of the outer electron have been described in § 7. Observed cross-sections are given in Table 18.25.

In the discussion of these results in § 7 it was pointed out that, for collisions with rare-gas atoms, spin flip must occur through spin-orbit interaction between the alkali-metal electron and the nucleus and electrons of the rare-gas atom.

Taking a coordinate system with the C.M. of the colliding atoms as origin, the spin-orbit interaction can be written as

$$V_{so} = V_{so}^{(n1)} + V_{so}^{(n2)} + V_{so}^{(e1)} + V_{so}^{(e2)},$$

where§
$$V_{so}^{(n1)} = \tfrac{1}{2}(\mu_b/c)Ze|\mathbf{r}-\mathbf{R}|^{-3}\{(\mathbf{r}-\mathbf{R})\times\tfrac{1}{2}\mathbf{v}\}.\mathbf{s},$$
$$V_{so}^{(n2)} = -\tfrac{1}{2}(\mu_b/c)Ze|\mathbf{r}-\mathbf{R}|^{-3}\{(\mathbf{r}-\mathbf{R})\times\mathbf{v}_N\}.\mathbf{s},$$
$$V_{so}^{(e1)} = -\tfrac{1}{2}(\mu_b/c)e\sum_j |\mathbf{r}-\mathbf{r}_j|^{-3}\{(\mathbf{r}-\mathbf{r}_j)\times\tfrac{1}{2}\mathbf{v}\}.\mathbf{s},$$
$$V_{so}^{(e2)} = \tfrac{1}{2}(\mu_b/c)e\sum_j |\mathbf{r}-\mathbf{r}_j|^{-3}\{(\mathbf{r}-\mathbf{r}_j)\times\mathbf{v}_j\}.\mathbf{s}.$$

Here μ_b is the Bohr magneton, Ze the charge on the nucleus of the rare-gas atom. $\mathbf{r}$ is the coordinate and $\mathbf{v}$ the velocity of the alkali-metal electron, $\mathbf{R}$, $\mathbf{v}_N$ the corresponding quantities for the nucleus of the rare-gas atom, and $\mathbf{r}_j$, $\mathbf{v}_j$ for the jth electron of this atom. $\tfrac{1}{2}\mathbf{s}\hbar$ is the spin angular momentum operator of the alkali-metal electron.

§ VAN VLECK, J. H., *Rev. mod. Phys.* **23** (1951) 213.

The four terms represent respectively the interaction of the spin of the alkali-metal electron with the electric field of the nuclear charge, the magnetic moment due to the nuclear motion, the electric field due to the rare-gas atomic electrons, and the magnetic moment due to the motion of these electrons.

In first order the only non-vanishing contribution of these interactions to the probability of spin flip per collision comes from $V_{\rm so}^{(n2)}$. The order of magnitude of this contribution can be readily estimated. The mean value of the nuclear velocity v_N is of order $(e^2/\hbar)(m/M)^{\frac{1}{2}}$, where m, M are the masses of the electron and rare-gas atom respectively. Taking the mean value of $|\mathbf{r}-\mathbf{R}|$ as of order a_0/Z, the probability per collision could be expected to be of order

$$Z^4(\hbar/mc)^2(e^2/\hbar c)^2(m/M)a_0^{-2} = Z^4(e^2/\hbar c)^4 \frac{m}{M}.$$

For helium this is less than 10^{-11}, which is somewhat smaller than the observed value which is about 10^{-9}. As we have only made a very rough estimate the discrepancy is not serious.

The electron–electron interactions make a contribution in second order which may be comparable with the first order contribution from $V_{\rm so}^{(n2)}$.

At the time of writing no detailed calculations have been carried out.

8.6. *Remarks on collisions involving molecules in which electronic transitions occur*

In dealing with collisions in which more than two atoms are involved the course of the reaction can no longer be traced out completely by following the relative motion in relation to a single set of potential energy curves. The potential energy for a given state of the polyatomic system will depend on more than one coordinate defining the configuration of the system. Even in the simplest case of three atoms, as, for example, in discussing the quenching of excited atoms by diatomic molecules, three such coordinates at least are required and, instead of a potential-energy curve for a given electronic state, it is necessary to consider a three-dimensional potential energy hypersurface.

The course of a reaction may be represented by the path of a representative point on the hypersurface,† starting from given initial conditions. On tracing out such paths the chance of a jump occurring to another hypersurface corresponding to a different electronic state may be ignored except where two such surfaces intersect. To examine the

† See GLADSTONE, S., LAIDLER, K., and EYRING, H., *The theory of rate processes*, Chap. iii (McGraw-Hill, New York, 1941).

consequence of a jump the Franck–Condon principle may be assumed—the configuration coordinates and velocities remain unchanged during the electronic transition. The chance of the jump taking place depends on the same formula as (455), the relevant slope of the functions V_i and V_f and component of velocity v being taken along the tangent to the path of the representative point at the hypersurface intersection concerned. Although it is usually out of the question to attempt to trace out the possible reaction paths in detail in all but the simplest cases, there are two general features distinguishing the polyatomic from the diatomic case which emerge, both due to the extra degrees of freedom.

In the diatomic case the probability of a reaction may be small due to the absence of any crossing or critical point between the corresponding potential-energy curves and of a third curve that interacts with both. The chance of a small probability in the polyatomic case from this cause is much less because of the greater number of degrees of freedom open to the representative point.

The second effect, which also tends to increase the probability of reaction, is that the chance of an electronic transition being reversed on the outward path is much reduced. So many paths are available that the chance that the path of the representative point will pass twice through the same intersection between two hypersurfaces is usually very small. The chance of an electronic transition persisting to the conclusion of the collision is then no longer given, in the notation of (455), by $2P(1-P)$ but more nearly by P. This is of order unity when the crossing point occurs with the atoms not too far apart, and there are no selection rules (such as forbidden change of multiplicity) operating. In the diatomic case the factor $1-P$, allowing for reversal of the process on the return path, very much reduces the chance of the transition persisting. For the polyatomic case the chance will remain high.

The chance of reversal may nevertheless be appreciable even in the polyatomic case if the separation of the interacting hypersurfaces is small over a wide region. Thus the chance of a transition involving a considerable change of electronic energy may be greater than that when a small change only is involved. This is in sharp contrast to the diatomic case where a resonance effect occurs as far as the change of electronic energy is concerned. It is not correct to suppose, however, that in the polyatomic case a large amount of electronic energy is transferred to energy of relative translation. Instead it will, in general, be partly or wholly taken up as energy of molecular vibration. A change of curvature of the path of the representative point corresponds to excitation of vibrational degrees of freedom and is almost certain to occur when a jump takes place to a different potential-energy hypersurface.

Summarizing these conclusions, we should expect that molecules would be much more effective than atoms in producing quenching of excited atoms and that large cross-sections are likely to be associated with large changes in electronic excitation. If the change is too large, however, the cross-section may be small even in the polyatomic case because of the absence of an intersection between the respective hypersurfaces or of a hypersurface intersecting both. The excess energy released by the quenching will be partly taken up as molecular vibration. There seems, however, to be no reason why all, or almost all, of this energy should be taken up by excitation of internal molecular motion. Nevertheless, although there are many exceptions it seems to be the case that in most reactions, including those between unexcited atoms and molecules (see Chap. 17, § 6.6), the amount of energy transferred to relative translation is small.

19

COLLISIONS UNDER GAS-KINETIC CONDITIONS—IONIC MOBILITIES AND IONIC REACTIONS

1. Introduction

WE now consider collisions under gas-kinetic conditions in which one of the colliding systems is ionized. In this connection a great deal of experimental and theoretical work has been devoted to the study of the mobility of ions in gases.† It was not at first realized that, during its passage through the gas, an ion may react with a gas molecule to produce an ion of a different species. In particular, the presence of polar impurities such as water vapour leads to clustering of the impurity molecules to the ions so that their mobility is greatly reduced. The importance of removing impurities of this kind was recognized before the last war but not much attention was paid, even at that time, to the possibility of reactions with other molecules, although Tyndall and his associates had observed the formation of compounds of alkali-metal ions with rare-gas atoms under certain conditions. It is now fully recognized that the occurrence of ionic reactions is the rule rather than the exception and attention is concentrated on the study of ionic reaction rates. For this purpose the experimental observation of mobilities remains of importance though new and powerful alternative techniques have been developed.

The first part of this chapter is concerned with the behaviour of positive ions. We begin by discussing the experimental and theoretical study of the mobility of alkali-metal ions in the rare gases, the cases which are least complicated by the occurrence of ionic reactions. The remainder of the first part then deals essentially with the study of ionic reactions, starting with a discussion of cluster formation and then proceeding to the simplest reactions involving rare-gas ions in their parent gases, the main features of which were revealed from observations of mobilities. We then proceed to describe further techniques for studying ionic reactions, including particularly the flowing afterglow method

† For a discussion of much of the earlier experimental work see TYNDALL, A. M., *The mobility of positive ions in gases* (Cambridge University Press, 1938).

which has proved so effective. The first part then concludes with a discussion of some of the results obtained. As in the preceding two chapters, we have been rather selective in the choice of reactions discussed because we would otherwise be led far into the field of ion chemistry. Our choice has been oriented towards reactions involving the simpler ions and molecules and those that are of interest in atmospheric physics.

In the second part of the chapter we deal with reactions involving negative ions, some of which may be studied by techniques that are not available for positive ions.

2. The behaviour of positive ions in gases—the mobility of positive ions

2.1. *Introductory remarks*

As pointed out above, cluster formation due to attractive forces between ions and polar molecules occurs so readily that, in order to measure the true mobility of unclustered ions, very special precautions must be taken to remove any polar impurities such as water vapour.† To reduce further the chance of clustering due to residual impurities, the measurement should be made as soon as possible after the ions enter the gas. It is also very desirable to have some means available for testing whether any clustering has occurred. In the early experiments, in which these precautions were not taken, low values of the mobility were found which were practically independent of the nature of the initial ion, depending only on the gas concerned.‡ Estimates of the mass of the ion made in terms of Langevin's theory (§ 2.4) indicated that cluster formation must have occurred. At this stage it was not clear whether the cluster was composed of atoms of the main gas or impurities. Experiments on the mobilities of electrons§ (see Chap. 2) threw light on this point. It was found that, in moist air, the mobilities of both electrons and positive ions are small and nearly the same, indicating that clusters had formed. On drying the air, the electron mobility increases to the normal value for electrons unaccompanied by molecular clusters. This shows that the molecules involved in the cluster are not those of the main gas but of the polar impurity, water. In these experiments

---

† TYNDALL, A. M. and PHILLIPS, L. R., *Proc. R. Soc.* **A111** (1926) 577; ERIKSON, H. A., *Phys. Rev.* **33** (1929), 403; **34** (1929) 635.
‡ See, for example, GRINDLEY, G. C. and TYNDALL, A. M., *Phil. Mag.* **48** (1924) 711.
§ See for example, FRANCK, J. and POHL, R. W., *Verh. dt. phys. Ges.* **12** (1910) 291, 613; TOWNSEND, J. S. and TIZARD, H. T., *Proc. R. Soc.* **A87** (1912) 357; ibid. **A88** (1913) 336.

the mobility of the positive ions was not increased by the drying, which must be effected to a much greater extent than for electrons—in diffusing a given distance positive ions take much longer than electrons and hence make many more collisions with impurity molecules. Evidence in favour of this was obtained from the experiments of Erikson† on the variation of mobility of positive ions with age. He was able to show that ions with age as short as 1/500 s had a mobility about 25 per cent higher than ions of the mobility characteristic of the situation in which clustering is complete.

The analysis of observed mobility data in terms of basic interactions between the ions and the atoms or molecules of the gas through which they are drifting under the action of a uniform electric field F is more complicated than for electrons. This is because it is not possible to obtain the velocity distribution function of the ions as for electrons in the form of a rapidly convergent expansion in velocity space (see Chap. 2) because the ratio M_1/M_2 of the mass M_1 of the ion to that M_2 of the gas molecule is not small.

However, it is true that, as for electrons, the drift velocity u, at a fixed temperature T, is a function of F/p where p is the gas pressure. Since the mobility μ is u/F it follows that μp is also a function of F/p. Further, it may be shown that, for small F/p, this function is a constant and equal to $eDp/\kappa T$, where D is the diffusion coefficient of the ions in the gas at pressure p and T the temperature of the gas. The zero-field mobility is thus inversely proportional to the pressure and it will be given in cm s^{-1} per V cm^{-1} at 1 atm pressure and a specified temperature, usually 0 °C. Measurements at other temperatures are usually reduced so as to refer to the pressure at this standard temperature‡ (see § 2.3 and Table 19.2 in which 18 °C is the standard temperature). It is to be noted that this is not the same convention as used for specifying the diffusion coefficient which is also inversely proportional to the pressure but given for a pressure of 1 torr, again at a specified temperature.

It has been shown by Wannier§ that, if the collisions between ions and molecules are essentially classical and lead to an isotropic distribution of scattering, and are such that the collision frequency is constant,

---

† loc. cit.

‡ To avoid confusion it is preferable to work in terms of F/N where N is the concentration of gas molecules under the experimental conditions but, as for electrons in Vols. I and II we retain F/p as the variable, where p refers to the standard temperature, because of earlier usage but often give F/N at the same time.

§ WANNIER, G. H., *Phys. Rev.* **83** (1951) 281.

then μp is constant for all F/p. Also the mean energy of a drifting ion is given by
$$\bar{\epsilon} = \tfrac{1}{2}M_1 u^2 + \tfrac{1}{2}M_2 u^2 + \tfrac{3}{2}\kappa T. \tag{1}$$

These simple results apply when the collisions between ions and molecules are determined to a dominating extent by polarization (see § 2.4). This will be the case at low temperatures and mean energies $\bar{\epsilon} \simeq \tfrac{3}{2}\kappa T$. At higher temperatures and higher F/p, for which $\bar{\epsilon} \gg \tfrac{3}{2}\kappa T$, the situation is more complicated.

As an indication of what is to be expected, consider ions drifting at such large values of F/p that $\bar{\epsilon} \gg \tfrac{3}{2}\kappa T$. Suppose that in each collision an ion loses the whole of its initial kinetic energy. The velocity acquired between two collisions by an ion of charge e is $(eFx/2M_1)^{\frac{1}{2}}$, where x is the free path between the collisions. Allowing for the distribution of free paths the mean velocity of motion of the ions in the gas in the direction of the field is

$$u = (eF/2M_1 l^2)^{\frac{1}{2}} \int_0^\infty x^{\frac{1}{2}} e^{-x/l} \, dx = \tfrac{1}{2}\{(\pi e F l)/2M_1\}^{\frac{1}{2}}, \tag{2}$$

provided the mean free path l is a constant. Since
$$l = 1/nQ,$$
where Q is the collision cross-section and n the concentration of gas molecules at pressure p, we have
$$u = (95\pi e/M_1 N Q)^{\frac{1}{2}}(F/p)^{\frac{1}{2}}, \tag{3}$$
where N is the concentration of gas molecules at a pressure of 760 torr.

This result was first obtained by Sena† who considered it to be a good approximation for cases in which charge-transfer collisions are predominant. In each such collision the transfer of charge from the ion to the gas atom effectively produces an ion with negligible kinetic energy. Wannier has shown, however, that if the collisions take place as between rigid spheres the drift velocity again comes out to be proportional to $(F/p)^{\frac{1}{2}}$, the formula differing from (3) only in the replacement of the numerical factor 95 by 318. He has also outlined a procedure for dealing with other types of collision but the elaborate calculations required have not been carried out.

Because of the difficulty of analysis of data obtained at high F/p it is important that observations are taken to low enough F/p to provide accurate interpolation to zero field.

† SENA, L., *J. Phys. USSR*, **10** (1946) 179.

2.2. Techniques for measuring mobilities of unclustered positive ions

2.2.1. *The electrical-shutter method.* A thorough study of the mobilities of unclustered ions was carried out by Tyndall and his associates[†] using an apparatus based on a method first suggested by van de Graaf.[‡]

The principle is the same as that described in Chapter 2 for measuring electron mobilities, but a different type of shutter is employed. Instead of an electron filter a pair of perforated electrodes such as CD in Fig. 19.1

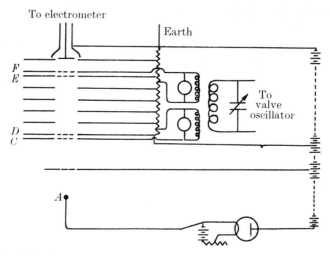

FIG. 19.1. Apparatus used by Tyndall and Powell for measuring ionic mobilities.

is used. If an alternating potential is applied between these, ions are only allowed to pass through both during one half-cycle. To reduce the proportional time during which ions may pass through, a steady potential of magnitude slightly less than the peak alternating potential, and in a sense to oppose the passage of the ions, is applied. Fig. 19.1 illustrates the electrode arrangement. A is the ion source. CD and EF are the two shutters between which the electric field is maintained uniform by four equidistant guard rings.

Owing to the variable field between the plates of the shutter an end correction is required in order to obtain an absolute value for the mobility. This correction does not affect relative values obtained with apparatus of given dimensions, provided the ratio of main field to peak shutter

---

[†] See TYNDALL, A. M., *Mobilities of positive ions in gases* (Cambridge University Press, 1938).
[‡] VAN DE GRAAF, R. J., *Phil. Mag.* **6** (1928) 210.

potential is unchanged. To determine the correction only one absolute determination is required. This was carried out for helium ions in helium by Tyndall and Powell.† They used an apparatus in which the distance DE could be varied and measured the electrometer current for different distances DE and constant values of the frequency and main electric field. The current is a maximum when the distance DE is covered in a whole number of cycles. Hence the change in DE between two successive current maxima gives the distance the ions travel in a cycle, the end correction being eliminated.

Special precautions were taken to avoid impurities. The whole apparatus was constructed entirely of metal and of Pyrex glass so that it could be thoroughly baked. Only specially purified gas was admitted, liquid-air traps being inserted between the apparatus and the last tap of the gas plant to exclude mercury and tap-grease vapours. In later experiments‡ tap-grease contamination was entirely eliminated by replacing the taps by mercury traps. To reduce further the absolute concentration of impurity the work was carried out at relatively low pressures (3–60 torr). The length of path DE was usually about 1·5 cm. The alternating frequency ranged from 1000 to 175 000 c/s so that drift velocities at values of F/p ranging from 2 to 50 V cm^{-1} at 3 torr could be studied. With this range available the age of the ions investigated was from 10^{-5} to 2×10^{-3} s, within which no evidence of any clustering was found.

It is an advantage of the shutter method that this can be checked by observing whether the successive peaks of electrometer currents occur at frequencies which are accurately in the ratio 1:2:3, etc. The homogeneity of the ions can also be tested from the electrometer current-frequency curve as it essentially provides a mass spectrum in first, second, etc., orders.

The electron filter used by Bradbury and Nielsen as a shutter for electron mobility investigations (see Chap. 2, § 3.1) has also been used for studying positive ion mobilities. It is rather less convenient for this purpose as it is necessary to employ very close spacing of the grid wires as well as high alternating potentials.

A shutter method has also been used by Samson and Weissler§ in conjunction with a photo-ionization source of positive ions. Using monochromatic radiation of a suitably chosen wavelength it is possible to

---

† TYNDALL, A. M. and POWELL, C. F., *Proc. R. Soc.* A**134** (1931) 125.
‡ Cf. MUNSON, R. J. and TYNDALL, A. M., ibid. **177** (1941) 187.
§ SAMSON, J. A. R. and WEISSLER, G. L., *Phys. Rev.* **137** (1965) A381.

produce unexcited ions in the presence of neutral molecules exclusively in the ground state (see also § 3.4.1). With a discharge source a variety of excited and dissociated species are present, and this complicates the interpretation of some of the experiments (see § 3.2.1). In their work Samson and Weissler used a high-voltage condensed spark discharge within a ceramic capillary in a low-pressure gas. This produced an intense line spectrum between 500 and 1000 Å with a duration of about 1 μs and repetition rate of 90 s^{-1}. Radiation with a 10 Å (0·25 eV) band pass was provided by a normal incidence, vacuum monochromator.

Experiments in O_2 were carried out in which the ions were produced by a line at 833 Å (14·88 eV) with frequency between the first (1226 Å) and second (770 Å) ionization thresholds. Thus only O_2^+ ions in their ground states were produced from the source.

2.2.2. *Hornbeck's method.* A time of flight method with high time resolution was introduced by Hornbeck.† This made it easier to identify separate ionic species by their arrival times at the collector. The method is, however, limited to the study of the mobilities of ions in their parent gases. Its use for the measurement of electron mobilities has already been described in Chapter 2, § 3.2.

The principle of the method (see Fig. 19.2) consists in observing the form of the transient current produced between two electrodes in the gas under investigation when electrons are released photoelectrically from one electrode, the cathode, by means of a light pulse of 0·1 μs duration. An electric field is applied between the electrodes so that the photoelectrons drift rapidly to the anode. In the course of their passage through the gas they produce positive ions which are collected at the cathode. Consider now the form of the current transient which will be observed, on the following assumptions.

(a) The electron drift velocity is so very much greater than that of any positive ions that it may be presumed infinite.
(b) All ions possess the same velocity as the drift velocity u.
(c) Each ion collected at the cathode releases γ_i secondary electrons.

Initially, through impact ionization, there will be an exponential distribution of positive ions between the electrodes determined by the ionization coefficient α_i for the electrons in the gas under the prescribed ratio of electric field F to gas pressure p. As these ions will all be drifting towards the cathode with velocity u they will give rise to an initial

† HORNBECK, J. A., *Phys. Rev.* **83** (1951) 374; **84** (1951) 615.

current in the tube proportional to u and the total number of ions present in the gas. In general, ions will not be produced in appreciable quantities until the photoelectrons have drifted a distance x_0 sufficient for them to have acquired, from the field, the necessary ionizing energy. No positive ions will reach the cathode until a time $t_0 = x_0/u$ elapses so that, until this time, the positive ion current will remain constant. It will thereafter decrease steadily with time until a time t_1 has elapsed where $t_1 = l/u$, l being the electrode separation. Following our assumption (b), no ion

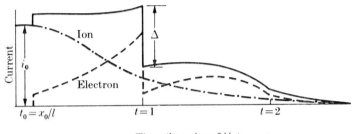

FIG. 19.2. Illustrating the idealized form of the transient current as a function of time following the formation, at $t = 0$, of an exponential distribution of positive ions between the electrodes. $-\cdot-\cdot$ ion current, $---$ electron current, ——— total current.

will take longer than t_1 to move from the neighbourhood of the anode to the cathode. The current due to the positive ions produced by the photoelectrons will therefore vanish at greater times, so its contribution to the total transient will be as shown in Fig. 19.2.

Account must now be taken of the contribution from electrons released at the cathode by positive ion impact. This will begin at a time $t_0 = x_0/u$ and increase, following the initial distribution of positive ions, until the time t_1 after which all these ions will have been collected. The shape of this contribution will therefore be as shown in Fig. 19.2, dropping suddenly to zero at t_1.

Actually, the total current will not vanish at t_1 because the secondary electrons will in turn produce more positive ions and hence a further set of electrons as these ions are collected. From such effects contributions as shown in Fig. 19.2 will arise for $t > t_1$. Adding all these, the final form of the transient will be as indicated, at least according to our assumptions. In fact, the ions will possess random velocities as well as the drift velocity so that the sharpness of the discontinuity at $t = t_1$, will be reduced. Fig. 19.3 illustrates a typical transient as actually observed. The oscillations in the initial stages are due to initial oscillations in the positive ion density. By measuring the time t_1 from an oscillogram such as in Fig. 19.3 the mobility of the ions concerned may be determined.

Effects due to release of electrons from the cathode by metastable atom impact are not observed because the time of diffusion of such atoms to the cathode is of the order 700 times larger than t_1. The only possible source of confusion could be effects due to resonance radiation from excited atoms. Not only would such radiation eject secondary

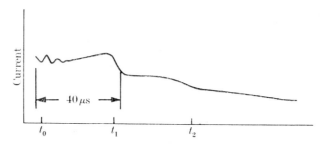

Fig. 19.3. Oscillogram showing the form of the transient signal observed by Hornbeck with argon as the working gas.

electrons from the cathode but the time required for it to diffuse across the space between the electrodes could be of the same order as t_1. The possibility of such effects arising should be recognized in analysing the experimental results.

Fig. 19.4 illustrates the general arrangement of the apparatus. The spark light source operated at a repetition frequency of 60 c/s. The light pulses were focused by a quartz lens system and passed through the perforated anode to fall on the cathode. By means of an external magnet the electrode separation, which was measured by a travelling microscope, could be varied. The tube current, of order $1\ \mu A$, was kept small so that any distortion by space charge was negligible. It was recorded through generation of a voltage between the ends of a resistor through which it flowed. This voltage was impressed across the vertical plates of a cathode-ray tube, the horizontal sweep of which was triggered by low capacitance coupling to the spark electrodes.

The electrodes were nickel discs, 2 inches in diameter. A uniform surface coating of mixed barium and strontium was given to the cathode by first coating with the carbonates, reducing to the oxides during outgassing, and then making uniform by high-frequency spark treatment.

The tube envelope was of 3-in outside diameter so the metal part could be heated by r.f. induction. To avoid effects due to charging up of the envelope walls the tube was coated inside with Aquadag. This was maintained at a potential midway between that of the anode and cathode. Before introduction of gas into the tube it was baked to 450 °C and the electrodes were heated to about 1000 °C.

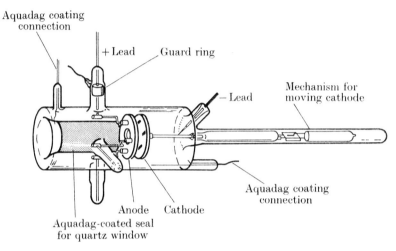

Fig. 19.4. General arrangement of apparatus used by Hornbeck for measuring mobilities of positive ions in gases.

As may be seen from the time scale on the typical record shown in Fig. 19.3 the ions observed have a very short lifetime and have little opportunity to form clusters with impurities. If the transit times of two ions differ by more than 10 per cent they may be recorded. On the other hand, the method is limited to ions produced in the gas through which they drift. While suitable for measurement at high values of F/p it is not satisfactory at such low values that the mobility is effectively independent of the field (see p. 1935).

A modification of the method to obtain data at low gas temperatures was used by Beaty.† Fig. 19.5 illustrates the experimental tube which could be immersed as a whole in a refrigerating bath. The electrodes were 2-in discs of molybdenum separated by 1 cm. Electrons were liberated by a flash of ultra-violet light which travelled down through the quartz windows and was reflected from the cathode into a light trap at the bottom of the tube.

† BEATY, E. C., *Phys. Rev.* **104** (1956) 17.

2.2.3. *The pulse method of Biondi and Chanin.* A time-of-flight method, particularly suitable for measurement of mobilities of ions which are nearly in thermal equilibrium with the gas, was introduced by Biondi and Chanin.† As with Hornbeck's method it is, however, limited to observations of ions in their parent gases.

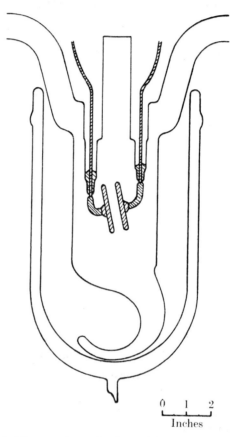

Fig. 19.5. The experimental tube used by Beaty for measuring mobilities of positive ions in gases at low temperatures.

Referring to Fig. 19.6, ions that are generated by application of a high-voltage pulse to the electrode A pass through a grid B into a drift space C throughout which there exists a constant and uniform electric field F. The ions drift across the space to a collector electrode D, and in so doing induce a current in an external circuit. This current suddenly decreases when the ions reach the collector. Hence, by recording the

† BIONDI, M. A. and CHANIN, L. M., *Phys. Rev.* **94** (1954) 910.

current-time relation, the time of transit across the drift space and thence the mobility may be determined. To eliminate errors due to the initial energies of the injected ions being greater than thermal, measurements of transit time are taken for two different drift distances. If t_1, t_2 are the transit times when the distances are l_1, l_2 respectively, then the mobility μ is given by

$$\mu = \frac{l_1 - l_2}{t_1 - t_2} \frac{1}{F}.$$

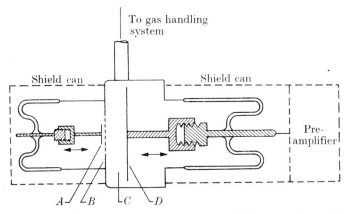

FIG. 19.6. Illustrating the general arrangement of the apparatus used by Biondi and Chanin.

Because the ions are generated outside the drift space it is not necessary that F/p should be large within that space as in Hornbeck's method, so making it possible to obtain data at very low F/p.

Fig. 19.6 illustrates the general form of the mobility tube which was of glass and Kovar. The discharge producing the ions was created by application of a pulse of 0·5 μs duration. The collector electrode D could be moved perpendicular to itself by a magnetic armature.

The ion current was measured, as in Hornbeck's method, by means of the voltage signal, of order 10^{-4} V, generated across a resistor through which the collector electrode was connected to the voltage supply. This was very small compared with the high-voltage discharge pulse (1000 V) so it was necessary to shield the discharge and drift regions very effectively from each other. This was done by means of an external shield screwed to the exterior of the metal mobility tube.

Before an experiment the tube was baked out at 420 °C for 15 hours to give a final background pressure $< 10^{-8}$ torr. The gas under investigation was then introduced through an ultra-high vacuum gas-handling system.

There is no difficulty in using this equipment for mobility measurements at low gas temperatures.

2.2.4. *Parallel-plate method.* A further variety of pulse method,† suitable for measurement of mobilities and of loss rates for positive ions in a gas, is one in which the ions are formed between plane parallel plates. Thus, referring to Fig. 19.9, suppose that ions are produced at a uniform volume rate q/cm^3 by some ionizing source. If a rectangular voltage-pulse

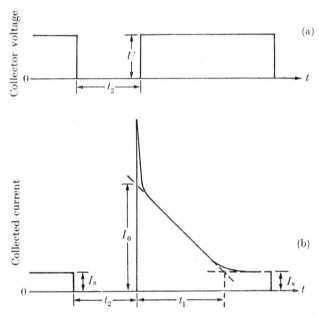

FIG. 19.7. Variation of (a) collector voltage and (b) transient current in the experiments of Young et al.

of the shape shown in Fig. 19.7 (a) is applied between the plates the terminal current collected when the voltage is on will have the form shown in Fig. 19.7 (b).

The sharp initial spike is due to electron collection. It is followed by a linearly decreasing current arising from collection of the positive ions that have accumulated between the plates while the collector voltage is off. This linearly decreasing region finally merges in a plateau of saturation current I_s which persists until the voltage shuts off once more.

Short extrapolations, indicated by the dotted lines in Fig. 19.7, enable the contributions from the accumulated positive ions and from the

† YOUNG, R. A., GATZ, C. R., SHARPLESS, R. L., and ABLOW, C. M., *Phys. Rev.* **A138** (1965) 359.

saturation current to be determined separately. As a first approximation the time t, from onset of the collector voltage pulse, taken to sweep the last of the accumulated positive ions to the collector electrode is given by

$$t = l/u = t_0, \text{ say,} \qquad (4)$$

where l is the plate separation and u the drift velocity of the ions. However, we must expect some correction to this result arising from the presence of a finite initial concentration of ions. To analyse this matter further we must consider the variation of ion concentration n^+ with time and position during the period in which there is no collector voltage.

If D is the diffusion coefficient of the ions and α the recombination coefficient for ions and electrons in the gas then

$$\frac{\partial n^+}{\partial t} = q - \alpha n^+ n_e + D\frac{\partial^2 n^+}{\partial x^2}. \qquad (5)$$

We may also write $n^+ = n_e$ if there is only one species of positive ion. At relatively high ion concentration the main loss will be by recombination (see Chap. 20); so, neglecting the diffusion term,

$$n^+(t) = (q/\alpha)^{\frac{1}{2}}[1-\exp\{-(4\alpha q)^{\frac{1}{2}}t\}]/[1+\exp\{-(4\alpha q)^{\frac{1}{2}}t\}]. \qquad (6\text{ a})$$

Hence, if q is independent of position between the plates, n^+ will also be uniform.

In the opposite extreme case of low ion concentration so that the recombination term may be neglected we have,† taking $n^+ = 0$ at each electrode at all times and throughout the region between the plates when $t = 0$,

$$n^+ = \frac{q}{2D}\left\{(\tfrac{1}{4}l^2 - x^2) - \frac{8l^2}{\pi^3}\sum_s(-1)^s(2s+1)^{-2}\exp\{-(2s+1)^2\pi^2 Dt/l^2\}\cos\{(2s+1)\pi x/l\}\right\}. \qquad (6\text{ b})$$

The first term in this series is much the most important, and gives rise to the spatial distributions of ionization at different times shown in Fig. 19.8. It will be seen that, for times t such that $4Dt/l^2 \ll 1$, the distribution is nearly uniform, whereas it is nearly parabolic when this condition is not satisfied.

The variation of the collected current with time from onset of a collected pulse may be calculated on the assumption of either a uniform or parabolic initial distribution of ionization, using the equations

$$\int_{-\frac{1}{2}l}^{\frac{1}{2}l} F\,dx = lF_0, \qquad \frac{\partial F}{\partial x} = -4\pi n^+ e,$$

$$\frac{\partial n^+}{\partial t} + \frac{\partial (n^+ u)}{\partial x} = 0, \qquad I = A\left(n^+ eu - \frac{1}{4\pi}\frac{\partial F}{\partial t}\right)_{x=\frac{1}{2}l}. \qquad (7)$$

Here lF_0 is the collector voltage and I the current collected, A being the area of a plate.

† CARSLAW, H. S. and JAEGER, J. C., *Conduction of heat in solids*, p. 131 (Oxford University Press, 1959).

If the initial distribution is uniform it is found that, provided
$$m^2 = 2\pi n_0 le/F_0 \ll 1,$$
$$2\pi Il/AuF_0 = m^2\{1-\tau+m^2\tau(1-2\tau+\tfrac{7}{3}\tau^2)\}+O(m^6), \tag{8}$$
where $\tau = t/t_0$, t_0 being as given in (4).

I vanishes when $\quad t = t_1 \simeq t_0/(1-m^2) = t_0/(1-CI_0) \tag{9}$

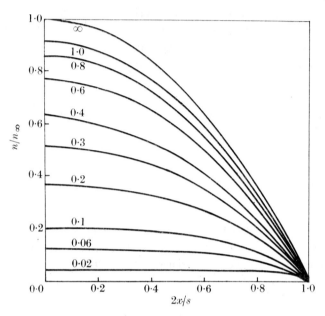

Fig. 19.8. Calculated spatial distribution of ions between plane parallel electrodes at $x = \pm\tfrac{1}{2}l$ resulting from a uniform and constant production rate when diffusion loss predominates. The different curves correspond to different values of $4Dt/l^2$, where D is the diffusion coefficient of the ions and t is the time after application of the voltage pulse. These values are indicated on each curve.

where I_0, the initial value of I (see Fig. 19.7 (b)), is given by
$$I_0 = AuF_0 m^2/2\pi l, \tag{10}$$
so that $\quad C = 2\pi l/AuF_0 = 2\pi t_0/AF_0. \tag{11}$

Fig. 19.10 (a) shows some calculated forms of the current transient for different values of m.

(9) may be rewritten in the form
$$t_1 = t_0 + Ct_1 I_0 \tag{12}$$
so that, to check whether the initial distribution is indeed nearly uniform, it is convenient to plot t_1 against $t_1 I_0$. A linear relation should result with intercept on the t_1 axis equal to t_0 and slope given by C.

If an initial ion distribution of parabolic shape is assumed (8) is replaced by
$$\frac{2\pi Il}{AuF_0} = \tfrac{2}{3}m^2(1-\tau)^2(1+2\tau)+\frac{m^4}{12}(1-\tau)p_6(\tau)+O(m^6), \tag{13}$$

p_6 being a polynomial of degree 6 in τ. It follows that I vanishes when $\tau = 1$, i.e. $t_1 = t_0$, up to times of order m^6. In this case, provided $m^2 \ll 1$, t_1 should not vary appreciably with I_0.

It is worth noting for future reference (see Chap. 20, § 4.7) that the total number of ions between the plates just before the collector pulse is applied is given by

$$N^+ = \tfrac{1}{2} \int_0^\infty I_a(t)\, dt, \tag{14}$$

where $I_a(t)$ is the contribution to the total current in the transient from the accumulated positive ions, at a time t after application of the collector pulse. If, further, the ion distribution is uniform, as will be the case where ion loss is predominantly by recombination, the concentration n^+ at the commencement of the collector pulse is given by

$$n^+ = N^+/Al. \tag{15}$$

In this way n^+ can be determined as a function of t_2, the time interval between collector pulses (see Fig. 19.7). This may be used to determine the recombination coefficient because $n^+(t_2)$ should have the form (6a) with t replaced by t_2. Application of this technique to measure the recombination of NO$^+$ ions with electrons is described in Chapter 20, § 4.7.

It is also worth noting that the saturation current I_s in the transient is related to the ion production rate q by

$$q = I_s/Ael. \tag{16}$$

Young, Gatz, Sharpless, and Ablow† have applied this technique to measure the zero-field mobility of NO$^+$ ions in helium, argon, nitrogen, and hydrogen.

The gas under investigation flowed from high-pressure tanks, through a copper line and a cold trap filled with glass beads, to a purification system composed of zirconium and titanium turnings heated to 850 °C, and thence to a second cold trap, more copper tubing, and a needle valve into a long glass tube leading to the electrode assembly. Provision was made for introduction of nitric oxide into the purified gas stream within this tube.

The gas pressure was measured with a McLeod gauge and the total flow rate monitored with a Tri-Flat ball-float flowmeter. The nitric oxide was stored in a reservoir of about 8-litre capacity which was pumped out and refilled each day. Its flow rate was measured by the rate of change of pressure in the reservoir.

† loc. cit., p. 1944.

As a check on the possible presence of contaminants containing carbon, oxygen, and sulphur, nitrogen was run through the system and excited by a microwave discharge, upstream from the electrode system. No sulphur bands were observed in the afterglow spectrum and the NO and CN bands were very weak. Further, from measurement of the rate of pressure rise of the system when closed off, and the flow rate, the contamination under operating conditions was estimated to be less than 10 p.p.m.

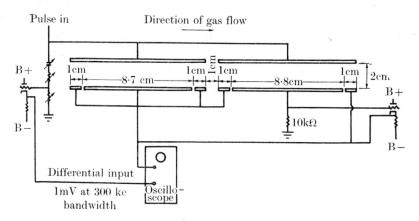

Fig. 19.9. Schematic diagram of the electrode arrangement in the experiments of Young *et al.*

The geometry of the electrode system is illustrated in Fig. 19.9. The collecting electrodes were 8·8 cm long, 4 cm wide, and separated by 2 cm. Guard electrodes 1 cm long were present on either side of a collector plate. The voltage pulses applied were of about 10 ms duration and of 10–200 V amplitude, separated by a time interval that could be varied from 0·1 ms to 1 s. The rise time of a pulse to a value constant to within 1 per cent was between 5 and 10 μs.

NO$^+$ ions were produced between the electrodes by irradiation with light from a sealed-off krypton resonance lamp with a lithium fluoride window. With low power microwave excitation the Kr resonance line at 1236 Å was excited with sufficient intensity to produce NO$^+$ by photo-ionization. The LiF window eliminated any radiation of sufficiently short wavelength to ionize any of the main gases. With 0·1 per cent of nitric oxide present the rate of photo-ionization greatly exceeded the rate of photoelectric emission of electrons from the surface of the electrodes.

Fig. 19.10 (b) illustrates a typical set of current transients. They are of the form expected for an initially uniform ion distribution between the electrodes (see Fig. 19.10 (a)). It is therefore to be expected that the time t_1 during which the accumulated charge is swept up will follow the form (12). Fig. 19.11 shows that this is indeed so for typical cases

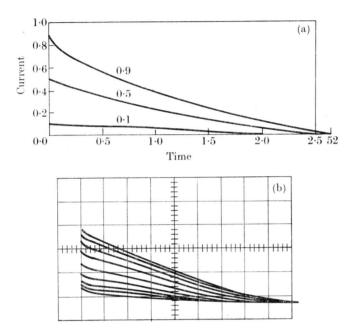

FIG. 19.10. Form of the transient curve as a function of time after initiation of the collection pulse in the experiments of Young et al. (a) Calculated presuming a uniform initial distribution of ions for $m^2 = 0.9$, 0.5, and 0.1 respectively (see (8)). (b) Observed for NO^+ ions in He at 3 torr for delay times of 7, 11, 16, 23, 31, 41, 64, 71, and 91 ms; $E_0 = 100$ V cm^{-1}, horizontal gain 0.02 ms cm^{-1}, vertical gain 1.0 μA cm^{-1}. NO partial pressure 0.004 torr, flow velocity 80 cm s^{-1}, $t_0 = 0.123$ ms, $\mu = 1.4 \times 10^4$ cm s^{-1}.

of NO in N_2. In this figure, $t_1 I_0$ (in the notation of (12)) is plotted against t_1 for three different collector voltage amplitudes. A linear plot is found in each case as given from (12). From the intercept on the t_1 axis the time t_0 is obtained and hence the drift velocity and mobility from (4). As a further check the slope of each linear plot gives the quantity C defined in (11). It was verified that the slope in each case is proportional to t_0/F_0 as required by (11) and that the constant of proportionality is of approximately the correct magnitude.

In the range of pressure p (from about 2 to 25 torr) covered in the experiments, the drift velocity of the ions was found to be proportional to F/p for values of F/p up to at least 7 V cm^{-1} torr^{-1} for He and H$_2$ and at least 15 V cm^{-1} torr^{-1} for Ar and N$_2$.

With the photo-ionizing source used the only molecules which could be ionized were those with ionization potential less than 10 V. The only possibilities, apart from NO, are free metals and complex hydrocarbons,

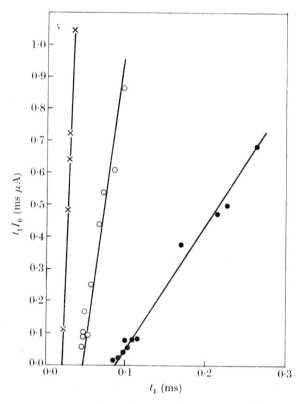

FIG. 19.11. Variation of the product $t_1 I_0$ with t_1, where t_1 is the ion transit time and I_0 the initial current (see Fig. 19.7 (b)), for NO$^+$ ions in N$_2$ at 3 torr total pressure for three values of the collection voltage pulse. × 200 V, ○ 100 V, ● 50 V.

the presence of which should certainly have been shown up in the spectroscopic tests with nitrogen. The absence of any appreciable distortion of the shape of the transients during the collection of accumulated ions provides evidence that only one ion was present. Finally, in earlier experiments, Gatz, Young, and Sharpless[†] used a similar technique to investigate the ions produced in mixtures of O and N in N$_2$. In these

[†] GATZ, C. R., YOUNG, R. A., and SHARPLESS, R. L., *J. chem. Phys.* **39** (1963) 1234.

experiments N_2 was the main gas and the microwave discharge upstream from the NO inlet was in operation. This discharge produced a concentration of N atoms which then partially reacted with the injected NO to form O atoms through the reaction (see pp. 1832, 2024)

$$NO+N \rightarrow N_2+O.$$

Ions were formed presumably as the result of reactions involving metastable species. Comparison of the shape of the usual transients produced when such ions are collected by a square-topped voltage pulse with the corresponding transients when the ions are produced by photo-ionization from a krypton resonance lamp as described above showed that the ion transit times are the same for both and that there is no evidence of the presence of more than one ion in either case. Mass-spectrometric analysis has confirmed that the ions formed by chemi-ionization are indeed NO^+.

The numerical results found for the mobilities are discussed in § 2.6 and further evidence found in support of the identification of the ions as NO^+.

2.2.5. *Mobilities derived from electron decay measurements in afterglows.* We have already discussed the application of electron concentration measurements in afterglows produced by microwave breakdown in gases to the study of electron attachment rates (Chap. 12, § 7.5.5 and Chap. 13, § 11.2) and of the rates of reactions which lead to the destruction of metastable atoms (Chap. 18, § 5.2). We now describe how, under suitable conditions, the positive ion mobility in the gas may be derived from such measurements.

The ions and electrons in the plasma move under the combined action of the concentration gradient and an electric field **F**. If n_e, n^+ are the respective concentrations of electrons and of positive ions and $\mathbf{j}_e$, $\mathbf{j}^+$ their corresponding particle current densities, then

$$\mathbf{j}_e = -D_e \operatorname{grad} n_e - \mu_e \mathbf{F} n_e,$$
$$\mathbf{j}^+ = -D^+ \operatorname{grad} n^+ + \mu^+ \mathbf{F} n^+, \qquad (17)$$

where D_e, μ_e, and D^+, μ^+ are the respective diffusion coefficients and mobilities of electrons and of positive ions. In a decaying plasma the only fields present are due to the space charge and, provided n_e and n^+ are not too small, the condition for preservation of a plasma,

$$|n^+ - n_e| \ll n_e,$$

is maintained. The currents $\mathbf{j}_e$, $\mathbf{j}^+$ may therefore be taken as equal and the field strength **F** eliminated from the equations (17) to give

$$\mathbf{j} = -D_a \operatorname{grad} n, \qquad (18)$$

where **j** and n now refer either to positive ions or electrons and D_a, the ambipolar diffusion coefficient, is given by

$$D_a = (D^+ \mu_e + D_e \mu^+)/(\mu_e + \mu^+). \quad (19)$$

In a decaying plasma, in which there is no source of ionization and no volume recombination,

$$\text{div}\,\mathbf{j} + \frac{\partial n}{\partial t} = 0,$$

so that

$$-D_a \nabla^2 n + \frac{\partial n}{\partial t} = 0.$$

Assuming an exponential rate of decay so that $n \propto e^{-t/\tau}$, we have

$$\nabla^2 n + n/D_a \tau = 0. \quad (20)$$

This equation must be solved subject to the conditions that n should be finite within the container and vanish at the walls. For a spherical container of radius R, a spherically symmetrical distribution of ion concentration must therefore satisfy these conditions and the equation

$$\frac{d^2}{dr^2}(rn) + rn/D_a \tau = 0.$$

This requires that

$$n = B r^{-1} \sin\{(1/D_a \tau)^{\frac{1}{2}} r\}, \quad (21\,a)$$

where

$$(1/D_a \tau)^{\frac{1}{2}} R = s\pi \quad (s = 1, 2, \ldots) \quad (21\,b)$$

and B is a constant.

The allowed values of τ correspond to different modes of diffusion. In general the concentrations n will be given by a linear combination of different modes determined by the distribution when the ionization is cut off. If it is assumed that at this time the distribution is that corresponding to the mode $s = 1$, then

$$\tau = R^2/\pi^2 D_a. \quad (22)$$

D_a could then be obtained if τ is measured. Similar considerations apply to containers of other shapes such as cylindrical containers (see also Chap. 5, § 3.1.2, Chap. 18, § 5.2, and Chap. 20, § 3.1.1).

To ensure that the lowest diffusion mode prevails, the exciting source is arranged as far as possible to give an ionization rate per electron which is constant throughout the container. This generates the lowest mode in the steady state and after the excitation is cut off it persists during the decay period. In any case as the lowest mode has the slowest decay rate this will be the ultimate rate of loss due to diffusion.

If the decay of electron density is observed after the electrons and ions have come into equilibrium with the gas and possess a Maxwellian distribution, then

$$D^+/\mu^+ = D_e/\mu_e = \kappa T/e, \qquad (23\text{a})$$

so

$$D_a = 2D^+\mu_e/(\mu_e+\mu^+) \simeq 2D^+,$$

as $\mu_e \gg \mu^+$. Hence

$$\mu^+ \simeq \tfrac{1}{2}eD_a/\kappa T. \qquad (23\text{b})$$

TABLE 19.1

Mobility of alkali-metal ions in rare gases in $cm^2 s^{-1} V^{-1}$ *at* 760 *torr and* 18 °C

Ion		He	Ne	Ar	Kr	Xe
Li$^+$	unclustered	25·8	11·85	4·97	3·97	3·04
	clustered	11·70	5·28	2·26	1·46	0·98
Na$^+$	unclustered	24·2	8·70	3·23	2·34	1·80
	clustered	11·15	5·25	2·25	1·43	0·94
K$^+$	unclustered	22·9	8·0	2·81	1·98	1·44
	clustered	11·85	5·26	2·19	—	0·92
Rb$^+$	unclustered	21·4	7·18	2·39	1·57	1·10
	clustered	12·8	5·38	2·10	1·37	0·87
Cs$^+$	unclustered	19·6	6·50	2·24	1·42	0·97
	clustered	13·9	5·48	2·18	—	0·83
Hg$^+$	unclustered	37·0				

Under these conditions D_a as observed should be inversely proportional to the gas pressure as in D^+. At very low pressures p a so-called 'electron-diffusion cooling' occurs which causes the electron temperature T_e to fall below the gas temperature to an increasing extent as p falls. This arises because of the selective effect of the potential barrier which resists diffusion of electrons to the walls. The slower electrons cannot penetrate the barrier so that, if the pressure is not high enough to maintain temperature equilibrium of the electrons with the gas, the mean energy of the electrons in the plasma will fall below that of the gas atoms. The lower the pressure the greater will be the effect. In terms of effective electron temperature T_e we must write, under these conditions,

$$D_e/\mu_e = \kappa T_e/e,$$

giving

$$D_a = \frac{\mu^+\mu_e}{\mu^++\mu_e}\frac{\kappa}{e}(T+T_e) \simeq \mu^+\kappa(T+T_e)/e = D^+(1+T_e/T). \qquad (24)$$

T_e/T is now a function of p which tends to zero with p. Thus when $p \to 0$, $D_a \to D^+$.

To allow for this effect it is necessary to observe pD_a as a function of p over a wide pressure range to ensure that the results are correctly interpreted.

Correction must be made for the effect of volume recombination. At pressures greater than around 20 torr this effect is predominant and measurement of the decay of electron density gives the recombination coefficient α (see Chap. 20, § 3). At pressures of 5 torr or so the recombination loss is relatively small and

$$\frac{dn_e}{dt} \simeq -(n_e/\tau) - \alpha n_e^2,$$

where τ is the decay time due to diffusion. To this approximation

$$n_e/(1+\alpha\tau n_e) = \{n_0/(1+\alpha\tau n_0)\}\exp(-t/\tau),$$

where n_0 is the electron concentration at $t = 0$. Assuming α to be independent of pressure, a rather doubtful assumption (see Chap. 20, § 3), this relation may be used to correct the data for recombination loss.

As a check on the extent to which the lowest diffusion mode has been excited, the value found for D_a should be independent of the shape of the container.

The decay time τ may be measured by observing the change in the resonant frequency of a microwave cavity enclosing the container with time during the afterglow (see Chap. 2, § 5.1).

This technique was first applied by Biondi and Brown[†] to study the mobility of helium ions in highly purified helium contained in a quartz bottle that could be thoroughly baked. The helium was ionized by a 250-μs microwave pulse from a 10-cm wavelength magnetron. During the pulse a stationary charge distribution of concentration between 10^{10} and 10^{11} per cm³ was produced. The magnetron was then turned off for 11 ms and measurements made during this interval of the change of electron concentration within the glass bottle.

The results obtained in these experiments as well as in later experiments are discussed in §§ 3.2 and 3.3.

2.2.6. *Cyclotron-resonance method.*[‡] In Chapter 2, § 6 we described the cyclotron-resonance method for measuring momentum-transfer cross-sections of atoms for impact of low energy electrons. An exactly similar method may be applied to collisions of positive and negative ions with atoms and molecules.

[†] BIONDI, M. A. and BROWN, S. C., *Phys. Rev.* **75** (1949) 1700.
[‡] WOBSCHALL, D., GRAHAM, J. R., and MALONE, D. P., *Phys. Rev.* **131** (1963) 1565.

Consider a concentration n/cm^3 of ions of charge e and mass M_1 in a gas of molecules of mass M_2. If the ions are subject to a uniform constant magnetic field H and an electric field $F\cos\omega t$ normal to H, the rate of power absorption by the ions is given by

$$\bar{P} = (e^2 F^2 n\nu/4M_1)/\{\nu^2+(\omega-\omega_H)^2\}, \qquad (25)$$

if the frequency, $\nu_c = \nu(M_1+M_2)/M_2$, of collisions between ions and molecules is independent of relative velocity and $\omega = eH/M_1$. This is the generalization of the formula for absorption by electrons given in Chapter 2, p. 75, equation (80). Under the assumed conditions the mobility μ of the ions is given by

$$\mu = e(M_1+M_2)/\nu_c M_1 M_2 = e/M_1\nu. \qquad (26)$$

It follows that, under these circumstances, μ may be obtained from measurement of the line width of the power absorption radio-frequency spectrum.

If the collision frequency is not a constant, the line shape will not be exactly of the form (25) but in most cases it may still be represented empirically by (25) with ν a mean collision frequency. Substitution of this mean value in (26) can be expected to lead to an effective mobility close to the true one.

These conclusions need reconsideration when the energy gained between collisions is comparable with or $> \kappa T$ so that appreciable heating of the ions occurs. The effective value of F/p, the ratio of electric field to gas pressure in d.c. mobility experiments, is $F_m/2p$ where F_m is the peak r.f. voltage, the factor $\frac{1}{2}$ arising because only one circularly polarized component of the r.f. field is effective. The range of F_m over which negligible ion heating occurs may then be estimated by reference to drift velocity measurements as a function of F/p and Wannier's formula (1).

If it is desired to obtain results for high equivalent F/p it is best to introduce an axial d.c. electric field. Heating through increase of F_m for the detector field or through an auxiliary r.f. field complicates the interpretation because the power absorption, and hence the temperature, varies with frequency. Any variation of collision frequency with temperature will then complicate the line shape.

In designing an ion cyclotron-resonance absorption spectrometer the mean free path of an ion must be long enough for the ion to make many revolutions around the magnetic field between collisions, either with the walls of the apparatus or with the gas molecules. Having chosen the operating pressure, the chamber length must be larger than the mean free path of the ions in the gas. The magnetic field is then set high enough to achieve the desired resolution.

Fig. 19.12 shows the general arrangement of a typical spectrometer designed by Wobschall.† The measurement chamber operates within a solenoid 102 cm long which provides a magnetic field of 1500 G. Ions

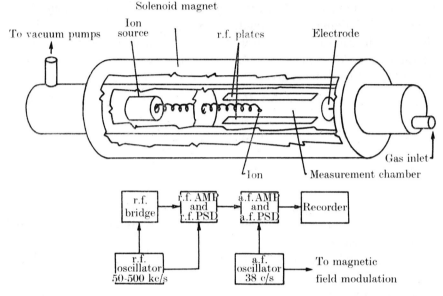

FIG. 19.12. Arrangement of the ion cyclotron-resonance spectrometer of Wobschall.

from a source follow the magnetic lines of force into the measurement chamber where they collide with the gas molecules between two parallel plates across which the r.f. electric field is applied. Alternatively, the ions, positive or negative, may be produced directly in the measurement chamber by firing an electron beam of suitable energy through it. To measure the power absorption the r.f. electrodes form one arm of a bridge that is initially balanced. Power absorption produces an imbalance because it is equivalent to a shunt resistance $F_m^2/\overline{P}\sqrt{2}$ where F_m is the peak r.f. voltage. $\overline{P}$ can therefore be measured by rebalancing the bridge. Any lack of balance of the bridge is amplified and detected by an r.f. phase-sensitive detector, the audio-frequency signal from which is further amplified and detected by a second phase-sensitive detector tuned to the modulation frequency (see Fig. 19.12). The output from this detector is displayed on a time base with the frequency swept with time so that the derivative of the power absorption is displayed as a

† WOBSCHALL, D., Rev. scient. Instrum. **36** (1965) 466.

function of frequency. A frequency—rather than magnetic—sweep is used because variation of magnetic field affects the diffusion of ions into the measurement chamber.

An advantage of this method is that it is relatively insensitive to impurities. An ion suffers at most 10–100 collisions before hitting the walls so that it has little chance to react with impurities. A further advantage is the positive identification of the mass of the ion concerned and the fact that measurements may be made even for ions which react rapidly in the gas.

Results obtained by this method, which agree well with those obtained by other methods, are given in §§ 3.2.2, 3.5.1–3, 4.1.

2.3. *Observed mobilities for the alkali-metal ions in rare gases*

A number of measurements[†] of the mobility of alkali metals in rare gases at 18 °C made using the shutter method of Tyndall and Powell are given in Table 19.1.

In addition the variation of mobility with temperature has also been observed (see Table 19.2).[‡] For such measurements the experimental chamber was surrounded by a Dewar flask filled with liquid hydrogen, nitrogen, oxygen, or ethylene, for low temperatures, and by a suitable bath at high temperatures. The electrical leads passing from outside into the apparatus were made very fine to reduce heat conduction. In these experiments the hot ion source was placed 7·5 cm above the shutter system, the ions being pulled down to it by an auxiliary field.

A value is also included for the mobility of Hg^+ ions in helium. This was derived by Biondi[§] from measurements of the rate of decay of electron concentration in an afterglow produced in a mixture of helium (1–2 torr pressure) and mercury vapour. In such an afterglow, while the ions are effectively all Hg^+, the electrons are rapidly brought to thermal equilibrium with the gas through collisions with the relatively light helium atoms—in an afterglow in pure mercury vapour, the electrons do not have time to come to thermal equilibrium with the main gas (see Chap. 18, § 5.2.4.2). At mercury vapour pressures of 10^{-6} torr the rate of decay is inversely proportional to the helium pressure and independent of the mercury vapour pressure and the main loss process is ambipolar diffusion of Hg^+ ions through helium. From such observations the mobility of Hg^+ ions in helium may be derived.

[†] TYNDALL, H. M., loc. cit., p. 1936.
[‡] Na^+ in He, Cs^+ in He: PEARCE, A. F., *Proc. R. Soc.* A**155** (1936) 490. Li^+–He, K^+–Ar, Rb^+–Kr, Cs^+–Xe: HOSELITZ, K., ibid. **177** (1941) 200.
[§] BIONDI, M. A., *Phys. Rev.* **90** (1953) 730.

TABLE 19.2

Variation of mobility, at constant gas density, with temperature

T (°K)	Mobility (cm² V⁻¹ s⁻¹ at 760 torr and 18 °C)					
	Li⁺ in He	Na⁺ in He	Cs⁺ in He	K⁺ in Ar	Rb⁺ in Kr	Cs⁺ in Xe
20·5	20·0	—	—	—	—	—
78	21·8	—	17·5	1·30	—	—
90	22·2	18·5	18·0	1·52	1·15	—
195	23·9	20·9	19·2	2·34	1·57	1·02
273	—	—	—	—	1·575	1·005
291	25·8	22·8	18·9	2·81	1·58	1·01
370	—	—	—	—	1·59	1·01
389	27·8	—	18·1	—	—	—
400	—	24·0	—	3·07	—	—
450	—	—	—	—	—	1·03
455	—	—	—	—	1·64	—
460	—	—	—	2·95	—	—
477	—	24·6	—	—	—	—
483	29·2	—	—	—	—	—
492	—	—	17·4	—	—	—

At mercury pressures between 2×10^{-5} and 10^{-4} torr the decay rate is inversely proportional to the mercury vapour pressure, showing that ambipolar diffusion of the ions through mercury vapour is dominant. This is discussed further in § 3.2.6.

2.4. *The theory of the mobility of ions in non-reacting gases*

If ions A⁺ drift through a gas of atoms B which have a higher ionization potential than A, the charge exchange process

$$A^+ + B \rightarrow A + B^+$$

is not energetically possible and does not complicate the interpretation of mobility data. Much attention has therefore been devoted to the study of the mobilities of alkali ions in various gases as in such cases charge transfer does not occur. A special case of interest arises when the ion drifts through its own gas. Such cases have been studied with the rare-gas ions He⁺, Ne⁺, Ar⁺, Kr⁺, Xe⁺. If the ion A⁺ is initially moving in a gas of atoms B with ionization potential less than that of A, the nature of the ions will change from A to B⁺ as the stream ages and the interpretation of the results will be complicated. The same considerations apply, of course, if any other reactions take place between the ions and gas atoms which are fast enough to change substantially the character of the ions.

We now discuss the theoretical calculations of the mobility of ions in gases of higher ionization potential assuming that no ionic reactions occur. Allowance for charge transfer when the ion is drifting in the parent gas will be discussed in § 3.2.

Under the conditions we are now assuming the classical formulae (9) and (42) of Chapter 16 may be applied.

The general nature of the interaction between an atom and an ion has been discussed in Chapter 16, § 2. At large separations the interaction has the form
$$V(r) \sim -\tfrac{1}{2}\alpha e^2/r^4,$$
where α, the polarizability of the atom, is given in terms of the dielectric constant K of the gas at s.t.p. by the relation
$$\alpha = (K-1)/4\pi n,$$
where n is the number of gas atoms per cm^3 at s.t.p. The interaction changes to a repulsion at a smaller distance. The simplest interaction including these features is the one used by Langevin,†
$$V(r) = -\tfrac{1}{2}\alpha e^2/r^4 \quad (r > r_0)$$
$$\to \infty \quad (r < r_0). \tag{27}$$

The mobility μ is given by $eD/\kappa T$, where D is the diffusion coefficient. Substituting the form (27) for $V(r)$ in the formulae (42), (44 a), and (44 c) of Chapter 16 and taking the suffix 1 to denote properties of the gas atoms and 2 of the ions we find, after some reduction, that

$$\mu = (1+M_1/M_2)^{\frac{1}{2}}\{\rho(K-1)\}^{-\frac{1}{2}}g(\lambda), \tag{28}$$

where
$$\lambda = \{8\pi n\kappa T r_0^4/(K-1)e^2\}^{\frac{1}{2}}, \tag{29}$$

$$g(\lambda) = \tfrac{3}{16}\lambda f(\lambda), \tag{30}$$

$$f(\lambda) = \int_0^\infty \int_0^\infty x^2 e^{-x} p' \cos^2 \tfrac{1}{2}\vartheta \, dp' dx, \tag{31}$$

$$\vartheta = 2p' \int_{r_1}^\infty \frac{dr'}{(r'^4 - p'^2 r'^2 + r_0^4/\lambda^2 x)^{\frac{1}{2}}}, \tag{32}$$

ρ is the density of the gas at n.t.p., and p the gas pressure. The lower limit r_1 in (32) is equal to the zero of the denominator or to r_0 whichever is the larger. The function $g(\lambda)$ has been calculated numerically by Langevin and by Hassè and Cook.‡ It is illustrated in Fig. 19.13. It will be seen that, for small values of λ, $g(\lambda)$ varies slowly with λ so that, for different ions in the same gas, the mobility should be nearly proportional to $(1+M_1/M_2)^{\frac{1}{2}}$. This is often useful in checking experimental results.

† LANGEVIN, P., *Annls Chim. Phys.* **5** (1905) 245.
‡ HASSÈ, H. R. and COOK, W. R., *Phil. Mag.* **12** (1931) 554.

To see the physical significance of a mobility independent of λ we note that (29) may be rewritten in the form

$$\lambda^2 = \kappa T / \tfrac{1}{2}\alpha e^2 r_0^{-4}. \tag{33}$$

When λ is small the mean energy of relative motion of an ion and an atom is small compared with the polarization energy at the hard sphere of radius r_0. Under these circumstances the dynamics of the collision will

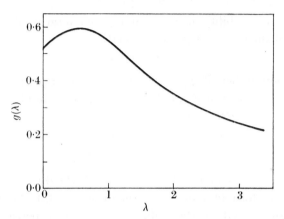

Fig. 19.13. The function $g(\lambda)$ appearing in Langevin's theory of ionic mobility.

be determined almost solely by the polarization—in general, deflexion will occur at distances great compared with r_0. Since

$$g(\lambda) \to 0 \cdot 510 \quad \text{as} \quad \lambda \to 0$$

the mobility, referred to a constant number density given by $2 \cdot 69 \times 10^{19}/$ cm^3 of gas atoms, is now given by

$$\mu = 35 \cdot 9/(\alpha M)^{\frac{1}{2}} \text{ cm}^2 \text{ V}^{-1} \text{ s}^{-1}, \tag{34}$$

where α is the polarizability of a gas atom measured in atomic units (a_0^3) and M is the reduced mass $M_1 M_2/(M_1+M_2)$ in units of the proton mass.

From the argument presented the formula (28) is likely to be valid no matter what form is taken by the repulsive potential, provided

$$\kappa T \ll \tfrac{1}{2}\alpha e^2 r_m^{-4}, \tag{35}$$

where r_m is the separation at which the interaction energy is a minimum. This has been confirmed by a detailed analysis carried out by Dalgarno, McDowell, and Williams.†

† DALGARNO, A., MCDOWELL, M. R. C., and WILLIAMS, A., *Phil. Trans. R. Soc.* A**250** (1958) 411.

It follows that it is convenient to start the discussion of the experimental data by testing whether (35) is valid. If not, it probably means that the repulsive interaction must be taken into account, in which case a better representation than the simple one (27) must be introduced.

To analyse the data for mobility in the rare gases it is convenient in view of (34) to examine the values of a reduced mobility

$$\mu' = \frac{n}{2 \cdot 69 \times 10^{19}} \mu M^{\frac{1}{2}}, \tag{36}$$

where n is the number density of the gas atoms at the operating temperature. If (34) is valid

$$\mu' = 35 \cdot 9 \alpha^{-\frac{1}{2}}, \tag{37}$$

independent of the temperature and of the mass of the ion.

TABLE 19.3

Comparison of observed and calculated values of the reduced mobility μ' (cm^2 V^{-1} s^{-1}) *for alkali-metal ions in rare gases at* 291 °K

Ion/Gas	He	Ne	Ar	Kr	Xe
			Obs.		
Li	38·6	25·2	11·4	9·4	7·3
Na	41·9	26·8	11·5	9·3	7·5
K	41·0	27·4	11·7	9·6	7·4
Rb	39·3	27·2	11·7	9·5	7·4
Cs	36·3	25·5	11·5	9·5	7·4
Hg	37·0	—	—	—	—
			Calc.		
$35 \cdot 9 \alpha^{-\frac{1}{2}}$	30·5	21·9	10·8	8·9	6·9
$38 \cdot 4 \alpha^{-\frac{1}{2}}$	32·8	23·5	11·6	9·6	7·4

In Table 19.3 values of μ' derived from the observations given in Table 19.1 are listed and compared with the values given by (37) using observed values of the polarizabilities. It will be seen that for Ar, Kr, and Xe the mobilities are indeed very closely independent of the nature of the ion. However, the absolute values are, for all these gases, about 1·08 times larger than given by (37). This may be due to an error in the absolute measurement of gas pressure. For neon there is an appreciable variation of μ' with the ion and this is more marked for helium. In these two cases the effect of the repulsion is presumably becoming appreciable.

Further evidence may be obtained from the observed temperature variation. In Table 19.4 values of μ' derived from the data of Table 19.2 are given. For Cs$^+$ in Xe, μ' is completely independent of temperature over the whole range investigated, showing that, in this case, the polarization force is responsible for the whole effect. The same applies over a more limited temperature range for Rb+ in Ar and even more so for K+ in Ar. In the other cases, although the temperature variation is not large, it is clearly present over the whole range studied.

TABLE 19.4

Variation with gas temperature T of the reduced mobility μ' (cm^{2}V^{-1}s^{-1}) for alkali-metal atoms in rare gases

T (°K)	Li+–He	Na+–He	Cs+–He	K+–Ar	Rb+–Ar	Cs+–Xe
20	29·9	—	—	—	—	—
78	32·5	—	—	5·4	—	—
79	—	—	32·3	—	—	—
90	33·2	—	—	6·5	6·9	—
92	—	32·1	33·3	—	—	—
195	35·7	36·2	35·5	9·7	9·5	7·9
290	—	39·5	—	—	—	—
291	38·5	—	35·0	11·5	—	7·6
389	41·5	—	—	—	9·6	—
392	—	—	33·5	—	—	7·7
405	—	41·6	—	12·7	—	—
477	—	42·6	—	12·0	9·9	7·8
483	43·7	—	—	—	—	—
492	—	—	32·2	—	—	—

To proceed further with the analysis of the mobilities in helium it is necessary to introduce an interaction allowing explicitly for the short-range repulsion. In 1944 Meyerott† carried out calculations of the mobility of Li+ in helium using the interaction

$$V(r) = Ae^{-\zeta r} - Br^{-4} - Cr^{-6}, \tag{38}$$

where $A = 1 \cdot 60 \times 10^{-9}$ erg, $\zeta = 5 \cdot 9 \times 10^8$ cm^{-1},

$B = \tfrac{1}{2}\alpha e^2 = 2 \cdot 37 \times 10^{-44}$ erg cm^4, $C = 1 \cdot 85 \times 10^{-60}$ erg cm^6,

α being the polarizability of helium. The repulsive term was chosen to fit the results of an approximate quantal calculation. His results are shown in Fig. 19.14. Also included in the figure are the results obtained by Dalgarno, McDowell, and Williams‡ for the same interaction except that C was taken as zero. The latter results agree with the observations at very low temperatures but deviate at higher temperatures. However, the same authors found that by changing ζ in (38) to $5 \cdot 26 \times 10^8$ cm^{-1}, keeping A and B unchanged and taking $C = 0$, good agreement as to the temperature variation was obtained (see Fig. 19.14).

† MEYEROTT, R., *Phys. Rev.* **66** (1944) 242.   ‡ loc. cit., p. 1960,

Mason, Schamp, and Vanderslice† have successfully fitted the observed behaviour and magnitude of the mobility of Li⁺ in He using an empirical interaction of the form

$$V(r) = Ar^{-12} - Br^{-4} - Cr^{-6}, \tag{39}$$

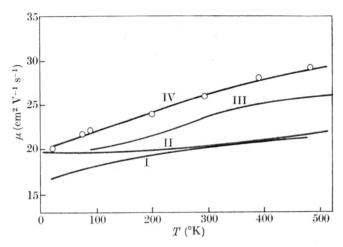

Fig. 19.14. Comparison of observed and calculated mobilities μ of Li⁺ ions in He. ○ observed. I, calculated by Meyerott. II, calculated by Dalgarno et al. with the Meyerott interaction but taking $C = 0$. III, best theoretical fit using the modified Meyerott interaction. IV, calculated by Mason et al. using the interaction (39).

which corresponds to the (12, 6) form used in analysing data on collisions between neutral atoms (see Chap. 16, § 6.1). The best fit was obtained, as shown in Fig. 19.14, with

$$A = 5{\cdot}80 \times 10^{-106} \text{ erg cm}^2, \qquad B = \tfrac{1}{2}\alpha e^2 = 2{\cdot}41 \times 10^{-44} \text{ erg cm}^4,$$
$$C = 1{\cdot}76 \times 10^{-60} \text{ erg cm}^6.$$

In this potential the characteristic parameters ϵ, r_m have the values 0·046 eV and $2{\cdot}22 \times 10^{-8}$ cm respectively. These are to be compared with the corresponding values 0·062 eV and $1{\cdot}95 \times 10^{-8}$ cm for the interaction used by Dalgarno, McDowell, and Williams which gave the best fit to the data. Both sets of values for ϵ are consistent with evidence derived from data on the clustering of Li⁺ ions in helium (see § 3.1).

Weber and Bernstein‡ have calculated differential and total elastic cross-sections for collisions between Li⁺ ions and He atoms over a range

† Mason, E. A., Schamp, H. W., and Vanderslice, J. T., Phys. Rev. 112 (1958) 445.
‡ Weber, G. G. and Bernstein, R. B., J. chem. Phys. 42 (1965) 2166.

of relative energies extending from 0·01 to 10^3 eV for various assumed interactions. These include the two empirical interactions which gave the best fit to the mobility data as well as others derived by approximate theoretical methods. The calculated cross-section is as large as 5×10^{-14} cm^2 at 0·01 eV. It falls smoothly (apart from 'glory' oscillations, see Chap. 16, § 5.4.3) to 3×10^{-15} cm^2 at 10 eV. Unfortunately, the experimental measurement of these cross-sections for very low energy ions of sufficiently homogeneous energy to show up the glory effects still presents a very difficult problem (see Chap. 22).

The reduced mobility μ' for Na$^+$ ions in He is so close to that of Li$^+$ ions at the same temperature over the range of observation that the assumption of the same interaction as for Li$^+$–He gives good agreement with the observations.

2.5. *Mobilities of alkali-metal ions in molecular gases*†

In principle, the problem of calculating mobilities for ions in molecular gases is much more complicated. The basic interactions are no longer isotropic but depend on the molecular orientation relative to the line joining the centres of mass of the ion and molecule. Allowance should also be made for the possibility of excitation of molecular rotation and vibration on impact. However, the anisotropic component of the interaction is usually small and it is often a good approximation to average over all orientations of the molecular axis to obtain an effective isotropic interaction. In that case the long-range term which remains is again of the form $-\tfrac{1}{2}\alpha e^2/r^4$ where α is some mean of the parallel and perpendicular polarizabilities of the molecule. It is therefore worth while to apply the same procedure as for the atomic gases to see whether there is evidence that the mobilities are indeed governed by an effective interaction of this form.

In Table 19.5 the reduced mobilities μ' are given for various ions in N$_2$, H$_2$, and CO. It will be seen that, for all ions except Li$^+$, μ' is remarkably constant in each gas. Indeed, for N$_2$, additional data for the mobilities of Al$^+$, Ga$^+$, Kr$^+$, In$^+$, Xe$^+$, Ba$^+$, Hg$^+$, and Tl$^+$ all give values between 10·1 and 10·3 for μ'. This strongly suggests that the formula (34) applies. Assuming this, the effective values of the polarizability come out to be, in atomic units, 11·3, 4·3, and 16·3 for N$_2$, H$_2$, and CO respectively. These are to be compared with the observed parallel and perpendicular polarizabilities which, in the same units, are respectively 16·1 and 9·8 for N$_2$, 6·3 and 4·8 for H$_2$, and 17·5 and 11·0 for CO. Except

† DALGARNO, A., McDOWELL, M. R. C., and WILLIAMS, A., loc. cit., p. 1960.

for H_2 these bracket the desired effective values of α. This supports the assumptions made and suggests that, at least for N_2 and CO, no complications arise due to excitation of inner molecular motion. It is also noteworthy that, for a heteronuclear molecule such as CO, the full calculation contains an anisotropic term which behaves asymptotically like r^{-3}. Despite its very long range, this term remains ineffective because it vanishes when averaged over all orientations.

TABLE 19.5

Reduced mobilities μ' ($cm^2 V^{-1} s^{-1}$) for alkali-metal ions in diatomic gases measured at 291 °K

Gas/Ion	Li	Na	K	Rb	Cs
N_2	9·3	10·1	10·2	10·3	10·3
H_2	15·6	17·3	17·4	17·5	17·6
CO	5·6	8·1	8·8	8·9	8·9

2.6. *Mobilities of NO^+ ions in He, Ar, H_2, and N_2*

As NO has a low ionization potential (9·6 V) NO^+ ions do not undergo charge-transfer reactions in monatomic and diatomic gases. Observations of the mobility of NO^+ in He, Ar, H_2, and N_2 made by Young, Gatz, Sharpless, and Ablow,† using the parallel-plate pulse collection method (see § 2.2.4), show no evidence of any other ionic reactions occurring under their experimental conditions, so that the results should not be too difficult to interpret.

TABLE 19.6

Reduced mobilities μ', for NO^+ and alkali-metal ions in He, Ar, H_2, and N_2, reduced to 291 °K

Gas/Ion	NO	Li	Na	K	Rb	Cs
He	34±4	38·6	41·9	41·0	39·3	36·3
Ar	14·2±2·6	11·4	11·5	11·7	11·7	11·5
H_2	20·8±1·6	15·6	17·3	17·4	17·5	17·6
N_2	11·7±0·8	9·3	10·1	10·2	10·3	10·3

Table 19.6 gives the reduced mobility μ' defined as in (36) for NO^+ ions in the different gases compared with corresponding values for the alkali-metal ions. If allowance is made for the uncertainty in the NO^+

† loc. cit., p. 1944.

data it does not fit too badly with that for the alkali-metal ions. In any case the trend of the agreement is sufficient to add further support to the identification of the ions used in the parallel-plate experiment as NO^+ (see Chap. 20, § 4.7).

3. Reactions involving positive ions

3.1. *Cluster formation*

The refinements of technique introduced by Tyndall and his collaborators made possible a quantitative study of certain aspects of cluster formation. They were able to observe the effect on the mobility of adding small measurable admixtures of polar substances to gases which were effectively pure.

3.1.1. *The clustering of water molecules.* The most thorough study of this kind was made by Munson and Tyndall[†] using water as the polar substance. The apparatus used was substantially the same as that with which the main mobility measurements were carried out except that, before the ions were allowed to pass through the regions between the shutters, which were 1 cm apart, they were aged by passage for 3 cm through an ageing field. In most of the experiments the ratio F_1/F_2 of the ageing field F_1 to the field F_2 between the shutters was maintained constant while both were varied. With this arrangement a study was made of the mobilities of the alkali ions in the rare gases containing measured admixtures of water vapour. Special techniques using optical magnification with a Rayleigh gauge[‡] had to be employed to measure the partial pressure of the added water vapour, which was, in certain experiments, as low as 4×10^{-4} torr.

It was found that only two groups of ions were obtained in each case, one having the full mobility and consisting of unclustered ions, the other with a lower mobility consisting of clustered ions. The mobility of this latter group was found, for a given F/p (see Fig. 19.15), to be smaller the smaller the percentage of water vapour. However, this effect became less marked as F/p was reduced so that, in the limit of vanishing F/p, the mobility of the clustered ion was independent of the percentage of water vapour. This shows that the size of the cluster formed is independent of the concentration of water vapour if the ions are very slow. As the water-vapour concentration was reduced, at small F/p, the current due to clustered ions became relatively smaller, but in no case were any ions of intermediate mobility observed. The measured values

[†] MUNSON, R. J. and TYNDALL, A. M., *Proc. R. Soc.* **A172** (1939) 28.
[‡] SCHRADER, J. E. and RYDER, H. M., *Phys. Rev.* **13** (1919) 321.

of the mobilities for the fully clustered ions (in the limit of very small F/p) are given in Table 19.1.

These results suggest that the chance of attaching the first water molecule is quite low, but once one such molecule is attached the chance of attaching others is so much increased that the final cluster builds up very rapidly. Presumably the first attachment can only take place in

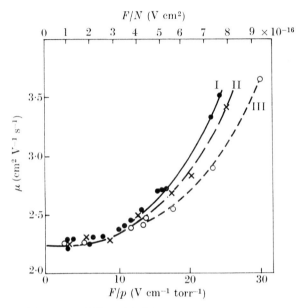

FIG. 19.15. Observed mobility of Li$^+$ ions in Ar in the presence of H$_2$O, as a function of F/p. I, 0·008–0·43 per cent H$_2$O, ●; II, 1·3 per cent H$_2$O, ×; III, 2·8 per cent H$_2$O, ○; where ●, ×, ○ are experimental points.

a three-body collision. Once this occurs the structure is so complex that the excess energy acquired in attaching a second molecule is redistributed among the many internal degrees of freedom and may be dissipated gradually in subsequent collisions. In support of these ideas the fraction of ions clustered for a given F/p is observed to increase with the pressure of the main gas. On the other hand, the chance of three-body encounters, under many conditions in which appreciable clustering was observed, was very small. Thus perceptible clustering occurred in xenon with 0·005 per cent water in which the chance of a three-body collision between an ion, a water molecule, and a gas molecule during the life of an ion was as low as 10^{-4}.

By applying Langevin's theory, assuming no increase in the effective diameter r_0, an upper limit of about 6 is found for the number of clustered molecules in the saturated condition. A closer estimate for Li⁺ suggests that the actual number is about 4, in agreement with the predictions of Bernal and Fowler† concerning hydration of ions in electrolytes. On the other hand, the latter authors predicted no attachment to Cs⁺, whereas clustering certainly occurred even for these ions.

The reason for the observed increase in mobility of the clustered group with F/p is not clear, but must be due in some way to the increased mean energy of the ions which corresponds to a temperature increase.

3.1.2. *The appearance of clustered alkali ions in pure rare gases.* It was found by Hoselitz,‡ in the course of an investigation of the mobility of Li⁺ in Xe, that a group of ions of smaller mobility than the main group was always present and did not disappear after the most careful purification. It was suspected that these ions really arose from addition of one or more xenon atoms to the lithium ion. Confirmation of this possibility was obtained by observation of a similar group when studying the mobility of Li⁺ in He at liquid-hydrogen temperatures at which all polar impurities would have been frozen out. A detailed study§ of the effect was therefore carried out on similar lines to the work described in § 3.1.1.

The fraction c of clustered ions was observed as a function of F/p and extrapolated to zero F/p. From the observed mobilities an estimate could be made, by Langevin's method, of the number of clustered atoms. Application of the usual formulae of statistical mechanics then gives an estimate of $\sum D_r/r$, where D_r is the energy required to remove an atom from a cluster containing r atoms. The results of such an analysis are given in Table 19.7 for lithium ions in various rare gases. They form a consistent set in that the dissociation energy increases with the mass of the rare-gas atom. The values found for helium are considerably smaller than would have been expected from the Li⁺–He interaction calculated by Meyerott (p. 1962).

Further discussion of these experimental results will be given in the next section in terms of a theory of cluster formation.

3.1.3. *Theory of ion clustering.* An interesting classical statistical theory of cluster formation has been given by Bloom and Margenau.∥ In most cases the interaction between an ion and a gas molecule is such that the separation of the vibrational levels is small and a large number

---

† BERNAL, J. D. and FOWLER, R. H., *J. chem. Phys.* **1** (1933) 515.
‡ HOSELITZ, K., *Proc. R. Soc.* **A177** (1941) 200.
§ MUNSON, R. J. and HOSELITZ, K., ibid. **172** (1939) 43.
∥ BLOOM, S. and MARGENAU, H., *Phys. Rev.* **85** (1952) 670; see also MAGEE, J. L. and FUNABASHI, K., *Radiat. Res.* **10** (1959) 622.

of such levels exist. Even for the lightest pair, Li⁺ in H_2, there will be at least 10, but in most cases it will be as large as 50. Correspondence-principle arguments show that under these conditions a classical treatment should be adequate.

TABLE 19.7

Effect of clustering on the mobility of Li⁺ *ions in the rare gases*

Gas	Temp. (°K)	μ/μ_c	$c/(1-c)$	n	D (eV)
He	20	1·10	3	1·8	0·016
	77	—	1	—	0·07
	90	—	1	—	0·08
Ne	90	1·98	10	—	0·13
	195	—	0·1	—	0·16
Ar	195	2·38	10	2·7	0·25
	290	—	0·33	—	0·29
Kr	290	3·09	10	2·6	0·31
	360	—	2	—	0·35
Xe	290	3·84	20	2·6	0·42

μ, μ_c = mobility of unclustered and clustered ion respectively,
c = fraction of clustered ions,
n = upper limit to average number of gas atoms attached to each ion,
D = mean energy to extract one atom from cluster.

Consider, for simplicity, a positive ion of effectively infinite mass surrounded by an atmosphere of neutral molecules of mass m within a spherical container of large radius R and at a temperature T. If $V(r)$ is the interaction energy of a molecule within the ion when at a distance r from it, the number of molecules with velocity between v and $v+dv$ which will be found within a distance between r and $r+dr$ from the ion is given, according to classical statistics, by

$$A \exp\{-(\tfrac{1}{2}mv^2+V)/\kappa T\}r^2v^2 \, dr dv. \tag{40}$$

A is a constant which is such that

$$A \int_0^R \exp(-V/\kappa T)r^2 \, dr \int_0^\infty \exp(-\tfrac{1}{2}mv^2/\kappa T)v^2 \, dv = N, \tag{41}$$

where N is the total number of molecules within the container. As V has the asymptotic form

$$V \sim -\tfrac{1}{2}\alpha e^2/r^4, \tag{42}$$

where α is the polarizability of a molecule, and as R is very large,

$$\tfrac{4}{3}\pi R^3 (2\pi)^{\frac{3}{2}}(m\kappa T)^{-\frac{3}{2}} A = N. \tag{43}$$

Since $N/\frac{4}{3}\pi R^3$ is the number density n of neutral molecules we have
$$A = 16n(m/2\kappa T)^{\frac{3}{2}}\pi^{\frac{1}{2}}. \qquad (44)$$
The number of molecules bound to the ion will be obtained by integrating (40) over all r and v for which
$$V(r) + \tfrac{1}{2}mv^2 \leqslant 0. \qquad (45)$$
$V(r)$ will have the general shape shown in Fig. 16.1 and the asymptotic form (42). For $V \leqslant 0$ (45) requires
$$v \leqslant (-2V/m)^{\frac{1}{2}} = v_\mathrm{m}(r), \text{ say,}$$
while for $V > 0$ there is no binding. Hence if $V(\sigma) = 0$, the number of molecules bound to the ion is given by
$$N_\mathrm{b} = 16n(m/2\kappa T)^{\frac{3}{2}}\pi^{\frac{1}{2}} \int_\sigma^\infty r^2 e^{-V/\kappa T}\left\{\int_0^{v_\mathrm{m}(r)} v^2 e^{-\frac{1}{2}mv^2/\kappa T}\, dv\right\} dr. \qquad (46)$$

Margenau and Bloom carried out detailed calculations for two interactions. The first combined a hard-sphere repulsion with polarization
$$V(r) = -\tfrac{1}{2}\alpha e^2 r^{-4} \quad (r > r_0),$$
$$\to \infty \qquad\qquad (r = r_0), \qquad (47)$$
while the second was of the form (see (39))
$$V(r) = Ar^{-12} - \tfrac{1}{2}\alpha e^2 r^{-4}. \qquad (48)$$
If the zero σ of this potential is chosen to be equal to r_0, the depth of the potential minimum is considerably smaller for (48) than for (47) so that the average size of a cluster for fixed n and T will be less.

Quantitative application of the formulae is difficult because N_b is sensitive to the force constants r_0 and A which are difficult to fix with any precision. Margenau and Bloom used for r_0 the sum of the radii of the ion and molecule obtained by Margenau.† These are given in Table 19.8 for interaction between alkali-metal ions and a number of neutral atoms and molecules. A was chosen so that the zero of the interaction (48) fell at r_0, so that $A = \tfrac{1}{2}\alpha e^2 r_0^8$.

We may write (46) in the form
$$N_\mathrm{b} = 2nr_0^3 \pi^{\frac{1}{2}} Q(b), \qquad (49)$$
where $b = \tfrac{1}{2}\alpha e^2/\kappa T r_0^4 = \lambda^{-2}$ with λ as in (33). $Q(b)$ may be calculated numerically and is illustrated as a function of b in Fig. 19.16 for the two interactions (47) and (48). It is clear that, because of the greater depth of the attractive well in (47), $Q(b)$, and hence the size of the cluster, is much larger for given b than for the more realistic form (48).

† MARGENAU, H., *Philosophy Sci.* **8** (1941) 603.

TABLE 19.8

Radii r_0 (in Å) of the equivalent short-range rigid-sphere interaction between alkali-metal ions and neutral atoms and molecules under thermal conditions

Neutral atom or molecule / Alkali-metal ion	Li$^+$	Na$^+$	K$^+$	Rb$^+$	Cs$^+$
Ar	2·21	2·41	2·76	2·92	3·08
Kr	2·35	2·55	2·90	3·06	3·22
Xe	2·49	2·69	3·04	3·20	3·36
H$_2$	2·16	2·36	2·71	2·87	3·03
O$_2$	2·23	2·43	2·78	2·94	3·10
CO$_2$	2·39	2·59	2·94	3·10	3·26

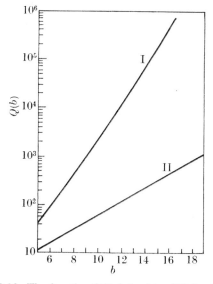

FIG. 19.16. The function $Q(b)$ defined by (49) for the two interactions (47) and (48). I, hard-sphere interaction (47); II, interaction (48).

Using $Q(b)$ for (48) some comparison may be made with the experiments of Munson and Hoselitz,† who found that only Li$^+$ would form clusters in the rare gases. According to the theory, the average number of atoms per cluster for Li$^+$ in argon at 195 °K and 19 torr is 1·7, which compares with the observed value of 2. Equally good agreement is found for Li$^+$ in Xe, theory predicting 2·3 and experiment showing 2 atoms per cluster. On the other hand, at the same temperature, the

† loc. cit., p. 1968.

cluster size in Kr was found to be also 2 whereas according to the theory it should be 0·2—it would only be expected to be as large as 2 for Kr at 225 °K, at 19 torr.

At 273 °K a highly polarizable molecule such as CO_2 would form a cluster about a Li^+ ion of about 28 molecules whereas for O_2 it would be about 2 and for H_2 0·08. Of the alkali ions Li^+ is the strongest clustering agent. For K^+ in O_2, for example, $N_b = 0·05$. These results are consistent with the failure of Munson and Hoselitz to observe cluster formation to any ion but Li^+ in the rare gases.

3.2. *The mobilities of ions in their parent gases—the effect of charge exchange and molecular ion cluster formation*

3.2.1. *The mobility of helium ions in helium—historical account.* The history of the study of the mobility of helium ions in helium is a remarkable one. It is worth describing in some detail as many of the complications are common to the development of mobility studies of most ions. Helium, being the simplest case, is especially instructive. We discuss first the mobility at room temperature.

The story begins with the observations by Tyndall and Powell[†] in 1930 of the mobility of helium ions derived from a radioactive source, which gave 13 cm^2 V^{-1} s^{-1} under standard conditions. Repetition of these experiments[‡] in 1931, using ions from a glow discharge source, gave 21·4 cm^2 V^{-1} s^{-1}. This was considered to be the most reliable value and the earlier measurements were disregarded. Part of the reason for this is the fact that the simple theory of § 2.4 gave 25·6 cm^2 V^{-1} s^{-1} when applied to He^+ ions in He. However, it was pointed out by Massey and Mohr[§] in 1933 that this theory was inapplicable because it failed to take account of the occurrence of charge transfer. They calculated the mobility allowing for this and obtained 11 cm^2 V^{-1} s^{-1}. The first reaction to this was to seek ways in which the theory could be seriously in error, particularly as, in 1935, Tyndall and Pearce[||] repeated the experiments with the glow discharge source and again found a high value 19·9 cm^2 V^{-1} s^{-1} for the mobility. However, in 1944, Meyerott[††] suggested that in the experiments carried out using a glow discharge source the main ion might well have been He_2^+. Although this ion is

---

[†] TYNDALL, A. M. and POWELL, C. F., *Proc. R. Soc.* A**129** (1930) 162.
[‡] ibid. **134** (1931) 125.
[§] MASSEY, H. S. W. and MOHR, C. B. O., ibid. **144** (1934) 188.
[||] TYNDALL, A. M. and PEARCE, A. F., ibid. **149** (1935) 426.
[††] MEYEROTT, R., *Phys. Rev.* **66** (1944) 242.

more massive than He^+ it would have a higher mobility because no charge transfer effects could arise with it. It was further suggested that, in the earlier experiments with the radioactive source, the ions might really have been He^+ so the observed mobility was not in serious disagreement with that calculated for He^+ by Massey and Mohr.

The matter rested until 1949 when Biondi and Brown† made the first mobility measurement in helium using the afterglow technique and obtained 12·5 cm² V⁻¹ s⁻¹, consistent with the earliest measurement by Tyndall and Powell using the radioactive source. This supported Meyerott's suggestion. Further support came from observations by Hornbeck,‡ using the pulsed Townsend discharge technique described in § 2.2.2. He found that, in helium, two ions were present simultaneously. Furthermore, Hornbeck and Molnar§ were able to show, using a mass spectrograph, that He_2^+ ions were formed in a Townsend discharge in helium.

A further step forward was made by Phelps and Brown‖ in 1952. They attached a mass spectrometer to their afterglow cavity and so were able to determine the ion composition at different times during the afterglow. In this way they verified the presence of He_2^+ and, by measuring the decay rates of electrons and He^+ and He_2^+ ions, were not only able to derive the mobility of the He^+ ions, but also produced evidence that the He_2^+ ions are formed in the three-body reaction

$$He^+ + 2He \rightarrow He_2^+ + He, \tag{50}$$

and determined the rate coefficient. The He^+ mobility they obtained was 13·0±0·5 cm² V⁻¹ s⁻¹, consistent with other measurements of the mobility of the above ion and with the theoretical value of Massey and Mohr.

A little later, in 1954, Biondi and Chanin,†† applying the method described in § 2.2.3, obtained values of 10·5 and 20·3 cm V⁻¹ s⁻¹ for the mobilities of the slower and faster ions, presumably He^+ and He_2^+. These were consistent with earlier measurements and at that time the situation seemed to be completely clear. More elaborate calculations of the mobility of He^+ improved the agreement with observation. It was not possible to be so definite about the mobility of He_2^+ but at any rate there seemed to be no serious difficulty, on theoretical grounds, in ascribing the higher mobility to this ion.

† BIONDI, M. A. and BROWN, S. C., ibid. **75** (1949) 1700; **76** (1949) 302.
‡ HORNBECK, J. A., ibid. **83** (1951) 374; **84** (1951) 615.
§ HORNBECK, J. A. and MOLNAR, J. P., ibid. **84** (1951) 621.
‖ PHELPS, A. V. and BROWN, S. C., ibid. **86** (1952) 102.
†† BIONDI, M. A. and CHANIN, L. M., ibid. **94** (1954) 910.

The first sign that all was still not clear came from the afterglow observations of Oskam,[†] who obtained a mobility of 16·2 cm² V⁻¹ s⁻¹ for the ions concerned, presumably He_2^+. This was in substantial disagreement with all earlier measurements of the higher mobility. Oskam and Mittelstadt[‡] followed this up with a more extensive afterglow investigation. They confirmed that the ion present at high pressure possessed an intermediate mobility 16·2 cm² V⁻¹ s⁻¹, while at low pressures they found evidence for the existence, as the major ion, of He⁺ with the mobility 10·7 cm² V⁻¹ s⁻¹.

Further evidence on the same lines was provided at about the same time by Kerr and Leffel[§] from a study of the time dependence of the intensity of the optical spectrum of the helium afterglow, giving mobilities for atomic and molecular ions 10·6 and 16·2 cm² V⁻¹ s⁻¹ consistent with those of Oskam and Mittelstadt.

In both these sets of experiments the purity of the helium was maintained by running a cataphoresis discharge.

Beaty and Patterson[∥] studied the problem further with a drift tube of the Tyndall and Powell type. Under most conditions they found two ions present which they identified as He⁺ and He_2^+ respectively, with mobilities 10·40±0·10 and 16·70±0·17 cm² V⁻¹ s⁻¹. From an analysis of the shape of the recorded pulse for each ion, they were able to show that He_2^+ was formed from He⁺ by a three-body reaction. The rate coefficient of this reaction was found to be $1·08 \times 10^{-31}$ cm⁶ s⁻¹. However, at low pressures and by sampling the ions very soon after breakdown of the gas, a third ion of mobility close to 20 cm² V⁻¹ s⁻¹ was also obtained.

It therefore appears that there are three ions, with mobilities around 10·6, 16·2, and 20 cm² V⁻¹ s⁻¹ respectively, the first being He⁺ and the latter two both He_2^+. The ion of highest mobility appears only when observations are made very soon after formation, while the one of intermediate mobility appears after ageing. This is consistent with all the earlier observations with the possible exception of those of Tyndall and Powell in 1931. It is not clear that in their work the ions were newly formed, although the observed mobility is close to 20 cm² V⁻¹ s⁻¹.

Soon after these observations were reported, Beaty, Browne, and Dalgarno[††] suggested that the fastest ion in helium is He_2^+ in the

[†] OSKAM, H. J., *Philips Res. Rep.* **13** (1958) 401.
[‡] OSKAM, H. J. and MITTELSTADT, V. R., *Phys. Rev.* **132** (1963) 1435.
[§] KERR, D. E. and LEFFEL, C. S., *Bull. Am. phys. Soc.* **7** (1962) 131.
[∥] BEATY, E. C. and PATTERSON, P. L., *Phys. Rev.* **137** (1965) A346.
[††] BEATY, E. C., BROWNE, J. C., and DALGARNO, A., *Phys. Rev. Lett.* **16** (1966) 723.

metastable $^4\Sigma_u^+$ state, so the ion of intermediate mobility would be ground-state He_2^+. If this is accepted no further difficulty remains as far as the observations at room temperature are concerned.

The situation at lower temperatures is more complicated. In 1935 Tyndall and Pearce† observed ions in helium at 77 °K of mobility near 19 cm² V⁻¹ s⁻¹. This was confirmed in 1957 by Chanin and Biondi,‡ who also observed ions with smaller mobility near 13 cm² V⁻¹ s⁻¹. The fast and slow ions were identified as He_2^+ and He^+. However, in 1967 Patterson,§ using the technique developed by Beaty and Patterson,‖ found evidence that below 200 °K the He_2^+ ions transform to He_3^+, which have a mobility close to that of the faster ion in the earlier experiments. A value was obtained for the dissociation energy of He_3^+.

It is clear that, even for the supposedly simple case of helium, the nature of the positive ions which are studied varies considerably with the experimental conditions. Even in the pure gas, ion-clustering occurs leading, in helium, to the production of He_2^+ and even He_3^+ ions. We must expect at least as complicated phenomena to occur in other cases. In particular, with diatomic gases such as N_2 it is to be expected that N_3^+ and even N_4^+ will be produced (see § 3.5.2). The reactions which lead to the formation of these complex ions are the simplest examples of ionic reactions.

We shall describe in some detail selected examples of the techniques and results obtained in the study of the mobility and ionic reactions of helium ions in helium. These serve as examples of some of the techniques for the study of ionic reactions in general. After discussing applications of these techniques to the study of other rare-gas ions in their parent gases we shall give an account of the theory of charge transfer and its application to the mobilities of atomic rare-gas ions. We shall then go on to discuss ionic reactions in general.

3.2.2. *The mobility of helium ions in helium—experimental techniques and results at* 300 °K—*drift-tube measurements.* Biondi and Chanin,†† using the technique described in § 2.2.3, measured the mobility of ions in helium as a function of the ratio F/p of electrical field strength to gas pressure, at 300 °K.

Fig. 19.17 illustrates their results at 300 °K. Comparison is made with the results of Hornbeck‡‡ obtained using the equipment described in § 2.2.2, and of the results for He^+ obtained by Wobschall, Fluegge,

---
† TYNDALL, A. M. and PEARCE, A. F., *Proc. R. Soc.* A **149** (1935) 426.
‡ CHANIN, L. M. and BIONDI, M. A., *Phys. Rev.* **106** (1957) 473.
§ PATTERSON, P. L., *J. chem. Phys.* **48** (1968) 3625.
‖ loc. cit., p. 1974.   †† loc. cit., p. 1942.   ‡‡ loc. cit., p. 1938.

and Graham,† using the cyclotron resonance method with d.c. axial field for heating the ions. The agreement is good for the slower He⁺ ion but not so satisfactory for the faster ion. It will be noticed from this figure that the results for the faster ion agree quite well with those obtained by Tyndall and Powell‡ using a glow discharge source.

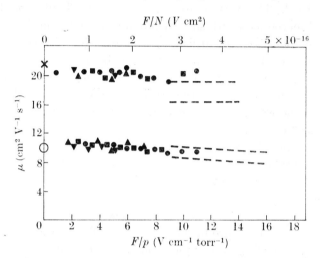

Fig. 19.17. Mobilities of ions in helium at 300 °K observed by pulse techniques. Observed by Biondi and Chanin at different pressures (in torr): ● 3·0, ■ 6·0, ▼ 9·0, ▲ 12·0. Observed by Hornbeck between ‒‒‒. Observed by Tyndall and Powell, ×; by Wobschall, Fluegge, and Graham, ○.

Extrapolation to zero F/p was made, for the atomic ion, by assuming that the mobility varies with F/p according to

$$\mu(F/p) = \mu(0)(1+aF/p)^{\frac{1}{2}},$$

where a is a constant. Results obtained for $\mu(0)$, the zero-field mobility, are given in Table 19.9.

Beaty and Patterson§ used a cylindrical double-shutter drift tube of similar design to that of Tyndall and Powell (§ 2.2.1). The electrode system is indicated in Fig. 19.18. Referring to the numbers in that figure, electrodes 14, 15 and 3, 4 constitute the shutters. Ions were produced by striking a discharge between a Kovar wire (electrode 18) and the cylindrical case (electrode 17). The discharge was pulsed and measurements taken during the afterglow period. This was done to eliminate the effect of noise generated in the discharge.

† WOBSCHALL, D. C., FLUEGGE, R. A., and GRAHAM, J. R., *J. appl. Phys.* **38** (1967) 3761.
‡ loc. cit., p. 1972. § loc. cit., p. 1974.

19.3 IONIC MOBILITIES AND IONIC REACTIONS

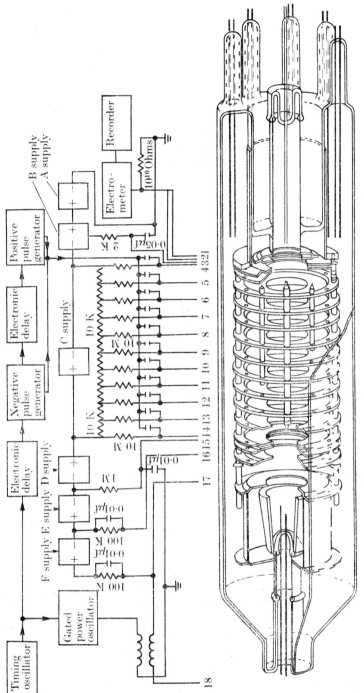

Fig. 19.18. Electrode system used in the experiments of Beaty and Patterson.

A cycle of operations commenced by triggering a power oscillator which drove the discharge. After a suitable electronically controlled delay the shutter nearest the ion source was opened during the afterglow. A second electronic delay, with calibrated control, opened the second shutter. The repetition rate was between 100 and 1000 c/s. The current received at the ion collector electrodes was measured as a function of time by an electrometer.

To eliminate background contamination of the helium the tube was evacuated to 10^{-9} torr and outgassed so that the rate of rise of background pressure after seal-off was around 10^{-6} torr/day. Reagent grade helium gas was admitted through a bakeable metal valve and purified by a cataphoresis discharge.

The raw data obtained with this equipment are the measurements of the charge $q(t_1)$ collected during the opening of the second shutter as a function of the time delay t_1 between the openings of the two shutters. To derive the mobilities and reaction rates from these data a considerable amount of analysis is required.

Suppose that we are dealing with two types of ion, distinguished by suffixes 1 and 2, which may be He^+ and He_2^+ respectively. We assume that, while drifting along the tube, ions of the first type are converted into those of the second type in collisions with the gas atoms. If D, μ, $\mathbf{F}$, q, and η are respectively the diffusion coefficient, mobility, electric field, ion concentration, and rate coefficient for the conversion reaction, then, provided the ions are close to thermal equilibrium with a large excess of gas atoms, we have

$$\frac{\partial q_1}{\partial t} = D_1 \nabla^2 q_1 - \mu_1 \operatorname{div}(\mathbf{F} q_1) - \eta q_1, \qquad (51\,\mathrm{a})$$

$$\frac{\partial q_2}{\partial t} = D_2 \nabla^2 q_2 - \mu_2 \operatorname{div}(\mathbf{F} q_2) + \eta q_1. \qquad (51\,\mathrm{b})$$

It is convenient, with the geometry used, to work in terms of cylindrical coordinates z, ρ, θ with z directed along the axis of the drift tube towards the entrance shutter. The electric field is constant along z and of magnitude F. As we require the total charge integrated over the cross-section of the ion swarm we need to change from q_1, q_2 to Z_1, Z_2 where

$$Z_{1,2} = \int_0^{\rho_0}\int_0^{2\pi} q_{1,2}\,\rho\,d\rho d\theta, \qquad (52)$$

ρ_0 being the radius of the swarm cross-section.

If we assume that any dependence of $q_{1,2}$ on θ is rapidly smoothed out by diffusion we find, by multiplying (51 a) and (51 b) by $2\pi\rho\,d\rho$ and integrating from 0 to ρ_0,

$$\frac{\partial Z_1}{\partial t} = 2\pi D_1 \left(\rho \frac{\partial q_1}{\partial \rho}\right)_{\rho=\rho_0} + D_1 \frac{\partial^2 Z_1}{\partial z^2} + u_1 \frac{\partial Z_1}{\partial z} - \eta Z_1, \qquad (53\,\mathrm{a})$$

$$\frac{\partial Z_2}{\partial t} = 2\pi D_2 \left(\rho \frac{\partial q_2}{\partial \rho}\right)_{\rho=\rho_0} + D_2 \frac{\partial^2 Z_2}{\partial z^2} + u_2 \frac{\partial Z_2}{\partial z} + \eta Z_1, \qquad (53\,\mathrm{b})$$

where u_1 and u_2 are the drift velocities $\mu_1 F$, $\mu_2 F$ respectively.

The charge collected when the exit shutter, of width δ_2, is open for a time τ_2 is

$$q_t(t_i) = \int_{t_i}^{t_i+\tau_2-\delta_2/u_1} D_1\left(\frac{\partial Z_1}{\partial z}\right)_{z=0} dt + \int_{t_i}^{t_i+\tau_2-\delta_2/u_2} D_2\left(\frac{\partial Z_2}{\partial z}\right)_{z=0} dt. \quad (54)$$

It is assumed that electrode 4 is a complete absorber of ions so

$$Z_1, Z_2 \to 0 \quad \text{as } z \to 0. \quad (55)$$

At $t = 0$, the time of opening the first shutter, the ion distribution is assumed to be given by

$$Z_1(z,0) = N_1/(u_1\tau_1-\delta_1), \quad d+\delta_1 < z < d+u_1\tau_1,$$
$$= 0 \quad \text{otherwise}, \quad (56\,\text{a})$$
$$Z_2(z,0) = N_2/(u_2\tau_1-\delta_1), \quad d+\delta_1 < z < d+u_2\tau_1,$$
$$= 0 \quad \text{otherwise}, \quad (56\,\text{b})$$

where τ_1 is the time for opening of the first shutter and d is the distance between electrodes 4 and 14.

To simplify the analysis it was also assumed that

$$2\pi\left[\rho \frac{\partial q_{1,2}}{\partial \rho}\right]_{\rho=\rho_0} = -\gamma Z_{1,2}, \quad (57)$$

where γ is independent of z and t. If the ion density is taken to vanish at $\rho = \rho_0$ and the diffusion takes place in the fundamental mode then (57) is valid with $\gamma = 2\cdot 93$ cm^{-2}. Most of the analysis was carried out taking this value of γ. It was verified in selected samples that changing to $\gamma = 0$ and $\gamma = 5\cdot 86$ cm^{-2} affected the final results by no more than a few per cent.

Fig. 19.19 (b) illustrates a set of observations of $q(t_1)$ under conditions in which it exhibits two peaks only, while the set shown in Fig. 19.19 (a) exhibits a third smaller peak due to faster ions.

It is relatively easy to obtain a good approximation to the mobilities from the location of the peaks but a more elaborate analysis is necessary to obtain the conversion rate coefficient η. The procedure adopted was to minimize a quantity ϵ given by

$$\epsilon = \sum_{i=1}^{j} \{q_t(t_i) - q_e(t_i)\}^2, \quad (58)$$

where $q_t(t_i)$ is the observed charge collected with a time delay t_i between the opening of the entrance and exit shutters, q_e that calculated with assumed parameters, and j is the total number of observations. According to the theory stated above, the variables to be adjusted to minimize ϵ are μ_1, μ_2, η, N_1, and N_2 because the diffusion coefficients D_1, D_2 should be related to μ_1 and μ_2 by (23 a). In practice it was found that, if D_1 and D_2 were also treated as variables, the values which minimized ϵ were somewhat larger than given by (23 a), unless the ion concentrations were kept below certain maximum values. This effect was probably due to space-charge spreading of the beam.

The results of carrying out an analysis of the data shown in Fig. 19.19 (a) and (b) are indicated in the form of continuous curves calculated for $q(t)$ using values of μ_1, μ_2, and η that minimized ϵ.

The pressure range covered was about 2–22 torr. Throughout this range He_2^+ was more abundant than He^+ and, at pressures above 10 torr,

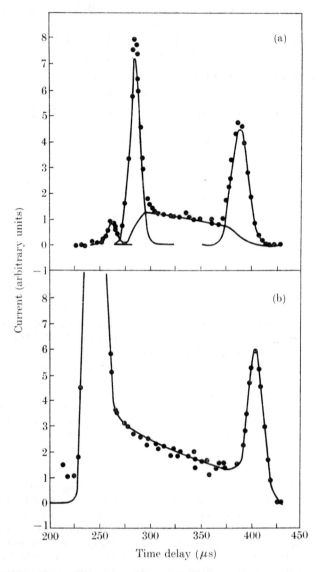

Fig. 19.19. Plots of the observed current to the ion collector in the experiments of Beaty and Patterson. (a) $p = 3\cdot 15$ torr, $F/p = 2\cdot 46$ V cm^{-1} torr^{-1}, $T = 305$ °K. (b) $p = 4\cdot 99$ torr, $F/p = 3\cdot 15$ V cm^{-1} torr^{-1}, $T = 304$ °K.

it was the only ion present. Under the latter conditions its mobility was independent of F/p and care was taken to check that it was unaffected by change of applied field, pulse width, age of the ions, power of the discharge, etc.

The best values found for these mobilities, extrapolated to zero field, were 10·4, 16·7, and 19·6 cm² V⁻¹ s⁻¹. The fastest ion was only observed at the lower end of the pressure range and, even then, only when the ions were sampled very shortly after breakdown.

Values of η were derived from observations in which conditions were chosen so that the fraction of He^+ ions which were converted to He_2^+ in

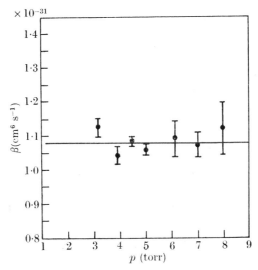

Fig. 19.20. Variation of the rate coefficient β ($= \eta/n^2$) with gas pressure in helium as observed by Beaty and Patterson.

flight was neither too large nor too small. If the He_2^+ ions are produced by the three-body reaction (50) η can be written as βn^2, where n is the neutral atom concentration and β is a constant. Fig. 19.20 shows a plot of β as a function of pressure verifying that it is indeed a constant and equal to $1·08 \pm 0·06 \times 10^{-31}$ cm⁶ s⁻¹. This corresponds to a value of 134 s⁻¹ torr⁻² for η.

Madson, Oskam, and Chanin† were the first to achieve a mass analysis in association with drift velocity measurements in helium, a technique that had already been applied to other gases by several investigators (see § 3.4.2). They used an ion source similar to that of Biondi and Chanin (§ 2.2.3) but, after passing through the drift tube, part of the ions entered a quadrupole mass spectrometer through an aperture of 10 μm diameter.

Fig. 19.21 shows typical results obtained in the pressure range 3–5 torr. The ion of lowest mobility is of atomic mass 4, definitely identified as

† MADSON, J. M., OSKAM, H. J., and CHANIN, L. M., *Phys. Rev. Lett.* **15** (1965) 1018.

He$^+$. It has a mobility close to 10·5 cm^2 V^{-1} s^{-1} which it has been assumed is that of the atomic ion. Two ions with distinct mobilities, one close to 16 and the other to 20 cm^2 V^{-1} s^{-1}, were observed, both with mass 8. Identification of either with O^{++} is ruled out by the absence of any ions of mass 16. It seems then that both the faster ions are He$_2^+$. Of these two, only the faster one was present when a narrow

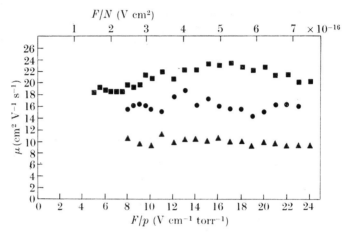

FIG. 19.21. Mobilities of He$^+$ and He$_2^+$ ions in helium as a function of F/p as observed by Madson, Oskam, and Chanin. ■ He$_2^+$ (fast ion), ● He$_2^+$ (slow ion), ▲ He$^+$.

discharge pulse of low amplitude was used. This suggests that the slower ion is only produced by some process of ageing, an interpretation consistent with the failure of Biondi and Chanin to observe it in their experiments and with evidence from Beaty and Patterson's experiments (see p. 1981). As mentioned on p. 1974 it is possible that the faster ion is metastable He$_2^+$ in the $^4\Sigma_u^+$ state, while the slower is normal He$_2^+$.

3.2.3. *The mobility of helium ions in helium—experimental techniques and results at* 300 °K—*afterglow observations.* The first investigations in a helium afterglow suffered from the absence of any means of identifying the ions. In 1952 Phelps and Brown† used a mass spectrometer in conjunction with a microwave cavity to identify He$^+$ and He$_2^+$ ions and observed the variation with time of their separate concentrations in the early stage of the afterglow. Fig. 19.22 illustrates the general arrangement of their experiment. The microwave equipment was the same as that used in the first microwave afterglow experiment by Biondi and Brown (see p. 1954). Ions diffusing to the cavity wall were

† loc. cit., p. 1973.

sampled through a hole of diameter 0·004″. The pressure in the analysing region was kept low by a 200-l/s oil diffusion pump. The ions were first accelerated in a uniform field region and then in a conical lens system which focused them on the entrance slit of a 60° sector-type mass spectrometer. To prevent ionization in the relatively high pressure region near the exit aperture from the cavity, the initial accelerating voltage was only 22·5 V. Ions leaving the spectrometer were detected by an electron multiplier.

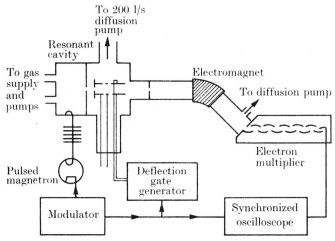

Fig. 19.22. General arrangement of apparatus used by Phelps and Brown for observing the positive ions present in an afterglow in helium.

Using commercial, spectroscopically pure helium supplied to the cavity through a liquid helium trap after baking the filling system at 270 °C for several days, no positive ion currents due to impurities were observed throughout the afterglow period provided the cavity pressure was below 1·5 torr.

During the afterglow we have the following equations for the concentrations n_m, n_1, and n_2 of the metastable atoms, He$^+$ ions, and He$_2^+$ ions respectively.

$$\frac{\partial n_\mathrm{m}}{\partial t} = D_\mathrm{m}\nabla^2 n_\mathrm{m} - \gamma n_\mathrm{m} - \beta n_\mathrm{m}^2, \qquad (59\,\mathrm{a})$$

$$\frac{\partial n_1}{\partial t} = D_\mathrm{a1}\nabla^2 n_1 - \eta n_1 + \beta n_\mathrm{m}^2, \qquad (59\,\mathrm{b})$$

$$\frac{\partial n_2}{\partial t} = D_\mathrm{a2}\nabla^2 n_2 + \eta n_1. \qquad (59\,\mathrm{c})$$

Here D_m is the diffusion coefficient of the metastable atoms, D_a1, D_a2

the ambipolar diffusion coefficients of He$^+$ and He$_2^+$ ions respectively, γ the rate coefficient for destruction of metastable atoms, η that for production of He$_2^+$ from He$^+$ in the gas, and β that for the process

$$\text{He}^* + \text{He}^* \to \text{He}^+ + \text{He} + e \qquad (60)$$

discussed in Chapter 18, § 5.2.4.

If we assume that diffusion loss occurs through the fundamental mode and that βn_m is small compared with γ, it is convenient to define three time constants

$$\tau_m^{-1} = \Lambda^{-2} D_m + \gamma, \qquad \tau_1^{-1} = \Lambda^{-2} D_{a1} + \eta, \qquad \tau_2^{-1} = \Lambda^{-2} D_{a2}, \qquad (61)$$

where Λ is the characteristic diffusion length of the cavity (see, for example, (22)). We then have

$$n_m = n_m^0 \exp(-t/\tau_m), \qquad (62\,\text{a})$$

$$n_1 = (n_1^0 + A)\exp(-t/\tau_1) - A\exp(-2t/\tau_m), \qquad (62\,\text{b})$$

$$n_2 = (n_2^0 + B n_1^0 + BC)\exp(-t/\tau_2) - B(n_1^0 + A)\exp(-t/\tau_1) + $$
$$+ B(A - C)\exp(-2t/\tau_m), \qquad (62\,\text{c})$$

where n_m^0, n_1^0, n_2^0 are initial concentrations and

$$A = \beta(n_m^0)^2 (2\tau_m^{-1} - \tau_1^{-1})^{-1}, \qquad B = \eta(\tau_1^{-1} - \tau_2^{-1})^{-1},$$
$$C = \beta(n_m^0)^2 (2\tau_m^{-1} - \tau_2^{-1})^{-1}.$$

The type of variation of n_1, n_2, and the electron concentration $n_e = n_1 + n_2$ expected from (62), using plausible values for the various parameters involved, is shown in Fig. 19.23. Observed variations showed quite good agreement for n_m and n_2 but for n_1 were considerably smaller than those calculated, at least in the first 4 ms.

At pressures below 2 torr, $\tau_1 > \tau_2 > \frac{1}{2}\tau_m$ so that the final decay rate for electrons should be as $\exp(-t/\tau_1)$. Because the conversion of He$^+$ to He$_2^+$ occurs in three-body collisions η must be proportional to the square of the main gas pressure p, so we may write

$$\tau_1^{-1} = (pD_{a1}/p\Lambda^2) + \zeta p^2, \qquad (63)$$

where pD_{a1} and ζ are independent of the pressure p. By plotting p/τ_1 against p^3 a straight line should be found with slope ζ and intercept pD_{a1}/Λ^2 on the pressure axis. Observed results were found to behave in this way and, from observations with $\Lambda^2 = 0.31$ and 0.85 cm^2, D_{a1} and η were obtained. From D_{a1} the mobility, derived as explained on p. 1953, came out to be 14 cm^2 V^{-1} s^{-1} and

$$\eta = 65 \pm 15 \text{ s}^{-1} \text{ torr}^{-2} \quad (0.52_5 \times 10^{-36} \text{ cm}^6 \text{ s}^{-1}).$$

The quantitative reliability of these pioneering observations was not high but the method was established and it was conclusively proved that both He$^+$ and He$_2^+$ were present in the afterglow.

We now describe a considerably later afterglow experiment by Oskam and Mittelstadt.† The frequency of the exciting source was 104 Mc/s,

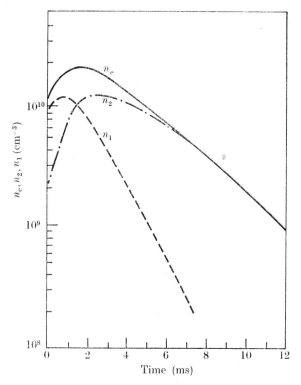

Fig. 19.23. Theoretical form of the variation with time of the concentrations of electrons (n_e), He$^+$ (n_1) and He$_2^+$ (n_2) ions in a helium afterglow. (Constants assumed: $D_{a1} = 540$ cm^2 s^{-1} at 1 torr, $D_{a2} = 840$ cm^2 s^{-1} at 1 torr, $D_m = 520$ cm^2 s^{-1} at 1 torr, $\gamma = 100p$ s^{-1} torr^{-1}, $\eta = 65p^2$ s^{-1} torr^{-1}.)

the maximum power 60 W, and pulse length 10 μs or longer. The microwave probing signal, of frequency 9000 Mc/s, was of very low power so as to avoid heating the afterglow electrons. To examine the possible effect of the shape of the electron distribution on the measurements, two cavities, one operating in the TM$_{010}$ and one in the TE$_{011}$ mode, were used. It was possible to measure the electron concentration as a function of time over a concentration range of 5000 to 1 (from 10^7 to 5×10^{10} cm^{-3}).

† loc. cit., p. 1974.

Special care was taken to ensure purity of the gas. The quartz bottle containing the gas was connected permanently to an ultra-high vacuum system and could be fired at 1000 °C. During bakeout the oil diffusion pump and associated molecular sieve traps were cut off from the system and a vac-ion pump substituted. The ultimate pressure obtained when cold was 10^{-9} torr with a rate of rise of 5×10^{-10} torr/min. The commercially pure gas was further purified by cataphoresis.† At helium pressures below 15 torr the efficiency of removal of neon impurities in this way is low so the helium was purified at high pressures and no cataphoresis discharge run at low pressure. When working at pressures above 15 torr the cataphoresis discharge was maintained throughout all the measurements.

The analysis follows on exactly the same lines as in that of the experiment of Phelps and Brown except that, as the measurements were taken at a stage in the afterglow at which all metastable atoms had effectively disappeared, $n_m = 0$. We then find from (62 b, c)

$$n_e = n_1 + n_2 = (1-B)n_1^0 \exp(-t/\tau_1) + (n_2^0 + Bn_1^0)\exp(-t/\tau_2), \quad (64)$$

where

$$\tau_1^{-1} = \Lambda^{-2}D_{a1} + \eta, \qquad \tau_2^{-1} = \Lambda^{-2}D_{a2}, \qquad B = \eta(\tau_1^{-1} - \tau_2^{-1})^{-1}.$$

Writing $\eta = \zeta p^2$, where p is the gas pressure, we have $\tau_1 \gtrless \tau_2$ according as

$$p^3 \gtrless (D_{a2} - D_{a1})p/\zeta\Lambda^2. \tag{65}$$

At high gas pressures, for which $\tau_2 > \tau_1$, $T_e = T_g$ and $p/\tau_2 = D_{a2}\,p/\Lambda^2$ should be independent of pressure. This was found to be the case at pressures greater than 5 torr. From this pressure region a mobility of 16·2 cm² V⁻¹ s⁻¹ was obtained from both cavities, agreeing closely with that expected for aged He_2^+ ions.

At sufficiently low pressures $\tau_1 > \tau_2$ since $D_{a2} > D_{a1}$ and the behaviour is essentially the same as that discussed above in analysing the experiment of Phelps and Brown (p. 1984). However, in that work no account was taken of electron-diffusion cooling (see p. 1953). Provided the pressure is not too small it should still be possible to determine the atom–molecule conversion coefficient η from a plot of p/τ_1 against p^3. This was found to be effective in the pressure range between 1·15 and 2 torr, the plot being linear in this region. From the slope of the line η was determined as 105 ± 10 s⁻¹ torr⁻² ($0·84_5 \times 10^{-36}$ cm⁶ s⁻¹) and, from the intercept on the p^3-axis, the mobility of He⁺ was determined at 10·7 cm² V⁻¹ s⁻¹. To bring out the effect of electron-diffusion cooling

† RIESZ, R. and DECKE, G. H., *J. appl. Phys.* **25** (1954) 196.

at lower pressures Oskam and Mittelstadt defined an effective mobility μ_{eff} by
$$\mu_{\text{eff}} = eD_{\text{eff}}/\kappa T, \qquad (66\,\text{a})$$
where
$$p/\tau_1 = (D_{\text{eff}}p/\Lambda^2) + \eta p^2. \qquad (66\,\text{b})$$

In the absence of diffusion cooling μ_{eff} would tend to $\mu(\text{He}^+)$ as $p \to 0$, as shown by the dotted curve of Fig. 19.24 which is calculated from (66 a, b) using the observed value for η. The observed results are

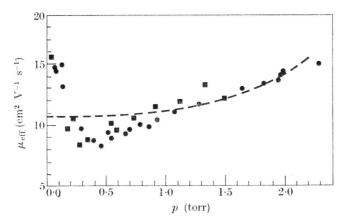

Fig. 19.24. Variation of μ_{eff} with pressure for He^+ ions in a helium afterglow calculated without allowance for diffusion cooling ---, observed by Oskam and Mittelstadt (TM_{00} mode) ●, (TE_{011} mode) ■.

seen to fall below this curve at pressures < 1.2 torr, which would be expected if diffusion cooling occurs. Thus, when T_e falls below T_g, D_{a1} falls below $2D^+$ until, when $T_e \ll T_g$, $D_{a1} = D^+$. The rise in the observed values of μ_{eff} at the lowest pressures is not understood—no such behaviour was found for Ne and Ar, both of which showed the diffusion cooling effect very clearly (see Fig. 19.28).

Very similar results were obtained by Mulcahy and Lennon,† the mobilities of the respective He^+ and He_2^+ ions being 10·6 and 16·7±0·5 cm² V⁻¹ s⁻¹ and η, the three-body rate coefficient, 115 s⁻¹ torr⁻² (0·93 cm⁶ s⁻¹). They did not work at pressures below 0·8 torr so found no evidence of diffusion cooling.

3.2.4. *The mobility of helium ions in helium—experimental techniques and results at 300 °K—determination of atomic to molecular ion conversion rate by measurements of the optical emission from an afterglow.* The theory of the electron–ion recombination processes in helium afterglows is described in Chapter 20, § 4. According to this theory, which has strong

† MULCAHY, M. J. and LENNON, J. J., *Proc. phys. Soc.* **80** (1962) 626.

observational support, the intensities of emission from the afterglow, of He_2^+ molecular bands and of He^+ atomic lines are proportional to the product of the electron concentration and that of the molecular or atomic ion, respectively. Hence, by measuring the intensity emitted in initially chosen wavelength ranges, at different times in the afterglow, it is possible to monitor the variations with time of the concentration of both He^+ and He_2^+ ions. The three-body conversion coefficient η can then be determined.

Since the intensity I_2 of the molecular light is proportional to $n_e n_2$, n_e, n_2 being the respective concentrations of electrons and He_2^+ ions,

$$\frac{d}{dt}(\ln I_2) = \frac{d}{dt}(\ln n_e) + \frac{d}{dt}(\ln n_2). \tag{67a}$$

At a later stage in the afterglow (of the order 1 ms or later) almost all the ions are He_2^+ so $n_e = n_2$ and (67a) becomes

$$\frac{d}{dt}(\ln I_2) = 2\frac{d}{dt}(\ln n_2). \tag{67b}$$

The intensity I_1 of the atomic light is proportional to $n_e n_1$ so that

$$\frac{d}{dt}(\ln I_1) = \frac{d}{dt}(\ln n_1) + \frac{d}{dt}(\ln n_e) = \frac{d}{dt}(\ln n_1) + \frac{1}{2}\frac{d}{dt}(\ln I_2). \tag{68}$$

Since, as in (62b) when $n_m = 0$,

$$n_1 = n_1^0 \exp(-t/\tau_1),$$

where
$$\tau_1^{-1} = \Lambda^{-2} D_{a1} + \zeta p^2,$$

we have
$$\zeta p^2 = -\frac{d}{dt}(\ln I_1) + \frac{1}{2}\frac{d}{dt}(\ln I_2) - \Lambda^{-2} D_{a1}. \tag{69}$$

In obtaining (69) we neglected the rate of loss of He^+ ions due to recombination but this is certainly justifiable (see Chap. 20, § 2). It follows from (69) that, if the long-time decay rates of I_1 and I_2 can be measured at different pressures, ζp^2 can be obtained.

Experiments on these principles have been carried out by Niles and Robertson.† The gas was enclosed in a Pyrex cell 17 cm long and of radius 2·35 cm. It was purified by passage through two liquid nitrogen cold traps, followed by two cataphoresis discharge tubes, one of which was in the bakeable system. The cell was baked out and evacuated below 10^{-8} torr before admitting the helium.

The breakdown pulse was of 120 μs duration. At a selected time in the afterglow period the photomultiplier was turned on, normally for

† NILES, F. E. and ROBERTSON, W. W., J. chem. Phys. **40** (1964) 3568.

10 μs, to measure the intensity of light from the afterglow after passage through one of a set of Bausch and Lomb interference filters. These were chosen to have peak transmission at 3870, 4475, 4650, 5080, and 5880 Å respectively.

The time variation of the molecular ion concentration was first observed using a monochromator to isolate the strong 4650 Å line of

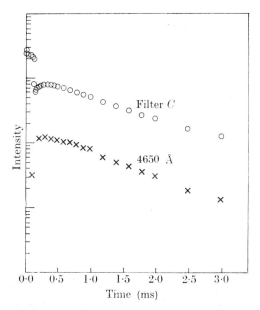

FIG. 19.25. Intensity of molecular emission in a helium afterglow, at a pressure of 5·17 torr, as a function of time since breakdown pulse cut-off. The upper set of observed points gives the emission transmitted through a filter with peak wavelength 4650 Å and a photomultiplier gate of 10 μs. The lower set of observed points gives that transmitted through a monochromator set at 4650 Å with a photomultiplier gate of 100 μs.

He_2 ($2s\,^3\Sigma_u^+ - 3p\,^3\Pi_g$) and a photomultiplier gating time of 100 μs. Fig. 19.25 illustrates the results obtained, showing clearly how n_2 at first builds up, due to conversion of He^+ ions, and then falls finally at an exponential rate. It was then verified that, if the monochromator were replaced by the 4650 Å filter, a similar result was obtained with an increased intensity, even when the gating time was reduced to 10 μs. This was particularly true of the final decay time. This filter was therefore used when it was desired to monitor $d(\ln I_2)/dt$ in the later stages of the afterglow.

Fig. 19.26 illustrates typical measurements from which $d(\ln I_1)/dt$ could be obtained. In this case the 3870 Å filter was used which transmits the strong $2\,^3S-3\,^3P$ He line at 3889 Å. From the total transmitted light the molecular contribution was subtracted. The decay rate of the difference curve shown then gives $d(\ln I_1)/dt$.

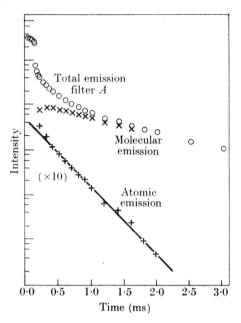

Fig. 19.26. Intensity of emission from a helium afterglow at a pressure of 5·17 torr, as a function of time since breakdown pulse cut-off. The upper set of observed points (○) gives the emission transmitted through a filter with peak wavelength 3870 Å with a photomultiplier gate of 10 μs. The middle set of observed points (×) gives that part of the total emission which is emitted by molecules. Subtraction of the middle from the upper set gives the atomic emission –+--+-.

From measurements of this kind, at pressures of 5·17, 9·93, and 20·3 torr, Niles and Robertson found from (69), using D_{a1} as given from earlier measurements,

$$\eta = 109 \pm 6 \text{ s}^{-1} \text{ torr}^{-2}\ (0.88 \times 10^{-31} \text{ cm}^6 \text{ s}^{-1}),$$

which is somewhat smaller than the value (134 s⁻¹ torr⁻²) obtained by Beaty and Patterson† using the drift tube technique, but agrees well with the results obtained from the later measurements of the decay of electron concentration in afterglows.

3.2.5. *The mobility of helium ions in helium—experimental results at temperatures below* 300 °K. Tyndall and Pearce,‡ using the shutter method described on p. 1936 with ions obtained from a helium glow

† loc. cit., p. 1974.   ‡ loc. cit., p. 1975.

discharge source, measured mobilities at 169 and 77 °K by immersing the mobility tube in dry ice and liquid nitrogen respectively.

In each case they observed one ion only with reduced mobility 21·0 and 19·3 cm² V⁻¹ s⁻¹ at the respective temperatures. Chanin and Biondi† also made observations at these temperatures using the technique described in § 2.2.3. They not only observed an ion with reduced mobility close to that observed by Tyndall and Pearce (21·7 and 18·0 cm² V⁻¹ s⁻¹ at 195 and 77 °K respectively), but also a slower ion with mobility 12·1 and 13·5 cm² V⁻¹ s⁻¹ at the respective temperatures.

It was naturally assumed that the faster and slower ions were He_2^+ and He^+ respectively but observations made a few years later by Patterson‡ have shown that the situation is more complicated. He used the technique of Beaty and Patterson described on p. 1976 and made observations over a series of temperatures between 300 and 77 °K. This was done, cooling the mobility tube by allowing cold N_2 vapour to accumulate around it and monitoring the output wave forms continuously as the temperature decreased. The temperature was measured by three copper–constantan thermocouples attached to the outer surface of the mobility tube. Checks were made at fixed temperatures of 196 and 76 °K.

At room temperature and gas concentrations N greater than 2×10^{17} cm⁻³ the only ion observed was He_2^+. The behaviour of the peak due to this ion was then observed as the temperature was reduced. At small F/N the time delay before arrival of this peak remained the same down to 190 °K, but it then became progressively smaller as the temperature was further reduced. This suggested that a reaction was taking place producing a partial conversion of He_2^+ into a faster ion. If an equilibrium is attained and n_1, n_2 are the concentrations of the two ions and μ_1, μ_2 their mobilities, the observed mobility will be given by

$$\mu = (n_1\mu_1 + n_2\mu_2)/(n_1+n_2). \tag{70}$$

Strong support for this explanation was provided by observations made at a fixed temperature and different values of F/N, the helium concentration being low enough for two ions to be present. This may be seen by references to Figs. 19.27 (a) and (b), which show plots of observed ion current received as a function of time delay for different values of F/N.

In (a) the shutter opening times were adjusted so that only the faster ion was admitted to the drift space. At low F/N a single peak is observed

† loc. cit., p. 1975.   ‡ loc. cit., p. 1975.

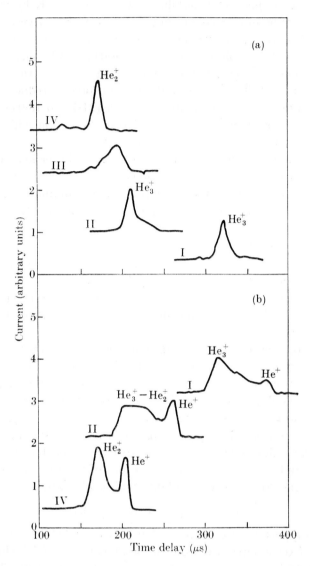

Fig. 19.27. Plots of ion currents as functions af time delay, observed by Patterson for ions in helium at 76 °K and a gas concentration N of 1.6×10^{17} cm^{-3} at different values of F/N as follows, in 10^{-17} V cm^2,

I:6;   II:10;   III:11;   IV:13.

In (a) only the faster and in (b) only the slower ion was allowed to enter the drift space. The likely identification of the ion peaks is indicated in each case.

but as F/N is increased the time delay falls and the shape of the peak becomes less simple. At still higher F/N it again becomes simple but arrives with still smaller time delay.

Fig. 19.27 (b) shows results obtained when the shutter opening times were adjusted so that only the slower (He$^+$) ion was admitted to the drift space. At high and low F/N only two peaks are present, one corresponding to He$^+$ and one to the faster ion observed at the particular F/N in Fig. 19.27 (a). On the other hand there is evidence of the presence of three ions at intermediate F/N.

These and other observations showed that the third ion, of high mobility at low F/N, is formed by reaction of He$_2^+$ with the gas. This ion could be destroyed by electrical heating, and this reproduced He$_2^+$ directly without passing through any intermediate stage. Furthermore, the rate of production of the ion from He$_2^+$, in such conditions as in Fig. 19.27 (b), must have been faster than that of He$_2^+$ from He$^+$ at low F/N. This follows from the absence of an intermediate peak at low F/N. Strong evidence that the third ion is not formed by reaction with impurities is provided from the fact that, before purification by cataphoresis, peaks due to impurities were observed. These disappeared after cataphoresis, leaving, at low pressures, He$_2^+$ ions which converted to the third ion at higher pressures in the pure gas.

The only likely possibility is that the ion is He$_3^+$. This is very difficult to verify by mass analysis because of the possible confusion with C$^+$. Assuming this identification, the equilibrium constant K for the reaction

$$\text{He}_2^+ + \text{He} + \text{He} \rightarrow \text{He}_3^+ + \text{He}$$

can be obtained from (70). Thus

$$K = n_1 N/n_2$$
$$= \{(\mu_2 - \mu)/(\mu - \mu_1)\}N.$$

Since K is given at temperature T by

$$K = (Z_1 Z/Z_2)\exp(-D/\kappa T),$$

where D is the dissociation energy of He$_3^+$ into He$_2^+$ and He and Z, Z_1, and Z_2 are the partition functions for He, He$_2^+$, and He$_3^+$ respectively, a plot of $\ln K$ against T^{-1} should be approximately linear. This was found to be so and, from the slope of the plot, D was found to be 0.17 ± 0.03 eV.

In view of this work it seems likely that the ion observed by Tyndall and Pearce and the faster ion observed by Chanin and Biondi was He$_3^+$. Values for the mobilities of the different helium ions in their parent gas on this interpretation are given in Table 19.9.

3.2.6. *The mobilities of other rare-gas ions in their parent gases.* Table 19.10 gives the results of observations of the mobilities of neon, argon, krypton, and xenon ions in their parent gases. The techniques used are the same as for helium and are indicated.

There is no evidence among these data of any serious inconsistencies, except for argon. For the slow ion, which is presumably Ar_2^+, Biondi and

TABLE 19.9

Observed mobilities of helium ions in helium

Ion	Temp. (°K)	Mobility (cm^2 V^{-1} s^{-1} reduced to s.t.p.)
He^+	300	10·5 (a), 10·4 (d), 10·7 (g)
	195	12·1 (b)
	77	13·5 (b)
He_2^+	300	20·3 (b), 16·7 (d), 19·6 (d), 16·3 (e), 16·2 (f), 16·2 (g), $\simeq$ 16 (h), 20 (h).
	195	$\simeq$ 16 (i)
	77	$\simeq$ 16 (i)
He_3^+	300	—
	195	21·7 (b), 21·0 (c), 20·5 (i)
	77	18·0 (b), 19·3 (c), $\simeq$ 19 (i)

(a) BIONDI, M. A. and CHANIN, L. M., *Phys. Rev.* **94** (1954) 910.
(b) CHANIN, L. M. and BIONDI, M. A., ibid. **106** (1957) 473.
(c) TYNDALL, A. M. and PEARCE, A. F., *Proc. R. Soc.* A **149** (1935) 426.
(d) BEATY, E. C. and PATTERSON, P., *Phys. Rev.* **137** (1965) A 346.
(e) OSKAM, H. J., *Philips Res. Rep.* **13** (1958) 335, 401.
(f) KERR, D. E. and LEFFEL, C. S., *Bull. Am. Phys. Soc.* **7** (1962) 131.
(g) OSKAM, H. J. and MITTELSTADT, V. R., *Phys. Rev.* **132** (1963) 1435.
(h) MADSON, J. M., OSKAM, H. J., and CHANIN, L. M., *Phys. Rev. Lett.* **15** (1965) 1018.
(i) PATTERSON, P. L., *J. chem. Phys.* **48** (1969) 3625.

Chanin using the drift tube method (§ 2.2.3) find a much greater mobility than Beaty, using essentially the shutter method of Tyndall and Powell, and Oskam and Mittelstadt, using the same afterglow technique as described in § 3.2.3.

Fig. 19.28 illustrates the observed variation with gas pressure of the effective mobility μ_{eff}, defined as in (66), for ions in an argon afterglow. The behaviour is very close to that expected. At high pressures ($>$ 1·5 torr) μ_{eff} is constant at 7·85 cm^2 V^{-1} s^{-1} which should then be the mobility of Ar_2^+. At lower pressures, μ_{eff} falls gradually to a limiting value of 0·8 cm^2 V^{-1} s^{-1}. If diffusion cooling is occurring then this should be one-half the mobility of Ar^+ which, according to Biondi and Chanin and to Hornbeck, is 1·6 cm^2 V^{-1} s^{-1}. This suggests that the afterglow observations are reliable and that the mobility of Ar_2^+ is well below that observed by Biondi and Chanin. The only doubt is that for the TE_{011} cavity there is some suggestion that the high-pressure behaviour may not lead to the same constant value of μ_{eff} as for the TM_{010}. This may be seen by reference to Fig. 19.28. Unfortunately the TE_{011} data stop at too low a pressure to check this.

TABLE 19.10

Observed mobilities of rare-gas ions in their parent gases

Ion	Gas	Temp. (°K)	Mobility ($cm^2\ V^{-1}\ s^{-1}$)					Munson and Tyndall
			Measured by					
			Drift tube				Afterglow	
			(a)	(b)	(c)	(d)	(e)	(f)
Ne^+	Ne	300	4·4	4·2	—	—	4·1	—
		195	—	4·5	—	—	—	—
		77	—	5·2	—	—	—	—
Ne_2^+	Ne	300	—	6·5	—	—	6·5	6·2
Ar^+	Ar	300	1·63	1·6	1·38	—	1·6	—
		195	—	1·95	—	—	—	—
		77	—	2·2	1·88	—	—	—
Ar_2^+	Ar	300	—	2·7	1·8	—	1·9	1·95
		195	—	2·9	—	—	—	—
		77	—	2·7	—	—	—	—
Kr^+	Kr	300	—	0·9	0·9	0·95	—	
Kr_2^+	Kr	300	—	1·21	1·00	—	—	0·94(?)
Xe^+	Xe	300	—	0·58	—	0·6–0·65	—	
Xe_2^+	Xe	300	—	0·79	—	—	—	0·65(?)

(a) HORNBECK, J. A., *Phys. Rev.* **84** (1951) 615.
(b) CHANIN, L. M. and BIONDI, M. A., ibid. **106** (1957) 473.
(c) BEATY, E. C., ibid. **104** (1956) 17.
(d) VARNEY, R. N., ibid. **88** (1952) 362.
(e) OSKAM, H. J. and MITTELSTADT, V. R., ibid. **132** (1963) 1435.
(f) MUNSON, R. J. and TYNDALL, A. M., *Proc. R. Soc.* **A177** (1941) 187.

It may be that, as in helium, Ar_2^+ possesses two mobilities depending upon age but this is purely speculative at present. Some further evidence is provided in § 3.3.3 from comparison with theory.

It is of interest to compare with the mobilities measured by Tyndall and his collaborators before 1941, using a glow discharge ion source with their shutter method. Referring to the last column of Table 19.10 we see that there is little doubt that the neon ion they studied was Ne_2^+. Their observations for argon then provide some support for the correctness of the mobility $1·9\ cm^2\ V^{-1}\ s^{-1}$ for Ar_2^+. On the other hand, there is no way of deciding whether their observations for krypton and xenon refer to the atomic or molecular ions. Taken at their face value the numbers suggest that, for these gases, they actually observed the mobilities of the atomic ions.

3.2.7. *The mobility and reactions of* Hg+ *ions in mercury vapour.* In § 3.2.6 we described how from measurements of the rate of decay of

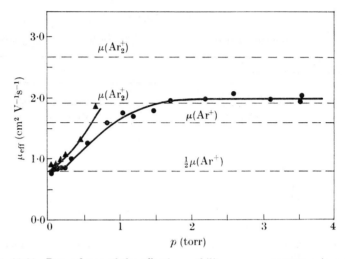

Fig. 19.28. Dependence of the effective mobility μ_{eff} on pressure p in argon between 0·025 and 4 torr, observed by Oskam and Mittelstadt. ● observed with cavity in TM_{010} mode, ▲ observed with cavity in TE_{011} mode. The mobilities of Ar^+ and Ar_2^+ ions observed by pulse methods are indicated by the dotted lines. For Ar_2^+ the larger mobility is that observed by Chanin and Biondi, the smaller by Beaty (see Table 19.10 for references).

electron concentration in an afterglow produced in a mixture of helium (1–2 torr pressure) and mercury vapour at about 10^{-6} torr pressure Biondi† was able to determine the mobility of Hg+ ions in helium under conditions in which the electrons were in thermal equilibrium with the gas.

From observations with the mixture containing greater relative concentrations of mercury vapour (between 2×10^{-5} and 10^{-4} torr) it was possible to attain conditions in which, while the electrons were in thermal equilibrium, the ambipolar diffusion of the Hg+ ions was determined by the neutral Hg, rather than He, atoms. In this way an ambipolar diffusion coefficient D_a of 10·1 cm² s⁻¹ at 1 torr was obtained for Hg+ ions in Hg when the electrons are in thermal equilibrium with the main gas at 350 °K. The corresponding mobility of Hg+ ions in Hg is then 0·2 cm² V⁻¹ s⁻¹. Measurements were also made in pure mercury

† BIONDI, M. A., *Phys. Rev.* **90** (1953) 730.

vapour at the same vapour temperature. These gave a value of D_a nearly four times larger. In terms of the formula (see (24))

$$D_a = D^+(1+T_e/T),$$

this gives $T_e \simeq 2400$ °K which agrees with observations made by Mierdel,† who determined the electron temperature as a function of time in a mercury afterglow, using a Langmuir probe technique. He found that T_e fell rapidly from 6200 °K, its value when the discharge was on, to about 2200 °K in the first few microseconds of the afterglow and then remained essentially constant. It is suggested that the first rapid fall is due to diffusion cooling but the reason for the near constancy thereafter is not clear. Mierdel also determined D_a with $T_e \simeq 2200$ °K and his value is within 15 per cent of that observed by Biondi.

Evidence of volume-loss processes of ions under conditions in which recombination is unimportant was also found by Biondi. In the mercury–helium mixture such a process was observed with a rate proportional to the Hg atom concentration. This was interpreted as

$$Hg^+ + Hg + He \rightarrow Hg_2^+ + He,$$

in which case the rate constant would be $1 \cdot 7 \times 10^{-31}$ cm^6 s^{-1}. Owing to the difficulty of carrying out experiments in which both the Hg and He pressures could be varied it was not possible to confirm this interpretation by showing that the rate is also proportional to the He atom concentration. At the higher Hg vapour densities, evidence was obtained of a loss process with rate depending on the square of the Hg concentration, presumably

$$Hg^+ + Hg + Hg \rightarrow Hg_2^+ + Hg.$$

The rate constant for this was estimated very roughly as 10^{-31} cm^6 s^{-1}.

3.3. *The theory of the effect of charge transfer on the mobilities of ions in their parent gases*

In Chapter 18, § 8 we discussed the theory of symmetrical resonance excitation transfer. An exactly similar theory may be applied to collisions involving symmetrical charge transfer. Thus a normal atom, possessing a complete outer shell of electrons, can interact in two different ways with an ion of the same kind, so the interaction energy at large separations is either $V_0(r) \pm U(r)$. The cross-section for charge transfer is then

$$Q_{tr} = (\pi/k^2) \sum (2l+1)\sin^2(\eta_l^g - \eta_l^u), \qquad (71)$$

just as in (314) of Chapter 18, $k\hbar$ being the momentum of relative motion and η_l^g, η_l^u the phase shifts for motion with relative angular momentum

† MIERDEL, G., *Z. Phys.* **121** (1943) 574.

$\{l(l+1)\}^{\frac{1}{2}}\hbar$ produced by the respective interactions $V_0 \pm U$, which we designate henceforth as $V_{g,u}$.

The semi-classical method described on p. 1874 may be used to give

$$Q_{\text{tr}}^{\text{sc}} = 2\pi\left\{\int_0^{p_0} p\sin^2\phi(p)\,\mathrm{d}p + \int_{p_0}^{\infty} p\sin^2\phi(p)\,\mathrm{d}p\right\}, \tag{72}$$

where
$$\phi(p) = (1/\hbar v)\int_{-\infty}^{\infty}\Delta E\{(z^2+p^2)^{\frac{1}{2}}\}\,\mathrm{d}z, \tag{73}$$

v being the relative velocity of impact and $\Delta E = V_g - V_u, = 2U$ at large r. p_0 is such that $\phi(p_0) = \frac{1}{4}\pi$, while for $p \gg p_0$, $\phi(p) \ll \frac{1}{4}\pi$ and for $p \ll p_0$, $\phi(p) \gg \pi$. Under these conditions

$$Q_{\text{tr}}^{\text{sc}} \simeq 2\pi\left\{\tfrac{1}{4}p_0^2 + \int_{p_0}^{\infty} p\phi^2(p)\,\mathrm{d}p\right\}. \tag{74}$$

For application to charge transfer ΔE is best represented by either

$$\Delta E = Ce^{-\lambda R}, \tag{75a}$$

or
$$\Delta E = ARe^{-\lambda R}. \tag{75b}$$

(73) gives
$$\phi(p) = (Ap/\hbar v)\{pK_0(\lambda p) + K_1(\lambda p)/\lambda\}, \tag{76}$$

where K_0 and K_1 are the usual Bessel functions (cf. Chap. 7, § 5.3.4, (137)). If p_0 is the largest value of p for which $\phi(p) = \frac{1}{4}\pi$, $\lambda p_0 \gg 1$ and

$$K_0(\lambda p) \sim (\pi/2\lambda p)^{\frac{1}{2}}e^{-\lambda p}, \qquad K_1(\lambda p) \sim -(\pi/2\lambda p)^{\frac{1}{2}}e^{-\lambda p},$$
$$\phi(p) \simeq (Ap^{\frac{3}{2}}/\hbar v)(\pi/2\lambda)^{\frac{1}{2}}e^{-\lambda p} \quad (p \gg p_0).$$

Hence
$$Q_{\text{tr}}^{\text{sc}} \simeq \tfrac{1}{2}\pi p_0^2 + 2\pi\int_{p_0}^{\infty} p\{\phi(p)\}^2\,\mathrm{d}p = \tfrac{1}{2}\pi p_0^2 + \pi^3 p_0/16\lambda, \tag{77}$$

with p_0 given by
$$p_0^3 e^{-2\lambda p_0} = \pi\lambda v^2\hbar^2/8A^2. \tag{78}$$

When v is very small
$$p_0 \simeq -\frac{1}{2\lambda}\ln(\pi\lambda v^2\hbar^2/8A^2\tilde{p}^3), \tag{79}$$

where $\tilde{p}$ is slowly varying, so as $v \to 0\dagger$ (cf. Chap. 18 (375)),

$$Q_{\text{tr}}^{\text{sc}} \to (\pi/2\lambda^2)\{\ln v + \tfrac{1}{2}\ln(\pi\lambda\hbar^2/8A^2)\}^2. \tag{80}$$

The semi-classical approximation will not be valid for very small v but there will be a considerable range of low velocities for which (80) will give a good representation (see Chaps. 23 and 24).

In connection with the calculation of mobilities the diffusion cross-section Q_d is required. Its definition when charge transfer occurs requires some care. The differential cross-section for scattering of the ions into the solid angle $\mathrm{d}\omega$ is given by

$$I(\theta)\,\mathrm{d}\omega = \{I_{\text{el}}(\theta) + I_{\text{tr}}(\pi - \theta)\}\,\mathrm{d}\omega \tag{81}$$

because the transfer process reverses the identity of ion and atom and

† DEMKOV, J. N., *Ann. leningr. Univ.* **146** (1952) 74.

interchange of nuclei changes θ to $\pi-\theta$. I_{el} and I_{tr} refer to direct elastic scattering and scattering with charge transfer respectively. Using the formulae (cf. Chap. 18, (322), (323), (329), and (330))

$$I_{el}(\theta) = \left| \frac{1}{2ik} \sum (2l+1)(e^{2i\eta_l^g} + e^{2i\eta_l^u} - 2) P_l(\cos\theta) \right|^2, \qquad (82\,\text{a})$$

$$I_{tr}(\theta) = \left| \frac{1}{2ik} \sum (2l+1)(e^{2i\eta_l^g} - e^{2i\eta_l^u}) P_l(\cos\theta) \right|^2, \qquad (82\,\text{b})$$

and substituting in the formula (6) of Chapter 16 for Q_d gives

$$Q_d = (4\pi/k^2) \sum (l+1) \sin^2(\chi_{l+1} - \chi_l), \qquad (83)$$

where
$$\chi_{l+1} = \eta_{l+1}^g, \qquad \chi_l = \eta_l^u \quad (l \text{ even})$$
$$= \eta_{l+1}^u, \qquad = \eta_l^g \quad (l \text{ odd}).$$

Using the semi-classical approximation, we have

$$Q_d \simeq 2\pi \int_0^\infty p\{\sin^2\zeta_A(p) + \sin^2\zeta_S(p)\}\, dp, \qquad (84)$$

where
$$\zeta_A(p) = \eta_l^g - \eta_{l+1}^u, \qquad \zeta_S = \eta_l^u - \eta_{l+1}^g,$$

so
$$\zeta_{A,S}(p) = \mp (1/\hbar v) \int_p^\infty \frac{(V_g - V_u)}{(1 - p^2/r^2)^{\frac{1}{2}}}\, dr +$$
$$+ (1/Mv^2) \frac{\partial}{\partial p} \int_p^\infty \frac{V_{g,u}}{(1 - p^2/r^2)^{\frac{1}{2}}}\, dr. \qquad (85)$$

As for Q_{tr} we may take $V_g - V_u$ as given by (75). For charge transfer

$$V_{g,u} \sim -\frac{1}{2} \frac{\alpha e^2}{r^4},$$

where α is the polarizability of the atom. Hence, in applying a similar method to evaluate Q_d as used for Q_{tr} we obtain

$$Q_d = \pi p_0^2 + (\pi A/v\hbar\lambda)^2 p_0^4 e^{-2\lambda p_0} + 3\pi^3 \alpha^2/8v^2 M^2 p_0^4, \qquad (86\,\text{a})$$

where
$$(4A^2/\hbar^2 v^2 \lambda) p_0^3 e^{-2\lambda p_0} + \frac{9}{8} \frac{\pi \alpha^2 e^4}{v^4 M^2 p_0^8} = \tfrac{1}{2}\pi. \qquad (86\,\text{b})$$

From (84) and (86) both Q_d and Q_{tr} may be obtained in terms of two parameters A and λ. These parameters may be fixed from observations of Q_{tr} made by ion beam experiments (see Chap. 24), at relative energies of the order 200 eV or so, and then used to derive Q_d and hence the mobility at ordinary temperatures from (24), or vice versa.

3.3.1. *Application to* H^+ *in* H *and* D^+ *in* D. The simplest application that can be made of this theory is to the mobility of H^+ in H and D^+

in D. Neither of these cases is of practical importance but they serve to illustrate the behaviour expected.

The interaction energies $V_{g,u}$ have been calculated without approximation by Bates, Ledsham, and Stewart.† A good fit to the calculated $\Delta E(R)$ can be made using the empirical form (75 b).‡ This may be checked by calculating Q_{tr} using (84) and comparing with the accurate calculations of Dalgarno and Yadav§ who used the actual interactions $V_{g,u}$. Applying the formula (86 b) the mobilities given in Table 19.11 may be derived.

TABLE 19.11

Calculated reduced mobilities of H+ *in* H *and* D+ *in* D

T (°K)	100	200	300	400	500	1000	2000	5000
Mobility (cm² V⁻¹ s⁻¹) of								
H+ in H	16·7	13·6	12·4	11·0	10·2	7·4	5·7	4·0
D+ in D	10·6	8·2	7·2	6·4	5·9	4·5	3·4	2·4

3.3.2. *Application to* He+. In 1934 Massey and Mohr‖ calculated V_g and V_u by quantum theory for He–He+ interactions, using approximate helium wave-functions available at the time which would be expected to be reasonably accurate at large distances from the nucleus. From these interactions they calculated the phase shifts $\eta_l^{g,u}$ in (71) and hence derived the mobility of He+ ions in He at 300 °K. They found a value 12·0 cm² V⁻¹ s⁻¹. Their calculations were later repeated by Lynn and Moiseiwitsch†† using more elaborate helium wave-functions and covering a considerable temperature range. Their results are given in Table 19.12 for comparison with the observations. The agreement is good and certainly within the range of experimental error and uncertainty in the theoretical calculation.

We also include in the table the results of a semi-empirical analysis by Dalgarno‡ using the formulae (84) and (86). He used the data on Q_{tr} from the experiments of Gilbody and Hasted‡‡ (see Chap. 24), obtained at relative energies > 250 eV, to derive the parameters A

---

† BATES, D. R., LEDSHAM, K., and STEWART, A. L., *Phil. Trans. R. Soc.* A**246** (1953) 215.
‡ DALGARNO, A., ibid. **250** (1958) 426.
§ DALGARNO, A. and YADAV, H. N., *Proc. phys. Soc.* A**66** (1953) 173.
‖ MASSEY, H. S. W. and MOHR, C. B. O., *Proc. R. Soc.* A**144** (1934) 188.
†† LYNN, N. and MOISEIWITSCH, B. L., *Proc. phys. Soc.* A**70** (1957) 474.
‡‡ GILBODY, H. B. and HASTED, J. B., *Proc. R. Soc.* A**238** (1957) 334.

and λ in (75). It was found that a good fit could be obtained with
$$A = 4\cdot92 \text{ a.u.}, \quad \lambda = 1\cdot40/a_0.$$
Using these values in (86) the mobility was calculated as a function of temperature, giving the results listed as semi-empirical in Table 19.12. The agreement with the observations and with the calculations of Lynn

TABLE 19.12

Comparison of observed and theoretical reduced mobilities for He^+ ions in He

Temp. (°K)	Mobility ($cm^2 V^{-1} s^{-1}$)		
	Obs.	Calc.	Semi-empirical
100	13·4	13·9	14·0
200	12·0	11·4	11·5
300	10·8	10·2	10·4
400	—	9·3	9·4
500	—	8·7	8·6
1000	—	6·8	6·5
2000	—	—	4·9
5000	—	—	3·4

and Moiseiwitsch is very good. It appears, however, that the charge transfer data below 250 eV do not agree with the predictions. This is probably due to the difficulty of the experiments and is further discussed in Chapter 24.

A striking feature of the effect of charge transfer is not only the reduction of the mobility but the changed temperature variation. Whereas for ions in unlike gases the mobility either falls steadily as the temperature falls below 200 °K or reaches a weak maximum with falling temperature and then decreases steadily (see § 2.3 and Table 19.4), for ions in their parent gases the mobility increases steadily as the temperature falls. This feature, which is of considerable generality (see Table 19.13), could be used to help in identifying the ion concerned in a particular set of experiments, a possibility pointed out by Dalgarno.†

3.3.3. *Application to other atomic ions in their parent gases.* For Ne^+ in Ne and Ar^+ in Ar, Dalgarno† applied the same semi-empirical procedure as for He^+–He, relating Q_d with Q_{tr}. Unfortunately the experimental data (Chap. 24) on Q_{tr} are less satisfactory and mutually consistent than for helium. Because of this A and λ were determined so as to give the observed mobility at 300 °K and reasonable agreement

† loc. cit., p. 2000.

with an average of the experimental data at several hundred eV energy. The results obtained are given in Table 19.13. The only observational check available under these circumstances is the temperature variation and this is satisfactory. As in other cases, the mobility increases steadily as the temperature falls.

TABLE 19.13

Semi-empirical and observed reduced mobilities of atomic ions in their parent gases

T (°K)	Mobilities in cm^2 V^{-1} s^{-1}							
	Ne		Ar		Kr		Xe	
	(a)	(b)	(a)	(b)	(a)	(b)	(a)	(b)
100	5·1	5·1	2·2	2·1	1·3	—	0·84	—
200	4·4	4·5	1·8	2·0	1·0	—	0·68	—
300	4·0	4·2	1·6	1·6	0·90	0·90	0·60	0·58
400	3·6	—	1·4	—	0·80	—	0·53	—
500	3·3	—	1·3	—	0·72	—	0·48	—
1000	2·5	—	0·95	—	0·53	—	0·35	—
2000	1·9	—	0·71	—	0·39	—	0·26	—
5000	1·3	—	0·47	—	0·26	—	0·17	—

In all cases the parameters in the semi-empirical analysis, (a), are adjusted to give agreement with the observed mobility, (b), at 300 °K.

For Kr$^+$–Kr and Xe$^+$–Xe the same procedure was applied, giving the results listed in Table 19.13, but here there is very little with which to check the semi-empirical theory as the only evidence about the temperature variation is the observation by Beaty† that between 300 and 90 °K the mobility of Kr$^+$ in Kr increases by a factor of 1·2.

There remains Hg$^+$–Hg but here the charge transfer data (Chap. 24) seem definitely inconsistent with the rather meagre mobility data. Thus the charge transfer measurements of Dillon, Sheridan, Edwards, and Ghosh‡ lead to a mobility about twice as high as observed.

3.3.4. *Mobilities of diatomic ions in their parent (atomic) gases.* The mobilities of ions such as Ne$_2^+$, Ar$_2^+$, Kr$_2^+$, and Xe$_2^+$ in their parent gases, over the observed temperature range, should be determined mainly by the polarization force and therefore should behave in much the same way as the alkali-metal ions in the corresponding gases.

In the limit of very low temperatures the mobility should be given by (34). For comparison we give in Table 19.14 the values calculated

† loc. cit., p. 1995.
‡ DILLON, J. A., SHERIDAN, W. F., EDWARDS, H. D., and GHOSH, S. N., *J. chem. Phys.* **23** (1955) 776.

in this way (0 °K), as well as the observations at higher temperatures, to see whether extrapolation of the latter to 0 °K would seem consistent with the theory.

TABLE 19.14

Reduced mobilities (cm^2 V^{-1} s^{-1}) *of diatomic rare-gas ions in their parent gases*

T (°K)	He$_2^+$		Ne$_2^+$	Ar$_2^+$			Kr$_2^+$		Xe$_2^+$	
0		18·6	5·98		2·10		1·18		0·74	
22		17·0	—		—		—		—	
77		18·0	6·7		2·7		—		—	
195		21·7	7·3		2·9		—		—	
300	16·6	20·3	6·5	1·8	1·9	2·7	1·21	1·00	0·94	0·79

Values for 0 °K are calculated. All others are observed.

For neon there seems to be no difficulty and the same applies to krypton and xenon though here there are fewer observations. For helium and argon there is the problem of understanding why there should be two ions involved with different mobilities. In each case the mobility of the faster ion is probably more consistent with the theoretical low-temperature limit, although, according to Beaty, Browne and Dalgarno,† the faster ion in helium is metastable He$_2^+$ ($^4\Sigma_u^+$).

3.4. *Mobilities of positive ions in their parent (diatomic) gases—ionic reactions in general—experimental methods*

The history of the study of the mobility of positive ions in the simplest atomic parent gas has been a chequered one as will have been clear from § 3.2.1. There is no difficulty in imagining how much more confusion has beset the study when the parent gas is diatomic. For the resolution of the puzzling data obtained it is even more necessary to have available means for definite identification of the ions. It is only recently that it has become usual to associate a mass spectrometer with a drift tube as in the experiments in helium by Madson, Oskam, and Chanin.‡ From observations made with such an arrangement it is possible not only to obtain information about the nature of the reactions occurring, but also about the rates of these reactions. For some time, information had been derived about ionic reactions occurring in an ion source by observing with a mass spectrograph the variation of the intensity of different ions issuing from the source under different pressure conditions (see § 3.4.1).

† loc. cit., p. 1974.   ‡ loc. cit., p. 1981.

Attachment of a mass spectrometer to sample the ions reaching the walls of the container of a discharge afterglow, as in the early experiments of Phelps and Brown (§ 2.2.5), opened up further possibilities of studying the rates of ionic reactions.

However, difficulties of interpretation remain in using these methods. The importance of the reactions of ions with neutral molecules in many contexts such as radiation chemistry, upper atmospheric and astrophysics, plasma physics, and so on, has presented a challenge to develop more direct methods of determining reaction rates. This has led to the introduction of the remarkable technique of the flowing afterglow and to the refinement of earlier techniques so that reliable information about many important reaction rates is now forthcoming. We begin by describing some of the techniques which are being used.

3.4.1. *Use of a mass spectrograph without a drift tube.* An account has been given in Chapter 12, § 7.3.2 of the technique involved in investigating the composition of the ions resulting from passage of an electron beam of homogeneous energy through a gas. In Chapter 12 the emphasis was placed on ions formed directly in a single electron impact. These may be recognized by the proportionality of the mass-analysed ion current to the gas pressure p and electron current i. Here we are concerned with the secondary ions resulting from reactions of the primary ions with neutral gas molecules. The current of such ions will still be proportional to i but will vary as p^2 or even p^3 if the process involves three-body collisions.

Consider a reaction $\quad A^+ + X \to B^+ + Y.$ \hfill (87)

If n_A^+, n_B^+, n_X are the respective concentrations of A^+, B^+, and X at the exit from the ion source the mean reaction cross-section $\bar{Q}$ will be given by
$$n_X \bar{Q} \bar{l} = \ln\{(n_A^+ + n_B^+)/n_A^+\} = \ln\{(i_A^+ + i_B^+)/i_A^+\}, \quad (88)$$
where $\bar{l}$ is the mean path length of a primary ion in the source before it reaches the exit slit. With a low-pressure ion source $i_B^+ \ll i_A^+$ and $\bar{l}$ is determined from the geometry of the source. Thus, if the ionizing agent is a well-collimated electron beam perpendicular to the extracting electric field F, $\bar{l}$ is simply the distance from the beam to the exit slit. Even in this case, however, $\bar{Q}$ will be a mean value over an ion kinetic energy ranging from 0 to eV_R where V_R, the repelling potential used on the 'pusher' electrode, will normally be around 10 V. In the absence of knowledge of the variation of the reaction cross-section with relative kinetic energy of the reactants it is difficult to extend the value of this cross-section to thermal energies from the data.

No reliable theory exists to predict the energy variation of the reaction cross-section. However, in many cases the following considerations are applicable. The condition for the onset of orbiting in a collision between an ion and a neutral atom or a molecule at an initial relative kinetic energy E is that (see Chap. 16, § 5.1.1)

$$\frac{Ep^2}{r^2} - \frac{\alpha e^2}{2r^4} - E = 0, \tag{89a}$$

where
$$-\frac{2Ep^2}{r^3} + \frac{2\alpha e^2}{r^5} = 0. \tag{89b}$$

Here p is the impact parameter for the collision and α is the polarizability of the target atom. From (89) it follows that orbiting will just occur when
$$p = p_c = (2\alpha e^2/E)^{\frac{1}{4}}. \tag{90}$$

This will be valid provided that, when $p = p_c$, the colliding systems never come close enough to experience any appreciable influence from the short-range forces. Gioumousis and Stevenson† suggest that, in many collisions, reactions take place when orbiting occurs. Then, if the chance of reaction is negligible when $p > p_c$ and unity for $p < p_c$, the reaction cross-section would be

$$\pi p_c^2 = 2\pi(\alpha e^2/2E)^{\frac{1}{2}}. \tag{91}$$

This gives a variation with relative kinetic energy as $E^{-\frac{1}{2}}$ in which case the cross-section for a particular kinetic energy could readily be derived from $\bar{Q}$ as observed with a low-pressure source. Thus Gioumousis and Stevenson have shown that, under these conditions, in which the rate coefficient k is a constant, independent of the relative kinetic energy of the reactants, we have

$$k = (eF\bar{l}/2M)^{\frac{1}{2}}\bar{Q}, \tag{92}$$

where M is the mass of an ion, provided $eF\bar{l} \gg \kappa T$. Since (91) may be evaluated absolutely from known values of α it is easy to check whether this picture applies to a particular observed reaction. When this fails it is not possible to obtain any idea of the thermal cross-section from observations with a low-pressure source unless new features are introduced into the experimental arrangement.

Apart from these problems it is, of course, essential that the collection efficiency of the mass spectrograph and of the detector should not depend on the nature of the ions. Furthermore, in some reactions, ions will be formed with considerable kinetic energy. For such cases it will be difficult to determine cross-sections with any accuracy because of the

† GIOUMOUSIS, G. and STEVENSON, D. P., *J. chem. Phys.* **29** (1958) 294.

well-known difficulty of collecting such ions effectively in a mass spectrometer (cf. Chap. 12, p. 853). It is possible, however, using a suitable analyser electrode between the exit slit of the ion source and the entrance slit of the spectrograph, to make a retarding potential analysis of the energy distribution of the ions.

One modification of the experimental procedure which is capable of yielding reactive cross-sections at thermal energies has been introduced by Tal'roze and Frankevich.†

Instead of producing the ions by bombardment with a steady electron beam they ionize the gas by short pulses of electrons. During the pulse there is no electric field in the source chamber, but at a time t_z after the termination of the electron pulse, a short-voltage pulse is applied to the repeller electrode at the back of the chamber to extract the ions for mass analysis.

Secondary ions will be formed during the electron pulse of duration t_i, the interval t_z between this pulse and the extraction pulse, and the duration t_E of this latter pulse.

Let n_A^+ be the concentration of primary ions at the termination of the ionizing pulse, n_0 the concentration of reactive neutral molecules, and k_t the rate constant for the production of secondary ions B$^+$ in two-body thermal collisions between A$^+$ and the neutral molecules. Then, during the interval between the pulses, the concentration n_B^+ of B$^+$ ions satisfies

$$\frac{dn_B^+}{dt} = k_t n_A^+ n_0, \tag{93}$$

so that the total number of B$^+$ ions produced per cm^3 throughout the interval t_z between the pulses is

$$n_{B,t}^+ = k_t n_A^+ n_0 t_z, \tag{94}$$

provided $n_B^+ \ll n_A^+$ throughout.

During the ionizing pulse the concentration of B$^+$ built up will be

$$n_{B,i}^+ = \tfrac{1}{2} k_i n_A^+ n_0 t_i, \tag{95}$$

where k_i is the appropriate rate constant which may not be equal to k because of space-charge effects.

Similarly, during the extraction phase,

$$n_{B,e}^+ = k_e n_A^+ n_0 t_e. \tag{96}$$

It follows that, if I_A^+, I_B^+ are the currents of A$^+$ and B$^+$ ions extracted,

$$I_B^+ / I_A^+ = n_0(\tfrac{1}{2} k_i t_i + k_t t_z + k_e t_e). \tag{97}$$

† TAL'ROZE, V. L. and FRANKEVICH, E. L., *Zh. fiz. Khim.* **34** (1960) 2709.

A plot of I_B^+/I_A^+ against t_z should therefore be linear with slope given by $n_0 k_t$. In this way the thermal reaction rate is isolated and all ambiguity of interpretation removed. The only obvious complication may arise from a dependence of the transmission of the spectrograph, or of the detector sensitivity, on the nature of the ion.

It would appear possible in principle to obtain information about reaction cross-sections near thermal energies by working with an ion source at such a high pressure that the ions drift under the action of the repeller field with a uniform drift velocity u. Thus, if $\bar{v}$ is the root-mean-square velocity of the ions, the path length $\bar{l}$ will be $l\bar{v}/u$, where l is the distance from the ionizing beam to the exit slit, so that

$$n_X \bar{Q} l \bar{v}/u = \ln\{(i_A^+ + i_B^+)/i_A^+\}. \tag{98}$$

$n_X \bar{Q} \bar{v}$ is the mean rate coefficient k for the reaction and l/u is the time τ which a primary ion spends in the source—the so-called residence time. Thus

$$k\tau n_X = \ln\{(i_A^+ + i_B^+)/i_A^+\}. \tag{99}$$

If τ can be measured by some pulse technique then k may be obtained.

The chief difficulty in applying this method to ions produced by electron impact ionization is the complexity of the secondary and higher order reactions which can occur in a mixture of gases at the pressures concerned. This problem is greatly reduced if the ionization is produced by a nearly monochromatic photon beam with suitably chosen quantum energy so that only one of the species present can be ionized (see also p. 1937). It is even possible to control the vibrational excitation of the primary ions, if molecular. An effective technique of this kind has been developed by Warneck,[†] following on discussion of its possibilities by Giese[‡] and by Koyano, Omura, and Tanaka.[§]

Fig. 19.29 shows the general arrangement of the apparatus used by Warneck. The light source was a pulsed nitrogen spark, of the type developed by Weissler, used in conjunction with a $\tfrac{1}{2}$-m Seya vacuum ultra-violet monochromator. The beam issuing from the exit slit passed through the cylindrical ion source in a direction perpendicular to the axis and could be monitored by a photomultiplier sensitized by a coating of sodium salicylate (see Chap. 14, p. 1081). Ions produced by photo-ionization were extracted by an electric field directed along the cylinder axis, analysed by a 180° magnetic analyser with wedge-shaped air gap,

---

[†] WARNECK, P., *J. chem. Phys.* **46** (1967) 502; for a detailed description of the equipment see POSCHENRIEDER, W. and WARNECK, P., *J. appl. Phys.* **37** (1966) 2812.
[‡] GIESE, G. F., *Adv. chem. Phys.* **10** (1966) 247.
[§] KOYANO, I., OMURA, I., and TANAKA, I., *J. chem. Phys.* **44** (1966) 3850.

and detected by a 20-stage electron multiplier together with a vibrating-reed electrometer and chart recorder. By means of differential pumping the pressure in the analyser could be maintained at 10^{-6} torr with the pressure in the ion source as high as 0·2 torr.

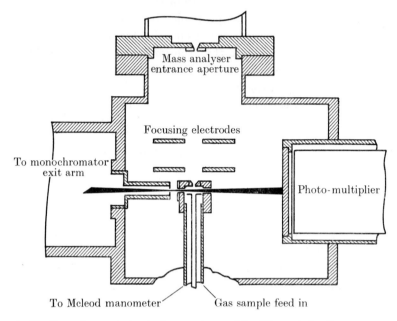

FIG. 19.29. General arrangement of apparatus used by Warneck for the measurement of ionic reaction rates.

The centre of the region in which ion formation occurs was 2 mm away from the extraction orifice which was 0·7 mm in diameter. The average width of the light beam inside the source was 0·6 mm with a spectral resolution of around 5 Å. Photo-electrons produced at the confining slit were prevented from entering the source by applying a suitable small electric field.

To determine the residence time under particular experimental conditions the light source was pulsed at a repetition rate of 120 pulses/s, each of duration about 0·5 μs. The residence time could then be derived from measurement of the total delay time between the formation of ions in the source by a light pulse and their arrival at the collector beyond the mass spectrometer. This was done by means of a calibrated Tektronix oscilloscope triggered by the photomultiplier signal.

In principle, the total delay time can be analysed into its two components, the residence time and the flight time through the mass spectrometer, by taking account of the fact that, if the repeller voltage is not too large, the flight time but not the residence time is independent of it. Observations of the time delay as a function of the inverse repeller voltage V_R^{-1}, when extrapolated to zero V_R^{-1}, should therefore yield the flight time, the residence time vanishing in this limit. This proved to be a practicable procedure, the extrapolation being effectively linear. Account had also to be taken of the pulse shape. At the high pressures used, the pulse is broadened by ion diffusion and the mean residence time should be determined from the half-rise time of the observed pulse. If τ_0 is the residence time derived from observations of the commencement of arrival of the pulse, it can be shown from diffusion theory that

$$\tau = \tau_0\{1-(a/l)(2D\tau_0)^{\frac{1}{2}}\}, \tag{100}$$

where D is the diffusion coefficient of the ions and a is a factor depending on the practical definition of τ_0. If the average practical detection limit is 10 per cent of the pulse height $a = 1\cdot 28$. Having determined D once and for all from measurements of both τ and τ_0 under suitable conditions it was convenient thenceforward to observe τ_0 and use (100) to obtain τ.

The drift velocity is now given by l/τ. From this the mobility μ is obtained. Since $\mu = eD/\kappa T_i$, where T_i is the ion temperature, this provides a means for determining T_i.

As an example of the practical application of this procedure Fig. 19.30 illustrates observations of τ and τ_0 as functions of pressure for N_2^+ ions in air while Fig. 19.31 illustrates the derived diffusion coefficients. These vary linearly with reciprocal pressure as they should. Fig. 19.32 illustrates the drift velocities derived from the residence times and reveals reasonably good agreement with observations made by other techniques (see § 3.5.2). Finally, from the diffusion coefficients shown in Fig. 19.31 and the drift velocities in Fig. 19.32, the derived ion temperatures are shown in Fig. 19.33.

A number of checks on the performance of the apparatus were made. Thus it was verified that a linear relation of ion current to gas pressure was obtained, when appropriate correction for light absorption in passing through the ion chamber was made, for repeller potentials and light intensities within the working range. It was checked also that, under working conditions, ion current ratios measured by the ratio of peak heights agreed with those measured from the ratio of the areas under the peaks.

Special attention was paid to checking that the sensitivity of the electron multiplier detector was independent of the nature of the ion

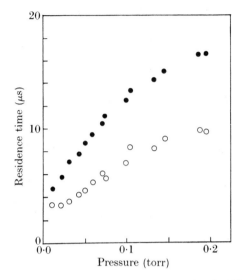

Fig. 19.30. Variation of the residence times τ and τ_0 for N_2^+ ions in a photo-ionization source in air as a function of pressure, observed by Warneck. ○ τ_0; ● τ.

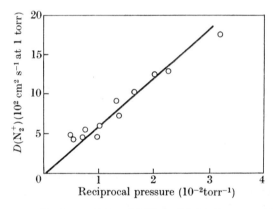

Fig. 19.31. Diffusion coefficient $D(N_2^+)$ for N_2^+ ions in air as a function of pressure derived from the residence times shown in Fig. 19.30.

observed. This was done by operating the source at such low pressures that only primary ion production occurred. The ion current may then be calculated from known photo-ionization cross-sections (see Chap. 14). Thus, using radiation at 685 Å, observations were made using Ar, N_2, O_2, CO, CO_2, H_2, and CH_4 as sample gases.

When working with a gaseous mixture in the ion source allowance must be made for the possibility that ions B⁺ are produced directly by

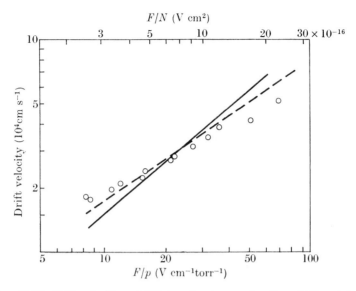

Fig. 19.32. Drift velocities of N_2^+ ions in air as a function of F/p. ○○○, derived from residence times shown in Fig. 19.30; ———, observed by Martin et al.; ---, observed by Dahlquist, both for ions in N_2.

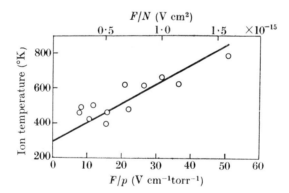

Fig. 19.33. Variation of N_2^+ ion temperature in air with F/p derived from the drift velocities and diffusion coefficients shown in Figs. 19.32 and 19.31 respectively.

the photon beam. If $i_A^+(0)$ and $i_B^+(0)$ are the currents of A^+ and B^+ produced respectively from the pure parent gases at the pressures equal to their partial pressures in an experiment then (99) becomes

$$k\tau n_X = \ln\{i_A^+(0)/(i_A^+(0)+i_B^+(0)-i_B^+)\}. \tag{101}$$

No attempt was made to overcome the difficult problem of the variation of the collection efficiency of the mass spectrometer with kinetic energy of the ions, so that results were not obtained for reactions in which the product ions possessed considerable kinetic energy. As an indication as to whether such ions were being produced comparison could be made between the reduction $i_A^+(0)-i_A^+$ of the current of A^+ and the increase of the current of B^+. If there is reason to believe that (87) is the predominant reaction these should be equal. If the observed increase for B^+ is less than the decrease of A^+ it is likely that the B^+ ions are being produced with considerable kinetic energy so that the observed i_B^+ is less than the true value.

Applications of this method are discussed in § 3.5.

3.4.2. *The associated drift tube and mass-spectrometer technique.* A typical example of equipment designed especially to investigate reactions of ions in their parent gases under near-thermal conditions is that of Albritton, Miller, Martin, and McDaniel.[†]

The general arrangement of the apparatus is shown in the perspective drawing in Fig. 19.34. The gas was contained within a drift tube at a pressure ranging from 0·02 to 1 torr or so. Primary ions were produced by an electron impact source. This could be placed with an accuracy of a few thousandths of an inch at any one of eight positions along the axis of the drift tube so that the drift length of the ions could be varied from 1 to 44 cm. The primary ions were produced by electron pulses of duration 0·5 to 20 μs, very short compared with the drift time of the ions down the tube.

The longitudinal electric field F along the drift space was maintained uniform by fourteen guard rings of 17·5 cm internal diameter, similar in design to that of Crompton, Elford, and Gascoigne.[‡] Surfaces exposed to the ions were gold-plated to reduce surface-charging effects.

The ions left the drift space through a grid aperture of diameter 0·079 cm and the core was selected by a conical skimmer for mass analysis by an r.f. quadrupole filter. Care was taken to avoid stray fields between the exit aperture and the skimmer so as to minimize the possibility of ionic reactions occurring in this region. For the same reason it was differentially pumped. Individual ions passing through the filter were detected by a 14-stage electron multiplier. The time interval between creation of an ion swarm and the detection of an ion was measured and stored by a 256-channel time-of-flight analyser triggered by the ionizing tube.

[†] ALBRITTON, D. L., MILLER, T. M., MARTIN, D. W., and McDANIEL, E. W., *Phys. Rev.* **171** (1968) 94; see also McDANIEL, E. W., MARTIN, D. W., and BARNES, W. S., *Rev. scient. Instrum.* **33** (1962) 2.

[‡] CROMPTON, R. W., ELFORD, M. T., and GASCOIGNE, J., *Aust. J. Phys.* **18** (1965) 409.

The chamber could be baked at 200 °C for 12 hours and achieved a background pressure as low as 10^{-9} torr while, after seal-off, it never exceeded 5×10^{-7} torr.

To obtain the mobility of a particular identified ion it is necessary to confirm that the ion has not spent part of its time as another species

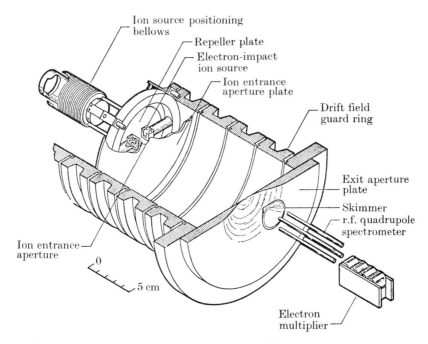

FIG. 19.34. General arrangement of the combined drift tube and mass spectrometric equipment used by McDaniel and his collaborators.

with a different drift time. As a test for this the shape of the histogram of recorded ion arrival times may be used. The flux of ions of thermal energy with diffusion coefficient D and drift velocity u which passes without undergoing reaction through an aperture of area A at a distance z from a source which, at time $t = 0$, creates a thin uniform disc of ions of radius r_0, in an infinite cylindrical volume, is given by

$$\phi(z,t) = \{A\sigma_0/4(\pi Dt)^{\frac{1}{2}}\}(u+z/t)\exp\{-(z-ut)^2/4Dt\} \times$$
$$\times \{1-\exp(-r_0^2/4Dt)\}, \quad (102)$$

where σ_0 is the initial surface ion density. If the observed shape of the histogram agrees with that expected from (102) with u the observed drift velocity and $D = \kappa T \mu/e$, where μ is the low-field mobility, then there is strong presumptive evidence that the ions concerned have drifted unchanged from the source to the exit aperture.

As an example of such a check Fig. 19.35 shows a histogram for H_3^+ ions in H_2 taken for $F/p = 0.52$ V cm^{-1} torr^{-1} at a pressure of 0·95 torr and temperature 304 °K. The shape fits very well with that expected from (102) with the drift velocity 0.43×10^4 cm s^{-1}. Other examples are

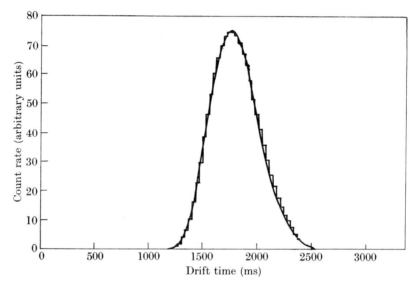

Fig. 19.35. Comparison of a histogram of arrival times of H_3^+ ions in H_2 observed in the experiment of Albritton, Miller, Martin, and McDaniel for F/p 0·52 V cm^{-1} torr^{-1} with that calculated assuming a drift velocity of 0.43×10^4 cm s^{-1} (full line curve).

given in § 3.5, including cases in which there is clear evidence that ionic reactions are occurring.

The nature and rates of such reactions may be investigated by using the variable drift distance to change the residence time of the ions in the gas and hence to determine the dependence of the intensity of a particular species on the residence time. By observations at different pressures effects due to loss by diffusion and by ionic reactions with rates depending in different ways on the pressure can be distinguished.

In another set of experiments Saporoschenko† combined an electrical shutter drift tube of the type developed by Tyndall and Powell‡ with a 90° sector mass spectrometer. The last grid of the drift tube (see Fig. 19.1) was replaced by a stainless steel to glass seal through which

---

† SAPOROSCHENKO, M., *Phys. Rev.* **139** (1965) A349.
‡ TYNDALL, A. M. and POWELL, C. F., *Proc. R. Soc.* **A129** (1930) 162.

the tube was connected with the spectrometer. Passage through the seal was closed by a platinum foil except for a pin-hole, of 2×10^{-3} cm radius, which allowed a sample of the ions passing the second shutter to continue on into the electrical accelerating field of the mass spectrometer. The current of ions which passed through the spectrometer was measured by a Bendix magnetic electron multiplier and a vibrating-reed electrometer.

By differential pumping the pressure close to the drift tube could be maintained at 2×10^{-4} torr with a pressure in the drift tube of 1·6 torr of hydrogen. The pressure in the magnetic analyser was 3×10^{-6} torr. A background pressure as low as 10^{-8} torr was achieved in the drift tube after long degassing at 200 °C.

3.4.3. *Drift tube combined with inlet and exit mass analysis.* One of the difficulties in interpreting data obtained by the techniques described above is that the source is located within the drift tube. Not only does this limit the investigation to ions in their parent gas but it also introduces into the drift tube excited species which may react in the gas to produce ions. Furthermore, it leaves uncertain the composition of the ions which enter the drift tube. The first two of these disadvantages may be eliminated by isolating the source but to remove the third the ions from the source must be mass-analysed before entering the drift space. Bloomfield and Hasted[†] were the first to carry out experiments on these lines. They relied upon the mobility measurements to discriminate between the ions arriving at the end of the drift tube, but it proved difficult to interpret the mobility spectrum unless it had a simple double-peaked form. To remove these difficulties Kaneko, Megill, and Hasted[‡] added a second mass spectrometer at the exit end.

Fig. 19.36 shows schematically the arrangement of their apparatus. The ions were produced in a Finkelstein ion source designed to produce a high ion current as the main difficulty seemed to be the low intensity of the final mass-analysed ion current. Unfortunately, the space charge due to the intense ionizing electron beam tended to hold positive ions within the source for such long periods that many of the ions left the source in excited states, thereby gravely complicating the interpretation of the observed data. The ion source was therefore operated with a smaller ionizing current than the maximum practicable.

Ions leaving the source were accelerated and focused by a lens system and mass-analysed by 90° deflection in a magnetic field. Selection of

† BLOOMFIELD, C. H. and HASTED, J. B., *Discuss. Faraday Soc.* **37** (1964) 176.
‡ KANEKO, Y., MEGILL, L. R., and HASTED, J. B., *J. chem. Phys.* **45** (1966) 3741.

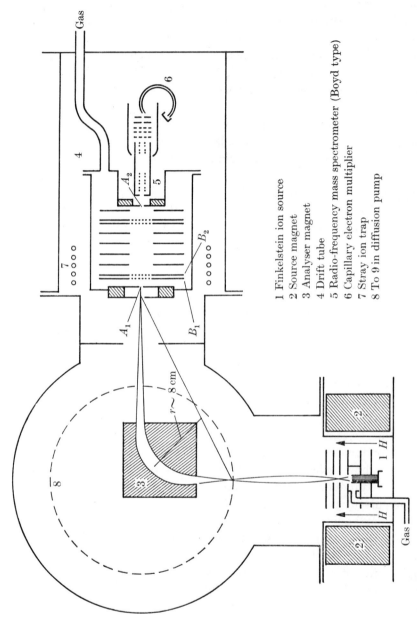

FIG. 19.36. General arrangement of the apparatus used by Kaneko, Megill, and Hasted in which a drift tube is combined with inlet and exit mass analysis.

1 Finkelstein ion source
2 Source magnet
3 Analyser magnet
4 Drift tube
5 Radio-frequency mass spectrometer (Boyd type)
6 Capillary electron multiplier
7 Stray ion trap
8 To 9 in diffusion pump

ions was carried out by varying the voltage on the ion source. With the analysis chamber pumped by a 9-in mercury diffusion pump the pressure in the analysing chamber could be maintained at less than 10^{-4} torr with pressures in the source and drift space of 10^{-2} and 1 torr respectively.

Ions were injected into the drift tube, which was of the conventional four-gauze-shutter type, through an entrance aperture A_1 of 1 mm diameter and were extracted through an aperture A_2 of 0·5 mm diameter.

As the ions entered the drift space at a relatively high energy of the order 100–400 eV they had to be brought to thermal equilibrium quickly before passing through the electrode system. To achieve this a gap was left between A_1 and the first gauze shutter electrode B_1 so that, at the working pressure in the drift space, the ion energies were degraded to thermal by the time they reached B_1. This was checked in the following way.

A drift voltage was applied so that ions arrived at the exit aperture A_2. It was then verified that the ion current vanished when the second gauze shutter B_2 was made slightly positive with respect to B_1.

The mass spectrometer at the exit aperture was an r.f. mass spectrometer, similar to that of Boyd and Morris,[†] but with a 24-grid system. This was chosen because of its high transmission and short path length for the ions.

The final ion current issuing from the spectrometer was detected by a Bendix capillary electron multiplier.

To make measurements of reaction rates the chosen primary ion was fired into the target gas and its drift velocity measured in the normal way. The system was then operated with steady potentials and the ion currents issuing from the second mass spectrometer measured. If there is negligible loss of ions by diffusion to the walls of the drift tube the rate coefficient for production of various secondary ions from the primary ion can be determined as follows.

For simplicity consider the case in which only one kind of secondary ion is produced. Distinguishing between the primary and secondary ions by suffixes 1 and 2 we have, if ν is the rate coefficient,

$$\frac{dn_1}{dt} = -\nu n_1, \tag{103}$$

so
$$n_1 = n_{10}\,e^{-\nu t}, \tag{104}$$

where n_{10} is the initial concentration of primary ions entering the gas. As the reaction continues for a time τ given by l/u_1 where l is the length

[†] BOYD, R. L. F. and MORRIS, D., *Proc. phys. Soc.* A**68** (1955) 1.

of the drift tube and u_1 the drift velocity of the primary ions, the concentration of primary ions arriving at the end of the drift tube is given by

$$n_1 = n_{10}\exp(-\nu l/u_1). \tag{105}$$

The concentration of secondary particles is similarly given by

$$n_2 = n_{10}\{1-\exp(-\nu l/u_1)\}u_1/u_2, \tag{106}$$

where u_2 is the drift velocity of the secondary ions. Hence the ratio of the currents of primary and secondary ions arriving at the end aperture is

$$i_2/i_1 = n_2 u_2/n_1 u_1 = \exp(\nu l/u_1)-1 \tag{107}$$

and the cross-section $\bar{Q}_{12}$ for the reaction, averaged over the distribution of relative kinetic energy of the ions and gas molecules, is given by

$$\bar{Q}_{12} = (u_1/v_r n_0 l)\ln(1+i_2/i_1), \tag{108}$$

where n_0 is the concentration of gas molecules and v_r is the mean relative velocity of an ion and a gas molecule.

If the rate of reaction is small then the correction due to the dependence of the diffusion coefficient D and the drift velocity u on the nature of the ion is to multiply (108) by

$$(1-\gamma_2/\gamma_1)/\ln(\gamma_1/\gamma_2), \tag{109}$$

where $\gamma_i = 4D_i/u_i l$. When the mean ion energy barely exceeds thermal this correction is very small because of the relation (24).

If the reaction cross-section is very large it may not be possible to work at a sufficiently high gas-pressure to degrade the incoming ion energies to thermal without converting all the primary into secondary ions. To overcome such a difficulty a non-reactive buffer gas may be used in the drift tube at sufficient pressure to produce rapid energy degradation. A suitably small partial pressure of the reacting gas may then be added to produce the reaction under investigation.

3.4.4. *Static-afterglow studies using mass spectrographs.* We have already given (§ 3.2.3) a brief account of the first experiments in which a mass spectrograph was used to sample, with high time resolution, the composition of the ions diffusing to the walls at different times in an afterglow. By studying the rates of decay of the He^+ and He_2^+ ions in a helium afterglow in this way Phelps and Brown[†] were able to obtain information about the rate of the ion–atom three-body reaction (50). It is clearly possible in principle to apply this technique to the study of ionic reactions in more complex afterglows to obtain information about their rates under thermal conditions. The main difficulty is one of

† loc. cit., p. 1973.

interpretation because of the presence of a variety of reactive excited and ionized species in an afterglow of this kind.

Consider a reaction in which a primary ion produces a secondary ion. If n_1, n_2 are their respective concentrations then

$$\frac{\partial n_1}{\partial t} = -\alpha_1 n_e n_1 - \beta n_0 n_1 - (\sum \gamma_i n_i) n_1 + D_{a1} \nabla^2 n_1, \qquad (110)$$

where α_1 is the coefficient of recombination of the primary ions with electrons, of concentration n_e, and β is that for the reaction with neutral molecules, of concentration n_0, which produces the secondary ions concerned. D_{a1} is the ambipolar diffusion coefficient of the primary ions. The third term on the right represents the rate of loss of primary ions through all other reactions which occur with excited or normal neutral molecules in the afterglow.

Similarly we may write for the secondary ions

$$\frac{\partial n_2}{\partial t} = -\alpha_2 n_e n_2 + \beta n_0 n_1 - (\sum \delta_i n_i) n_2 + D_{a2} \nabla^2 n_2. \qquad (111)$$

To obtain β from observation of $\partial n_1/\partial t$ and $\partial n_2/\partial t$ we require not only the recombination and ambipolar diffusion coefficients α_1 and α_2 and D_{a1}, D_{a2} respectively but also the $\gamma_i n_i$ and $\delta_i n_i$. There is some hope of determining the $\alpha_{1,2}$ and D_{a1}, D_{a2} but we cannot expect to be able to include all the important contributions to $\sum \gamma_i n_i$ and $\sum \delta_i n_i$.

In an attempt to minimize the importance of these terms Fite, Rutherford, Snow, and van Lint[†] considered the ratio $R = n_2/n_1$ which, according to (110) and (111), satisfies

$$\frac{\partial R}{\partial t} = \{(\alpha_1 - \alpha_2) n_e + \gamma - \delta + D_{a2} n_2^{-1} \nabla^2 n_2 - D_{a1} n_1^{-1} \nabla^2 n_1\} R + \beta n_0 (1+R), \qquad (112)$$

where $$\gamma = \sum \gamma_i n_i, \qquad \delta = \sum \delta_i n_i. \qquad (113)$$

At sufficiently small R $$\frac{\partial R}{\partial t} \simeq \beta n_0, \qquad (114)$$

but this may only be valid for such small R as to defy accurate measurement of $\partial R/\partial t$. There is always the hope that, since the terms in curly brackets all involve differences between properties of the two ions, the differences might be quite small if the ions are fairly similar.

In practice what is measured is not the concentration of a particular ion but the current density which reaches the walls of the afterglow

[†] FITE, W. L., RUTHERFORD, J. A., SNOW, W. R., and VAN LINT, V. A. J., *Discuss. Faraday Soc.* **33** (1962) 264.

container. This is usually taken to be proportional to the mean concentration as the error introduced by this assumption is likely to be much smaller than those that arise from other sources. The only test, necessary but not sufficient, of the applicability of the relation (114) is that the rates of the secondary and primary ion currents should be small and increase linearly with time in the afterglow.

Three sets of experiments on these lines have been carried out. Fite, Rutherford, Snow, and van Lint used an instrument composed of stainless steel with copper gaskets throughout to allow baking. Gas in the afterglow chamber, also of stainless steel, could be ionized either by an r.f. pulse in the usual way, or by a pulse of 20-MeV electrons from an electron linear accelerator. In the latter case the high-energy beam entered and left the chamber through windows of stainless steel foil. The ions were sampled through a hole in the wall of the afterglow chamber, analysed in a 90° magnetic sector mass spectrograph, and detected with an electron multiplier.

Apart from the advantage of checking whether the initial distribution of ionization had any serious effect on the results, the use of the beam from the linear accelerator made it possible to work with short pulse intervals, at such wide separation in time that the pumping system was able to prevent accumulation of impurities generated during each pulse. With the r.f. excitation, the pulse separation could not be too long as residual ionization from the previous pulse was found necessary to act as pre-ionization for the next.

Langstroth and Hasted† used an r.f. mass spectrometer of the Boyd–Morris type which, requiring no magnet, has the advantage that it may be inserted as a probe into the afterglow, which was generated in a Pyrex glass tube 90 cm long and of 10 cm diameter. The spectrometer was introduced through a short side-arm. The discharge was initiated by d.c. pulses of 2 μs duration, with a repetition rate of 10 per second.

Sayers and his collaborators‡ also used an r.f. mass spectrometer. The afterglow was generated in a glass tube which, in the later experiments, was 30 cm long and 14·5 cm diameter. Breakdown of the gas was produced by 100 kW pulses of r.f. power of 7 Mc/s frequency. The pulses were of about 20 μs duration with a repetition frequency of 50 per second.

Even if the conditions in an afterglow happen to be such as to yield a good value for the rate constant of a particular ionic reaction, there

† LANGSTROTH, G. F. O. and HASTED, J. B., *Discuss. Faraday Soc.* **33** (1962) 298.
‡ DICKINSON, P. H. G. and SAYERS, J., *Proc. phys. Soc.* **A76** (1960) 137.

is no control over the state of excitation, electronic or vibrational, of the reacting species and it is not possible to study reactions with neutral radicals such as O or N atoms. Most of the difficulties arise from the fact that the reactants are all exposed to the discharge excitation. We now describe a method of afterglow experiment in which the afterglow is spread over a considerable distance in space, so making it possible to introduce reactive species and sample the products in the afterglow where no discharge excitation is present. This overcomes almost all the difficulties.

3.4.5. *The flowing-afterglow method.* This remarkable method has proved very effective in determining reliably the rates of ionic reactions under well-defined thermal conditions. It originated from the work of Schmeltekopf and Broida† in 1963. Their attention was drawn to an observation of Kunkel and Hurlbut‡ of a very short duration afterglow in helium in a low-density wind tunnel. They allowed helium to expand through a single de Lune converging–diverging nozzle into a Pyrex tube 100 cm long and 9 cm in diameter attached to a high-speed (500 l/s) pump. A discharge was produced at the nozzle by application of a 2450 Mc/s, 125 W, magnetron power supply. An afterglow was observed some centimetres down the tube from the discharge, provided sufficiently pure helium was used. It was found that small traces of added impurities caused large changes in the afterglow. Addition of nitrogen downstream produced bright blue luminosity due to emission by N_2^+. Similarly, emission from O_2^+ bands, due to addition of O_2, produced a green 'flare'. It was realized by Schmeltekopf and Broida that the production of an afterglow flowing at high speed in this way opened up new possibilities for investigating the rates of thermal reactions. The first application to ionic reactions was made in 1964 by Ferguson, Fehsenfeld, Dunkin, Schmeltekopf, and Schiff.§

The principle of the method is illustrated in Fig. 19.37. Helium is passed at a flow rate of about 2×10^4 cm s^{-1} through a quartz tube. Near the front end of the tube it is subjected to an r.f. electric field which produces a discharge in the gas. The afterglow from this discharge travels down the tube with the main gas. Downstream from the discharge a neutral gas is fed into the flowing gas, through a fine nozzle, at an adjustable rate. Ions produced by the reaction of He$^+$ ions in the afterglow are detected by a quadrupole mass spectrometer connected

---

† SCHMELTEKOPF, A. L. and BROIDA, H. P., *J. chem. Phys.* **39** (1963) 1261.
‡ KUNKEL, W. B. and HURLBUT, F. C., *J. appl. Phys.* **28** (1957) 827.
§ FERGUSON, E. E., FEHSENFELD, F. C., DUNKIN, D. B., SCHMELTEKOPF, A. L., and SCHIFF, H. I., *Planet. Space Sci.* **12** (1964) 1169.

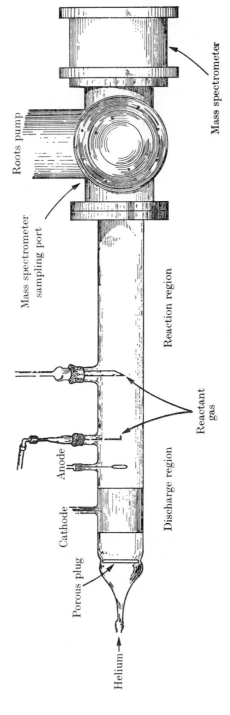

FIG. 19.37. Arrangement of the apparatus used in flowing-afterglow experiments.

to a fine aperture at the end of the quartz tube, the gas stream being pumped out at the side.

Let $[\text{He}^+]_0$ be the concentration of He^+ ions immediately behind the injection nozzle. In the absence of any injected molecules the concentration of He^+ ions will decay with a time constant τ_0 as they flow down the tube. The concentration at the end of the tube will then be given by

$$[\text{He}^+]_1 = [\text{He}^+]_0 \, e^{-T/\tau_0}, \tag{115}$$

where T is the flow time from the nozzle.

Suppose now that a gas is injected which reacts with the He^+ ions so that there is an additional rate of loss of He^+ concentration given by $\lambda[\text{M}][\text{He}^+]$, where $[\text{M}]$ is the concentration of the injected molecules. This has the effect of changing the time constant for decay of He^+ from τ_0 to τ, where

$$\tau^{-1} = \tau_0^{-1} + \lambda[\text{M}]. \tag{116}$$

The concentration of He^+ at the end of the tube is now given by

$$[\text{He}^+]_2 = [\text{He}^+]_0 \, e^{-T/\tau}, \tag{117}$$

so

$$\ln[\text{He}^+]_2 = \ln[\text{He}^+]_0 - \tau_0^{-1} T - \lambda[\text{M}] T. \tag{118}$$

Since the ion current $i(\text{He}^+)$ issuing from the spectrograph is proportional to $[\text{He}^+]_2$ a plot of $i(\text{He}^+)$ against $[\text{M}]$ will give a straight line with slope λT.

A major advantage of this method is the timing through flow rate so that the time during which the added molecules react is known. With flow rates of order 10^4 cm s^{-1} and tube lengths of the order of some tens of centimetres or more, T is of the order of milliseconds, which is generally very suitable when the gas pressure is around 0·25 torr.

It is possible to extend the technique to reactions with other ions by using two injectors. Thus, suppose it were wished to investigate reactions between N^+ ions and some other molecules M. All that is necessary is first to convert He^+ ions to N^+ by injecting nitrogen gas through the first nozzle, the reaction being

$$\text{He}^+ + \text{N}_2 \rightarrow \text{He} + \text{N} + \text{N}^+ + 0\cdot3 \text{ eV}. \tag{119}$$

If the molecules M are now injected from a second nozzle, so far downstream that all the He^+ ions have reacted, it is now possible, by observing the current of N^+ ions issuing from the spectrometer as a function of $[\text{M}]$, to determine the reaction rate between those ions and molecules M. Alternatively, the ions whose reactions are to be investigated may be produced by adding the parent gas to the helium stream before entering the discharge.

Many precautions have to be taken in any particular case to ensure that the conditions under which a relation of the form (118) is obtained are valid. Suppose, for example, the reactions of ions X^+ with molecules M are being studied by injection of X into the helium flow. Then it is essential that any reactions which produce X^+ should have ceased to be important at the point of injection of the molecules M. In taking this into account with helium as the buffer gas, allowance must be made for the high internal energy of $2\,^3S$ metastable helium atoms, which are capable of ionizing almost all other atoms and molecules. Such reactions producing ions X^+ must also have proceeded to completion before the nozzle injecting the reactive neutral species is reached. A second complication with helium is the possibility of production of X^+ downstream of the final nozzle through photo-ionization of molecules X by helium resonance radiation. This effect may be eliminated by operating the flowing afterglow on a pulsed instead of a continuous regime. The resonance photons travel faster than the ionized and neutral atoms and molecules so their effects are not detected if observations are made after a suitable time delay from the cessation of the exciting pulse.

It is possible to apply the flowing-afterglow technique to reactions of ions with radicals, such as atomic nitrogen and oxygen. Thus N atoms may be produced in a subsidiary flow tube by a microwave discharge in N_2 and added downstream in the usual way. To measure the production rate of N atoms use is made of the reaction

$$N+NO \to N_2+O. \tag{120}$$

This has the advantage of giving a visible indication of the completion of the reaction (see also Chap. 18, § 5.5) so that by measuring the flow rate of NO which brings this about the flow rate of N may be obtained.

The same reaction could be used to provide a supply of O atoms at a known rate, it being assumed that the O flow rate was the same as that of the NO. To check that in fact recombination of the O atoms does not occur experiments were carried out with two quite different arrangements for introducing the atomic oxygen. In the first the atoms entered from the NO titration chamber through a nozzle with a constriction, a right-angled bend, and a number of exit holes about 0·5 mm in diameter, in the second, through a straight glass tube of 11 mm internal diameter and 30 cm length. No difference was found in the rates of the ionic reaction with atomic oxygen under study, namely

$$N_2^+ + O \to NO^+ + N. \tag{121}$$

An unexpected difficulty occurred in introducing atomic oxygen. This led to a great decrease in the efficiency of ion sampling by the spectrometer. It seems likely that this is due to some charging up of the surface of the sampling plate, as the effects were ameliorated by painting the plate with colloidal graphite dispersed in alcohol.

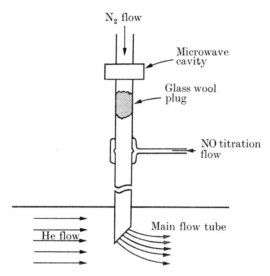

Fig. 19.38. A typical inlet system for N and O atoms in flowing-afterglow experiments.

Fig. 19.38 shows a typical inlet system for N and O atoms. N_2 flowing in at about 0·25 atmos cm³ s⁻¹ passes through a microwave discharge. Excited and ionized species then pass through a glass-wool plug which removes excited neutral species.† This is important because otherwise excited N_2 molecules seriously confuse the interpretation of reaction rates.

It is important to verify that serious error is not introduced due to the hydrodynamic effects arising from the high-speed flow. To check this, experiments have been carried out using different inlet nozzles with apertures ranging from pin-hole size to 1 cm diameter, introducing the neutral reactant at the tube wall or at the tube axis, and varying the reaction distance. Results obtained agree within 30 per cent. Theoretical estimates‡ of the errors due to diffusion and mixing effects have been

---

† SCHIFF, H. I. and MORGAN, J. E., Can. J. Chem. 41 (1963) 903.
‡ GOLDAN, P. D., SCHMELTEKOPF, A. L., FEHSENFELD, F. C., SCHIFF, H. I., and FERGUSON, E. E., J. chem. Phys. 44 (1966) 4095.

made and again suggest that these are likely to be less than 30 per cent. By use of the theory it is possible to apply corrections which should reduce the uncertainty very considerably.

Another source of error is the determination of the flow velocity. The value obtained from absolute flow-rate and density measurements is less than that determined from timing the passage of the afterglow plasma down the tube. The latter is probably to be preferred for the following reason. The ion density will be greater near the axis, where it is sampled, so greater weight should be given to the velocity of the ions in this region. Owing to the parabolic velocity distribution across the tube section the velocity is greatest on the axis. As the light emitted from the plasma comes also largely from the axial region where the electron concentration is greatest, the velocity determined from the passage of the glow is more nearly that of the ions which contribute most to the observations.

In general, a particular ion can react with a particular molecule in more than one way and it is important to be able to identify which reactions are making the main contribution to the reaction rate observed. Evidence about this can be obtained by comparing the rate of decrease of concentration of the reacting ion with the growth-rates of the concentrations of various other ions as sampled by the mass spectrometer.

A great advantage of the flowing-afterglow method is that the ionized and neutral species will be in their ground electronic states. There is a sufficient concentration of free electrons in the afterglow to ensure that deactivation of any electronically excited molecules occurs in a short travel distance in the flow tube. Vibrational deactivation may not be so rapid but it is usually possible to determine the vibrational state involved in any particular case. It is also possible to inject molecules, neutral or ionized, with a specific distribution of vibrational states so that the dependence of the reaction rate on the vibrational excitation can be determined.

We may sum up the advantages of the flowing afterglow, compared with static time-resolved afterglow technique, under the following heads.

(a) The fact that neutral reactants may be introduced at a controlled time or position along the flow tube, without their being subject to the electrical discharge which produces the afterglow, makes it possible to select a particular reaction and determine the time during which it has been occurring. Complications of all sorts arising from the presence of a neutral species added to the main gas in a static afterglow are avoided.

(b) The fact that the reacting species will normally be in their ground electronic states. In addition it is possible either to control or determine the vibrational excitation.

(c) The possibility of studying ionic reactions with neutral radicals.

(d) As we shall discuss in § 4, the technique may be applied without difficulty to study reactions of negative ions.

Before discussing the results obtained by this and other methods we give some further details of the apparatus which has been used so effectively up to the present. Referring to Fig. 19.37, the flow tubes have been made from Pyrex, quartz, and stainless steel, about 1 m long and 8 cm internal diameter. The helium is pumped through the tube by a large Roots-type blower, backed by a large mechanical forepump, at a flow rate of around 190 atmos $cm^3 s^{-1}$ and a flow velocity of 10^4 cm s^{-1}. The discharge produces about 10^{11} ions/cm^3 and a comparable concentration of metastable atoms. He_2^+ ions are formed from the He^+ in the tube so that, in very pure He at 0·3 torr, their concentration at the mass spectrometer is about 0·1 of that of the atomic ion. This is consistent with the rate for the reaction

$$He^+ + 2He \rightarrow He_2^+ + He, \qquad (122)$$

measured in static afterglow studies (see § 3.2.3).

The sampling orifice into the mass spectrometer is of diameter 0·020 in and is at a potential of 0·2 V, a drawout potential of about 4·5 V being applied between the plate and the mass spectrometer which is differentially pumped by a 6-in oil diffusion pump.

When operated in the pulsed mode to eliminate effects due to ions produced by photo-ionization, as explained on p. 2024, the ion detection system is gated on at a time equal to the plasma flow time down the tube ($\simeq$ 6 ms). The pulse width may be varied from 1 μs to 1 ms and the excitation produced either by a 2700-Mc magnetron discharge of 2·5 kW peak power or a variable voltage source of up to 5 kV with average power 2·5 kW.

Pulse-counting techniques are used, particularly when the system is operated in the pulsed mode.

Although we have been concerned particularly with the use of helium as the main gas, there is no difficulty in using any of the heavier inert gases in place of helium, although the cost of the large volumes of gas required is likely to be prohibitive except for helium and argon. We have only dealt with the application of the flowing-afterglow technique for studying ionic reactions but it is also applicable to the study of reactions involving metastable atoms and in fact to any reactions which have been studied by the static-afterglow method (see Chap. 18, § 5 and § 3.2.3 of this chapter).

3.5. *Results of mobility and ionic reaction observations*

3.5.1. *Hydrogen ions in hydrogen.* A great number of measurements have been made of the mobilities of positive ions in hydrogen, the values found ranging from 7·5 to 17·5 cm² V⁻¹ s⁻¹. It was difficult to compare and interpret the results obtained until identification of the ions through mass analysis was possible. In 1965 Saporoschenko[†] carried out measurements for values of F/p ranging from 5·5 to 100 V cm⁻¹ torr⁻¹, using the equipment described on p. 2014. He identified three ions, H^+, H_3^+, and H_5^+, the relative abundance of which varied with F/p. At all observed values of F/p, H_3^+ was much the most abundant ion, contributing 74 per cent of the whole at $F/p = 6·45$ V cm⁻¹ torr⁻¹ and 95 per cent at a value of F/p four times larger. The H^+ percentage abundance increased with F/p but was never greater than 10 per cent of the whole. H_5^+ behaved in the inverse way. At the lowest observed values of F/p it contributed 25 per cent of the total ion current, but by $F/p = 25·6$ V cm⁻¹ torr⁻¹ this had fallen to 0·5 per cent. Barnes, Martin, and McDaniel,[‡] using the equipment described on p. 2012, found H_3^+ and H^+ but no H_5^+ ions. They did find H_2^+ in addition at low values of pd where p is the gas pressure and d the drift distance. H_5^+ ions had been identified earlier by Dawson and Tickner[§] and by Kirchner[∥] in a beam of ions extracted from a glow discharge in hydrogen.

Albritton, Miller, Martin, and McDaniel,[††] using the equipment described on p. 2012, made definitive measurements of the low-field mobilities for H^+ and H_3^+ ions. Fig. 19.39 (a) shows results which they obtained for H_3^+ as a function of F/p for a range of gas pressures extending from 0·025 to 0·950 torr. Over this range the shape of the histogram (see Fig. 19.35) of ion transit times was as expected for ions drifting from the source without undergoing reaction. In Fig. 19.39 (b) comparison is made with results obtained for identified H_3^+ ions by Saporoschenko,[†] and by Wobschall, Fluegge, and Graham[‡‡] using the cyclotron resonance method, and for unidentified ions by a number of other

---

[†] SAPOROSCHENKO, M., *Phys. Rev.* **139** (1965) A349.
[‡] BARNES, W. S., MARTIN, D. W., and MCDANIEL, E. W., *Phys. Rev. Lett.* **6** (1961) 110.
[§] DAWSON, P. H. and TICKNER, A. W., *J. chem. Phys.* **37** (1962) 672.
[∥] KIRCHNER, F., *Z. Naturf.* **18a** (1963) 879.
[††] loc. cit., p. 2012.
[‡‡] WOBSCHALL, D., FLUEGGE, R. A., and GRAHAM, J. R., *J. chem. Phys.* **47** (1967) 4091.

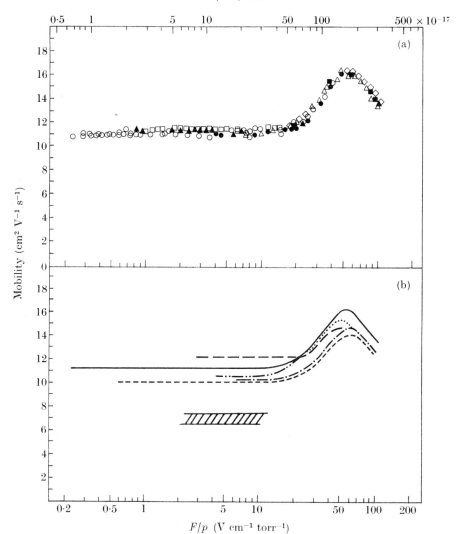

FIG. 19.39. (a) Mobilities of H_3^+ ions in H_2 at different values of F/p observed by Albritton, Miller, Martin, and McDaniel at different pressures; ○ 0·95, △ 0·65, □ 0·35, ● 0·15, ▲ 0·05, ■ 0·035, and ◇ 0·025 torr. (b) Mobilities of H_3^+ and unidentified hydrogen ions in H_2 as a function of F/p. H_3^+ : ——— Albritton *et al.*; – · – · Saporoschenko; ///// Wobschall, Fluegge, and Graham. Unidentified : — — — Chanin; – · · – · · Dutton *et al.*; · · · Jaeger and Otto; – – – Sinnott.

observers.† There seems little doubt that in the latter experiments H_3^+ was the ion studied. It will be seen that the mobility, defined by the ratio of the drift velocity to the electric field strength as measured by Albritton et al., is effectively constant at small F/p so that the zero-field mobility may be taken to be 11·1 cm³ V⁻¹ s⁻¹.

At pressures above 0·2 torr there was some evidence for the occurrence of a reaction converting H_3^+ to H^+. This is shown in Fig. 19.40 (a) which compares histograms observed for H_3^+ and H^+ ions. A small 'tail' is perceptible on the short transit-time side of the H_3^+ curve and its position relative to the H^+ histogram suggests that it arises from conversion of H^+, which are the faster ions at the particular values of F/p, through the reaction

$$H^+ + 2H_2 \to H_3^+ + H_2.$$

The amount of the conversion is so small as not to affect seriously the determination of the mobility of the H_3^+ ions.

Fig. 19.41 shows the results obtained by Albritton et al. for the mobility of H^+ ions. For $F/p < 54$ V cm⁻¹ torr⁻¹ the histograms for the transit times of these ions were of the shape (see Fig. 19.35) expected for non-reacting ions from the source but for higher F/p there was evidence of production from H_3^+ during flight. Thus in Fig. 19.40 (b), taken at 56 V cm⁻¹ torr⁻¹, a pressure of 0·05 torr and temperature of 302 °K, the H^+ histogram clearly shows a tail on the short transit-time side, the location of which clearly indicates conversion from H_3^+ which are the faster ions at these values of F/p (see Figs. 19.39 (a) and 19.41). Comparison of the mobility data for H^+ with those of Saporoschenko,‡ available at higher F/p, reveals good agreement as may be seen from Fig. 19.41. At lower F/p the mobility has become independent of F/p giving a zero-field mobility for H^+ in H_2 of 16·0 cm² V⁻¹ s⁻¹. Wobschall, Fluegge, and Graham§ obtain 14·7 cm² V⁻¹ s⁻¹ with the cyclotron-resonance technique.

H_5^+ ions were detected only at pressures above 0·65 torr. Typically the histograms for the arrival times for these ions were of the form shown in Fig. 19.42. Comparison with the corresponding histograms for H_3^+ shows that a large fraction of the H_5^+ ions is produced from H_3^+ in transit, close to the exit aperture. The fact that the H_5^+ histogram

---

† CHANIN, L. M., Phys. Rev. **123** (1961) 526; JAEGER, G. and OTTO, W., Z. Phys. **169** (1962) 517; SINNOTT, G., Phys. Rev. **136** (1964) A370; DUTTON, J., LLEWYLLYN JONES, F., REES, W. D., and WILLIAMS, E. M., Phil. Trans. R. Soc. **A259** (1966) 299.

‡ loc. cit., p. 2028.   § loc. cit., p. 2028.

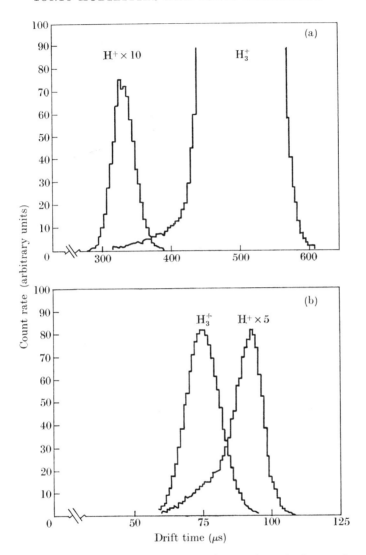

Fig. 19.40. Histograms of arrival times of H^+ and H_3^+ ions in the experiment of Albritton et al. (a) F/p 5·0 V cm^{-1} torr^{-1}, p 0·35 torr, drift distance 18·8 cm. (b) F/p 56 V cm^{-1} torr^{-1}, p 0·05 torr, drift distance 44·8 cm.

extends beyond that for H_3^+ on the long transit-time side indicates that the H_5^+ ions have the lower mobility.

Wobschall, Fluegge, and Graham,[†] using the cyclotron-resonance technique, obtained a low-field mobility for H_2^+ ions in H_2 of 7·2 cm^2 V^{-1} s^{-1}.

[†] loc. cit., p. 2028.

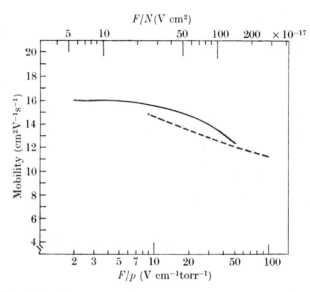

Fig. 19.41. Mobilities of H⁺ ions in H_2 at different values of F/p. —— observed by Albritton et al.; --- observed by Saporoschenko.

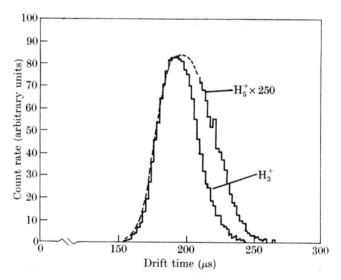

Fig. 19.42. Histograms of arrival times of H_3^+ and H_5^+ ions in the experiments of Albritton et al. F/p 5 V cm⁻¹ torr⁻¹, $p = 0.65$ torr, and drift distance 6·3 cm.

Mason and Vanderslice† have made theoretical estimates of the mobilities of H^+, H_2^+, and H_3^+ ions in H_2. The effective interaction between H^+ and H_2 involves a strong chemical attraction at short distances in addition to the polarization. Assuming that the short-range interaction averaged over all orientations can be represented by a Morse function (Chap. 10, p. 677), the three parameters involved were determined so as to give good agreement with observed data on the scattering of low-energy proton beams in hydrogen gas (Chap. 22). In practice, these data were not sufficient to determine all three parameters so the equilibrium separation was taken as 1·5 Å, the value given by quantal calculations.

For the mobility calculations an interaction of the form

$$V(r) = \tfrac{1}{2}\epsilon\{(1+\gamma)(r_m/r)^{12} - 3(1-\gamma)(r_m/r)^4 - 4\gamma(r_m/r)^6\} \tag{123}$$

was used, the parameters being chosen to give the correct coefficient of the polarization term,‡ and the same position and depth of the minimum as for the Morse function discussed above. This gives $\epsilon = 2\cdot7$ eV, $r_m = 1\cdot5$ Å, and $\gamma = 0\cdot72$. Carrying out the calculation of the mobility in the usual way (see § 2.4) the value obtained for a temperature of 300 °K is 18·3 cm² V⁻¹ s⁻¹ which is a little larger than that, 16·0 cm² V⁻¹ s⁻¹, found by Albritton et al. (see Fig. 19.41).

A somewhat similar procedure was used for H_3^+. In this case there is no chemical attraction and the short-range repulsion, assumed to be of exponential form, was determined from elastic scattering measurements§ as for H^+. Polarization and van der Waals terms were then added and the whole represented empirically by the form

$$V(r) = \epsilon\{(r_m/r)^8 - 2(r_m/r)^4\} \tag{124}$$

with $\epsilon = 0\cdot0366$ eV, $r_m = 2\cdot97$ Å. The calculated mobility at 300 °K is 22 cm² V⁻¹ s⁻¹ which is much higher than the observed value of 11·1 cm² V⁻¹ s⁻¹. It is hard to see how such a large discrepancy could be due to error in the interaction as the polarization term is the most important and it is quite well known. One possibility is the suggestion by Varney‖ that exchange of protons occurs in thermal collisions between

---

† MASON, E. A. and VANDERSLICE, J. T., *Phys. Rev.* **114** (1959) 497.

‡ The effective polarizability $\bar{\alpha}$ is some weighted mean of the respective polarizabilities $\alpha_\parallel$ and $\alpha_\perp$ for fields parallel and perpendicular to the nuclear axis. The mean was calculated assuming all orientations of this axis to have equal weight. In that case

$$\bar{\alpha} = \tfrac{1}{3}\alpha_\parallel + \tfrac{2}{3}\alpha_\perp.$$

The observed values of $\alpha_\parallel$ and $\alpha_\perp$ are $0\cdot934$ and $0\cdot718 \times 10^{-24}$ cm³.

§ SIMONS, J. H., FONTANA, C. M., MUSCHLITZ, E. E., and JACKSON, S. R., *J. chem. Phys.* **11** (1943) 307.

‖ VARNEY, R. N., *Phys. Rev. Lett.* **5** (1960) 559.

H_3^+ and H_2, thereby producing a drag effect rather like that due to charge exchange.

The mobility of H_2^+ in H_2 was calculated, allowing for charge exchange, by using the relation between mobility and charge transfer cross-sections discussed in § 3.3. At 300 °K this gave a mobility of 13·9 cm² V⁻¹ s⁻¹, considerably larger than the value observed by the cyclotron-resonance technique.

It seems that, under most conditions, the ion studied in drift experiments in hydrogen will be H_3^+. This ion is known to be very stable. The dissociation energy calculated from quantum theory is at least as high as 8·3 eV towards dissociation into H⁺+H+H. It follows that the reaction
$$H_2^+ + H_2 \to H_3^+ + H \tag{125}$$
is exothermic to at least 1·2 eV. The rate of this reaction under thermal conditions has been measured by Harrison, Ivko, and Shannon,† using a pulse technique similar to that of Tal'roze and Frankevich‡ described on p. 2006. They obtained a rate constant of $5·9 \times 10^{-10}$ cm³ s⁻¹ which is to be compared with $20·8 \times 10^{-10}$ cm³ s⁻¹ according to the theory of Gioumousis and Stevenson§ (p. 2005). Observations were also made without the use of the pulse method so the ions were extracted through the ion source by an electric field of 10·5 V cm⁻¹. For these ions the rate was measured as $13·3 \times 10^{-10}$ cm³ s⁻¹ suggesting that the conditions assumed by Gioumousis and Stevenson of unit probability of reaction when orbiting occurs are approached as the mean kinetic energy of relative motion increases.

On the other hand Warneck,∥ applying the technique described on p. 2007, in which the ions are produced by photo-ionization, obtained a rate constant of $18·5 \times 10^{-10}$ cm³ s⁻¹, not far below that expected from the theory. In these experiments, at a pressure in the ion source of 0·05 torr, the ions followed an effectively collision-free path through the ion chamber and as the repeller potential was 1 V their mean energy was considerably above thermal.

Weingartshofer and Clarke†† have investigated the dependence of the reaction rate on the initial vibrational state of the H_2^+ ion. The measurements were carried out with a mass spectrometer attached to an ion source in which ionization was produced by electrons of closely controlled energy from an electron selector of the Clarke type (see Chap. 3, p. 112). The electron energy spread was less than 0·05 eV at half peak and the

† HARRISON, A. G., IVKO, A., and SHANNON, T. W., *Can. J. Chem.* **44** (1966) 1351.
‡ loc. cit., p. 2006. § loc. cit., p. 2005. ∥ loc. cit., p. 2007.
†† WEINGARTSHOFER, A. and CLARKE, E. M., *Phys. Rev. Lett.* **12** (1964), 591.

H_2^+ ions drifted through the ion chamber with nearly thermal energies. Fig. 19.43 shows the appearance potential curves for H_2^+ and H_3^+ plotted as a function of electron energy above threshold.

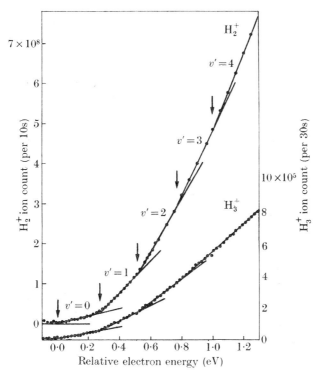

Fig. 19.43. Variation with electron energy above threshold of the intensity of production of H_2^+ and H_3^+ ions as observed by Weingartshofer and Clarke. Thresholds for excitation of different vibrational states are indicated.

The locations of the different vibrational levels of the H_2^+ ion are indicated and it appears that the number N_v of H_2^+ ions formed in a particular vibrational level v is given by

$$N_v = \alpha_v(V-V_v), \qquad (126)$$

where V_v is the appearance potential for ions with this degree of vibrational excitation. It will be seen that the yield curve for H_3^+ can similarly be built up from contributions $\alpha'_v(V-V_v)$ from each vibrational level of H_2^+. This suggests that the relative cross-section for production of H_3^+ from H_2^+ ions in a particular vibrational state is given by α'_v/α_v. This assumes that the collection efficiency of the spectrometer is the same for both H_3^+ and H_2^+ ions. If this is valid then, taking α_0'/α_0 as unity, the values obtained for $v = 1, 2, 3, 4$, and 5 respectively are 0·5, 0·4, 0·5, 0·0, and 0·0.

Miller, Moseley, Martin, and McDaniel† measured the rates of the reactions
$$H^+ + 2H_2 \to H_3^+ + H_2,$$
$$D^+ + 2D_2 \to D_3^+ + D_2,$$

using the apparatus and method described on pp. 2012–4. They found three-body rate coefficients of $3\cdot 2 \pm 0\cdot 3$ and $3\cdot 0 \pm 0\cdot 4 \times 10^{-29}$ cm^6 s^{-1} respectively.

3.5.2. *Nitrogen ions in nitrogen.* Four ions N^+, N_2^+, N_3^+, and N_4^+ are observed in nitrogen, the appearance potentials observed in different experiments being given in Table 19.15.

TABLE 19.15

Observed appearance potentials in V for ions in nitrogen

Observers	N^+	N_2^+	N_3^+	N_4^+
Saporoschenko‡	24·2±0·4	15·5±0·2	22·1±0·5	15·8±0·3
Kaul and Fuchs§	24·2	15·4	21·7±0·5	—
Curran‖	—	—	21·04±0·05	15·04±0·5
Tekhomirov, Komarov, and Tumitskii††	—	—	20·4±1·3	—
Čermak and Herman‡‡	—	—	21·2±0·5	—
Asundi, Schulz, and Chantry§§	24·2±0·1	15·6	21·1±0·1	15·1±0·1

Moseley, Snuggs, Martin, McDaniel, and Miller‖‖ have studied the mobilities of the ions, using the same apparatus as that of Albritton *et al.* described on pp. 2012–4, except that more precise control of the ion swarm entering the drift tube was provided by an electrical shutter. They obtained zero-field mobilities for N^+, N_2^+, N_3^+, and N_4^+ under conditions

---

† MILLER, T. M., MOSELEY, J. T., MARTIN, D. W., and McDANIEL, E. W., *Phys. Rev.* **173** (1968) 115.
‡ SAPOROSCHENKO, M., ibid. **139** (1965) A352.
§ KAUL, V. W. and FUCHS, R., *Z. Naturf.* **15a** (1960) 326.
‖ CURRAN, R. K., *J. chem. Phys.* **38** (1963) 2974.
†† TIKHOMIROV, M. V., KOMAROV, V. N., and TUNITSKII, N. N., *Soviet J. phys. Chem.* **38** (1964) 515.
‡‡ ČERMAK, V. and HERMAN, Z., *Colln Czech. chem. Commun. Engl. Edn.* **30** (1965) 1343.
§§ ASUNDI, R. K., SCHULZ, G. J., and CHANTRY, P. J., *J. Chem. Phys.* **47** (1967) 1584.
‖‖ MOSELEY, J. T., SNUGGS, R. M., MARTIN, D. W., McDANIEL, E. W., and MILLER, T. M., Gaseous Electronics Conference, Boulder, Oct. 1968.

in which they drifted without reaction, as checked from the shape of the histogram of ion transit times (see p. 2031) of 2·97, 1·87, 2·26, and 2·33 cm^3 V^{-1} s^{-1} respectively (cf. McKnight, McAfee, and Sipler†). For the N$^+$ and N$_2^+$ ions the measurements were made for F/p ranging from 2 to 250 V cm^{-1} torr^{-1} at pressures from 0·02 to 1·00 torr while for N$_3^+$ and N$_4^+$ ions the F/p range was from 0·6 to 13 V cm^{-1} torr^{-1}.

Fig. 19.44 compares these low-field mobilities with the observations of Saporoschenko,‡ using the equipment described on p. 2014, and of Wobschall, Fluegge, and Graham, using the cyclotron resonance technique, as well as with mobilities for unidentified ions observed by other investigators.§ It will be seen that, for N$_4^+$, there is good agreement with Saporoschenko, whose observations extend to sufficiently small values of F/p to have attained independence of that parameter. The ions observed by Huber at low F/p seem definitely to be N$_4^+$. In other cases the data are obviously complicated by the occurrence of ionic reactions. Thus, for $F/p > 60$ V cm^{-1} torr^{-1} the effective mobility of N$_4^+$ as observed by Saporoschenko falls very rapidly to that of N$_2^+$.

Further light is thrown on the nature of the reactions involved from the shapes of the pulses of the N$_2^+$ and N$_4^+$ ions when arriving at the end of the drift tube, observed by Saporoschenko. Records taken at four values of F/p are shown in Fig. 19.45. At the lowest value of F/p (44·3 V cm^{-1} torr^{-1}) the pulses for the two ions are quite distinct and N$_4^+$ is much more abundant. The symmetrical shape of the pulses shows that both ions have formed near the entrance to the tube. A quite different situation is apparent when F/p is increased to 60 V cm^{-1} torr^{-1}, the critical value beyond which the effective mobility of N$_4^+$ tends rapidly to that of N$_2^+$. Referring to Fig. 19.45 (b) we see that now a considerable fraction of the N$_4^+$ ions have taken longer to reach the end of the tube while equally well a considerable fraction of N$_2^+$ ions have been quicker. This suggests at once that a substantial fraction of N$_4^+$ ions have spent their time delayed as N$_2^+$ ions while for the N$_2^+$ the reverse is the case. As F/p increases still further we eventually reach the situation in Fig. 19.45 (d) ($F/p = 101·5$ V cm^{-1} torr^{-1}) in which both ions travel with the slower mobility and the N$_2^+$ is the more abundant.

† McKnight, L. G., McAfee, K. B., and Sipler, D. P., *Phys. Rev.* **164** (1967) 62.
‡ loc. cit., p. 2036.
§ Vogel, J. K., *Z. Phys.* **148** (1957) 355; Frommhold, L., ibid. **160** (1960) 554; Bradbury, N. E., *Phys. Rev.* **40** (1932) 508; Huber, E. L., ibid. **97** (1955) 267; Mitchell, J. H. and Ridler, K. I. W., *Proc. R. Soc.* A**146** (1934) 911; Davies, P. G., Dutton, J., and Llewellyn-Jones, F., *Proc. Fifth int. Conf. Ioniz. Phenom. Gases*, 1961, vol. ii (North Holland, Amsterdam, 1962), p. 1326; Samson, J. A. R. and Weissler, G. L., *Phys. Rev.* **137** (1965) A381.

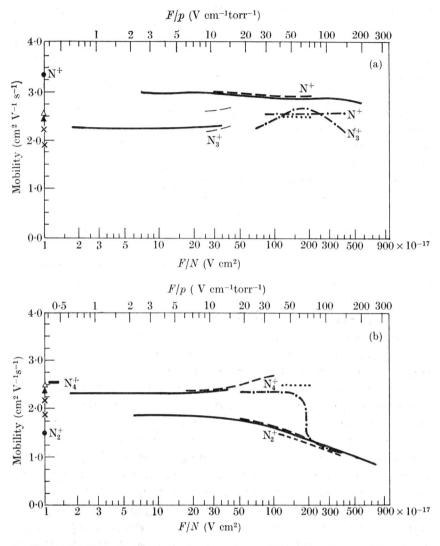

FIG. 19.44. Mobilities of nitrogen ions in nitrogen. —— Moseley *et al.*; — — —McKnight, McAfee, and Sipler; — · — · Saporoschenko; ● Wobschall, Fluegge, and Graham; ▬ Samson and Weissler. Low-field mobilities observed for unidentified ions are indicated: △ Davies *et al.*, Mitchell and Ridler; ▲ Huber; × Bradbury; · · · · Vogel, Frommhold. (a) N^+ and N_3^+, (b) N_2^+ and N_4^+.

This strongly supports the original suggestion made by Varney† that the N_2^+ and N_4^+ ion concentrations are related through the reactions

$$N_2^+ + N_2 \to N_4^+, \quad (127\,a)$$

$$N_4^+ + N_2 \to N_2^+ + 2N_2. \quad (127\,b)$$

He proceeded to apply thermochemical equilibrium theory to the dissociation reaction

$$N_4^+ \rightleftharpoons N_2^+ + N_2. \quad (128)$$

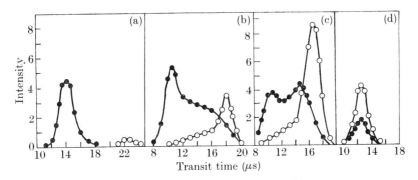

Fig. 19.45. Pulses of ion transients observed by Saporoschenko. ● N_4^+, ○ N_2^+, for the following values of F/p in V cm^{-1} torr^{-1}: (a) 44·3, (b) 60, (c) 67, (d) 101·5.

If ϵ is the fraction of N_4^+ dissociated at a temperature T then the equilibrium constant K is given by

$$K = \epsilon p/(1-\epsilon), \quad (129)$$

where p is the gas pressure. The Nernst equation then gives

$$\ln K = -(\Delta H/RT) + R^{-1} \int \left\{ \int (C_{p2} + C_{p3} - C_{p1})\,dT \right\} T^{-2}\,dT + \Delta S_0/R, \quad (130)$$

where ΔH is the heat of dissociation, C_{p1}, C_{p2}, C_{p3} are the respective molar heat capacities at constant pressure of N_4^+, N_2^+, and N_2, and ΔS_0 is the difference between the molar entropy constant of N_4^+ and that of the dissociation products. R is the gas constant.

Varney‡ then based a detailed analysis on observations made by Kovar, Beaty, and Varney,§ using a four-shutter drift tube of the type discussed on p. 1941, of the drift velocity of the main ion in N_2 at 77, 300, and 450 °K as a function of F/p. These observations were made at pressures in the range 0·1–32 torr and provided the data shown in smoothed form in Fig. 19.46. In terms of the ion identifications the

† Varney, R. N., *Phys. Rev.* **89** (1953) 708.
‡ Varney, R. N., *J. chem. Phys.* **31** (1959) 1314.
§ Kovar, F. R., Beaty, E. C., and Varney, R. N., *Phys. Rev.* **107** (1957) 1490.

drift velocities of N_2^+ and N_4^+ are shown as dashed lines. From data such as these ϵ, and hence K, may be derived. The temperatures of the ions must next be estimated. Varney assumed on the basis of Wannier's theory (p. 1935) that the ion 'temperature' could be written

$$T = \Theta + T_g, \tag{131}$$

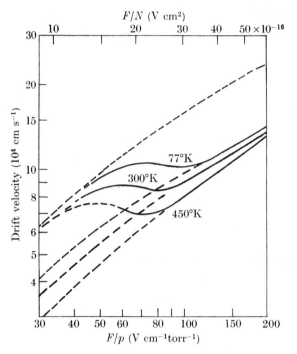

Fig. 19.46. Drift velocities of ions in nitrogen at various temperatures including extrapolation of velocities to high and low values of F/p. —— experimental (various investigators). - - - expected behaviour.

where T_g is the gas temperature and Θ depends on F/p. He plotted $\ln K$ as a function of p/F for each gas temperature and obtained three parallel linear plots just as would be obtained by plotting $\ln K$ against $1/\kappa T$ in a case in which the temperature T is well defined. He therefore assumed that

$$\Theta = \frac{1}{G}\frac{F}{p}, \tag{132}$$

where G is a constant, so that

$$\ln K = -\{G\,\Delta H/R(GT_g + F/p)\} + C, \tag{133}$$

C being a constant if the variation of the specific heats appearing in (130) can be ignored. G was then chosen so that a plot of $\ln K$ against

$1/T$, with T defined by (131) with Θ as in (132), gave a single straight line. With F/p in the usual units G came out to be 8×10^{-2} giving, for $F/p = 100$ V cm^{-1} torr^{-1}, $\Theta = 1250$ °K.

Having obtained the linear plot ΔH may be read off from the slope. It was found to be 0·50 eV, a not unreasonable value for the dissociation energy of N_4^+.

In a later paper, Varney† examined the significance of the absolute values found for K in terms of the degrees of freedom of the reactants. He found that the value of K derived from the observations at 300 °K would not agree with that expected from (130) unless there is a strong degree of activation of vibration in N_4^+ at 300 °K.

Asundi, Schulz, and Chantry‡ have studied the formation of N_3^+ and N_4^+ ions by electron impact in N_2 at pressures p up to 0·1 torr using a 90° magnetic mass-spectrometer. They found that the abundance of the N_4^+ ion varies as p^2 at electron energies close to the threshold (15·0 eV, see Table 19.15). At electron energies greater than 17·5 eV the production varied as p^3. These results are consistent with the assumption that, close to the threshold, the ions are produced through collisions of impact-excited molecules N_2^* with normal molecules,

$$N_2^* + N_2 \to N_4^+ + e, \tag{134}$$

whereas at higher electron energies they are formed by the three-body reaction

$$N_2^+ + 2N_2 \to N_4^+ + N_2. \tag{135}$$

If the ion-repeller field is large, sufficient kinetic energy is supplied to the N_4^+ for N_2^+ ions to be produced by the inverse of (135) at an electron-impact energy below the threshold for direct production of N_2^+.

For N_3^+ ions, at not too high pressures, the production rate varies as p^2. The appearance potential is consistent with the formation process being

$$N_2^+(^4\Sigma_u^+) + N_2 \to N_3^+ + N. \tag{136}$$

Moseley, Snuggs, Martin, and McDaniel§ have used the combined drift tube and mass spectrometer described on p. 2012 to measure rates for the reaction (135) and for

$$N^+ + 2N_2 \to N_3^+ + N_2. \tag{137}$$

It was found that, at low F/p, both N^+ and N_2^+ ions were lost by three-body reactions, the reaction rates for N_2^+ (through (135)) and N^+ (through

---

† VARNEY, R. N., *J. chem. Phys.* **33** (1960) 1709.
‡ loc. cit., p. 2036.
§ MOSELEY, J. T., SNUGGS, R. M., MARTIN, D. W., and MCDANIEL, E. W., *Phys. Rev. Lett.* **21** (1968) 873.

(137)) being 5.0×10^{-29} and 1.8×10^{-29} cm^6 s^{-1} respectively. These rates decreased slowly as F/p increased.

Warneck† has also observed the reaction (135) using his combined photo-ionization source and mass spectrometer technique (see p. 2007). He worked with nitrogen pressures in the source ranging from 0·09 to 0·2 torr and ionizing radiation at 764 Å. Despite the inclusion of a liquid-nitrogen-cooled trap, water vapour remained a noticeable impurity

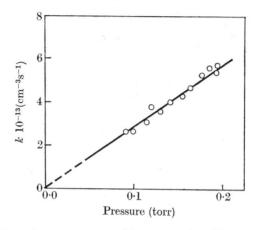

Fig. 19.47. Dependence on pressure of the rate constant k for the reaction in N$_2$ of N$_2^+$ ions which converts them to N$_4^+$, as observed by Warneck.

and H$_2$O$^+$ ions were formed by fast charge transfer from nitrogen ions. To allow for this, the intensity of these ions observed in the mass analyser was added to the sum of N$_2^+$ and N$_4^+$ currents to determine the initial N$_2^+$ currents required for the determination of rate constants (see p. 2011).

In agreement with the observations of Asundi et al. and of Moseley et al., the apparent two-body rate constant was found to be proportional to the nitrogen pressure, showing that the reaction is a three-body one. This is shown in Fig. 19.47. The average value for the three-body coefficient was found to be 8.5×10^{-29} cm^6 s^{-1}, which is a little larger than, but not inconsistent with, that found by Moseley et al. Under the experimental conditions, which correspond to mean ion energies up to 700 °C, the chance that N$_4^+$ ions, when formed, should be reconverted to N$_2^+$ before leaving the ion chamber is negligible.

There seems little doubt that the reaction leading to N$_4^+$ production from N$_2^+$ is a three-body one though in some earlier experiments‡ evidence was produced that it is a two-body reaction.

† loc. cit., p. 2007. ‡ Fite, W. L. et al., loc. cit., p. 2019.

3.5.3. *Ions in oxygen and in helium–oxygen mixtures.* Whereas in hydrogen the charge-exchange reaction

$$H^+ + H_2 \rightarrow H + H_2^+ \qquad (138)$$

is endothermic and the reaction

$$H_2^+ + H_2 \rightarrow H_3^+ + H \qquad (139)$$

is strongly exothermic the reverse situation applies in oxygen. Thus the reaction

$$O^+ + O_2 \rightarrow O + O_2^+ \qquad (140)$$

is exothermic by 1·3 eV while

$$O_2^+ + O_2 \rightarrow O_3^+ + O, \qquad (141)$$
$$\rightarrow O_3 + O^+ \qquad (142)$$

are endothermic by 4·8 and 5·5 eV respectively. We would therefore expect that, in mobility measurements, O_2^+ would be the dominant ion. There may well be circumstances in which more complex ions would be formed, as in nitrogen. While in mass analysis of the ions formed in a glow discharge in oxygen only O^+ and O_2^+ were identified, O_3^+ ions have been observed in addition to the lighter ions in afterglows in oxygen (see Chap. 20, § 4.6).

Varney,[†] in mobility studies by the method of § 2.2.1, observed only one ion, with a mobility of 2·25 cm² V⁻¹ s⁻¹. This agrees well with the later observations of Eiber[‡] (2·22±0·15 cm² V⁻¹ s⁻¹) and is consistent with observations of Frommhold[§] using a pulse technique, of Burch and Geballe[∥] using Hornbeck's method (§ 2.2.2), and of Samson and Weissler[††] using a photo-ionization source of the ions (p. 1937). Furthermore, Wobschall, Fluegge, and Graham,[‡‡] using the cyclotron resonance technique, found 2·0 cm² V⁻¹ s⁻¹ for O_2^+ ions in O_2. According to the Langevin theory (§ 2.4) the mobility should be about 2·9 cm² V⁻¹ s⁻¹, so there is evidence of a significant contribution from charge transfer. There is a difficulty which arises in this connection through the relation between the mobilities of O_2^+ and O_2^-. At 300 °K these mobilities are about equal (see § 4.1) but the effect of charge transfer on such a diffuse structure as O_2^- would be expected to be considerably greater than on O_2^+.

The rate of the exothermic charge-transfer reaction (140) under thermal conditions has been measured by the flowing-afterglow and

---

[†] Varney, R. N., *Phys. Rev.* **89** (1953) 705.
[‡] Eiber, H., *Z. angew. Phys.* **15** (1963) 103.
[§] Frommhold, L. *Atomic Coll. Proc.* ed. McDowell, M. R. C., p. 556 (North Holland, Amsterdam, 1964).
[∥] Burch, D. S. and Geballe, R., *Phys. Rev.* **106** (1957) 183.
[††] Samson, J. A. R. and Weissler, G. L., ibid. **137** (1965) A381.
[‡‡] loc. cit., p. 2028.

time-resolved afterglow methods and by the use of a mass spectrograph with a photo-ionization source.

In the flowing-afterglow experiments (see § 3.4.5) oxygen was added to the flowing helium, downstream of the discharge, to produce O^+ and O_2^+ ions by the reactions

$$He^+ + O_2 \to He + O + O^+ \tag{143}$$

and (see Chap. 18, § 5.3)

$$He\,(2\,^3S) + O_2 \to He + O + O^+ + e, \tag{144}$$

$$He\,(2\,^3S) + O_2 \to He + O_2^+ + e, \tag{145}$$

and by

$$O^+ + O_2 \to O_2^+ + O. \tag{146}$$

Additional oxygen was then added so far further downstream (nearly 30 cm from the discharge or 3 ms in time at the helium flow rate of 10^4 cm s^{-1}) that the reactions (143), (144), and (145) had proceeded to completion. By observing the variation, with concentration of added O_2, of the O^+ ion current sampled by the mass spectrometer the rate of (146) was determined as $4\cdot 0 \pm 1 \times 10^{-11}$ cm^3 s^{-1} at 300 °K. It was verified that decrease of the O^+ current was quantitatively matched by the increase of the O_2^+ current.

One of the major experimental difficulties in the use of the time-resolved afterglow technique is the 'clean-up' of oxygen that occurs through successive pulses of discharge operation. In the earlier experiments all observations were taken as quickly as possible to minimize this effect. Thus Langstroth and Hasted used exciting pulses of 2 μs duration with a repetition frequency of 10 s^{-1} and carried out all observations, including mass spectrometer adjustments, in less than 2 min. Sayers and his collaborators used pulses of 10 μs duration at a repetition frequency of 50 s^{-1} and completed the observations in less than 40 s. In later experiments they observed the decay signals at different times in the pulse sequence and extrapolated the decay rate to zero time. Finally they used a single-pulse mode of operation—they found that the measured decay constant was reduced by 15 per cent after a discharge pulse.

In the latest experiments of Copsey, Smith, and Sayers[†] a fully bakeable vacuum system was used. The sampling orifice of the mass spectrometer was of 0·02 cm diameter, in the centre of a disc maintained at the same d.c. potential as the internal metal electrodes. It was placed midway between the discharge electrodes and protruded 2 cm into the cylindrical discharge vessel, of 2·5 l active volume. The discharge power

[†] COPSEY, M. J., SMITH, D., and SAYERS, J., *Planet. Space Sci.* **14** (1966) 1047.

was around 10 kVA and the single discharge pulses were of 10 μs duration and radio-frequency 10 Mc/s. The vacuum system was baked at 350 °C and discharge-cleaned in pure helium so that background pressures were reduced to 10^{-7} torr.

Fig. 19.48 illustrates typical logarithmic plots at different oxygen partial pressures of the decay during the afterglow of O^+ concentration as given by the rate of decay of the mass-spectrometer current. From such curves the total decay constant λ can be obtained. This can be written
$$\lambda = \lambda_d + \lambda_o,$$
where λ_d is the contribution from diffusion to the walls and λ_o from the reaction (146). To determine λ_d observations were made at different helium pressures in mixtures containing such little oxygen that λ_o is negligible. Since $\lambda_d p$ is a constant, where p is the helium pressure, this constant could be determined from these observations and hence λ_o at any helium pressure.

From the derived values of λ_o the rate coefficient was found to be 2.0×10^{-11} cm^3 s^{-1}. This is to be compared with the earlier results of Dickinson and Sayers† (2.5×10^{-11} cm^3 s^{-1}) and Batey, Court, and Sayers‡ (1.64×10^{-11} cm^3 s^{-1}). It is not seriously in conflict with the flowing-afterglow results but disagrees with the much smaller rate coefficient (0.18×10^{-11} cm^3 s^{-1}) found by Langstroth and Hasted.§

Fite, Rutherford, Snow, and van Lint‖ made some observations, using their technique (see p. 2020), in a pure oxygen afterglow. Because of the rapid decay rate of O^+ ions in such an afterglow they were only able to place a lower limit of 1×10^{-11} cm^3 s^{-1}, which supports the higher values rather than that of Langstroth and Hasted. With a 10:1 mixture of helium and oxygen they obtained an even higher apparent rate of 1.5×10^{-10} cm^3 s^{-1}.

Warneck†† measured the rate of the reaction (146), using his photoionization technique, for O^+ ion temperatures from 700 to 1400 °K. In one set of experiments he used pure oxygen as the working substance and ionizing radiation at 585 Å. No systematic variation of the rate coefficient with ion temperature was found, the average value being $2.2_5 \times 10^{-11}$ cm^3 s^{-1}, agreeing well with the results of Copsey, Smith, and Sayers. Further measurements were made using air as the working

---

† DICKINSON, P. H. G. and SAYERS, J., *Proc. phys. Soc.* **76** (1960) 137.
‡ BATEY, P. H., COURT, G. R., and SAYERS, J., *Planet. Space Sci.* **13** (1965) 911.
§ LANGSTROTH, G. F. O. and HASTED, J. B., *Discuss. Faraday Soc.* **33** (1963) 298.
‖ FITE, W. L., RUTHERFORD, J. A., SNOW, W. R., and VAN LINT, V. A. J., ibid. **33** (1963) 518.
†† WARNECK, P., *Planet. Space Sci.* **15** (1967) 1349.

substance and ionizing radiation at 585 and at 630 Å. In these experiments the reaction (146) competes with

$$O^+ + N_2 \to NO^+ + N. \tag{147}$$

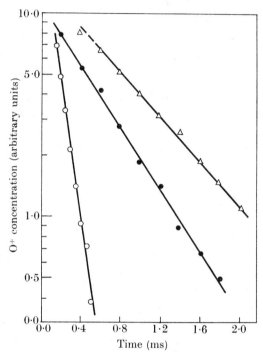

FIG. 19.48. Typical variations of O^+ concentrations with time in the afterglow as observed by Copsey, Smith, and Sayers, the O_2 partial pressures being: ○ 10.26×10^{-3}, ● 1.58×10^{-3}, △ 0.80×10^{-3} torr.

If we denote the concentration of a species M by [M] we have

$$\frac{d[O^+]}{dt} = -k_1[O^+][O_2] - k_2[O^+][N_2], \tag{148}$$

$$\frac{d[NO^+]}{dt} = k_2[O^+][N_2] = k_1[O^+][O_2]\epsilon, \text{ say.} \tag{149}$$

Integrating these equations with the appropriate initial conditions we find, for an ion residence time τ,

$$k_1 = \frac{\ln\{[O^+]_0/[O^+]\}}{(1+\epsilon)[O_2]\tau}, \quad k_2 = \frac{\epsilon k_1[O_2]}{[N_2]},$$

$$\epsilon = k_2[N_2]/k_1[O_2] = \{[O^+]_0 - [O^+]/[NO^+]\} - 1. \tag{150}$$

$[O^+]_0$ is the primary ion concentration, $[O^+]$, $[NO^+]$ the respective concentrations when reactions occur.

Fig. 19.49 shows the observed variation of the currents of O+ and NO+ ions with the air pressure when the gas is irradiated by 630 Å and by 585 Å radiation. From such observations k_1 and k_2 may be obtained. Again, no systematic variation of k_1 with ion energy is found, the mean values being $1·90 \times 10^{-11}$ cm³ s⁻¹ and $2·04 \times 10^{-11}$ cm³ s⁻¹ respectively

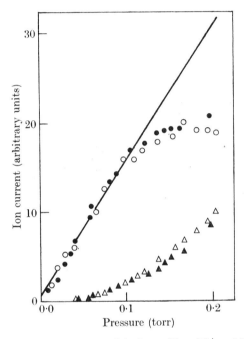

FIG. 19.49. Variation with pressure of the intensities of O+ and NO+ ions in air irradiated by 630 Å and 585 Å radiation. ● O+ (585 Å), ○ O+ (630 Å), ▲ NO+ (585 Å), △ NO+ (630 Å). —— O+ intensity before reaction.

when 585 and 630 Å ionizing radiations are used. These are quite consistent with the mean value obtained using pure oxygen. Results for k_2 are discussed in § 3.5.6.

Smith and Fouracre,† using equipment very similar to that of Copsey, Smith, and Sayers, have observed the variation of k_1 with gas temperature in an afterglow. The temperature range, 300–576 °K, was covered using an oven surrounding the experimental tube, while temperatures of 190 and 250 °K were obtained by using freezing mixtures of solid CO_2 and of ice–sodium chloride respectively. It was assumed that the electron, ion, and gas temperatures in the afterglow were the same as the wall temperature of the containing vessel, which was directly measured.

† SMITH, D. and FOURACRE, R. A., *Planet. Space Sci.* **16** (1968) 243.

There is some evidence† that this assumption is questionable. Fig. 19.50 shows the observed variation of the rate coefficient with wall temperature. These results can be represented by writing

$$k = 3 \cdot 4 \pm 0 \cdot 5 \times 10^{-10} T^{-(0 \cdot 48 \pm 0 \cdot 05)} \text{ cm}^3 \text{ s}^{-1}. \tag{151}$$

This does not conflict with Warneck's observations, which are not precise enough to show clearly a temperature dependence of this kind. Kaneko, Megill, and Hasted,‡ using the combined drift tube mass

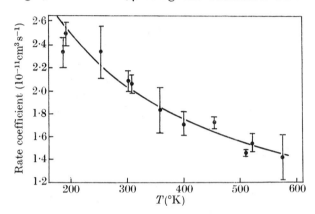

FIG. 19.50. Variation of rate coefficient with temperature for reaction of O^+ ions with O_2 molecules, obtained by Smith and Fouracre.

spectrometer technique described in § 3.4.3, found a similar variation of rate coefficient with mean ion energy in the range 0·06–0·4 eV. However, the absolute magnitudes of the coefficients they observed are three times larger than those of Smith and Fouracre, which are consistent with the other measurements made by Sayers and his associates and by Warneck.

Although it seems now well established that the rate coefficient at thermal energies is around $2 \cdot 5 \times 10^{-11}$ cm^3 s^{-1} it is still uncertain by at least 50 per cent. For many initial applications this is already very useful but there is room for considerable improvement.

The rate for the charge transfer reaction (143) is of interest in connection with the interpretation of the composition of the topside ionosphere of the earth and its variation with time of day, height, etc (see Chap. 20, § 6). Sayers and Smith§ obtained a value for the rate at 308 °K

† GUSINOW, M. A., GERARDO, J. B., and VERDEYEN, J. T., *Phys. Rev.* **149** (1966) 91; GERARDO, J. B., VERDEYEN, J. T., and GUSINOW, M. A., *J. appl. Phys.* **36** (1965) 3526; BORN, G. K. and BUSER, R. G., ibid. **37** (1966) 4918.
‡ KANEKO, Y., MEGILL, L. R., and HASTED, J. B., *J. chem. Phys.* **45** (1966) 3741.
§ SAYERS, J. and SMITH, D., *Discuss. Faraday Soc.* **37** (1964) 167.

from observations of the decay rate of He⁺ ions in the mixed oxygen–helium afterglow. They found that this rate varied with the time from initiation of the first discharge, presumably due to oxygen clean-up. To overcome this difficulty they photographed the signals from the spectrograph on a cine-camera and determined decay constants from various parts of the film. The decay rate used in the final analysis was obtained by extrapolating such data to zero time from initiation of the discharge. In this way a rate coefficient of $1.05 \pm 0.08 \times 10^{-9}$ cm³ s⁻¹ was found corresponding to a mean cross-section of $7.7 \pm 0.6 \times 10^{-15}$ cm². This refers to the sum of the rates of the two reactions

$$He^+ + O_2 \rightarrow O^+ + O + He, \qquad (152)$$
$$\rightarrow O_2^+ + He, \qquad (153)$$

but as the main oxygen ion present during these observations was O⁺ it was very likely that the reaction (152) was predominant. Similar conclusions were arrived at by Fite, Smith, Stebbings, and Rutherford† who obtained a rate coefficient around 5×10^{-10} cm³ s⁻¹.

Measurements have also been made by the flowing-afterglow technique‡ yielding a combined rate of 1.5×10^{-9} cm³ s⁻¹ for (152) and (153).

Fig. 19.51 shows the observed variations (on a logarithmic scale) of the currents $i(He^+)$, $i(O_2^+)$, $i(O^+)$ of He⁺, O_2^+, and O⁺ ions respectively with the flow rate of injected O_2. The variation of $\ln i(He^+)$ is linear except at the largest flow rates. Deviations from linearity at these rates are probably due to production of He⁺ by collisions between pairs of metastable He atoms (see Chap. 18, § 5.2.3.1), which will be most effective when the He⁺ concentration is low.

Evidence was obtained in these experiments that (152) is the dominant reaction. Thus the rate of production of O⁺ agreed with that of loss of He⁺ at low O_2 flow rates. Furthermore, the He⁺ concentration was decreased, without altering the flow conditions, by local microwave heating upstream of the O_2 injector. This increased the loss of He⁺ to the walls by ambipolar diffusion. It was found that this decrease led to a proportionate decrease in O⁺ current in the spectrometer but had no effect on the O_2^+. In addition, the O_2^+ current observed was of the magnitude expected from an origin through the Penning process using rates measured for this process by methods described in Chapter 18, § 5.3.

† FITE, W. L., SMITH, A. C. H., STEBBINGS, R. F., and RUTHERFORD, J. A., *J. geophys. Res.* **68** (1963) 3225.
‡ FEHSENFELD, F. C., SCHMELTEKOPF, A. L., GOLDAN, P. D., SCHIFF, H. I., and FERGUSON, E. E., *J. chem. Phys.* **44** (1966) 4087.

Warneck† has applied his method, described on p. 2007, by observing with a mass spectrometer the ions produced in an He–O_2 mixture by a pulsed beam of 492 Å radiation. He observed only O^+ ions and obtained a rate coefficient of $1 \cdot 2 \times 10^9$ cm³ s⁻¹ consistent with the values obtained by the other methods. However, the increase in O^+ production did not balance the loss of He^+. A possible explanation of this behaviour, which did not occur in comparable observations with He–N_2 mixtures, is that

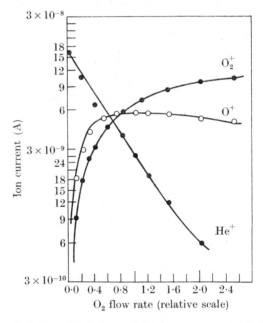

FIG. 19.51. Variation of He^+, O^+, and O_2^+ ion currents with O_2 flow rate in flowing-afterglow experiments.

O^+ ions are formed with considerable kinetic energy and for such ions the collecting efficiency of the spectrometer is reduced. Evidence in support of this was found by making a retarding-potential analysis of the O^+ ions emerging from the source. This confirmed that ions with kinetic energy around 1 eV were indeed being produced with considerable probability. The existence of such ions, produced from the same reaction, was also observed by Moran and Friedman‡ using an electron impact ion source with a mass spectrograph. Considerable interest attaches to the fact that a substantial fraction of the energy released by the reaction is converted to kinetic energy of relative translation of the

† WARNECK, P., J. chem. Phys. **47** (1967) 4279.
‡ MORAN, T. F. and FRIEDMAN, L., ibid. **45** (1966) 3837.

products and is not almost all absorbed as internal energy (contrast the chemical reactions discussed in Chapter 17, § 6). This departure from near energy-resonance (see Chap. 17, § 6 and Chap. 18, §§ 8.4–8.5) may be due to the formation of an intermediate complex HeO^+. Moran and Friedman obtained some evidence for this by comparing the kinetic energy distributions of the O^+ produced by reactions with $^4He^+$ and $^3He^+$ ions. If the kinetic energy of the O^+ results from breakup of the HeO^+ complex it should be reduced when ^4He is replaced by ^3He. Such a reduction was definitely observed.

Aquilanti and Volpi,[†] using a mass spectrometer in conjunction with a tritium beta-ray ionization source, have obtained a rate coefficient of 0.52×10^{-9} cm^3 s^{-1}. Allowing for the spread of the observations it is likely that the true rate coefficient is within 50 per cent of 1.1×10^{-9} cm^3 s^{-1}.

3.5.4. *Ions in nitrogen–helium mixtures.* The reactions

$$He^+ + N_2 \rightarrow He + N + N^+ + 0.3 \text{ eV}, \tag{154}$$

$$\rightarrow He + N_2^+ + 9.0 \text{ eV} \tag{155}$$

have been investigated by the same techniques as for O_2 and are of equal if not even greater importance in the interpretation of the properties of the topside ionosphere (Chap. 20, § 6).

Warneck,[‡] using his photo-ionization technique, obtained a coefficient of 1.5×10^{-9} cm^3 s^{-1} for the over-all rate of both reactions. In addition, he determined the ratio of the N^+ to N_2^+ ions produced as 48:52 so that, in contrast to the reaction with O_2, the branching ratio for the two reactions is almost unity, a somewhat surprising result in view of the large amount of energy released in reaction (155). The combined increase of N^+ and N_2^+ currents was found to balance the decrease of He^+ currents, indicating that the ions are formed with little kinetic energy. This is in agreement with the observations of Moran and Friedman§ using their electron impact source. However, those same authors found a rate coefficient of 1.69×10^{-9} cm^3 s^{-1} for (154) alone, effectively independent of the energy of the incident ion between 0.3 and 12 eV. This is very close to the value 1.66×10^{-9} cm^3 s^{-1} expected from the theory of Gioumousis and Stevenson‖ and would suggest that (155) is unimportant, in contrast to the conclusions of Warneck.

The flowing-afterglow method†† gives very nearly the same value,

---

† AQUILANTI, V. and VOLPI, G. G., *Ric. Sci. Rend.* **36** (1966) 359.
‡ loc. cit., p. 2050.   § loc. cit.   ‖ loc. cit., p. 2005.
†† FEHSENFELD, F. C. et al., loc. cit., p. 2049.

1.7×10^{-9} cm^3 s^{-1}, for the over-all rate constant but tends to support Warneck's results about the branching ratio.

Sayers and Smith,[†] using the static-afterglow method, observed the over-all reaction rate as a function of temperature, obtaining the rate coefficients of 1·75, 1·45, 1·40, and 1.03×10^{-9} cm^3 s^{-1} for temperatures of 195, 293, 408, and 503 °K respectively.

3.5.5. *Other reactions of He$^+$ ions.* The reactions of He$^+$ ions with NO, CO, and CO$_2$ have been studied by the flowing-afterglow method.[‡]

With NO the reaction seems to be mainly

$$\text{He}^+ + \text{NO} \rightarrow \text{He} + \text{O} + \text{N}^+ + 3\cdot6 \text{ eV}, \qquad (156)$$

NO$^+$ ions observed being due to Penning ionization of NO by metastable He atoms. The rate coefficient is found to be 1.5×10^{-9} cm^3 s^{-1}. This is not far from the value 1.62×10^{-9} cm^3 s^{-1} predicted from the theory of Gioumousis and Stevenson.[§] Moran and Friedman,[||] using their electron impact ion source+mass analyser, found a rate coefficient approximately constant over the ion energy range 0·3–12 eV with a value (2.1×10^{-9} cm^3 s^{-1}) larger than given by the theory. They also found that the N$^+$ ions were produced with kinetic energy of around 0·8 eV and that this energy was reduced when ^{3}He was substituted for ^{4}He (see the similar result for O$_2$ on p. 2051).

For CO the dominant reaction seems to be

$$\text{He}^+ + \text{CO} \rightarrow \text{He} + \text{O} + \text{C}^+ + 2\cdot2 \text{ eV} \qquad (157)$$

with a rate constant of 1.7×10^{-9} cm^3 s^{-1}. This is to be compared with the prediction of 1.75×10^{-9} cm^3 s^{-1} from the theory and the nearly constant value 1.63×10^{-9} cm^3 s^{-1} observed by Moran and Friedman[||] over the ion energy range 0·3–12 eV. The latter authors found that the C$^+$ ions were produced with a kinetic energy around 0·6 eV.

The over-all rate coefficient for the reactions

$$\text{He}^+ + \text{CO}_2 \rightarrow \text{He} + \text{O}^+ + \text{CO} + 4\cdot5 \text{ eV}, \qquad (158)$$

$$\rightarrow \text{He} + \text{CO}^+ + \text{O} + 4\cdot1 \text{ eV} \qquad (159)$$

was measured by the flowing-afterglow method as 1.2×10^{-9} cm^3 s^{-1}.

3.5.6. *Ions in nitrogen–oxygen mixtures.* One of the most interesting ionic reactions in a mixture of nitrogen and oxygen is

$$\text{O}^+ + \text{N}_2 \rightarrow \text{NO}^+ + \text{N}. \qquad (160)$$

This is because of the importance of this reaction in the interpretation

---

[†] SAYERS, J. and SMITH, D., loc. cit., p. 2048.
[‡] FEHSENFELD et al., loc. cit., p. 2049.
[§] loc. cit., p. 2005. [||] loc. cit., p. 2050.

of the behaviour of the ionosphere at altitudes up to a few hundred kilometres (see Chap. 20, § 6).

The flowing-afterglow observations were made by generating O^+ ions downstream from the discharge in a helium flow system. This was done by injection of O_2 from which the ions were produced by the reaction (143). Nitrogen was injected further downstream and the intensity of NO^+ ions monitored as a function of N_2 concentration. Complications may ensue because of the possibility of production of O^+ by the Penning reaction (144) with metastable helium atoms. To eliminate this, argon was injected into the stream before the addition of nitrogen so as to eliminate the metastable atoms.

There is also the important question as to whether the O^+ ions are in their ground states when they react. It seems likely that, under the experimental conditions, superelastic collisions with electrons will have reduced any O^+ ions in one of the metastable ^2P or ^2D states to the ground state. This is based on the measured electron concentration in the region where ions are formed, $\simeq 10^{11}/cm^3$, and the cross-sections for deactivation of the metastable states of O^+ calculated by Seaton (see Chap. 8, § 9.3). These are of the order 5×10^{-15} cm^2, giving a lifetime of $O^+(^2D)$ of order 10^{-4} s, whereas the flow time to the N_2 nozzle is about 4×10^{-3} s. Similar considerations apply to $O^+(^2P)$, which goes to the ground state via the ^2D state at almost the same rate. The measured reaction rate for (160) is $1 \cdot 8 \times 10^{-12}$ cm^3 s^{-1}.

Warneck† has measured the rate for O^+ ion temperatures from 700 to 1400 °K by applying his photo-ion source+mass spectrometer technique using dry air as the working gas. Observations were made using ionizing radiation of 585 and 630 Å wavelength. The reaction (160) was sorted out from the alternative one

$$O^+ + O_2 \to O_2^+ + O, \qquad (161)$$

discussed in § 3.5.3, by observing the variation with pressure of the intensities of both O^+ and NO^+ ions. In this way the rate constant both for the production of NO^+ and for the reaction (161) could be derived. The latter results were compared with those obtained using pure O_2 as the working gas and found to agree within the expected uncertainty (see Table 19.16). The rate coefficient for NO^+ production was found to show no systematic variation with pressure, and hence with O^+ ion temperature, and the average value found is $4 \cdot 6 \pm 1 \cdot 1 \times 10^{-12}$ cm^3 s^{-1}. This is considerably higher than the values derived from the afterglow observations.

† loc. cit., p. 2045.

Schmeltekopf, Fehsenfeld, Gilman, and Ferguson† have investigated the variation of the reaction rate with the distribution of vibrational states of the N_2 molecules. To do this, they subjected the N_2 stream to a microwave discharge prior to entry into the main flow tube. This led to a marked decrease in the intensity of the O^+ signal. To interpret this result we note that the effect of the discharge could be to produce vibrationally excited N_2 molecules, dissociation of N_2 into N atoms, ionization of N_2, and electronic excitation of N_2.

The possibility that the effects are due to electronically excited or ionized molecules can be ruled out because, to bring about the observed decrease, a concentration of excited or ionized species of about 5 per cent of the whole would be required, assuming unit probability of reaction on collision. This is several orders of magnitude higher than the possible concentration.

As to the N atoms, their concentration can rise to as much as 1 per cent of the whole, but it was verified that removal of these atoms by titration through the reaction (120) had no effect on the O^+ signal. In any case N atoms would hardly be expected to affect the O^+ concentration as the charge transfer reaction

$$N+O^+ \to N^++O \tag{162}$$

is endothermic by 2 eV.

Finally, no impurities could be present in sufficient concentration to produce the observed effects. Moreover, it was checked from the mass spectrum that no unexpected impurities were present to a significant extent.

The only remaining explanation is that vibrational excitation of the N_2 by the microwave discharge was responsible. An ingenious method was then used to determine the vibrational distribution. Advantage was taken of certain features of the Penning ionization of N_2 by metastable He ($2\,^3S$) atoms. As discussed in Chapter 18, § 5.3 about one-half of the collisions that lead to ionization from N_2 molecules in the ground vibrational state produce excited $B\,^2\Sigma_u^+$ molecular ions (see Chap. 13, p. 958) in which the ratio of the population of the $v' = 0$ to $v' = 1$ vibrational states is exactly as expected from the Franck–Condon principle (Chap. 13, p. 977). It was then assumed that, if the N_2 molecules were initially in excited vibrational states, the final distribution in the $B\,^2\Sigma_u^+$ state of N_2^+ produced by Penning ionization would also be as given from the Franck–Condon principle. This being so, the initial

---

† SCHMELTEKOPF, A. L., FEHSENFELD, F. C., GILMAN, G. L., and FERGUSON, E. E., *Planet. Space Sci.* **15** (1967) 401.

vibrational distribution in the N_2 molecules could be inferred from the intensity distribution in the first negative bands $(B\,^2\Sigma_u^+ \to X\,^2\Sigma_g^+)$ emitted by the excited N_2^+ ions. To check the validity of these assumptions, this spectrum was compared with one excited by bombardment of the vibrationally excited N_2 molecules by 7-keV electrons. In this case the vibrational distribution in the N_2^+ $(B\,^2\Sigma_u^+)$ molecules must certainly be that given by the Franck–Condon principle. Little difference was observed between the two spectra.

The effective vibrational temperature T_v could be changed by varying the distance between the exciting discharge and the inlet orifice to the main flow tube. Increasing this distance decreased T_v because of vibrational deactivation in collisions with the glass wall.

Fig. 19.52 shows the observed variation of reaction rate with vibrational temperature, the values given being relative to the rate for $T_v = 300\,°K$.

Time-resolved afterglow experiments have been carried out, usually with small partial pressures of oxygen and nitrogen in helium. In such experiments, the possibility that the N_2^+ ions possess some vibrational excitation cannot be ruled out and it is true that the results obtained exhibit a rather wide variation, at least part of which may be ascribed to this source. Thus Batey, Court, and Sayers,† using the technique already described on p. 2044 for the O^+–O_2 reaction, found a rate constant as high as 28×10^{-12} cm^3 s^{-1} for (160).

On the other hand, in later experiments Copsey, Smith, and Sayers,‡ using single-pulse operation (see p. 2044), obtained a value $2.4 \pm 0.4 \times 10^{-12}$ cm^3 s^{-1}, which is not inconsistent with the results from the flowing-afterglow and photoionization techniques. Langstroth and Hasted§ also obtained a consistent value: $4.7 \pm 0.05 \times 10^{-12}$ cm^3 s^{-1}.

The reaction $\qquad N_2^+ + O_2 \to N_2 + O_2^+ \qquad\qquad (163)$

has also been studied by both afterglow techniques and by the photoionization method of Warneck. In the flowing-afterglow experiments,∥ the N_2^+ ions are produced from injected N_2 by the Penning reaction

$$\mathrm{He}\,(2\,^3S) + N_2 \to \mathrm{He} + N_2^+ \qquad (164)$$

and react with O_2 injected from a nozzle further downstream. Two other reactions which may occur are

$$N_2^+ + O_2 \to NO^+\,(^3\Sigma) + NO\,(^2\Pi) - 0.1\ \mathrm{eV}, \qquad (165)$$
$$\to NO^+\,(^1\Sigma) + NO\,(^2\Pi) + 4.5\ \mathrm{eV}. \qquad (166)$$

---
† loc. cit., p. 2045. ‡ loc. cit., p. 2044. § loc. cit., p. 2045.
∥ GOLDAN, P. D., SCHMELTEKOPF, A. L., FEHSENFELD, F. C., SCHIFF, H. I., and FERGUSON, E. E., *J. chem. Phys.* **44** (1966) 4095.

Under conditions in which N_2^+ is the dominant ion it was found that addition of O_2 leads mainly to production of O_2^+ and it is estimated that the rate of (163) is greater than 10 times the combined rates of (165) and (166).

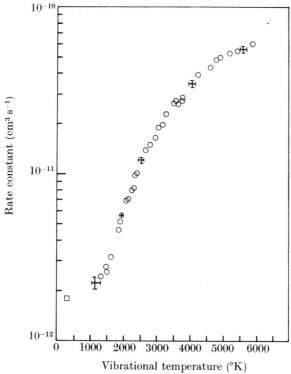

FIG. 19.52. Observed variation of the rate of the reaction
$$O^+ + N_2 \rightarrow NO^+ + N$$
with vibrational temperature of the target N_2 molecules.

O_2^+ ions could also be formed from the Penning reaction (145). To eliminate any confusion from this source the injected concentration of N_2 was sufficiently large to ensure that all metastable atoms were quenched by (164) before the oxygen was added.

At first† it was not realized that N_2^+ ions were being formed by photo-ionization of N_2 by helium resonance radiation and much too small values were found for the reaction rate. When the afterglow was pulsed and the mass-spectrometer observations made at suitable intervals after each pulse, this effect was eliminated and a rate of $1 \cdot 0 \pm 0 \cdot 5 \times 10^{-10}$ cm³ s⁻¹

---

† FEHSENFELD, F. C., SCHMELTEKOPF, A. L., and FERGUSON, E. E., *Planet. Space Sci.* **13** (1965) 219; ibid. 919.

found at 300 °K. This agrees reasonably well with the value $2 \cdot 0 \times 10^{-10}$ cm³ s¹ obtained by Fite, Rutherford, Snow, and van Lint† from time-resolved mass analysis of the ions in a static afterglow in a 100:1 mixture of N_2 and O_2. In fact it was the large discrepancy with the early flowing-afterglow results (which gave a rate of 4×10^{-13} cm³ s⁻¹) that drew attention to the effects of photo-ionization.

Warneck‡ applied his photo-ionization method using air as working substance. The reaction rate was measured using 760, 630, and 585 Å radiation. Whereas the first produces N_2^+ ions in the $X\,^2\Sigma_g^+$ ground state the latter two produce a proportion of ions in excited $A\,^2\Pi_u$ and $B\,^2\Sigma_u$ states (see Chap. 14, Fig. 14.52). These excited ions will revert to the ground state by radiative transitions in a time short compared with the residence time in the ion chamber but, in so doing, it is likely that vibrational levels of the $X\,^2\Sigma_g^+$ ground state will be populated in different proportions to those produced directly by the 760 Å radiation. If the reaction rate depends on the vibrational distribution in the ion, different results should be obtained when the different ionizing radiations are employed. No significant difference was obtained and the result for the rate coefficient ($1 \cdot 1 \times 10^{-10}$ cm³ s⁻¹) agrees well with those obtained by the pulsed flowing-afterglow method. No evidence was obtained of any systematic variation with initial ion temperature in the range 400–800 °K as determined from measurement at different pressures.

No NO⁺ ions were observed in these experiments and an upper limit of 3×10^{-14} cm³ s⁻¹ was derived for the rate coefficients for the reactions (165) and (166).

The rates of reactions of N⁺ with O_2 have been measured by the flowing-afterglow and photo-ionization methods. The possible exothermic reactions are

$$N^+(^3P)+O_2(^3\Sigma) \rightarrow NO^+(^3\Sigma)+O(^3P)+2 \cdot 1 \text{ eV}, \qquad (167)$$

$$NO^+(^1\Sigma)+O(^3P)+6 \cdot 7 \text{ eV}, \qquad (168)$$

$$N(^4S)+O_2^+(^2\Pi)+2 \cdot 5 \text{ eV}, \qquad (169)$$

$$N(^2D)+O_2^+(^2\Pi)+0 \cdot 1 \text{ eV}. \qquad (170)$$

In the flowing-afterglow experiments,§ N⁺ ions are produced by the fast reaction (154). The sum of the rates for all the reactions (167)–(170) at 300 °K was found to be $1 \cdot 0 \pm 0 \cdot 5 \times 10^{-9}$ cm³ s⁻¹. NO⁺ and O_2^+ ions

---
† loc. cit., p. 2019.
‡ WARNECK, P., *J. chem. Phys.* **46** (1967) 502 and *Planet. Space Sci.* **15** (1967) 1349.
§ GOLDAN, P. D., SCHMELTEKOPF, A. L., FEHSENFELD, F. C., SCHIFF, H. I., and FERGUSON, E. E., *J. chem. Phys.* **44** (1966) 4095.

were observed to be produced in approximately equal proportions, suggesting that (167)–(168) and (169)–(170) each proceed at a rate of about 0.5×10^{-9} cm^3 s^{-1}. There may have been some production of NO$^+$ through the reactions (169)–(170) followed by

$$O_2^+ + N \to NO^+ + O, \qquad (171)$$

so that it is likely that (169)–(170) are in fact faster than (167)–(168).

Warneck,† in his experiments with the photo-ionization technique, produced the N$^+$ ions by photo-ionization of air with 482 Å radiation, just shorter than the threshold for N$^+$ production from N$_2$ (501 Å). At high pressures, NO$^+$ production from the reaction

$$O^+ + N_2 \to NO^+ + N \qquad (172)$$

complicated the interpretation of the results. However, at pressures below 0.06 torr, the wanted reaction with N$^+$ was dominant. It was not possible to determine the residence time of the N$^+$ ions in the ion chamber directly by time-of-flight measurements because the presence of a considerable excess of N$_2^+$ ions rendered the oscilloscope traces of the N$^+$ pulses too ragged. The N$^+$ residence time was assumed to be $1/\sqrt{2}$ times that of the N$_2^+$, which could be measured in the usual way. This is really only justified at pressures lower than those actually used in the observations, in which the ions follow effectively collision-free trajectories in the ionization chamber. Nevertheless, the observed over-all rate coefficient ($6.5 \pm 2.5 \times 10^{-10}$ cm^3 s^{-1}) is consistent with the flowing-afterglow observations, particularly when it is noted that, because of the low working pressure, the mean energy of the N$^+$ ions (0.15 eV) is considerably above thermal.

The observed ratio of NO$^+$ production to N$^+$ consumption was 0.27, suggesting individual rate coefficients of 4.8×10^{-10} and 1.8×10^{-10} cm^3 s^{-1} for the charge transfer reactions (169)–(170) and the ion interchange reactions (167)–(168) respectively. This assumes that the NO$^+$ ions are produced with such small kinetic energy that no discrimination against them occurs in the mass spectrometer.

Reactions of ions with atomic nitrogen and oxygen have been measured by the flowing-afterglow method, the atomic radicals being introduced at known rates using the technique described on p. 2024.

The reaction $\qquad O_2^+ + N \to NO^+ + O + 4$ eV $\qquad (173)$

is found‡ to have a rate at 300 °K of $1.8 \pm 0.5 \times 10^{-10}$ cm^3 s^{-1}. For the

† WARNECK, P., *Planet. Space Sci.* **15** (1967) 1349.
‡ GOLDAN, P. D., SCHMELTEKOPF, A. L., FEHSENFELD, F. C., SCHIFF, H. I., and FERGUSON, E. E., *J. chem. Phys.* **44** (1966) 4095.

corresponding reactions of N_2^+ with O there are two possibilities:

$$N_2^+(X\,^2\Sigma_g^+)+O\,(^3P) \to NO^+(X\,^1\Sigma^+)+N+3\cdot05\text{ eV}, \qquad (174)$$
$$\to N_2(X\,^1\Sigma_g^+)+O^+(^4S)+1\cdot96\text{ eV}. \qquad (175)$$

Measurements were made† with the helium discharge pulsed so that, by suitable gating of the mass spectrograph, photo-ionization by helium resonance radiation did not affect the results.

Fig. 19.53 illustrates the variation of the ion current recorded by the mass spectrometer as a function of added concentration of atomic

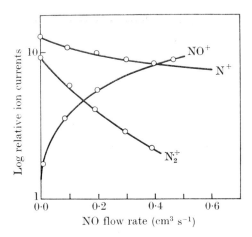

Fig. 19.53. Variation of ion currents resulting when O atoms are injected into a flowing afterglow containing N_2^+ and N^+ ions. The O atom concentration is proportional to the NO flow rate (see p. 2024).

oxygen. It will be seen that there is no evidence of the formation of O^+ ions. Even though such ions could be converted to NO^+ through the reaction (172) it can be estimated that this would not proceed to near completion under the experimental conditions. This seems to indicate that the rate of (175) is less than 0·1 of that for (174).

It was also necessary to verify that the reaction

$$N_2^+ + N \to N_2 + N^+ \qquad (176)$$

is much slower than (175), because the concentration of N atoms varies with the amount of NO added to the nitrogen afterglow to produce the O (see (120)). This was done by turning off the discharge which produces the N atoms and checking that this did not produce any appreciable change in the N_2^+ signal.

† FERGUSON, E. E., FEHSENFELD, F. C., GOLDAN, P. D., SCHMELTEKOPF, A. L., and SCHIFF, H. I., *Planet. Space Sci.* **13** (1965) 823.

The observed reaction rate for (174) is $2 \cdot 5 \times 10^{-10}$ cm³ s⁻¹. That for (175) is less than 10^{-11} cm³ s⁻¹.

3.5.7. *Reactions of nitrogen and of oxygen ions with nitric oxide.* The flowing-afterglow method has been applied to measure the rates of reactions of N_2^+, N^+, O_2^+, and O^+ ions with NO.

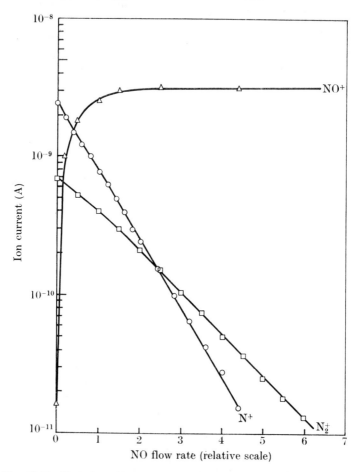

Fig. 19.54. Variation of ion currents with the inflow rate of NO added to a flowing afterglow containing N_2^+ and N^+ ions. ○ N^+, □ N_2^+, △ NO^+.

Fig. 19.54 shows the observed variation of N_2^+ and N^+ current with NO flow rate. From such observations as these,† over-all rate coefficients of 5×10^{-10} cm³ s⁻¹ and 8×10^{-10} cm³ s⁻¹ have been derived for

† GOLDAN, P. D., SCHMELTEKOPF, A. L., FEHSENFELD, F. C., SCHIFF, H. I., and FERGUSON, E. E., *J. chem. Phys.* **44** (1966) 4095; see also FEHSENFELD, F. C., SCHMELTEKOPF, A. L., and FERGUSON, E. E., ibid. **46** (1967) 2019.

the reactions with N_2^+ and N^+ respectively. With N_2^+ the possible reactions are

$$N_2^+(X\,^2\Sigma_g^+) + NO(X\,^2\Pi) \to N_2(X\,^1\Sigma_g^+) + NO^+(X\,^1\Sigma^+) + 6.3 \text{ eV}, \quad (177)$$
$$\to N_2(X\,^1\Sigma_g^+) + NO^+(a\,^3\Sigma^+) + 1.3 \text{ eV}, \quad (178)$$
$$\to N_2(A\,^3\Sigma_u^+) + NO^+(X\,^1\Sigma^+) + 0.1 \text{ eV}. \quad (179)$$

There is no evidence as to the branching ratio among these. Similar ignorance persists about the branching ratio for the reactions with N^+,

$$N^+(^3P) + NO(X\,^2\Pi) \to NO^+(X\,^1\Sigma^+) + N(^4S) + 5.3 \text{ eV}, \quad (180)$$
$$\to NO^+(a\,^3\Sigma^+) + N(^4S) + 0.35 \text{ eV}, \quad (181)$$
$$\to NO^+(X\,^1\Sigma^+) + N(^2D) + 2.9 \text{ eV}, \quad (182)$$
$$\to NO^+(X\,^1\Sigma^+) + N(^2P) + 1.7 \text{ eV}. \quad (183)$$

The over-all rate of the charge transfer reactions of O_2^+ with NO was measured by Warneck,† using his photo-ionization method with ionizing radiation at 980 Å. As for the N^+–O_2 reaction the measurements were carried out at low pressures so that the mean energy of the O^+ ions was around 0.15 eV. He obtained a rate coefficient of $7.7 \pm 1.5 \times 10^{-10}$ cm^3 s^{-1} which agrees well with the value 8×10^{-10} cm^3 s^{-1} measured in the flowing-afterglow experiments. In this case the possible reaction modes are

$$O_2^+(X\,^2\Pi_g) + NO(X\,^2\Pi) \to NO^+(X\,^1\Sigma^+) + O_2(X\,^3\Sigma_g^-) + 2.82 \text{ eV}, \quad (184)$$
$$\to NO^+(X\,^1\Sigma^+) + O_2(a\,^1\Delta_g) + 1.87 \text{ eV}, \quad (185)$$
$$\to NO^+(X\,^1\Sigma^+) + O_2(b\,^1\Sigma_g^+) + 1.22 \text{ eV}. \quad (186)$$

For the O^+–NO reaction the possibilities are

$$O^+(^4S) + NO(X\,^2\Pi) \to NO^+(X\,^1\Sigma^+) + O(^3P) + 4.35 \text{ eV}, \quad (187)$$
$$\to NO^+(X\,^1\Sigma^+) + O(^1D) + 2.39 \text{ eV}, \quad (188)$$
$$\to NO^+(X\,^1\Sigma^+) + O(^1S) + 0.18 \text{ eV}, \quad (189)$$

and the over-all rate coefficient according to the flowing-afterglow method is $< 2.4 \times 10^{-11}$ cm^3 s^{-1}.

3.5.8. *Summary of results for reactions of helium, nitrogen, and oxygen ions with* N_2, O_2, *NO, N, and O*. In Table 19.16 we summarize the results of measurements of the rates of the reactions we have discussed in §§ 3.5.2–3.5.7. For comparison the reaction rate (92), calculated on the assumption that the reaction takes place when the relative impact parameter is less than the critical value p_c for orbiting to occur, is also given in each case. For a number of reactions the rate estimated in this way gives good results but for others the observed rate is much smaller.

† loc. cit., p. 2045.

The rates given in the table refer to near-thermal conditions and do not include the results of observations taken at different temperatures, which are fully described in §§ 3.5.2–3.5.7.

3.5.9. *Reactions of nitrogen and oxygen ions with CO and CO_2.* The rates of the reactions of N_2^+, N^+ with CO and CO_2 and of O^+ with CO_2, which are of some interest in planetary atmospheric physics, observed by the flowing-afterglow method are given in Table 19.17. The N_2^+–CO_2 reaction has also been studied by Warneck using the photo-ionization

TABLE 19.16

Rate coefficients for thermal or near-thermal reactions of He^+, N^+, O^+, N_2^+, and O_2^+ ions with N_2, O_2, NO, N, and O

Reaction	ΔE (eV)	Rate coefficient (10^{-9} cm^3 s^{-1}) Obs.	Calc. (Giou-mousis and Stevenson)	Section reference
$He^+ + N_2 \to N^+ + N + He$	0.3	1.7†	1.7	3.5.4
$\to N_2^+ + He$	9.0	1.5,† 1.5§		
$He^+ + NO \to N^+ + O + He$	3.6	1.5†	1.6	3.5.5
$He^+ + O_2 \to O^+ + O + He$	5.9	1.5†, 1.1,‡ 1.2§	1.5	3.5.3
$N^+ + NO \to NO^+ + N$	5.3	0.8†	0.95	3.5.7
$N^+ + O_2 \to O_2^+ + N$	2.47	0.5,† 0.45§	0.9	3.5.6
$\to NO^+ + O$	6.70	0.5,† 0.5,‡ 0.16§		
$O^+ + N_2 \to NO^+ + N$	1.10	1.8×10^{-3},† 2.4×10^{-3},‡ 4.6×10^{-3}§	1.0	3.5.6
$O^+ + NO \to NO^+ + O$	4.36	≤ 0.025†	1.00	3.5.7
$O^+ + O_2 \to O_2^+ + O$	1.30	0.04,† 0.02,§ 0.02‡	0.9	3.5.3
$N_2^+ + N \to N^+ + N_2$	1.04	< 0.01†	0.82	
$N_2^+ + O \to NO^+ + N$	3.05	0.25†	0.65	3.5.6
$\to O^+ + N_2$	1.94	< 0.01†		
$N_2^+ + O_2 \to O_2^+ + N_2$	3.5	0.10,† 0.11,§ 0.20‡	0.76	3.5.6
$\to NO^+ + NO$	4.5	$< 3 \times 10^{-5}$§		
$N_2^+ + NO \to NO^+ + N_2$	6.3	0.50,† 0.48§	0.80	3.5.7
$O_2^+ + N \to NO^+ + O$	4.0	0.18†	0.8	3.5.6
$O_2^+ + NO \to NO^+ + O_2$	2.8	0.80,† 0.77§	0.75	3.5.7
$O_2^+ + N_2 \to NO^+ + NO$		$< 10^{-6}$,† 3×10^{-6}§	0.80	3.5.6

† Flowing-afterglow method.   ‡ Static-afterglow method.
§ Mass spectrograph with photo-ionization source.

For detailed references see the sections of the text as indicated in the last column.

method. His result for the over-all rate coefficient is given in Table 19.17 and agrees quite well with that obtained by the flowing-afterglow method. The O^+–CO_2 reaction

$$O^+(^4S) + CO_2(^1\Sigma) \to O_2^+(^2\Pi) + CO(^1\Sigma) + 1.5 \text{ eV} \qquad (190)$$

is of interest because it violates the Wigner spin conservation rule (Chap. 18, § 4.2.3) and yet has a very high rate, corresponding to a mean cross-section of 1.8×10^{-14} cm^2.

3.5.10. *Reactions of* C^+, CO^+, *and* CO_2^+ *with* O_2 *and* CO_2. These reactions are also of some interest in planetary atmospheric physics. Over-all rates determined by the flowing-afterglow method are given in Table 19.17.

TABLE 19.17

Rate coefficients for thermal reactions of N^+, N_2^+, O^+ *with* CO *and* CO_2 *and of* C^+, CO^+, *and* CO_2^+ *with* O_2 *and* CO_2

Reaction	$N_2^+ + CO \rightarrow CO^+ + N_2 + 1.57$ eV	$N_2^+ + CO_2 \rightarrow CO_2^+ + N_2 + 1.79$ eV
Rate coefficient 10^{-10} cm^3 s^{-1}	0.7†	9†
Reaction	$N^+ + CO \rightarrow CO^+ + N + 0.53$ eV	$N^+ + CO_2 \rightarrow CO_2^+ + N + 0.75$ eV
Rate coefficient 10^{-10} cm^3 s^{-1}	5†	13†
Reaction	$O^+ + CO_2 \rightarrow O_2^+ + CO + 1.5$ eV	
Rate coefficient 10^{-10} cm^3 s^{-1}	12‡	
Reaction	$C^+ + O_2 \rightarrow CO^+ + O + 3.27$ eV	$C^+ + CO_2 \rightarrow CO^+ + CO + 2.96$ eV
Rate coefficient 10^{-10} cm^3 s^{-1}	11.0§	19.0§
Reaction	$CO^+ + O_2 \rightarrow O_2^+ + CO + 1.94$ eV	$CO^+ + CO_2 \rightarrow CO_2^+ + CO + 0.22$ eV
Rate coefficient 10^{-10} cm^3 s^{-1}	2.0§	11.0§
Reaction	$CO_2^+ + O_2 \rightarrow O_2^+ + CO_2 + 1.72$ eV	
Rate coefficient 10^{-10} cm^3 s^{-1}	6.8‖	

3.5.11. *Reactions of* Ar^+ *ions*. We conclude with a comparison of results obtained by the flowing-afterglow†† and photo-ionization techniques‡‡ for the rate coefficient of a number of reactions involving Ar^+ ions. To make such measurements by the former method the helium is replaced by argon. Any Ar^+ ions produced in the $^2P_{\frac{1}{2}}$ state will be de-excited to the $^2P_{\frac{3}{2}}$ ground state by electron impact before reaction so that the flowing-afterglow results refer to reactions with $Ar^+(^2P_{\frac{3}{2}})$.

The photo-ionization measurements were carried out using 780 Å radiation, which is just above the threshold wavelength (778 Å) for

---

† FEHSENFELD, F. C., SCHMELTEKOPF, A. L., and FERGUSON, E. E., *J. chem. Phys.* **44** (1966) 4537.
‡ FEHSENFELD, F. C., FERGUSON, E. E., and SCHMELTEKOPF, A. L., ibid. **44** (1966) 3022.
§ FEHSENFELD, F. C., SCHMELTEKOPF, A. L., and FERGUSON, E. E., ibid. **45** (1966) 23.
‖ Idem, unpublished results.
†† FEHSENFELD, F. C., FERGUSON, E. E., and SCHMELTEKOPF, A. L., ibid. **45** (1966) 404.
‡‡ WARNECK, P., ibid. **46** (1967) 513.

production of $Ar^+(^2P_{\frac{1}{2}})$ so that again the results should refer to reactions with $Ar^+(^2P_{\frac{3}{2}})$. The ion source was operated at low pressures so the motion of the ions is mainly one of acceleration by the repeller field, collisions being negligible. With this arrangement the average energies of the reacting ions were between 0·11 and 0·17 eV.

TABLE 19.18

Observed rates of reactions of Ar^+ ions with various molecules

Reaction Rate coefficient (10^{-10} cm³ s⁻¹)	$Ar^+ + N_2 \rightarrow N_2^+ + Ar + 0·18$ eV		$Ar^+ + NO \rightarrow Ar + NO^+$
Flowing afterglow	6×10^{-2}		—
Photo-ionization	0·66		3·9
Theoretical	7·60		7·8
	$Ar^+ + CO \rightarrow CO^+ + Ar + 1·75$ eV		$Ar^+ + CO_2 \rightarrow CO_2^+ + Ar + 1·97$ eV
Flowing afterglow	0·9		7·6
Photo-ionization	1·25		7·0
Theoretical	8·10		8·7₅
	$Ar^+ + O_2 \rightarrow O_2^+ + Ar + 3·68$ eV		$Ar^+ + H_2 \rightarrow ArH^+ + H$
Flowing afterglow	1·1		11
Photo-ionization	1·1		—
Theoretical	6·9		15

It will be seen that, apart from the reaction with N_2, there is remarkably good agreement between the results obtained by the two methods, particularly when it is remembered that the flowing-afterglow measurements refer to thermal energies, those with the photo-ionization technique to somewhat higher energies.

The other striking feature is that in almost all cases the observed rate coefficient is much smaller than that expected on the theory of Gioumousis and Stevenson,† as may be seen from Table 19.18. The reason for this is not clear.

4. Reactions involving negative ions

We now consider the mobilities and reactions of negative ions in gases. Of special interest in this regard are the ions O^- and O_2^- as these are the primary negative ions formed in the upper atmosphere by attachment of free electrons to gas atoms and molecules. The ionic composition of the atmosphere at any altitude and epoch therefore depends strongly on the rates of reactions of O^- and O_2^- ions with the neutral molecules present. This is one of the reasons why attention has been concentrated

† loc. cit., p. 2005.

on these ions, but it is also partly due to the fact that, while oxygen is not the easiest substance to work with in ionic experiments, it is certainly less troublesome than the halogens. From the experimental point of view, since oxygen forms negative ions quite readily, it is the natural gas to work with, at least initially. Because of this we shall be mainly, but not exclusively, concerned with the negative ions of oxygen.

We discuss first experiments that have been carried out on the mobilities of negative ions formed in oxygen. As with positive ions this leads naturally into the study of ionic reactions. In addition to the techniques available for positive ions, such as the drift tube plus mass spectrometer and the flowing afterglow, the pulse technique of Chanin, Phelps, and Biondi[†] used for studying attachment, discussed in Chapter 12, § 7.5.2, may be extended to study certain types of reactions involving negative ions. We shall conclude with an account of the application of these methods to the measurement of reaction rates.

4.1. *Mobility of negative ions in oxygen*

The mobilities of negative ions may be measured by similar techniques to those used for positive ions, such as the shutter methods of Tyndall and Powell[‡] and of Bradbury and Nielsen[§] and the pulse method of Hornbeck.[||] McDaniel and Crane[††] have used an interesting technique to measure the mobilities of a number of negative ions in pure oxygen and oxygen mixtures. This work was carried out to provide data likely to assist in the interpretation of the behaviour of a Geiger counter filled with carbon dioxide.

Fig. 19.55 shows a cross-section through the experimental chamber. The central feature is a drift tube consisting of a series of metal rings to which potentials can be applied to give a uniform longitudinal field. Ions are produced in a slot between two neighbouring rings by an alpha-particle source. Passage of the alpha particle is recorded by the proportional counter 1. The alpha-particle source is movable in a direction parallel to the axis of the drift tube so that the ions may be formed in any chosen slot. Once formed, the negative ions drift along the tube to be recorded by the proportional counter 2. To record the times involved, the horizontal sweep of a synchroscope is triggered by the passage of the alpha particle across the drift tube, while pulses from the proportional counter 2 are impressed on the grid of the cathode-ray tube, thereby brightening the trace at points indicating the time of arrival of the ions. Due to such effects as straggling, ionic reactions, etc., all of the ions produced by one alpha particle do not arrive simultaneously at the counter. A camera is mounted in front of the synchroscope with shutter open throughout a run so as to integrate over a large number

---

† CHANIN, L. M., PHELPS, A. V., and BIONDI, M. A., *Phys. Rev. Lett.* **2** (1959) 344.
‡ loc. cit., p. 1937.   § loc. cit., p. 58. || loc. cit., p. 1938.
†† McDANIEL, E. W. and CRANE, H. R., *Rev. scient. Instrum.* **28** (1957) 684.

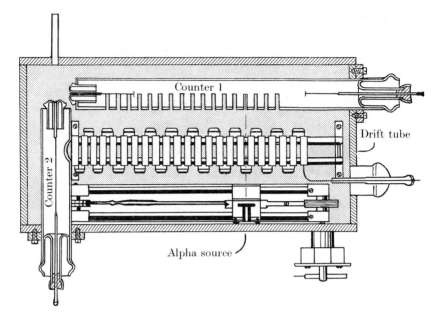

Fig. 19.55. Sectional drawing of the mobility chamber of the apparatus used by McDaniel and Crane.

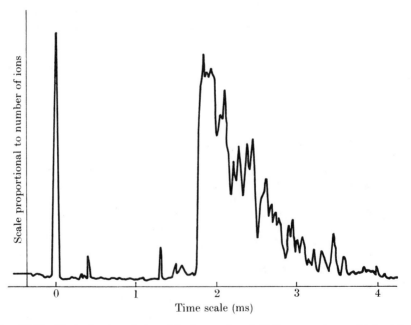

Fig. 19.56. Typical microphotometric trace of a drift-time spectrum obtained by McDaniel and Crane.

of events. Fig. 19.56 shows a typical microphotometric trace which is obtained in this way. The time of arrival of the fastest ion of appreciable abundance is read off from the interval between the record of the passage of the alpha particle and the front edge of the drift-time 'spectrum'.

The drift tube consisted of twenty-three brass cylinders, 1 inch in internal diameter and $\frac{1}{4}$ in long, spaced $\frac{1}{8}$ in apart, while the alpha-particle source consisted of 10 μC of polonium on a platinum disc 7 mm in diameter covered with mica at a surface density of 1 mg cm^{-2}.

Although it was not possible to bake the apparatus, the metal assembly was cleaned with a mixture of nitric and sulphuric acid, washed with distilled water, and pumped for several days. The range of gas pressures used was from 100 to 250 torr and the range of F/p from 0·11 to 0·79 V cm^{-1} torr^{-1}.

Measurements were carried out for mixtures of oxygen with the rare gases, hydrogen, nitrogen, and carbon dioxide. In all cases the fastest ion was the most abundant. The mobility μ_f for a mixture of oxygen with a gas X in which the percentage of oxygen is 100f, is given by

$$\frac{1}{\mu_f} = \frac{f}{\mu_{O_2}} + \frac{1-f}{\mu_X}, \qquad (191)$$

where μ_{O_2} and μ_X are the respective mobilities in pure oxygen and in the pure gas X. Hence, if the reciprocal of the observed mobility is plotted as a function of f, a straight line should result. If this is done for various gases X all these lines should extrapolate to the same point as f tends to unity since this limit is the reciprocal of the mobility in pure oxygen. Fig. 19.57 shows how well the observed data for mixtures with rare gases follow these expectations. Similar results were obtained for mixtures with H_2, N_2, and CO_2.

The mobility obtained for the fastest ion in pure oxygen was 2·46±0·05 cm^2 V^{-1} s^{-1}. No information about the nature of this ion is forthcoming from these experiments. The linearity of the reciprocal mobility plots and the fact that they all tend to the same limit when f tends to 1 strongly indicate that it is the same ion in all cases. As we shall see below it is almost certainly O_3^- (see Fig. 19.68). Further evidence against interpretation of the ion as one associated with an impurity is provided by the fact that the same results were obtained using oxygen from quite different sources.

Table 19.19 gives the mobility of the ion in pure gases X, determined by extrapolation of reciprocal mobility plots to zero f.

TABLE 19.19

Mobilities of oxygen negative ions (O_3^-) in various gases

Gas	He	Ne	Ar	Kr	Xe	H_2	N_2	CO_2
Mobility (cm^2 V^{-1} s^{-1})	11·8 ±1·2	5·14 ±0·5	2·06 ±0·04	1·39 ±0·07	0·88 ±0·10	11·4 ±1	2·22 ±0·06	1·08 ±0·03

We now turn to consider results obtained for the mobilities of negative oxygen ions obtained by other observers. It is to be expected from the shape of the drift-time spectrum shown in Fig. 19.56 that ionic reactions take place during drift and that more than one kind of ion exists.

At room temperature, three ions with different mobilities have been observed in oxygen. Fig. 19.58 summarizes results obtained in a number of different experiments. From this figure it seems that there are three types of ions, which were referred to as A, B, C respectively.

It is now well established that the ions should be identified as O^-, O_3^-, and O_2^-, respectively, the corresponding mobilities being 3·0, 2·5, and 2·4 cm² V^{-1} s^{-1}. The first direct evidence came from experiments

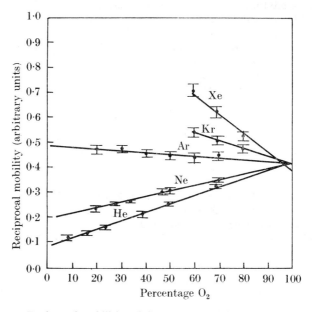

FIG. 19.57. Reciprocal mobilities of the oxygen negative ion in mixtures of O_2 with each of the rare gases, as functions of the percentage of O_2 in the mixture.

carried out by Eiber in which rough estimates of e/M for each set of ions could be made (see p. 2072). Further evidence came from experiments on electron detachment from negative oxygen ions (see p. 2088), and finally, by operating a mass spectrometer in conjunction with a drift tube, Moruzzi and Phelps† were able to confirm the identification of the ions and provide new detailed information about the reactions in which they are involved. Wobschall, Fluegge, and Graham,‡ using the cyclotron resonance technique, found 4·3 and 2·1 cm² V^{-1} s^{-1} for O^- and O_2^- respectively.

Before discussing this interpretation we shall describe briefly the techniques used in the different experiments.

† loc. cit., p. 2077. ‡ loc. cit., p. 2028.

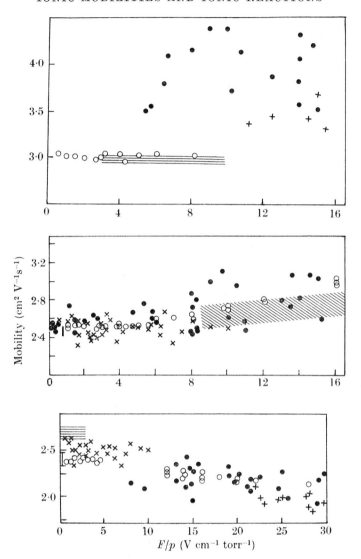

FIG. 19.58. Mobilities of the three types A, B, C of negative ions as functions of F/p observed in oxygen by ○ Rees, + (type A and C) Burch and Geballe, ▨ (type B) Burch and Geballe, ● Eiber, ≡ Chanin, Phelps, and Biondi, × Doehring, | McDaniel and Crane, ⊥ Voshall, Pack, and Phelps.

Burch and Geballe† used a method similar to that of Hornbeck (§ 2.2.2). They generated the oxygen from $KMnO_4$ and, from mass spectrographic analysis, estimated the partial impurity content to be less than 4×10^{-5}.

† BURCH, D. S. and GEBALLE, R., *Phys. Rev.* **106** (1957) 183.

The pulse methods described in Chapter 12, § 7.5.2 primarily for determining the attachment coefficient for a swarm of electrons drifting in a gas under the action of a uniform electric field may also be used for measuring the mobilities of the negative ions formed. Thus Doehring[†] used the method described in Chapter 12, p. 864 while Chanin, Phelps, and Biondi[‡] used that described in Chapter 12, p. 866, in which only one shutter electrode is used. Voshall, Pack, and Phelps[§] repeated the latter experiment using a two-shutter system similar in principle to that of Doehring. The grid shutters were spaced at distances 5·1 and 10·2 cm from the photo-cathode and the delay time between light and grid pulses could be varied from 3 to 300 ms. The mobility tube, manifold, and associated valves were baked out at 400 °C until a pressure of less than 10^{-7} torr could be attained on cooling. Under these circumstances the pressure rose by less than 10^{-8} torr/min. The oxygen was admitted through a small-diameter copper and stainless steel line which could be baked out, a metal needle valve, and a bakeable high-vacuum valve. As supplied the most abundant impurity was 0·04 per cent of nitrogen. By immersing the tube in different temperature baths, measurements of mobility could be made at temperatures of 77·4, 87·6, 195, and 373 °K as well as at room temperature.

With the geometry used, electrons attached close to the photo-cathode and measurements were confined to such low values of F/p that detachment was unimportant (see § 4.2.2). Under these conditions the form of the pulses of ion current received at the collector when voltage pulses were applied to the respective grids took the simple form shown in Fig. 19.59. The drift time between the grids is then taken as the interval between the current minima associated with the pulses applied to the respective grids. This eliminates end-effects that occur when a single grid is used. Comparison with the earlier results of Chanin, Phelps, and Biondi, who used a single grid, shows that the latter authors obtained somewhat too high mobilities at room temperature. We shall discuss results obtained at other temperatures in § 4.2 below.

Rees[||] used both the Doehring pulse method and the method of Bradbury and Nielsen (p. 1937). Fig. 19.60 shows a typical record of the collected ion current as a function of the frequency of the sinusoidal voltage applied to the shutters (see Chap. 2, § 3.1). The presence of four ions is indicated. One of these, that of lowest mobility, is ascribed by

[†] DOEHRING, A., *Z. Naturf.* **7a** (1952) 253.
[‡] CHANIN, L. M., PHELPS, A. V., and BIONDI, M. A., *Phys. Rev.* **128** (1962) 219.
[§] VOSHALL, R. E., PACK, J. L., and PHELPS, A. V., *J. chem. Phys.* **43** (1965) 1990.
[||] REES, J. A., *Aust. J. Phys.* **18** (1965) 41.

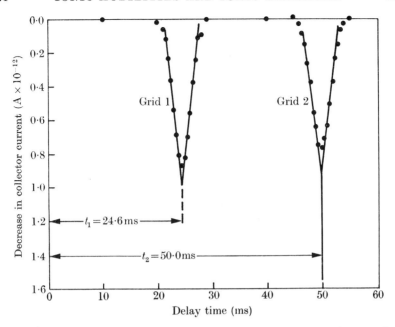

Fig. 19.59. Typical ion current wave-form observed in the experiments of Voshall, Pack, and Phelps for $F/p = 0.1$ V cm^{-1} torr^{-1} ($F/N = 3.1 \times 10^{-18}$ V cm^2), $p = 228$ torr.

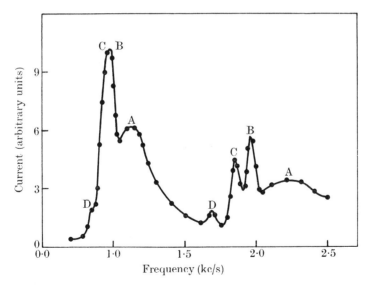

Fig. 19.60. Typical ion current–frequency curve observed by Rees for negative oxygen ions in O_2 using the Bradbury–Nielsen technique ($F/p = 4.7$ V cm^{-1} torr^{-1}, $p = 41.3$ torr). Contributions from ions of types A, B, and C are indicated as well as from a further ion labelled D, probably due to an impurity.

Rees to an impurity. The other three are assignable to the types A, B, and C.

Eiber carried out two different experiments. In the first† he measured the mobilities of the ions in two ways while in the second‡ he determined the nature of the ions by an ingenious method. For the mobility measurement he used the electrode system shown in Fig. 19.61. The ions entered the drift tube proper through the grid G_0 and diffused through the equally-spaced grids G_1–G_6 to be collected at K. In the first mode of operation, similar in many ways to the method of Biondi and Chanin (see § 2.2.3), the ion source was pulsed. Breakdown was produced between the platinum ring electrodes S_1, S_2 by pulses of either 0·6 or 10 μs duration at a repetition rate of 200 s^{-1}. The ions were extracted by a strong electric field between S_2 and G_0 and drifted under the action of the uniform electric fields between G_1 and G_6. Passage of the ion cloud through each grating was recorded from the induced current pulse, using an integrating circuit and an oscillograph with time base triggered by the breakdown pulse on the ion source. Fig. 19.62 shows a superposition of typical records of pulses induced at each successive grating. From the time intervals between successive maxima the mobility could be determined.

The second mode of operation was basically similar to the filter method of Bradbury and Nielsen (see Chap. 2, § 3.1) or even more closely to that of Doehring. The gratings G_1 and G_2 were connected to one phase and the grating G_3 to the other phase of a symmetrical square-wave voltage generator with an adjustable repetition frequency. Ions of a particular drift velocity were unable to pass from G_1 through G_5 unless their time of passage between G_1 and G_3 was synchronized to half the repetition frequency. For these experiments the ion source was a cylindrical ring of platinum foil with some inward point projections from which a corona discharge was produced by application of a high d.c. voltage.

Measurements were carried out over a pressure range from 0·7 to 545 torr, the results being as shown in Fig. 19.63.

The second experiments carried out by Eiber‡ were directed towards the identification of the ions with the three different mobilities. For this purpose he used a development of the electron filter first introduced by Loeb and Cravath, and described in Chapter 2, § 3.1 (see also Chap. 12, § 7.5.1). This filter was devised so that electrons could be collected from a mixed swarm of electrons and negative ions without at

† EIBER, H., Z. angew. Phys. 15 (1963) 103.
‡ ibid. 461.

the same time collecting the ions. It consisted of a grid of copper between alternate wires of which a high-frequency voltage could be applied. By choosing the amplitude and frequency of this voltage, electrons could be collected while the much more massive ions passed through without

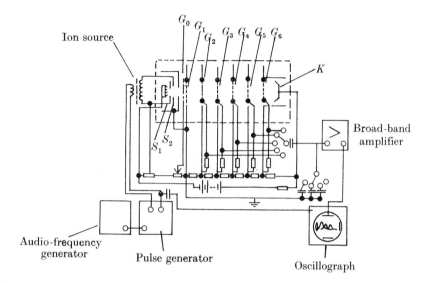

Fig. 19.61. Arrangement of equipment in Eiber's measurements, by the pulse method, of the mobilities of negative oxygen ions in O_2.

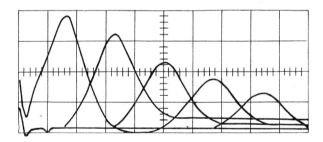

Fig. 19.62. Typical form of the pulses recorded at each grating in Eiber's experiments (see Fig. 19.61).

suffering sufficient sideways deviation on the average to reach some of the grid wires. Eiber designed a filter capable of distinguishing in its collecting power between ions of different mass, a much more difficult problem.

The grid wires were chosen to be of such cross-sectional diameter D and separation d that $l \gg D > d$, where l is the mean free path of the ions in the gas. This is in contrast to the Loeb electron filter for which

d was as large as 2 mm, considerably greater than l. Thus, whereas in this filter the collecting power depends on the amplitude of the field strength F and the mobility, in Eiber's grid it depends on the field strength F, the angular frequency ω of the alternating field, and the mass M of the ion. The motion of an ion through the effective slit of

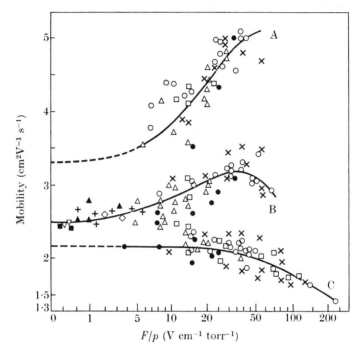

FIG. 19.63. Mobilities of the three types of negative ions observed in O_2, as functions of F/p, measured at different pressures by Eiber. $\times$ 0·7, $\square$ 0·8, $\bigcirc$ 2·4, $\bullet$ 4·9, $\triangle$ 6·4, $\diamondsuit$ 9·6, $+$ 19, $\blacktriangle$ 47, $\triangledown$ 260, $\blacksquare$ 545 torr.

width d between a pair of wires is unaffected by collisions, so the amplitude of the transverse displacement suffered by the ions is given by

$$A = eF/M\omega^2, \qquad (192)$$

where e is the charge on the ion. We can expect that, as F increases, the current of those ions transmitted through the grid will fall linearly to zero when $F = \frac{1}{2}M\omega^2 d/e = F_\mathrm{m}$, say. If there are two or more types of ions present the variation of total transmitted current with the amplitude of the applied field will be as shown in Fig. 19.64. At a field strength for which $F = F_\mathrm{m}$ there will be a discontinuity of slope of the current–field strength characteristic. From observation of this discontinuity the corresponding mass M may be derived. The choice of geometry of the grid system and of the frequency ω should be such as to make the location of the discontinuity as definite as possible.

For this reason ω must be sufficiently large for the wavelength λ of the transverse oscillations in the path of an ion to be smaller than about $2d$. This ensures that at least one maximum of sideways displacement occurs in the region between the wires in which field inhomogeneities are not marked. On the other hand, the frequency must not be so high

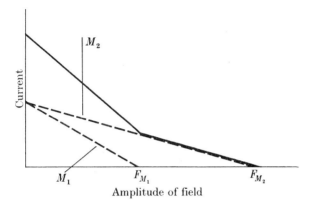

Fig. 19.64. Idealized form of the total transmitted current as a function of the amplitude of the applied high-frequency electric field, in Eiber's experiments, when ions of masses M_1, $M_2 = 2M_1$ are present. The broken lines are the contributions from ions of mass M_1, M_2 respectively.

that the total path of an ion in the slit region becomes comparable with the mean free path.

For application to ions in oxygen the filter grid was constructed of tungsten wires 0·031 mm in diameter with axial separation 0·04 mm so the minimum distance between wires was only 0·009 mm. If the ion energies are effectively thermal the minimum frequency to secure $\lambda \leqslant 0 \cdot 02$ mm is 68 Mc/s. With this choice the amplitude of the voltage oscillations between successive grid wires to reduce the O⁻ current to zero will be 1·23 V.

Fig. 19.65 illustrates the arrangement of electrodes used in the oxygen experiments. The ion source S was a glow discharge between a thin wire and a small platinum plate. The ions drifted through the gas under the action of the uniform electric field between the grids G_1, G_2, and G_3 to reach the filter grid G_4. Those ions which penetrated G_4 were collected on the electrode C.

Typical collector current–alternating voltage characteristics are shown in Fig. 19.66. The first of these refers to positive ions and indicates that

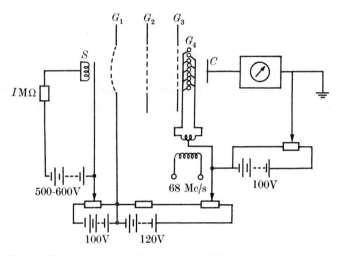

FIG. 19.65. Arrangement of electrodes in Eiber's experiment on the identification of the negative oxygen ions in O_2.

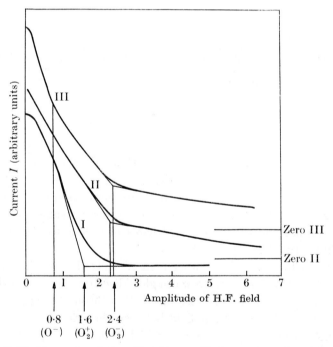

FIG. 19.66. Observed variation of total transmitted current as a function of the amplitude of the applied high-frequency electric field in Eiber's experiments on the ions in O_2. I, Positive ions, $F/p = 20$ V cm^{-1} torr^{-1}. II, Negative ions, $F/p = 5$ V cm^{-1} torr^{-1}. III, Negative ions, $F/p = 10$ V cm^{-1} torr^{-1}.

these are definitely O^+ (see § 3.5.3). The other two characteristics refer to negative ions. At the higher value of F/p (curve III in Fig. 19.66) two discontinuities are discernible which are located as expected for O^- and O_3^- ions respectively. There is no evidence for the presence of O_2^-. Curve II taken at a lower F/p clearly shows the presence of O_3^- but only very slight evidence of O^-.

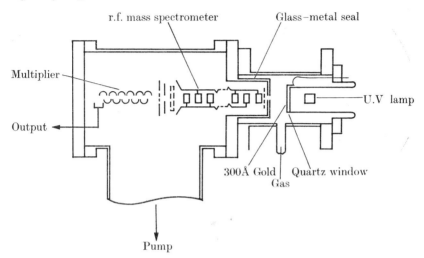

FIG. 19.67. General arrangement of the drift tube and mass spectrometer used by Moruzzi and Phelps.

In later investigations, with a higher resolving power obtained by using finer grid wires, smaller and more precisely equal separations between the wires, and a higher frequency, evidence of the presence of O_2^- at high F/p was found.

These experiments went far towards confirming definitely the nature of the negative ions of type A, B, and C as suggested by Burch and Geballe.[†] The final stage in this type of experiment was reached when Moruzzi and Phelps[‡] were able to operate successfully a mass spectrometer with a drift tube so as to study the passage of negative ions through a gas in which they were produced from electrons by attachment. Fig. 19.67 shows the general arrangement of their apparatus, the principle of which is essentially the same as for experiments with positive ions (see § 3.4.2).

The most satisfactory electron source for operation in oxygen was a photoelectric one constructed by depositing a thin layer (about 300 Å

[†] loc. cit., p. 2069.
[‡] MORUZZI, J. L. and PHELPS, A. V., *J. chem. Phys.* **45** (1966) 4617.

thick) of gold or palladium on a quartz window. By illuminating this window from the rear with a 10–20-W u.v. lamp, electron emission currents of 10^{-8} A were obtained over periods of weeks when gases such as oxygen were present in the drift tube. The drift distance, between the source and the anode, between which a uniform electric field could be maintained, was 2·35 cm. A hole 0·01 inch in diameter in the centre of the anode allowed drifting ions to pass through to the r.f. mass

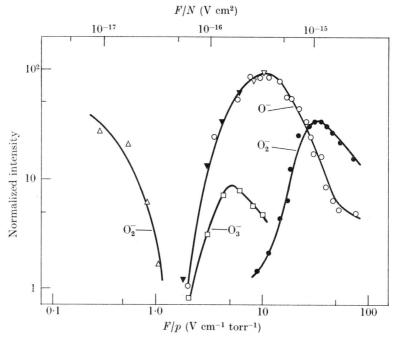

FIG. 19.68. Relative intensity of different negative ions in O_2 as a function of F/p, as observed by Moruzzi and Phelps. $p = 1\cdot76$ torr ○ ● ; $p = 3\cdot37$ torr □ △ ▼.

spectrometer via a cylindrical accelerating and focusing lens. Ions selected by the spectrometer were detected by an electron multiplier, the currents at the output of which were of the order 10^{-12} A. The mass spectrum was scanned by mechanically sweeping the radio-frequency. In these experiments it was assumed that the collecting power of the spectrometer and detection efficiency of the multiplier were the same for all ions.

The entire system could be baked at 400 °C so that the background pressure was less than 10^{-8} torr.

Fig. 19.68 shows typical observations of the relative intensity of different ions as a function of F/p in pure oxygen.

The ions of type A, the O^- ions, are relatively abundant at low pressures and drift distances and low to intermediate values of F/p. This is to be expected if these ions are primary ions formed by dissociative attachment of electrons to O_2 molecules (see Chap. 13, § 4.4.2).

Ions of type B, the O_3^- ions, must be formed by a secondary process. As they tend to be abundant at high pressures and intermediate F/p it is likely that they are produced in some exothermic or only slightly endothermic reaction. One such possibility is

$$O^- + 2O_2 \rightarrow O_3^- + O_2. \tag{193}$$

If $A(O_3)$, the electron affinity of O_3, is 0·4 eV this will be exothermic.

Under the conditions in which O_3^- ions are prominent we can, in analysing the relative abundance of O^- and O_3^- ions as observed in the experiments of Moruzzi and Phelps,† ignore all processes but the dissociative attachment and the reaction (193).

If n_e, n_1, n_3 are the concentrations of electrons, O^-, and O_3^- at a point x from the source along the axis of the drift tube, and u_e, u_1, u_3 their corresponding drift velocities, we have

$$u_e \frac{dn_e}{dx} = -k_a N n_e, \tag{194}$$

$$u_1 \frac{dn_1}{dx} = k_a N n_e - k_t N^2 n_1, \tag{195}$$

$$u_3 \frac{dn_3}{dx} = k_t N^2 n_3. \tag{196}$$

Here k_a, k_t are the rate coefficients for dissociative attachment and (193) respectively, and N is the concentration of oxygen molecules. Solving these equations subject to the boundary conditions $n_e = n_0$, $n_1 = n_3 = 0$, $x = 0$, we find that at the anode, which is at a distance l from the source,

$$\frac{u_3 n_3}{u_1 n_1} = \frac{\alpha_t\{1-\exp(-\alpha_a l)\} - \alpha_a\{1-\exp(-\alpha_t l)\}}{\alpha_a\{\exp(-\alpha_a l) - \exp(-\alpha_t l)\}}, \tag{197}$$

where $\alpha_a = k_a N/u_e$, $\alpha_t = k_t N^2/u_1$. At low pressures p,

$$\frac{u_3 n_3}{u_1 n_1} = \tfrac{1}{2} k_t N^2 l/u_1 \tag{198}$$

and so should be proportional to p^2.

Fig. 19.69 illustrates the observed pressure dependence of i_3/i_1, the mass-analysed ion current ratio, for $F/p = 4\cdot8$ V cm^{-1} torr^{-1}. For

† loc. cit., p. 2077.

pressures below 3 torr it varies as p^2 as expected. A good fit over the whole pressure range studied (up to 6 torr) is obtained if α_t/N^2 is taken as $1\cdot0\times10^{-35}$ cm^5 in conjunction with measured values of k_a and u_e in (198) as may be seen from Fig. 19.69. Taking u_1 at the particular value

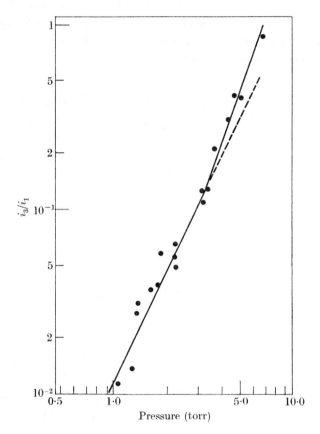

Fig. 19.69. Variation with O_2 pressure p of the ratio i_3/i_1 of the currents i_3, i_1 of O_3^- and O^- ions observed by Moruzzi and Phelps for $F/p = 4\cdot8$ V cm^{-1} torr^{-1} ($F/N = 1\cdot49\times10^{-16}$ V cm^2). ———— calculated from (197) with appropriate choice of rate constants. – – – p^2 dependence of the ratio, valid at low pressures.

of F/p as given by Burch and Geballe[†] k_t is found to be $1\cdot4\times10^{-31}$ cm^6 s^{-1}. This is considerably smaller than the value found by Beaty, Branscomb, and Patterson,[‡] 9×10^{-31} cm^6 s^{-1}. The reason for this discrepancy is not clear.

[†] loc. cit., p. 2069.
[‡] BEATY, E. C., BRANSCOMB, L. M., and PATTERSON, P. L., Bull. Am. phys. Soc. 9 (1964) 535.

The O_2^- ions, type C, which are observed at low F/p, are probably primary ions produced by the three-body process discussed in Chapter 13, § 4.4.3. Secondary O_2^- ions could be produced through the charge-transfer process
$$O^- + O_2 \to O + O_2^-. \tag{199}$$

This is endothermic by an amount equal to the difference of the electron affinities $A(O)$ and $A(O_2)$ of O and O_2 respectively, which is likely to be as high as 1 eV (see Chap. 13, § 4.4.1 and pp. 1015–16, and § 4.2.2.3 of this chapter). We would therefore expect O_2^- ions, produced in this way, to appear only at values of F/p high enough for the mean energy of the O^- ion to be comparable with 1 eV. According to Wannier's formula (1) this requires F/p to be comparable with 50 V cm^{-1} torr^{-1}. Actually O_2^- ions were observed by Moruzzi and Phelps† for $F/p > 10$ V cm^{-1} torr^{-1} (see Fig. 19.68). Furthermore, they found that the ratio $i(O_2^-)/i(O^-)$ of the currents of O_2^- and O^- ions at low pressures was consistent with the linear dependence on pressure required if (199) is the dominant source of O_2^-. Assuming that these ions do arise in this way Moruzzi and Phelps analysed the results of their observations at different pressures, allowing for electron production by ionization‡ and by impact detachment from O^-§ (see § 4.2.2.4). Using the observed mobilities of O^- they found rate constants for (199) of $1\cdot 8 \times 10^{-13}$ and $5\cdot 5 \times 10^{-13}$ cm^2 for $F/p = 25$ and 30 V cm^{-1} torr^{-1} respectively. This shows a strong dependence of the rate on the mean ion energy, as would be expected for an endothermic reaction. In fact, Moruzzi and Phelps showed that an even faster increase of the rate constant with F/p would be expected on the assumption of a constant cross-section for (199) above the energy threshold and a Maxwellian energy distribution of the O^- ions about the mean value given by Wannier's formula (1).

According to the simple Langevin theory (§ 2.4) the zero-field mobility of O^- ions should be $3\cdot 35$ cm^2 V^{-1} s^{-1}, which is not too different from the observed value which is close to $3\cdot 0$ cm^2 V^{-1} s^{-1}. Even better agreement is found for O_3^-, the Langevin theory giving $2\cdot 49$ cm^2 V^{-1} s^{-1}, which is not significantly different from the mean of the observed values. For O_2^- there is a substantial discrepancy, Langevin's theory giving $2\cdot 73$ as compared with $2\cdot 40$ cm^2 V^{-1} s^{-1} as observed. This is not unexpected because of the possibility of resonance charge transfer
$$O_2^- + O_2 \to O_2 + O_2^-, \tag{200}$$

† loc. cit., p. 2077.
‡ PRASAD, A. N. and CRAGGS, J. D., *Proc. phys. Soc.* **77** (1961) 385.
§ FROMMHOLD, L., *Fortschr. Phys.* **12** (1964) 597.

which will certainly reduce the mobility. As pointed out on p. 2043 the observed mobility of O_2^+ is very close to that of O_2^-, which is a little surprising in that the relatively diffuse charge distribution of the negative ion should lead to a contribution from charge transfer appreciably different from that for O_2^+.

The observed temperature variation of the mobility of O_2^- is shown in Fig. 19.70. At first, as the temperature falls, the mobility increases in the manner characteristic of an ion which may suffer resonant charge exchange (see Table 19.11). However, a maximum is reached at about 370 °K and thereafter the mobility falls with the temperature.

At temperatures T above 195 °K the product μp of the mobility and gas pressure is independent of p over a range 20–700 torr in pressure and 0·04 to 1·2 V cm^{-1} torr^{-1} in F/p. However, at the lower temperatures, 77·4 and 87·7 °K, for which observations were carried out it was found that μp, while independent of F/p (in the range 0·025 to 3·5 V cm^{-1} torr^{-1}) for a fixed p and T, is greater at given T the smaller the value of p. This is illustrated in Fig. 19.71 for $T = 87·6$ °K.

To interpret these results Voshall, Pack, and Phelps† assume that, under these conditions, a statistical equilibrium is set up between O_2^- and O_4^- ions through the reaction

$$O_2^- + O_2 \rightleftharpoons O_4^-. \tag{201}$$

This is a similar situation to that discussed for N_2^+ and N_4^+ ions by Varney (see p. 2041). However, whereas Varney and others were able to determine the mobilities of both N_2^+ and N_4^+ under conditions in which they were present alone, Voshall, Pack, and Phelps were unable to work at sufficiently high densities for complete conversion to O_4^-, because of condensation of the oxygen. Nevertheless, at each temperature the mobility of O_4^- as well as the equilibrium constant K were treated as adjustable variables to obtain the best fit to the observed variation of μp with $1/p$ (see Fig. 19.71). The values found for the normalized mobility of O_4^- are given in Fig. 19.70. They are somewhat lower than would be expected from the Langevin theory but this may be due to the low temperature.

From the values of the equilibrium constant at the two temperatures, the Nernst equation (130) may be used to obtain the energy E of dissociation of O_4^- into $O_2^- + O_2$. In contrast to the nitrogen positive ion equilibrium discussed by Varney, the ion temperature in the oxygen case is effectively equal to the gas temperature. Thus, whereas Varney

† loc. cit., p. 2070.

used the observed mobilities as functions of F/p at three different temperatures and the same pressure, Voshall, Pack, and Phelps used the mobilities observed at different pressures at two different temperatures, these being independent of F/p in the range considered.

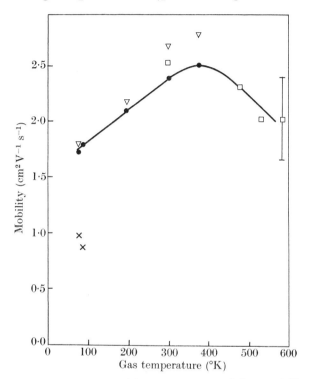

FIG. 19.70. Observed variation with gas temperature of the zero-field mobility of O_2^- and O_4^- ions in O_2. O_2^-: ▽ Chanin, Phelps, and Biondi (loc. cit., p. 2070); □ Phelps and Pack (loc. cit., p. 2085); ● Voshall, Pack, and Phelps (loc. cit., p. 2070). O_4^-: × (see p. 2082).

Assuming completely free internal rotation but no vibration in the molecules concerned, the dissociation energy of O_4^- which gives a good fit to the derived values of the equilibrium constant is 0·06 eV. The formation of complex negative ions in CO_2, H_2O, and in O_2–CO_2 and O_2–H_2O mixtures is discussed in § 4.4.

4.2. *Detachment of electrons from negative ions in collisions with gas molecules*

4.2.1. *Introduction.* Electrons may be directly detached from negative ions in collisions with gas molecules by reactions such as

$$X^- + Y \rightarrow X + Y + e, \tag{202}$$

in which the energy required to detach the electron is obtained by transfer from kinetic energy of relative motion. At ordinary temperatures T the fraction of gas molecules possessing sufficient translational energy will be small, unless the electron affinity $A(X)$ of X is comparable with $\tfrac{3}{2}\kappa T$ or the electric field accelerating the ions is large enough for them to acquire a mean kinetic energy comparable with $A(X)$.

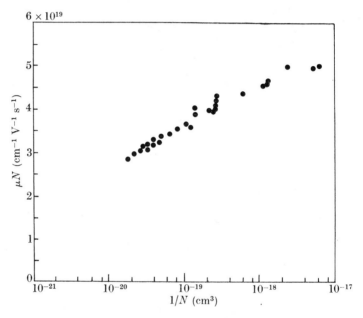

FIG. 19.71. Variation of the product μN with $1/N$ for negative oxygen ions in O_2 at 87·6 °K, μ being the zero-field mobility and N the concentration of oxygen molecules.

A second type of detachment reaction which may occur is that of associative detachment
$$X^- + Y \rightarrow XY + e, \qquad (203)$$
in which the energy required to detach the electron comes from that released by formation of the molecule XY from X and Y. If $D(XY)$, the dissociation energy of XY, is greater than $A(X)$ the reaction will be exothermic even at the lowest temperatures, releasing electrons with considerable kinetic energy.

The first direct quantitative observations of detachment reactions were made by Pack and Phelps[†] while extending the observations on attachment coefficients and negative ion mobilities in oxygen by the

† PACK, J. L. and PHELPS, A. V., *J. chem. Phys.* **44** (1966) 1870; **45** (1966) 4316.

pulse method (Chap. 12, § 7.5.2) to high gas-pressures and temperatures. As they worked at low values of F/p, the ratio of field strength to gas pressure, they were concerned with O_2^- ions (see Chap. 13, § 4.4.3), for which the electron affinity is low enough for reactions of the type (202) to give an appreciable yield of detached electrons under thermal conditions.

Somewhat later, Frommhold[†] worked at such high values of F/p that not only were O^- ions predominantly formed (see Chap. 13, § 4.4.2) but they were accelerated to sufficient energies for detachment through a reaction of the type (202) to be observed.

In the course of their later experiments using the combined drift tube and mass spectrometer described on p. 2077, Moruzzi and Phelps[‡] observed detachment reactions of O^- ions with CO at thermal energies which almost certainly were of the associative type (203). Moruzzi, Ekin, and Phelps[§] then found, using the pulse technique of Pack and Phelps at values of F/p for which O^- ions are predominantly formed, that rapid detachment occurs at thermal energies in NO, CO, and H_2, again almost certainly due to the associative process.

Meanwhile, the flowing-afterglow technique described in § 3.4.5, which has proved so effective for positive ion reactions, was applied to reactions of negative ions and the rates of associative detachment reactions of O^- ions with various neutral species measured. Quite good agreement has been obtained between reaction rates measured by the three methods when all are applicable and the flowing-afterglow technique has been applied to measure a number of further rates of reactions involving negative ions which are of interest in upper atmospheric physics.

4.2.2. *Adaptation of the pulse method.* The possibility of determining rates of detachment of electrons from negative oxygen ions by collisions with gas molecules was first realized by Phelps and Pack[||] in the course of extending their observations on attachment coefficients and negative ion mobilities by the pulse method to high gas-pressures and temperatures.

In Fig. 19.72 the arrangement of their apparatus, which is essentially the same as that described in Chapter 12, p. 866, is shown diagrammatically. The mode of operation is as follows. A pulse of electrons is produced photoelectrically from a cathode by a pulse of ultra-violet light of about 120 μs duration. Just as in the earlier attachment experi-

[†] FROMMHOLD, L., *Fortschr. Phys.* **12** (1964) 597.
[‡] MORUZZI, J. L. and PHELPS, A. V., loc. cit., p. 2077.
[§] MORUZZI, J. L., EKIN, J. W., and PHELPS, A. V., *J. chem. Phys.* **48** (1968) 3070.
[||] PHELPS, A. V. and PACK, J. L., *Phys. Rev. Lett.* **6** (1961) 111.

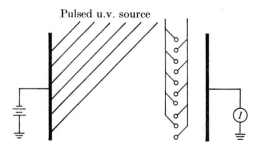

Fig. 19.72. Arrangement of apparatus used by Phelps and Pack for observing attachment and detachment of electrons in pure oxygen and in oxygen mixtures.

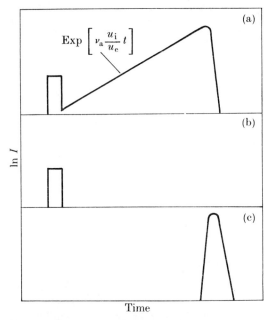

Fig. 19.73. Idealized form of the variation with time of the change I of the collector current, in pulse experiments in O_2 at room temperature. (a) Low pressure with d.c. voltage applied to the grid. (b) Low pressure with H.F. a.c. voltage applied to the grid. (c) High pressure with d.c. voltage applied to the grid.

ments, a pulse of voltage, also of about 120 μs duration, is applied after an adjustable time delay τ to collect negative ions and electrons in the neighbourhood of the grid, so reducing the current to the collector by an amount proportional to the concentration of both ions and electrons near the grid. Fig. 19.73 (a) shows the change of collector current as a function of the time delay τ.

At room temperature and low pressures it takes the form already discussed in Chapter 12, § 7.5.2 from which $\nu_a u_i/u_e$ may be obtained, ν_a being the attachment frequency ($\alpha_a u_e$ where α_a is the attachment coefficient), u_i and u_e the ion and electron drift velocities respectively. If instead a pulse of high-frequency alternating voltage (~ 1 Mc/s) is

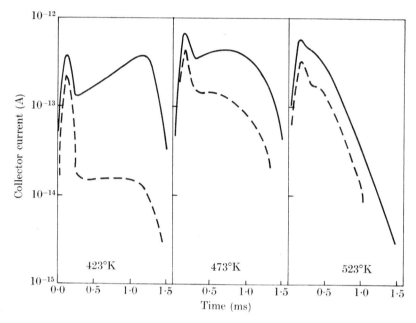

Fig. 19.74. Observed change I in collector current with time in the pulse experiment of Pack and Phelps in O_2, at three temperatures. O_2 concentration 10^{18} cm^{-3}, $F/N = 3.1 \times 10^{-17}$ V cm². ——— current due to electrons and negative ions (d.c. pulse), − − − current due to electrons only (high-frequency pulse).

applied the negative ions are unable to follow the field and only electrons are collected by the grid. The dependence of the collector current on the time delay τ in this case, under the same conditions of temperature and pressure, is as shown in Fig. 19.73 (b). This shows that there is no contribution to the collector current change when an a.c. pulse is applied, except for small time delays.

However, if the experiments are repeated at higher temperatures and pressures a delayed electron current is observed. At low pressures ($\simeq 30$ torr), but a temperature of 423 °K, the form of the collector current change when a high-frequency pulse is applied is shown in Fig. 19.74. The presence of the delayed electrons is clear but the current wave-form when d.c. pulses are applied is still mainly of the same form as in Fig. 19.73. As the temperature rises the relative importance of the delayed electrons increases while the d.c. wave-form becomes more and more strongly distorted.

At much higher pressures (a few hundred torr) and not too large F/p, electrons attach to form negative ions close to the cathode. The wave-form of the pulse collected at temperatures below 373 °K is indicated in Fig. 19.73 (c). The peak occurs at a time determined by the drift velocity of the negative ions and no delayed current is observed. Under these conditions detachment is unimportant and interest is directed mainly on the determination of the composition of the negative ions (see § 4.2.2.2).

At higher temperatures the peak shape of the collected current remains generally similar but the maximum current appears at an earlier time. The electron component behaves in exactly the same way, apart from being somewhat smaller in magnitude. This behaviour is interpreted in terms of an equilibrium between attachment and detachment being set up during the time of drift from cathode to collector—an electron on the average attaches and is detached many times during this period. The ions and electrons therefore drift together as a mixed cloud.

Fig. 19.75 illustrates the variation of the shape and time of arrival of the pulse with oxygen pressure at 477 °K and $F/N = 3 \cdot 1 \times 10^{-17}$ V cm². It will be seen that the apparent mobility and the pulse width increase as the pressure decreases.

We now consider how, from an analysis of these phenomena, it is possible to derive both attachment and detachment rates, and thence, from a consideration of statistical equilibrium between them at different temperatures, derive the electron affinity of O_2^-.

4.2.2.1. *Analysis of the low-pressure data in* O_2. The continuity equations satisfied by the respective concentrations n_e, n_i of electrons and of negative ions in the drift region are given by

$$\frac{\partial n_e}{\partial t} + u_e \frac{\partial n_e}{\partial x} = -\nu_a n_e + \nu_d n_i, \tag{204}$$

$$\frac{\partial n_i}{\partial t} + u_i \frac{\partial n_i}{\partial x} = -\nu_d n_i + \nu_a n_e, \tag{205}$$

it being assumed that ionization and diffusion are unimportant and that only one kind of negative ion, O_2^-, is present. ν_a, ν_d are the respective attachment and detachment rates per electron and per negative ion while u_e, u_i are the corresponding drift velocities. Solution of the equations (204)–(205) gives for the electron current density at a distance L from the cathode at a time t after the pulse

$$J_e/Q_0 = \delta(t-t_e)e^{-a} + e^{-a}z(t-t_e)^{-1}\exp\{\tau(a-d)\}I_1(2z), \tag{206}$$

where $t_e = L/u_e$, $t_i = L/u_i$ are the transit times for electrons and for negative ions respectively. $a = \nu_a L/u_e$ is the probability that an electron will form a negative ion during its transit time, while $d = \nu_d L/u_i$ is the corresponding probability that a negative ion will lose its electron during its transit time. Of the remaining symbols

$$\tau = (t-t_e)/(t_i-t_e), \tag{207}$$
$$z = \{ad(1-\tau)\tau\}^{\frac{1}{2}}, \tag{208}$$

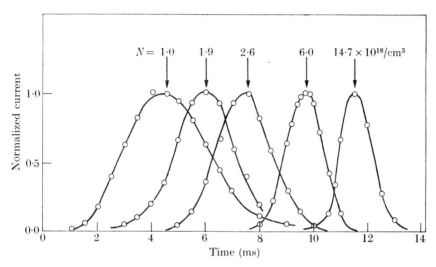

FIG. 19.75. Observed change in collector current with time in the pulse experiments of Pack and Phelps in O_2 at 477 °K for $F/N = 3 \cdot 1 \times 10^{-17}$ V cm² and different O_2 concentrations N as indicated, showing the apparent decrease of mobility as N increases.

I_1 is the usual Bessel function, Q_0 is the charge emitted in the electron pulse per unit area of cathode, and $\delta(t)$ is the unit impulse function such that

$$\int \delta(t)\,dt = 1. \tag{209}$$

The first term in the right-hand side of (206) is the current due to electrons that have travelled the distance L without attaching, while the remaining terms arise from electrons that have been detached from the negative ions which they formed by attachment at some stage in their passage. Fig. 19.76 shows the form of J_e for a typical case. However, for the analysis of data to obtain ν_d a simplified solution may be used.

We have $t_e \ll t_i$. In addition, we shall assume that $t \ll t_i$ so $\tau \ll 1$. For small values of $ad\tau$
$$I_1(2z) \simeq z(1+\tfrac{1}{2}z^2),$$
and so
$$J_e/Q_0 = \delta(t-t_e)e^{-a}+(ad/t_i)e^{-a}\{1+(\tfrac{1}{2}ad+a-d)\tau+...\}$$
$$\simeq \delta(t-t_e)e^{-a}+(ad/t_i)e^{-a}, \tag{210}$$
since we may take
$$|(\tfrac{1}{2}ad+a-d)|t \ll t_i. \tag{211}$$

In practice the light pulse that produces the electrons will be of a finite duration Δt_L. We then have from (210) for $t_e < t < t_e + \Delta t_L$
$$J_e/q_0 = e^{-a}+\{ad(t-t_e)/t_i\}e^{-a}, \tag{212}$$
where $q_0 = Q_0/\Delta t_L$.

Thus the current collected by the grid due to the electrons that have passed directly across without attachment has the constant value $J_0 = q_0 e^{-a}$, while that due to electrons that have been detached from negative ions increases linearly with the time.

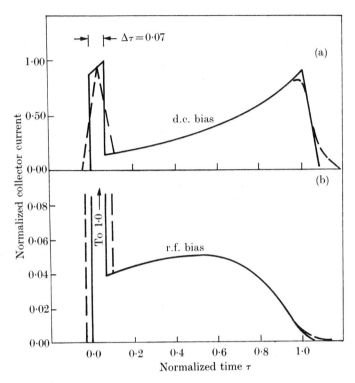

Fig. 19.76. Calculated form of the variation with time of the change I of the collector current in the pulse experiment of Pack and Phelps for values $a = 2$, $d = 0.2$, and an electron emission pulse width of 0.07. These assumed values correspond roughly to the case $F/N = 3.1 \times 10^{-17}$ V cm^2, $N = 10^{18}$ cm^{-3}, $T = 410$ °K. —— for a very narrow control grid pulse. – – – for a control grid pulse of width equal to that of the pulse of ultra-violet light: (a) with d.c. voltage applied to the grid, (b) with H.F. a.c. voltage applied to the grid.

For $t > t_e + \Delta t_L$ but $\ll t_i$ the current due to electrons that have passed across without attachment vanishes, while that due to detached electrons remains constant at
$$J_d = q_0 (ad\, \Delta t_L / t_i) e^{-a}. \tag{213}$$
Thus
$$J_d / J_0 = ad\, \Delta t_L / t_i = (v_a L / u_e) v_d\, \Delta t_L. \tag{214}$$

The physical significance of (213) is clear. A contribution to the 'initial' delayed current is made from electrons which have attached anywhere along the drift path, provided detachment occurs in a time Δt_L thereafter.

So far we have neglected the effect of a finite duration Δt_g of the collecting pulse applied to the grid. If $\Delta t_g = \Delta t_L$ then
$$J_d / J_0 = J_d / (J_m - \tfrac{1}{2} J_d), \tag{215}$$

where J_m is the maximum current collected at the grid. If $\Delta t_L \neq \Delta t_g$ then J_m is the current at a time $\tfrac{1}{2}|\Delta t_L - \Delta t_g|$ earlier than that at which the current is a maximum. Moreover in (214) Δt_L must be taken as the smaller of Δt_L and Δt_g.

Having determined the product $\nu_a \nu_d / u_e$ from (214) we may obtain ν_d if ν_a/u_e can be measured under the same conditions. This cannot be done by the method of Chapter 12, § 7.5.2 unless ν_d is very small because the wave-form of the total current pulse collected by the grid departs markedly from the form shown in Fig. 19.73. However, if J_{es} is the current collected by the grid when all electrons that cross the drift tube without attachment are collected by it and J_t the total current of electrons and negative ions, then

$$J_{es} = J_t \exp(-\nu_a L/u_e). \tag{216}$$

To measure J_{es} the voltage of the r.f. pulse is increased until the electron current collected by the grid is saturated. J_t is obtained by measuring the current received on the collector when no voltages are applied to the grid. To check the validity of this method it may be applied under conditions in which detachment is negligible so that ν_a/u_e may also be obtained by the method described in Chapter 12, § 7.5.2. It was found by Pack and Phelps that the two methods gave the same results within 10 per cent.

4.2.2.2. *Analysis of the high-pressure data in O_2.* At high pressures, as described above, attachment and detachment occur many times during drift from cathode to grid. The relative concentrations n_e, n_i of electrons and negative ions remain fixed throughout and the mean drift velocity u of the mixed cloud of electrons and negative ions is given by

$$u = (u_e n_e + u_i n_i)/(n_e + n_i). \tag{217}$$

As $u_e \gg u_i$ but $u < 10 u_i$ it follows that $n_e \ll n_i$ so (217) becomes

$$u = u_i + u_e n_e/n_i. \tag{218}$$

If μ, μ_i, μ_e are the corresponding mobilities we then have

$$\mu p = \mu_i p + \mu_e p n_e/n_i, \tag{219}$$

where p is the gas pressure. Since $\mu_i p$ and $\mu_e p$ are expected to be independent of p for fixed F/p in the temperature range concerned any variation of μp with p must be due to variation of n_e/n_i.

It is found empirically that

$$n_e/n_i = K'N, \tag{220}$$

where N is the concentration of neutral O_2 molecules, so

$$\mu p = \mu_i p + \mu_e p K'N. \tag{221}$$

Hence, by plotting μp against N, a linear relation is found. Extrapolation to zero N then gives $\mu_i p$ and hence μ_i. In this way the mobilities shown in Fig. 19.70 at high temperatures were obtained.

We also have, in view of the dynamical equilibrium in the mixed electron–ion cloud,
$$\nu_a/\nu_d = K'N. \tag{222}$$

To obtain ν_a and ν_d separately, use is made of the fact that $\nu_a \nu_d$ varies as N^3, ν_a varying as N^2 (see Chap. 13, § 4.4.3) and ν_d as N, for fixed F/p and gas temperature T. This result may be established by measurements, in the low-pressure range, of $\nu_a \nu_d/u_e$ and may then be used to obtain $\nu_a \nu_d$ at higher pressures.

4.2.2.3. *Experimental results in O_2 and their interpretation.* In the experiments of Pack and Phelps the pulsed light source was a hot-cathode hydrogen lamp. This was chosen because of the short decay time, about $0.2\ \mu s$, of the lamp when turned off, so that very short light pulses could be applied.

The whole system was baked out at about 400 °C for about 14 hours before each set of measurements. At room temperature, after bakeout, the pressure in the system was less than 10^{-8} torr and the rate of rise of background pressure less than 10^{-9} torr/min. When operating at 310 °C this rate of rise was about 10^4 times faster, leading to changes in the ion current wave-form within an hour. At higher temperatures the system was flushed after each run and it was thereby possible to obtain wave-forms without appreciable peaks due to impurities.

Fig. 19.77 illustrates results obtained in the low-pressure range for the product $\nu_a \nu_d/u_e$ as a function of N^3 at a fixed value of F/p and different gas temperatures. The linearity of the relation is apparent, so lending support to the soundness of the analysis technique.

By determining ν_a/u_e separately, as described above, ν_d is obtained. ν_d/N is constant for fixed F/p and T and the values obtained as a function of T for the same F/p are shown in Fig. 19.78. In the same figure the values of ν_a/N^2, which is again a constant for fixed F/p and T, are shown as a function of T for the same F/p. To obtain these, use must be made of the values of u_e measured as indicated in Chapter 11, Fig. 11.43. The mean electron energy at the F/p value concerned is about 0.6 eV and the electron energy distribution is determined by the field rather than the gas temperature.

At high gas pressures we show first in Fig. 19.79 some of the results which justify the empirical relationship (220). In this figure measured values of μN at 529 °K are plotted as functions of $1/N$ for a number of values of F/N. In all cases a linear relation is found from which $\mu_i N$ and $\mu_e N K'$ may be obtained. Since $\mu_e N$ is known from other measurements K' and hence from (222) the ratio ν_a/ν_d may be derived. Finally,

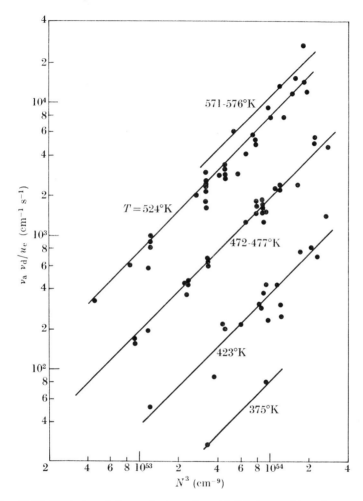

FIG. 19.77. Variation of the product $\nu_a \nu_d/u_e$ with N^3 observed in pulse experiments in O_2 for $F/N = 3 \cdot 1 \times 10^{-17}$ V cm² and different gas temperatures as indicated.

from the observed constancy of $\nu_a \nu_d/u_e N^3$ for different N at fixed F/p and temperature T obtained from the low-pressure measurements, further values of ν_d/N may be obtained. These are shown in Fig. 19.78 and it is seen that they fit well with the results obtained at low pressures. This seems to verify not only that ν_d is proportional to N over the entire

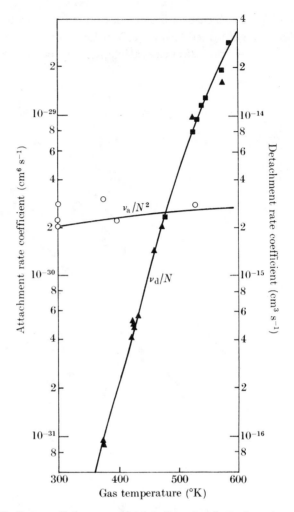

Fig. 19.78. Rate coefficients ν_d, ν_a for detachment and attachment respectively of electrons in pure O_2, for $F/N = 3 \cdot 1 \times 10^{-17}$ V cm^2, as a function of temperature, derived from pulse experiments. ▲ ν_d/N at low pressures. ■ ν_d/N at high pressures. ○ ν_a/N^2.

pressure range but that ν_a is proportional to N^2 over the same range—otherwise consistency could not be expected.

All of these results are in accord with the assumption that the detachment reaction is
$$O_2^- + O_2 \rightarrow 2O_2 + e, \tag{223}$$

while attachment occurs through
$$2O_2 + e \rightarrow O_2^- + O_2, \tag{224}$$

19.4 IONIC MOBILITIES AND IONIC REACTIONS

in other words, that we are dealing with the equilibrium

$$2O_2 + e \rightleftharpoons O_2^- + O_2. \tag{225}$$

Further support for the assumption (224) is forthcoming from the observation of attachment in mixtures of O_2 with other gases (see Chap. 13, § 4.4.4).

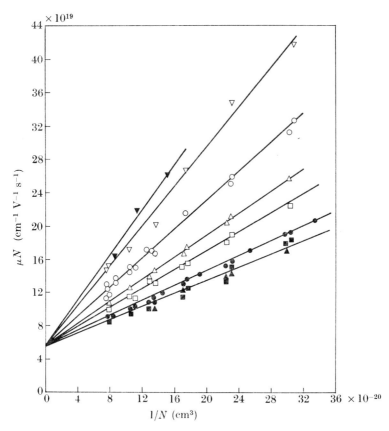

FIG. 19.79. Variation of μN with $1/N$ observed using the pulse method in O_2 at 529 °K, for different values of F/N, N being the O_2 concentration and μ the apparent mobility. Observed results for F/N, in V cm², ▲, 3·1 ▼, 4·65 ▽, 9·3 ○, 18·6 △, 31 □, 93 ●, 186 ■, 310.

We may now apply the method of statistical mechanics to calculate the equilibrium constant K for (225) in terms of the electron affinity $A(O_2)$ of O_2 just as was done for other ionic reactions on pp. 2039 and 2082. Since $K = n_e N/n_i$ it is identical with the empirical constant K' introduced in (220). Assuming that O_2 and O_2^- have identical vibrational and rotational properties and that all O_2^- ions are in the ground $^2\Pi_g$

electronic state, it is found that the best fit between observed K' and calculated K is obtained if the electron affinity of O_2 is given by

$$A(O_2) = 0\cdot43 \pm 0\cdot02 \text{ eV}. \tag{226}$$

This is to be compared with $0\cdot15\pm0\cdot05$ eV from photo-detachment data (see Chap. 15, § 6.3), $0\cdot58$ eV from electron impact data in O_3 (see Chap. 13, § 10), and $0\cdot15$–$1\cdot0$ eV from thermochemical data (see p. 1016).

Assuming that (226) is correct it is possible to obtain ν_d for O_2^- ions in thermal equilibrium with the gas at lower temperatures. The attachment coefficient ν_a for thermal electrons at such temperatures may be obtained from observations in mixtures of O_2 with CO_2, and ν_d then follows from the relation

$$\nu_d/\nu_a = K(T), \tag{227}$$

where $K(T)$ is the equilibrium constant for the temperature T in question, calculated using (130). Thus at 230 °K, a temperature of interest for atmospheric applications, we find

$$\nu_d/N = 7 \times 10^{-21} \text{ cm}^3 \text{ s}^{-1}, \tag{228}$$

where N is the concentration of neutral O_2 molecules.

4.2.2.4. *Attachment and detachment rates in oxygen at high F/p—detachment from O^-.* An essentially similar investigation has been carried out by Frommhold† covering the pressure range 1–50 torr and a range from 40 to 150 V cm^{-1} torr^{-1} of F/p. The higher mean energy of the electrons involved leads to production of O^- instead of O_2^- but the higher mean energy of the ions leads to an appreciable detachment rate, even though the electron affinity of O^- is considerably greater than that of O_2^-.

The method of analysis of the data is similar to that described above. However, at these higher electron energies it is necessary to allow for production of additional electrons by ionization so that on the right-hand side of (204) the additional term $\nu_i n_e$ must be added, ν_i being the ionization rate $\alpha_i u_e$ where α_i is the ionization coefficient. The electron current received as a function of t is then given in terms of ν_i, ν_e, u_e, u_i, ν_a, and ν_d. ν_i and u_e were determined directly as described in Chapters 5, § 3.1.1 and 2, § 3 respectively while u_i was taken from the observations of Burch and Geballe (see § 4.1 and Fig. 19.58). A first approximation to ν_a was obtained as in the experiments of Pack and Phelps

† loc. cit., p. 2081.

from the ratio r of the current carried by electrons moving with the full drift velocity u_e to the total current. To a close approximation

$$r = \exp\{-\nu_a L/u_e\}, \tag{229}$$

where L is the distance between cathode and anode.

To obtain a first approximation to ν_d a mean delay time $\bar{t}$ of the delayed electron current was defined as shown in Fig. 19.80. This

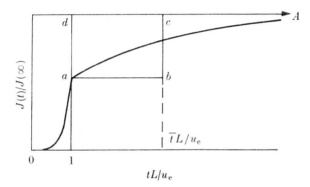

FIG. 19.80. Typical form of the integrated current pulse $J(t)$ in the experiment of Frommhold in O_2 at high F/p, showing how the mean delay time $\bar{t}$ of the detached electron current is defined. Thus

$$(L/u_e)\bar{t} \times ad = \text{area } adA,$$

A being at a time $t \geqslant t$.

illustrates the typical form of an integrated electron current pulse. It may be shown that
$$\bar{t} \simeq \nu_a L/\nu_d u_e,$$
so that, if $\bar{t}$ is measured, ν_d may be estimated.

Having determined approximate values for ν_a and ν_d these were used together with observed values of ν_i, u_e, and u_i to calculate the form of the integrated current pulse. If this did not agree sufficiently well with the observed form, modified values of ν_a and ν_d were substituted the procedure repeated until good agreement was found.

The variation of ν_a/p with F/p obtained in this way by Frommhold is illustrated in Fig. 19.81. It was found that, in contrast to the measurements of Pack and Phelps at lower F/p, ν_a/p is independent of p, indicating a two-body attachment process. This is almost certainly the dissociative attachment reaction (Chap. 13, § 4.4.2)

$$O_2 + e \rightarrow O^- + O. \tag{230}$$

To provide further confirmation of this it is necessary to know the mean energy of the electrons as a function of F/p. In the absence of consistent observations Frommhold adopted an indirect procedure based on the measured ionization frequency ν_i. Assuming a Maxwellian distribution of electrons, a theoretical value of ν_i was calculated as a function of mean electron energy $\bar{\epsilon}$ from the ionization cross-section of O_2 as a function of electron energy observed by Tate and Smith (see Chap. 13, Fig. 13.73). $\bar{\epsilon}$ was then adjusted so that the theoretical and observed

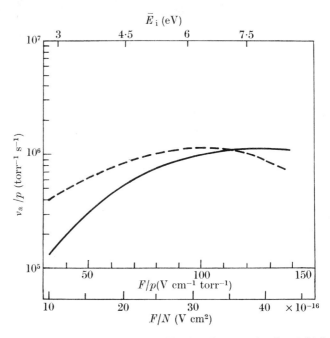

Fig. 19.81. Variation of ν_a/p with F/p for electrons in O_2 at high F/p. —— observed by Frommhold. – – – calculated from observed attachment cross-section assuming a Maxwellian energy distribution for the electrons, at each F/p, with mean energy $\bar{E}_i$, indicated on the upper scale, which gives good agreement between the observed ionization coefficient and that calculated from measured ionization cross-sections.

values of ν_i were in agreement. In this way the mean energy scale shown in Fig. 19.81 was obtained. It is now possible to check whether the negative ions produced are actually O^- by calculating what ν_a should be if the observed cross-section for the reaction (230) as a function of electron energy is taken in conjunction with the Maxwellian energy distribution at the appropriate mean energy. Using the cross-section for (230) measured by Buchel'nikova (Chap. 12, Table 12.7), Frommhold obtained a 'theoretical' curve for ν_a as a function of F/p, shown in Fig. 19.81. Allowing for the inaccuracy of the assumption of a Maxwellian distribution of electron energies, the agreement of this curve with the observed

19.4 IONIC MOBILITIES AND IONIC REACTIONS

is not unsatisfactory and strongly supports the assumption that (230) is indeed the attachment process involved.

The ratio v_d/p was found to be independent of p for fixed F/p showing that a two-body process is again involved. Fig. 19.82 illustrates the

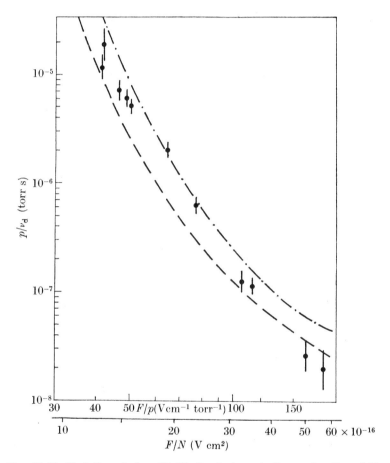

FIG. 19.82. Variation of p/v_d with F/p for electrons in O_2 as a function of F/p at high F/p. ⬥ observed by Frommhold. — — — calculated assuming the detachment cross-sections (a) of Fig. 19.83. — · — · — calculated assuming the detachment cross-sections (b) of Fig. 19.83.

observed variation of p/v_d with F/p. To examine the plausibility of these results and derive information about the detachment cross-section as a function of electron energy, Frommhold extrapolated observed cross-sections for the reaction

$$O^- + O_2 \rightarrow O + O_2 + e, \tag{231}$$

measured by beam methods for ion energies of 10 eV and greater (see Chap. 24), down to the threshold (2·25 eV = $\frac{3}{2} \times$ electron affinity of O) by two methods. In the first, referred to as (*a*), he assumed the cross-section to remain constant for ion energies between the threshold and 10 eV, while in the second, referred to as (*b*), linear extrapolation to zero at the threshold was adopted. The resulting detachment cross-sections are shown in Fig. 19.83. The ion energy distribution was assumed to

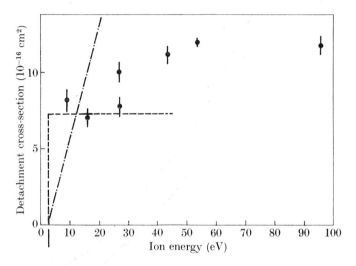

FIG. 19.83. Cross-section for the detachment reaction
$$O^- + O_2 \rightarrow O + O_2 + e.$$
⦁, observed by Hasted and Smith (R. A.). Extrapolation to threshold: (*a*) ― ― ― ―, (*b*) ―·―·―

be Maxwellian about a mean energy $\bar{E}_i$ given by Wannier's formula (1),

$$\bar{E}_i = \tfrac{1}{2}(M_1 + M_2)u_i^2 + \tfrac{3}{2}\kappa T, \tag{232}$$

where T is the gas temperature and M_1 and M_2 are the respective masses of O^- and O_2. The drift velocity u_i of the O^- ions has not been measured at the high values of F/p involved. A constant mobility of 4·2 cm² V⁻¹ s⁻¹ was assumed. This is a very rough approximation (see Fig. 19.58) based on observations made by Eiber (see p. 2072) for F/p in the range 7–30 V cm⁻¹ torr⁻¹. Values of $\bar{E}_i$ derived in this way are shown in Fig. 19.81 as well as calculated values of p/v_d under the assumptions (*a*) and (*b*).

In view of the crudity of the approximations made the agreement between the observed and calculated p/v_d is not unsatisfactory.

It is of interest to note that, because the mobility of O_2^- falls quite rapidly at high F/p as compared with that of O^-, the mean energy of

O_2^- ions would be much smaller for given F/p in this region than for O^-, so much so as to make up for the smaller energy required to detach an electron from O_2^- than from O^-—the detachment frequency for O_2^- might well be less than for O^-.

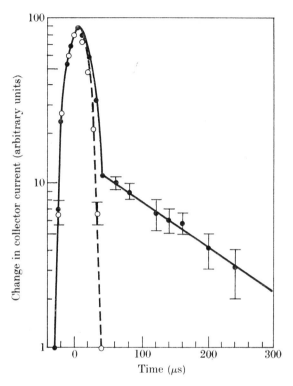

FIG. 19.84. Observed change of collector electron current with time in the pulse experiments of Moruzzi, Ekin, and Phelps for $F/p = 5$ V cm^{-1} torr^{-1}, –o–o– for pure O_2 at 5 torr pressure, –●–●– for O_2 at 5 torr containing 0·001 torr partial pressure of H_2.

4.2.3. *The rates of associative detachment reactions.* The pulsed method which we have been discussing in § 4.2.2 is applicable to the study of associative as well as other detachment processes. Thus in Fig. 19.84 the form of the electron current pulse obtained in pure O_2 at a pressure of 5 torr for $F/p = 5$ V cm^{-1} torr^{-1} shows no sign of a delayed electron current. Addition of a fractional concentration of 2×10^{-4} of H_2 has a dramatic effect in producing a delayed electron current, due presumably to the associative detachment reaction

$$H_2 + O^- \rightarrow H_2O + e, \tag{233}$$

which is exothermic by 3·5 eV. There is no difficulty in applying the analysis of § 4.2.2.1 to obtain the associative detachment rate coefficient, which is shown for a number of values of F/p in Fig. 19.86.† The corresponding mean ion energy values are also included.

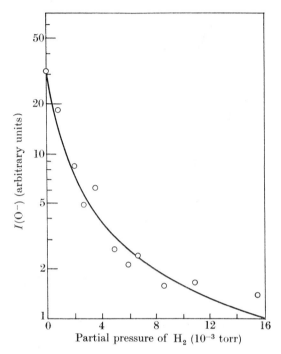

Fig. 19.85. Observed variation of the current $I(O^-)$ of O^- ions, in the experiments of Moruzzi and Phelps, with the partial pressure of H_2 in an O_2–H_2 mixture at a total pressure of 1·2 torr and $F/p = 10$ V cm^{-1} torr^{-1} ($F/N = 3·1 \times 10^{-16}$ V cm^2). ○ experimental points. —— curve giving the best fit to the data.

The reaction (233) can also be investigated by the combined drift tube and mass spectrometer technique. Fig. 19.85 shows the dependence of O^- current on the partial pressure of H_2 in such an apparatus, observed by Moruzzi and Phelps.‡ Using observed values of O^- mobility, the rate constant for the reaction (233), assuming it to be responsible for the reduction of O^- current, is found to be $1·0 \times 10^{-9}$ cm^3 s^{-1} for $F/p = 10$ V cm^{-1} torr^{-1}. This value, which is independent of any variation of detecting efficiency between different ions, agrees well with that obtained by the pulse method.

† Moruzzi, J. L., Ekin, J. W., and Phelps, A. V., loc. cit., p. 2085.
‡ loc. cit., p. 2077.

Finally, the flowing-afterglow method (see § 3.4.5) may be applied without difficulty to obtain reaction rates for negative ions of thermal energy.† O^- ions are produced by adding small fractional concentrations of O_2 or CO_2 to the flowing buffer gas which is either helium or argon. When this mixture is subject to a pulsed discharge, O^- is produced by the usual dissociative attachment reactions (see Chap. 13, § 4.4.2 and p. 2097 of this chapter). The observed loss of O^- ions in the presence of admixed hydrogen may be due either to (233) or to the dissociative charge transfer reaction

$$O^- + H_2 \to OH^- + H + 0 \cdot 16 \text{ eV}. \tag{234}$$

However, very little OH^- production was observed to accompany the loss of O^- so that (233) must be dominant. The observed rate coefficient, $1 \cdot 5 \times 10^{-9}$ cm^3 s^{-1}, which did not depend on whether the O^- ions were generated from admixed O_2 or CO_2 or whether the buffer gas was argon or helium, is consistent with the results obtained by the other methods and there seems little doubt that the rate of the associative detachment reaction has been measured in all cases.

Similar results obtained for the reactions

$$O^- + CO \to CO_2 + e, \tag{235}$$

$$O^- + NO \to NO_2 + e \tag{236}$$

are shown in Fig. 19.86. For the first, all three methods have been used and once again the results are reasonably consistent. The rapid fall in the rate coefficient with increasing F/p for the reaction with NO is consistent with observations made using a flowing mixture of O_2 and NO from which negative ions, produced by electron impact, were drawn into a mass spectrometer.‡

Table 19.20 gives the rate constants for a number of associative detachment reactions observed by the flowing-afterglow technique. This makes it possible to observe these reactions with radicals such as O and N atoms (see p. 2024). O_2^- ions were produced by two methods. One involved some attachment process in the helium afterglow. This had the disadvantage that O_2^- ions were being produced down the flow tube into the reaction zone. An alternative method is to use the fast charge-transfer reaction

$$NO^- + O_2 \to O_2^- + NO, \tag{237}$$

---

† FEHSENFELD, F. C., FERGUSON, E. E., and SCHMELTEKOPF, A. L., *J. chem. Phys.* **45** (1966) 1844.
‡ MORUZZI, J. L. and PHELPS, A. V., ibid. **45** (1966) 4617.

the NO⁻ ions being produced by flowing a mixture of argon and NO_2 through the discharge.

OH⁻ ions were produced by adding H_2O to the discharge while Cl⁻ ions were present as common impurities.

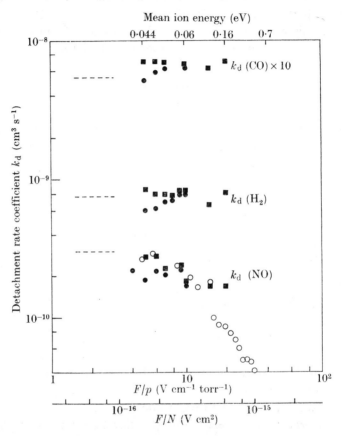

FIG. 19.86. Observed rate coefficients for associative detachment of electrons from O⁻ ions by collisions with CO, H_2, and NO molecules. ■ from pulse detachment experiments using attachment coefficient data of Huxley, Crompton, and Bagot. ● from pulse detachment experiments using attachment coefficient data of Chanin, Phelps, and Biondi. ○ from combined drift tube and mass spectrometer observations of Moruzzi and Phelps. – – – from flowing-afterglow experiments.

For the reaction of O⁻ with N_2 the energy release is uncertain because the dissociation energy of N_2O into NO and O is uncertain. The two values given in the table are obtained assuming dissociation energies of 1·66 and 1·2 eV respectively. The failure to observe the reaction either in the flowing-afterglow or the pulse experiments provides some support for the lower value.

The yield of O^- associated with the loss of O_2^- in the presence of O and N is so small that, even though the O^- may have partly been removed through associative detachment reactions listed above, there seems little

TABLE 19.20

Rate constants for detachment reactions at thermal energies

Reaction	Energy release ΔE (eV)	Rate constant (10^{-10} cm³ s⁻¹)	
		Obs.	Calc. from orbiting radius
$O^- + H_2 \rightarrow H_2O + e$	3·5	15	16
$\rightarrow OH^- + H$	0·16	—	—
$O^- + NO \rightarrow NO_2 + e$	1·6	5	10
$O^- + CO \rightarrow CO_2 + e$	4·0	5	10
$O^- + O \rightarrow O_2 + e$	3·6	1·9	7
$O^- + N \rightarrow NO + e$	5·1	2·2	7
$O^- + N_2 \rightarrow N_2O + e$	$\begin{cases} -0·25 \\ +0·21 \end{cases}$	0·01	—
$O^- + O_2 \rightarrow O_3 + e$	−0·4	0·01	—
$O^- + CO \rightarrow CO + O + e$		10^{-4}	—
$Cl^- + H \rightarrow HCl + e$	+0·8	Fast	—
$Cl^- + O \rightarrow ClO + e$	−0·9	0·1	—
$Cl^- + N \rightarrow ClN + e$	−0·9	0·1	—
$O_2^- + O \rightarrow O_3 + e$	0·6	3·3	6
$\rightarrow O^- + O_2$	1·0	—	—
$O_2^- + N \rightarrow NO_2 + e$	4·1	5	8
$\rightarrow NO + O + e$	1·0	—	—
$\rightarrow O^- + NO$	2·5	—	—
$OH^- + O \rightarrow HO_2 + e$	1·0	2	7
$OH^- + N \rightarrow HNO + e$	2·4	0·1	9

doubt that the dominant reaction leading to loss of O_2^- is the associative detachment one. For the O_2^-–N reaction it was not possible to discriminate between the two reactions leading exclusively to electrons and neutral products.

Comparison of the reactive rates with those obtained from the 'orbiting' model of Gioumousis and Stevenson (see (92)) shows that in some cases there is good agreement but in others the observed rate is considerably smaller.

4.3. *Charge transfer reactions involving negative ions at thermal energies*

The flowing-afterglow method has been applied to measure the rates of charge transfer and other reactions involving ions at thermal energies.†
Results obtained are given in Table 19.21.

† FERGUSON, E. E., *Rev. Geophys.* **5** (1967) 305; *Can. J. Chem.* **47** (1969) 1815.

TABLE 19.21

Rate constants for charge transfer reactions involving negative ions at thermal energies

Reaction	Rate constant (10^{-10} cm^3 s^{-1})	
	Obs.	Calc. from orbiting radius
$O^- + O_3 \rightarrow O_3^- + O$	5·3	—
$O^- + NO_2 \rightarrow NO_2^- + O$	12·0	—
$O_2^- + O_3 \rightarrow O_3^- + O_2$	4·0	—
$O_2^- + NO_2 \rightarrow NO_2^- + O_2$	8·0	—
$O_3^- + CO_2 \rightarrow CO_3^- + O_2$	0·4	—
$O_3^- + NO \rightarrow NO_3^- + O$	0·1	7·5
$CO_3^- + O \rightarrow O_2^- + CO_2$	0·8	5·7
$CO_3^- + NO \rightarrow NO_2^- + CO_2$	0·09	7·2
$CO_3^- + NO_2 \rightarrow NO_3^- + CO_2$	0·8	—
$NO^- + O_2 \rightarrow O_2^- + NO$	9	7·6
$NO_2^- + O_3 \rightarrow NO_3^- + O_2$	0·18	—
$H^- + NO_2 \rightarrow NO_2^- + H$	29	—
$OH^- + NO_2 \rightarrow NO_2^- + OH$	10	—
$NH_2^- + NO_2 \rightarrow NO_2^- + NH_2$	10	—
$F^- + NO_2 \rightarrow NO_2^- + F$	< 0·25	—
$Cl^- + NO_2 \rightarrow NO_2^- + Cl$	< 0·06	—

4.4. *The formation of complex negative ions*

We have already discussed the formation of O_3^- and O_4^- ions in oxygen. These are examples of the building up of complex ions by associative reactions of one kind or other. Such complex formation is often of importance in many circumstances, as for example in the D region of the earth's ionosphere where the pressure is high enough and the chemical composition varied enough for a wide variety of complex ions to be produced. We conclude the discussion of reactions involving negative ions by considering the information available about the rates of formation and some of the properties of a few complex ions of importance.

4.4.1. *Negative ions in CO_2 and in O_2–CO_2 mixtures.* The observations of Moruzzi and Phelps,[†] using their combined drift tube and mass spectrometer, show that in CO_2 no negative ions are formed at low F/p, in agreement with earlier observations (see Fig. 13.117 b), but that at higher field strengths O^- and CO_3^- are present as shown in Fig. 19.87. Fite and Rutherford[‡] also found CO_3^- ions in a discharge afterglow in an

---

[†] loc. cit., p. 2077.
[‡] FITE, W. L. and RUTHERFORD, J. A., *Discuss. Faraday Soc.* **37** (1964) 192.

O_2–CO_2 mixture. It is probable that the reactions responsible for the production of these ions are

$$CO_2 + e \to CO + O^-, \qquad (238)$$

$$O^- + CO_2 + CO_2 \to CO_3^- + CO_2. \qquad (239)$$

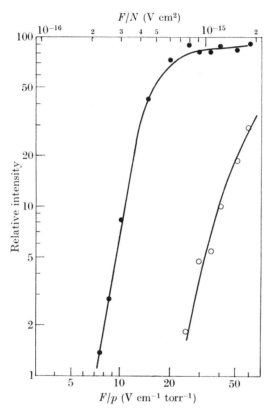

Fig. 19.87. Relative intensity of different negative ions observed as functions of F/p by Moruzzi and Phelps in pure CO_2 at 1·74 torr. ● CO_3^-, ○ O^-.

From the observed variation of the ratio of the CO_3^- to O^- current, assuming a drift velocity of $1·3 \times 10^5$ cm s^{-1} for O^- in CO_2 at $F/p = 50$ V cm^{-1} torr^{-1} and observed values of the attachment and ionization coefficients for CO_2,† the rate coefficient for (239) is found to be 4×10^{-29} cm^6 s^{-1}.

In O_2–CO_2 mixtures the relative intensity of different ions observed is as shown in Fig. 19.88. In addition to O^- and CO_3^-, O_2^- and CO_4^- are

† BHALLA, M. S. and CRAGGS, J. D., *Proc. phys. Soc.* **76** (1960) 369.

also present, mainly at low F/p. The likely reactions are now

$$O_2 + e + M \rightarrow O_2^- + M, \qquad (240)$$
$$O_2^- + CO_2 + M \rightarrow CO_4^- + M, \qquad (241)$$

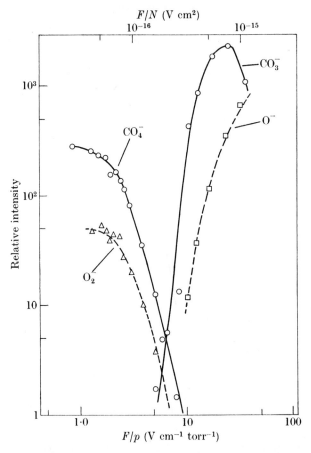

Fig. 19.88. Relative intensity of different negative ions observed as a function of F/p by Moruzzi and Phelps in a mixture of 0.8 torr CO_2 with 0.5 torr O_2.

where the third body, M, may be either O_2 or CO_2. From observations of the ratio of the O_2^- and CO_4^- currents, using electron drift velocities for pure CO_2† (Chap. 11, Fig. 11.47) and the observed rate constants for the attachment reaction for O_2–CO_2 mixtures (Chap. 13, § 4.4.4),‡ together with an estimated drift velocity of 7.7×10^3 cm s^{-1} for O_2^- in CO_2 at $F/p = 3$ V cm^{-1} torr^{-1}, the rate coefficients for (241) are found

† PACK, J. L., VOSHALL, R. E., and PHELPS, A. V., *Phys. Rev.* **127** (1962) 2084.
‡ PACK, J. L. and PHELPS, A. V., *J. chem. Phys.* **44** (1966) 1870.

to be 9×10^{-30} and 2×10^{-29} cm^6 s^{-1} for M = CO_2 and M = O_2 respectively.

Pack and Phelps† have applied the pulse technique described above to study attachment and detachment in mixtures of O_2 and CO_2.

Fig. 19.89 shows the effect of addition of a small admixture of CO_2 under conditions in which the transient wave-form is similar to that

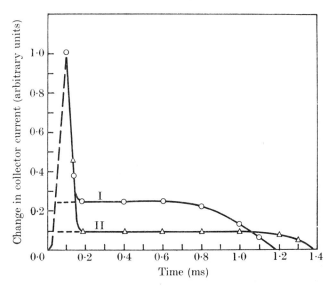

Fig. 19.89. Observed change of collector electron current with time in the pulse experiment of Phelps and Pack at 461 °K, and $F/N = 3.1 \times 10^{-17}$ V cm^2. I for pure O_2 at a concentration of 1.1×10^{18} cm^{-3}, II for O_2 containing 0.23 per cent of CO_2.

shown in Fig. 19.84 for pure O_2. The current of delayed electrons is markedly reduced, even though the CO_2 concentration is only about 1/400 of that of the O_2. One possible explanation of this result is that, in the presence of the CO_2, O_2^- is converted to a more stable ion from which detachment occurs with reduced probability. For a fixed relative composition an equilibrium would exist between the rates of production and loss of the new ion from O_2^-. Suppose further, for simplicity, that the new ion does not suffer detachment. From the observations on attachment it is clear that the presence of such a small concentration of CO_2 can have no appreciable effect on the rate of attachment. Assuming that the new ion, X^-, say, does not suffer detachment and that an equilibrium is set up between O_2^- and X^-, the detached current II

† Pack, J. L. and Phelps, A. V., *J. Chem. Phys.* **45** (1966) 4316.

in Fig. 19.89 would be proportional to the O_2^- concentration $n(O_2^-)$ and I to the sum $n(O_2^-)+n(X^-)$. From observations of I and II at different partial pressures of CO_2 and a fixed temperature it is found that $n(X^-)/n(O_2^-)$ is proportional to $n(CO_2)$. This suggests that X^- is CO_4^-, the equilibrium being
$$O_2^- + CO_2 \rightleftharpoons CO_4^-. \tag{242}$$

Further evidence concerning the ion was obtained from observations at high gas pressure and low F/p. Under these conditions, attachment and detachment occur many times while an ion traverses the drift space. Following a similar analysis to that of § 4.2.2.2 it is found that the ratio $n_e/n(X^-)$ is proportional to $1/n(O_2)n(CO_2)$, which would be so if the new ion were indeed CO_4^- and we were concerned with the equilibrium
$$e + O_2 + CO_2 \rightleftharpoons CO_4^- + \Delta E. \tag{243}$$

This being so, an estimate may be made of the energy ΔE from the observed value of the equilibrium constant
$$K = n_e n(O_2) n(CO_2) / n(CO_4^-). \tag{244}$$

Assuming complete freedom of internal motion in CO_4^- but no vibrational activation of O_2 or CO_2, ΔE is found to be $1\cdot2\pm0\cdot1$ eV. This would give an energy of dissociation of CO_4^- into O_2^- and CO_2 as high as $0\cdot8\pm0\cdot1$ eV. The observation of CO_4^- by Moruzzi and Phelps under similar conditions supports this interpretation.

4.4.2. *Ions in H_2O and in O_2–H_2O mixtures.* Moruzzi and Phelps† observed seven negative ions in triply distilled water vapour for F/p 10 V cm^{-1} torr^{-1} as shown in Fig. 19.90. Apart from H^- and OH^- all the other ions are water molecules attached to OH^-. Moruzzi and Phelps suggest that the reactions that lead to production of these ions are as follows. H^- is first formed by the reaction discussed in Chapter 13, § 7.2
$$H_2O + e \rightarrow H^- + OH. \tag{245}$$

OH^- is then formed through the reaction
$$H^- + H_2O \rightarrow OH^- + H_2, \tag{246}$$
which is exothermic by at least $0\cdot3$ eV. Clustering then occurs through three-body reactions
$$OH^- + H_2O + H_2O \rightarrow OH^-.H_2O + H_2O, \tag{247}$$
and so on.

It will be seen that, as F/p increases, the degree of hydration falls until, for $F/p = 100$ V cm^{-1} torr^{-1}, only H^-, OH^- and $OH^-.H_2O$ remain.

† loc. cit., p. 2077.

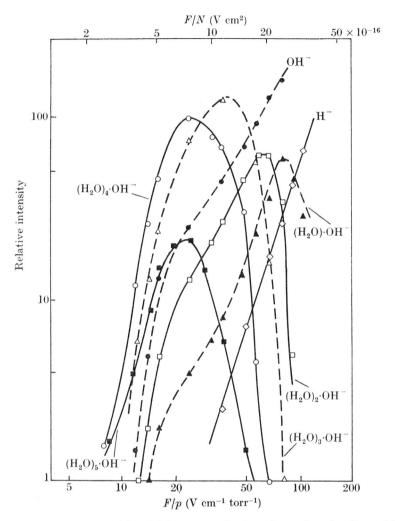

Fig. 19.90. Relative intensities of different negative ions observed as functions of F/p by Moruzzi and Phelps in triply distilled H_2O.

For O_2–H_2O mixtures† an even greater variety of complex ions is found. Fig. 19.91 shows the relative intensities in a mixture for $F/p = 5 \cdot 0$ V cm⁻¹ torr⁻¹. O_2^- ions with up to five clustered water molecules are observed. Even when the partial pressure of H_2O was reduced to 10^{-4} torr, $O_2^-\cdot H_2O$ and $O_2^-\cdot 2H_2O$ persisted. The rate coefficient for the three-body collision is certainly large (cf. the discussion on

† Moruzzi, J. L. and Phelps, A. V., loc. cit.

positive ion clustering in § 3.1.1), probably at least as high as 10^{-28} cm^6 s^{-1}, but no even semiquantitative value could be found for it.

At higher F/p clustered O$^-$ ions up to O$^-$.(H$_2$O)$_5$ were observed.

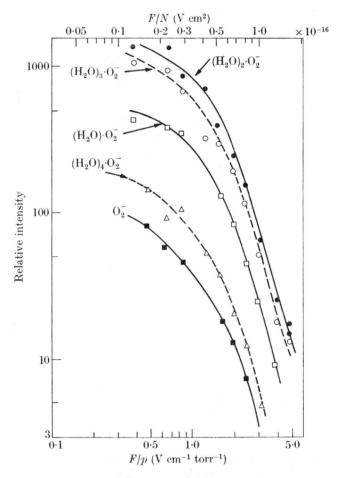

Fig. 19.91. Relative intensity of different negative ions observed as functions of F/p for $F/p < 5$ V cm^{-1} torr^{-1} by Moruzzi and Phelps in a mixture of O$_2$ at 2·16 torr with H$_2$O at 0·44 torr.

Pulse experiments carried out in O$_2$† containing 0·2 to 2·5 per cent of H$_2$O show that the presence of the H$_2$O reduces the detachment rates for the same reason as for O$_2$–CO$_2$ mixtures, i.e. through conversion of the primary O$_2^-$ ion into more stable complex ions such as those discussed above.

† PACK, J. L. and PHELPS, A. V., loc. cit., p. 2084.

4.4.3. *Ions in O_2–N_2 mixtures.* No evidence of the formation of complex ions involving nitrogen has been found, in O_2–N_2 mixtures, either by the combined drift tube and mass spectrometer technique or by pulse methods.

4.5. Notes on the theory of detachment reactions

It has been pointed out that the rate coefficients for ionic reactions do not always follow the simple behaviour expected according to the theory of Gioumousis and Stevenson† in which the reaction cross-section is given by πp_c^2, where p_c is the impact parameter at which orbiting occurs. There has not been much progress towards the development of a more detailed theory. However, Dalgarno and Browne‡ and Herzenberg§ have discussed the detachment reactions involving negative ions along similar lines to the theoretical analysis of the inverse reaction of dissociative attachment discussed in Chapter 12, § 3.6.1 and Chapter 13, § 1.6.

We consider specifically the reactions

$$H+H^- \rightarrow H_2+e \tag{248a}$$
$$\rightarrow H+H+e. \tag{248b}$$

Referring to Fig. 13.36 we see that the interaction energy between an H and H⁻ ion in the lowest $^2\Sigma_u^+$ state of H_2^- is greater than that for the ground $^1\Sigma_g$ state of H_2 for nuclear separations $< R_s$ (see Chap. 13, § 1.6.3). Under these circumstances the state is metastable towards autodetachment

$$H_2^-(\Sigma_u^{2+}) \rightarrow H_2+e, \tag{249}$$

the lifetime of the ion when the nuclear separation is R being $1/\Gamma(R)$.

The situation is very similar to that discussed in Chapter 18, § 8.1.3.1 for radiative collisions between $He(1\,^1S)$ and $He(2\,^1S)$ atoms. In a collision between H and H⁻ interacting in the $^2\Sigma_u^+$ state with impact parameter p and initial relative velocity v the probability that autodetachment will occur is given by

$$P(p) = 1-\exp\left\{-\int_{-\infty}^{\infty} \Gamma(R)\,dt\right\}$$
$$= 1-\exp\left\{2\int_{R_c}^{R_s} \Gamma(R)\frac{dt}{dR}\,dR\right\}, \tag{250}$$

---

† loc. cit., p. 2005.
‡ DALGARNO, A. and BROWNE, J. C., *Astrophys. J.* **149** (1967) 231.
§ HERZENBERG, A., *Phys Rev.* **160** (1967) 80.

where
$$\frac{dR}{dt} = \{v^2 - p^2v^2R^{-2} - 2V(R)/M\}^{\frac{1}{2}},$$

$V(R)$ being the interaction energy, or more correctly the real part of it (see Chap. 12, § 2.5 and Chap. 13, § 1.6.3), assuming the imaginary part to be small.

The cross-section for the associative detachment process is then given by
$$Q(v) = 2\pi g \int_0^\infty P(p)p\,dp,$$

and the reaction rate is the average of $vQ(v)$ over the Maxwellian distribution of velocities of the atoms and ions. g is the probability, $\frac{1}{2}$, that an H atom and H$^-$ ion will interact in the $^2\Sigma_u^+$ state.

Using values for $V(R)$ and $\Gamma(R)$ calculated by Herzenberg and Mandl in their analysis of the inverse process (see Chap. 13, § 1.6.3) Dalgarno and Browne obtained in this way a reaction rate of $1\cdot 9\times 10^{-9}$ cm^3 s^{-1} at 300 °K which varies only slowly with temperature. This is not far from the value 3×10^{-9} cm^3 s^{-1}, independent of temperature, given by the orbiting model of Gioumousis and Stevenson. This is not surprising because the rate at which autodetachment occurs is high. Thus $1/\Gamma(R)$ is of the order 10^{-12} s, which is at least four orders of magnitude smaller than the radiative lifetime associated with an optically allowed transition and is comparable with the time of collision when orbiting occurs. It is not clear why, in some other cases (see Table 19.20), the observed rate constant falls so far below that expected under these circumstances.

If the H atom and H$^-$ ion interact in the $^2\Sigma_g^+$ state, which is repulsive and presumably lies above the $^1\Sigma_u$ repulsive state of H$_2$ (see Fig. 13.36), autodetachment, following the mechanism discussed above, will lead to the reaction (248 b). As this is endothermic by 0·7 eV the reaction rate is very low at ordinary temperatures. Dalgarno and Browne† estimate the rate coefficients as $1\cdot 1\times 10^{-21}$, $2\cdot 9\times 10^{-17}$, $1\cdot 8\times 10^{-12}$, and $4\cdot 9\times 10^{-11}$ cm^3 s^{-1} at temperatures of 500, 1000, 4000, and 16 000 °K respectively.

† loc. cit.

AUTHOR INDEX

Ablow, C. M., 1944, 1947, 1965.
Abragam, A., 1616, 1617.
Ackerman, M., 1448, 1633, 1634.
Adams, C. E., 1473, 1578.
Airey, J. R., 1448, 1633, 1634.
Albritton, D. L., 2012, 2014, 2028, 2029, 2030, 2031, 2032, 2033.
Alder, K., 1591.
Alkemade, C. T. J., 1656, 1657, 1658, 1661, 1662, 1693, 1695.
Alleman, R. S., 1473.
Allison, A. C., 1573, 1576, 1589.
Allison, D. C., 1880, 1881.
Alpert, D., 1663, 1664.
Amdur, I., 1426.
Andersen, W. H., 1488.
Anderson, E. M., 1748.
Anderson, J. M., 1780.
Anderson, L. W., 1842, 1857, 1865, 1867.
Anderson, R. A., 1751, 1754.
Angona, F. A., 1531.
Aquilanti, V., 2051.
Arnold, J. W., 1527.
Asundi, R. K., 2036, 2041, 2042.
Aybar, S., 1531.

Bagot, C. H., 2104.
Baird, J. C., 1842, 1857.
Barnes, W. S., 2012, 2025.
Barrat, J. P., 1693, 1700, 1701, 1702, 1703, 1704, 1708, 1712, 1715, 1717, 1718, 1733, 1905, 1907.
Barrat, M., 1717, 1720.
Barth, C. A., 1857, 1865.
Barwig, P., 1390, 1394, 1397.
Bates, D. R., 1748, 1753, 1920, 1924, 2000.
Bates, J. R., 1676, 1693.
Batey, P. H., 2045, 2055.
Bauer, E., 1920, 1926.
Bauer, H. J., 1522, 1741.
Baumann, M., 1715, 1716, 1733.
Baumann, R. B., 1454.
Beach, J. Y., 1419.
Beahn, T. J., 1739, 1740.
Bean, D. T. W., 1453, 1456.
Beaty, E. C., 1941, 1942, 1974, 1975, 1976, 1977, 1980, 1981, 1982, 1990, 1991, 1994, 1995, 1996, 2002, 2039, 2080.
Beck, D., 1363, 1364, 1376, 1378, 1380, 1381, 1382, 1388, 1389, 1390, 1397, 1629, 1631, 1633.
Becker, E. W., 1401, 1403, 1404, 1454, 1455, 1458, 1459, 1460.

Becker, R., 1479, 1521, 1526, 1531, 1536.
Becker, R. L., 1635, 1637.
Bederson, B., 1355, 1448, 1773, 1774.
Beenakker, J. J. M., 1401, 1403.
Beier, H. J., 1345, 1346, 1411, 1413, 1414, 1415, 1416, 1417.
Bendt, P. J., 1405, 1406, 1456, 1457.
Benedict, W. S., 1500.
Bennett, W. R., 1797.
Bennewitz, H. G., 1384, 1439, 1594, 1595, 1596, 1599, 1613.
Benton, E. E., 1792, 1794, 1797, 1821.
Berg, H. C., 1862.
Berkling, K., 1347, 1363, 1385, 1437, 1439, 1441, 1595.
Bernal, J. D., 1968.
Bernheim, R. A., 1867.
Bernstein, R. B., 1326, 1332, 1336, 1337, 1338, 1353, 1354, 1356, 1357, 1358, 1359, 1368, 1382, 1383, 1384, 1385, 1390, 1393, 1397, 1398, 1412, 1413, 1414, 1423, 1424, 1445, 1446, 1447, 1448, 1449, 1450, 1588, 1589, 1591, 1592, 1593, 1613, 1635, 1638, 1639, 1641, 1643, 1963.
Beterov, I. M., 1797, 1798.
Bethke, G. W., 1508.
Beutler, H., 1744, 1745, 1757.
Bhalla, M. S., 2107.
Bigeon, M. C., 1692.
Binkele, H. E., 1454.
Biondi, M.A., 1801, 1803, 1804, 1806, 1807, 1809, 1822, 1942, 1943, 1954, 1957, 1973, 1975, 1976, 1981, 1982, 1991, 1993, 1994, 1995, 1996, 1997, 2065, 2069, 2070, 2072, 2083, 2104.
Bird, R. B., 1408, 1416, 1417, 1427.
Birely, J. H., 1635, 1638, 1639, 1641, 1643.
Bitter, F., 1712, 1714.
Black, G., 1832, 1833, 1838, 1839.
Blackman, V., 1482, 1483, 1484, 1520, 1521.
Blais, N. C., 1582.
Bloom, M., 1616, 1618, 1619, 1621.
Bloom, S., 1858, 1968, 1970.
Bloomfield, C. H., 2015.
Bochkova, O. P., 1746.
Boers, A. L., 1658.
Boks, J. O. A., 1410, 1411.
Booth, F. B., 1857, 1865.
Born, G. K., 2048.
Bouchiat, M. A., 1845, 1850, 1865, 1867.
Bourgin, D. G., 1467.
Boyd, R. L. F., 2017

Boyer, R. A., 1578.
Bradbury, N. E., 1937, 2037, 2038, 2065, 2070, 2072.
Branscomb, L. M., 2080.
Bresalon, A., 1529.
Brewer, R. G., 1867.
Brickl, D., 1484.
Britton, F. R., 1453, 1456.
Broadway, L. F., 1351.
Brody, J. K., 1741.
Broida, H. P., 1522, 1583, 1585, 1586, 1587, 2021.
Bromley, L. A., 1581.
Brossel, J., 1712, 1714, 1715, 1722, 1733, 1865, 1905.
Brown, H. H., 1355, 1448.
Brown, S. C., 1954, 1973, 1982, 1983, 1986 2004, 2018.
Browne, J. C., 1880, 1881, 1974, 2113, 2114.
Browning, R., 1426, 1428.
Bruch, L. W., 1410.
Buchel'nikova, I. S., 2098.
Buck, U., 1390, 1397, 1429, 1430, 1431.
Buckingham, R. A., 1335, 1400, 1401, 1403, 1404, 1405, 1406, 1409, 1413, 1419, 1424, 1425, 1427, 1428, 1454, 1455, 1459, 1460, 1771, 1876, 1877, 1878, 1879, 1882.
Burch, D. S., 2043, 2069, 2077, 2080, 2096.
Burhop, E. H. S., 1880.
Busala, A., 1531.
Buschmann, K. F., 1527.
Buser, R. G., 2048.
Busko, Z. A., 1748.
Buttner, H., 1801.
Byron, F. W., 1733, 1901, 1902.

Cahn, J. H., 1799, 1800, 1801.
Callaway, J., 1920, 1926.
Callear, A. B., 1503, 1507, 1508, 1522, 1534, 1535, 1538, 1667, 1670, 1677, 1678, 1680, 1689, 1693, 1694, 1807.
Camac, M., 1524.
Cario, G., 1743, 1750.
Carrington, T., 1522, 1583, 1585, 1586, 1587.
Carslaw, H. S., 1945.
Carver, T. R., 1865, 1867.
Casalta, D., 1693, 1700, 1701, 1702, 1703, 1704, 1733.
Cashion, J. K., 1647.
Čermak, V., 1817, 1818, 2036.
Chamberlain, G. E., 1773.
Chanin, L. M., 1942, 1943, 1973, 1975, 1976, 1981, 1982, 1991, 1993, 1994, 1995, 1996, 2003, 2029, 2030, 2065, 2069, 2070, 2072, 2083, 2104.
Chantry, P. J., 2036, 2041.

Chapman, G. D., 1736.
Chapman, S., 1302, 1303, 1304, 1579.
Chebotaev, V. P., 1797, 1798.
Chen, F. M., 1617, 1621.
Claes, A., 1454.
Clarke, E. M., 2034, 2035.
Clifton, D. G., 1427.
Clouston, J. G., 1487.
Cohen, E. G. D., 1455, 1459, 1460.
Cohen, V. W., 1355.
Cojan, J. L., 1691, 1692, 1693, 1700, 1701, 1702, 1703, 1704, 1717, 1718, 1720, 1733, 1907.
Colegrove, F. D., 1823, 1825, 1827, 1878.
Collins, C. B., 1800.
Colombo, L., 1798, 1799.
Compaan, K., 1404.
Condell, W. J., 1739, 1740.
Connor, J. V., 1578.
Cook, W. R., 1959.
Coolidge, A. S., 1420.
Copsey, M. J., 2044, 2045, 2046, 2047, 2055.
Coremans, J. M. J., 1401, 1403.
Corner, J., 1401, 1413.
Corran, P. G., 1531.
Cottrell, T. L., 1514, 1515, 1516, 1525, 1528, 1530, 1531, 1532, 1536, 1562.
Coulliette, J. H., 1681, 1807, 1808.
Court, G. R., 2045, 2055.
Cowan, G. R., 1486.
Cowling, T. G., 1579.
Craggs, J. D., 2081, 2107.
Crampton, S. B., 1862.
Crane, H. R., 2065, 2066, 2069.
Cravath, A. M., 2072.
Crawford, B. L., 1490.
Crompton, R. W., 2012, 2104.
Cross, R. J., 1350, 1450, 1607, 1611, 1612, 1614.
Curnutte, B., 1691, 1693, 1751, 1752, 1754.
Curran, R. K., 2036.
Curtiss, C. F., 1333, 1416, 1417, 1427.
Curtiss, C. W., 1484.
Čvetanovic, P. J., 1693.
Czajkowski, M., 1736.

Dahlquist, J. A., 2011.
Dalgarno, A., 1385, 1386, 1387, 1416, 1417, 1419, 1420, 1421, 1422, 1423, 1424, 1425, 1433, 1445, 1573, 1576, 1588, 1589, 1771, 1773, 1774, 1876, 1877, 1878, 1879, 1880, 1881, 1889, 1890, 1960, 1962, 1963, 1964, 1974, 2000, 2001, 2113, 2114.
Damgaard, A., 1748.
Dash, J. G., 1356, 1358.
Datz, S., 1411, 1413, 1461, 1622, 1635, 1637, 1647, 1648, 1649.

AUTHOR INDEX

Davidovits, P., 1865.
Davidson, N., 1489.
Davies, A. E., 1409, 1454, 1455.
Davies, A. R., 1409, 1454, 1455, 1459, 1460.
Davies, P. G., 2037, 2038.
Davison, W. D., 1385, 1386, 1387, 1410, 1416, 1417, 1571, 1572.
Dawson, P. H., 2028.
Decius, J. C., 1490, 1514.
Decomps, B., 1725, 1726, 1727, 1728, 1733.
Dehmelt, H. G., 1842.
Delany, M. E., 1513, 1517, 1525.
Demkov, J. N., 1998.
Demtröder, W., 1662, 1663, 1693.
Devonshire, A. F., 1548.
de Boer, J., 1403, 1404, 1408, 1455, 1459, 1460.
de Breyn, N. G., 1919.
de Bruyn, P., 1526.
de Bruyn Ouboter, R., 1401, 1403, 1404.
de Haas, W. J., 1401, 1403.
DeMore, W., 1841.
de Vries, A. E., 1583.
Dickinson, P. H. G., 2020, 2045.
Dieke, G. H., 1986.
Dillon, J. A., 2002.
Dinsmore, H. L., 1490.
Dixon, J. R., 1791, 1792.
Dobbie, R. C., 1532, 1562.
Doehring, A., 2069, 2070, 2072.
Dohmann, H. D., 1384.
Donat, K., 1744, 1749.
Donovan, R. J., 1670.
Druyvesteyn, M. J., 1778.
Duffendack, O. S., 1755.
Duffie, J. A. H., 1410, 1411.
Dummel, H., 1390, 1397.
Dumont, M., 1725, 1726, 1727, 1728, 1733.
Dunkin, D. B., 2021.
Dunoyer, L., 1342.
Durand, E., 1506, 1538.
Düren, R., 1326, 1337, 1413, 1414, 1415, 1416, 1417.
Dushman, S., 1794.
Dutton, J., 2029, 2030, 2037.
Dwyer, R. J., 1496.
D'yakonov, M. L., 1898.

Ebbinghaus, E., 1780.
Eden, D. C., 1528.
Edmonds, P. D., 1475.
Edwards, A. J., 1533, 1534.
Edwards, H. D., 2002.
Eiber, H., 2043, 2068, 2069, 2072, 2073, 2074, 2075, 2076, 2100.
Eisenschimmel, W., 1757.
Ekin, J. W., 2085, 2101, 2102.

Elford, M. T., 2012.
Ellett, A., 1355, 1703.
Emrich, R. J., 1484.
Ener, C., 1578.
Enskog, D., 1302, 1303.
Erikson, H. A., 1933, 1934.
Estermann, I., 1354, 1355, 1433.
Eucken, A., 1521, 1526, 1527, 1531, 1536, 1561, 1579, 1582.
Evans, M. G., 1676, 1693.
Evans, R. B., 1582.
Evett, A. A., 1453.
Eyring, H., 1929.

Faroux, J. P., 1721, 1722, 1733, 1905.
Fehsenfeld, F. C., 2021, 2025, 2049, 2051, 2052, 2054, 2055, 2056, 2057, 2058, 2059, 2060, 2063, 2103.
Feltgen, R., 1413, 1415, 1462.
Ferguson, E. E., 1792, 1794, 1797, 1821, 2021, 2025, 2049, 2054, 2055, 2056, 2057, 2058, 2059, 2060, 2063, 2103, 2105.
Ferguson, M. G., 1532.
Fine, J., 1681, 1682, 1683, 1693, 1694.
Fite, W. L., 2019, 2020, 2045, 2049, 2057, 2106.
Florin, H. 1385.
Fluegge, R. A., 1975, 1976, 2028, 2029, 2030, 2031, 2037, 2038, 2043, 2068.
Fluendy, M. A. D., 1355, 1361, 1362, 1452.
Flynn, G. W., 1508.
Fogg, P. G. T., 1531.
Fokkens, K., 1401, 1403, 1404.
Foley, H. M., 1901, 1902.
Foner, S. N., 1354, 1355, 1433.
Fontana, C. M., 2033.
Ford, K. W., 1311, 1319, 1333.
Fouracre, R. A., 2047, 2048.
Fowler, R. G., 1759, 1762, 1767.
Fowler, R. H., 1968.
Fox, J. W., 1419, 1421, 1424, 1425, 1426, 1427, 1428, 1886, 1887, 1888.
Fox, R. E., 1452.
Francis, W. E., 1923.
Franck, E. U., 1581, 1582.
Franck, J., 1743, 1750, 1933.
Franken, P. A., 1823, 1842, 1865.
Frankevich, E. L., 2006, 2034.
Franz, F. A., 1729, 1733.
Franzen, W., 1845, 1866, 1867.
Fred, M., 1741.
Fricke, E., 1536.
Fricke, J., 1729, 1731, 1733.
Friedman, L., 2050, 2051, 2052.
Frish, S. E., 1746.
Frommhold, L., 2037, 2038, 2043, 2081, 2085, 2096, 2097, 2098, 2099.
Frost, L. S., 1817.

Fuchs, R., 2036.
Fujii, Y., 1528.
Funabashi, K., 1968.

Gabriel, A. H., 1760.
Gabrysh, A. F., 1578.
Gaide, W., 1413, 1462.
Gal, E., 1419, 1421, 1424, 1425, 1427, 1428, 1886, 1887, 1888.
Gallagher, A., 1729, 1733, 1913.
Gascoigne, J., 2012.
Gasilevich, E. S., 1517.
Gatz, C. R., 1944, 1947, 1950, 1965.
Gaydon, A. G., 1487, 1488, 1518, 1520, 1521, 1670.
Geballe, R., 2043, 2069, 2077, 2080, 2096.
Generalov, N. A., 1520, 1521.
Gerardo, J. B., 2048.
Ghosh, S. N., 2002.
Gibbs, H. M., 1845, 1846, 1848, 1854, 1865.
Giese, G. F., 2007.
Gilbody, H. B., 2000.
Gilles, D. C., 1454, 1459, 1460.
Gilman, G. L., 2054.
Gioumousis, G., 2005, 2034, 2051, 2052, 2062, 2064, 2105, 2113, 2114.
Gislason, E. A., 1350, 1450, 1611, 1612.
Gladstone, S., 1929.
Glass, I. I., 1487.
Glotov, I. I., 1778.
Goldan, P. D., 2025, 2049, 2055, 2057, 2058, 2059, 2060.
Goldenberg, M., 1861, 1862.
Gordon, R., 1623.
Gordon, R. G., 1607, 1614.
Gorelik, G., 1511.
Grad, H., 1479.
Graffunder, W., 1780.
Graham, J. R., 1954, 1976, 2028, 2029, 2030, 2031, 2037, 2038, 2043, 2068.
Gran, W. H., 1755.
Grant, F. A., 1788, 1789, 1791, 1792.
Greenawalt, E. M., 1881.
Greene, E. F., 1356, 1448, 1486, 1565, 1567, 1629, 1631, 1633, 1634.
Greenspan, M., 1578.
Grey, J., 1344.
Griffith, W., 1484, 1565, 1567.
Grilly, E. R., 1409.
Grinberg, R. O., 1748.
Grindley, G. C., 1933.
Groblicki, P. J., 1393, 1398.
Grosser, A. E., 1635, 1638, 1639, 1641, 1643.
Grossetête, F., 1715, 1733, 1845, 1850, 1865.
Gulley, E. R., 1579.
Gusinow, M. A., 2048.

Gutowski, F. A., 1536.

Haas, J., 1729, 1733.
Hamel, J., 1693, 1700, 1701, 1702, 1703, 1704, 1733.
Hamermesh, M., 1741.
Hamilton, J., 1400.
Hammel, E. F., 1404.
Hanks, P. A., 1531.
Hanle, W., 1703.
Hanson, H. G., 1675, 1693.
Happer, W., 1723, 1733.
Harrison, A. G., 2034.
Harrison, H., 1411, 1412, 1413, 1452, 1453, 1457, 1458, 1461.
Harteck, P., 1426, 1456, 1457.
Hartland, A., 1618.
Hassè, H. R., 1959.
Hasted, J. B., 2000, 2015, 2016, 2020, 2044, 2045, 2048, 2055, 2100.
Heath, H. R., 1456, 1457.
Heberling, R., 1454.
Heddle, D. W. O., 1760.
Helbing, R., 1325, 1347, 1363, 1369, 1371, 1372, 1373, 1383, 1413, 1414, 1448, 1462.
Hellwig, H., 1861, 1864, 1865, 1886, 1887.
Henkel, U., 1390, 1397.
Herm, R. R., 1623, 1640, 1643.
Herman, R., 1500.
Herman, Z., 2036.
Herriott, D. R., 1797.
Herschbach, D. R., 1350, 1355, 1361, 1362, 1450, 1452, 1611, 1612, 1623, 1630, 1631, 1635, 1638, 1639, 1640, 1641, 1643, 1644, 1645, 1646.
Herzenberg, A., 2113, 2114.
Herzfeld, K. F., 1467, 1530, 1549, 1550, 1551, 1552, 1554, 1563.
Hildebrandt, A. F., 1857, 1865.
Hill, D. L., 1333.
Hilsenrath, J., 1581, 1582.
Hirschfelder, J. O., 1416, 1417, 1427, 1436, 1437, 1552, 1576.
Hobart, J., 1842, 1865.
Hocker, L. C., 1508.
Holborn, L., 1410, 1411.
Holmes, R., 1520, 1521, 1524.
Holstein, T., 1654, 1663, 1664, 1785.
Hooker, W. J., 1489, 1491, 1520, 1521, 1524.
Hooymayers, H. P., 1656, 1657, 1658, 1661, 1662, 1693, 1695.
Hornbeck, J. A., 1938, 1940, 1941, 1973, 1975, 1976, 1994, 1995, 2043, 2065.
Hornig, D. F., 1486, 1488, 1565, 1567, 1578, 1579.
Hoselitz, K., 1957, 1968, 1971, 1972.
Hostettler, H. U., 1356, 1357, 1358.

Howard, A. J., 1406, 1407.
Hubbard, J. C., 1472, 1529, 1531, 1565, 1578.
Huber, E. L., 2037, 2038.
Huber, P. W., 1521, 1528, 1565, 1567.
Hudson, B. C., 1691, 1693, 1751, 1752, 1754.
Huet, M., 1691.
Huggins, R. W., 1799, 1800, 1801.
Hughes, R. H., 1761, 1768.
Hull, R. J., 1845, 1846, 1854, 1865.
Hundhausen, E., 1626, 1627, 1634.
Hunten, D. M., 1839.
Hurlbut, F. C., 2021.
Hurle, I. R., 1518, 1520, 1521, 1670.
Hurt, W. B., 1805.
Husain, D., 1670.
Husseini, I., 1409, 1454.
Huxley, L. G. H., 2104.

Ibbs, T. L., 1456, 1457.
Ilün, B., 1410, 1411.
Ivko, A., 2034.

Jaacks, H., 1527.
Jackson, J. M., 1541, 1545.
Jackson, S. R., 2033.
Jaeger, G., 2029, 2030.
Jaeger, J. C., 1945.
James, H. M., 1420.
Jarrett, S. M., 1851, 1865.
Javan, A., 1508, 1797, 1798.
Jean, P., 1733.
Jenkins, D. R., 1656, 1657, 1658, 1659, 1660, 1661, 1693.
Jesse, W. P., 1811, 1812, 1821, 1822.
Johnson, B. R., 1446.
Johnston, H. L., 1409, 1454, 1455.
Jordan, J. A., 1740.
Josephy, B., 1744, 1745.

Kaneko, Y., 2015, 2016, 2048.
Kantrowitz, A., 1344, 1492, 1493, 1494, 1496, 1521, 1528, 1565, 1567.
Karl, G., 1683, 1689, 1690.
Karplus, M., 1649.
Kastler, A., 1712.
Kaul, V. W., 2036.
Kay, R. B., 1761, 1768.
Keesom, P. H., 1401, 1403, 1454.
Keesom, W. H., 1401, 1403, 1410, 1411, 1454, 1559, 1560, 1606, 1612, 1613.
Keller, W. E., 1403, 1404.
Kellogg, J. M. B., 1617.
Kennard, E. H., 1303, 1315, 1347.
Kerr, D. E., 1974, 1994.
Khaikin, A. S., 1746, 1747, 1748, 1749.
Kiefer, J. H., 1484, 1518.

Kilpatrick, J. E., 1404.
Kingston, A. E., 1385, 1387, 1416, 1417, 1433, 1771, 1773, 1774.
Kinsey, J. L., 1620, 1646.
Kirchner, F., 2028.
Kirkwood, J. G., 1387.
Kisilbach, B., 1693.
Klein, O., 1435.
Kleinberg, A. V., 1583.
Kleppner, D., 1861, 1862.
Knaap, H. F. P., 1401, 1403.
Knable, N., 1865.
Kneser, H. O., 1467, 1475, 1522, 1741.
Knipp, J. K., 1434.
Knötzel, H., 1536.
Knötzel, L., 1536.
Knudsen, V. O., 1475, 1536.
Kodera, K., 1448, 1633, 1634.
Kolb, A. C., 1657.
Kolos, W., 1420, 1421, 1422, 1423.
Komarov, V. N., 2036.
Kondratiev, V., 1693.
Konowalow, D. D., 1436, 1437.
Kovacs, M. A., 1508.
Kovar, F. R., 2039.
Koyano, I., 2007.
Kramer, K., 1347, 1363.
Kramer, K. H., 1591, 1592, 1593, 1595.
Kraulinya, E. K., 1746, 1747, 1748, 1751, 1754.
Krause, L., 1736, 1737, 1739, 1740.
Kruger, P. G., 1879.
Kruithof, A. A., 1778.
Krumbein, A. D., 1788, 1789.
Kruus, P., 1683, 1689, 1690.
Küchler, L., 1526, 1527.
Kunkel, W. B., 2021.
Kusch, P., 1351, 1356.
Kvifte, G., 1831, 1836, 1840.
Kwei, G. H., 1630, 1631, 1643, 1644, 1645, 1646.

Lacroix-Desmazes, F., 1720.
Laidler, K., 1929.
Lalita, K., 1621.
Lamb, H., 1800.
Lamb, J., 1475.
Lambert, J. D., 1528, 1529, 1530, 1531, 1532, 1533, 1534, 1535.
Landau, L. D., 1324, 1326, 1345, 1383, 1916.
Landman, D. A., 1743.
Landorf, R. W., 1365, 1366, 1413, 1415, 1462.
Langevin, P., 1959.
Langmuir, I., 1432.
Langstroth, G. F. O., 2020, 2044, 2045, 2055.

AUTHOR INDEX

Laniepce, B., 1718, 1733, 1907.
Lecler, D., 1733.
Lecluse, Y., 1717, 1718, 1733, 1906, 1907, 1909.
Ledsham, K., 2000.
Lee, K. P., 1520, 1521.
Lees, J. H., 1757, 1758.
Leffel, C. S., 1974, 1994.
Legvold, S., 1527, 1531, 1533, 1535.
Leipunsky, A., 1693.
Lennon, J. J., 1987.
Lester, W. A., 1589.
Levine, J., 1774.
Levine, R. D., 1466.
Lezdin, A. E., 1751, 1754.
Libby, P. A., 1479.
Lifshitz, E. M., 1324, 1326, 1383.
Lin, C. C., 1760.
Linder, B., 1436, 1437.
Lindsay, R. B., 1528.
Lipscomb, F. J., 1504, 1538.
Lipsicas, M., 1456, 1457, 1616, 1618, 1619.
Lipson, H. C., 1693.
Litovitz, T. A., 1549, 1550, 1551, 1563.
Llewyllyn Jones, F., 2030, 2037.
Lochte-Holtgreven, W., 1734.
Loeb, L. B., 2072.
Loesch, H. J., 1363, 1364, 1376, 1378, 1380, 1381, 1382, 1390.
Loria, S., 1744.
Los, J., 1583.
Losev, S. A., 1520, 1521.
Lukasik, S. J., 1521.
Lulla, K., 1355, 1448.
Luscher, E., 1729, 1733.
Luszczynski, K., 1406, 1407.
Lutz, R. W., 1484, 1518.
Lyman, T., 1879.
Lynn, N., 1419, 1420, 1421, 1422, 1423, 1424, 1924, 2000, 2001.

McAfee, K. B., 2037, 2038.
McCloskey, K. E., 1454, 1455.
McCoubrey, A. O., 1663, 1664, 1809.
McCoubrey, J. C., 1527, 1536.
McDaniel, E. W., 2012, 2013, 2014, 2025, 2028, 2029, 2036, 2041, 2065, 2066, 2069.
McDermott, N. M., 1733.
McDowell, M. R. C., 1960, 1962, 1963, 1964.
McElroy, M. B., 1839.
McFarland, R. H., 1751, 1754.
Macfarlane, I. M., 1514, 1515, 1525.
McGee, I. J., 1410.
Mack, J., 1484.
McKnight, L. G., 2037, 2038.
McLain, J., 1532, 1562.
McNeal, R. J., 1867.

Madson, J. M., 1974, 1981, 1982, 1994, 2003.
Magee, J. L., 1968.
Mais, W., 1383.
Maisch, W. G., 1435, 1436.
Malone, D. P., 1954.
Mandelberg, H. I., 1739, 1740, 1926, 1927, 1928.
Mandl, F., 2114.
Mann, J. B., 1582.
Margenau, H., 1386, 1427, 1453, 1968, 1970.
Mariens, P., 1526, 1536, 1565.
Marino, L. L., 1414, 1415, 1416, 1417.
Markovic, B., 1798.
Marriott, R., 1553, 1554, 1555, 1556, 1557, 1558, 1559, 1561, 1563, 1589, 1880.
Martin, D. W., 2011, 2012, 2014, 2028, 2029, 2036, 2041.
Martin, M., 1733.
Martin, P. E., 1531.
Martin, R. M., 1355, 1361, 1362, 1452.
Mason, E. A., 1322, 1323, 1325, 1339, 1403, 1404, 1405, 1406, 1427, 1435, 1436, 1437, 1455, 1576, 1580, 1581, 1582, 1615, 1963, 2033.
Massey, H. S. W., 1306, 1307, 1308, 1309, 1324, 1327, 1383, 1400, 1403, 1445, 1588, 1589, 1624, 1760, 1972, 1973, 2000.
Matheson, A. J., 1528, 1530, 1532, 1562.
Matland, C. G., 1664, 1683.
Matsen, F. A., 1792, 1794, 1797, 1821, 1877, 1878, 1881.
Matsuzawa, M., 1881, 1882, 1883.
Matthews, D. L., 1520, 1521.
Mazo, R. M., 1857.
Megill, L. R., 2015, 2016, 2048.
Meinke, C., 1412.
Meissner, K. W., 1780.
Messenger, H. A., 1679.
Metropolis, N., 1404.
Meunier, J., 1724, 1725, 1733, 1898, 1899, 1904, 1912.
Meyer, E., 1578.
Meyerott, R., 1962, 1963, 1968, 1972, 1973.
Michels, A., 1403, 1410, 1411.
Mierdel, G., 1807, 1997.
Miller, F. L., 1760.
Miller, N. E., 1406, 1407.
Miller, R. C., 1356.
Miller, T. M., 2012, 2014, 2028, 2029, 2036.
Millikan, R. C., 1489, 1491, 1498, 1499, 1500, 1519, 1520, 1521, 1524, 1532, 1534, 1535, 1537.
Milne, E. A., 1654.
Minten, A., 1457.
Minturn, R. E., 1635, 1637.

Misenta, R., 1401, 1403, 1404, 1454, 1455, 1459.
Mitchell, A. C. G., 1676, 1693, 1849.
Mitchell, J. H., 2037, 2038.
Mittelstadt, V. R., 1974, 1985, 1987, 1994, 1995, 1996.
Mohler, F., 1734.
Mohr, C. B. O., 1307, 1324, 1327, 1383, 1400, 1760, 1972, 1973, 2000.
Moiseiwitsch, B. L., 2000, 2001.
Molnar, J. P., 1780, 1788, 1789, 1792, 1804, 1973.
Monchick, L., 1322, 1323, 1403, 1404, 1405, 1406, 1580, 1581, 1582, 1615.
Montroll, E. W., 1490.
Moore, G. E., 1411, 1413, 1461, 1500.
Moos, H. W., 1854, 1855, 1865.
Moran, T. F., 2050, 2051, 2052.
Morduchow, M., 1479.
Morgan, J. E., 2025.
Morris, D., 2017.
Morse, F. A. 1368, 1390, 1397, 1412, 1413.
Moruzzi, J. L., 2068, 2077, 2078, 2079, 2080, 2081, 2085, 2101, 2102, 2103, 2104, 2106, 2107, 2108, 2110, 2111, 2112.
Moseley, J. T., 2036, 2041, 2042.
Mott, N. F., 1298, 1306, 1308, 1309, 1541, 1545, 1624.
Mott-Smith, H. M., 1480.
Moursund, A. L., 1448, 1633, 1634.
Muckerman, J. T., 1446.
Mueller, C. R., 1365, 1366, 1413, 1415, 1462.
Mulcahy, M. J., 1987.
Müller, F., 1409, 1454, 1455.
Mulliken, R. S., 1891.
Munn, R. J., 1403, 1404, 1405, 1406.
Munson, R. J., 1937, 1966, 1968, 1971, 1972, 1995.
Muschlitz, E. E., 1355, 1361, 1362, 1452, 1774, 1775, 1815, 1816, 1821, 1822, 1823, 1828, 2033.

Nakamura, H., 1881, 1882, 1883.
Nee Tsu-wei, 1760, 1767, 1768.
Neynaber, R. H., 1359, 1375, 1378, 1379, 1380, 1382, 1384, 1385, 1411, 1413, 1414, 1417, 1769, 1770, 1771, 1772, 1773, 1776.
Niblett, P. D., 1459, 1460.
Nickerson, J. L., 1879.
Nielsen, R. A., 1937, 2065, 2070, 2072.
Nijhoff, G. P., 1410, 1411.
Niles, F. E., 1988, 1990.
Noble, J. D., 1621.
Norberg, R. E., 1406, 1407.
Norris, J. A., 1643, 1644, 1645, 1646.
Norrish, R. G. W., 1504, 1538, 1654, 1655, 1670, 1677, 1678, 1693, 1694.

Novick, R., 1733, 1842, 1865.
Nümann, E., 1526, 1527, 1531, 1536, 1561.

O'Brien, T. J. P., 1338.
O'Connor, C. L., 1531.
Offerhaus, M. J., 1455, 1459, 1460.
Omont, A., 1697, 1719, 1724, 1725, 1733, 1892, 1898, 1899, 1902, 1904, 1908, 1911, 1912.
Omura, I., 2007.
Onnes, H. K., 1410, 1411.
Opfer, J. E., 1406, 1407.
Oppenheim, I., 1616, 1619.
Osberghaus, O., 1457.
Osburg, L. A., 1532, 1537.
Oskam, H. J., 1974, 1981, 1982, 1985, 1987, 1994, 1995, 1996, 2003.
Otto, J., 1410, 1411.
Otto, W., 2029, 2030.

Pack, J. L., 1781, 2069, 2070, 2071, 2082, 2083, 2084, 2085, 2086, 2087, 2089, 2090, 2091, 2092, 2097, 2108, 2109, 2112.
Padley, P. J., 1658.
Parbrook, H. D., 1473, 1565, 1578.
Parker, J. G., 1473, 1474, 1524, 1578, 1580.
Parks-Smith, D. G., 1534, 1535.
Patterson, P. L., 1974, 1975, 1976, 1977, 1980, 1981, 1982, 1990, 1991, 1992, 1994, 2080.
Paul, W., 1594, 1595.
Pauling, L., 1419.
Pauly, H., 1325, 1326, 1337, 1347, 1350, 1363, 1369, 1371, 1372, 1373, 1383, 1390, 1397, 1413, 1414, 1429, 1430, 1431, 1447, 1448, 1462, 1626, 1634.
Pavlović, Z., 1798.
Payman, W., 1478.
Pearce, A. F., 1957, 1972, 1975, 1990, 1991, 1993, 1994.
Pemberton, D., 1533, 1534.
Penner, S. S., 1508.
Penning, F. M., 1778.
Percival, I. C., 1445, 1589.
Perel, V. I., 1898.
Peršin, A., 1798.
Peters, H. E., 1842, 1862, 1865.
Petralia, S., 1578.
Phelps, A. V., 1780, 1781, 1783, 1788, 1789, 1792, 1795, 1797, 1803, 1804, 1806, 1814, 1817, 1821, 1822, 1879, 1880, 1973, 1982, 1983, 1986, 2004, 2018, 2065, 2068, 2069, 2070, 2071, 2077, 2078, 2079, 2080, 2081, 2082, 2083, 2084, 2085, 2086, 2087, 2089, 2090, 2091, 2092, 2097, 2101, 2102, 2103, 2104, 2106, 2107, 2108, 2109, 2110, 2111, 2112.
Phillips, L. R., **1933**.

Pierce, G. W., 1470, 1523.
Piketty-Rives, C. A., 1715, 1733.
Pipkin, F. M., 1842, 1857.
Platzman, R. L., 1815.
Pohl, R. W., 1933.
Polanyi, J. C., 1647, 1683, 1689, 1690.
Pollack, E., 1773.
Porter, R. N., 1649.
Poschenrieder, W., 2007.
Poshusta, R. D., 1877, 1878.
Powell, C. F., 1936, 1937, 1957, 1972, 1973, 1974, 1976, 1994, 2014, 2065.
Powers, R. S., 1333.
Prasad, A. N., 2081.
Present, R. D., 1420.
Presnyakov, L., 1919.
Prileschaweja, N., 1675.
Pringsheim, P., 1751.

Rabi, I., 1351, 1352, 1355, 1383.
Rae, A. G. A., 1736.
Raff, L. M., 1649.
Rali, I. I., 1617.
Ramsauer, C., 1342.
Ramsey, A. T., 1865, 1867.
Ramsey, N. F., 1342, 1346, 1617, 1861, 1862.
Raper, O. F., 1841.
Rapp, D., 1923.
Rautian, S. G., 1746, 1747, 1748, 1749.
Raw, C. J. G., 1325.
Rayleigh, Lord, 1703.
Read, A. W., 1514, 1515, 1525, 1532, 1562.
Reck, G. P., 1448, 1633, 1634.
Rees, A. L. G., 1435.
Rees, J. A., 2069, 2070, 2071, 2072.
Rees, W. D., 2030.
Reich, G., 1412.
Rhodes, C. K., 1508.
Rhodes, J. E., 1565, 1573.
Rice, F. O., 1467.
Rice, W. E., 1455.
Richards, H. L., 1774, 1775, 1815, 1822, 1823, 1828.
Richardson, E. G., 1521.
Ridler, K. I. W., 2037, 2038.
Riehl, J. W., 1620.
Riesz, R., 1986.
Rietveld, A. O., 1409, 1459.
Robben, F., 1522.
Roberts, C. S., 1570, 1571, 1572, 1576.
Roberts, R. W., 1356.
Robertson, W. W., 1759, 1792, 1794, 1797, 1800, 1821, 1988, 1990.
Robinson, E., 1774.
Roessler, F., 1505, 1506, 1507, 1538.

Rol, P. K., 1359, 1375, 1378, 1379, 1380, 1382, 1384, 1385, 1414, 1417.
Rontgen, W. C., 1511.
Roos, B. W., 1455, 1459, 1460.
Rose, M. E., 1566.
Rosen, N., 1920.
Rosenberg, P., 1353, 1383.
Rosenfeld, J. L. J., 1629, 1630.
Rosin, S., 1351, 1352, 1355, 1383.
Ross, J., 1356, 1448, 1629, 1630, 1631, 1633, 1634.
Rossing, T. D., 1527, 1531.
Rothe, E. W., 1353, 1354, 1359, 1375, 1378, 1379, 1380, 1382, 1383, 1384, 1385, 1411, 1413, 1414, 1415, 1416, 1417, 1447, 1448, 1449, 1613, 1769, 1770, 1771, 1772, 1773, 1776.
Rothman, A. J., 1581.
Roy, A. S., 1566.
Rudge, M. R. H., 1889, 1890.
Rutgers, A. J., 1467.
Rutherford, J. A., 2019, 2020, 2045, 2049, 2057, 2106.
Rydberg, R., 1435.
Ryder, H. M., 1966.

Sadauskis, J., 1811, 1812, 1821, 1822.
Sagalyn, P., 1712.
St. John, R. M., 1759, 1760, 1762, 1766, 1768.
Saloman, E. B., 1723, 1733.
Salop, A., 1773.
Salter, P., 1452.
Salter, R., 1528, 1529, 1530, 1531, 1532.
Samson, E. W., 1679, 1680.
Samson, J. A. R., 1937, 1938, 2037, 2038, 2043.
Sands, R. H., 1842, 1854, 1855, 1865.
Saporoschenko, M., 2014, 2028, 2029, 2030, 2032, 2036, 2037, 2038.
Saulgozka, A. K., 1748.
Sayers, J., 2020, 2044, 2045, 2046, 2047, 2048, 2052, 2055.
Schade, R., 1801.
Schafer, K., 1527.
Schamp, H. W., 1963.
Schearer, L. D., 1742, 1743, 1823, 1825, 1827, 1878.
Scheer, M. D., 1681, 1682, 1683, 1693, 1694.
Schiff, H. I., 2021, 2025, 2049, 2055, 2057, 2058, 2059, 2060.
Schiff, L. I., 1325, 1438, 1591.
Schillbach, H., 1791.
Schlier, C., 1339, 1347, 1359, 1363, 1378, 1380, 1381, 1382, 1385, 1392, 1437, 1439, 1441, 1594, 1595.
Schmeissner, F., 1401, 1403, **1404**.

Schmeltekopf, A. L., 2021, 2025, 2049, 2054, 2055, 2056, 2057, 2058, 2059, 2060, 2063, 2103.
Schmidt, H. W., 1456, 1457.
Schmidt, T. W., 1647, 1648, 1649.
Schneider, W. G., 1410, 1411.
Schrader, J. E., 1966.
Schulz, G. J., 1794, 2036, 2041.
Schumacher, H., 1449, 1613.
Schutz, W., 1655, 1791.
Schwartz, R. N., 1551, 1552.
Schweitzer, W. G., 1720.
Scott, B. W., 1380, 1382.
Scott, D. R., 1881.
Scriven, R. A., 1404, 1405, 1406.
Seaton, M. J., 2053.
Seiwert, R., 1740.
Sena, L., 1935.
Sessler, G., 1578.
Sette, D., 1529, 1531.
Shannon, T. W., 2034.
Sharma, R. D., 1649.
Sharpless, R. L., 1832, 1944, 1947, 1950, 1965.
Shepherd, W. C. F., 1478.
Sheridan, W. F., 2002.
Sherman, F. S., 1488, 1579.
Shields, F. D., 1520, 1521, 1526.
Shiffrin, R. M., 1406, 1407.
Sholette, W. P., 1816, 1821.
Shortley, G. H., 1791.
Shuler, K. E., 1490.
Silin, Y. A., 1751.
Silverman, S., 1500.
Simons, J. H., 2033.
Simpson, O. C., 1354.
Sinnott, G., 2029, 2030.
Sipler, D. P., 2037, 2038.
Siskin, M., 1693.
Sittig, E., 1522, 1741.
Skinner, B. G., 1920.
Skinner, H. W. B., 1757, 1758.
Slater, J. C., 1387, 1400.
Slawsky, Z. I., 1551, 1552.
Slobodskaya, P. V., 1517, 1525.
Smiley, E. F., 1521, 1526.
Smit, J. A., 1658.
Smith, A. C. H., 1426, 2049.
Smith, D., 2044, 2045, 2046, 2047, 2048, 2052, 2055.
Smith, F. A., 1520, 1521, 1524.
Smith, F. J., 1403, 1404, 1405, 1406, 1424, 1425, 1433, 1436, 1437, 1885, 1891, 1892, 1893.
Smith, I. W. M., 1690.
Smith, P. T., 2098.
Smith, R. A., 2100.
Smith, W. M., 1654, 1655, 1693.

Sneddon, I. N., 1298.
Snider, R. F., 1617, 1621.
Snow, W. R., 2019, 2020, 2045, 2057.
Snuggs, R. M., 2036, 2041.
Sobel'man, I., 1919.
Sommer, L. A., 1879.
Sommers, H. S., 1356, 1358.
Srivastava, B. N., 1415.
Srivastava, K. P., 1415.
Stavseth, R. M., 1473, 1578.
Stebbings, R. F., 1816, 2049.
Stehl, O., 1458, 1460.
Stern, O., 1354, 1355, 1433.
Stevenson, D. P., 2005, 2034, 2051, 2052, 2062, 2064, 2105, 2113, 2114.
Stewart, A. L., 2000.
Stewart, E. S., 1472, 1565, 1567.
Stewart, J. L., 1472, 1565, 1567.
Streed, E. R., 1657.
Stretton, J. L., 1533, 1534, 1535.
Strunck, H. J., 1359, 1378, 1380, 1381, 1382, 1392.
Stueckelberg, E. C. G., 1917, 1921.
Sugden, T. M., 1658.

Taconis, K. W., 1401, 1403, 1404.
Takayanagi, K., 1459, 1460, 1546, 1550, 1568, 1570, 1571, 1572, 1573.
Talbot, L., 1488.
Tal'roze, V. L., 2006, 2034.
Tanaka, I., 2007.
Tanaka, Y., 1879.
Tate, J. T., 2098.
Taylor, E. H., 1622.
Taylor, G. I., 1478.
Taylor, J. B., 1432.
Taylor, R. L., 1489.
Teller, E., 1545.
Tempest, W., 1473, 1520, 1521, 1524, 1565, 1578.
Terenin, A. N., 1583, 1675.
Teter, M. P., 1759.
Thaler, W. J., 1578.
Thomson, K., 1755.
Thrush, B. A., 1504, 1538.
Thys, L., 1565.
Tickner, A. W., 2028.
Tikhomirov, M. V., 2036.
Tip, A., 1583.
Tisza, L., 1483, 1578.
Tittel, K., 1715, 1717, 1733.
Tizard, H. T., 1933.
Toennies, J. P., 1595, 1599, 1601.
Tomkins, F. S., 1741.
Toschek, P., 1347, 1363, 1385, 1437, 1439, 1441, 1595.
Touw, T. R., 1622.
Townsend, J. S., 1933.

Trautz, M., 1409, 1454, 1455.
Trischka, J. W., 1622.
Trujillo, S. M., 1359, 1375, 1378, 1379, 1380, 1382, 1414, 1417, 1769, 1771, 1772.
Tsuchiya, S., 1671, 1693.
Tunitskii, N. N., 2036.
Tyerman, W. J. R., 1667.
Tyndall, A. M., 1932, 1933, 1936, 1937, 1957, 1966, 1972, 1973, 1974, 1975, 1976, 1990, 1991, 1993, 1994, 1995, 2014, 2065.
Tyndall, J., 1511.

Ubbelohde, A. R., 1527.
Ubbink, J. B., 1401, 1403.
Uhlenbeck, G. E., 1580.
Ung, A. Y.-M., 1839.
Urushihara, K., 1528.

Vainshtein, L., 1919.
Valley, L. M., 1533, 1535.
van de Graaf, R. J., 1936.
van den Berg, G. J., 1409.
Vanderslice, J. T., 1325, 1435, 1436, 1963, 2033.
van der Valk, F., 1406, 1407, 1411, 1412, 1461.
Vanier, J., 1860, 1862, 1865.
van Itterbeek, A., 1401, 1403, 1409, 1454, 1459, 1526, 1536, 1565, 1567.
van Kranendonk, J., 1404, 1608.
van Leeuwen, J. M. J., 1455, 1459, 1460.
van Lint, V. A. J., 2019, 2020, 2045, 2057.
van Paemel, O., 1401, 1403, 1454.
Van Vleck, J. H., 1928.
Varney, R. N., 1995, 2033, 2039, 2041, 2043, 2082.
Varnum, C. M., 1867.
Vegard, L., 1831, 1836, 1840.
Veneklasen, L. H., 1382.
Verdeyen, J. T., 2048.
Verhaegen, L., 1565, 1567.
Vessot, R. F. C., 1862.
Vines, R. G., 1581.
Vogel, J. K., 2037, 2038.
Volpi, G. G., 2051.
von Busch, F., 1350, 1359, 1378, 1380, 1381, 1382, 1392.
von Hundhausen, E., 1390, 1397, 1448, 1626, 1627, 1634.
von Keussler, V., 1703.
Voshall, R. E., 2069, 2070, 2071, 2082, 2083, 2108.

Wade, C. G., 1616, 1619.
Waech, T. G., 1423, 1424.
Wakano, M., 1333.

Waldmann, L., 1456, 1457.
Walker, R. A., 1527.
Wallace, L., 1839.
Walters, G. K., 1823, 1825, 1827, 1878.
Wang Chang, G. K., 1580.
Wannier, G. H., 1934, 1935.
Warburton, B., 1531.
Warneck, P., 2007, 2008, 2010, 2034, 2042, 2045, 2048, 2050, 2051, 2052, 2053, 2055, 2057, 2058, 2061, 2062, 2063.
Warnock, T. T., 1641.
Watanabe, T., 1898.
Watson, G. M., 1582.
Watson, W. W., 1406, 1407.
Waugh, J. S., 1620.
Webb, H. W., 1679, 1680.
Weber, D., 1508.
Weber, G. G., 1963.
Weingartshofer, A., 2034, 2035.
Weissler, G. L., 1937, 1938, 2007, 2037, 2038, 2043.
Weissman, S., 1427.
Wergeland, A., 1307.
Wheeler, J. A., 1311, 1319, 1333.
White, D. R., 1518, 1519, 1520, 1521, 1524, 1536, 1537.
White, J. U., 1496.
Wight, H. M., 1527.
Wigner, E., 1757.
Wild, N. E., 1456, 1457.
Williams, D. A., 1576.
Williams, E. M., 2030.
Williams, G. J., 1670, 1677, 1678, 1680, 1689, 1693, 1694, 1807.
Williams, S. A., 1960, 1962, 1963, 1964.
Wilson, K. R., 1643, 1646.
Winans, J. G., 1675, 1693.
Winckler, J., 1484.
Windsor, M. W., 1489.
Winkler, E. H., 1521, 1526.
Winter, A., 1591.
Winter, T. G., 1559.
Wobschall, D., 1954, 1956, 1975, 2028, 2029, 2030, 2031, 2037, 2038, 2043, 2068.
Wobser, R., 1409, 1454, 1455.
Wolfhard, H. G., 1488.
Wolniewicz, L., 1420, 1421, 1422, 1423.
Wood, R. W., 1703, 1734.
Woodmansee, W. E., 1514.
Wouters, H., 1410, 1411.

Yadav, H. N., 2000.
Yarnell, C. F., 1616, 1619.
Yntema, J. L., 1410, 1411.
Yoshino, K., 1879.
Young, A. H. (1), 1514, 1515, 1525.
Young, A. H. (2), 1794.
Young, J. E., 1521.

Young, R. A., 1832, 1833, 1838, 1839, 1944, 1947, 1948, 1949, 1950, 1965.
Yun, K. S., 1436, 1437.

Zacharias, J. R., 1617.
Zandbergen, P., 1401, 1403.
Zemansky, M. W., 1676, 1680, 1693, 1849.
Zener, C., 1545, 1916, 1920.
Zink, H., 1528.
Zink, R., 1409.
Zmuda, A. J., 1578.
Zorn, J. C., 1773.

SUBJECT INDEX

Absorption:
  band, of I_2 and vibrationa relaxation, 1496.
  line, coefficient integrated over line width, 1848.
  of high frequency sound:
    effect of vibrational relaxation on, 1466–9.
    methods of measurement of, 1470–5.
    variation of, with pressure, temperature, and impurity content, 1470.
  of nuclear magnetic resonance radiation, 1854.
  of optical pumping radiation, 1845, 1852.
  of resonance radiation, 1499, 1653, 1654.
Acoustic impedance, 1471.
Affinity, electron, of O_2, 2096.
Air:
  diffusion coefficient of N_2^+ in, 2010.
  ionic reactions in, 2046.
  rotational relaxation in, 1578.
Airy function, and theory of rainbow scattering, 1321.
Alignment of electron angular momentum, 1696.
  by absorption of resonance radiation, 1700.
  of metastable (2^3S) ^{4}He atoms, 1823.
  of normal ^{3}He atoms, 1824.
  of ^{87}Rb and ^{85}Rb, 1851.
Analysis, electric, of rotational states of a molecular beam, 1593, 1641.
Angle of deflexion
  in classical scattering theory, 1309.
  relative to phase shifts, 1317.
Angular distribution of elastic scattering:
  at small angles:
    classical theory of, 1315.
    experimental study of, 1369.
    for Cs–He, Ar and K–He, Ar, Xe collisions, 1372.
    quantal theory of, 1326.
  by rigid spheres, 1306.
  classical theory of, 1309.
  effect of chemical attraction on, 1631.
  in C.M. and laboratory systems, 1367.
  in glory scattering, 1312, 1321.
  in H–H collisions, 1422.
  in reactive collisions:
    of K scattered by CCl_4, CH_3I, HCl, HAr, HI, $SiCl_4$, 1631; SnI_4, 1627.

  of Na scattered by $C_2H_4Br_2$ and HBr, 1627.
  relation of elastic to reactive scattering, 1629.
  theory of, and complex phase shifts, 1626.
  measurement of, 1387.
  near the rainbow angle, 1312, 1321;
    of K scattered by Hg, 1391; Kr, 1388.
    of Li scattered by Hg, 1398.
    of metastable He scattered by He, Ne, 1774, 1777.
    of Na scattered by Hg, 1391; Kr, 1392, 1395; Xe, 1394, 1396.
  semi-classical theory of, 1321.
  of products in reactive collisions:
    Cs halides from Cs–Br, ICl, IBr, and I_2 collisions, 1644.
    DBr from D–Br_2 collisions, 1647, 1649.
    KBr from K–Br_2 collisions, 1637, 1641, 1643.
    K, Rb, and Cs halides from collisions between alkali-metal atoms and alkyl-halide molecules, 1645, 1646.
Angular resolution, in single beam measurements of total cross-sections, 1350.
  Kusch criterion for, 1351.
Appearance potentials, of ions:
  in H_2, 2035.
  in N_2, 2036.
Ar:
  cluster formation in, by Li^+, 1969, 1971.
  collisions of:
    with Cs, 1378, 1384; D_2 and H_2, 1462; Ga, 1442; He, 1416; He (metastable), 1769, 1770; Na, 1352; Ne, Kr, and Xe, 1417; K, 1353, 1363, 1369, 1376; TlF, 1605.
  deactivation by, on impact:
    of $Cs(6^2P)$, $K(4^2P)$, $Rb(5^2P)$, and $Tl(7^2S)$, 1693, 1732; He (metastable), 1825; $Na(3^2P)$, 1660, 1671, 1674, 1693; $O(^1S)$, 1837; $Se(4^3P_0)$, 1668.
  depolarization by, on impact:
    of $Cd(5^3P_1)$, 1905; $Cd(6^3S_1)$, 1907; $Hg(6^3P_1)$, 1712, 1742, 1905, 1911; Hg (other states), 1732, 1907–9; Rb and Cs states, 1733, 1913.
  effect of:
    in quenching fluorescence of γ-bands of NO, 1507.

SUBJECT INDEX

on vibrational relaxation in CO and O_2, 1524.
enhancement by, of ionization by α-particles in He and Ne, 1815.
excitation of fine-structure levels of alkali-metal atoms by, 1740–1, 1926.
interaction of:
  with Ar, Kr and Xe, 1417; D_2 and H_2, 1462; Ga, 1442; He, 1416; K, 1377, 1382, 1384; Li and Na, 1382; Ne, 1417.
ionization of, by α-particles, enhancement of, 1815.
mobilities in:
  of alkali-metal ions, 1953, 1958, 1961; Ar^+, 1994, 2001, 2002; Ar_2^+, 1994, 1995, 2002; NO^+, 1965; O_3^-, 2067.
Penning ionization of, 1778.
$^3P_2$ state of:
  diffusion coefficient of, in Ar, 1792.
  loss processes for, in Ar, 1792.

Ar^+:
mobility of, in Ar, 1994, 1995, 2001.
effect of charge transfer on, 2002.
reaction of, with CO, CO_2, H_2, N_2, NO, and O_2, 2004.

Ar_2^+:
mobility of, in Ar, 1994, 1995, 2002.

AsD_3 and AsH_3:
vibrational relaxation in, 1530–2.
theory of, 1562.

Atoms:
interaction between:
  analytical representation of, 1333–5.
  asymptotic form of, determination from scattering measurements, 1325.
  determination of, from experimental data, 1340.
  general nature of, 1298.
  number of bound states associated with, and glory effects, 1337, 1382.
  interaction of:
    with ions, 1300.
    with like excited atoms or ions, 1301.
  rigid spherical, classical and quantal cross-sections for, 1304.

Attachment, dissociative, of electrons to O_2, 2079, 2097.

Autodetachment, and theory of detachment reactions, 2113.

Beams, molecular:
detection of, 1344–6.
production of, 1343–4.

Br, $4^2P_{\frac{1}{2}}$ state of, impact deactivation of, 1670, 1741.

Br_2:
reactive collisions of:
  with Cs, 1644; D, 1642–8; K, 1635.
vibrational relaxation in, 1520–1.

Buckingham potential, 1335.

C^+:
production of, from He^+–CO reaction, 2052.
reactions of, with O_2 and CO_2, 2063.

Cataphoresis, 1986, 1988.

CCl_4:
collisions, reactive of:
  with K, 1634; other alkali-metal atoms, 1646.
effect of, on vibrational relaxation of O_2, 1536.
interaction of, with K, 1448.

Cd:
5^3P_1 state of:
  deactivation of, by various gases, 1693.
  depolarization of, by various gases, 1733, 1905.
  self-depolarization of, 1902, 1903.
6^3P_1 state of:
  depolarization of, by various gases, 1733.
sensitized fluorescence of, in presence of $Hg(6^3P_1)$, 1743.

CD_4:
vibrational relaxation in 1530, 1532.
theory of, 1562.

CF_4:
collisions of, with CsCl, 1450.
interaction of, with Cs, 1449; K, 1449.

CH_4:
bulk viscosity of, 1578.
collisions of:
  with CsCl, 1450; TlF, 1605.
deactivation by, on impact of $Cd(5^3P_1)$, $Hg(6^3P_0)$, $Hg(6^3P_1)$, and $Na(3^2P)$, 1693.
interaction of:
  with Cs, 1449; K, 1448.
rotational relaxation in, 1583.
spin disorientation of Rb by, 1867.
vibrational relaxation in, 1517, 1530, 1532.
theory of, 1562.

C_2H_2, enhancement by, of ionization by α-particles in Ar, 1815.

C_2H_4:
deactivation of $Hg(6^3P_1)$ by, 1677.
destruction of He (metastable) by, 1821.
enhancement by, of ionization by α-particles in He and Ar, 1815.
spin disorientation of Rb by, 1867.

C_2H_6:
deactivation of $Hg(6^3P_1)$ by, 1677.
enhancement by, of ionization by α-particles in Ar, 1815.
spin disorientation of Rb by, 1867.
vibrational relaxation in, 1529.

$C_2H_4Br_2$, reactive collisions of, with Na, 1627.

$CHCl_3$:
reactive collisions of, with Na, 1646.
thermal conductivity of, 1615.

CH_2Cl_2:
collisions of, with CsCl, 1613–14.
vibrational relaxation in, 1529.

CH_2F_2 and CHF_3, collisions of, with CsCl, 1613–14.

$(CH_3)_2CO$:
destruction:
of He (metastable) by, 1821; of $O(^1S)$ by, 1837.
interaction of:
with Cs, 1449; K, 1448.

CH_3Cl and CH_3OH, thermal conductivity of, 1615.

CH_3I:
elastic collisions of, with CsBr, CsCl, and KCl, 1613.
reactive collisions of:
with Na, Rb, and Cs, 1646; K, 1631–4, 1645, 1649.

Cl^-, 2104.
associative detachment of electrons from, in H, O, N, 2105.
charge transfer reaction of, with NO_2, rate of, 2106.

Cl_2:
bulk viscosity of, 1578.
rotational relaxation in, 1581.
thermal conductivity of, 1580–1.
vibrational relaxation in, 1520–1.
theory of, 1552.

Classical theory:
of diffusion and viscosity coefficients, 1315.
of scattering:
at small angles, 1315, 1369.
by extended range interactions, 1309.
'total' cross-section for, 1312.
for rigid spherical atoms, 1304–8.
glory effects in, 1312.
orbiting effects in, 1310.
rainbow effects in, 1312.

ClO_2, vibrational relaxation in, 1538.

Close-coupling (truncated eigenfunction) approximation, application to:
resonance effects in collisions between molecules, 1346.

rotational relaxation:
in D_2 and H_2, 1573; in H_2–He mixtures, 1576; in more general cases, 1589.
transfer of electronic excitation, 1918.
vibrational relaxation, 1553.
of CO, 1556; CO_2, 1557.

Cluster formation:
by alkali-metal ions in rare gases, 1968.
by D_2^- in H_2O, 2111.
by OH^- in H_2O, 2110.
by positive ions in general, 1966.

CN, vibrational relaxation in, 1496.

CO:
bulk viscosity of, 1578.
deactivation by, on impact:
of $Cd(5^3P_1)$ and resonance levels of K, Rb, and Tl, 1693; $Hg(6^3P_1)$, 1674–83; $Hg(6^3P_0)$, 1679–83; $Na(3^2P)$, 1661, 1671, 1674, 1693.
depolarization by, on impact, of $Hg(6^3P_1)$, 1712.
destruction of He (metastable) by, 1821.
detachment of electrons from O^- in, 2105.
effect of, on vibrational relaxation of NO, 1522.
interaction of, with K, 1448.
mobilities in, of alkali-metal ions, 1965.
Penning ionization of, 1819–20.
quenching of infra-red fluorescence in, 1498–1501.
reactions of:
with Ar^+, 2064; He^+, 2052; N^+ and N_2^+, 2063.
rotational relaxation in, 1583.
transfer of vibration:
to D_2, 1536; N_2, 1537.
vibrational excitation of, by $Hg(6^3P_1)$ and $Hg(6^3P_0)$, 1683.
vibrational relaxation in, 1520–1.
effect of Ar, H_2, and He, 1524; of D_2, 1536–7.
theory of, 1556.

CO^+:
production of:
from He^+–CO_2 reaction, 2052; from C^+–CO_2, C^+–O_2, N^+–CO, and N_2^+–CO reactions, 2063.
reaction of, with CO_2 and O_2, 2063.

CO_2:
bulk viscosity of, 1587.
charge transfer reaction of, with O_3^-, 2106.
deactivation by, on impact:
of Hg (6^3P_1), 1677, 1693; Hg (6^3P_0), 1693; $Na(3^2P)$, 1661, 1693; resonance levels of Li, K, Rb, and Tl, 1693.

depolarization by, on impact of $Hg(6^3P_1)$, 1712.
enhancement by, of ionization by α-particles in He, 1815.
interaction of:
with Cs, 1449; K, 1448.
low-lying vibrational levels of, 1510.
mobility in, of O_3^-, 2067.
reactions of:
with Ar^+, 2064; CO^+, N^+, and O^+, 2063; He^+, 2052.
rotational relaxation in, 1580, 1583.
thermal conductivity of, 1580–1.
vibrational relaxation in, 1489, 1494, 1508–11, 1517, 1525–6.
effect of:
He on, 1526; H_2 on, 1558; H_2O on, 1536, 1559.
theory of:
semi-empirical, 1552.
three-dimensional, 1557.
vibrational temperature in, 1558.

CO_2^+:
production of from CO^+–CO_2, N^+–CO_2, and N^+–CO_2 reactions, 2063.
reaction of, with O_2, 2063.

CO_3^-:
charge transfer reactions of, with NO, NO_2, and O, 2106.
formation of:
in O_2–CO_2 mixtures, 2106.
from O^-, 2107.

CO_4^-:
dissociation energy of, 2160.
formation of:
in O_2–CO_2 mixtures, 2107.
from O_2^-, 2108.

Collision parameters, reduced, in theory of gas-kinetic collisions, 1335.

Conductivity, thermal, of gases:
effect of rotational relaxation on, 1579.
relation to viscosity, 1303.

COS, vibrational relaxation in, 1517.

Cross-sections for:
associative detachment, 2114.
charge transfer collisions, 2004–5, 2007.
deactivation of electronic excitation and quenching of resonance radiation, 1653.
measurement of, 1653–96.
destruction of metastable atoms, 1782, 1793, 1815, 1821, 1836, 1879.
disalignment and disorientation:
of electronic angular momentum, 1696.
measurement of:
by double resonance, 1711–15.
by Hanle effect, 1703–10.
theory of:
for atoms with J half integral, 1912–14.
for atoms with J integral:
for dissimilar atoms, 1901, 1905.
for similar atoms, 1892.
of rotation, in molecular collisions, 1617.
elastic scattering, in collisions between gas atoms, 1301.
differential, measurement of, 1366–9.
effect of symmetry on, 1308.
for extended range interactions, 1309.
for rigid spherical atoms, 1304–8.
in terms of reduced parameters, 1337.
low and high velocity regions for, 1338.
relation between classical and quantal formulae for, 1317.
total:
classical, 1317.
dependence of, on molecular orientation, 1595.
measurement of, 1342.
angular resolution in, 1350.
by double-beam method, 1359.
by single-beam method, 1346.
excitation and deactivation:
of molecular rotation, 1569–76, 1587, 1592, 1599–1605, 1607–13.
of molecular vibration, 1469, 1550, 1553.
excitation of fine-structure transitions:
measurement of, 1734, 1739.
theory of, 1925.
hyperfine exchange, 1719.
ionic reactions, 2004–5, 2007.
reactive collisions, 1631.
and complex phase shift, 1624.
spin-exchange collisions:
observations of, 1841–65.
theory of, 1884, 1888, 1890.
spin-reversal collisions:
measurement of, 1865–8.
theory of, 1928.
transfer of electronic excitation, 1748, 1754, 1761, 1763, 1796, 1822, 1828.
theory of:
for collisions between different atoms, 1915, 1918.
for collisions between similar atoms, 1871, 1876, 1881.
semi-classical approximation for, 1873.
in accidental resonance case, 1923.

Crossed-beam method:
for measurement:
of differential cross-sections, 1366–9.
of total cross-sections, 1361, 1365.
for studying reactive collisions, 1622.
Crossing point, of potential energy curves, and theory of excitation transfer, 1915.
Cs:
beams of, deflexion by gravity, 1354.
collisions:
elastic of:
with Ar, Kr, Xe, 1378, 1384; Cs, 1355, 1430–3; He, 1355, 1384; Hg, 1363, 1368, 1376; Na, K, and Rb, 1430–3; Ne, 1384; various molecules, 1447–9.
reactive:
with alkyl halides, 1646; Br_2, I_2, IBr, and ICl, 1643–4.
spin-exchange:
with Cs, 1845, 1857, 1890; Rb, 1845, 1890.
depolarization of resonance radiation of, 1729, 1733, 1913.
excitation of transitions between fine-structure levels of, 1737, 1740, 1926.
interaction of:
with K, Na, Rb, 1433; Kr, 1382; various molecules, 1449.
resonance level of, impact deactivation of, 1661, 1693.
Cs^+, mobility of:
in CO, H_2, N_2, 1965; rare gases, 1953, 1958, 1961, 1962.
CS_2, interaction of with K, 1448.
CsBr, collisions of:
with CH_3I, 1613; Xe, 1451.
CsCl, collisions of:
with CH_4, CF_4, D_2, H_2, He, Kr, Ne, SF_6, $SiCl_4$, Xe, 1450–2; with CHF_3, CH_2F_2, cis-$C_2H_4Cl_2$, H_2S, NH_3, and NO, 1614; with CH_3I and HBr, 1613.
Cyclotron resonance method, for measurement of mobilities of positive ions in gases, 1954.

D:
mobility of D^+ in, 2000.
reactive collisions of, with Br_2, 1646–8.
D^+:
mobility of, in D, 2000.
reaction of with D_2, 2036.
D_2:
collision of:
with Ar, Kr, Ne, and Xe, 1462; CsCl, 1450; ^{3}He, 1461.

deactivation of various excited atoms by, 1693.
diffusion of, into H_2, 1457.
effect of, on vibrational relaxation of CO, 1536.
interaction of:
with Cs, 1449; K, 1448.
reaction of, with D^+, 2036.
rotational excitation of, by H and He, 1576.
rotational relaxation in:
observed, 1565.
theory of, 1567.
spin disorientation of Rb by, 1867.
transfer of vibration from, to CO, 1537.
vibrational relaxation in, 1518.
DBr, from D–Br_2 reaction, internal excitation of, 1647–8.
DCl, vibrational relaxation in, 1532.
Depolarization:
cross-sections for:
observed values of, 1732–3.
theory of:
for atoms of different structure, 1901.
for atoms of same structure, 1894.
of resonance radiation:
theory of, 1700–3.
use of, for measurement of disorientation and disalignment cross-sections, 1703, 1712, 1729, 1732, 1733.
Detachment of electrons from negative ions:
by impact, 2083.
for O^-, 2085, 2096, 2100; O_2^-, 2085, 2092, 2096.
associative, 2084.
for Cl^-, in H, N, and O, 2105; O^- in CO, H_2, and NO, 2102, 2105; O_2^- in N and O, 2105; OH^- in N and O, 2105.
theory of, 2113.
Detectors for molecular beams, 1344–6, 1622.
by electron impact ionization, 1345.
by surface ionization, 1345.
Diffusion:
ambipolar, 1952.
coefficient:
formula for, 1302.
classical approximation to, 1315.
of Ar (metastable) in Ar, 1792; D_2 in H_2, 1457; H in H_2, 1428–9; He (metastable) in He, 1782, 1879; ^{3}He in ^{4}He, 1405–6; Hg (metastable) in Hg, 1682, 1807–8; N_2^+ in air, 2010; Na in He, Ne and

SUBJECT INDEX

Rb in rare gases, hydrocarbons, H_2, and N_2, 1867; ortho in para H_2, 1457.
cross-section, 1302.
   expansion of, under near-classical conditions, 1408.
   for H–H collisions, 1419.
   for rigid spherical atoms, 1307–8.
   in presence of charge transfer, 1998.
   thermal:
     definition of, 1304.
     of ^{3}He in ^{4}He, 1407.
Dispersion of high frequency sound:
   double, 1533.
   effect of vibrational relaxation on, 1467–9.
   variation of, with pressure, temperature, and impurity content, 1470.
Dissociation energy:
   of CO_4^-, 2110; N_4^+, 2041; O_4^-, 2083.
Distorted-wave method, application to:
   rotational depolarization collisions:
     in H_2, 1619; in H_2–He mixture, 1620.
   rotational excitation:
     in H_2, 1567–72; of H_2 and D_2 by H and He, 1576.
   transfer of vibration, 1545, 1547.
   vibrational relaxation, in head-on collisions, 1545.
     three-dimensional empirical theory, 1547.
D_2O, deactivation of $Hg(6^3P_0)$ and $Hg(6^3P_1)$ by, 1693.
Double-beam method, for measuring total cross-sections, 1359.
Double resonance, 1711.
Drift velocity, of positive ions in gases, 1934.
   variation of, with F/p, 1935.

Electron concentration in He and Ne afterglows, measurement of, 1801.
Electron-diffusion cooling, 1953, 1987, 1994.
Electron filter, 1937, 2072.
Energy distribution of electrons from ionization by He (metastable):
   of CO, 1819–20; N_2, 1818–20, 2054, 2055; NO, 1821.
Excitation, impact:
   of rotation in molecules, effect of on total cross-section, 1445.
   of transitions between fine structure levels, 1734.
   of upper level of 21-cm line of H, 1844.
Excitation temperature, measurement of, from line reversal, 1670.

F^-, charge transfer reaction of, with NO_2, 2106.
F_2, vibrational relaxation in, 1520–1.
Fine-structure levels, transitions between:
   excitation of, 1734.
   theory of, 1925.
Flash photolysis and spectroscopy, use of:
   for observing vibrational relaxation, 1502–4.
   for quenching measurements, 1666.
Fluorescent yield, 1656.
Forces:
   between atoms and ions, 1300.
   chemical, between atoms, 1299.
   polarization, 1300.
   van der Waals, 1298.

Ga:
   collisions of, with rare-gas atoms, 1439–43.
   van der Waals interaction of, with rare-gas atoms, anisotropy of, 1442.
Glory effect in gas-kinetic collisions:
   classical theory of, 1312.
   in terms of reduced parameters, 1337.
   observation of, 1377–8.
   quantum theory of, 1326–9.
   semi-classical theory of, 1320.

H:
   collisions, elastic:
     with H:
        differential cross-sections for, 1421.
        diffusion and viscosity cross-sections for, 1419.
        total cross-section for, 1421.
     with H_2, 1427, 1452.
   detachment, associative, of electrons from Cl^- in, 2105.
   diffusion coefficient of, in H_2, 1428.
   interaction of:
     with H, 1300, 1419.
        resonance levels in, 1422–3.
     with $H(2s)$, 1882.
     with $H(2p)$, 1883.
     with H^+, 1301.
     with H_2, 1427, 1452.
   mobility of H^+ in, 2000.
   polarization of, 1842.
   reaction of, with H_2, theory of, 1649.
   rotational excitation of H_2 by, 1576.
   $2s$ and $2p$ states of, impact deactivation of, by normal H, 1881.
   spin-exchange collisions in:
     with H_2, 1842, 1844, 1857, 1861, 1865, 1885, 1890.
     with O, 1891.

SUBJECT INDEX

H^+:
  interaction of, with H, 1301.
  mobility of:
    in H, 2000; in H_2, 2028, 2030, 2032–3.
  reaction of, with H_2, 2030, 2036.

H^-:
  charge transfer reaction with NO_2, 2106.
  detachment reactions with H, theory of, 2113.

H_2:
  collisions, elastic of:
    with CsCl, 1450; H, 1452; H_2, 1457; ^{3}He and ^{4}He, 1461; other rare gases, 1462.
  deactivation by, on impact, of resonance levels:
    of Cd, Li, K, Rb, Cs, Tl, 1693; He (metastable), 1821; $Hg(6^3P_0)$, 1693; $Hg(6^3P_1)$, 1677, 1693; $Na(3^2P)$, 1660, 1693; $O(^1S)$, 1837.
  depolarization by, on impact:
    of $Hg(6^3P_1)$, 1712, 1732; other Hg states, 1732.
  detachment, associated, of electrons from O^- in, 2105.
  effect of, on vibrational relaxation:
    in CO and O_2, 1524; CO_2, 1558.
  enhancement by, of ionization by α-particles in He and Ne, 1815.
  interaction of:
    anisotropic component of, 1457, 1619.
    with Cs, 1449; H, 1427, 1452; H_2, 1453, 1455.
    with rare-gas atoms, 1461, 1462.
  mobility in:
    of alkali-metal ions, 1965; H^+, 2028, 2030, 2033; H_2^+ and H_3^+, 2028, 2033; H_5^+, 2028; hydrogen ions, 2028; NO^+, 1965; O_3^-, 2067.
  ortho:
    diffusion of into para H_2, 1456.
    effect of, on vibrational relaxation in CO, 1501.
    quenching of infra-red fluorescence of CO by, 1501.
    rotational disorientation of:
      by ortho H_2, 1620.
      by para H_2, 1620.
    viscosity cross-section of, 1459.
  para:
    effect of, on vibrational relaxation in CO, 1501.
    interaction of, with para H_2 and ortho H_2, 1453.
    quenching of infra-red fluorescence of CO by, 1501.
    viscosity cross-section of, 1459.

  quenching of infra-red fluorescence of CO by, 1501.
  reaction of:
    with H, theory of, 1649; H^+, 2030, 2036; H_2^+, 2034; dependence on vibrational state of H_2^+, 2035.
  rotational disorientation of:
    by H_2, 1618; He, 1620.
  rotational excitation of, by H and He, 1576.
  rotational relaxation in:
    observed, 1565.
    theory of, 1567, 1577.
  spin disorientation of Rb by, 1867.
  spin–orbit and spin–spin interaction of, with H_2, 1616.
  vibrational relaxation in, 1518.
  viscosity of, 1426–7, 1454.

H_2^+:
  mobility of, in H_2, 2028, 2031.
  theory of, and charge transfer, 2034.
  reaction of, with H_2, 2034.
  effect of initial vibrational state of H_2^+ on, 2035.

H_2^-, electronic states of, 2113.

H_3^+:
  mobility of, in H_2, 2028.
  theory of, 2033.
  reaction of, with H_2, 2030.

H_5^+:
  mobility of, in H_2, 2028.
  production of, from H_3^+ in H_2, 2030.

Hanle effect, 1703.
  and nuclear spin, 1722.

HBr:
  collisions of, with CsCl and KCl, 1613.
  interaction of, with K, 1448.
  reaction of:
    with Na, 1627; K, 1631–4.
  thermal conductivity of, 1615.

HCl:
  bulk viscosity of, 1578.
  collisions, reactive of with K, 1631–4.
  interaction of, with K, 1448.
  thermal conductivity of, 1615.
  vibrational relaxation in, 1530–2.

H_2CO, interaction of:
  with Cs, 1449; K, 1448.

HD, rotational excitation of, theory of, 1573.

^{3}**He**:
  coefficient:
    for diffusion of, in He, 1405–6.
    for self-diffusion, 1406.
  polarization of, from Rb, 1824, 1867.
  second virial coefficient of, 1404.

thermal conductivity of, 1403.
thermal diffusion of, in ^{4}He, 1407.
viscosity of, 1402.
^{4}He:
cluster formation in, by Li$^+$, 1968, 1969.
collisions of:
 with Cs, 1384; CsCl, 1450; Ga, 1442; He, 1411–13; K, 1353; 1379, 1384; other rare gases, 1413–15; TlF, 1605.
deactivation, impact:
 of excited atoms, 1693; O(^1S), 1837.
depolarization:
 of Cd(5^3P_1), 1905, 1906–9; Cd(6^3S_1), 1907; Cs and Rb levels, 1732, 1913; Hg(6^3P_1), 1712, 1732, 1905, 1911; Hg (other states), 1732; Ne($2p_4$), 1733.
effect of, on vibrational relaxation:
 of CO and O$_2$, 1523; CO$_2$, 1526.
excitation of transitions between fine-structure levels of alkali-metal atoms, 1740–1, 1926.
interaction of:
 empirical forms for:
  Buckingham–Corner, 1401.
  Slater, 1400.
  van der Waals, 1401, 1410.
 with Ga, 1442.
 with He, at large and intermediate separations, 1399.
 with other rare-gas atoms, 1416.
metastable:
 collisions of:
  differential cross-sections for, in gases, 1724.
  total cross-sections for, in He, Ar, and Kr, 1769, 1770–3.
 destruction of in collisions:
  with Ar, C$_2$H$_4$, CO$_2$, H$_2$, N$_2$, Kr, and Xe, 1815; metastable He, 1801, 1804, 1806.
 diffusion coefficient of, in He, 1782, 1795, 1800, 1804.
 ionization of atoms by, 1816.
 loss processes of, in He afterglows, 1781.
  effect of impurities on, 1792.
2^1S state:
 transfer of excitation from:
  to He, 1822, 1828.
   theory of, 1876.
  to Ne, 1796, 1799.
 two-body destruction of, 1782, 1804, 1814, 1821.
  theory of, 1879.
2^3S state:
 excitation of O$_2^+$ bands by, 1800.

in flowing afterglows, 1799.
 production of, by α-particle impact, 1814.
 three-body destruction of, 1782, 1795, 1800, 1814, 1821.
 transfer of excitation from:
  to He, 1822, 1828.
   theory of, 1876.
  to Ne, 1796, 1799.
mobilities in:
 of alkali metal ions, 1953, 1958, 1961, 1962.
 of helium ions, 1976, 1969, 1981.
  at low temperatures, 1991.
  theory of, 2000.
 He$^+$, 1954, 1972, 1982, 1986, 1987, 1994.
 He$_2^+$, 1973, 1981, 1982, 1984, 1986, 1987, 1994, 2003.
 He$_3^+$, 1975, 1993, 1994.
 NO$^+$, 1965.
 O$_3^-$, 2067.
2^3P_1 state, impact excitation of transitions between fine structure levels of, 1741.
quenching of Na resonance radiation by, 1693.
rotational excitation of H$_2$ and D$_2$ by, theory of, 1576.
second virial coefficient of:
 at low temperatures, 1400.
 at medium to high temperatures, 1409.
shock front thickness in, 1487.
spin disorientation of Na and Rb by, 1867, 1868.
thermal conductivity of, at low temperatures, 1400, 1403.
thermal diffusion in, of ^{3}He, 1407.
transfer of charge to, from He$^+$, 2000.
transfer of excitation in:
 and apparent breakdown of Wigner rule, 1760.
 effect of on optical excitation functions, 1758, 1759–61, 1766–7.
 from 2^3P_0 state to Ne, 1799.
 importance of F states in, 1758.
viscosity of:
 at low temperatures, 1400, 1403.
 at medium to high temperatures, 1409.
He$^+$:
conversion of, to He$_2^+$, 1973, 2027.
mobility of, in He, 1954, 1972, 1982, 1986, 1987, 1994, 2000.
reactions of:
 in flowing-afterglow method, 2021.
 with CO and CO$_2$, 2052; N$_2$, 2051,

He⁺ (cont.):
 2052, 2062; NO, 2052, 2062; O_2, 2049.
 transfer of charge from, to He, 2000.

He₂:
 $A\Sigma_u$ state of, and deactivation of He (2^1S), 1879.
 metastable ($2^3\Sigma_u$) state of, 1782, 1800, 1814.
 diffusion coefficient of, in He, 1783.
 natural lifetime of, 1782.

He₂⁺:
 conversion of, to He_3^+, 1993.
 formation of, from He⁺, 1973, 1981, 1984, 1986, 1990, 2027.
 metastable ($^4\Sigma_u^+$) state of, 1982.
 mobility of, in He, 1973, 1981, 1982, 1984, 1986, 1987, 1994, 2003.

He₃⁺:
 dissociation energy of, 1993.
 mobility of, in He, 1975, 1993, 1994.

HeO⁺, as intermediate complex, 2051.

Hg:
 collisions of:
  with Cs, 1363, 1368, 1376; K, 1363, 1368, 1376, 1390, 1391; Li, 1393.
 6D states of, depolarization of, 1718, 1732.
 destruction of He (metastable) by, 1821.
 enhancement by, of ionization by α-particles in He and Ne, 1815.
 interactions of:
  with K, 1378, 1382, 1390, 1397; Li, 1382; Na and Rb, 1390, 1397.
 metastable atoms of:
  destruction of, in collisions with normal Hg and Hg(6^3P_2), 1807.
  diffusion coefficient of, in Hg, 1807.
 mobility of Hg⁺ ions in, 1996.
 and charge transfer, 2002.
 6^3P_0 state of, deactivation of, 1676, 1691.
  by various gases, 1693, 1694.
 6^3P_1 state of:
  deactivation of, 1670, 1676.
   by CO, 1678, 1679–83; Hg(6^1S), 1678, 1693; N_2, 1678, 1679, 1683.
  depolarization of, by various gases, 1712, 1732, 1733, 1905.
  self-depolarization of, 1724, 1733, 1902, 1903, 1911.
 6^3P_2 state of, depolarization of, 1715, 1732.
 7^3S_1 and 7^3D_1 states of, destruction of by N_2, 1691.

Hg⁺:
 conversion of, to Hg_2^+, 1997.
 mobility of:
  in He, 1957; Hg, 1996.
 relation to charge transfer, 2002.

Hg₂, band fluorescence of, 1809.

Hg₂⁺, formation of, from Hg⁺, 1997.

HI, reactive collisions of, with K, 1631–4.

High-velocity region, for gas-kinetic collisions, 1338.

H₂O:
 clustering of:
  to negative ions, 2106, 2110.
  to positive ions, 1966.
 collisions of, with TlF, 1605, 1613.
 deactivation by, on impact:
  of resonance levels of Cs, Li, K, Rb, Cs, and Tl, 1693; Hg(6^3P_0) and Hg(6^3P_1), 1677, 1693; Na(3^2P), 1660, 1693.
 effect of, on vibrational relaxation:
  of CO_2, 1536, 1559; NO, 1522; O_2, 1536.
 interaction of:
  with Cs, 1449; K, 1448.
 vibrational relaxation of, 1528.

H₂S:
 collisions of, with CsCl, 1613–14.
 effect of, on vibrational relaxation of O_2, 1536.
 interaction of, with K, 1448.

I, $5^2P_\frac{1}{2}$ state of, deactivation of, 1670, 1741.

I₂:
 band absorption in, and vibrational relaxation, 1496.
 quenching of Na resonance fluorescence by, 1675–6.
 reactive collisions of, with Cs, 1644.
 sensitized band fluorescence of, and vibrational relaxation, 1505, 1538.
 vibrational relaxation in, 1520–1.

IBr and ICl, reactive collisions of, with Cs, 1644.

In, Hg-sensitized fluorescence of, 1749.

Infra-red emission:
 behind shock front, and vibrational relaxation, 1489.
 fluorescent, quenching of, 1497.
 measurement of:
  in CO excited by Hg(6^3P_1) and Hg(6^3P_0), 1683.
  in shock-excited CO, 1498–1502.

Interaction between:
 alkali-metal ions and neutral atoms and molecules, 1971.
 atoms:
  analytical representation of, 1333–5.
  determination of, from experimental data on gas-kinetic collisions, 1341.
  in general, 1298.
  in ^2S states, 1418.
 atoms and ions, 1300.

atoms and like excited atoms or ions, 1301, 1870, 1997.
H–H, 1420; H–H($2s$) and H–H($2p$), 1882; H–H$^+$, 2000; H–H$^-$, 2113; H$_2$–H$_2$, 1453, 1455, 1457, 1619; H$_2$—H$^+$, H$_2$–H$_3^+$, 2033; He–He 1399–1401, 1410; He–He(2^1S) and He(2^3S), 1876; He–He$^+$, 2000; He with other rare-gas atoms, 1416; He–Li$^+$, 1962.
chemical, effect of on elastic differential cross-sections, 1630.
experimental form of, use of in theory of vibrational relaxation, 1548.
imaginary, and theory of reactive scattering, 1626.
long range, between molecules, 1444.
polar molecules, 1606.
Morse form of, use in theory of vibrational relaxation, 1548.
van der Waals, see van der Waals constant.

Interferometer:
scanning, Fabry–Perot, 1847.
ultrasonic, 1470–3.
use of for observing shock-front thickness, 1484.
application to D$_2$, 1518.
Inversion, in ND$_3$, NH$_3$, 1528.
Ionic reactions:
methods for measuring rates of, 2003.
theory of the rates of, 2005.
Ionization coefficient, observed, effect of Penning ionization on, in Ar and Ne, 1778.

K:
collisions of:
with Ar, 1353, 1363, 1369, 1376, 1384; Cs, Na, Rb, 1430–3; He, 1353, 1379; Hg, 1363, 1368, 1376, 1391; Kr, 1363, 1376, 1378, 1384, 1388; Ne, 1379, 1384; various molecules 1446–8; Xe, 1359, 1363, 1369, 1375.
interaction of:
with Ar and Kr, 1382; Cs, Na, and Rb, 1433; Hg, 1378, 1382, 1390, 1397; various molecules, 1448; Xe, 1382, 1390, 1397.
4^2P state,
excitation of transitions between fine structure levels of, 1740, 1926.
impact deactivation of, 1661, 1662, 1693.
reactive collisions:
with Br$_2$, 1634–43:
angular distribution of KBr from, 1638.
rotational energy distribution of KBr from, 1641.
total reaction cross-section for, 1643.
velocity distribution of KBr from, 1640.
with CCl$_4$, CH$_3$I, HBr, HCl, HI, SF$_6$, SiCl$_4$, SnCl$_4$, 1631–4.
with CH$_3$I:
angular distribution of products from, 1645.
theory of, 1649.
with other alkyl halides, 1646.
spin-exchange collisions of, with Na, 1865.
theory of, 1890.
K$^+$, mobility of:
in CO, H$_2$, and N$_2$, 1965; in rare gases, 1953, 1958, 1961, 1962.
KBr, resulting from K–Br$_2$ reaction:
angular distribution of, 1638.
rotational energy distribution of, 1641.
velocity distribution of, 1640.
KCl, collisions of:
with CH$_3$I and HBr, 1613; Kr and Xe, 1451.
Kihara potential, 1335.
Kinematics, of reactive collisions, 1635.
Kr:
collisions of:
with Ar, Kr, 1417; Cs, 1378, 1384; CsCl, 1450; Ga, 1442; H$_2$ and D$_2$, 1462; He, 1416; He(2^1S) and He(2^3S), 1769; K, 1363, 1376, 1378, 1384; KCl, 1452.
deactivation by, on impact:
of He (metastable), 1821; O(^1S), 1837.
depolarization by, on impact:
of Cd(5^3P_1), 1905; Cs and Rb states, 1733; Hg(6^3P_1), 1712, 1732, 1905, 1911; other Hg states, 1732.
effect of, on vibrational relaxation of NO, 1522.
enhancement by, of ionization by α-particles in He and Ne, 1815.
excitation of fine-structure levels of alkali-metal atoms by, 1926.
interaction of:
with Ar and Kr, 1417; Cs, Li, K, and Rb, 1382; Ga, 1442; He, 1416; Na, 1382, 1397.
mobility in:
of alkali-metal ions, 1953, 1958, 1961, 1962; krypton ions, 1995, 2002; Kr$_2^+$, 2003; O$_3^-$, 2067.

Kr (*cont.*):
  spin disorientation of Na and Rb by, 1867.
Kr⁺, mobility of, in Kr, 2003.
  and charge transfer, 2002.
Kr$_2^+$, mobility of, in Kr, 2003.

Langevin theory of ionic mobilities, 1958.
Laser, use of:
  for measuring:
    depolarizing cross-section for Ne, 1725.
    rates of transfer reactions from He (metastable) to Ne, 1797.
    shock-front thickness, 1484–5.
  for observing vibrational relaxation in CO_2, 1508–11.
Lennard-Jones potential, 1334.
  use in theory of vibrational relaxation, 1549.
Li:
  collisions of:
    with Ar, He, and Kr, 1771; Hg, 1393; including isotope effects, 1394; Xe 1359, 1378.
    interaction of with Ar, Hg, Kr, and Xe, 1382.
  2^2S state, deactivation of, 1661.
    by various gases, 1693.
  spin-exchange collisions of, with alkali metal atoms, theory of, 1890.
Li⁺:
  cluster formation by, in rare gases, 1968, 1969, 1971.
  mobility of:
    in CO, H_2, and N_2, 1965; rare gases, 1953, 1958, 1961, 1962.
Lorentz broadening, 1658.
Low-velocity region, for gas-kinetic collisions, 1338.

Maser, hydrogen, 1861.
Mean energy:
  of a drifting ion, 1935, 2040.
  of O^- in O_2, 2100.
Metastable atoms:
  loss processes of:
    in discharge afterglows, 1780, 1781, 1801, 1807.
    in flowing afterglows, 1799.
  Penning ionization by, 1778.
Method, experimental,
  for identification of negative ions in O_2:
    by Eiber's method, 2072.
    by mass analysis, 2077.
  for measurement of:
    conversion rate of He^+ to He_2^+, from optical emission from afterglows, 1987.
  cross-sections, for deactivation of excited atoms:
    from lifetime measurements, 1662.
    from the decay rate of imprisoned resonance radiation, 1663.
    from the quenching of resonance radiation, 1654; using flames, 1656.
    observing infra-red radiation from vibrationally excited molecules, 1683–9.
    using flash photolysis, 1666, 1677.
    using optical dissociation, 1676.
    using shock-wave excitation, 1670.
  cross-sections, for disalignment and disorientation of electronic angular momentum:
    for atoms excited by stepwise excitation, 1717.
    for $Hg(6^3P_2)$, 1715.
    for self-depolarization, 1722.
    using Hanle and related effects, 1703.
    using laser excitation, 1725.
    using optical pumping, for Rb and Cs, 1729.
    when nuclear spin is involved, 1720.
  cross-sections, for disalignment and disorientation of molecular rotation, from nuclear magnetic relaxation, 1616.
  cross-sections, for excitation of rotation in polar molecules, 1599.
  cross-sections, for excitation of transitions between fine structure levels by sensitized fluorescence, 1734.
  cross-sections, for mixing collisions between levels of $He(2^3P)$, 1740.
  cross-sections, for spin-exchange:
    from relaxation of a hyperfine population, 1844.
    using a hydrogen maser, 1861.
    using nuclear magnetic resonance, 1854.
    using optical pumping to Zeeman levels, 1851.
    using stimulated emission, 1858.
  cross-sections, for spin reversal, 1865.
  cross-sections, for transfer of excitation between $He(2^3S)$ and $He(1^1S)$, 1822.
    by optical pumping, 1823.
  differential cross-sections:
    for gas-kinetic collisions, 1366:
      kinematics of, 1367.

for collisions of metastable atoms in gases, 1774.
mobilities and reaction rates, of negative ions:
by pulse methods, 2070.
of detachment reactions:
associative, 2012.
by pulse methods, 2085.
using flowing afterglow, 2103.
using drift tube and mass analysis, 2079.
using Geiger counter, 2065.
mobilities and reaction rates of positive ions:
from electron decay in afterglows, 1951.
using cyclotron resonance, 1954.
using drift tube:
without mass analysis, 1976.
with mass analysis, 1981.
using electrical shutter, 1936.
using mass spectrograph without drift tube, 2004.
in pulsed operation, 2006.
photoionization source for, 2007.
using time of flight,
Hornbeck's method, 1938.
parallel-plate method, 1944.
pulse method, 1942.
rates of loss processes for metastable atoms:
in discharge afterglows 1780, 1801, 1805.
from observations of electron decay, 1801.
in He, 1781; Hg, 1807; Ne, 1784; Ne–He, 1791.
involving Penning ionization, 1816.
energy distribution of electrons produced in, 1817, CO, 1819–20; N_2, 1818–20; NO 1821.
rotational relaxation rates:
from thermal conductivity, 1579.
from thermal transpiration, 1582.
from ultrasonic absorption and dispersion, 1564.
using shock waves, 1577.
using spectroscopic methods, 1583.
total cross-sections in gas-kinetic collisions:
between alkali-metal atoms, 1429.
for metastable atoms, 1769.
for polar molecules in gases, 1593.
using crossed-beam method, 1361.
using double-beam method, 1359.
using single-beam method, 1346, 1351.
vibrational relaxation rates, 1466:
by spectroscopic methods, 1496:
for molecules in excited electronic states, 1504.
using flash spectroscopy, 1502.
using laser-induced fluorescence, 1508.
using quenching of infra-red fluorescence, 1497.
from persistence of vibration in gas dynamics, 1491–6.
from shock waves, 1482, 1483–90.
from ultrasonic absorption and dispersion, 1466, 1473.
the ultrasonic interferometer, 1470.
using spectrophone, 1511.
for observing:
cluster formation by positive ions, 1966.
collisions involving transfer of electronic excitation in He, 1759.
using time-resolved spectroscopy, 1761.
effect of anisotropy of van der Waals forces on total cross-sections, 1436, 1439.
enhancement of ionization by α-particles due to metastable atoms, 1810.
Hg-sensitized fluorescence:
of Na, 1744, 1746; Tl, 1751.
quenching of $O(^1S)$, 1831.
for studying reactive collisions between molecular beams, 1631, 1634–43, 1647.

Mobility of ions in gases:
in non-reacting gases, theory of, 1958.
methods of measurement of, 1936–57.
of alkali-metal ions:
in molecular gases, 1964.
in rare gases, 1957.
of helium ions in helium, 1976, 1979, 1981.
at low temperatures, 1992.
historical account of, 1972.
theory of, 2000, 2003.
of Hg^+ ions in helium, 1957.
of hydrogen ions in H_2, 2028, 2030, 2032, 2033.
of negative ions in O_2, 2065.
of nitrogen ions in N_2, 2036.
of NO^+ in various gases, 1947.
of oxygen ions in O_2, 2043.
of rare-gas ions in parent gases, 1994, 2002, 2003.
relation of, to diffusion coefficient, 1303.

Momentum-transfer cross-section, *see* diffusion cross-section.

Monte Carlo method, and theory of reactive collisions, 1649.

N:
associative detachment of electrons from O^-, O_2^-, and OH^- by, 2105.
charge transfer reactions of:
  with N_2^+, 2059, 2062; O^+, 2054.
exchange of, between N and N_2, on impact, 1857.
reactions of:
  in flowing afterglows, 2024.
  with O_2^+, 2058, 2062.
titration of, 2024, 2054.

N^+:
charge transfer reaction with O_2, 2057, 2062.
formation of from reaction:
  between He^+ and N_2, 2051; He^+ and NO, 2052, 2062.
mobility of, in N_2, 2037.
reactions of:
  in flowing afterglows, 2023.
  with CO and CO_2, 2063; N_2, 2041; NO, 2061, 2062.

N_2:
appearance potentials for ions formed in, 2036.
bulk viscosity of, 1578.
collisions of with K, 1447.
deactivation by, on impact:
  of $Cd(5^3P_1)$ and resonance levels of Li, K, Rb, Cs, and Tl, 1693; $Hg(6^3P_1)$, 1677, 1679–83, 1693; $Hg(6^3P_0)$, 1678–83, 1690; $Hg(7^3S_1)$ and $Hg(6^3D_1)$, 1691; $Na(3^2P)$, 1661, 1670, 1671, 1674, 1693.
depolarization by, on impact:
  of $Hg(6^3P_1)$, 1712, 1732; of other Hg states, 1733.
destruction of He (metastable) by, 1821.
detachment of electrons from O^- in, 2105.
effect of:
  in quenching fluorescence of γ-bands of NO, 1507, 1538.
  on vibrational relaxation of NO, 1522, 1552.
enhancement by, of ionization by α-particles, in He and Ne, 1815.
interaction of:
  with Cs, 1449; K, 1448.
mobilities in:
  of alkali-metal ions, 1965; of nitrogen ions, 2036–7; NO^+, 1965; O_2^-, 2067.
Penning ionization of, 1818–21, 2054.
reactions of:
  with Ar^+, 2064; He^+, 2051, 2052, 2062; O^+, 2046, 2052, 2058; O_2^+, 2062.
rotational relaxation in, 1577–8, 1580, 1581, 1583.
spin disorientation of Rb by, 1867.
thermal conductivity of, 1580–1.
transfer of vibration with CO, 1537.
vibrational relaxation in, 1520–1, 1670.

N_2^+:
charge transfer reactions of:
  with N, 2059, 2062; O_2, 2055–7, 2062.
diffusion coefficient of, in air, 2010.
formation of, from He^+–N_2 reaction, 2051.
mobility of, in N_2, 2037.
reactions of:
  in air, 2009.
  with CO and CO_2, 2063; N_2, 2039, 2041, 2042; NO, 2061, 2062; O, 2024, 2059, 2062.

N_3^+:
formation of, from N^+ and N_2^+, 2041.
mobility of, in N_2, 2037.

N_4^+:
dissociation energy of, 2041.
formation of:
  from N_2 (metastable), 2041; N_2^+, 2039, 2041.
mobility of, in N_2, 2037.

Na:
collisions of:
  with Ar, 1352; Cs, K, and Rb, 1432; Hg, 1351, 1390–2; Kr, 1293; Xe, 1394.
interaction of:
  with Ar, 1382; Cs, K, and Rb, 1433; Hg, 1382, 1390, 1397; Kr, 1382, 1397; Xe, 1382, 1390, 1397.
Hg-sensitized fluorescence of, 1744.
3^2P state of:
  excitation of transitions between fine structure levels of, 1740, 1926.
  impact deactivation of, 1654, 1656, 1660, 1662, 1663, 1670, 1671, 1674, 1675, 1693, 1694.
polarization of, 1842.
reactive collisions of, with CH_3I, 1646; $C_2H_4Br_2$, 1627; HBr, 1627.
spin-disorientation collisions of, with He, Ne, 1867.
spin-exchange collisions of, with Na, K, and Rb, 1865.
theory of, 1890.

Na^+, mobility of:
in CO, H_2, N_2, 1965; in rare gases, 1953, 1958, 1961.

NaI, photodissociation of, 1675.

ND$_3$:
  inelastic collisions of, with TlF, 1606.
  vibrational relaxation of, 1528, 1530–2.

Ne:
  cluster formation in, by Li$^+$, 1969.
  collisions of:
    with Ar, 1417; Cs, 1384; CsCl, 1450; D$_2$ and H$_2$, 1462; Ga, 1442; He (metastable), 1773; K, 1379, 1384; TlF, 1605.
  deactivation by, on impact, of O(^1S), 1831.
  depolarization by, on impact:
    of Cd(5^3P_1), 1905; Cd(6^3S_1), 1907; Hg(6^3P_1), 1712, 1732, 1905, 1911; Hg (other states), 1732, 1907–9; Rb and Cs states, 1735, 1913.
  enhancement by, of Al$^+$, Cu$^+$, and Pb$^+$ spectral lines, 1755.
  excitation, of fine-structure levels of alkali-metal atoms by, 1740–1, 1926.
  excitation-transfer collisions:
    with He (metastable), 1796, 1799; He(2^3P_0), 1799.
  interaction of:
    with Ar, 1417; Ga, 1442; He, 1416.
  metastable:
    loss processes for:
      in Ne afterglow, 1738; Ne–He, 1791.
    $^3P_0$ state:
      destruction of, in collisions with normal Ne, 1793, 1804.
      diffusion coefficient of, in Ne and He, 1793, 1804.
    $^3P_1$ state, destruction of, in collisions with normal Ne, 1793, 1804.
    $^3P_2$ state:
      diffusion coefficient of, in Ne and He, 1793, 1804.
      three-body destruction rate of, in Ne, 1793.
    Penning ionization by, 1778.
  mobility in:
    of alkali-metal ions, 1953, 1958, 1961, 1962; Ne$^+$, 1995, 2001; Ne$_2^+$, 1995, 2002; O$_3^-$, 2067.
  $2p_4$ state of, depolarization of, in Ne and He, 1725, 1733.
  spin disorientation of Na and Rb by, 1867.

Ne$^+$, mobility of, in Ne, 1995.
  theory of, and charge transfer, 2001, 2002.

Ne$_2^+$, mobility of, in Ne, 1995.
  theory of, 2002.

Negative ions:
  charge transfer reactions of, 2105.
  complex, formation of, 2106.
    in CO$_2$ and CO$_2$–O$_2$ mixtures, 2106.
    in H$_2$O and H$_2$O–O$_2$ mixtures, 2110.
  detachment of electrons from, on collisions with gas molecules, 2083.
    associative, 2084.
    methods for measuring rates of, 2084–8.
  mobility of, 2065.
  see also Cl$^-$, CO$_3^-$, CO$_4^-$, F$^-$, H$^-$, H$_2^-$, NH$_2^-$, NO$^-$, NO$_2^-$, O$^-$, O$_2^-$, O$_3^-$, O$_4^-$, and OH$^-$.

NH$_2^-$, charge transfer reaction with NO$_2$, 2106.

NH$_3$:
  bulk viscosity of, 1578.
  collisions of:
    with CsCl, 1614; TlF, 1605, 1606, 1613.
  deactivation by, on impact:
    of Cd(5^3P_1), Hg(6^3P_1), Hg(6^3P_0), 1693.
  effect of, on vibrational relaxation of O$_2$, 1536.
  interaction of, with K, 1448.
  vibrational relaxation in, 1528, 1530–2.

NO:
  $A\,^2\Sigma^+$ state of, 1507, 1508.
  γ-bands of, quenching of fluorescence of, 1507.
  charge transfer reactions of, with O$_3^-$, CO$_3^-$, 2106.
  collisions of, with CsCl, 1613–14.
  deactivation by, on impact:
    of Hg(6^3P_1), 1677, 1693; Hg(6^3P_0), 1693.
  detachment of electrons from O$^-$ in, 2105.
  excitation of transitions between fine-structure levels of, 1584, 1784.
  interaction of, with K, 1448.
  Penning ionization of, by metastable He, 1821.
  reactions of:
    with Ar$^+$, 2064; He$^+$, 2052, 2062; N$^+$ and N$_2^+$, O$^+$, and O$_2^+$, 2061, 2062.
  rotational relaxation in, 1583–7.
  use of, in titration of N, 2024, 2054.
  vibrational relaxation in, 1522.
    in mixtures with CO, H$_2$O, N$_2$, Kr, 1522.

NO$^+$:
  formation of:
    from N$^+$–NO, N$_2^+$–NO, O$^+$–NO, O$_2^+$–NO reactions, 2061, 2062; N$^+$–O$_2$, 2057, 2062; N$_2^+$–O, 2059, 2062;

NO⁺ (cont.):
  N_2^+–O_2, 2055, 2057; O^+–N_2, 2046, 2052, 2062; O_2^+–N, 2058, 2062.
  mobility of, in various gases, 1947, 1965.
NO⁻, charge transfer reaction of, with O_2, 2103, 2106.
N_2O:
  bulk viscosity of, 1577.
  collisions of:
    with CsCl, 1613; TlF, 1605, 1613.
  deactivation by, on impact:
    of $Hg(6^3P_1)$, 1677, 1693; $Hg(6^3P_0)$, 1693; $O(^1S)$, 1836.
  vibrational relaxation of, 1517, 1525–6.
    effect of He in, 1527; H_2O, 1536.
NO_2:
  charge transfer reactions of, with Cl^-, CO_3^-, F^-, H^-, NH_2^-, O^-, O_2^-, and OH^- 2106.
  use of, in titration of O, 1833.
  vibrational relaxation from highly excited states of, 1538.
NO_2^-, reaction of, with O_3, 2106.
Nozzle, supersonic, as molecular beam source, 1344.
Nuclear spin:
  effect of on impact depolarization, 1719.
  of $Hg(6^3P_1)$, 1909.

O:
  associative detachment of O^-, O_2^-, and OH^- in, 2105.
  ^1D state of, impact deactivation of, 1839.
    in liquid N_2, 1841.
  interaction of:
    quadrupole–quadrupole, 1435.
    van der Waals, 1435.
    with O, 1434–6.
  metastable, reactions of, 1828.
  reactions of:
    in flowing afterglows 2024–5.
    with CO_3^-, 2106; N_2^+, 2024, 2059, 2062.
  red line of, in atmospheric emission, 1839.
  ^1S state of:
    excitation of, in three-body recombination, 1838.
    impact deactivation of:
      by Ar, CO_2, H_2, He, and Kr, 1836; Ne, 1831; O and O_2, 1831, 1836.
  spin-exchange collisions of, with H, theory of, 1890.
  thermal conductivity of, 1433–6.
  viscosity of, 1433–6.
O⁺:
  charge transfer reactions of, with O_2, 2043, 2044, 2045, 2053, 2062.
  dependence on vibrational excitation:
    of N_2, 2054; NO, 2061, 2062.
  formation of:
    from He^+–CO_2 reaction, 2056; He^+–O_2, 2049, 2062; N_2^+–O, 2059, 2062.
  reaction of:
    with CO_2, 2062, 2063; N_2, 2046, 2052, 2058, 2062.
O⁻:
  charge transfer reactions of, with O_3, NO_2, 2106.
  cluster formation by, in O_2–H_2O mixtures, 2112.
  detachment of electrons from, on impact, associative:
    in CO, H_2, NO, 2103, 2105; N, N_2, O, O_2, 2105.
  mobility of, in O_2, 2068.
    theory of, 2081.
  reactions of:
    with CO_2, 2107; O_2, 2079, 2080.
O_2:
  bulk viscosity of, 1578.
  charge transfer reactions of:
    with N_2^+, 2055, 2062; NO^-, 2106; O^+, 2044, 2053, 2062.
  collisions of, with TlF, 1605.
  deactivation by, on impact:
    of $Hg(6^3P_1)$, 1677; $Na(3^2P)$, 1660.
  depolarization by, on impact, of $Hg(6^3P_1)$, 1712, 1732.
  destruction by, on impact:
    of He (metastable), 1821; $O(^1S)$, 1831, 1836.
  detachment of electrons from O_2^- in, 2092, 2105.
  interaction of, with K, 1448.
  mobility in:
    of O_2^+, 2043; O_2^-, 2043, 2068; O^- and O_3^+, 2068.
  reaction of:
    with Ar^+, 2064; C^+, CO^+, CO_2^+, 2063; He (metastable), 2044; He^+, 2044, 2062; N^+, 2057, 2062; NO^-, 2103; O^-, 2079, 2080, 2081; O_2^-, 2081.
  rotational relaxation in, 1577–8, 1580, 1583.
  thermal conductivity of, 1580–1.
  vibrational relaxation in, 1504, 1520–1, 1552.
    effect of CCl_4, C_2H_2, H_2S, H_2O, and NH_3, 1536; H_2 and He, 1524.
O_2^+:
  charge transfer reactions of, with O_2, 2043.
  formation of:
    from CO^+–O_2, CO_2^+–O_2, and O^+–CO_2

SUBJECT INDEX

reactions, 2063; He^+-O_2, 2044, 2050; N^+-O_2, 2057, 2062.
mobility of, in O_2, 2043.
reactions of:
  with N, 2058, 2062; N_2, 2062; NO, 2061, 2062; O_2, 2043.
second negative bands of, excitation of, by $He(2^3S)$, 1800.

O_2^-:
charge transfer reactions of:
  with NO_2 and O_3, 2106; O_2, 2081.
cluster formation by, in H_2O-O_2 mixtures, 2111.
detachment of electrons from:
  direct, in O_2, 2094.
  on impact, associative, with N and O_2, 2105.
formation of:
  from NO^--O_2 reaction, 2103; O^--O_2, 2081.
  in O_2-CO_2 mixtures, 2107; O_2-H_2O mixtures, 2111.
mobility of, in O_2, 2068.
  effect of charge transfer on, 2081.
  temperature variation of, 2082.
reaction of, with CO_2, 2108.

O_3, charge transfer reactions in, of NO^-, NO_2^-, O^-, and O_2^-, 2106.

O_3^+, observation of, in oxygen afterglow, 2043.

O_3^-:
charge transfer reactions of, with CO_2 and NO, 2106.
formation of, from O^--O_2 reaction, 2079.
mobility of:
  in CO_2, H_2, N_2, and rare gases, 2067; O_2, 2068.
theory of, 2081.

O_4^-:
dissociation energy of, 2083.
formation of, from O_2^- in O_2, 2081.
mobility of, in O_2, 2083.

OH, rotational relaxation in, 1587.

OH^-, 2104.
associative detachment of electrons from, in N, O, 2105.
charge transfer reactions of, with NO_2, 2106.
cluster formation by, in H_2O, 2111.
formation of, from H^--H_2O reaction, 2110.

Optic-acoustic effect, 1511.
Optical excitation function, apparent, for D states of He, 1758, 1766–7.
Optical pumping:
and depolarization of Cs and Rb resonance radiation, 1729.

and hyperfine relaxation, 1845, 1846.
and transfer collisions between He (metastable) and He, 1823.
to Zeeman levels, and spin exchange, 1851.

Orbiting in gas-kinetic collisions:
and theory of ionic reaction rates, 2005.
conditions for, in terms of reduced parameters, 1340.
theory of:
  classical, 1310.
  quantal, 1329–30.

Orientation, of electronic angular momentum, 1696.
production of, by absorption of resonance radiation, 1701.

Paraflint, 1849.
Parameters, reduced:
for gas-kinetic collision theory, 1335–6.
for interaction of unlike atoms, 1415–16.

Pb, $^3P_1$ state of, self-depolarization of, 1733.
theory of, 1902, 1903.

Pb^+, enhancement of spark lines of, by Ne, 1755.

PD_3, vibrational relaxation in, 1530–2, 1562.

Penning ionization, 1778, 1801, 1810, 1816, 2024.
energy distribution of electrons produced by, 1817.

PH_3, vibrational relaxation in, 1530–2, 1562.

Phase shifts, in scattering:
behaviour of, near resonance levels, 1330.
complex, and reactive scattering, 1624.
relation of, to angle of deflexion, 1317.

Photodissociation, of NaI, 1675.
Polarizability, of atoms:
and ion-atom interactions, 1300.
and van der Waals interaction, 1386–7.

Polarization:
of alkali-metal atoms and H atoms, 1842.
of resonance radiation, 1700.

Polar molecules:
collisions between, theory of, 1606.
comparison of observed and calculated cross-sections for, 1613.

Predissociation, 1332.
Production of molecular beams, 1342–4.
Proton exchange, in $H_3^+-H_2$ collisions, 2033.

Pumping effect, errors in pressure measurement due to, 1349, 1383, 1384, 1414, 1770.

Quantum theory:
of collisions between rigid spherical atoms, 1304–8.
of orbiting, 1329.
of scattering, at small angles, 1325.
by extended range interactions, 1322, 1323.
by r^{-s} potentials, 1323.

Quenching:
by flame gases, 1656.
in CO, 1498, 1501.
by ortho and para H_2, 1501–2.
of forbidden lines of O, 1829.
of infra-red fluorescence, 1497.
of radiation in general, 1653.
experiments on, 1653.

Rainbow angle, in gas-kinetic collisions:
and analysis of reactive scattering, 1626.
observation of:
at high resolution, 1390–4.
for K–Kr collisions, 1388; Na and K with Hg, 1390.
theory of:
classical, 1317.
in terms of reduced parameters, 1339.
semi-classical, 1321.

Rare-gas atoms, collisions of:
with alkali-metal atoms, 1351; H_2, 1461; TlF, dependence on orientation, 1595–1600.

Rb:
collisions of, with Cs, Na, K, 1430–3.
depolarization of resonance radiation of, 1729, 1733, 1913.
excitation of transitions between fine-structure levels of, 1740, 1926.
interaction of:
with Cs, Na, K, 1433; Hg, 1397; Kr, 1382.
polarization of, 1842.
reactive collisions of, with alkyl halides, 1646.
resonance level of, deactivation of, 1661, 1693.
spin-disorientation of, on impact with various gases, 1867, 1868.
spin-exchange collisions of:
theory of, 1888.
with Cs, 1845, 1850, 1865; Na, 1865; Rb, 1845, 1850, 1851, 1861, 1865.

Rb$^+$, mobility of:
in CO, H_2, and N_2, 1965; in rare gases, 1953, 1958, 1961, 1962.

Rebound reactions, 1644.
Recombination, 1954.
Resonance, symmetrical and accidental in inelastic collisions between atoms, 1869, 1923.

Resonance effects:
effect of, on thermal conductivity of polar gases, 1615.
in collisions between molecules, 1445–6.
polar molecules, 1609.
in enhancement of Pb$^+$ spark lines, 1755.
in quenching collisions, 1694.
in sensitized fluorescence, 1744.
of Na, 1766; In, 1749; Tl, 1750.

Resonance fluorescence, quenching of, 1653.

Resonance levels, in atom–atom interactions, 1330.
and predissociation, 1332.
for H–H interaction, 1422–3.

Resonance radiation:
imprisonment of, 1654.
Holstein's theory of, 1664.
use of, in measuring quenching cross-section, 1663.
polarization of, 1701.

Ringing, molecular, 1858.

Rotational excitation:
effect of, on interpretation of molecular-beam experiments, 1588.
of DBr from D–Br$_2$ reactions, 1647–8.
of KBr, from K–Br$_2$, 1641.
of TlF, by impact of various molecules, 1599–1606.
resonance transfer of, and thermal conductivity of polar gases, 1615.
theory of:
close-coupling (truncated eigenfunction expansion), 1588–9.
sudden approximation, 1590.
semi-classical form of, 1591.

Rotational relaxation:
effect of:
on shock-front thickness, 1483.
on thermal conductivity of molecular gases, 1579.
on thermal transpiration, 1582.
in D_2 and H_2, 1565; N_2 and O_2, 1577–8.
spectroscopic evidence on:
in NO, 1583–7; OH, 1587.

S$_2$, sensitized band fluorescence of, and vibrational relaxation, 1506–7, 1538.

Scattering, in gas-kinetic collisions:
at small angles:
classical theory for, 1315.
relation to quantal formula, 1317–19.

semi-classical theory for, 1319.
by long-range potentials:
classical theory of, 1322.
quantal theory of, 1322.
by r^{-s} potentials, 1323.
at small angles, 1325.
reactive, 1622.
$Na-C_2H_4Br_2$, Na–HBr, K–SnI$_4$, 1627.
Schlieren method, for measuring shock-front thickness, 1484–5.
application to D_2, 1518.
Se, deactivation of 4^3P_0 state of, 1667, 1741.
Semi-classical theory of:
diffusion cross-section for ions in gases, 1999.
glory effects, 1320.
molecular ringing, 1858.
rainbow effects, 1321.
rotational excitation (sudden approximation), 1591.
scattering, 1319.
symmetric charge transfer, 1988.
symmetric excitation transfer, 1873.
Sensitized fluorescence, 1731.
and excitation of transitions between fine-structure levels, 1734.
and excitation transfer, 1743.
and vibrational excitation of S_2, 1506, 1538.
SF$_6$:
collisions of:
with CsCl, 1450; TlF, 1605.
interaction of:
with Cs, 1449; K, 1448.
reactive collisions of, with K, 1631–4.
Shock front:
effect on:
of rotational relaxation, 1483.
vibrational relaxation, 1481–3.
infra-red emission behind, 1489.
reflectivity of, 1486.
temperature distribution behind, optical technique for measurement of, 1487.
thickness of, 1479.
observation of, 1484–7.
in Ar and He, 1487.
Shock waves:
generation of, 1470–1.
theory of, 1477–9.
use of, in quenching experiments, 1671.
SiCl$_4$:
collisions of, with CsCl, 1450.
interaction of, with K, 1448.
SiD$_4$, vibrational relaxation in, 1530–2.
SiF$_4$, interaction of, with K, 1448.
SiH$_4$, vibrational relaxation in, 1530–2.

SnCl$_4$:
interaction of, with K, 1448.
reactive collisions of, with K, 1631–4.
SnI$_4$, analysis of reactive scattering of K by, 1627.
SO$_2$:
interaction of with K, 1448.
vibrational relaxation of, 1528.
Spark lines, enhancement of, 1755.
Spectrophone, 1511.
use of for vibrational relaxation, 1512–17.
Spin-conservation rule:
in transfer collisions, 1756.
in He, 1760.
in O^+–CO_2 reaction, 2062.
Spin-density matrix, 1697.
Spin-exchange collisions, 1841.
methods for measurement of, 1844–64.
theory of, 1884.
Spin Hamiltonian, for two H$_2$ molecules, 1616.
Spin-lattice relaxation time, 1616.
effect of rotational transitions on, 1621.
for H$_2$, 1617; for ortho H$_2$–para H$_2$, 1618; for ortho H$_2$–ortho H$_2$, 1618.
Spin scattering matrix, 1699.
Spin temperature, 1843.
Stripping reactions, 1644.
Sudden approximation, application to theory of rotational excitation, 1590.
Symmetry, effect of:
on gas-kinetic cross-sections, 1308.
for H–H collision, 1418–19.
on transport properties of ^{3}He and ^{4}He, 1403.
on viscosity and thermal conductivity of H, 1425.

Thermal conductivity of gases:
effect of:
resonance transfer of rotation on, for polar gases, 1615.
rotational relaxation on, 1579.
of ^{4}He, 1400; ^{3}He, 1403; O, 1433–6.
relation to viscosity, 1303.
Thermal diffusion:
definition of, 1304.
of ^{3}He in ^{4}He, 1407.
Thermal transpiration, and rotational relaxation, 1582.
Tl:
deactivation by, on impact of Hg(6^3P_1), 1691, 1693.
resonance level of, deactivation of, 1661, 1675, 1693.
sensitized fluorescence of, 1743, 1750, 1754.

TlF:
  cross-section for elastic scattering, by rare gases, dependence on orientation, 1595–1600.
  rotational excitation of, by various molecules, observed cross-sections for, 1599–1605.

Transfer, on impact:
  of charge:
    between D^+–D and H^+–H, 2000; O^+–O_2, 2043; O_2^+–O_2, 2043; O_2^-–O_2, 2043.
    between ions and parent atoms, theory of, 1907.
    involving negative ions, 2105.
  of electronic excitation:
    and sensitized fluorescence, 1743.
    between He atoms, 1757; He (metastable) and He, 1822; He (metastable) and Ne, 1796.
    between molecules, 1929.
    theory of:
      for symmetrical case, 1869.
      for unsymmetrical case, 1914.
        when crossing point occurs, 1915.
        when no crossing point exists, 1918.
  of electron spin, 1841, 1844.
    theory of, 1884.
  of energy between vibrational and electronic excitation:
    for CO–Hg, 1683.
      absolute cross-section for, 1689.
      dependence on vibrational state, 1688.
    for N_2–Na, 1670; NO–Hg, 1683.
  of energy between vibrational and rotational excitation, theory of, 1561.
  of vibration, 1533, 1535.
    theory of, for head-on collisions, 1545.

Transformation from laboratory to C.M. coordinates:
  in crossed-beam experiments, 1367.
  in reactive scattering experiments, 1635, 1638.

Truncated eigenfunction expansion, see close-coupling approximation.

van der Waals attraction, 1298.
  anisotropy of, effect of, on collision cross-section, 1437–8.
  observation of, for Ga–rare gas interactions, 1439–42.

van der Waals constant:
  calculation of, 1386.
  Slater–Kirkwood formula for, 1387.

  determination of, from gas-kinetic collision experiments, 1341, 1381.
  for interactions:
    between alkali-metal atoms, 1433; and rare-gas atoms, 1385; Ar and H_2, 1462; and rare-gas atoms 1417; Cs, K and various molecules, 1448–9; CsCl, KCl, and various atoms and molecules, 1450–1; Ga and rare-gas atoms, 1442; H_2 and H_2, 1453, 1455–6; H and O, 1891; He and He, 1401, 1409, 1410; and rare-gas atoms, 1416; He (metastable) and He, 1876; and rare-gas atoms, 1773; Kr and Kr, 1417; Ne (metastable) and rare-gas atoms, 1774; O and O, 1435.

Velocity:
  of high-frequency sound:
    effect of vibrational relaxation on, 1466–9.
    measurement of, 1470–3.
  of shock waves:
    measurement of, 1483.
    theory of, 1477–9.

Velocity distribution, of KBr resulting from K–Br_2 reaction, 1640, 1643.

Velocity selector:
  for observing scattering of H atoms in gases, 1361.
  in molecular-beam scattering experiments, 1355.
  typical example of, 1356–8.

Vibrational relaxation (deactivation), 1465.
  and band absorption:
    in CN, 1496–7; I_2, 1496, 1506.
  effect of:
    in gas dynamics, 1491–6.
    on dispersion and absorption of high frequency sound, 1466–9.
    on shock-front thickness, 1483.
  of electronically excited molecules, 1504–8.
  of first vibrational level of NO ($X^2\Pi$), 1503.
  of gas mixtures, dependence on molar composition, 1533–7.
  of hydrides, as compared with deuterides, 1530.
  of polyatomic molecules, dependence of, on H-atom content, 1530–1.
  theory of:
    for head-on collisions:
      by distorted wave method, 1540.
      by semi-classical method, 1544.
      with more realistic interaction, 1547.

three-dimensional:
 close-coupling method, 1557.
 semi-empirical, 1550.
Vibrational temperature:
 in CO_2, theory of, 1558.
 measurement of, behind shock fronts:
  from infra-red emission, 1489.
  from line reversal, 1487.
Virial coefficient, second:
 of ^{3}He at low temperatures, 1402, 1404.
 of ^{4}He, at low temperatures, 1400–2;
  medium to high temperatures, 1411.
Virtual levels, *see* resonance levels.
Viscosity of a gas:
 bulk, 1483.
  in CH_4, Cl_2, CO_2, HCl, N_2, N_2O, and O_2, 1578.
 shear:
  for gas mixtures, 1416; of H, calculated, 1424–5; H_2, 1426; as a function of ortho-para composition, 1458; H–H_2 mixture, 1426; ^{3}He and ^{4}He at low temperatures, 1400–2; ^{4}He at medium to high temperatures, 1409; O, 1433–6.
 theory of, 1303.
  classical formula for, 1315.
  cross-section for gas-atom collisions, 1303.
   expansion of, under near-classical conditions, 1408.
   for H–H collisions, 1419.
   for rigid spherical atoms, 1307–8.

Xe:
 cluster formation by Li^+ ions in, 1968, 1969, 1971.
 collisions of:
  with Ar, 1417; Cs, 1378, 1384; CsBr, 1452; CsCl, 1450, 1452; D_2 and H_2, 1462; Ga, 1442; He, 1416; K, 1359, 1363, 1369, 1375, 1378, 1384; Li, 1359.
 deactivation by, on impact of He (metastable), 1878.
 depolarization by, on impact:
  of $Cd(5^3P_1)$, 1905; Cs states, 1733, 1913; $Hg(6^3P_1)$, 1712, 1732, 1905, 1911; other Hg states, 1732.
 enhancement by, of ionization by α-particles in He and Ne, 1815.
 excitation of fine-structure levels of alkali atoms by, 1926.
 interaction of:
  with Ar and Kr, 1417; Ga, 1442; He, 1416; Li, 1382; Na and K, 1382, 1390, 1397.
 mobility in:
  of alkali-metal ions, 1953, 1958, 1961, 1962; O_3^-, 2067; xenon ions, 1995, 2002, 2003.
 spin disorientation of Rb by, 1867.
Xe^+, mobility of:
 in Xe, 1995.
 relation of, to charge transfer, 2002.
Xe_2:
 mobility of in Xe, 1995, 2003.

PRINTED IN GREAT BRITAIN
AT THE UNIVERSITY PRESS, OXFORD
BY VIVIAN RIDLER
PRINTER TO THE UNIVERSITY